Electromagnetic and Electrical Engineering II

WITPRESS
WIT Press publishes leading books in Science and Technology.
Visit our website for the current list of titles.
www.witpress.com

WITeLibrary
Home of the Transactions of the Wessex Institute.
Papers contained in this title are archived in the WIT elibrary in volume 107 of
WIT Transactions on Engineering Sciences (ISSN 1743-3533).
The WIT electronic-library provides the international scientific community with immediate
and permanent access to individual papers presented at WIT conferences.
http://library.witpress.com

WIT Transactions

Electromagnetic and Electronics Engineering II

Editors

Zhang Jun
Hebei University of Science and Technology, China

&

Jeffrey
The Scientific Research and Technical Service Centre, China

Editors:

Zhang Jun
Hebei University of Science and Technology, China

Jeffrey
The Scientific Research and Technical Service Centre, China

Published by

WIT Press
Ashurst Lodge, Ashurst, Southampton, SO40 7AA, UK
Tel: 44 (0) 238 029 3223; Fax: 44 (0) 238 029 2853
E-Mail: witpress@witpress.com
http://www.witpress.com

For USA, Canada and Mexico

Computational Mechanics International
25 Bridge Street, Billerica, MA 01821, USA
Tel: 978 667 5841; Fax: 978 667 7582
E-Mail: infousa@witpress.com
http://www.witpress.com

British Library Cataloguing-in-Publication Data

A Catalogue record for this book is available
from the British Library

ISBN: 978-1-78466-117-5
eISBN: 978-1-78466-118-2
ISSN: (print) 1746-4471
ISSN: (on-line) 1743-3533

The texts of the papers in this volume were set individually by the authors or under their supervision. Only minor corrections to the text may have been carried out by the publisher.

No responsibility is assumed by the Publisher, the Editors and Authors for any injury and/or damage to persons or property as a matter of products liability, negligence or otherwise, or from any use or operation of any methods, products, instructions or ideas contained in the material herein. The Publisher does not necessarily endorse the ideas held, or views expressed by the Editors or Authors of the material contained in its publications.

Printed in Great Britain by Lightning Source, UK.

Preface

The 2014 Second Symposia on Electromagnetic and Electronic Engineering (SEEE 2014 II) was held by The Scientific Research and Technical Service Center and Hebei University of Science and Technology. The goal of SEEE 2014 II is to provide a platform for researchers, engineers, academics as well as industrial professionals from all over the world to present their research results and development activities in Electromagnetic and Electronic Engineering. It also allow the attendees both in industry and academia to share their state-of-art results, to explore new areas of research and development.

SEEE 2014 II focuses on the Electromagnetic and Electrical Engineering field. Such as Electromagnetic field and microwave technology, EMC, environment effect, Electromagnetic materials, Electromagnetic Modeling and Simulation, Electromagnetic emission, Microwave and antenna, Electromagnetic signal processing, Electrical Power Systems and automation, Electric motor and control, Electrical theory and technology, Circuit and system, Fault diagnosis theory, Mechanical and electrical integration and so on. SEEE 2014 II received more than 130 submissions in all. After receiving papers which reviewed by two specialists in this field, we collected 81 papers into the proceedings.

There is a long list of individuals we would like to thank for their support in organizing this conference. We would like to take this opportunity to express our thanks to the individuals and organizations for their efforts to serve the conference. We would also like to extend our thanks to the conference academic committee, technical committee and organizing committee for their hard work. Finally, we would like to express our appreciation to each participant.

The organizing committee of SEEE 2014 II
Shijiazhuang China 2015

Contents

Section 1 Electromagnetic

Section 2 Electrical engineering

Section 1
Electromagnetic

A speech recognition method based on sub-frequency band analysis and running spectrum filter

Zhang Yuxin, Ding Yan
School of Computer Science and Technology,
Changchun University of Science and Technology, Changchun, China

Abstract

In this paper, we propose a new algorithm based on the sub-frequency band of spectrum entropy. The whole spectrum is divided into sub-frequency bands limited. Then, the spectrum entropy of each sub-band is calculated, respectively. Some sub-frequency band is extracted to detect, because the major changes of voice happen in these band and it expresses the degree of signal change. The experiment shows that the method improved VOD performance in low SNR environment. On the other hand, we modified the conventional RSF. The band-pass filter to reduce the noise for RSF. And then, CMS method is used to reduce the remanent noise in whole frequency domain. The calculation cost is far lower than that of the conventional RSF. Recognition accuracy of ASR is improved much more.
Keywords: spectrum entropy, running spectrum analysis, speech recognition.

1 Introduction

Recently, the automatic speech recognition technology (ASR) is a popular research. So, the machine can understand human language and communicate with humans, which become our study goal [1, 2]. Voice activity detection (VAD) and noise reduction are two important problems in field of speech recognition to be solved [3–5].

VAD based on short-time energy and zero-crossing rate is only suitable for high SNR environment. In the steady low SNR environment, the spectrum entropy method (SE) has strong anti-interference [6, 7]. When environmental noise deteriorates, the accuracy of VAD will be a sharp decline. Since the

WIT Transactions on Engineering Sciences, Vol. 107, © 2015 WIT Press
www.witpress.com, ISSN 1743-3533 (on-line)
doi:10.2495/SEEEII140011

environmental noise is varied, different spectrum entropies have big difference in each other. White noise and babble noise are relatively stable, its entropy is larger. Entropy factor noise is very small. So the SE method cannot be well adapted to changes of all noises.

Running spectrum filtering (RSF) is a noise reduction method that exploits the difference of temporal variability between the spectra of speech and noise signals to remove the noise [8]. However, RSF cannot reduce the noise, which is in speech spectrum. On the other hand, we know that the calculation cost of RSF algorithm is high, since the high order (240) is used for FIR, and RSF is used for reducing noise twice.

In this paper, we propose a new algorithm based on the sub-frequency band of spectrum entropy. The entire spectrum is divided into sub-bands limited, then, respectively, the entropy of each sub-band is calculated, so that we can judge the degree of signal change. Because the major changes of voice happen in some scope, though extracting sub-band entropy that has big changes for endpoint detection, in our experiment, we find that the method improves VOD performance in low SNR environment. On the other hand, we modified the conventional RSF and use RSF with band-pass filter to reduce the noise. And then, CMS method is used to reduce the remanent noise in whole frequency domain. The calculation cost is far lower than that of conventional RSF. Recognition accuracy of ASR is improved much more.

This paper is organized as follows. In Section 2, we propose VAD algorithm based on spectrum entropy of sub-frequency band analysis. In Section 3, we analyze the shortcoming and propose modified RSF. Section 4 shows the experiment result. Finally, we conclude our paper in Section 5.

2 Sub-frequency band analysis algorithm (STE)

The frequency distribution of the noise is different with speech signal. When a speech signal is pronounced, the energy of some frequencies is bound to enhance. However, the energy of the start and end parts of the speech signal is low. Thus, it is difficult how to detect the boundary between the noise and the speech signal. Through the analysis of the speech signal frequency energy diagram, we found that the energy of the speech signal is not evenly distributed in whole frequency spectrum, the energy of some frequencies has big changes and has small changes in others, the energy variance of frequency whose energy change is larger than the whole frame. Thus, these frequencies are more for detection. The whole frequency can be divided into a few sub-frequency bands. We analyzed that the sub-frequency band is easier than whole spectrum. Therefore, we proposed a VAD algorithm based on sub-frequency band entropy analysis (SFE).

Usually, the transform point of FFT is 512 point. The whole frequency is divided into 15 sub-frequency bands. Each sub-frequency band is 34 points. Therefore, the energy of the ith sub-frequency band is defined as

$$e_x(j) = \sum_{k=34j-33}^{34j} |v_i(f_k)|^2, 1 \le j \le 15 \quad (1)$$

where $v_i(f_k)$ is the *k*th frequency component of the *i*th frame.

The spectrum entropy represents the change of energy spectrum of a frame. The spectrum entropy of sub-frequency band is protean with the frame number. It does not have the well reflected changes in the characteristics of the current signal. Furthermore, the change of speech signal is short time. Thus, the sliding windows are proposed. The sub-frequency band is only calculated, which is in windows, then their spectrum entropies well reflect the instantaneous change for speech. The frequency of noise is stationary distribution. The change of spectrum entropy is gently. When speech signal comes into window, the change of spectrum entropy of sub-frequency band is acute than the whole frame, which belongs to speech. Therefore, this apparent change is beneficial voice endpoint detection.

Algorithm description:

Step 1: Divide noisy speech signal into frames, then FFT transform for each frame.

Step 2: The spectrum is divided into 15 sub-bands for each frame, and then calculate the energy of each sub-frequency band.

Step 3: Set the value of the sliding window, w = 10. Usually, the first some frames are considered as non-speech, so the first 10 frames are in the window. And then calculate the spectrum entropy of each sub-frequency band. The short sub-frequency band spectrum entropy is defined as

$$p_x(j) = \frac{e_x(j)}{\sum_{x=1}^{w} e_x(j)} \tag{2}$$

$$H(j) = \sum_{x=1}^{w} |p_x(j)| \cdot \log p_x(j) \tag{3}$$

where $ex\ (j)$ is the energy of the *j*th sub-frequency band of the *x*th frame, $p_x\ (j)$ is the normalized probability density, and $H(j)$ is spectrum entropy.

Step 4: Set sub-frequency band spectrum entropy threshold $\overline{H(j)} = \omega H(j)$, ω is experience value. The frame is considered as speech signal, whose spectrum entropy is more than $\overline{H(j)}$.

Step 5: Detect the endpoint method: The sliding window moves forward frame by frame. The spectrum entropy of each sub-frequency band in sliding window is calculated, when a frame is in sliding window.

The value of the sliding window W_i is defined: The number of which sub-frequency band spectrum entropy is more than the threshold value $H(j) > \overline{H(j)}$. Initialize W_0 = 0. If W_i < 2, the window slides forward. If $W_i \geq 10$, then the endpoint is in the window. If $2 \leq W_i \leq 10$ for continuous 5 windows, then the endpoint is in the first window of them. Otherwise, windows move forward.

Step 6: After the windows are found that where endpoint is, the spectrum entropy of each frame is calculated which are in sliding window. In each band, find the first frame, which is $e_x(j) > \frac{1}{w}\sum_{x=1}^{w} e_x(j)$. The frame number is as $x(j)$ the starting frame of speech is arg min $x(j)$.

Step 7: Detecting the end point is similar to the method described above. After finding the window, a look back from the first sub-band energy continuously decreasing, the last frame is the end.

3 Noise reduction

3.1 Running spectrum filtering algorithm

In the modulation spectrum, we have found that the noise spectrum is concentrated in the direct component (DC). Most of the noise energy is distributed in the low-frequency band of the modulation spectrum. Figures 1–4 show the power spectrum of speech and noise. This shows that the significant constituent of speech is in the band [0, 16] Hz. The additivity noise on frequency band [0, 1] Hz exerts such tremendous effect on speech signal. To speech recognition, the important information of speech is about in the frequency band [0, 16] Hz. The multiplication noise exerts such tremendous influence on frequency band, which is close to 0 Hz.

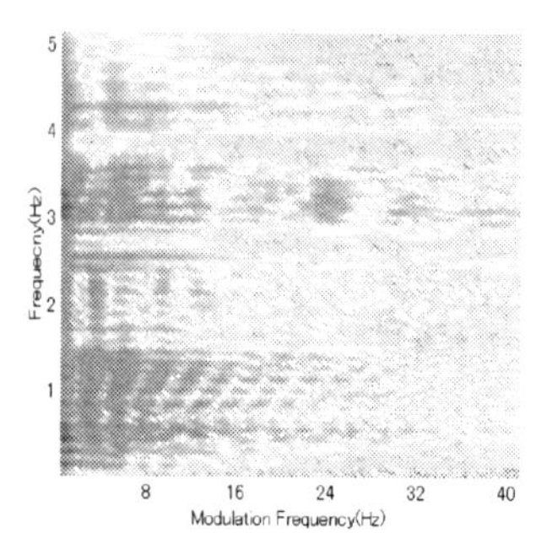

Figure 1: The power spectrum of clean speech in the modulation spectra.

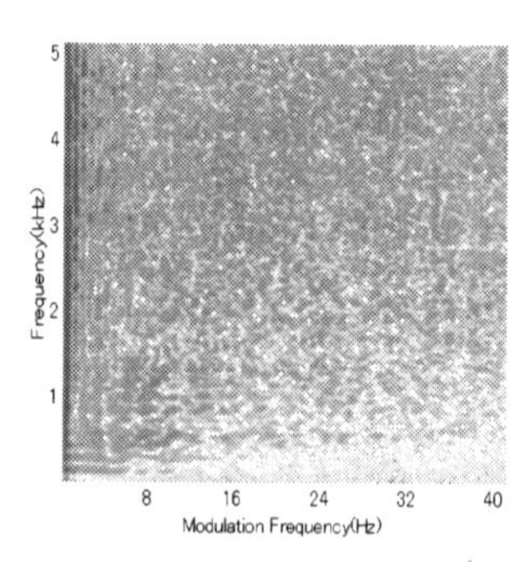

Figure 2: The power spectrum of 5 dB white noise in the modulation spectra.

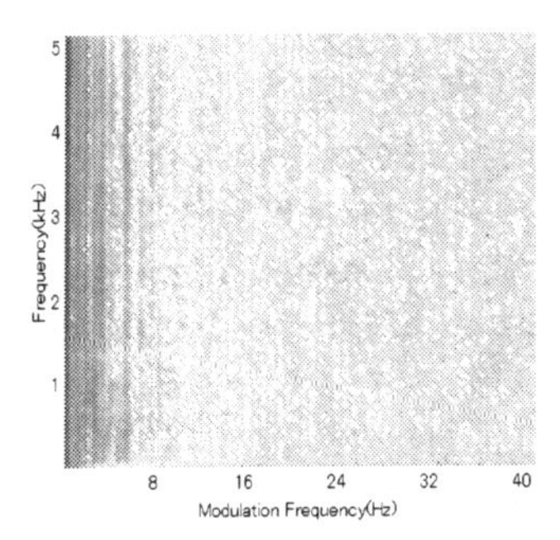

Figure 3: The power spectrum of noisy speech with 0 dB white noise in the modulation spectra.

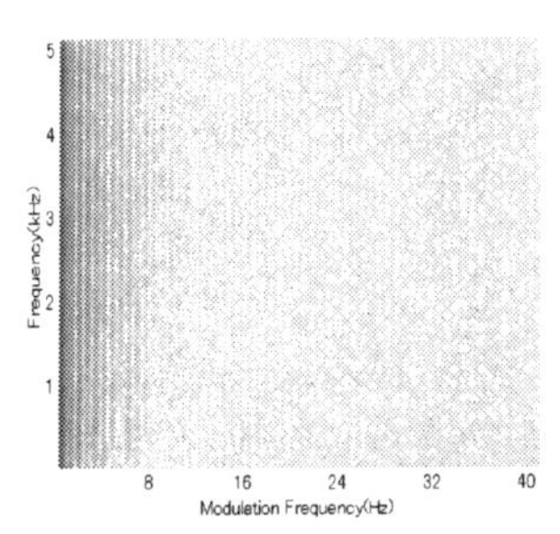

Figure 4: The power spectrum of noisy speech with 5 dB white noise in the logarithm spectra.

Thus, removing low-frequency components with a high-pass filter can reduce the noise. On the other hand, we can use a bandpass filter to separate speech from noise. The additive noise is reduced in the power spectra and the multiplicative noise is reduced in the logarithm spectra by RSF.

3.2 Cepstrum mean subtraction algorithm (CMS)

CMS is often reduced noise [9]. Because the white noise is uniformly distributed in a spectrum, then the time-invariant noise features in such a range are considered as almost constant. Thus, the average of whole noisy speech signal is subtracted from noisy speech features of every frame, the energy of noise components is almost zero. We assume that a speech waveform is divided into h short frames. $f_i(t)$ is the tth component of the ith frame. Noise reduction is then executed as

$$f_i^x(t) = f_i(\mathrm{t}) - \frac{1}{h}\sum_{j=1}^{h} f_i(t) \tag{3}$$

3.3 Proposed noise reduction method

In real environment, the additivity and multiplicative noises are simultaneous. Hence, the mixed superimposed speech waveform is as follows in time domain [1].

$$x(t) = s(t) \otimes h(t) + n(t) \tag{5}$$

where $x(t)$ is the noisy speech signal, $s(t)$ is the speech signal, $h(t)$ is the multiplicative noise, and $n(t)$ is the additivity noise. Eq. (5) is Fourier transformed on both sides. In frequency and power spectrums, the equation is followed, which is effected by the additivity and multiplicative noises:

$$X(t,i) = S(t,i)H(t,i) + N(t,i) \tag{6}$$

$$\begin{aligned} |X(t,i)|^2 &= |S(t,i)H(t,i) + N(t,i)|^2 \\ &= |S(t,i)H(t,i)|^2 + |N(t,i)|^2 + 2Re[S(t,i)H(t,i)N(t,i)] \\ &= |S(t,i)|^2 |H(t,i)|^2 + |N(t,i)|^2 \\ &= +2|S(t,i)||H(t,i)||N(t,i)| cos(\theta(t,i)) \end{aligned} \tag{7}$$

where $\theta(t, i)$ is the phase separation between speech signal and additivity noise on the ith point Because the speech and noise can be supposed as mutually independent zero-mean distribution, the desired value of last item is zero in Eq. (7). Although instantaneous value of each frame is not zero in this item, the output value of each filter unit is equal to weighted sum of energies of all points when computing Mel-filter. Hence, Mel-energy of noisy speech signal is approximately equal to

$$P_x(t,i) \approx P_s(t,i)P_h(t,i) + P_n(t,i) \tag{8}$$

where $P_x(\cdot)$, $P_s(\cdot)$, $P_h(\cdot)$, and $P_n(\cdot)$ are Mel-energy of noisy speech, clean speech, additivity noise, and multiplicative noise.

In logarithm spectrum, we defined $X^{\log}$, $S^{\log}$, $N^{\log}$, and $H^{\log}$ as values of vector for noisy speech, clean speech, additivity noise, and multiplicative noise. So

$$X^{\log} = S^{\log} + H^{\log} + \log\left(I + e^{\left(N^{\log} - S^{\log} - H^{\log}\right)}\right) \tag{9}$$

Similarly, we defined X^{cep}, S^{cep}, N^{cep}, and H^{cep} as values of cepstrum feature vector for noisy speech, clean speech, additivity noise, and multiplicative noise in cestrum spectrum. So

$$X^{cep} = S^{cep} + H^{cep} + D\log\left(I + e^{D^{-1}\left(N^{cep} - S^{cep} - H^{cep}\right)}\right) \tag{10}$$

where D is the discrete cosine transformation (DCT) matrix.

In Eq. (10), the H^{cep} can be almost removed by RSF, but the effect of $D\log\left(I + e^{D^{-1}\left(N^{cep} - S^{cep} - H^{cep}\right)}\right)$ is in the whole modulation frequency domain. On the other hand, we know the calculation cost of RSF algorithm is high, since the high order (240) is used. Moreover, the conventional RSF algorithm is used for reducing noise twice. One is in power spectrum, the other is in logarithm spectrum. Hence, the calculation time of ASR system with RSF is relatively high. In order to improve the performance of ASR system, we remove the RSF for noise reduction in power spectra. After cepstrum computing, we use RSF with band-pass filter to reduce the noise. And then, CMS method is used to reduce the remanent noise in whole frequency domain. CMS is simpler than RSF. The calculation cost is far lower than that of RSF.

4 Experiment result analysis

Our ASR system is implemented in MATLAB. The dynamic time (DTW) with nonlinear media filter method (NMF) is used to match, which has been proposed in paper [10]. Further details of the experimental settings and parameters appear in Table 1.

Table 2 shows the detection accuracy of STE and ZCR is very low in low SNR. That of SE is better than STE and ZCR from 0 to 20 dB. That of SFE is best among them. The accuracy is more than 80% in two noises. The tables indicate that detection accuracy improves greatly with SFE.

Furthermore, Table 3 shows the recognition accuracy for 10 and 20 dB SNR white and babble noises with VAD. These tables indicate that recognition accuracy improves greatly with modified RSF. Moreover, it shows that modified RSF is better than conventional RSF approach.

Table 1: The experimental settings and parameters.

Parameter	Value
Recognition task	100 isolated Japanese words
Sampling	11.025 kHz, 16 bits
Window length	23.2 ms (256 samples)
Shift length	11.6 ms (128 samples)
FFT	512 point

Table 2: The detection accuracy of VAD (%).

VAD	White			Babble		
	20 dB	10 dB	0 dB	20 dB	10 dB	0 dB
STE and ZCR	91	72	40	90	70	32
SE	94	83	71	94	82	69
SFE	96	89	83	95	89	81

Table 3: Recognition accuracy of ASR (%).

Noise		Noise Reduction Method			
Name	SNR	CMS	RSF	Modified RSF	Nothing
White	10 dB	85.3	86.78	86.9	28.6
	20 dB	97.9	98.06	98.12	84.42
Babble	10 dB	83.22	83.6	84.1	37.68
	20 dB	97.2	96.7	97.82	87.92
Clean		98.96			

5 Conclusion

Spectrum entropy only considers the difference of each band of internal frame. And it did not take into the difference between adjacent frames. If the noise spectrum is very similar to the spectrum of the speech, it is difficult to detect between voice and non-voice. Therefore, by dividing the band, we can analyze energy change of the main band of voice. Setting sliding window reflects the continuity of speech signal, which can avoid identification errors of instantaneous increase of entropy that the sudden emergence of short-term noise leading. And we only calculate the frequency bands which have large energy changes. On the other hand, the modified approach combines the advantages of RSF and CMS. The recognition accuracy is better than that of RSF, as well as the calculation cost is lower than that of RSF.

References

[1] X. Huang, A. Acero, A. Acero, et al, *Spoken Language Processing: A Guide to Theory, Algorithm, and System Development*. Upper Saddle River, Prentice Hall, NJ, 2001.

[2] D. Jurafsky, J.H. Martin, *Speech and Language Processing: An Introduction to Natural Language Processing, Computational Linguistics, and Speech Recognition*. Prentice Hall, 2009.

[3] Z. Lu, B. Liu, L. Shen, Speech endpoint detection in strong noisy environment based on the Hilbert-Huang transform. In: *International Conference on Mechatronics and Automation*, pp. 4322–4326, 2009.

[4] X.U. Dawei, W.U. Bian, Z. Jianwei, et al., A real time algorithm for voice activity detect on in noisy environment. *Computer Engineering and Application*, **24(1)**, pp. 115–117, 2003.

[5] H. Ghaemmaghami, R. Vogt, S. Sridharan, Speech endpoint detection using gradient based edge detection techniques. In: *Second International Conference on Signal Processing and Communications System, ICSPCS*, pp. 01–08, 2008.

[6] C. Jia, B. Xu, An improved entropy-based endpoint detection algorithm. In: *International Conference on Spoken Language Processing, ICSLP-2002*, pp. 285–288, 2002.

[7] J.L. Shen, J.W. Hung, L.S. Lee, Robust entropy-based endpoint detection for speech recognition in noisy environments. In: *International Conference on Spoken Language Processing, ICSLP-98*, pp. 232–238, 1998.

[8] N. Hayasaka, Y. Miyanaga, N. Wada, Running spectrum filtering in speech recognition. In: *International Conference on Soft Computing and Intelligent Systems*, pp. 154–157, 2002.

[9] D. Naik, Pole-filtered cepstral mean subtraction. In: *International Conference on Acoustics, Speech, and Signal Processing* (Vol. 1), pp. 157–160, 1995.

[10] Y. Zhang, Y. Miyanaga, C. Siriteanu, Robust speech recognition with dynamic time warping and nonlinear median filter. *Journal of Signal Processing,* **16(2)**, pp. 147–157, 2012.

Numerical dispersion analysis of higher-order LOD-FDTD method for micro-scale structures

Min Su, Pei-Guo Liu
Department of Electronic Science and Engineering, National University of Defense Technology, Changsha, China

Abstract

Numerical dispersion analysis of three-dimensional locally one-dimensional finite difference time domain, when analyzing for radio frequency micro-electro-mechanical-system, structures was proposed. It is the key issue for saving simulation time and assuring simulation accuracy. Several factors (such as spatial deviation, aspect ratio) which affect the numerical dispersion were discussed in detail. The basic setup principles of parameters were suggested.
Keywords: RF MEMS, locally one-dimensional finite-difference time-domain, higher-order numerical dispersion, micro-scale structures.

1 Introduction

RF MEMS was widely used in micro-electromagnetic system because of its superior electrical properties at high frequency [1]. The traditional finite-different time-domain (FDTD) method has been one of the most popular and attractive means to solve radio frequency micro-electro-mechanical-system (RF MEMS) structures of electromagnetic problems [2]. However, the traditional FDTD method is an explicit method, and the time step size is constrained by the Courant–Friedrichs levy (CFL) condition, which affects its computational efficiency when fine meshes are required. Recently, to overcome the CFL condition, some unconditionally stable FDTD methods such as alternating direction implicit (ADI) and locally-one-dimensional (LOD) FDTD methods were developed. However, numerical dispersion error is larger than condition FDTD method. Several methods have been proposed to improve the accuracy of the numerical dispersion [3, 4], for example, higher-order schemes are a usual method to reduce the numerical dispersion error [5, 6].

www.witpress.com, ISSN 1743-3533 (on-line)
doi:10.2495/SEEEI140021

Some researches which using ADI-FDTD method to solve the RF MEMS structures problems have been reported [7]. These researches' basic setup principles of parameters were suggested under the routine size. Numerical dispersion analysis of higher-order LOD-FDTD method for micro-scale structures has not been reported.

In this paper, numerical dispersion analysis of LOD-FDTD method for micro-scale structures was studied. Spatial deviation, aspect ratio, wave propagation angle and time step, which affect numerical dispersion were discussed in detail.

2 The formulation of high-order LOD-FDTD

In this section, formulations of the proposed method will be derived. For simplicity, the 3D wave propagation is considered in a linear, lossless isotropic medium with permittivity ε and permeability μ. Throughout the paper, our finite difference approximation schemes for Maxwell's equation possess second-order accuracy in time domain.

The Maxwell's equations can be written in a matrix form as

$$\nabla\times\vec{H}=\varepsilon\frac{\partial\vec{E}}{\partial t}\qquad \nabla\times\vec{E}=-\mu\frac{\partial\vec{H}}{\partial t} \tag{1}$$

According to X, Y and Z directions, (1) is approximated with the following equation:

$$\phi^{n+1}=\frac{([I]+\frac{\Delta t}{2}[A])([I]+\frac{\Delta t}{2}[B])([I]+\frac{\Delta t}{2}[C])}{([I]-\frac{\Delta t}{2}[A])([I]-\frac{\Delta t}{2}[B])([I]-\frac{\Delta t}{2}[C])}\phi^{n} \tag{2}$$

Here, $\phi=\left[E_x,E_y,E_z,H_x,H_y,H_z\right]^{\mathrm{T}}$, Δt is the time step and ϕ^{n+1} represents the field values at $t=n\Delta t$, and

$$[A]=\begin{bmatrix}0&0&0&0&0&0\\0&0&0&0&0&-\frac{1}{\varepsilon}\frac{\partial}{\partial x}\\0&0&0&0&\frac{1}{\varepsilon}\frac{\partial}{\partial x}&0\\0&0&0&0&0&0\\0&0&\frac{1}{\mu}\frac{\partial}{\partial x}&0&0&0\\0&-\frac{1}{\mu}\frac{\partial}{\partial x}&0&0&0&0\end{bmatrix}\quad [B]=\begin{bmatrix}0&0&0&0&0&\frac{1}{\varepsilon}\frac{\partial}{\partial y}\\0&0&0&0&0&0\\0&0&0&-\frac{1}{\varepsilon}\frac{\partial}{\partial y}&0&0\\0&0&-\frac{1}{\mu}\frac{\partial}{\partial y}&0&0&0\\0&0&0&0&0&0\\\frac{1}{\mu}\frac{\partial}{\partial y}&0&0&0&0&0\end{bmatrix}\quad [C]=\begin{bmatrix}0&0&0&0&-\frac{1}{\varepsilon}\frac{\partial}{\partial z}&0\\0&0&0&\frac{1}{\varepsilon}\frac{\partial}{\partial z}&0&0\\0&0&0&0&0&0\\0&\frac{1}{\mu}\frac{\partial}{\partial z}&0&0&0&0\\-\frac{1}{\mu}\frac{\partial}{\partial z}&0&0&0&0&0\\0&0&0&0&0&0\end{bmatrix}$$

The time advancement is made from n to n + 1/3 and then from n + 1/3 to n + 2/3, then from n + 2/3 to n + 1. (2) can be split into and computed into three steps:

Combining (3a)–(3c), $\phi^{n+1}=[\Lambda]\phi^{n}$ can be derived where $[\Lambda]$ is a matrix of the whole time step.

Using Matlab software, we can get six eigenvalues of the matrix $[\Lambda]$:

$$\text{Step 1: } \phi^{n+1/3} = \frac{([I]+\frac{\Delta t}{2}[A])}{([I]-\frac{\Delta t}{2}[A])}\phi^{n} = [\Lambda_1]\phi^{n} \qquad n \to n+1/3 \tag{3a}$$

$$\text{Step 2: } \phi^{n+2/3} = \frac{([I]+\frac{\Delta t}{2}[B])}{([I]-\frac{\Delta t}{2}[B])}\phi^{n+1/3} = [\Lambda_2]\phi^{n+1/3} \qquad n+1/3 \to n+2/3 \tag{3b}$$

$$\text{Step 3: } \phi^{n+1} = \frac{([I]+\frac{\Delta t}{2}[C])}{([I]-\frac{\Delta t}{2}[C])}\phi^{n+2/3} = [\Lambda_3]\phi^{n+2/3} \qquad n+2/3 \to n+1 \tag{3c}$$

$$\lambda_1 = \lambda_2 = 1, \lambda_3 = \xi_1 + \mathrm{j}\sqrt{1-\xi_1^2}, \lambda_4 = \xi_1 - \mathrm{j}\sqrt{1-\xi_1^2}, \lambda_5 = \xi_2 + \mathrm{j}\sqrt{1-\xi_2^2}, \lambda_6 = \xi_2 - \mathrm{j}\sqrt{1-\xi_2^2} \tag{4}$$

Here,

$$\xi_1 = \frac{\left[1+b^3d^3P_x^2P_y^2P_z^2 - dbP_x^2 - dbP_y^2 - dbP_z^2 - b^2d^2P_x^2P_y^2 - b^2d^2P_x^2P_z^2 - b^2d^2P_y^2P_z^2 + 4db\sqrt{db}P_xP_yP_z\right]}{A_xA_yA_z}$$

$$\xi_2 = \frac{\left[1+b^3d^3P_x^2P_y^2P_z^2 - dbP_x^2 - dbP_y^2 - dbP_z^2 - b^2d^2P_x^2P_y^2 - b^2d^2P_x^2P_z^2 - b^2d^2P_y^2P_z^2 - 4db\sqrt{db}P_xP_yP_z\right]}{A_xA_yA_z}$$

$A_\alpha = 1+bdP_\alpha^2$, $B_\alpha = 1-bdP_\alpha^2$, $d = \Delta t/(2\mu)$, $b = \Delta t/(2\varepsilon)$, $\alpha = x, y, z$,

$$P_a = \frac{-2\sin(\frac{k_\alpha \Delta\alpha}{2})}{\Delta\alpha}, \partial/\partial\alpha = \mathrm{j}P_\alpha, k_x = k\sin\theta\cos\varphi, k_y = k\sin\theta\sin\varphi,$$

$$k_z = k\cos\theta.$$

Numerical dispersion value is given as

$$\tan^2(\frac{\omega\Delta t}{2}) = \frac{\left[bdP_x^2 + bdP_y^2 + bdP_z^2 + b^2d^2P_x^2P_y^2 + b^2d^2P_x^2P_z^2 + b^2d^2P_y^2P_z^2\right]}{1+b^3d^3P_x^2P_y^2P_z^2} \tag{5}$$

First-order space partial differential operator is approximated as follows:

$$\frac{\partial(\phi|_{i,j,k}^{n})}{\partial\alpha} = \frac{1}{\Delta\alpha}\sum_{l=-L+1}^{L} d_{(2l-1)/2}\phi|_{i+(2l-1)/2,j,k}^{n} \tag{6}$$

From Eq. (5), many factors that affect the numerical dispersion, such as spatial deviation, aspect ratio, wave propagation angle and time step can be derived. In this section, according to investigating the characteristic of microwave device structure, the relationship between mentioned above factors and the numerical dispersion analysis of high-order LOD-FDTD method is discussed by using Matlab software.

Here, $d_{(2l-1)/2} = -d_{-(2l-1)/2} = \dfrac{(-1)^{l-1}(2L-1)!!^2}{2^{2L-2}(L+l-1)!(L-l)!(2l-1)^2}$,

$\dfrac{\partial}{\partial\alpha} = jP_\alpha = -2j\dfrac{1}{\Delta\alpha}\sum_{l=1}^{L} d_{(2l-1)/2}\sin\left(\dfrac{2l-1}{2}k_\alpha\Delta\alpha\right)$, L = 1, 2, 3, …, and coefficient d are exponent number, N = 2, 4, 6, 8, 10.

3 Numerical dispersion result

(1) *Effect of spatial deviation on numerical dispersion*

Figure 1 presents the normalized numerical phase velocity versus wave propagation angle φ (in degrees) for the second-, fourth-, sixth-, eighth- and tenth-order schemes with CFLN = 1 and $\lambda = 1\,\text{m}$. Here, $\Delta t = \lambda/\left(20\sqrt{3}c\right)$ is equal to the CFL limit corresponding to the minimum mesh size of $\lambda/20$ and θ is selected as 45° throughout this section.

It can be seen that the dispersion error can be reduced when the order is increased. The dispersion error of the second-order and higher-order schemes is lower than 2%. Furthermore, the dispersion error of the fourth-order and higher-order schemes is lower than 1.8%. But for the sixth-, eighth- and 10th-order

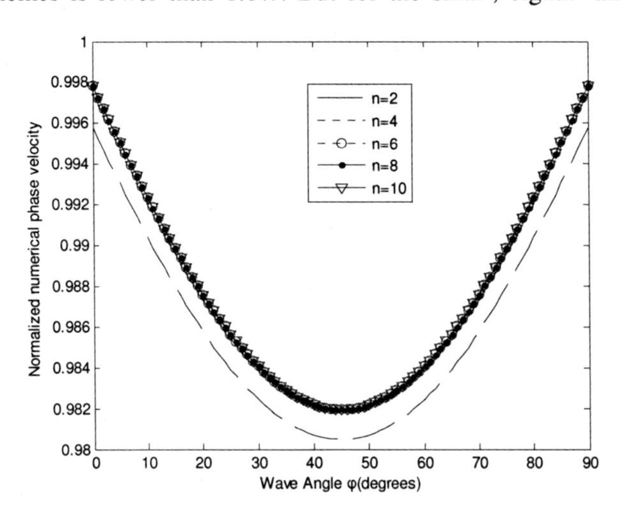

Figure 1: Normalized numerical phase velocity error (%) versus wave propagation angle φ for the second-, fourth-, sixth-, eighth- and 10th-order schemes with CFLN = 1, CPW = 20, Ry = 1, Rz = 1, $\lambda = 1$.

schemes, the dispersion errors are almost the same as that of the fourth-order scheme. Generally speaking, fourth-order scheme can satisfy the precision requirement of almost applications.

(2) *Effect of wave propagation angle on numerical dispersion*
In Figure 2, the maximum normalized numerical phase velocity error of the fourth-order scheme is 0.9666 under the given conditions (Ry = 5, Rz = 10, CFLN = 3, CPW = 20).

It can be seen that the maximum normalized numerical phase velocity is presented at $\theta = 45^{\circ}$. The result shows that numerical dispersion error is maximal when electromagnetic wave propagates along the direction. But micro-structures of transmission direction have little impact on algorithm of numerical dispersion from the whole transmission direction. The numerical dispersion of electromagnetic wave discrete maximum along space direction is the unique considered factor when loading excitation source.

(3) *Effect of CFLN on numerical dispersion*
Figure 3 presents the normalized numerical phase velocity versus wave propagation angle φ (in degrees) for the fourth-order scheme with CPW = 40 and $\lambda = 1$. Here, $\Delta t = \text{CFLN} \times \lambda / \left(\text{CPW} \times \sqrt{3}c\right)$ is equal to the CFL limit corresponding to the minimum mesh size of $\lambda / 40$ and θ is selected as 45° throughout this section.

It can be seen that the dispersion error can be increased when the CFLN is increased. The drawing shows that CFLN obviously affect the normalized numerical phase velocity.

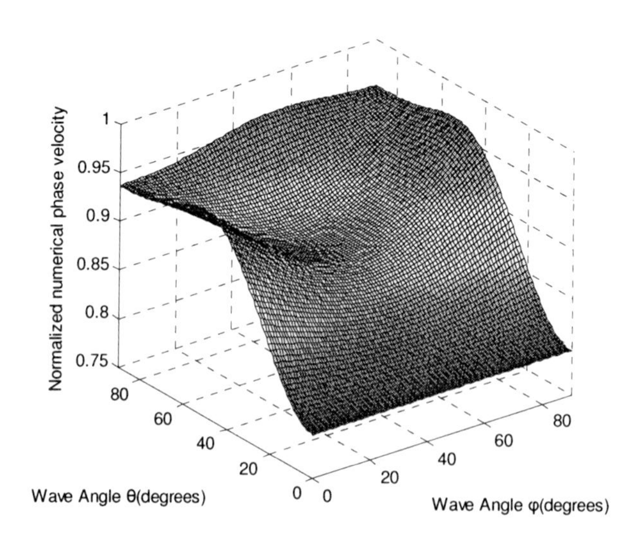

Figure 2: Normalized numerical phase velocity error (%) versus wave propagation angles (φ and θ) for the second-order schemes with CFLN = 3, CPW = 20, Ry = 5, Rz = 10, $\lambda = 1$.

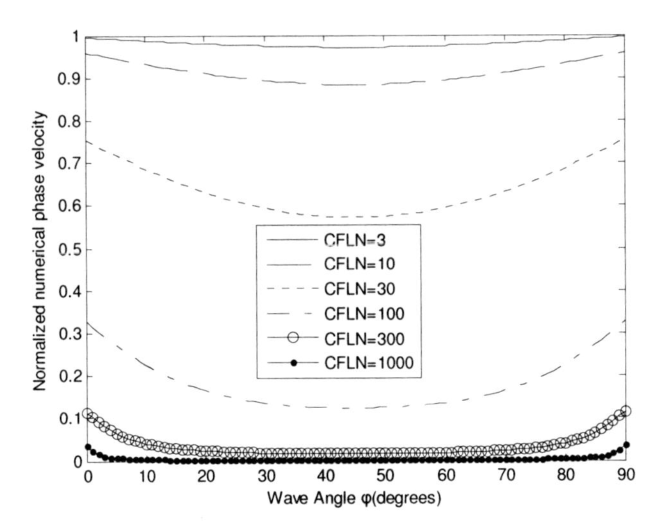

Figure 3: Normalized numerical phase velocity error (%) versus wave propagation angle for the different CFLN with CPW = 40, Ry = 1, Rz = 1, $\lambda = 1$.

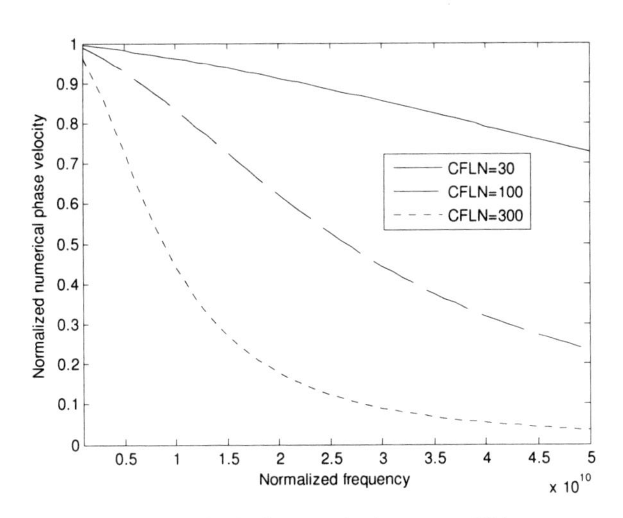

Figure 4: Normalized numerical phase velocity error (%) versus scan frequency for the second-order scheme with CPW = 4 0, Ry = 1, Rz = 1, $\theta = 45°$, $\varphi = 45°$.

(4) *Effect of scan frequency on numerical dispersion*
Figure 3 presents the normalized numerical phase velocity versus scan frequencies for the second-order scheme with CPW = 40 and $\lambda = 1$. Here, $\Delta t = \text{CFLN} \times \lambda / \left(\text{CPW} \times \sqrt{3}c\right)$, the minimum mesh size of $\lambda / 40$ is selected, meanwhile, φ and θ is selected as 45° throughout this section.

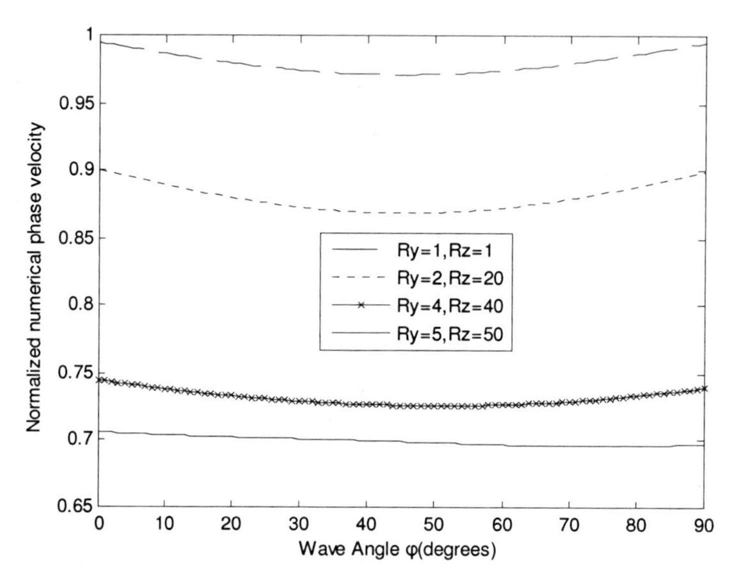

Figure 5: Normalized numerical phase velocity error (%) versus wave propagation angle for the second-order scheme with CFLN = 3, CPW = 40, $\lambda = 1$.

It is clear that the dispersion error can be increased when the frequency is increased. Meanwhile, the dispersion error can be increased when CFLN is increased. Therefore, micro-device tries to choose a narrow side allowed the step length when choosing space grid size. Determine the time step length when to scan the high part of the spectrum as the spatial sampling rate of reference point.

(5) *Effect of grid depth-to-width on numerical dispersion*
Figure 5 shows the normalized numerical phase velocity versus wave propagation angle φ (in degrees) for the second-order scheme with CPW = 40, CFLN = 3 and $\lambda = 1$. Here, θ is selected as 45° throughout this section.

It can be seen that the dispersion error can be increased when the depth-to-width is increased. In order to enhance the computational efficiency, choose the maximum space step in sufficient accuracy.

4 Conclusion

Through the generalized analysis of micro-electro-mechanical-system (RF MEMS) structures, we have proved the spatial deviation, wave propagation angle, CFLN, scan frequency and depth-to-width which affect numerical dispersion.

Analyzing micro-structures of LOD-FDTD method, the parameters are selected by the following principles: (1) High-order LOD-FDTD method has little influence on the precision, generally speaking, fourth-order scheme can satisfy the precision requirement of almost applications. (2) The grid depth-to-width has little effect on numerical dispersion. (3) Full wave analysis has a great impact on

the algorithm of numerical dispersion band, general admission high frequency point as a reference point.

Acknowledgment

This research work was supported by the National Natural Science Foundation of China under Grant No. 61372027.

References

[1] G.M. Rebeiz, *RF MEMS Theory, Design, and Technology*. John Wiley & Sons, New York, USA, 2003.

[2] L. Vietzorreck, EM modeling of RF MEMS. In: *Proceedings of the IEEE 7th International Conference on Thermal, Mechanial and multiphysics Simulation and Experiments in Micro-Electronics and Micro-system, EuroSimE*, IEEE, USA, pp. 1–4, 2006.

[3] Erping Li, Numerical dispersion analysis with an improved LOD-FDTD method. *IEEE Microwave Wireless Components Letters,* **17(5)**, pp. 319–321, 2007.

[4] Weiming Fu, Stability and dispersion analysis for higher order 3-D ADI-FDTD method. *IEEE Transactions on Antennas and Propagation,* **53(11)**, pp. 3691–3696, 2005.

[5] Yong-Dan Kong, High-order split-step unconditionally-stable FDTD methods and numerical analysis. *IEEE Transactions on Antennas and Propagation,* **59(9)**, pp. 3280–3289, 2011.

[6] F. Zheng, Z. Chen, J. Zhang, Toward the development of a three dimensional unconditionally stable finite-different time-domain method. *IEEE Transactions on Microwave Theory and Techniques*, pp. 1550–1558, 2000.

[7] Yu Wenge, Zhong Xianxin, Wu Zhengzhong, et al., Novel stackshorted microstrip bluetooth antenna. *Optics and Precision Engineering*, **11(4)**, pp. 23–27, 2003.

Full-circuit pre-evaluation method for conducted noise based on precise component model

Jiajia Zhang, Shangbin Ye
Department of Electronic and Information Engineering,
Tongji University, Shanghai, China

Abstract

A full-circuit pre-evaluation method for conducted noise is proposed based on converter topology and structure, which are described by component models. In this method, many aspects are taken into consideration including dynamic characteristic of active components, stray parameters of passive components, distributed structure parameters of high-power bus and near-filed coupling in the converter. Thus pre-evaluation of conducted noise is realized. A high-power DC/DC converter used in electric vehicle is chosen as the modeling object, with which the procedure for this pre-evaluation method is explained. Moreover, the influence of the component-level models on the conducted noise of system is also analyzed.

Keywords: conducted noise pre-evaluation, component-level model, DC/DC converter, electric vehicle.

1 Introduction

High-power energy conversion equipment is the crucial part of power system of electric vehicle (EV), which can produce strong electromagnetic interference (EMI). Not only the devices near the car, but also other sensitive electric control unit in the car will be influenced, threatening personal and vehicle safety. Therefore, solution of electromagnetic compatibility (EMC) problem of high-power energy conversion equipment used in EV is an important premise of safety promotion and overall performance breakthrough of EV. In order to provide the scientific basis for solution of EMC problem, it is necessary to predict the characteristic of noise emission at the early stage of converter design [1].

In the power system platform of EV, tougher requirement is put forward for simulation for noise evaluation in the special electromagnetic environment of

WIT Transactions on Engineering Sciences, Vol. 107, © 2015 WIT Press
www.witpress.com, ISSN 1743-3533 (on-line)
doi:10.2495/SEEEI140031

power converter. Considering the sensitive frequency band of on-vehicle receiver, the frequency band of on-vehicle devices conducted EMI is at 150 kHz to 108 MHz, while the frequency band of industrial/scientific/medical devices conducted EMI is at 150 kHz to 30 MHz. In order to evaluate the EMI in on-vehicle environment, the accuracy requirement of device model is higher at high frequency and more stray effect and coupling effect is considered in system level simulation.

A high-power DC–DC converter used in EV is chosen for research in the paper. A full-circuit conducted noise evaluation platform is built based on topology and geometric structure of converter. The dynamic characteristic of active device, stray parameter of passive device, distributed structure parameter of high-power bus and near-field coupling effect in the converter are considered. Conducted noise pre-evaluation of the converter is realized. On the basis, conducted interference measurement platform of on-vehicle devices is built for noise measurement. And feasibility and accuracy of simulation platform for the full-circuit evaluation are verified by contrast of measured result and simulated result.

2 Structure of pre-evaluation platform

Based on the differences of modeling method, the models of system level conducted EMI in converters can be divided as terminal EMI behavior model [2], CM/DM equivalent circuit model [3] and full circuit model [4].

Terminal EMI behavior model regards the device as a whole terminal network which is described by lumped circuits. The circuit for this model is the simplest and relatively high accuracy can be achieved. But internal interference source and noise coupling cannot be analyzed. CM/DM equivalent circuit model is established based on the difference of the noise coupling path in the converter. Unlike the terminal behavior model, the noise sources in this model are the dynamic voltage waveform of switching devices. Internal coupling mechanism can be described. But it is difficult to implement EMI pre-evaluation because the characteristic of noise source is often obtained by measured result of prototype.

Full circuit simulation model is directly established by theoretical circuit of converters. Both nonlinear characteristic of devices and noise coupling effect are taken into consideration. This model is intuitive and describes the internal generation and coupling mechanism, making noise pre-evaluation possible.

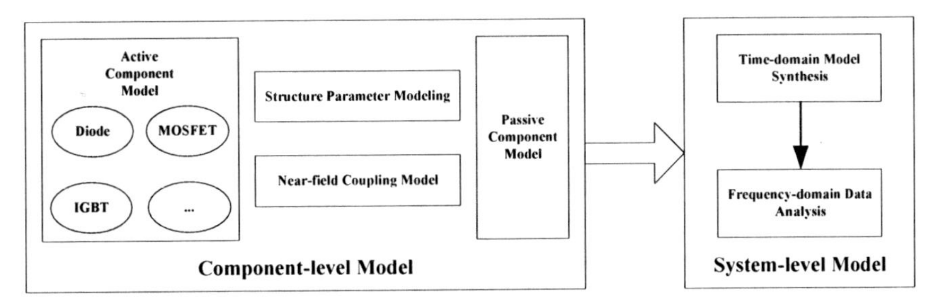

Figure 1: Diagram of full-circuit pre-evaluation platform for conducted EMI.

Full circuit simulation platform established in this paper is divided into two parts: component-level model and system-level simulation, which is shown in Figure 1. Component-level contains not only active component (such as diode and MOSFET) and passive component, but also generalized structure parameter model and near-field coupling model. Active component model is capable of describing its nonlinear dynamic process [5], such as forward and reverse recovery of diode, turn-on/turn-off procedure of MOSFET. Passive component model is usually equivalent impedance model in which noise coupling path is provided by equivalent stray parameters. Structure parameter can be obtained by software simulation, while extraction of structure parameter of high-power on-vehicle bus structure needs additional research. In addition, near-field effect model is more complex in bus structure of converters. System level simulation is established by the time-domain combination of component-level model, which is based on the circuit topology and parameter coupling of converters. After vehicle-level linear impedance stabilization network (LISN) model is added, conducted EMI of terminal can be extracted according to the standard and frequency spectrum can be analyzed.

3 Component-level model

3.1 Diode model

A PIN diode is chosen for this paper. The carrier concentration varies with time and space, which decides dynamic characteristic of PIN diode. Micro-model is more used in EMI pre-evaluation which is expressed in mathematical and circuit form by solvation of semiconductor internal bipolar-diffusion equation. An analytical micro-model based on Laplace-transformation is used in this paper [6]. The diode current i_d can be expressed as sum of carrier injection current in base region i_1 and compound effect current of emitter. Based on AWE principle and high-order truncation, expression of i_1 is equivalent to the same order RC network after Laplace-transformation, which is shown in dashed box in Figure 2. Emitter compound current is expressed by the controlled source between node 2 and ground. Nonlinear static module of diode is expressed by the sub-circuit consisted by node 30, 31 and ground. The voltage drop of diode is displayed by the network between node 10 and 12. The shunt resister between node 3 and ground should be set up as nonlinear device for description of boundary moving effect. The model parameters for PIN MUR8100 diode in the paper are: $T_{au} = 4.08*10^{-6}$ s, $T_o = 7.84*10^{-6}$ s, $I_E = 23.8$ A, $I_{KF} = 0.0577$ A, $N = 1.496$, $I_S = 1.96*10^{-6}$ A, $V_M = 0.453$ V, $R_{lim} = 0.046$ Ω, $R_{epi} = 25.007$ Ω.

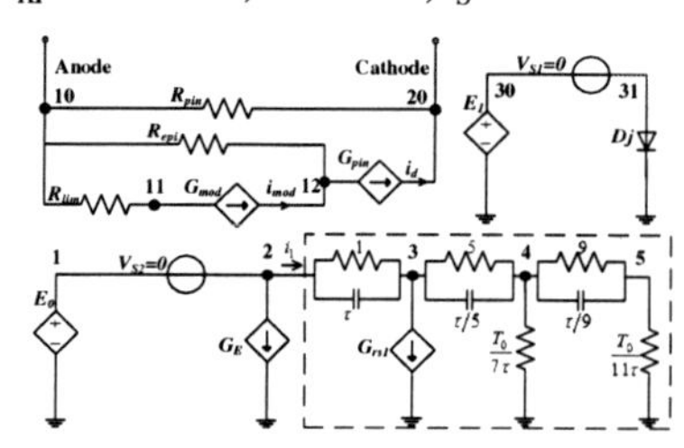

Figure 2: PIN diode model.

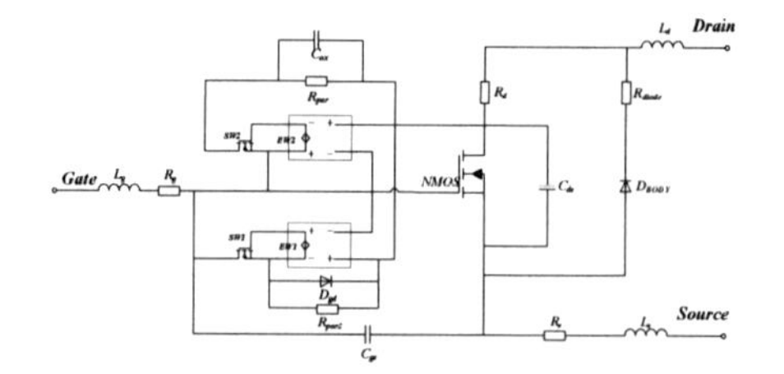

Figure 3: MOSFET sub-circuit model.

3.2 MOSFET model

The high-frequency MOSFET model is needed to describe three nonlinear inter-electrode capacitors C_{gs}, C_{gd} and C_{ds} and reverse recovery of tagma stray diode. Regular MOSFET high-frequency model includes sub-circuit model and lumped charge-based analytical model, and the former one is suitable for circuit simulation due to its faster calculation speed and higher accuracy [7].

The sub-circuit consists of core NMOS and peripheral circuit shown in Figure 3. The NMOS core can describe basic static characteristic. D_{BODY} and R_{diode} are tagma diodes. SPW16N50C3 from Infineon COOLMOS series is chosen and its main parameters are valued by information in the datasheet [8]. The parameters of D_{BODY} are extracted by characteristic of reverse diode, and results are: $I_S = 0.2$ pA, $T_M = 520$ ns, $N = 1$. Series resistance of the diode is estimated as 6 mΩ. C_{gs} and C_{ds} are calculated by input capacitance, output capacitance and transmission capacitance measurement, results are C_{gs} = 1474 pF and C_{ds} = 806 pF. C_{gd} is described by sub-circuit model controlled by switching of ideal MOSFET SW1, SW2, which contributes to the miller effect. The oxide layer capacitance is estimated to be 3.5 nF. Other inductance and resistance of lead wire are also estimated: $L_g = 8$ nH, $R_g = 1.5\ \Omega$, $L_d = 3$ nH, $L_s = 10$ nH, $R_s = 2$ mΩ.

3.3 Passive component model

Impedance characteristic of passive components are measured by Agilent 4395A at 150 kHz to 108 MHz and are described by proper lump parameter model.

The impedance model of output capacitor C_{out} is equivalent to an RLC series model, whose parameters are $C_{s1} = 9.312$ μF, $L_{s1} = 26.974$ nH, $R_{s1} = 66.869$ mΩ. In Figure 4, the impedance measured result is shown in black solid line, model fitting result is shown in red dashed line, and the two results are in good agreement.

The impedance model for energy storage inductor L_{main} is relatively more difficult. The conventional inductor is modeled by an RLC parallel model, which only has one resonance peak of impedance (blue dashed line in Figure 5), and does not agree with the measured result with multiple resonant peak (black solid line in Figure 5) at high frequency, especially at frequency upper than 30 MHz which is crucial in the vehicle environment. Thus, more RLC parallel model series need to be combined. In this paper, four series of RLC parallel model are combined with the parameters of: $L_{p1} = 0.032$ mH, $C_{p1} = 41.2$ pF, $R_{p1} = 3.6$ kΩ, $L_{p2} = 750$ nH, $C_{p2} = 395$ pF, $R_{p2} = 360\ \Omega$, $L_{p3} = 48$ nH, $C_{p3} = 61.5$ pF, $R_{p3} = 95\ \Omega$, $L_{p4} = 43.5$ nH, $C_{p4} = 10$ pF, $R_{p4} = 800\ \Omega$. The impedance of this new model (red dashed line in Figure 5) is in better agreement with the measured results at the whole frequency band.

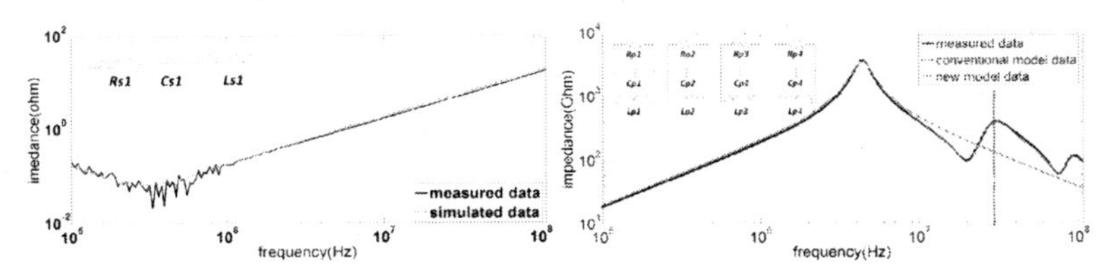

Figure 4: Impedance of C_{out}.

Figure 5: Impedance of L_{main}.

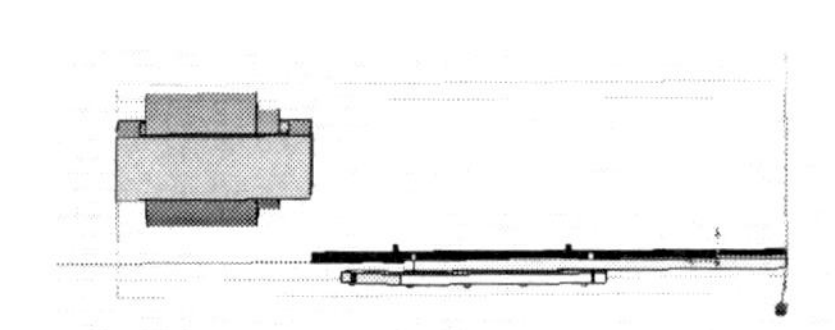

Figure 6: Side view of 3D model of bus.

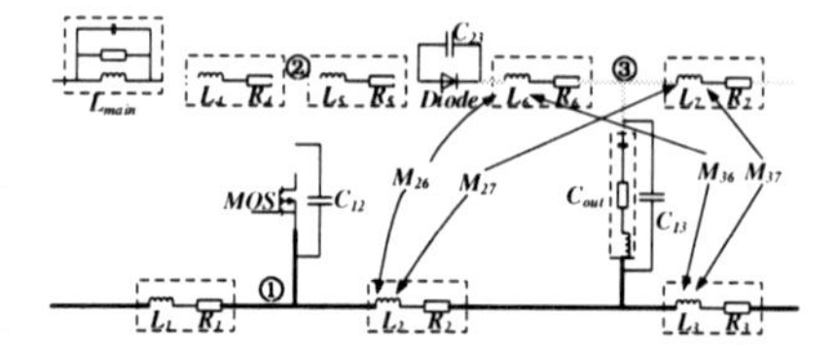

Figure 7: Diagram of structure and coupling parameter.

3.4 Structure parameter and near-field coupling model

The 3D model of a BOOST converter in Maxwell FEM software is shown in Figure 6. There are three buses in this converter. No. 1 bus (black) is the ground bus for both input and output terminal of the converter. No. 2 bus (yellow) connects the main inductor, drain pole of MOSFET and anode node of diode. No. 3 bus (blue) connects cathode node of diode to the output capacitor.

The distributed structure parameters are equivalent AC resistance and self-inductance of the bus denoted as R_1~R_7 and L_1~L_7. They may exist at different locations in the circuit as shown in Figure 7. As the diode and MOSFET are located very close to the main inductor, L_1, R_1, L_4, R_4, L_5, R_5 are small enough to be neglected. R_3 and R_7 can also be neglected as they are in series of R_{load}. The equivalent AC resistance is calculated by empirical formula and self-inductance of bus is extracted by Maxwell. The results are $R_2 = 0.249$ mΩ, $R_6 = 0.228$ mΩ, $L_2 = 27.038$ nH, $L_3 = 18.823$ nH, $L_6 = 28.404$ nH, $L_7 = 18.390$ nH.

The near-field couplings are divided into electric field coupling and magnetic field coupling, which are described by coupling capacitance of different nodes and mutual inductance of different loops, respectively. The specific kind and extraction method of near-field coupling effect is shown as follows:

(i) *The coupling capacitance between different buses.* Each two of the three buses create a coupling capacitance, denoted as C_{12}, C_{13} and C_{23}. Their location in the equivalent circuit is also shown in Figure 7. The parameters extracted by Maxwell are: $C_{12} = 1.014$ pF, $C_{23} = 1.022$ pF, $C_{13} = 0.871$ pF.

(ii) *The coupling capacitance between the active component and heat sink.* The equivalent capacitances are denoted as C_{mh} between drain pole of MOSFET and ground and C_{dh} between anode of diode and ground. There is a layer of heat dissipation pad between the active component and heat sink with heat conduct glue uniformed smeared at both sides. The stray capacitance is difficult to calculate and is measured directly: $C_{mh} = 10.19$ pF, $C_{dh} = 8.08$ pF.

(iii) *The mutual inductor of different bus made by magnetic coupling of bus.* The mutual parameter should be divided into different groups between different self-inductance, which is shown in Figure 7. The results from simulation of Maxwell are $M_{26} = -0.909$ nH, $M_{27} = 9.941$ pH, $M_{36} = 9.2967$ pH,

$M_{37} = -1.010$ nH. M_{27} and M_{36} can be neglected as they are much smaller than self-inductance of bus.

(iv) *Near-field magnetic coupling between main inductor and bus.* These parameters can be neglected as the simulation results are in large difference in order of magnitude with L_{main} and self-inductance of bus.

4 System-level simulation and analysis

As shown in Figure 8, the system-level simulation platform is the combination of component-level models which are explained in Section 3. A 1 GΩ resistance is added in parallel of the near-field capacitors except for C_{13}, which guarantees the existence of DC path in the circuit and has no influence on the result. For the CE simulation of vehicle-level, a pair of vehicle LISN is added at the power terminal.

The input voltage for the boost converter is 12 V, the load is 10 Ω, switching frequency is 100 kHz, duty cycle is 0.4, driving resistance is 10 Ω and driving voltage is $V_{pulse} = 15$ V. The voltage on the sampling resistor (V_{lp} and V_{ln}) in LISN is the noise of positive/negative line in the time domain, and the spectrum of noise by FFT is shown in Figure 9. The peaks of the conducted noise occur at the frequency-doubling of the switching frequency.

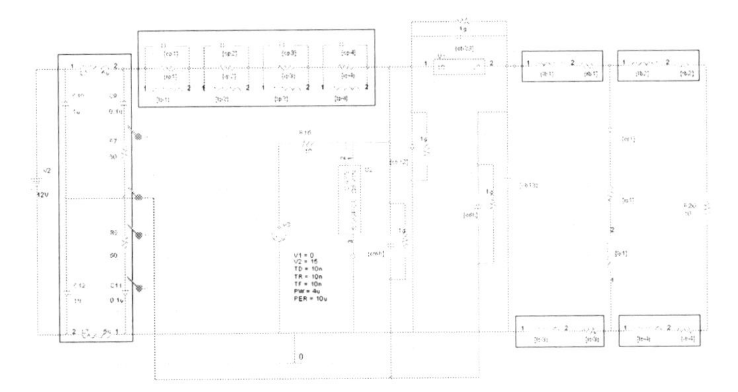

Figure 8: Diagram of full system simulation platform.

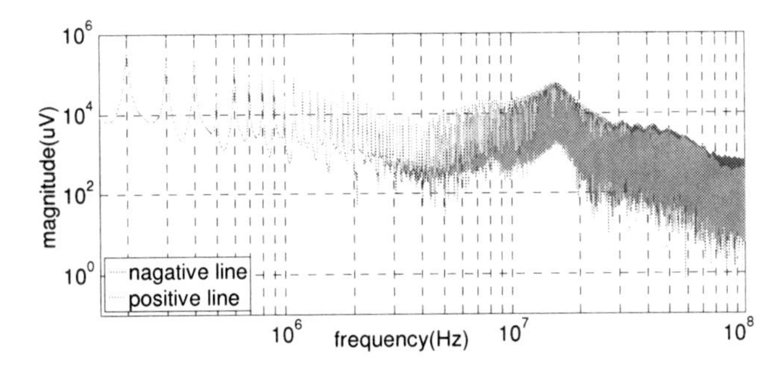

Figure 9: Pre-evaluation result of power terminal noise spectrum.

Noise of positive/negative line is converted to DM/CM noise displayed as $(V_{lp}+V_{ln})/2$ and $(V_{lp}+V_{ln})/2$. Effects of component-level model on the system noise are analyzed by contrast of different simulation situations. And the five simulation situations are:

(i) Only nonlinear effects of active components is considered.
(ii) High-frequency characteristic of active and passive components are considered, while bus structure parameters and near-field coupling are neglected.
(iii) All nonlinear parameters except near-field coupling are considered.
(iv) All parameters except C_{mh} and C_{dh} are considered.
(v) Complete system-level simulation model.

The DM noise spectrum of situations 1, 2 and 3 are shown in Figure 10. There is no significant change at low frequency, because noise characteristic in this range is mainly affected by circuit topology and working condition. After the stray effect of passive component is considered, DM noise increases in large scale in the middle and high frequency. The increase at high frequency above 10 MHz is more than 30 dB. This is because the high-frequency impedance of inductor decreased significantly in the accurate model. It is found that there is some increase above 5 MHz after bus structure parameters are added and there exists a peak of envelope between 10 and 20 MHz. It can be concluded that bus structure parameters have a great influence on the coupling path of DM noise. As a result, accurate models of passive components and bus structure parameters are needed in conducted EMI pre-evaluation.

The contrast of DM noise spectrum from situations 3 to 5 is shown in Figure 11. The results show that coupling parameter has little effect on noise amplitude. There are two reasons for this phenomenon: the coupling parameter is about 1 nH, which is relatively small in this example and less than 7% of self-inductance of bus. On the other hand, C_{mh} and C_{dh} are in a typical CM noise loop which would not affect the DM noise in theoretical analysis.

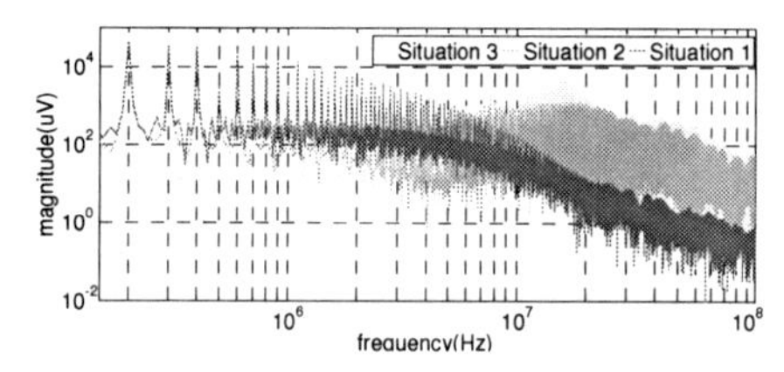

Figure 10: DM spectrum contrast between situations 1, 2 and 3.

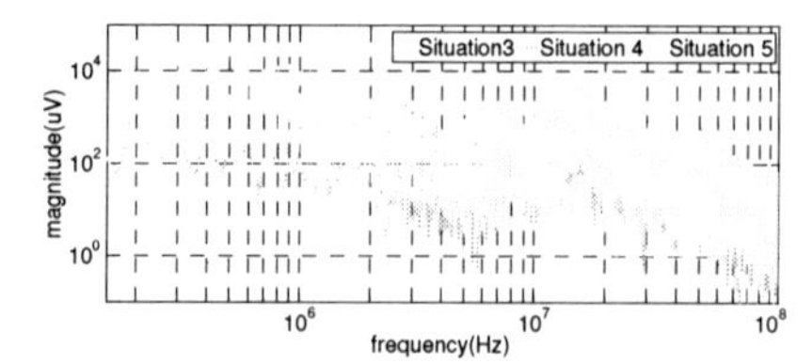

Figure 11: DM spectrum contrast between situations 3, 4 and 5.

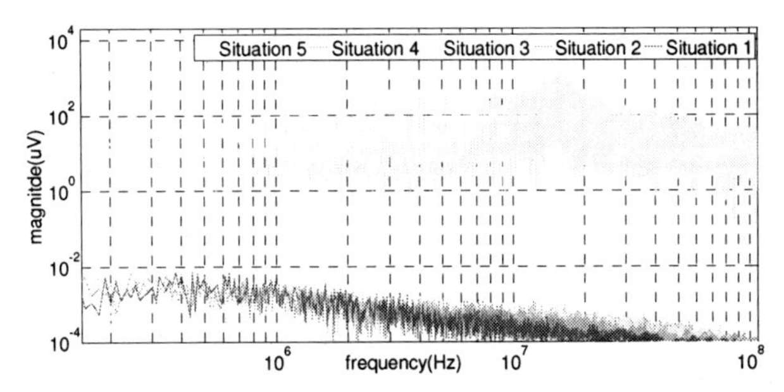

Figure 12: CM spectrum contrast of five situations.

The changes of CM noise shown in Figure 12 are quite different. There is no significant difference between spectrum of CM noise from situations 1 to 4, and the amplitude is very low. This is because there is no effective CM noise loop in these situations. Changes of passive component model and bus model are only effective to equivalent DM noise loop. C_{mh} and C_{dh} are added in situation 5, even though the value is lower (pF level), these two capacitors offer effective loop for emission of CM noise, thus there is significant difference for CM noise.

5 Conclusions

A full-circuit pre-evaluation method for conducted noise is proposed in this paper, which is divided into component-level model and system-level simulation. Dynamic characteristic of active components, stray parameters of passive components, distributed structure parameters of high-power bus and near-filed coupling effect inside the converter are taken into consideration by precise component-level modeling. The time-domain integrated platform for system-level simulation is built based on the real circuit topology and coupling relations of parameters. A pair of vehicle-level LISN is added for extraction of terminal conducted noise regulated by standard.

A DC/DC converter for electric vehicle is used for accurate modeling of the component-level models and establishment of system-level pre-evaluation platform. Based on the platform, influence of component-level model on the DM/CM noise is analyzed. The results show that high-frequency characteristic of main inductor has a significant effect on DM noise, which results in an increase of more than 30 dB at frequency above 10 MHz. Structure parameter has an influence on the coupling path of DM noise as well, which makes a local envelope peak for DM noise between 10 and 20 MHz. The effect of near-field coupling between buses on DM noise is not significant. The change of CM noise is quite different, passive component model and bus model are only effective to equivalent DM noise path instead of CM noise. However, the capacitance ceased by near-field coupling between component and heat-sink offers effective loop for CM noise which makes CM noise much higher, though the value of capacitance may be small.

By changing the component model and topology of circuit, this method can be used for pre-evaluation of conducted noise of other devices (such as DC/AC inverter) as well.

Acknowledgment

This research work was supported by the National High-tech Research and Development Program (863 Program) under Grant No. 2011AA11A265.

References

[1] D. Boroyevich, X. Zhang, H. Bishinoi, et al., Conducted EMI and systems integration. In: *Proceedings of the 8th International Conference on Integrated Power Systems*, Nuremberg, Germany, 2014.
[2] H. Bishnoi, P. Mattavelli, R. Burgo, et al., EMI behavioral models of DC-fed three-phase motor drive systems. *IEEE Transactions on Power Electronics*, **29(9)**, pp. 4633–4645, 2014.
[3] M. Moreau, N. Idir, P.L. Moigne, Modeling of conducted EMI in adjustable speed drives, *IEEE Transactions on Electromagnetic Compatibility*, **51(3)**, pp. 665–672, 2009.
[4] X. Gong, I. Josifovic, J.A. Ferreira, Modeling and reduction of conducted EMI of inverters with SiC JFETs on insulated metal substrate. *IEEE Transactions on Power Electronics*, **28(7)**, pp. 3138–3146, 2013.
[5] Zhang Yi-Cheng, Han Xin-chun, Shen Yu-zuo, et al. Electromagnetic interference and suppression of DC/DC converter used in electric vehicle. *Journal of Tongji University (natural science)*, **33(1)**, pp. 108–111, 2005.
[6] A.G. Strollo, A new spice model of power PIN diode based on asymptotic waveform evaluation. *IEEE Transactions on Power Electronics*, **12(1)**, pp. 12–20, 1997.
[7] Sun Kai, Lu Jue-jing, et al., Modeling of SIC MOSFET with temperature dependent parameters. *Proceedings of CSEE*, **33(3)**, pp. 37–43, 2013.
[8] Zhao Ya-xing, *PSPICE and Electric Component Model*. Press of Beijing University of Post and Telecommunications, Beijing, 2003.

Research on induction current of bridge wire of industrial electric detonator using FDTD arithmetic

Du Bin[1], Liu Yahui[1], Yao Hong-zhi[2], Ji Xiangfei[2]
[1]*Department of Chemical Engineering,*
Anhui University of Science and Technology, Huainan, Anhui, China
[2]*The 213 Research Institute of China North Industries Group Corporation, Xi'an, China*

Abstract

This paper introduces the principles and characteristics of the electromagnetic FDTD method. It has, based on the orthogonal FDTD arithmetic, deduced the non-orthogonal algorithm in arbitrary medium, and calculated the induced current of industrial electric detonator in continuous electromagnetic radiation. The results show that the industrial electric detonator could be probably ignited by induction current flowing through the bridge wire under continuous electromagnetic radiation.

Keywords: orthogonal FDTD arithmetic, non-orthogonal FDTD arithmetic, stability conditions, industrial electric detonator, induction current, electromagnetic radiation.

1 Introduction

First put forward by Yee in 1996, FDTD arithmetic is a numerical computation method based on Maxwell equation for solving the problem of electromagnetic field; it directly converts two curl equations in Maxwell equation to the difference form, discretizes the electromagnetic field in space and time, and gets the iterative equations, so as to realize the data sampling and compressing of the continuous electromagnetic field within a certain volume and a certain time. At present, FDTD arithmetic has been widely applied to the electromagnetic scattering, electromagnetic compatibility, microwave circuit, analysis of time domain of light path, and biological electromagnetic metrology.

WIT Transactions on Engineering Sciences, Vol. 107, © 2015 WIT Press
www.witpress.com, ISSN 1743-3533 (on-line)
doi:10.2495/SEEEI140041

FDTD arithmetic stars from Maxwell equation relying on the time variable, and converts the differential equation satisfied by constants to the difference equation by utilizing the centered difference scheme with the two-order precision in the cyber space alternatively arranged by electric fields and magnetic fields. FDTD arithmetic disposes any problem in terms of the initial-value problem, calculates in terms of the time step, and calculates the electric field and the magnetic field of each discrete point in the alternation of each time step. Though FDTD arithmetic is the method of calculation of time domain in nature, it also can be directly used for calculation of steady electromagnetic field.

Viewed from the above principle, the application of FDTD arithmetic first needs to discretize the studied computational domain to form a gridding body. In order to reach the enough computational accuracy, a single grid needs to be very fine, and the total of grids in the computational domain is often big. On each grid cell, variables related are six in all, respectively, the electric-field strength and magnetic induction in the three-dimensional coordinate, and the medium parameter related with the spatial location includes the electric conductivity, magnetic conductivity, dielectric constant, magnetic loss, and so on with the number related with the medium attribute.

With an increasingly expanded application of FDTD arithmetic, various types of unconventional FDTD arithmetics are gradually developed. Wherein, the arithmetic in different coordinate systems and the arithmetic by adopting different grid sizes are included. When the studied object or medium has the curved surface (circular wave guide, and surface of aircraft, etc.), the "ladder" boundary is formed if the traditional rectangular grid cell is used to fit the surface, and the use of the "ladder" boundary will not only stimulate the transmission of surface wave and arouse the additional numerical dispersion, but also need reducing the size of grid cell and correspondingly reducing the time step, and thereby greatly increasing the computational storage content and lengthening the computation time, so as to fit the medium interface in small curvature radius or the multi-layer medium interface. For the purpose of overcoming the problem aroused by the "ladder" surface, the better method is to use the common curvilinear coordinate system, making the coordinate line fit the geometrical surface of the medium. Thus, Holland puts forward FDTD arithmetic in the common curvilinear coordinate system [1], and Lee improves the discrete iterative equations [2–6] to be more intuitional and easier to program on the basis of Holland' research. Whereas, discrete iterative equations given by them all direct at the single medium. This paper deduces the non-orthogonal FDTD arithmetic suitable for arbitrary medium and tries to apply the arithmetic to the calculation of the induction current of bridge wire of industrial electric detonator on the basis of reference of predecessors' related literatures.

2 Vector relation in common curvilinear coordinate system

With expanded application of FDTD arithmetic, various types of FDTD arithmetics are gradually developed, and in the common three-dimensional non-orthogonal coordinate system (u^1, u^2, u^3), the base vector A_i can be used to define any differential vector dr (see Figure 1)

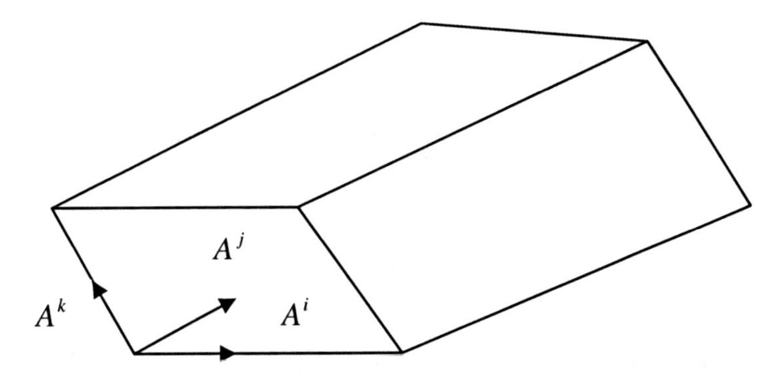

Figure 1: Base vector and dual base vectors in non-orthogonal coordinate system.

$$\mathrm{d}r = \sum_{i-1}^{3} \frac{\partial \vec{r}}{\partial r^i} \mathrm{d}u^i = \sum_{i-1}^{3} A_i \mathrm{d}u^i \tag{1}$$

Viewed from Figure 1, the base vector A_i (i = 1,2,3) defines the base vector of the unit grid related with the point A, and the three base vectors define a parallelepiped, of which the volume is

$$V = A_1(A_2 \times A_3) = A_2(A_1 \times A_3) = A_3(A_2 \times A_1) \tag{2}$$

In accordance with the definition of base vector in Figure 1, the corresponding dual base vectors are as follows:

$$\begin{aligned} A^1 &= \frac{A_2 \times A_3}{\sqrt{g}} \\ A^2 &= \frac{A_3 \times A_1}{\sqrt{g}} \\ A^3 &= \frac{A_1 \times A_2}{\sqrt{g}} \end{aligned} \tag{3}$$

wherein $\sqrt{g}$ is defined as the volume of the unit body, and the vector direction of the base vector A_i is defined as the normal direction of the plane determined by vectors A_j and A_k (i, j and k cyclically represent). Therefore, the base vector A_i and the dual base vectors A_i meet the following mutual relation:

$$A^i \cdot A_j = \delta_{ij} \tag{4}$$

wherein δ_{ij} is defined as the Kronecker function. Therefore, any vector can be represented by the two base vectors, for example, the electric field $\overline{E}$ can be represented as the following equations by the two base vectors:

$$\vec{E} = \sum_i E^i A_i \qquad \dot{E} = \sum_i A^i E_i \tag{5}$$

In the equations, E^1 and E_i are, respectively, defined as the contravariant component and the covariant component of the electric field $\dot{E}$. From their physical significances, we can know the covariant component E_i represents the "flow" of the electric field E along the coordinate line **i**, and from the mutual relation, we can also get the relation,

$$\vec{E} \cdot A_i = \left(\sum_i E^j A_j\right) \cdot A_i = E_i \tag{6}$$

From Eq. (6), we can get that the covariant component E_i represents the "flow" of the electric-field component $\dot{E}$ along the boundary **I**. In short, the following equation can be obtained by utilizing the mutual relation in Eq. (4) if viewed from the dot product A^i in Eq. (5):

$$\vec{E} \cdot A_i = \left(\sum_j E^j A_j\right) \cdot A^i = E^i \tag{7}$$

From Eqs. (3) and (7), the contravariant component E^i represents the quotient of total electric field passing through the surface I and the unit volume $\sqrt{g}$.

In order to expand FDTD arithmetic to the non-orthogonal grid, it is necessary to approximately represent Maxwell equation as the form of contravariant component and covariant component. As for arbitrary medium, the integral form of Maxwell equation can be represented as the following equations:

$$\oint \left(s\vec{H} + \mu \frac{\partial \vec{H}}{\partial t} \right) \mathrm{d}s = \oint \overline{E} \mathrm{d}l \tag{8}$$

$$\oint \left(\sigma \vec{E} + \varepsilon \frac{\partial \vec{E}}{\partial t} \right) \mathrm{d}s = \oint \overline{H} \mathrm{d}l \tag{9}$$

The left-hand side of Eq. (8) shows the rate of change of the electric field and the magnetic field passing through the surface Ω, while the right-hand side of

Eq. (8) shows the "circulation" of the electric field and the magnetic field on the corresponding boundary $\partial\Omega$, so the finite difference scheme of Faraday's law can be deduced from the above two equations and shown as follows:

$$\begin{aligned} &E^1(i,j,k)^{n+1} \\ &= C_1(i,j,k)\Big\{-D_1(i,j,k)E_1(i,j,k)n + \frac{1}{\sqrt{g}}\Big[H_3(i,j+\tfrac{1}{2},k) - H_3(i,j-\tfrac{1}{2},k) \\ &\quad - H_2(i,j,k+\tfrac{1}{2}) + H_2(i,j,k-\tfrac{1}{2})\Big]^{n+\frac{1}{2}}\Big\} \end{aligned} \tag{10}$$

Other two contravariant components E_2 and E_3 can be obtained by the mutual substitution, and likewise, we can get the following finite difference scheme of Ampere's theorem:

$$\begin{aligned} &H^1(i,j,k)^{n+\frac{1}{2}} \\ &= C_1'(i,j,k)\Big\{-D_1'(i,j,k)H^1(i,j,k)^{n-\frac{1}{2}} - \frac{1}{\sqrt{g}}\Big[E_3(i,j+\tfrac{1}{2},k) - E_3(i,j-\tfrac{1}{2},k) \\ &\quad - E_2(i,j,k+\tfrac{1}{2}) + E_2(i,j,k-\tfrac{1}{2})\Big]^{n}\Big\} \end{aligned} \tag{11}$$

wherein

$$\begin{aligned} C_\eta(i,j,k) &= \left\{\sqrt{\frac{\mu_0}{\varepsilon_0}}\left[\frac{\sigma_\eta(i,j,k)}{2}\right] + \frac{\varepsilon_\eta(i,j,k)}{\Delta t}\right\}^{-1} \\ D_\eta(i,j,k) &= \sqrt{\frac{\mu_0}{\varepsilon_0}}\left[\frac{\sigma_\eta(i,j,k)}{2}\right] + \frac{\varepsilon_\eta(i,j,k)}{\Delta t} \\ C_\eta^1(i,j,k) &= \left\{\sqrt{\frac{\sigma_0}{\mu_0}}\left[\frac{s_\eta(i,j,k)}{2}\right] + \frac{\mu_\eta(i,j,k)}{\Delta t}\right\}^{-1} \\ D_\eta(i,j,k) &= \sqrt{\frac{\varepsilon_0}{\mu_0}}\left[\frac{s_\eta(i,j,k)}{2}\right] + \frac{\varepsilon_\eta(i,j,k)}{\Delta t} \end{aligned} \tag{12}$$

The subscript η represents the non-orthogonal coordinate system i, j and k.

Likewise, we can get expressions of H^2 and H^3 through the superscript substitution.

It is noteworthy that the contravariant component and the covariant component of electric field and magnetic field are used in the paper, and the notable advantage of the method is that this difference scheme is quite similar to the difference scheme in the sense[7-9].

Finally, the covariant component g_{ij} can be calculated by using contravariant components of electric fields $\overline{E}$ and $\overline{H}$,

$$A_i = \sum_j g_{ij} A^j, \quad g_{ij} A_i \cdot A_j \tag{13}$$

Thus, the covariant component H_i and the contravariant component of the magnetic field can meet the following relation:

$$H_i = \sum_j g_{ij} H^j \tag{14}$$

Through the average value of adjacent four points, we can get the approximate finite difference scheme of Eq. (14), the contravariant component can be converted to the covariant component, and H_1 can be represented as:

$$\begin{aligned} H_1 &= g_{11} H^1(i, j, k) \\ &+ g^{12}/_4 \left[H^2(i + 1/2, j - 1/2, k) + H^2(i - 1/2, j - 1/2, k) \right. \\ &\left. + H^2(i + 1/2, j + 1/2, k) + H^2(i - 1/2, j + 1/2, k) \right] \\ &+ g^{13}/_4 \left[H^3(i + 1/2, j, k - 1/2) + H^3(i - 1/2, j, k - 1/2) \right. \\ &\left. + H^3(i + 1/2, j, k + 1/2) + H^3(i - 1/2, j, k + 1/2) \right] \end{aligned} \tag{15}$$

E_1 can be represented as:

$$\begin{aligned} E_1 &= g_{11} E^1(i, j, k) \\ &+ g^{12}/_4 \left[E^2(i + 1/2, j - 1/2, k) + E^2(i - 1/2, j - 1/2, k) \right. \\ &\left. + E^2(i + 1/2, j + 1/2, k) + E^2(i - 1/2, j + 1/2, k) \right] \\ &+ g^{13}/_4 \left[E^3(i + 1/2, j, k - 1/2) + E^3(i - 1/2, j, k - 1/2) \right. \\ &\left. + E^3(i + 1/2, j, k + 1/2) + E^3(i - 1/2, j, k + 1/2) \right] \end{aligned} \tag{16}$$

Other covariant components E_2 and E_3, H_2 and E_3, as well as H_2 and H_3 of the electric field $\overline{E}$ and the magnetic field $\overline{H}$ can be obtained through the same method.

3 Stability conditions of non-orthogonal FDTD arithmetic

The deduction of stability conditions of the FDTD formula of non-orthogonal curve grids is basically similar to that under Cartesian grids and it is a more

general form to be used to come to a conclusion under special circumstances conveniently[10][11]. At first, we assume that it is the three-dimensional space filled with the homogeneous medium, the field in the passive area shall comply with the wave equation as follows:

$$\nabla \times \nabla \times \overline{E} \frac{-1}{c^2} \frac{\partial^2 E}{\partial t^2} \tag{17}$$

wherein c is the speed of light in the homogeneous medium. If we apply it to Lorenz conditions, we can get:

$$\nabla^2 \overline{E} = \frac{1}{c^2} \frac{\partial^2 E}{\partial t^2} \tag{18}$$

The electric field conforming to the equation as above can be expanded to be the vector sum of the eigenmode or the composition of a series of plane waves. To guarantee the stability of the iterative formula of FDTD, every component in the field will be stable. At the same time, if any plane wave is stable, the whole field will be stable. Then, we will show the electric field in the form of spectral domain of the plane wave,

$$\overline{E}(u^1, u^2, u^3, t) = \overline{e}(t) \cdot \exp(-j(k_1 u^1 + k_2 u^2 + k_3 u^3)) \tag{19}$$

wherein

$$k_i = \dot{k} \cdot \dot{A}_t \tag{20}$$

In the non-orthogonal coordinate system,

$$\nabla = \dot{A}_1 \frac{\partial}{\partial u^1} + \dot{A}_2 \frac{\partial}{\partial u^3} + \dot{A}_3 \frac{\partial}{\partial u^3} \tag{21}$$

Putting Eq. (5.19) into the above equation, and then applying the centered difference, we can get,

$$\nabla \overline{E} = 2je(t) \cdot \exp\left\{ -j\overline{k} \cdot \overline{r} \left[\sum_1^3 \frac{\overline{A1}}{\Delta u^t} \sin\left(\frac{\Delta [k_j u^t]}{2} \right) \right] \right\} = 2j \left[\sum_1^3 \overline{A}^1 \sin\left(\frac{\Delta [k_1 u^t]}{2} \right) \right] \overline{E} \tag{22}$$

In the above equation, $u^i = 1$ is adopted. Putting the above equation into Eq. (18) and conducting the centered difference on the time item, we can get,

$$
\begin{aligned}
\nabla^2 \dot{E} &= (\nabla\cdot\nabla)\dot{E} - 4\left[\sum_i^3 \dot{A}^j \sin\left(\frac{\nabla[k_i u_i]}{2}\right)\right]\cdot\left[\sum_j^3 \dot{A}^j \sin\left(\frac{\nabla[k_i u_i]}{2}\right)\right]\cdot \dot{E}^n \\
&= \frac{1}{c^2}\frac{\overline{E}^{n+1} - 2\overline{E}^{n} + \overline{E}^{n-1}}{\Delta t^2} \\
&= \frac{1}{c^2}\left(\frac{\lambda^2 - 2\lambda + 1}{\lambda(\gamma\Delta t^2)}\right)\cdot \dot{E}^n
\end{aligned} \tag{23}
$$

In the above equation, $\overline{E}^{n+1} = \lambda\overline{E}^{n}$ is adopted and then λ is obtained,

$$
\lambda = \left[1 - 2s^2(\Delta t)^2\right] \pm 2s(\Delta t)\sqrt{s^2(\Delta t)^2 - 1} \tag{24}
$$

wherein:

$$
s^2 = c^2\left[\sum_i^3 \dot{A}^i \sin\left(\frac{\Delta[k_i u^i]}{2}\right)\right]\cdot\left[\sum_i^3 \dot{A}^j \sin\left(\frac{\Delta[k_j ui]}{2}\right)\right] \tag{25}
$$

In order to guarantee the stability of the iterative algorithm, $|\lambda| \le 1$ must be satisfied, which means λ must be in the unit circle or circle on the complex plane. By analysis, we can get the answer of the inequality:

$$
s^2(\Delta t)^2 \le 1 \tag{26}
$$

Expand s^2 in the inequality (26)

$$
\begin{aligned}
s^2 &= c^2\sum_i^3\left\{\sum_j^3 \overline{A}^i\cdot\overline{A}^j \sin\left(\frac{\Delta[k_i u^i]}{2}\right)\sin\left(\frac{\Delta[k_j u^j]}{2}\right)\right\} \\
&= c^2\sum_i^3\left\{\sum_j^3 g^{i\cdot j} \sin\left(\frac{\Delta[k_i u^i]}{2}\right)\sin\left(\frac{\Delta[k_j u^i]}{2}\right)\right\} \\
&\le c^2\sum_i^3\sum_j^3 g^{i\cdot j}
\end{aligned} \tag{27}
$$

From Eqs. (37) and (38), we can get that the conditions for long time stability of the FDTD formula of non-orthogonal grids:

$$
\Delta t \le \frac{1}{c\sqrt{\sum_i^3\sum_j^3 g^{i\cdot j}}} \tag{28}
$$

The stability condition in the homogeneous medium is in inequality (28). As for arbitrary medium, only the value of c can be selected to be the maximum value in the medium. Seen from the above inequality, the time step length to meet the requirement of the stability condition is related to the length of grids and the basic included angle.

4 Computational analysis of the radiation current of bridge wire of the industrial electric detonator

Due to the geometric structure and the complexity of the selected materials of the industrial electric detonator, there are many factors influencing the electromagnetic compatibility. It is very hard to conduct the overall modeling analysis on the industrial electric detonator with the current calculation method and the computer level. So, some simplified treatments have to be conducted when analyzing the electromagnetic compatibility of the industrial electric detonator. Based on the simplification, safety criterions are proposed to make it possible for the analysis and calculations.

The ideal simplified model of the industrial electric detonator is shown in Figure 2. The shield and attenuation effects of the case of the industrial electric detonator are ignored in the analysis. Only the worst electromagnetic radiation is considered in such treatment.

Generally, the resistance of the bridge wire is about 2Ω. The resistance of the bridge wire can be very small comparing with the resistance of the lead, so it can be ignored. The lead of the bridge wire is 10 cm with the diameter of 1 mm.

Under the exposure of the outer-electromagnetic field, the industrial electric detonator can produce the electromagnetic scattering as well as induction current in the bridge wire. In order to guarantee that the calculation of values is conducted in the finite region, the corresponding absorption boundary conditions must be adopted on the truncated boundary. PML absorption boundary condition is adopted in this calculation.

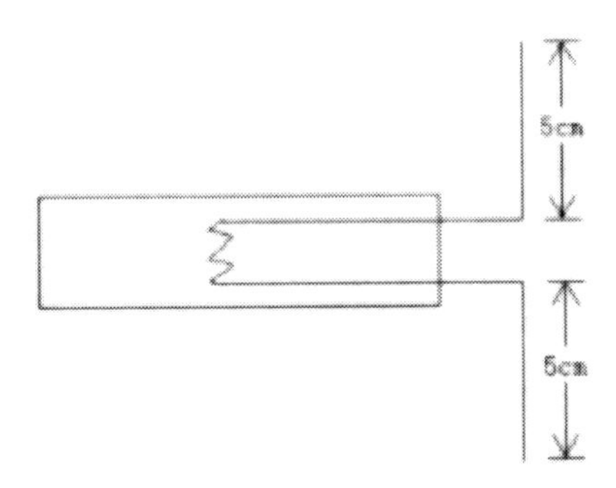

Figure 2: Simplified model of the industrial electric detonator of bridge wire.

The sine electromagnetic wave of the high power microwave under the gauss impulse modulation undergoes approximate treatment, and the electric field of the incident wave is

$$E^{inc} = E_0 \sin(2\pi f_0 t)\exp\left[\frac{-(t-t_0)^2}{T_0^2}\right] \tag{29}$$

wherein the direction of $E_0 = 1.0\times10^5 V/m$, $f_0 = 6GHz, t_0 2ns$, $T_0 = 0.6ns, E^{inc}(t)$ is parallel with the axis of the bridge wire.

The large-equivalent nuclear explosion in high altitude can produce very strong electromagnetic impulse. When the electromagnetic impulse spreading to the ground is simplified to one plane wave, the normalized nuclear electromagnetic impulse in high altitude can be approximately described by a double-index single impulse,

$$E(t) = 5.25\times10^4\left[\exp\left(-4\times10^6 t\right) - \exp\left(-4.67\times10^8 t\right)\right] \tag{30}$$

where $E(t)$ is the intensity of the electric field (v/m) and t is the time (s).

The induction current of the bridge wire can be obtained by Eq. (31). The integral path can be the circle around the bridge wire,

$$I = \oint_c H\bullet \mathrm{d}l \tag{31}$$

Results of the induction current flowing through the bridge wire under the radiation of the high-power microwave and the nuclear electromagnetic impulse calculated by Feco software are, respectively, shown in Figures 3 and 4.

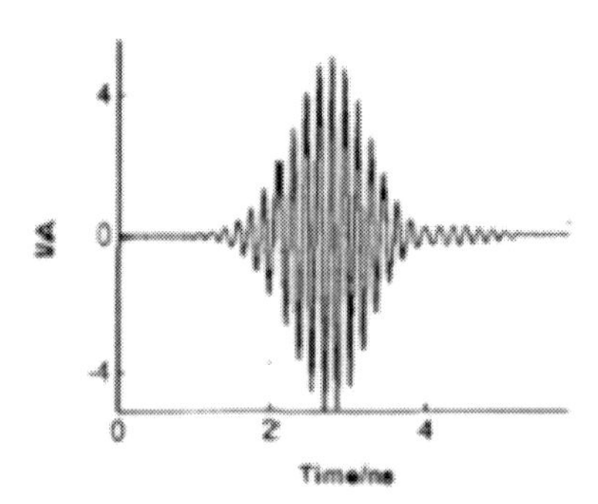

Figure 3: Induction current of the bridge wire under the radiation of the high power microwave.

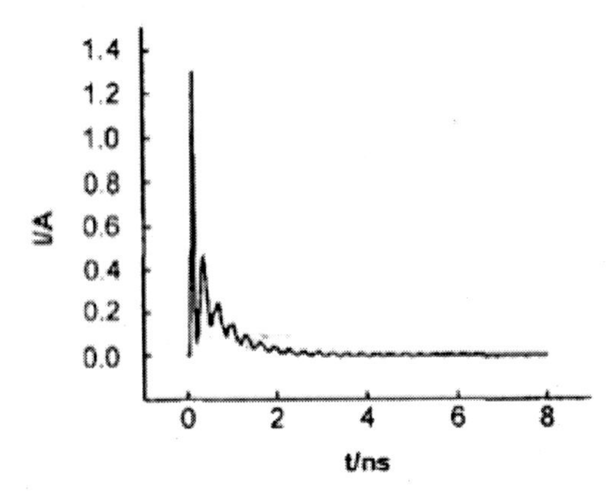

Figure 4: Induction current of the bridge wire under the radiation of the nuclear electromagnetic impulse.

Seeing from Figure 4, we can know that the incident wave used for numerical calculation has impact on the wave form of the induction current in the conditions of frequency $\oint_0$ or field strength (or power), but has a little impact on the peak value of the current. In addition, seen from Figures 3 and 4, peak values of the induction current are about 5 and 1.4 A, far higher than the striking current of the industrial electric detonator. So it is confirmed that the industrial electric detonator can be detonated by the induction current flowing through the bridge wire under the continual radiation of the high power microwave or the nuclear electromagnetic impulse.

4 Conclusions

This paper firstly introduces the principles and characteristics of the electromagnetic FDTD method. Then, it has, based on the orthogonal FDTD arithmetic, deduced the non-orthogonal algorithm in arbitrary medium, and discussed the stability conditions for calculation of values. Thereafter, this paper calculates the induced current of industrial electric detonator in continuous electromagnetic radiation and comes to the conclusion theoretically that the industrial electric detonator could be probably ignited by induction current flowing through the bridge wire under continuous electromagnetic radiation.

References

[1] K.S. Yee, Numerical solution of initial boundary value problems involving Maxwell's equation in isotropic media. *IEEE Transactions on OD AP*, **14**, pp. 302–307, 1966.
[2] Gao Benqing, *FDTD Method*. National Defense Industry Press, p. 3, 1995.

[3] Wang Changqing, Zhu Xili, *FDTD Method to Calculate Electromagnetic Field*. Peking University Press, 2014.
[4] Jia-Ying Wang, Ben-Qing Gao, A method of sub-region connection in FDTD algorithm. *Microwave and Optical Technical Letters*, **19(3)**, 1998.
[5] A. Taflove, M.E. Brodwin, Numerical solution of steady state electromagnetic scattering problems using the time domain Maxwell's equations. *IEEE Transactions on MTT*, **23(8)**, pp. 623–630, 1975.
[6] G. Mur, Absorbing boundary conditions for the finite-difference approximation of time domain electromagnetic field equations. *IEEE Transactions on EMC*, **23**, pp. 377–382, 1981.
[7] Jia Ying, Time Domain Electromagnetic Field Calculation Technology and its Application. Ph.D. and other doctoral theses, Beijing Institute of Technology, p. 12, 1998.
[8] Li Chaowei, Application and Research of FDTD Method in the Planar Circuit and Antenna. Degree paper, Beijing University of Aeronautics and Astronautics, p. 9, 2000.
[9] Ma Jinping, Mao Naihong, Research of twisted-pair electromagnetic compatibility. *Chinese Journal of Radio Science*, p. 3, 1998.
[10] Feng Qingmei, Hu Xinyi, The Harmful Effects of Electromagnetic Radiation to Electric Initiating Explosive Device and its Preventive Measures. 213 Institute of Chinese Weapons Industry.
[11] Wang Zhitian, et al., *Electromagnetics Calculation Theory*. Peking University Press, 2013.

Self-stability analysis of eddy current magnetic suspension

Zhengnan Hou, Xiaolong Li
Key Laboratory of Maglev Technology and Vehicle, Ministry of Education, Chengdu, China
School of Electrical Engineering, Southwest Jiaotong University, Chengdu, China

Abstract

The eddy current magnetic suspension levitates the body by the repulsive force generated by the interaction of the magnetic flux with eddy currents which are generated in a conducting plate when the plate is subjected to a time-varying magnetic flux. Unlike the attractive force, when the object deviates from the setting position, the sum of the repulsive force and gravitational force directs to the setting position, so that the eddy current suspension can achieve stable levitation without feedback loop. In this paper, mathematical model of the coil eddy current suspension is derived to theoretically prove the self-stability. Finally, through the comparison between the results of theoretical calculation, Ansoft simulation and experiment, the correctness of the mathematical model and the self-stability theory of the eddy current suspension are validated.

Keywords: magnetic suspension, eddy current, repulsive force, self-stability analysis.

1 Introduction

There are two kinds of magnetic levitation: attractive and repulsive [1, 2]. In attractive levitation, a ferromagnetic body is attracted to a source of magnetic flux, as a piece of steel is attracted to a permanent magnet [3]. This type of levitation is unstable without feedback control, therefore numerous analog and digital control techniques are available [4, 5]. Eddy current magnetic suspension is one of the many kinds of repulsive levitation, which using the repulsive force generated by the interaction of the magnetic flux with eddy currents which are

www.witpress.com, ISSN 1743-3533 (on-line)
doi:10.2495/SEEEI140051

generated in a conducting body when the body is subjected to a time-varying magnetic flux [6, 7]. Unlike the attractive force [8], when the object deviates from the setting position, the sum of the repulsive force and gravitational force directs back to the setting position, so that the eddy current suspension can achieve stable levitation without feedback loop.

In this paper, mathematical model of the coil eddy current suspension is derived to theoretically prove the self-stability. Finally, through the comparison between the results of theoretical calculation, Ansoft simulation and experiment, the correctness of the mathematical dynamic model and the self-stability theory of the eddy current suspension are validated.

This paper is organized as follows. In the next section, the principle of the eddy current suspension is presented. In Section 3, we propose the math model that we research in this paper. Section 4 presents the method and the result of theoretical calculation, Ansoft simulation and experiment. Finally, we conclude our paper in Section 5.

2 Eddy current suspension theory

One typical experiment of eddy current magnetic suspension is coil levitation. The geometry of the levitation experiment is shown in Figure 1. A circular copper coil was placed on top of a conducting plate that is much wider than the coil.

The Ampere's law shows a flowing current creates a magnetic field as $\Delta \times \vec{H} = \bar{J}$. From here, we know the φ-directed current in the coils generates a time-varying magnetic flux.

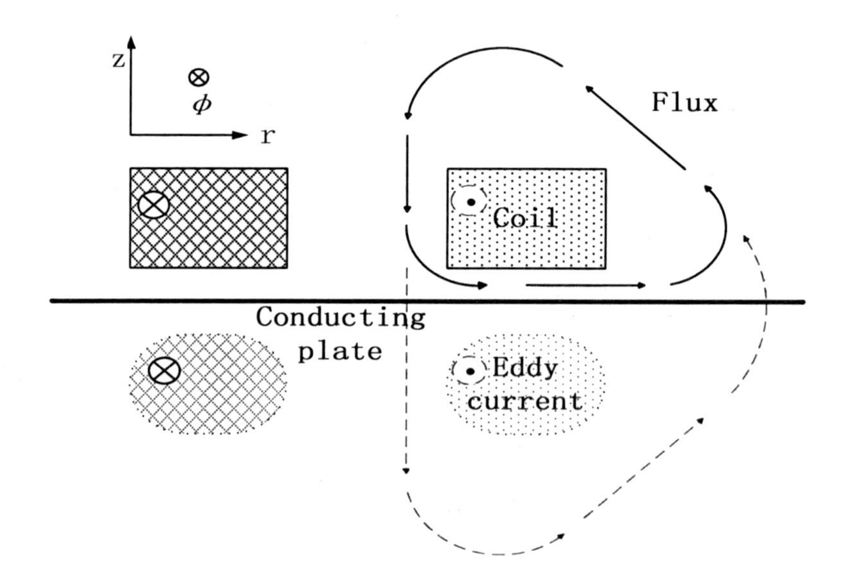

Figure 1: Principle of eddy current suspension.

Faraday's law shows the mechanisms by which a changing magnetic flux generates eddy current in a conducting material as $\vec{J} = \sigma\vec{E}$ which can be derived as $\frac{1}{\sigma}\Delta \times \vec{J} = \frac{-\partial\vec{E}}{\partial t}$. And we know the changing magnetic flux impinging on the plate induces an electric field (eddy current) in the conducting plate.

Furthermore, the Lorentz force law states that a magnetic force is created if there is a current flow in a region where there is magnetic flux, by $\vec{F} = \vec{J} \times \vec{B}$. So the φ-direction current component in the plate interacts with the component of the magnetic field to generate force to levitate the coil.

3 Self-stability analysis

3.1 Stable suspension

The suspension stability analysis can be given by comparing the gravity with repulsive force which is calculated by constructing a function of air gap inductance and setting up an energy model. The analysis of experimental data (the experiment will be elaborated later in this paper) shows that the air gap inductance (L) is a function of the distance (z) between the coil and the conducting plate, and decreases as the distance decreases. A possible functional dependence by fitting the experimental data through the software of Matlab is

$$L_{(z)} \approx L_{\infty} - L_r \mathrm{e}^{-z/r} \tag{1}$$

The term L_{∞} is the air gap inductance of the coil when it is far away from the plate. The terms L_r and r are two character constants that depend on the material, thickness of the plane, the size of the coil and frequency of the AC current.

For AC current source (effective value is I) as excitation source, the magnetic energy of the whole system is

$$E_m = \frac{1}{2} L_{(z)} I^2 \tag{2}$$

From here we can see that the energy changes with the air gap. Thus, the electromagnetic force is the derivative field energy on air gap as

$$f_z = \frac{\mathrm{d}}{\mathrm{d}z} E_m = \frac{I^2}{2} \frac{\mathrm{d}L_{(z)}}{\mathrm{d}z} = \frac{I^2 L_r}{2r} \mathrm{e}^{-z/r} \tag{3}$$

For the coil mass (m), at the setting position (z_0), the electromagnetic repulsive force is

$$f_{z_0} = \frac{I_{z_0}^2 L_r}{2r} \mathrm{e}^{-z_0/r} = mg \tag{4}$$

So the effective value of required excitation source current corresponding to the setting position (z_0) is

$$I_{z_0} = \sqrt{\frac{2rmg}{L_r \mathrm{e}^{-z_0/r}}} \tag{5}$$

3.2 Self-stabilization process

When the coil is deviated from the setting position z_0 to $z = z_0 + \Delta z$, by formula (3), the repulsive force can be expressed as

$$f_z = \frac{I_{z_0}^2 L_r}{2r} \mathrm{e}^{-\frac{z}{r}} = \frac{I_{z_0}^2 L_r}{2r} \mathrm{e}^{-\frac{z_0+\Delta z}{r}} = \frac{I_{z_0}^2 L_r}{2r} \mathrm{e}^{-\frac{z_0}{r}} \bullet \mathrm{e}^{-\frac{\Delta z}{r}} = mg \bullet \mathrm{e}^{-\Delta z/r} \tag{6}$$

Then the kinematic equation of the coil is

$$\frac{\mathrm{d}z}{\mathrm{d}t} = \frac{\Delta f}{m} = \frac{fz - mg}{m} = g \bullet (\mathrm{e}^{-\frac{\Delta z}{r}} - 1) \tag{7}$$

It follows that:

(1) when $\Delta z > 0$, that is, the coil is above the setting position, the repulsion is smaller than the gravity, the direction of the acceleration is downward.
(2) when $\Delta z < 0$, that is, the coil is below the setting position, the repulsion is greater than the gravity, the direction of the acceleration is upward.

Thus, in the eddy current suspension, the coil can automatically find the setting position.

It is worth mentioning that, when $\Delta z \ll r$, we have $\mathrm{e}^{-\frac{\Delta z}{r}} \approx 1 - \frac{\Delta z}{r}$, the formula (5) can be derived as

$$\frac{\mathrm{d}z}{\mathrm{d}t} = -\frac{g}{r} \bullet \Delta z \tag{8}$$

This is a typical like-Hooke's law expression, which is the basic principle of the displacement feedback loop commonly used in the attractive levitation.

3.3 Lift-off requirement

For given coil and plate, the values of m, r and L_r are fixed. According to the formula (2), the excitation current (I_{z_0}) is a monotone increasing function about the setting position (z_0). Then the lift-off critical current is

$$I_0 = \sqrt{\frac{2rmg}{L_r e^{-z_0/r}}}\Bigg|_{z_0=0} = \sqrt{\frac{2rmg}{L_r}} \tag{9}$$

As the formula (7) shows, as long as the setting position $z_0 > 0$, then the corresponding excitation current $I_{z_0} > I_0$, the coil will lift off, and reach the stable position z_0 according to the kinematics law shown in the formula (5).

4 Ansoft simulation and experiment

In order to verify the previous theory, this paper builds a simple coil levitation experiment platform. The coil parameters are as shown in Table 1.

4.1 Fitting expression of the air gap inductance

In order to verify the function form of the air gap inductance and get the value of the parameter L_0, L_r and r, we need to measure the air gap inductance value of different air gap value, and then use the Matlab function fitting to get the concrete function.

The air gap inductance was measured using a Tonghui TH2816A Precision LCR Meter at 0, 1.5, 3, 5, 7.5 and 10 mm above the conducting plate at 50 Hz. The measured results and the fitting curve are shown in Figure 2.

Figure 2 shows that the fitting effect of the function is good, which has verified the correctness of the air gap inductance function form. In the mean time, we got the fitted parameter values result as: $L_0 = 1020\ \mu H$. $L_r = 260\ \mu H$ and $r = 15\ mm$. Thus, the air gap inductance function of experiment system is

$$L_{(z)} = 1020 - 260\, e^{-z/25}\ (\mu H) \tag{10}$$

Table 1: Coil parameters.

Coil Turns	Coil Mass	Coil Resistance (300 K)	Coil Inductance (in free space)
$n = 100$	$m = 0.4$ kg	$R = 0.64\ \Omega$	$L_\infty = 1020\ \mu H$

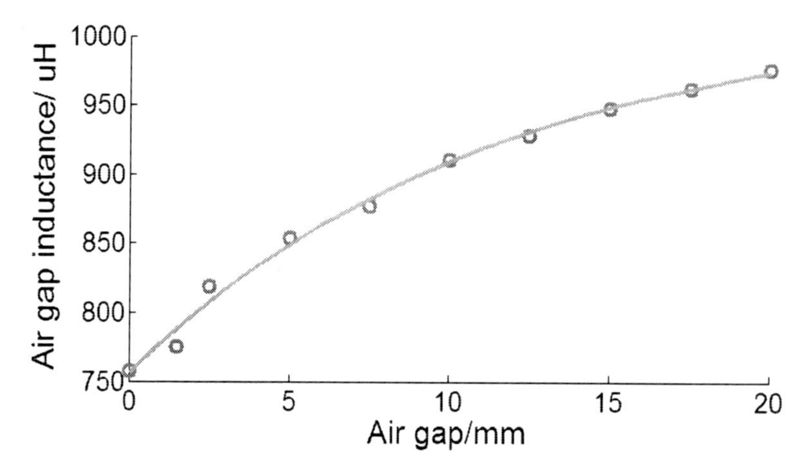

Figure 2: Fitting of air gap inductance.

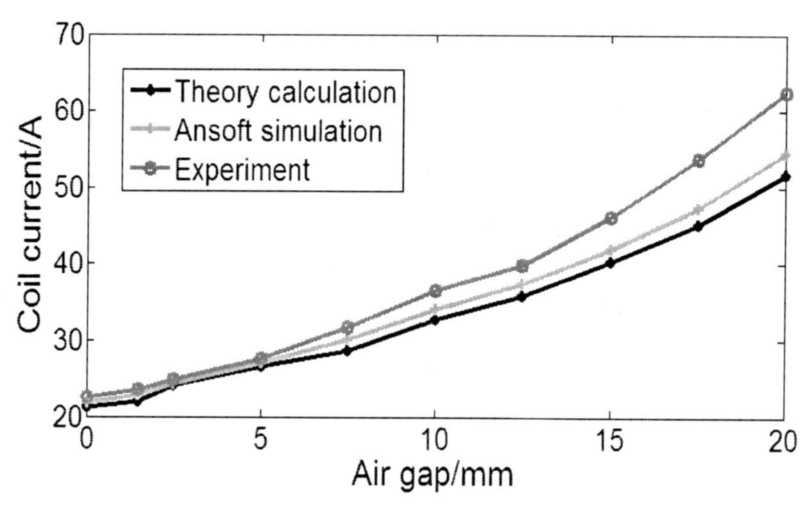

Figure 3: Comparison between the results of theoretical calculation, Ansoft simulation and experiment.

4.2 Analysis of the coil levitation

Simulation calculation is done by using Ansoft. During the simulation, after setting air gap value, the value of current through into the coil is regulated, when the repulsive force is equal to gravity of the coil, the corresponding current value is recorded.

During the experiment, the value of current is regulated, when the coil reaches the setting position, the corresponding current value is recorded.

The coil current value is recorded at the air gap of 0, 1.5, 3, 5, 7.5 and 10 mm in theory calculation, Ansoft simulation and experiment. The comparison of the three group data is shown in Figure 3.

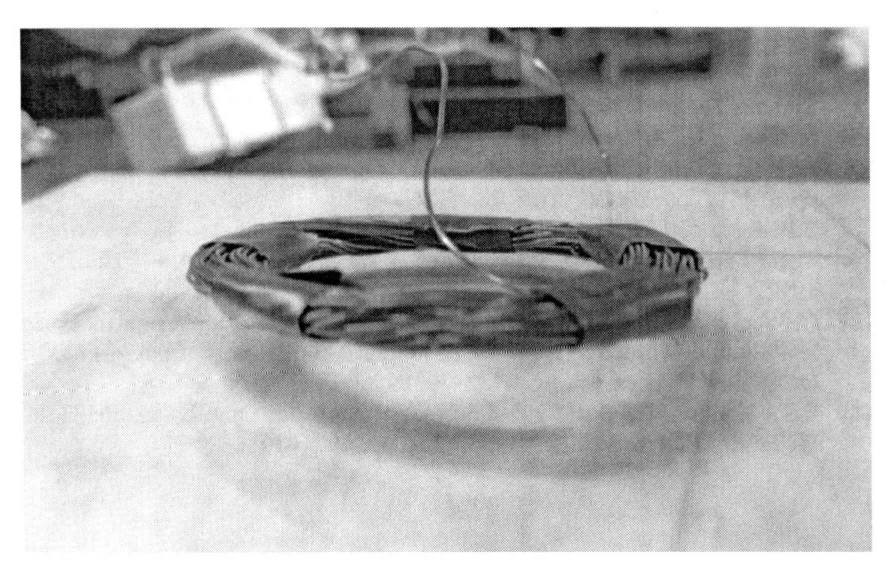

Figure 4: Coil suspended 5 mm above aluminum plate.

Figure 3 shows the results of the calculation, Ansoft simulation and experiment generally align, which verified the correctness of the mathematical model established in the paper. At the same time, the coil can achieve the setting position without feedback loop, which verified the self-stability of eddy current suspension. Figure 4 shows that the coil suspends stably at 5 mm.

5 Conclusions

In this paper, a mathematical model of the coil eddy current suspension is established, which theoretically proves that the eddy current suspension can automatically reach the setting position without feedback loop. Through the comparison and analysis of the results of the theoretical calculation, the Ansoft software simulation and experiment, the correctness of the model and the self-stability of eddy current suspension are verified. These works lay theoretical and experimental base for the future research of eddy current suspension.

References

[1] P.K. Sinha, *Electromagnetic Suspension Dynamics and Control*. Peter Peregrinus Ltd., London, 1987.

[2] M.T. Thompson, Electrodynamic magnetic suspension—models, scaling laws, and experimental results. *IEEE Transactions on Education*, **43(3)**, pp. 336–343, 2000.

[3] J.U. Jeon, T. Higuchi, Electrostatic suspension of dielectrics. *IEEE Transactions on Industrial Electronics*, **46(6)**, pp. 938–946, 1998.

[4] A. Charara, J. De Miras, B. Caron, Nonlinear control of a magnetic levitation system without premagnetization. *IEEE Transactions on Control Systems Technology*, **4(5)**, pp. 513–523, 1996.
[5] S. Sivriog lu, Adaptive backstepping for switching control active magnetic bearing system with vibrating base. *IET Control Theory Applications*, **1(4)**, pp. 1054–1059, 2007.
[6] E.E. Kriezis, T.D. Tsiboukis, S.M. Panas, J.A. Tegopoulos, Eddy currents: Theory and applications. *Proceedings of the IEEE*, **80(10)**, pp. 1559–1590, 1992.
[7] C. Elbuken, M.B. Khamesee, M. Yavuz, Eddy current damping for magnetic levitation: downscaling from macro to micro-levitation. Journal of Physics D: Applied Physics, **39**, pp. 3932–3938, 2006.
[8] B.Z. Kaplan, G. Sarafian, Control strategy for stabilising magnetic levitation. *Electronics Letters*, **33(23)**, pp. 1961–1962, 1997.

Magnetic properties of octacyano-based magnet $Mn_2[M(CN)_8] \cdot 8H_2O$ (M = Mo, W)

Qing Lin[1,2], Jinpei Lin[1], Yun He[1], ShaoHong Chen[1], Jianmei Xu[2]
[1]*College of Physics and Technology,*
Guangxi Normal University, Guilin, China
[2]*College of Medical Informatics, Hainan Medical University,*
Haikou, China

Abstract

The octacyano complexes $Mn_2[M(CN)_8] \cdot 8H_2O$ (M = Mo, W) have been synthesized. In the compound $Mn_2[Mo(CN)_8] \cdot 8H_2O$, the magnetic susceptibility obeys the Curie–Weiss law with $\Theta = -6.052$ K, $C = 1.307$ cm^3 K mol^{-1}, $\chi_m T$ decrease at lower temperature. The effective $\mu_{\rm eff}$ sharply decreases at lower temperature, indicating anti-ferromagnetic interaction between the paramagnetic centers. In the compound $Mn_2[W(CN)_8] \cdot 8H_2O$, the magnetic susceptibility obeys the Curie–Weiss law with $\Theta = -5.143$ K, $C = 1.199$ cm^3 K mol^{-1}, the negative Weiss constant (θ) indicates anti-ferromagnetic interactions. The magnetic measurements reveal that ferromagnetic and anti-ferromagnetic couplings occur between Mn and Mo, W centers through the CN bridges.

Keywords: molecule-based magnet, octacyano, molybdenum, tungsten.

1 Introduction

In recent years, molecule-based magnetic materials have attracted extensive attention because of their potential applications in magnetic devices, and octacyano complexes of molybdenum (IV,V) and tungsten (IV,V) have been the subject of numerous studies on their magnetic [1, 2]. The first octacyanometalate-based magnet $Cu_3[W(CN)_8]_2 \cdot 3.4H_2O$ was reported by Garde et al. [3]. The octacyano-bridged bimetallic complexes $Cs^{I}Cu^{II}[W^{V}(CN)_8] \cdot 0.5H_2O$ were reported by Shen et al. [4]. $[M(CN)_8]^{n-}$ as building blocks have been extensively studied. Molecular structures of $[M(CN)_8]^{4-}$ (as shown in Figure 1) show

WIT Transactions on Engineering Sciences, Vol. 107, © 2015 WIT Press
www.witpress.com, ISSN 1743-3533 (on-line)
doi:10.2495/SEEEI140061

various magnetic properties depending on their transition metal ion. Thus, cyano-bridged systems based upon octacyanomolybdate or octacyanotungstate building blocks attract considerable interest and a few bimetallic complexes based on $[M(CN)_8]^{4-}$ (M=Mo or W) and high spin Mn^{2+} have been structurally and magnetically characterized [4]. In this context, we have prepared octacyano-based complexes $Mn_2[Mo(CN)_8] \cdot 8H_2O$ and $Mn_2[W(CN)_8] \cdot 8H_2O$, and the magnetic properties of the compounds have been studied.

2 Experimental

2.1 Materials and physical measurements

Material. $MnCl_2 \cdot 4H_2O$, $Mn(ClO_4)_2 \cdot 6H_2O$ are reagent grade. Thermal gravimetric analysis (TG) was performed on Perkin Elmer TGA detector. The magnetic properties were determined using an MPMS-7 Quantum Design SQUID (Superconducting QUantum Interference Device) magnetometer operating within the range of 2–300 K.

2.2 Preparation process of $K_4[Mo(CN)_8] \cdot 2H_2O$ and $K_4[W(CN)_8] \cdot 2H_2O$

The precursors of $K_4[Mo(CN)_8] \cdot 2H_2O$ were synthesized according to the published procedures [5–7]. The precursors of $K_4[W(CN)_8] \cdot 2H_2O$ were synthesized according to the published procedures [5, 8, 9]. All the other reagents were purchased as reagent grade chemicals and used without further purification.

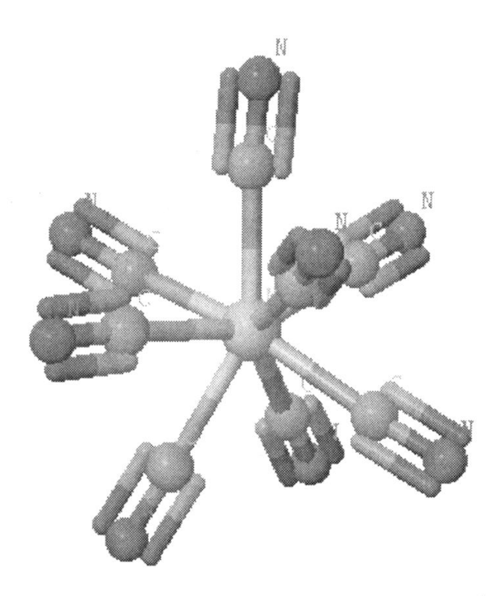

Figure 1: Molecular structures of $[M(CN)_8]^{4-}$.

2.3 Synthesis of $Mn_2[Mo(CN)_8] \cdot 8H_2O$

The precursors of $Mn_2[Mo(CN)_8]{\cdot}2H_2O$ were synthesized according to the literature method [10]. Polycrystalline compound was prepared by co-precipitation method. To a solution of 100 ml $K_4[Mo(CN)_8] \cdot 2H_2O$ (1 mmol) 100 ml solution of $MnCl_2 \cdot 4H_2O$ (2 mmol) was slowly added, and a solid was precipitated immediately. In the case of $Mn_2[Mo(CN)_8] \cdot 8H_2O$ the process started in about 10 min. The solids were collected by filtration, washed twice with 10 ml of water and dried above KOH, and dried at 45 °C. Thermal gravimetric analysis (TG) has given weight ratio of the crystal water in molecular formula weight, the compound contains 8.0 crystal waters by calculation.

2.4 Synthesis of $Mn_2[W(CN)_8] \cdot 8H_2O$

The precursors of $Mn_2[W(CN)_8] \cdot 2H_2O$ were synthesized according to the published procedures [11]. The yellow crystals of $Mn_2[W(CN)_8] \cdot 8H_2O$ were obtained by adding 30 ml of an aqueous solution of $Mn(ClO_4)_2 \cdot 6H_2O$ (0.4 mmol) and 30 ml of an aqueous solution of $K_4[W(CN)_8] \cdot 2H_2O$ (0.2 mmol) into the two shoulders of an H-shaped tube with 10 ml isopropyl alcohol. Thermal gravimetric analysis (TG) has given weight ratio of the crystal water in molecular formula weight, the compound contains 8.0 crystal waters by calculation.

3 Results and discussion

3.1 DC magnetic susceptibility of $Mn_2[Mo(CN)_8] \cdot 8H_2O$

The magnetic susceptibility of the compound $Mn_2[Mo(CN)_8] \cdot 8H_2O$ was measured from 2 to 300 K in a 1 kOe field. The inverse susceptibility as a function of temperature in the paramagnetic state is shown in Figure 2. The values of χ_m gradually increase, and then sharply increase after 25 K with a further decrease of the temperature. The χ_m shows a sharp maximum at 7 K, which is a characteristic of a ferromagnet [11–13]. According to χ_m^{-1} vs. T curve (as shown in Figure 2),

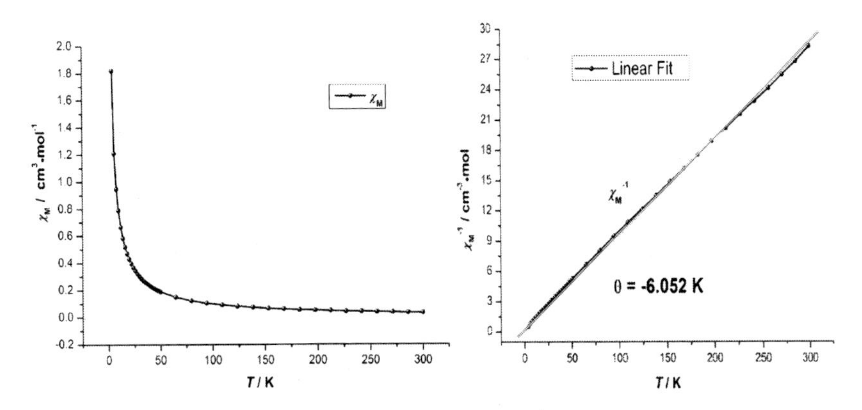

Figure 2: Temperature dependence of χ_m and χ_m^{-1} for $Mn_2[Mo(CN)_8] \cdot 8H_2O$.

the χ_m^{-1} vs. T above 2 K obey the Curie–Weiss law with a Curie constant of C = 1.307 cm^3 K mol^{-1} and Weiss paramagnetic Curie temperature of θ = −6.052 K. These results indicate the presence of a weak magnetic interaction in complexes, which can be attributed to anti-ferromagnetic coupling between Mn^{II} ions through the diamagnetic $[Mo^{IV}(CN)_8]^{4-}$ bridges.

The magnetic properties of compound $Mn_2[Mo(CN)_8] \cdot 8H_2O$ under the form of $\chi_m T$ vs. T plot are shown in Figure 3. In the low-temperature range (5–40 K), the $\chi_m T$ product decreases with a decrease of temperature, reaching a value of 7.12 cm^3 K mol^{-1} at 4 K. Then $\chi_m T$ decreases abruptly on further lowering the

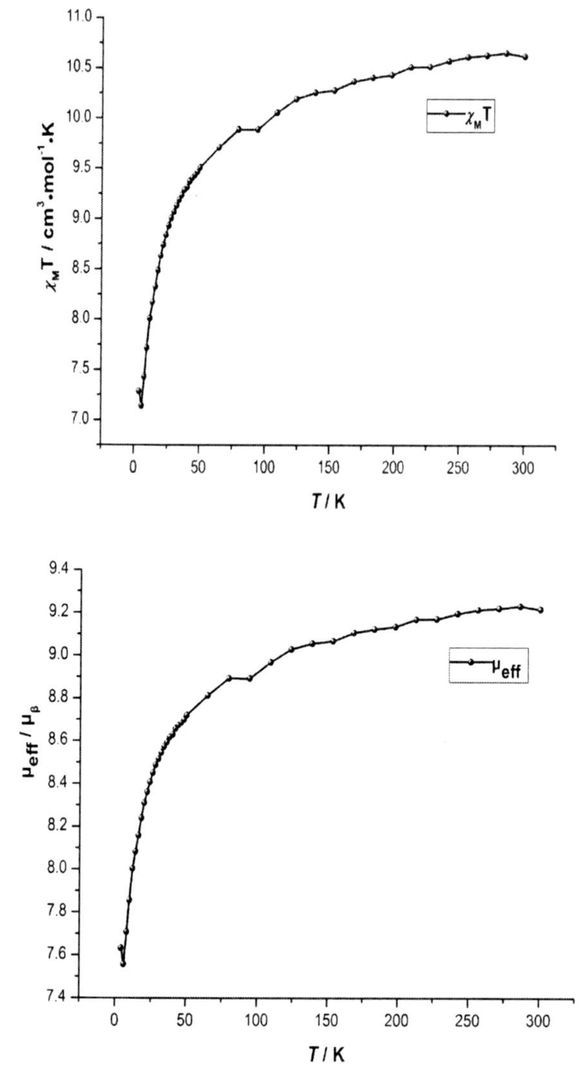

Figure 3: Temperature dependence of $\chi_m T$ and μ_{eff} for $Mn_2[Mo(CN)_8] \cdot 8H_2O$.

temperature due to zero-field splitting, suggesting the presence of an anti-ferromagnetic interaction for the complexes. A curve of μ_{eff} vs. T is shown in Figure 3, the μ_{eff} value at 300 K is $9.21\mu_{\beta}$. The effective moment μ_{eff} sharply decrease at lower temperature, indicating anti-ferromagnetic interaction between the paramagnetic centers. The behavior is a characteristic of a ferrimagnet [14–16]. These values correspond to what is expected for the anti-ferromagnetic interactions between the adjacent Mo^{IV} and Mn^{II} ions [17, 18].

3.2 DC magnetic susceptibility of $Mn_2[W(CN)_8] \cdot 8H_2O$

The magnetic susceptibility of the compound $Mn_2[W(CN)_8] \cdot 8H_2O$ was measured from 2 to 300 K in a 1 kOe field. The inverse susceptibility as a function of temperature in the paramagnetic state is shown in Figure 4.

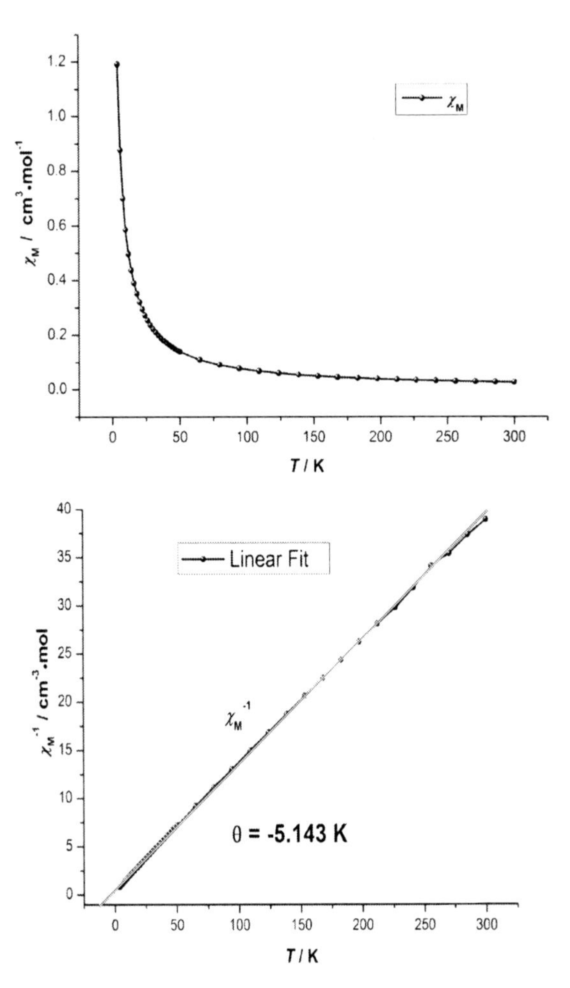

Figure 4: Temperature dependence of χ_m and χ_m^{-1} for $Mn_2[W(CN)_8] \cdot 8H_2O$.

The values of χ_m gradually increase, and then sharply increase after 25 K with a further decrease of the temperature (as shown in Figure 4). The χ_m shows a sharp maximum at 2 K, which is in consistent with the weak anti-ferromagnetic interaction between Mn^{II} and W^{IV} through the long-range cyano bridges [11–13]. According to χ_m^{-1} vs. T curve (as shown in Figure 4), the χ_m^{-1} vs. T above 2 K obey the Curie–Weiss law with a Curie constant of C=1.199 cm^3 K mol^{-1} and Weiss paramagnetic Curie temperature of θ = −5.143 K. The negative Weiss constant (θ) indicates anti-ferromagnetic interactions. These values correspond to what is expected for the anti-ferromagnetic interactions between the adjacent W^{IV} and Mn^{II} ions [17, 18].

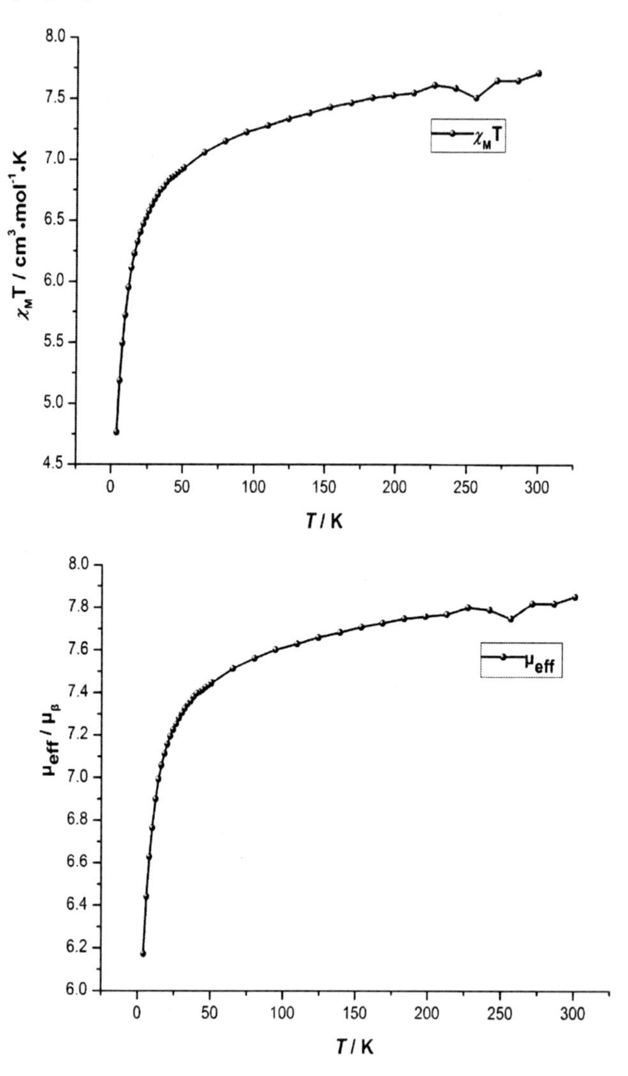

Figure 5: Temperature dependence of $\chi_m T$ and μ_{eff} for $Mn_2[W(CN)_8] \cdot 8H_2O$.

The magnetic properties of compound $Mn_2[W(CN)_8] \cdot 8H_2O$ under the form of $\chi_m T$ vs. T plot are shown in Figure 5, the $\chi_m T$ at room temperature is 7.75 cm^3 K mol^{-1}. Upon cooling, $\chi_m T$ decrease at lower temperature. This behavior is typical of anti-ferromagnetic couplings between the paramagnetic centers. A curve of μ_{eff} vs. T is shown in Figure 5, the μ_{eff} value at 300 K is 7.9 μ_B. The effective μ_{eff} sharply decreases at lower temperature, indicating anti-ferromagnetic interaction between the paramagnetic centers [14–16]. Such magnetic behavior is characteristic of Mn^{2+} in an axially distorted octahedral surrounding and a weak anti-ferromagnetic interaction between the Mn ions through the diamagnetic bridges [17, 18].

4 Conclusions

Bimetallic cyanides have attracted considerable interest in the molecular chemistry community because of their unusual electronic and magnetic properties. The octacyano complexes $Mn_2[M(CN)_8] \cdot 8H_2O$ (M=Mo, W) have been synthesized. In the compound $Mn_2[Mo(CN)_8] \cdot 8H_2O$, the magnetic susceptibility obeys the Curie–Weiss law with $\Theta = -6.052$ K, C=1.307 cm^3 K mol^{-1}, $\chi_m T$ decreases at lower temperature. The effective μ_{eff} sharply decreases at lower temperature, indicating anti-ferromagnetic interaction between the paramagnetic centers. In the compound $Mn_2[W(CN)_8] \cdot 8H_2O$, the magnetic susceptibility obeys the Curie–Weiss law with $\Theta = -5.143$ K, C=1.199 cm^3 K mol^{-1}, the negative Weiss constant (θ) indicates anti-ferromagnetic interactions. The magnetic measurements reveal that ferromagnetic and anti-ferromagnetic couplings occur between Mn and Mo, W centers through the CN bridges.

Acknowledgments

This work was financially supported by the National Natural Science Foundation of China (Nos. 11164002 and 11364004) and NSF of Guangxi Province (No. 0991092).

References

[1] R. Kania, K. Lewijski, B. Sieklucka, *Dalton Transactions*, p. 1033, 2003.
[2] J.M. Herrera, D. Armentano, G. de Munno, et al., *New Journal of Chemistry*, **27**, p. 128, 2003.
[3] R. Garde, C. Desplanches, A. Bleuzen, et al., *Molecular Crystals and Liquid Crystals*, **334**, p. 587, 1999.
[4] L. Shen, Y. Zhang, J. Uiu, *Journal of Coordination Chemistry*, **59(6)**, pp. 629–635, 2006.
[5] J.G. Leipoldt, L.D. Bok, P.J. Cilliers, *Zeitschrift für Anorganische und Allgemeine Chemie*, **409**, pp. 343–344, 1974.
[6] N.H. Furman, C.O. Miller, P.G. Arvan, *Inorganic Syntheses*, **3**, pp. 160–162, 1950.

[7] N.W. Alcock, A. Samotus, J. Szklarzewicz, *Journal of Chemical Society, Dalton Transactions*, p. 885, 1993.

[8] `L.D. Bok, J.G. Leipoldt, S.S. Basson, *Zeitschrift für Anorganische und Allgemeine Chemie*, **415**, pp. 81–83, 1975.

[9] P.J. Cilliers, J.G. Leipoldt, L.D.C. Bok, *Zeitschrift für Anorganische und Allgemeine Chemie*, **407,** p. 350, 1974.

[10] Z. Mitróová, M. Mihalik, A. Zentko, et al., *Ceramics – Silikáty*, **49(3)**, pp. 181–187, 2005.

[11] W. Dong, Y.-Q. Sun, L.-N. Zhu, et al., *New Journal of Chemistry*, **27**, pp. 1760–1764, 2003.

[12] S. Ohkoshi, T. Iyoda, A. Fujishima, et al., *Physical Review B*, **56**, pp. 11642–11652, 1997.

[13] S. Ohkoshi, Y. Abe, A. Fujishima, et al., *Review Letter*, **82**, pp. 1285–1288, 1999.

[14] M. Ohba, H. Okawa, *Coordination Chemistry Reviews*, **198**, pp. 313–328, 2000.

[15] A. Kumar, S.M. Yusuf, L. Keller, *Physical Review B*, **71(5)**, p. 054414, 2005.

[16] A.K. Sra, G. Rombaut, F. Lahit^ete, et al., *New Journal of Chemistry*, **24**, 871, 2000.

[17] A. Kumar, S.M. Yusuf, *Physica B*, **362**, pp. 278–285, 2005.

[18] S.I. Ohkoshi, N. Machida, Z.J. Zhong, et al., *Synthetic Metals*, **122**, p. 523, 2001.

Synthesis and magnetic properties of octacyanotungstate-based magnet $Mn^{II}Fe^{II}[W^{IV}(CN)_8]\cdot 8H_2O$

Yun He[1], Qing Lin[1,2], Jinpei Lin[1], Henian Chen[2], Ruijun Wang[2]
[1]*College of Physics and Technology,*
Guangxi Normal University, Guilin, China
[2]*College of Medical Informatics,*
Hainan Medical University, Haikou, China

Abstract

The octacyano complex $MnFe[W(CN)_8]\cdot 8H_2O$ has been synthesized, the magnetic susceptibility obeys the Curie-Weiss law with $\Theta = -3.165$ K, $C = 1.103$ $cm^3 \cdot k \cdot mol^{-1}$, $\chi_m T$ decreases abruptly on further lowering the temperature due to zero-field splitting, suggesting the presence of an antiferromagnetic interaction for the complex. The effective moment μ_{eff} sharply decreases at lower temperature, which also indicate an intramolecular antiferromagnetic coupling between Mn^{II} and W^{IV} ions.

Keywords: octacyano, tungsten, molecule-based magnet, magnetic interaction.

1 Introduction

Recently, the octacyanometalates $[M(CN)_8]^{3-/4-}$ (M = Mo,W) complexes have attracted much attention [1, 2]. Cyanometalates are excellent building blocks for constructing molecule-based clusters and networks, because they show abundant topologies and can potentially be used as functional materials [3, 4]. 3D molecules structures of $[W(CN)_8]^{4-}$ (as shown in Figure 1) show various magnetic properties depending on cyano groups and exhibit interesting properties like electric field-induced conductance switching, photo-induced magnetization, and so on. Octacyanometalates $[M(CN)_8]^{n-}$ (M = Mo, W) have been successfully employed as useful building blocks to achieve diverse bimetallic magnetic systems. In this context, we have prepared octacyano-based complex MnFe $[W(CN)_8]\cdot 8H_2O$, and the magnetic properties of this compound have been studied.

WIT Transactions on Engineering Sciences, Vol. 107, © 2015 WIT Press
www.witpress.com, ISSN 1743-3533 (on-line)
doi:10.2495/SEEEII140071

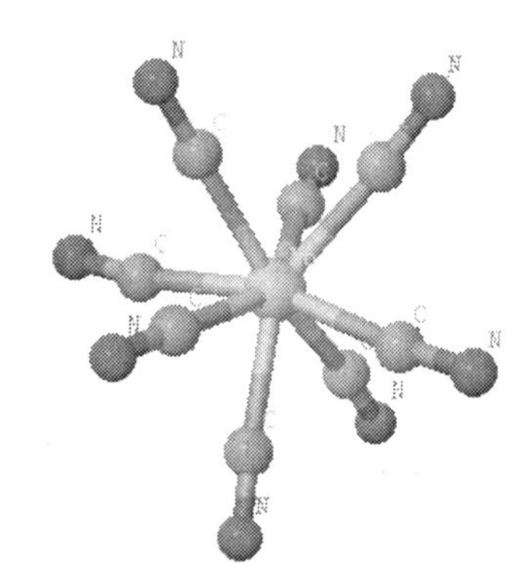

Figure 1: Molecular structures of $[W(CN)_8]^{4-}$.

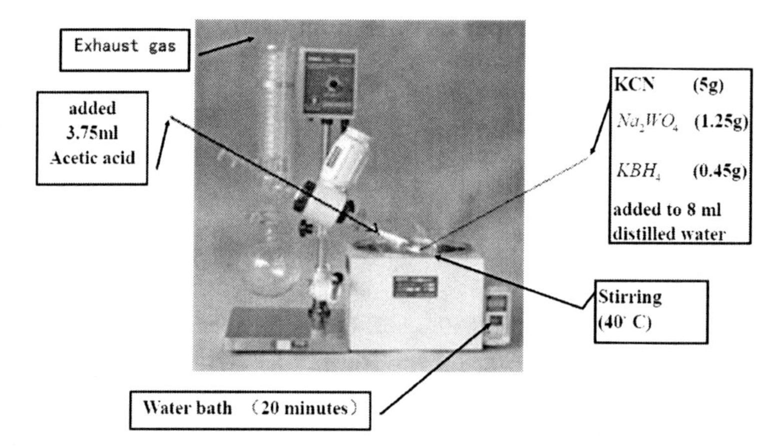

Figure 2: Preparation process of $K_4[W(CN)_8]\cdot 2H_2O$.

2 Experimental

2.1 Materials and physical measurements

Material. $MnCl_2\cdot 4H_2O$, $FeSO_4\cdot 7H_2O$ are reagent grade. Thermal gravimetric analyses (TG) were performed on Perkin Elmer TGA detector. The magnetic properties were determined using an MPMS-7 Quantum Design SQUID (Superconducting QUantum Interference Device) magnetometer operating within the range of 2–300 K.

2.2 Preparation process of $K_4[W(CN)_8]\cdot 2H_2O$

The precursors of $K_4[W(CN)_8]\cdot 2H_2O$ were synthesized according to published procedures [5–7].

Preparation process of $K_4[W(CN)_8]\cdot 2H_2O$ is shown in Figure 2. All the other reagents were purchased as reagent grade chemicals and used without further purification.

2.3 Synthesis of $MnFe[W(CN)_8]\cdot 8H_2O$

Polycrystalline compound was prepared by coprecipitation method [8]. A solution of 100 ml $K_4[W(CN)_8]\cdot 2H_2O$ (1 mmol) was slowly added to 100 ml mixed solution of $MnCl_2\cdot 4H_2O$ (1 mmol) and $FeSO_4\cdot 7H_2O$ (1 mmol), and a solid was precipitated immediately. In the case of $MnFe[W(CN)_8]\cdot 8H_2O$ the process started in about 10 min. The solids were collected by filtration, washed twice with 10 ml of water and dried above KOH, and dried at 45 °C. Thermal gravimetric analysis (TG) has given weight ratio of the crystal water in molecular formula weight, the compound contains 8.0 crystal waters by calculation.

3 Results and discussion

3.1 DC magnetic susceptibility

The magnetic susceptibility of the compound $MnFe[W(CN)_8]\cdot 8H_2O$ was measured from 2 to 300 K in a 1 kOe field, which corresponds to the steepest rise of magnetization with decreasing temperature. The inverse susceptibility as a function of temperature in the paramagnetic state is shown in Figure 3, the values of χ_m gradually increase, and then sharply increase after 20 K with a further decrease of the temperature [9, 10]. The χ_m shows a sharp maximum at 2 K, which is a characteristic of the ferromagnet. According to χ_m^{-1} versus T curve (as shown in Figure 3), the χ_m^{-1} vs T above 2 K obeys the Curie–Weiss law with a Curie constant of C = 1.103 $cm^3\cdot K\cdot mol^{-1}$ and Weiss paramagnetic Curie temperature of $\theta = -3.165$ K, while the negative θ value confirms the presence of antiferromagnetic interaction between Mn^{II} and Fe^{II} ions in $MnFe[W(CN)_8]\cdot 8H_2O$.

The magnetic properties of compound $MnFe[W(CN)_8]\cdot 8H_2O$ under the form of χ_mT vs. T plot are shown in Figure 4, the χ_mT at room temperature is 7.01 $cm^3\cdot K\cdot mol^{-1}$. Upon cooling, χ_mT decreases at lower temperature, indicating antiferromagnetic interaction [11–13]. This value is consistent with antiferromagnetic Mo–Mn interactions in line with the previous observations.

A curve of μ_{eff} vs T is shown in Figure 4, the μ_{eff} value at 300 K is 7.46 μ_B. The effective moment μ_{eff} sharply decreases at lower temperature, indicating antiferromagnetic interaction between the paramagnetic centers. The behavior is a characteristic of a ferrimagnet [14–16]. A sharp drop below the cusp temperature is observed, designating that prevalent antiferromagnetic couplings in both complexes are operative.

3.2 Field-dependent of magnetization and hysteresis behavior

The zero-field-cooled (ZFC) and field-cooled (FC) magnetization for the sample $MnFe[W(CN)_8]\cdot 8H_2O$ was carried out in the temperature range 2–60 K at the external field of 200 Oe. The result is presented in Figure 5. Magnetic studies reveal that the complexes confirm the occurrence of significant zero-field splitting in the Mn^{II} ion at low temperature. The cyano group has been shown to be an

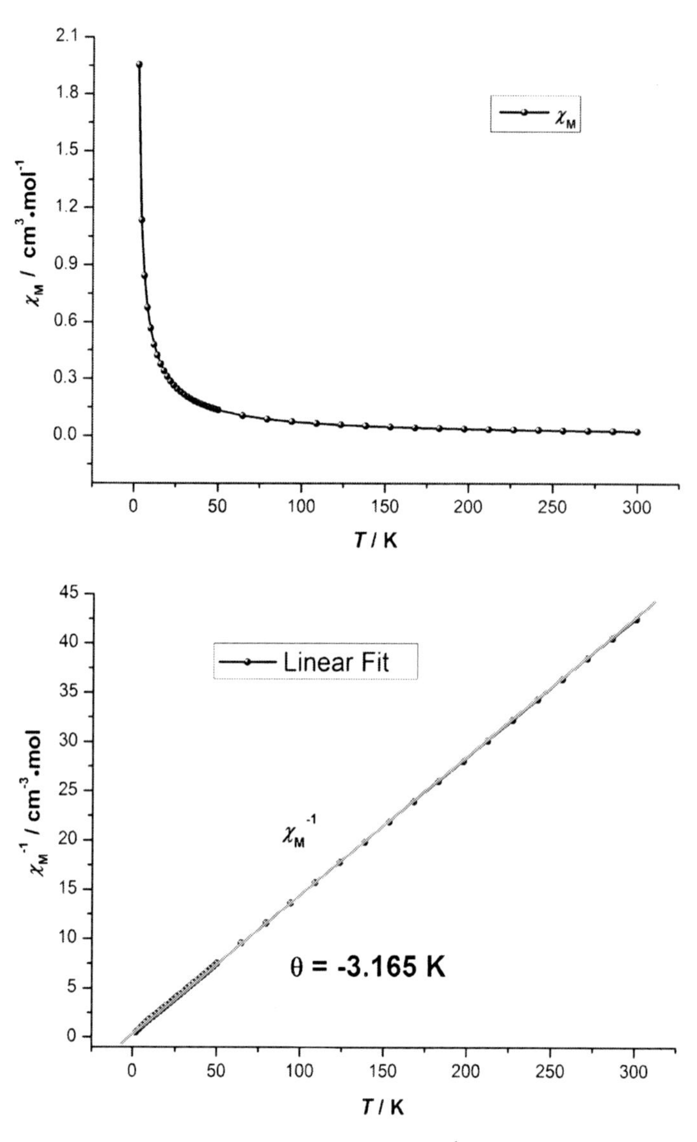

Figure 3: Temperature dependence of χ_m *and* χ_m^{-1} for $MnFe[W(CN)_8]\cdot 8H_2O$.

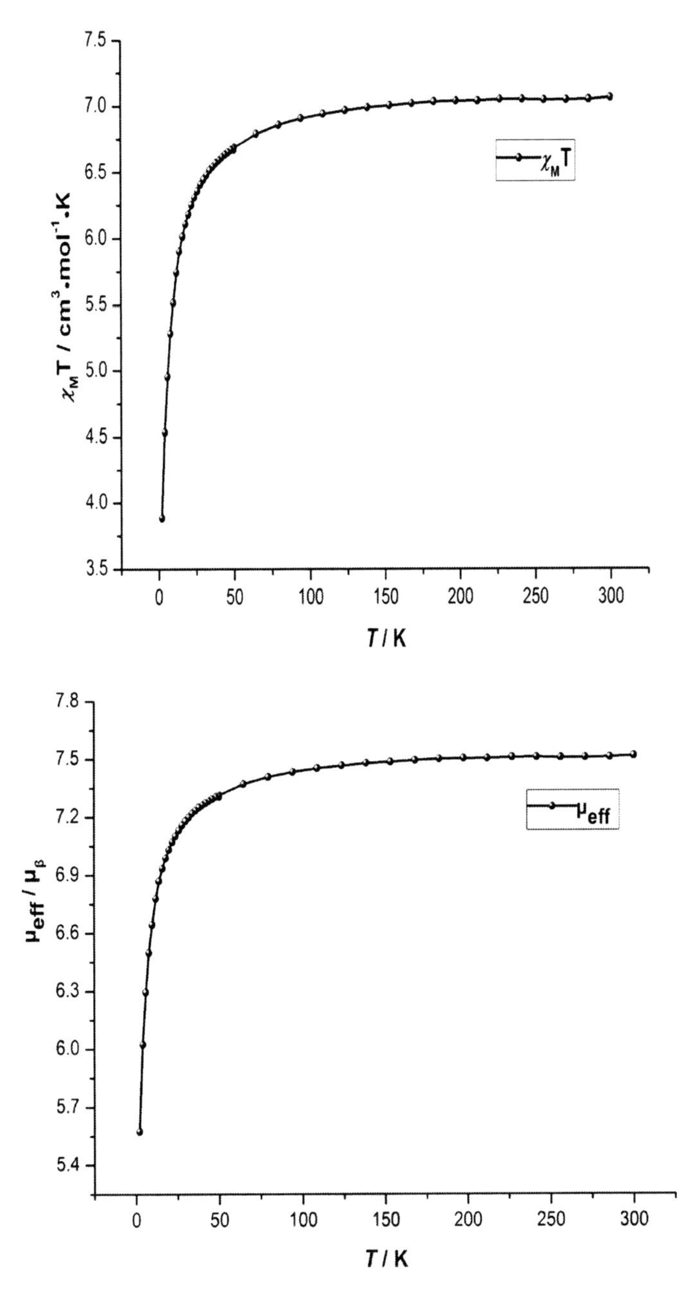

Figure 4: Temperature dependence of $\chi_m T$ and μ_{eff} for $MnFe[W(CN)_8]\cdot 8H_2O$.

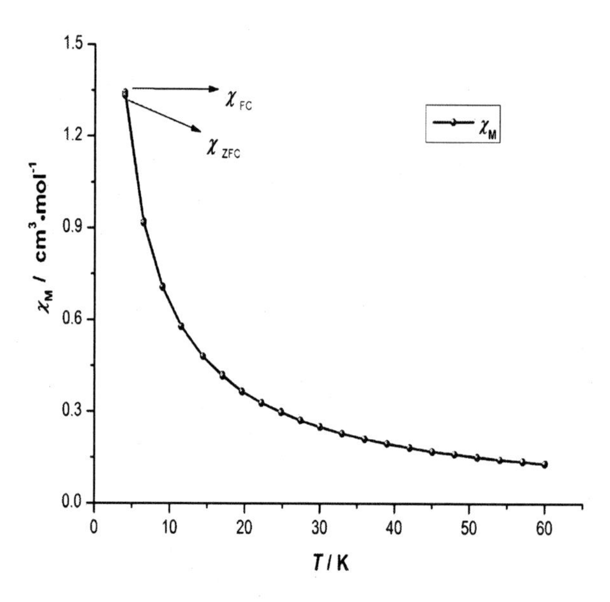

Figure 5: M_{FC} and M_{ZFC} vs T for the sample.

excellent bridging ligand in supporting the magnetic exchange interaction, while the FC curve at high temperature coincides with ZFC curve. Plots of the magnetization versus external magnetic field (M vs H, T = 4 K) for the compound $MnFe[W(CN)_8]\cdot 8H_2O$ are shown in Figure 6, in the magnetic field rang of 0–50 kOe, magnetization is 6.5 μ_B. The shape of the $M(H)$ dependence strongly suggests an antiferromagnetic interaction.

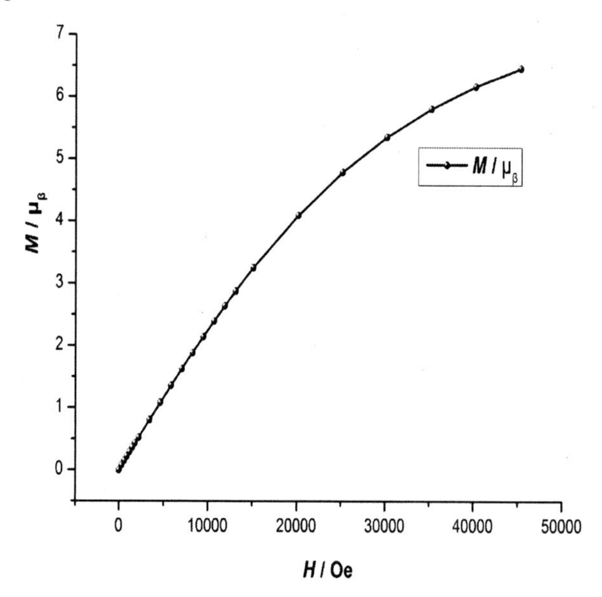

Figure 6: M vs H for the sample.

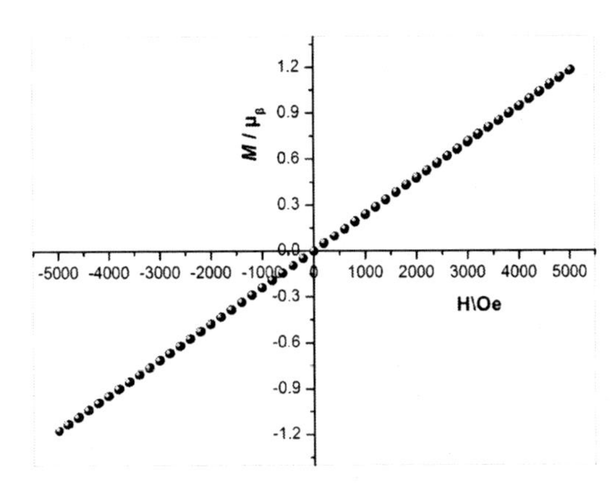

Figure 7: Hysteresis curve of the sample at 4 K.

The ferrimagnetic behavior is further characterized by the measurements of hysteresis behavior as shown in Figure 7. The shape of the *M*(*H*) dependence strongly suggests an antiferromagnetic interaction, which also indicates an intramolecular antiferromagnetic coupling between the adjacent Mn^{II} and W^{IV} ions through the cyano bridges.

4 Conclusions

Recently, octacyano metalate-based magnets have drawn much attention. In order to investigate the nature of Mn(II)–W(V) magnetic coupling and to search a new family of molecule-based magnets, the manganese(II) octacyanotungstate(V)-based magnet $MnFe[W(CN)_8]\cdot 8H_2O$ has been synthesized, the magnetic susceptibility obeys the Curie-Weiss law with $\Theta = -6.052$ K, $C = 1.307$ $cm^3\cdot k\cdot mol^{-1}$, $\chi_m T$ decreases abruptly on further lowering the temperature due to zero-field splitting, suggesting the presence of an antiferromagnetic interaction for the complexes. The effective moment μ_{eff} sharply decreases at lower temperature, which also indicates an intramolecular antiferromagnetic coupling between Mn^{II} and W^{IV} ions, while the FC curve at high temperature coincides with ZFC curve.

Acknowledgments

This work was financially supported by the National Natural Science Foundation of China (No. 11164002, 11364004) and NSF of Guangxi Province (No. 0991092).

References

[1] J.M. Herrera, D. Armentano, G. de Munno, et al., New Journal Of Chemistry, 27, p. 128, 2003.

[2] Guillaume Rombaut, Corine Mathonie`re, Philippe Guionneau, Inorganica Chimica Acta, 326, pp. 27–36, 2001.
[3] R. Garde, C. Desplanches, A. Bleuzen, et al., Molecular Crystallography Liquid Cryst., 334, p. 587, 1999.
[4] Ryo Watanabe, Hiroki Ishiyama, Goro Maruta, et al., Polyhedron, 24, pp. 2599–2606, 2005.
[5] J.G. Leipoldt, L.D. Bok, P.J.Z. Cilliers, Anorg. Allg. Chem., 409, pp. 343–344, 1974.
[6] N.H. Furman, C.O. Miller, P.G. Arvan, et al., Inorg. Synth., 3, pp 160–162, 1950.
[7] N.W. Alcock, A. Samotus, J. Szklarzewicz, Journal of Chemical Society, Dalton Transactions, p. 885, 1993.
[8] Zuzana Mitróová, Marián Mihalik, Anton Zentko, et al., Ceramics – Silikáty, 49(3), pp. 181–187, 2005.
[9] Juan Manuel Herrera, Donatella Armentano, Giovanni de Munno, et al., New Journal of Chemistry, 27, pp. 128–133, 2003.
[10] Wen Dong, Ya-Qiu Sun, Li-Na Zhu, et al., New Journal of Chemistry, 27, pp. 1760–1764, 2003.
[11] Zhuang Jin Zhong, Hidetake Seino, Yasushi Mizobe, et al., Inorganic Chemistry, 39, pp. 5095–5101, 2000.
[12] S. Ohkoshi, Y. Abe, A. Fujishima, et al., Review Letters, 82, pp. 1285–1288, 1999.
[13] Liang Shen, Yijian Zhang, Jin Uiu, Journal of Coordination Chemistry, 41(6), pp. 1760–1764, 2006.
[14] Hu Zhou, Ying-Ying Chen, Ai-Hua Yuan, et al., Inorganic Chemistry Communications, 11, pp. 363–366, 2008.
[15] Hyun Hee Ko, Jeong Hak Lim, Houng Sik Yoo, et al., Dalton Transactions, pp. 2070–2076, 2007.
[16] Robert Podgajny, Ce´dric Desplanches, Barbara Sieklucka, et al., Inorganic Chemistry, 41(5), p. 1323, 2002.

A discriminant method of whether lightning flashover occur based on integral area of discretized zero-sequence current

Zhangke[1], Zhangling[1], Zhaoyu[1], Xutao[1], Fang Xia[2], Suhong Chun[2]
[1]*Dali Power Supply Bureau, Dali, China*
[2]*Kungming University of Science and Technology, Kunming, China*

Abstract

The application of high-voltage transmission lines protection based on traveling-waves or transient components suffers maloperation in case of lightning strokes on transmission lines. Lighting stroke is one of the important causes of the accidents that occur on transmission lines. With the development of power system, the proportion of outages on transmission lines because of lightning stroke also increases. And according to the lightning accidents results, the lightning stroke characteristics is related to the time factors tightly. In order to analyze the correlativity between the lightning flashover amount and the time factors. The correlativity between the lightning stroke accidents investigation records, the lightning flashover amount of transmission lines increases from the recently years. According to the lightning flashovers mostly happen in the month of June, July, and August. Similarly in each day the flashover amount also varies with the time of day obviously. These lightning flashovers mainly occur during the afternoon. The analysis results in this paper have a good agreement with the meteorological observations and lightning detection data of lightning location system (LLS). And these results provide good reference for the lightning protection work in power system. According to the differences, the paper proposes a method identify lightning flashover based on zero sequence currents. AutoCAD simulation tests prove the method is feasible.

Keywords: index terms-identification, zero sequence current, lightning flashover.

1 Introduction

In order to distinguish whether lightning flashover occurs in the electric transmission line, this paper presents the discriminated method that based on integral area

WIT Transactions on Engineering Sciences, Vol. 107, © 2015 WIT Press
www.witpress.com, ISSN 1743-3533 (on-line)
doi:10.2495/SEEEI140081

of discredited zero-sequence current. This measure belongs to relay protection of electrical power system, extracts zero-sequence current from the lightning area. Zero-sequence current was recorded, and calculated to present time domain waveform in short-time window [1, 2]. Integral area of discredited zero-sequence current will reflect the relationship between flashover or not. The experimental results indicate that there are different data result after comparing integral area that belongs to two situations. According to these different results, whether lightning flashover occurs in the electric transmission line is clear [3, 4]. Thunderstorm static will misdirect the judgments from traveling wave protection and transient protection and will lead to malfunction. Thunderstorm static usually is a unipolar pulse wave belongs to high frequency signal [5, 6]. Because of the high frequency interference signals caused by lightning will be filtered, it is not a big interference for traditional protection based on main frequency parameters. But the transient signals caused by thunderstorm static and lightning stroke fault both are high frequency signals, it is necessary to distinguish whether lightning flashover is occur to avoid malfunction. So, distinguish whether lightning flashover is occur in the electric transmission line is a key problem for traveling wave protection and transient protection.

2 The method of discrimination

The zero sequence current will be extracted in transmission lines, in which lightning struck was happened, and the device will immediately start. The time domain waveform extracted of zero sequence current after lightning occur within a short time windows [7, 8]. The criterion is integral calculated to obtain the area on both sides of the 0 axis, and the method of discrimination is depending on comparisons of zero sequence current waveform's areas of the positive and negative side to 0 axis. The method of discrimination in accordance with the principle of criterion of transmission line lightning flashover or not. The processes are as follows:

(1) When transmission lines are struck by lightning, the device will immediately start, and three currents I_A, I_B, I_C will be extracted.

(2) The zero sequence current in transmission struck by lightning is calculated by:

$$i_0(j) = i_A(j) + i_B(j) + i_C(j) \tag{1}$$

$j = 1, 2, \quad , n$. $i_A(j)$, $i_B(j)$, $i_C(j)$ are the discrete signal points of the three-phase line currents I_A, I_B, I_C, respectively. n is the total extracting numbers of the discrete signal.

(3) The time domain waveform extracted of zero sequence current after lightning occur within a short time windows, and it will be integrated by:

$$A_1 = \sum_{p=1}^{m} i_0^{+}(p) \tag{2}$$

And

$$A_2 = \sum_{q=1}^{n-m} \left| i_0^-(q) \right| . \tag{3}$$

$i_0^+(p)$ and $i_0^-(q)$ are the discrete signal of the zero sequence current, $i_0^+(p) \geq 0$, $i_0^-(q) < 0$, $p = 1, 2, \ldots, m$, $q = 1, 2, \ldots, n-m$;

m is the numbers of discrete signal $i_0^+(p)$.

(4) Comparisons parameter of zero sequence current waveform's areas of thc positive and negative side to 0 axis is calculated by:

$$k = \frac{\max(A_1, A_2)}{\min(A_1, A_2)} \tag{4}$$

$\max(A_1, A_2)$ means the bigger one in A_1, A_2. $\min(A_1, A_2)$ means the smaller one in A_1, A_2.

(5) Threshold value Φ of k is determined by simulation model PSCAD. The method of discrimination is as follows:

If $k > \Phi$, the discrimination flashover occurs.
If $k \leq \Phi$, the discrimination flashover does not occur.

3 Models and parameters of transmission line

As shown in Figure 1, PSCAD software is used to simulate a typical 500KV EHV transmission system. The simulation takes into account the integral calculated to obtain the area on both sides of the 0 axis, and the method of discrimination is depending on comparison of zero sequence current waveform's areas of the positive and negative side to 0 axis. Figure 2 shows the flow chart of discriminate method of lightning flashover occur or not in transmission line. The bus stray capacitance is 0.05μf and the fault angle is 40°. The signal-sampling rate is 1MHz. The insulator is represented by current-controlled switch, and its lightning current is 25KA.

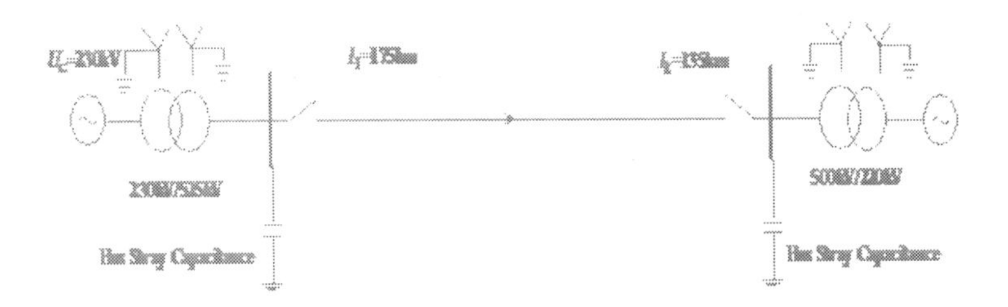

Figure 1: PSCAD software work flow.

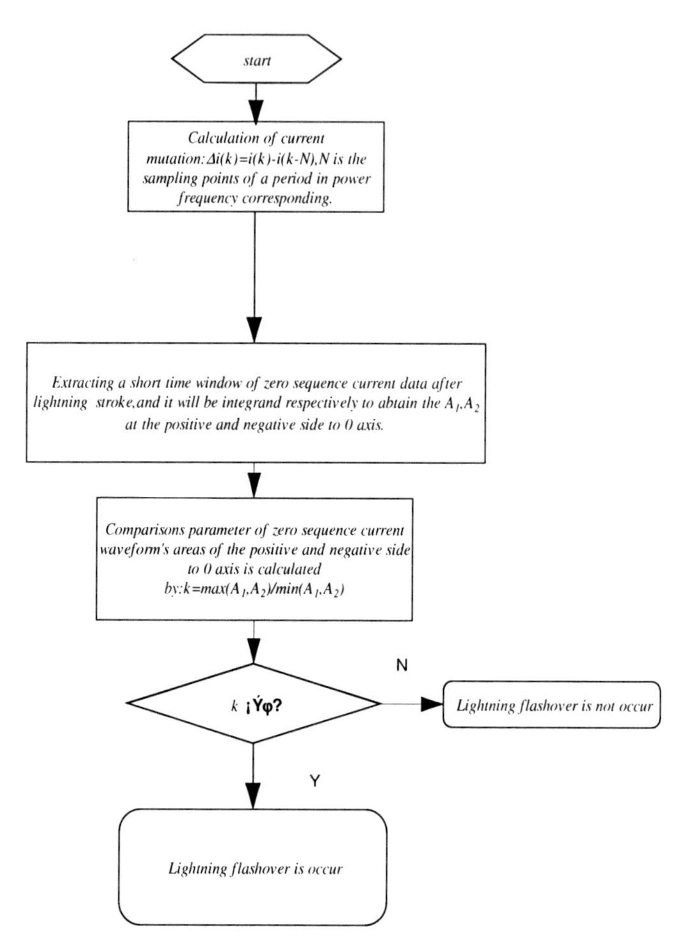

Figure 2: The flow chart of discriminate method of lightning flashover occur or not in transmission line.

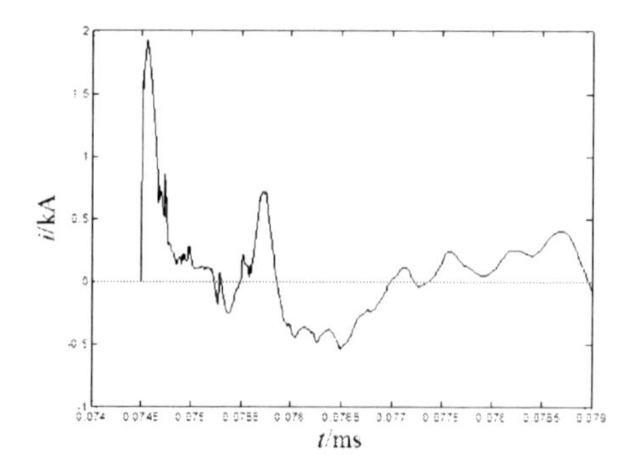

Figure 3: Zero sequence without lightning flashover.

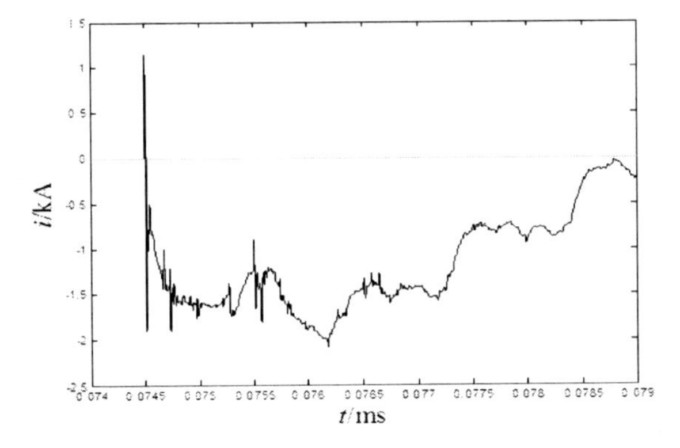

Figure 4: Zero sequence when lightning flashover occur within 5ms.

The zero sequence current is influenced by some stochastic factors, has uncertain amplitude. Figure 3 shows the chart of zero sequence without lightning flashover. Figure 4 shows the chart of zero sequence when lightning flashover occur within 5ms.

4 Analysis of the simulation process and results

In simulation, the lightning stroke occur at the 175KM to the bus in the model of Figure 1. The range of power transmission is 100–500MW, and the range of length is 100–300KM while this simulation is 300KM of the whole transmission with the voltage classes 230KV. And the flashover occurs. The bus stray capacitance is 0.05μf and the fault angle is 40°. The signal-sampling rate is 1MHz. According to the formulas:

$$A_1 = \sum_{p=1}^{n} i_0^+(p) \qquad i_0^+(p) > 0,\ p = 1, 2, \ldots, m \tag{5}$$

$$A_2 = \sum_{q=1}^{n-m} \left| i_0^-(q) \right| \quad i_0^-(q) \le 0, q = 1, 2, \ldots, n-m \tag{6}$$

$$k = \frac{\max(A_1, A_2)}{\min(A_1, A_2)} \tag{7}$$

Comparison parameters of zero sequence current waveform's areas of the positive and negative side to 0 axis calculated results are A_1=847.6440, A_2=411.3540, k=2.0606, respectively. Under the large number of simulation experiments, set the threshold value $\varphi = 50$ of comparison of zero sequence current waveform's areas of the positive and negative side to 0 axis. The result of comparison:

$$k = 2.0606 < \varphi = 50$$

Thus, the discrimination of flashover occurs.

Table 1 shows the different results influenced by different lightning currents.

Table 1: Different results of lightning currents.

Lightning Currents	2.5×10^4 A	2.0×10^4 A	1.5×10^4 A	1.0×10^4 A
A1	847.6440	1253.5230	1852.5231	2243.7253
A2	411.3540	241.4235	125.1345	41.4586
k	2.0606	5.1922	14.8043	54.1203
Results	Flash-over occur	Flash-over occur	Flash-over occur	Flash-over not occur

5 Conclusions

In recent years, a hot wave of round of power grid construction has been raised in YunNan. And the correlativity's between lightning flashover amount with the year, month, and the time of day are analyzed in this paper. Also the results are compared with the lightning detection data of lightning location system (LLS), and some relationships are found out by the comparison. Based on the analysis, whether the lightning flashover is occurred can be much more accurately discriminate. The discrimination of lightning flashover can be completed by the sampling data directly without the switch between time domain and frequency domain and only need a short window of zero sequence current sampling data.

References

[1] Huang Xiaqian, Xion Xilin, Yang Xuequan, Yunnan power grid transmission lines around the hit anti-technology research and application, May 2008.

[2] Ren Jin-feng, Duan Jian-dong, Zhang Bao-hui, Identification of lightning disturbance in ultra-high-speed transmission line protection, 2005.

[3] Li Xiao-lan, Chen Jia-hong, Gu Shan-qiang, Tong Xue-fang, Statistics and analysis of lightning flashovers of transmission lines during 2000–2007, March 2009.

[4] Yi Hui, The analysis of lightning-protection performance and the improvement measure for the shared tower double-circuit 500 kV transmission lines. *High Voltage Engineering*, **24(2)**, pp. 52–55, 1998.

[5] Ge Dong, Du Shu-chun, Zhang Cui-xia, Lightning protection of AC 1000 kV UHV transmission line. *Electric Power*, **39(10)**, pp. 24–28, 2006.

[6] Mei Zhen, Chen Shui-ming, Gu Qin-wei, et al., Statistics of lightning accidents during 1998–2004 in China. *High Voltage Engineering*, **33(12)**, pp. 173–176, 2007.

[7] Jiang Xing-liang, Yuan Ji-he, Sun Cai-xin, et al., External insulation of 800 kV UHV DC power transmission lines in China. *Power System Technology*, 30(9), pp. 1–9, 2006.

[8] Hu Yi, Analysis on operation faults of transmission line and countermeasures. *High Voltage Engineering*, **33(3)**, pp. 1–8, 2007.

Design and simulation of multilayer energy selective surfaces

Bo Yi, Cheng Yang, Peiguo Liu, Yanfei Dong
College of Electronic Science and Engineering,
National University of Defense Technology, Hunan, Changsha, China

Abstract

Electromagnetic weapons have shown their force on the battlefield in Iraq and pose a threat to the survival of information equipment. This paper presents a broad band multilayer energy selective surface to protect important equipment from electromagnetic weapon attack. A simulation model is built up in CST. The influence of different parameters to multilayer energy selective surface is studied. The transmission attenuation of multilayer energy selective surface can easily be less than $2\,\mathrm{dB}$ in low power level and greater than $20\,\mathrm{dB}$ in high power level at broad band.
Keywords: multilayer energy selective surface, strong electromagnetic protection, broad band defending.

1 Introduction

Microwave receivers are very sensitive devices in order to detect small signals [1]. Electromagnetic weapons develop rapidly and threaten to survival of electronic equipment especially broadband microwave receivers [2]. Currently FSS as spatial filter is the main protection method implemented at front-door and used to filter the energy out of band [3]. But they can't provide protection from high power wave within their pass-band [4]. The concept of energy selective surface has been presented in Ref. [5]. It moves limiter in the circuit to space and has good shielding effectiveness at lower frequency. Changeable surface impendence is the key to realize self-actuation surface. The surface impendence is high in normal state, and when attack waves come, it decreases rapidly and reflects most of energy. But its effective working band of surface is narrow. The energy selective surface performance influence factors are analyzed in Ref. [6]. The

WIT Transactions on Engineering Sciences, Vol. 107, © 2015 WIT Press
www.witpress.com, ISSN 1743-3533 (on-line)
doi:10.2495/SEEEI140091

analysis results show that the smaller the structure, the better the performance. But too small surface is difficult to process.

In this contribution, a 3D multilayer energy selective surface equipped with PIN diode is presented for improving defending band. It is composed by two or more layers energy selective surface and substrate. Compared to the tuning FSS [7], the 3D multilayer energy selective surface mainly relies on intensity of incident wave.

The paper is organized as follows. Section 2 introduces traditional energy selective surface. In Section 3 multilayer energy selective surface is designed and its working principle is explained through its equivalent circuit. In Section 4, multilayer energy selective surface is simulated in CST to illustrate its performance. Conclusions are drawn in Section 5.

2 Design and analysis of multilayer energy selective surface

The structure of multilayer energy selective surface is shown in Figure 1. The sandwich-like structure contains at least two layer energy selective surfaces and one substrate. The width l of substrate has a great impact on insertion loss [8]. When l is set to a proper value, the protection bandwidth of multilayer energy selective surface will become wider. That is to say, the insertion loss is low and shielding effectiveness is high in broad band. But thicker substrate may bring larger loss. In order to reduce loss of substrate, high dielectric constant material is employed. PIN diode is chosen as variable impendence device owning to its low junction capacitance and inductance.

The incident electromagnetic waves propagate in multilayer energy selective surface as shown in Figure 2.

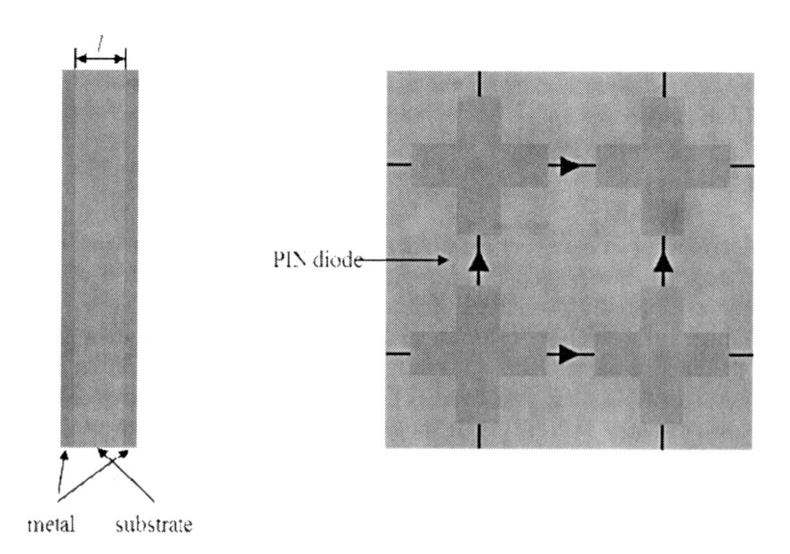

Figure 1: Geometry of two layers energy selective surface.

In Figure 2, E_{1i}, E_{1r}, and E_{1t} represent the incident wave, reflective wave and transmission wave respectively at left interface. E_{2i}, E_{2r}, and E_{2t} represent the incident wave, reflective wave and transmission wave respectively at right interface. As shown in Figure 2, when the incident field E_{1i} reaches left surface, some part of field E_{1t} penetrates into the substrate and the other E_{1r} reflects. The field E_{1t} goes on transmission in substrate. When the field E_{2i} reaches right surface, the field E_{2t} penetrates multilayer energy selective surface and goes on transmission in free space, and the other field E_{2r} reflects. The field E_{2r} arrives at the left face, part of the energy reflects and the other penetrates the surface into free space. The progress goes on like the one above. When diodes is on most of energy reflect to left space and when diodes is off most of energy transmit to right space.

According to diode state, multilayer energy selective surface has two different equivalent circuits. The equivalent circuits of multilayer energy selective surface are shown in Figure 3 [9, 10].

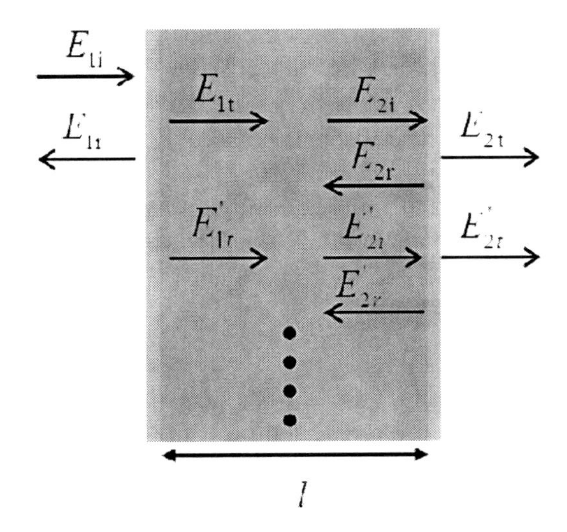

Figure 2: The propagation field of multilayer energy selective surface.

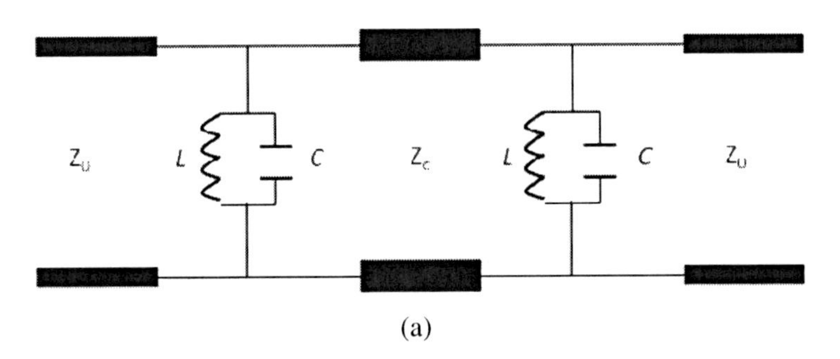

(a)

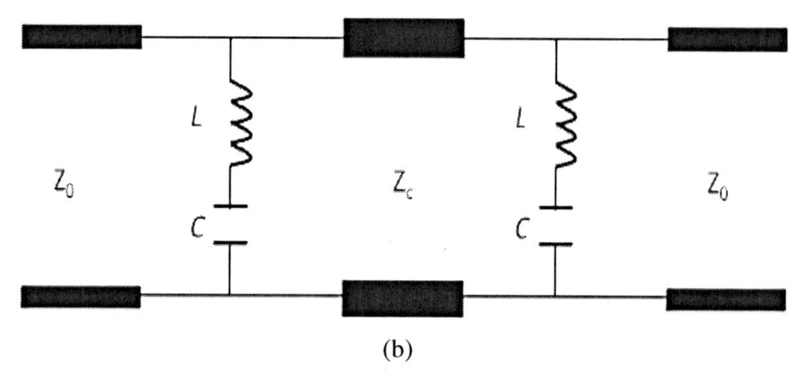

(b)

Figure 3: The equivalent circuit of multilayer energy selective surface. (a) the equivalent circuit when diode on. (b) the equivalent circuit when diode off.

When diode is on, the left surface and right surface can be seen as metal grid. The substrate can be seen as transmission line with characteristic impedance Z_C.Its equivalent circuit is shown in Figure 3(a). Transmission line with characteristic impedance Z_0 represents free space. When diode is off, the left surface and right surface can be seen as patches array. The substrate also can be seen as transmission line with characteristic impedance Z_C. Its equivalent circuit is shown in Figure 3(b).

Compared to single layer, multilayer energy selective surface introduces extra transmission pole when diode is off. When the value of l is chosen properly, defending band will become wider.

3 Simulation of multilayer energy selective surface

In this chapter, a structure is designed and modelled using CST. In order to understand how different parameters influence the performance of multilayer ESS, we keep parameters unchangeable expect one. In simulation, diode is replaced by its equivalent circuit. Bap 6302 Silicon PIN diode of NXP semiconductors is chosen as in section two.

The substrate thickness l of multilayer energy selective surface, the number of layers for energy selective surface and period length for cell are three important factors to multilayer energy selective surface. Figures 4–6 show simulation result of multilayer energy selective surface transmission attenuation at different substrate thickness l, number of layers and period size respectively. In simulation, the energy selective surface parameters are same as Table 1. The substrate thickness is 15 mm expect in Figure 4. For Figures 4 and 6, two layers energy selective surface is simulated. Working band referred to the band where insertion loss is less than 2 dB and defending band referred to the band where shielding effectiveness is more than 15 dB are used to evaluate the performance of multilayer energy selective surface.

Table 1: List of parameters for energy selective surface.

Parameter Definition	Dimensions (mm)
Thickness of substrate, t_1	0.5
Thickness of metal, t_2	0.035
Width of geometrical cell, W	3
Width of metal patch, w_1	2.4
Length of geometrical cell, D	3
Length of metal patch, d_1	2.2

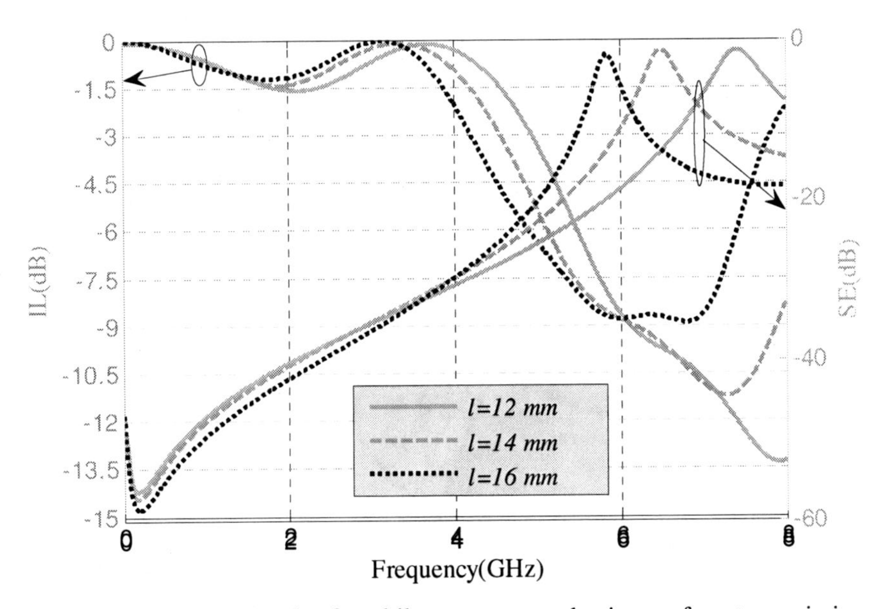

Figure 4: Simulation result of multilayer energy selective surface transmission attenuation for different substrate thickness l.

Figure 4 shows that when diode is off, working band of multilayer energy selective surface decreases as substrate thickness l increases. But at lower frequency, insertion loss is smaller for thicker substrate. When diode is on, defending band of multilayer energy selective surface decreases as substrate thickness l increases. There are two transmission poles as diode is off and the substrate thickness is about a quarter of wavelength of second pole. This can explain the reason why the thicker the substrate, the wider the working band.

Figure 5 shows that when diode is off, the working band of multilayer energy selective surface increases as the number of layers increases. But at lower frequency, the insertion loss is smaller for less layers energy selective surfaces. When diode is on, the shielding effectiveness is higher for more layers energy selective surfaces. At this case, energy selective surface can be seen as metal grid. The more layers the metal grid, the higher shielding effectiveness.

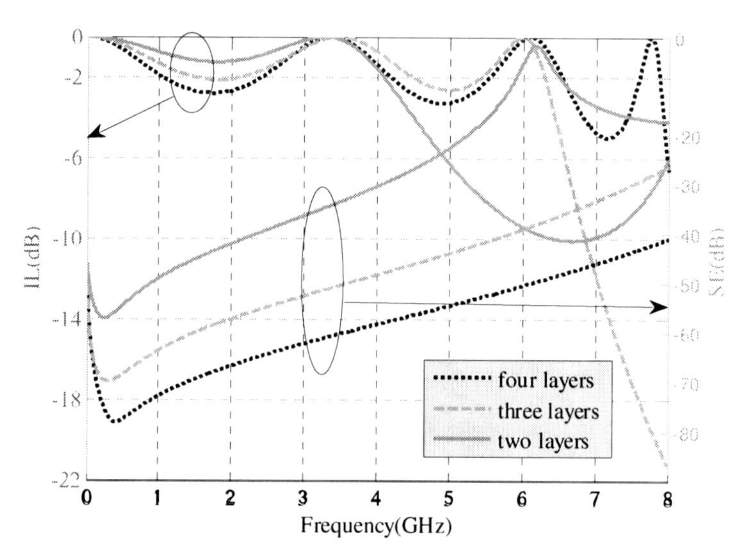

Figure 5: Simulation result of multilayer energy selective surface transmission attenuation for different number of layers.

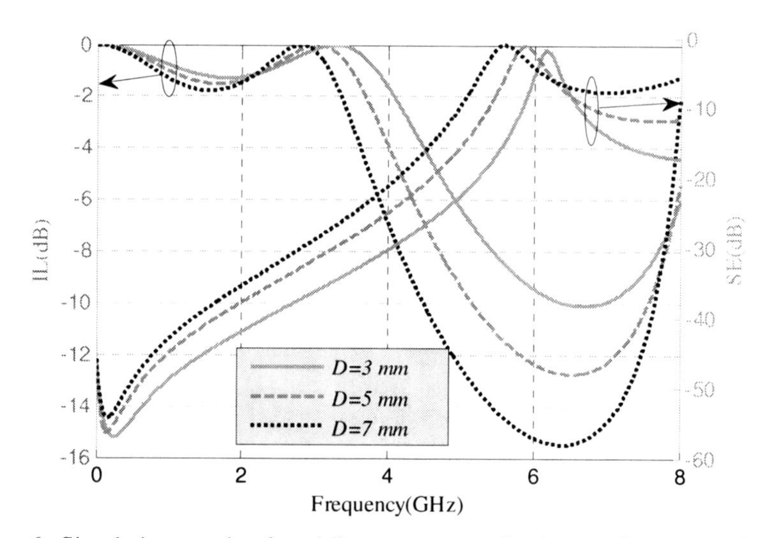

Figure 6: Simulation result of multilayer energy selective surface transmission attenuation for different period *D*.

Figure 6 shows that when diode is off, working band of multilayers energy selective surface decreases as period *D* increases. When diode is on, defending band of multilayers energy selective surface decreases as period *D* increases. Based on the simulation result, we know that the smaller the cell, the better the performance of multilayer energy selective surface.

4 Conclusions

A multilayer energy selective surface is presented in this paper, which is constructed by cascading two or more energy selective surfaces and substrates. Compared with traditional 2D energy selective surface, multilayer energy selective surface provide wider working band when diode is off and defending band when diode is on. Simulations for multilayer energy selective surface show that defending band can meet the requirement easily. When diode is off, insertion loss of multilayer energy selective surface can be less than 2dB. When diode is on, shielding effectiveness can gain more than 20 dB at broad band.

References

[1] S. Monni, D.J. Bekers, M. vanWanum, R. vanDijk, A. Neto, G. Gerini, F.E. van Vliet, Limiting frequency selective surfaces. In: *Proceedings of EuMC*, pp. 606–609, October 2009.

[2] Libor Palisek, Lubos Suchy, High power microwave effects on computer networks. In: *Proceedings of 10th International Symposium on Electromagnetic Compatibility*, pp. 18–21, September 2011.

[3] B.A. Munk, *Frequency Selective surface: Theory and Design*. Wiley Interscience, USA, 2000.

[4] Sean Scott, Christopher D. Nordquist, Michael J. Cich, et al., A frequency selective surface with integrated limiter for receiver protection. In: *IEEE Antennas and Propagation Society International Symposium* (APSURSI), pp. 1–2, 2012.

[5] C. Yang, P.L. Liu, X.J. Huang, A novel method of energy selective surface for adaptive HPM/EMP protection. *IEEE Antennas and Wireless Propagation Letters*, **12**, pp. 112–115, March 2013.

[6] Jianfeng Tan, Peiguo Liu, Cheng Yang, Bo Yi, Xianjun Huang, EM simulation analysis of the energy selective surface. *High Voltage Engineering*, **39(10)**, pp. 30897–30903, October 2013.

[7] Yangjun Zhang, Mengqing Yuan, Qinghuo Liu, Ultra wide band response of an electromagnetic wave shield based on a diode grid. *Progress in Electromagnetic Research*, **141**, pp. 591–606, 2013.

[8] Ping Lu, Guang Hua, Chen Yang, Wei Hong, A wideband bandstop FSS with tripole loop. In: *Proceedings of International Symposium on Antennas and Propagation*, pp. 1291–1294, October 2013.

[9] Abbas Abbaspour Tamijani, Kamal Sarabandi, Gabriel M. Rebeiz, Antenna-filter-antenna arrays as a class of bandpass frequency selective surface. *IEEE Transactions on Microwave Theory and Techniques*, **52(8)**, pp. 1781–1789, August 2004.

[10] Dubrovka, Donnan, Equivalent circuit of FSS loaded with lumped elements using modal decomposition equivalent circuit method. In: *Proceedings of the 5th European Conference on Antennas and Propagation*, pp. 2250–2253, April 2011.

Dynamic analysis and numerical simulation of multipole field electromagnetic launcher

Yingwei Zhu[1], David Thomas[2]
[1]*School of Electronical Engineering and Information, Sichuan University, Chengdu, China*
[2]*Department of Electrical and Electronic Engineering, University of Nottingham, Nottingham, UK*

Abstract

Based on eddy current problem with moving conductor, this paper presents the eddy current calculational formulation with magnetic vector and scalar potentials and the boundary condition equations. The electromagnetic force and torque is expressed by dynamic analysis formulations. An optimizational experimental three-stage launch model with arced saddle coils is proposed and numerical simulated. The results indicate that a 227g projectile is accelerated to an exit velocity of 164.66m/s with three-stages of significant pulsed electromagnetic force and magnetic torque.
Keywords: multipole field electromagnetic launcher, eddy current, moving conductor, dynamic analysis, FEM simulation.

1 Introduction

The multipole field electromagnetic launcher is characterized by the interaction of the radial magnetic field with azimuthal eddy current. It has a better ratio of the axial thrust force to the radial compression force. In the recent papers [1, 2], the operation principle and launch model of the multipole field electromagnetic launcher are introduced and analyzed in theory, and the launch performance of an improved twisty launch model is simulated with transient motion Finite Element Method in multiple degrees of freedom. The paper [3] derives an analytical solution approach based on the second-order vector potential formulation. The paper [4] analyses the different coil connection patterns and presents the experimental performances.

www.witpress.com, ISSN 1743-3533 (on-line)
doi:10.2495/SEEEI140101

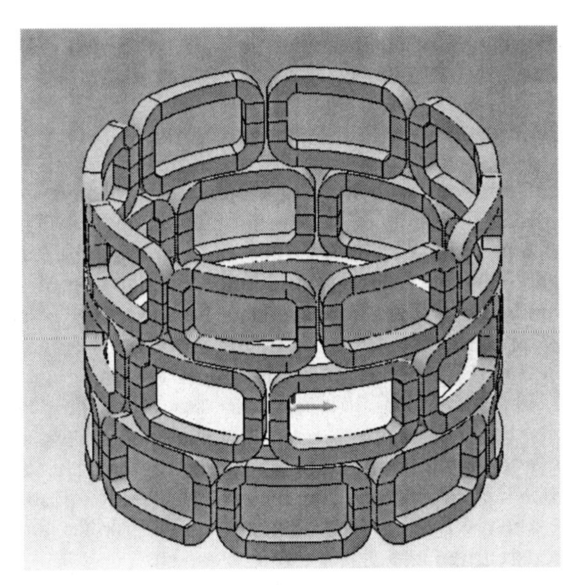

Figure 1: An optimizational configuration of three-stage twisty octapole electromagnetic launcher with arced saddle coils.

In this paper, an optimizational configuration of three-stage twisty octapole electromagnetic launcher with arced saddle coils is proposed shown as in Figure 1. The projectile is aluminum cylinder sleeve. This proposed launch model has better reasonable mechanical structure. As the multi-stages launch coils are excited with sequential pulsed power currents, they could generate a better helicoidal travelling magnetic wave. The transient multipole magnetic field induces the moving projectile and produces the mainly axial acceleration force and azimuthally restoring torsion. The projectile could be continuously accelerated to high speed and helically flies to a long distance.

2 Dynamic analysis

The basic principle of multipole electromagnetic launcher is Faraday's Law of Induction. As the projectile has a good conductivity with fast moving speed, the analysis model of the launcher lies in identifying the current distribution on the projectile, thereby complicating any calculations in which electromagnetic force and energy conversion efficiency are involved. There are two effective analysis models for the launch model [5]: (I) a simple model based on an assumed two-dimensional current distribution in the conductors; (II) a filamentary model in which both the drivecoil and the projectile are divided into a number of separate but interacting thin currents rings. The two methods focus on calculating the mutual inductance and electromagnetic force between drivecoil and projectile. The great difficulty of dynamic analysis for the multipole electromagnetic launcher is the eddy current calculation. We adopt the magnetic vector potential formulation with FEM (Finite Element Method) and BEM (Boundary Element

Method). The calculation of electromagnetic force is carried out by cylindrical-coordinate system.

2.1 Eddy current problem with moving conductor

The eddy current problem with moving conductor is always one of the most complex and difficult problems in electro-mechanical devices. The review paper [6] has outlined four typical techniques for eddy current problem with moving conductor: air-gap element, Lagrange multiplier, overlapping mesh, and moving band. Some formulations [7] are now available for modeling the electromagnetic behavior of moving conductor devices. The 3D formulation combining a Lagrangian description and FEM-BEM coupling [8] is an important progress. The composite grid method (CGM) [9] based on the FEM for determining eddy currents in moving conductors is evolutionally developed.

Figure 2 shows a simple configuration of three regions causing eddy current problem with moving conductor. The region Ω_1 is moving conductor with a relative velocity $\boldsymbol{u}$ to the fixed power supply current region Ω_2, and the region Ω_3 is air. $\boldsymbol{J}_e$ is the eddy current and $\boldsymbol{J}_s$ is the source current.

In the laboratory reference frame and coordinate system, the electric field $\boldsymbol{E}$ on the moving conductor could be expressed as three components with magnetic vector potential $\boldsymbol{A}$ and electric scalar potential V.

$$\boldsymbol{E} = \boldsymbol{u} \times \boldsymbol{B} - \frac{\partial \boldsymbol{A}}{\partial t} - \nabla V \tag{1}$$

The eddy current on the moving conductor could be calculated by

$$\boldsymbol{J}_e = \sigma \boldsymbol{u} \times \nabla \times \boldsymbol{A} - \sigma \frac{\partial \boldsymbol{A}}{\partial t} - \sigma \nabla V \tag{2}$$

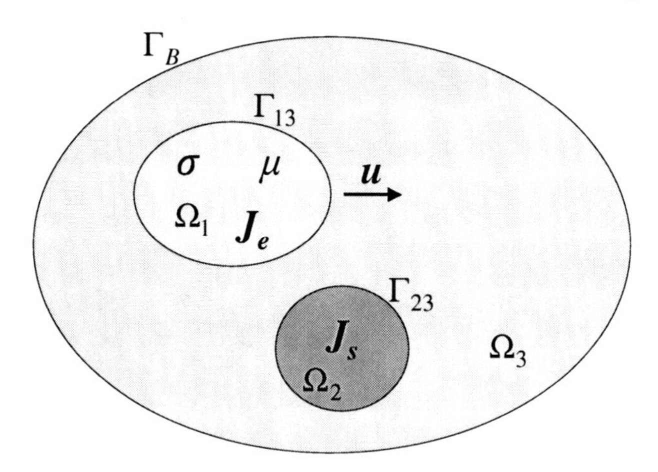

Figure 2: Eddy current problem with moving conductor.

In the moving conductor region Ω_1, deriving from Maxwell's equations, we could equivalently obtain:

$$\left.\begin{aligned}\nabla^2 \boldsymbol{A}_1 - \mu\sigma\boldsymbol{u}\times\nabla\times\boldsymbol{A}_1 + \mu\sigma\frac{\partial \boldsymbol{A}_1}{\partial t} + \mu\sigma\nabla V_1 = 0\\ \nabla\cdot(\sigma\boldsymbol{u}\times\nabla\times\boldsymbol{A}_1 - \sigma\frac{\partial \boldsymbol{A}_1}{\partial t} - \sigma\nabla V_1) = 0\end{aligned}\right\} \tag{3}$$

For a unique system, Coulomb gauge $\nabla\cdot\boldsymbol{A} = 0$ should be chosen to impose to Eq. (3).
In the fixed power supply current region Ω_2, source current should satisfy

$$\nabla^2 \boldsymbol{A}_2 - \boldsymbol{J}_S = 0 \tag{4}$$

In the air region Ω_3, magnetic vector potential should satisfy

$$\nabla^2 \boldsymbol{A}_3 = 0 \tag{5}$$

Eqs. (3)–(5) could be solved by adding the boundary conditions [10]

$$\left.\begin{aligned}\boldsymbol{A}_1 - \boldsymbol{A}_3 = 0\\ \boldsymbol{n}_{13}\cdot(\sigma\boldsymbol{u}\times\nabla\times\boldsymbol{A}_1 - \sigma\frac{\partial \boldsymbol{A}_1}{\partial t} - \sigma\nabla V_1) = 0\end{aligned}\right\} \quad \text{on } \Gamma_{13} \tag{6}$$

$$\boldsymbol{A}_2 - \boldsymbol{A}_3 = 0 \quad \text{on } \Gamma_{23} \tag{7}$$

$$\left.\begin{aligned}\nabla\cdot\boldsymbol{A}_3 = 0\\ \nabla\times\boldsymbol{A}_3 = 0\end{aligned}\right\} \quad \text{on } \Gamma_{B} \tag{8}$$

The most powerful numerical method for solving Eqs. (3)–(5) should be the ordinary Galerkin Finite Element Method.

2.2 Electromagnetic force calculational formulation

The distribution of magnetic field generated by multipole drivecoil is shown as sectional view in Figure 3. The major component of magnetic field in the space is the radial direction B_r. The eddy current distributes in the skin depth of the projectile. The major components of eddy current are axial J_z and azimuthal J_φ.

Thus, the eddy current $\boldsymbol{J}_e$ on the projectile acting with magnetic field $\boldsymbol{B}$ produces the magnetic force density as

$$\begin{aligned}\boldsymbol{f} &= \boldsymbol{J}_e\times\boldsymbol{B} = f_r\boldsymbol{e}_r + f_\varphi\boldsymbol{e}_\varphi + f_z\boldsymbol{e}_z\\ &= (J_\varphi B_z - J_z B_\varphi)\boldsymbol{e}_r + (J_z B_r - J_r B_z)\boldsymbol{e}_\varphi + (J_r B_\varphi - J_\varphi B_r)\boldsymbol{e}_z\end{aligned} \tag{9}$$

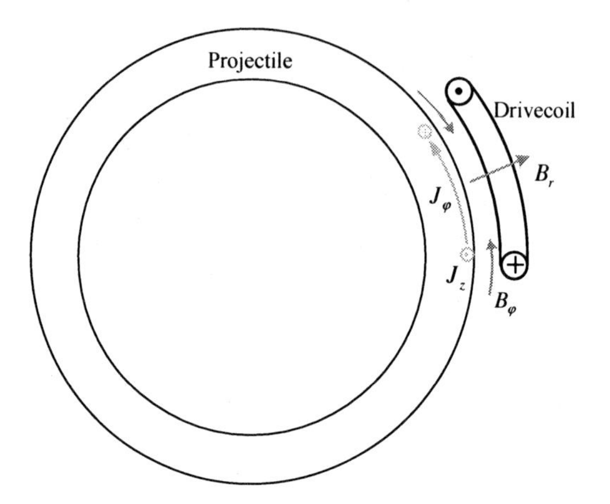

Figure 3: The distribution of magnetic field and eddy current.

Under the hypothesis of thin projectile sleeve, it is possible to assume the radial eddy current J_r is close to zero. Then the axial acceleration force of projectile is approximate

$$f_z = -J_\varphi B_r \qquad F_z = \int_V f_z \mathrm{d}V \tag{10}$$

The azimuthal torsional force of projectile is approximate

$$f_\varphi = J_z B_r \qquad F_\varphi = \int_V f_\varphi \mathrm{d}V \tag{11}$$

The projectile has an axial torque as

$$m_z = rf_\varphi = rJ_z B_r \qquad M_z = \int_V m_z \mathrm{d}V \tag{12}$$

The radial components force of projectile is

$$f_r = J_\varphi B_z - J_z B_\varphi \qquad F_r = \int_V f_r \mathrm{d}V \tag{13}$$

2.3 Motion equations and energy conversion efficiency

The axial acceleration force F_z is to accelerate the projectile to high speed. The axial translational motion equations of projectile are

$$a_z = F_z / m = \int_V J_\varphi B_r \mathrm{d}V / m \tag{14}$$

$$v_z = v_0 + \int_0^T a_z \mathrm{d}t \qquad s_z = s_0 + \int_0^T v_z \mathrm{d}t \tag{15}$$

The axial torque makes the projectile have a rotational motion about its axis, which makes the projectile stabilized as if from a rifle. The rotational motion equations of projectile are

$$\alpha_z = M_z / I_z = \int_V r J_z B_r \mathrm{d}V / I_z \tag{16}$$

$$\omega_z = \omega_0 + \int_0^T \alpha_z \mathrm{d}t \qquad \theta_z = \theta_0 + \int_0^T \omega_z \mathrm{d}t \tag{17}$$

The multipole electromagnetic field launcher is powered by the pulsed capacitor banks, the launch energy conversion efficiency is the ratio of the projectile's kinetic energy increment to the electrical energy initially stored in the capacitors. The efficiency is expressed by

$$\eta = \frac{\Delta E_k}{E_C} = \frac{\frac{1}{2}m(v_z^2 - v_0^2) + \frac{1}{2}I_z(\omega_z^2 - \omega_0^2)}{\sum_N \frac{1}{2} C_i U_{i0}^2} \tag{18}$$

where m is the mass of the projectile, v_0 is the initial speed of the projectile, v_z is the final speed, N is the launch stage number, C_i is the capacitance of the pulsed capacitors, U_{i0} is the initial voltage of the capacitors.

3 Numerical simulation evaluation

The transient with motion simulation model of three-stage twisty octapole electromagnetic launcher is established as shown in Figure 4. The multipole coils is optimizational designed with arced saddle coils. Every stage coils in series are fed by pulsed power capacitors. The projectile is aluminum cylinder sleeve with a bottom. The discharge sequence of the capacitors to multipole coils is determined by the presupposition moving location of projectile.

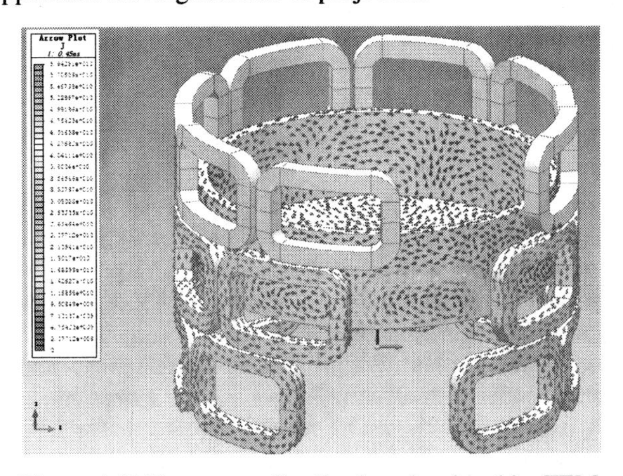

Figure 4: Eddy current distribution simulated by FEM.

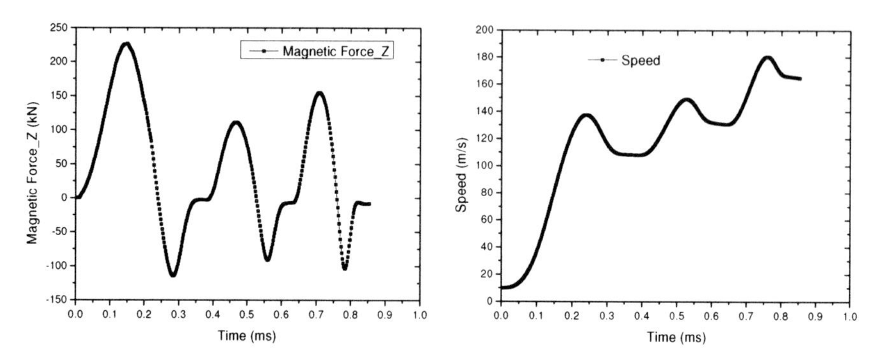

Figure 5: The axial acceleration force and translational speed of projectile.

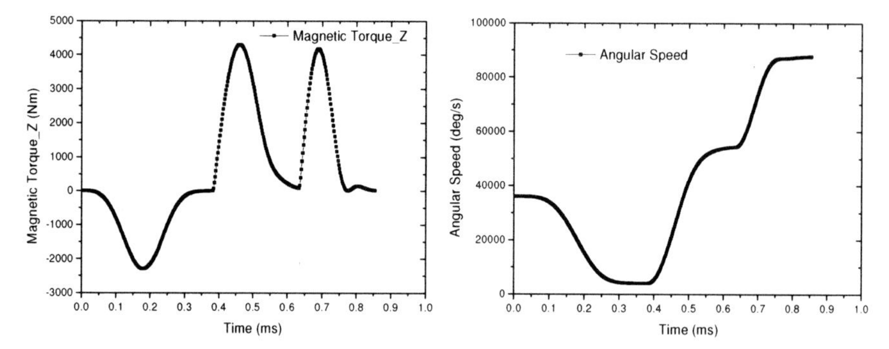

Figure 6: The azimuthal torsional force and angular speed of projectile.

Figures 5 and 6 show the FEM simulation results of the optimizational configuration of three-stage twisty octapole electromagnetic launcher with arced saddle coils. The results indicate that a 227g projectile is accelerated to an exit velocity of 164.66m/s with three-stage significant pulsed electromagnetic force. The magnetic torque and angular speed curves of the projectile show an impressive demonstration of spinning motion.

4 Conclusions

The essential principle of all kinds of electromagnetic launcher is the interaction of magnetic field with material. The multipole electromagnetic launcher belongs to the eddy current problem with moving conductor. This paper presents the eddy current calculational formulation with the scheme of AV-A. The electromagnetic force and torque and motion equations of the multipole launcher is expressed by vector dynamic analysis method. We propose an optimizational and experimental three-stage launch model with arced saddle coils. The launch model is numerical simulated with transient moving. The results indicate that a 227g projectile is accelerated to an exit velocity of 164.66m/s with significant electromagnetic force and torque.

Acknowledgement

The research work was supported by National Natural Science Foundation of China under Grant No. 51207097 and Supported by Doctoral Fund of Ministry of Education of China under Grant No. 20120181120100.

References

[1] Zhu Ying-Wei, Wang Yu, Yan Zhong-Ming, Dong Liang, Xie Xiao-Fang, Li Hai-Tao, Multipole field electromagnetic launcher. *IEEE Transactions on Magnetics*, **46(7)**, pp. 2622–2627, 2010.
[2] Zhu Ying-Wei, Wang Yu, Chen Wei-Rong, Yan Zhong-Ming, Li Hai-Tao, Analysis and evaluation of three-stage twisty octapole field electromagnetic launcher. *IEEE Transactions on Plasma Science*, **40(5)**, pp. 1399–1406, 2012.
[3] Musolino A., Rizzo R., Tripodi E., Travelling wave multipole field electromagnetic launcher: an SOVP analytical model. *IEEE Transactions on Plasma Science*, **41(5)**, pp. 1201–1208, 2013.
[4] Luo Wen-Bo, Wang Yu, Gui Zhi-Xing, Yan Zhong-Ming, Chen Wei-Rong, Connection pattern research and experimental realization of single stage multipole field electromagnetic launcher. *IEEE Transactions on Plasma Science*, **41(11)**, pp. 3173–3179, 2013.
[5] Novac B., Smith I., Enache M., Studies of a very high efficiency electromagnetic launcher. *Journal of Physics D: Applied Physics*, **35(12)**, pp. 1447–1457, 2002.
[6] Trowbridge C.W., Sykulski J.K., Some key developments in computational electromagnetics and their attribution. *IEEE Transactions on Magnetics*, **42(4)**, pp. 503–508, 2006.
[7] Rodger D., Lai H.C., A comparison of formulations for 3D finite element modeling of electromagnetic launchers. *IEEE Transactions on Magnetics*, **37(1)**, pp. 135–138, 2001.
[8] Kurz S., Fetzer J., Lehner G., A novel formulation for 3D eddy current problems with moving bodies using a Lagrangian description and BEM-FEM coupling. *IEEE Transactions on Magnetics*, **34(5)**, pp. 3068–3073, 1998.
[9] Peng Ying, Ruan Jiang-Jun, Zhang Yu, Gan Yan, A composite grid method for moving conductor eddy-current problem. *IEEE Transactions on Magnetics*, **43(7)**, pp. 3259–3265, 2007.
[10] Albertz D., Dappen S., Henneberger G., Calculation of the 3D nonlinear eddy current field in moving conductors and its application to braking systems. *IEEE Transactions on Magnetics*, **32(3)**, pp. 768–771, 1996.

Wireless power grid monitoring system based on ZigBee

Xudong Wu, Weidong Jin
Department of Electrical Engineering,
Southwest Jiaotong University, Chengdu, China

Abstract

This article represents a kind of design and implementation of wireless power grid monitoring system based on ZigBee for some factories, schools, and buildings. The aim of this system is to effectively and efficiently provide electrical power supply and management, especially save the energy of how many electrical power it used by LED of public schools. This system use power monitoring module, and realize the real time wireless communication based on ZigBee. The test confirms the correctness of the design, and the results meet the requirements. Some procedures and block diagrams have also been presented in this paper as well. The design has a certain novelty and practicality.
Keywords: smart meter, ZigBee, wireless network, monitoring.

1 Introduction

The "smart grid" is rapid development that brings a new modern-day revolution to people's life. Energy demand is urgent needed all over the world that is likely to beyond our ability to generate it [1-3].

According to a survey, the entire energy consumption will increase by 30% at a high speed to 5000 billion kWh in 2035 in the United States. However, the new energy resource may not keep up the pace of energy demand, just estimate to grow 22% in this period [4]. In Singapore, the largest consumers of fuel are power generation, transportation and factory while electricity is mainly consumed by the school, flat, and industry [5]. Therefore, wireless monitoring and management the electrical power of public buildings power is urgently required.

WIT Transactions on Engineering Sciences, Vol. 107, © 2015 WIT Press
www.witpress.com, ISSN 1743-3533 (on-line)
doi:10.2495/SEEEI140111

This article uses the ZigBee technology to design Electric Power Usage Monitoring System, it can easily achieve the functions such as monitoring, controlling, and protecting substations and related facilities.

2 System description

Wireless Power Grid Monitoring System consists of two parts, section one is Power Monitoring Module, section two is Wireless Communication Module. The first part, make a smart power meter based on wireless sensor network, and install in every building, the second part, the entire network and communication protocol will be set up. The main system includes function blocks of power sampling, electricity metering part and wireless transmission. The general block diagram is shown in Figure 1.

Town council office need to monitor and control a unit or location, show the data on the server PC online. The entire system is made up of the following parts: data collection module get information from smart meter, transmit it into central control system by wireless sensor network.

2.1 Smart meter module

Purpose of smart meter module(Fig.2): to measure and calculate power and energy, programming to realize its function and show it on the meter after data collection. So user can get information through smart meter about how many electricity they have used.

It can directly drive the LCD and communicates via UART with wireless device(Fig.3), offering a connection for meter calibration and access to the device power calculations, the system can calculates active and reactive energy ; active, reactive and apparent power; power factor; RMS current; RMS voltage, and line frequency.

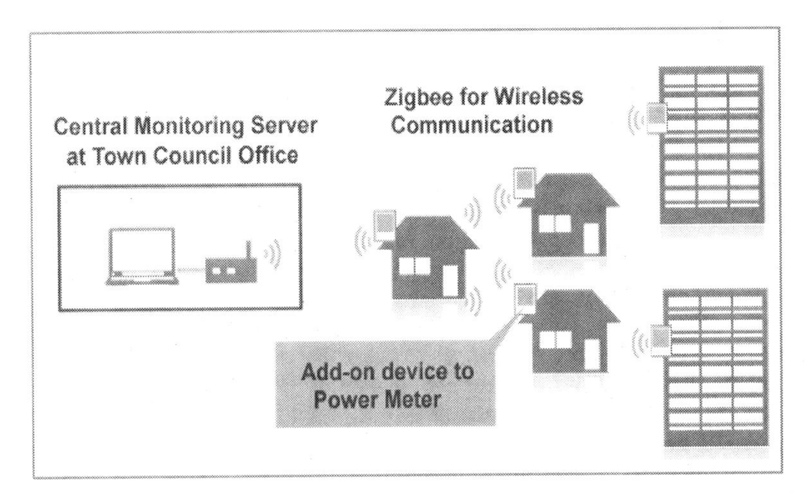

Figure 1: General block diagram.

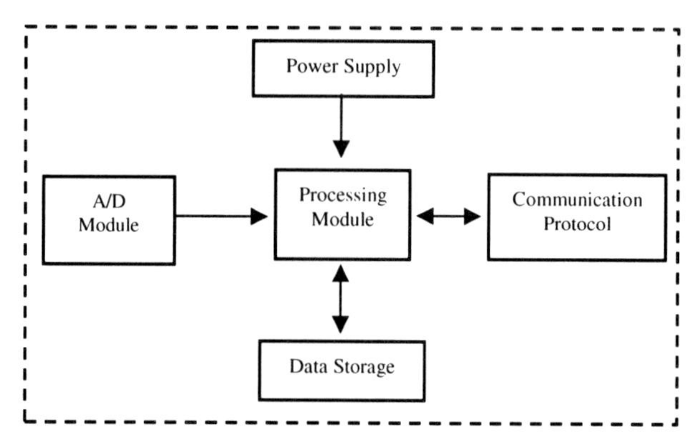

Figure 2: Smart meter module.

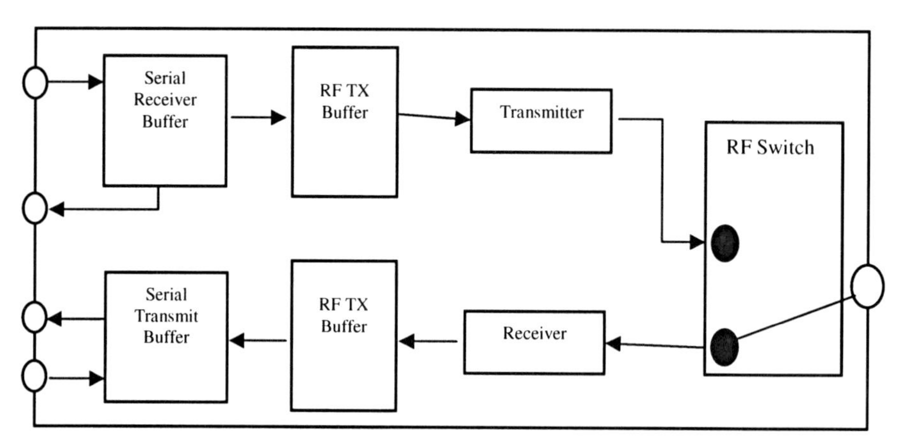

Figure 3: Wireless TX–RX.

2.2 Wireless module

Purpose of wireless module: to transmit the data via wireless device from smart meter into central control system in PC, to realize two-way communication instead of traditional USB module.

The model we're using is wireless device. It has a high level communication protocol using small, low-power digital radios [6].

2.3 Central control system

Purpose of central control system: to monitor the entire power grid system that covered ZigBee device by reading and analyzing data. On the other hand, feed-back the electricity information to user (Fig.4).

The model we're using is central control system, one PC and a set of monitoring system, the monitoring software is designed for building a variety of applications and user interface [7].

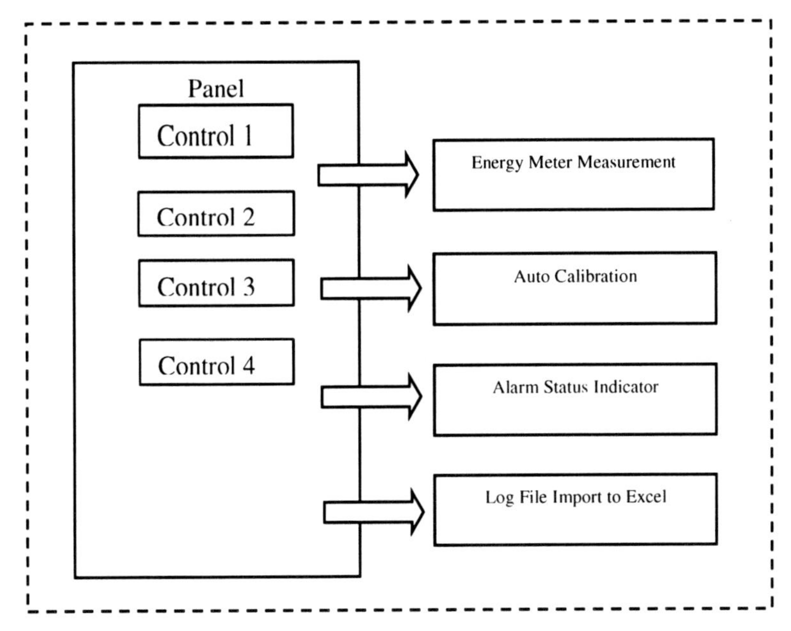

Figure 4: User interface.

3 Details of implementation

The method we are going to adopt is the one provided by Microchip [8]. The reason is it's a fully functional single-phase meter that can measure lots of data, current, power factor, active energy, active power, etc. On the other hand, it's easy to learn with 8 bit chip, with low-cost design.

The energy delivered by a power source into a load can be expressed as:

$$E=\int_0^t V(t)*I(t)\mathrm{d}t \tag{1}$$

The outputs of an energy meter, power and RMS values are obtained by multiplying two AC signals, computing the average value and then multiplying it with a calibration gain.

$$\begin{aligned} S_1(t) &= A_1\cos(wt) \\ S_2(t) &= A_2\cos(wt+\phi) \end{aligned} \tag{2}$$

The two signals (S1 and S2) can be the voltage and/or the current waveform. The instantaneous power value is obtained by multiplication:

$$P(t) = S_1(t)\times S_2(t) \tag{3}$$

4 Experiments and results

In this project, we are going to do is test the LED and fluorescent parameter, meanwhile, how many active energy the LED can save than fluorescent.

After 24 hours test, we record one set of data in every 15min, and get 96 data, is the power usage information. The RMS current/active power/PF/active energy, as Figures 5–8 show.

Figure 5 shows the current difference between LED and fluorescent, LED is better than fluorescent in less current consumption at 0.06A, while the fluorescent is about 0.38A.

Figure 6 shows the fluorescent consume more active power at about 36W than LEDs 10W, so it obviously consume more energy, makes more energy waste.

Figure 7 shows that LED lighting tube's power factor is much higher at above 0.9, while the fluorescent is just about 0.5, the reason is the lighting tube itself circuit internal structure, the chip and driver which one is made by Philips, the other is designed by Temasek Micro-electronics center.

Figure 8 shows that the active energy between LED and fluorescent, during 24 hours, LED lighting tube totally consume just 0.26KWh while fluorescent is about 0.87KWh, so for one day's test, it can save 70% energy for using one LED. It can obvious prove its good quality with low power consumption.

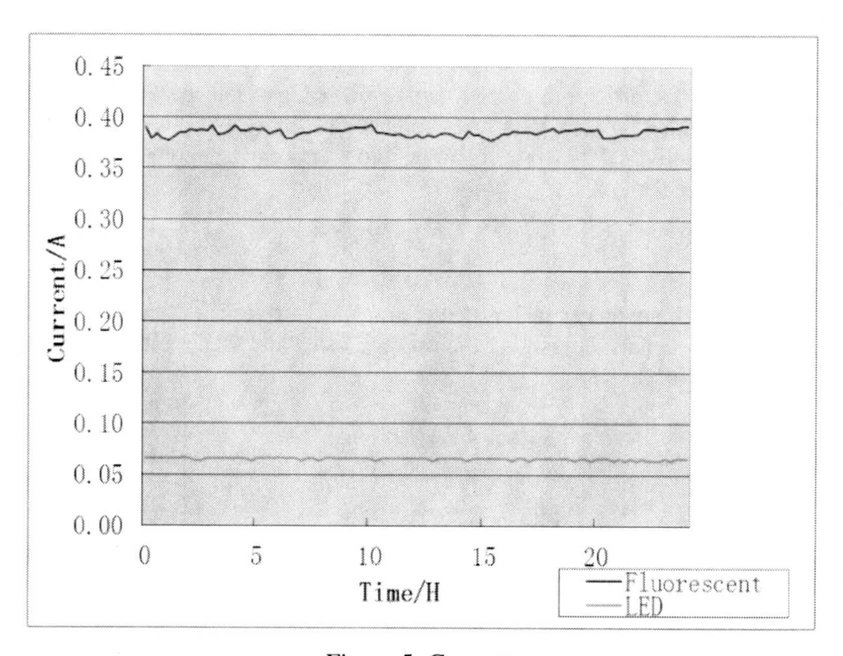

Figure 5: Current.

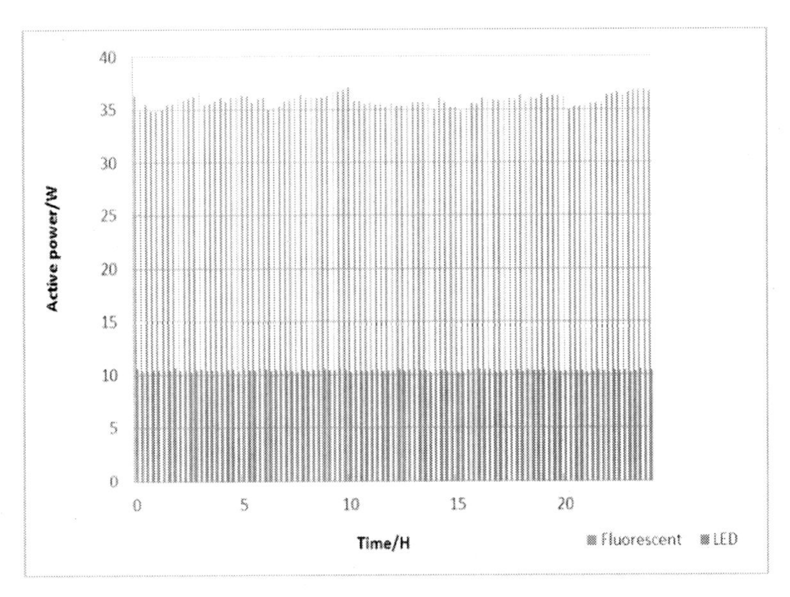

Figure 6: Active power.

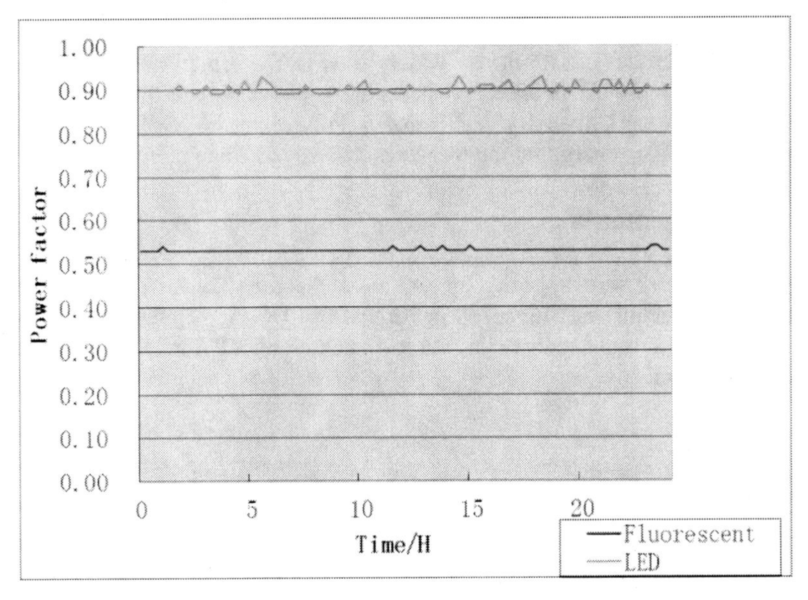

Figure 7: Power factor.

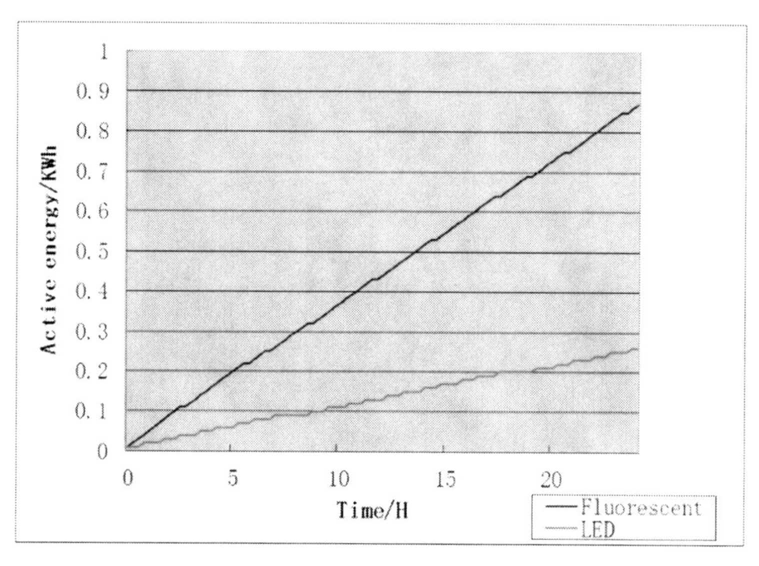

Figure 8: Active energy.

5 Conclusions

In this paper, we have described this wireless power grid monitoring system based on ZigBee device, it will be widely used in the school and HDB houses. Meanwhile, this project is accomplished during the exchange program between Southwest Jiaotong University and Temasek Polytechnic of Singapore. A future project is to develop energy saving measures by analysis some more data resources.

Acknowledgements

The research work was supported by Southwest Jiaotong University of China and Temasek Polytechnic of Singapore in August 2013 to September 2014. Thanks for all the staffs and teachers in Micro-E center of TP for this international exchange project.

References

[1] Richard Henry Douglin, *An Investigation of the Utilization of Smart Meter Data to Adapt Overcurrent Protection for Radial Distribution Systems with a High Penetration of Distributed Generation*. Texas A&M University, 2012.

[2] Nasim Beigi Mohammadi, *An Intrusion Detection System for Smart Grid Neighborhood Area Network*. Ryerson University, 2013.

[3] Mustafa Amir Faisal, *Securing Advanced Metering Infrastructure in Smart Grid using Intrusion Detection System*. Masdar Institute of Science and Technology, 2012.

[4] Gan Kejiang, Xu Pingping, Application and implement of wireless sensor networks in electric system. *Telecommunications for Electric Power System*, **28(1)**, pp. 11–14, 2007.
[5] Fawzi Al-Naima and Bahaa Jalil. Building a prototype prepaid electricity metering system based on RFID. *International Journal of Electronics and Electrical Engineering*, pp. 20–36, 2011.
[6] Zhao Yan, Yue Bingliang, Gao Dawei, Studying of ZigBee network layer. Computer Measurement & Control, **15(5)**, pp. 689–691, 2007.
[7] Rappaport TS, *Wireless Communications Principle and Practice*. Prentice Hall Inc., Upper Saddle River, NJ, 1996.
[8] IEEE std. 802.15.4-2003: Wireless Medium Access Control (MAC) and Physical Lay (PHY) specifications for Low Rate Wireless Personal Area Networks (LR-WPANs)[S], 2003.

The influence of containment on efficiency in electromagnetic railgun system

Yutao Lou, Gang Wan, Yong Jin, Baoming Li
National Key Laboratory of Transient Physics,
Nanjing University of Science and Technology, Nanjing, Jiangsu, China

Abstract

Eddy current in the containment of electromagnetic railgun (EMG) is induced by the pulse current, and it will badly weaken the propulsion of armature and efficiency of EMG system. In order to investigate the influence of containment on efficiency, a full circuit model in combination with finite element method was employed. Firstly, the launch efficiency and eddy-current loss were calculated and analysed. Results show that a stainless steel containment will reduce the launch efficiency very seriously. And the eddy-current loss in containment is more than half of armature muzzle kinetic energy. Thus, containment cannot be ignored in EMG system simulation. In addition, the efficiency of EMG system with different containments were discussed. As a result, efficiency with laminated silicon steel containment is four times more than stainless steel containment. And the efficiency of laminated stainless steel containment is close to without containment.

Keywords: efficiency, containment, electromagnetic railgun, eddy-current loss.

1 Introduction

For supporting the entire railgun and resisting the huge transverse forces of rails due to the pulse current, containment is an integral part of EMG. In order to guarantee the stiffness and strength of barrel, metal containment is often used [1]. Pulse current in rails not only produce complicated electromagnetic field, but also induce large eddy current in the containment. And the eddy current will generate ohmic losses, weaken the magnetic field around the armature, leading to decrease the armature propulsion and launch efficiency.

Based on 2D static model, Elham, S.M. and Galanin, M.P. analyzed the influence of different containment materials, structures and sizes on inductance gradient of

WIT Transactions on Engineering Sciences, Vol. 107, © 2015 WIT Press
www.witpress.com, ISSN 1743-3533 (on-line)
doi:10.2495/SEEEII140121

railgun [2, 3]. Results show that the higher conductivity, and closer of containment and rails, the smaller inductance gradient. Through repeated experiments, J.V. Parker found that there was a 20% reduction of inductance gradient with laminated stainless steel containment, compared with without containment condition [4]. Dwight Landen discussed the static and dynamic distributions of eddy current in containments by EMAP3D, and concluded that the thickness of lamination was the most sensitive factor for inductance gradient [5].

Most researches paid more attentions on the influence of containment on inductance gradient, while ignored the influence of containment on efficiency of EMG system. And the containment eddy-current loss was always not considered when discussing EMG system energy distribution [6]. In this paper, efficiency and eddy-current loss were calculated and analysed with a full circuit model in combination with finite element method. And the efficiency with different containments were also discussed.

2 Basic principles of circuit and field model

2.1 Full circuit model of EMG system.

Figure 1 shows the full circuit model of EMG system. In this model, railgun is modeled as a variable resistance in series with a variable inductance changing as a function of the armature position [7]. R_s and L_s are the pulse-forming resistance and inductance, respectively. R_c and L_c are the resistance and inductance of coaxial cables, respectively. R_g and L_g are the resistance and inductance of railgun, respectively.

When the switch K is closed, the diode is disconnection. If all modules charge synchronously, the mathematical model can be written as follows:

$$\begin{cases} C\dfrac{\mathrm{d}U_c}{\mathrm{d}t} = -i \\ U_c = (R_s + R_c)i + (L_s + L_c)\dfrac{\mathrm{d}i}{\mathrm{d}t} + R_g \cdot I + L_g\dfrac{\mathrm{d}I}{\mathrm{d}t} + L_g' \cdot v \cdot I \end{cases} \tag{1}$$

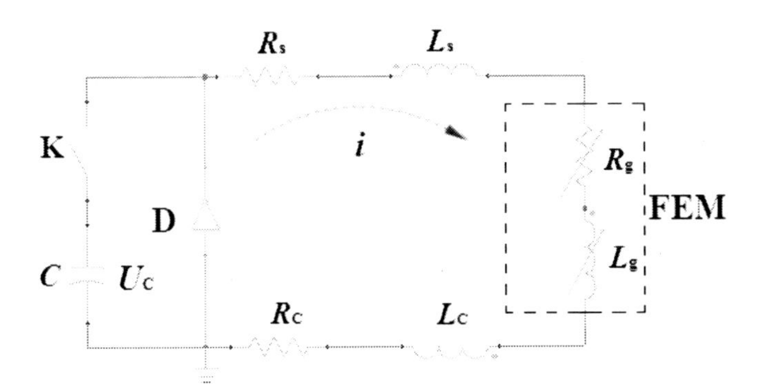

Figure 1: Full circuit model of EMG system.

where v is the velocity of armature, L_g' is the inductance gradient of railgun, and $I=\sum_{k=1}^{n} i_k$. Where n is the number of modules. The armature propulsive force can be given by

$$F = m \cdot a = \frac{1}{2} L_g ' \cdot I^2 \tag{2}$$

Because $v = v_0 + \int_0^t a\mathrm{d}t$ and $s = s_0 + \int_0^t v\mathrm{d}t$, hence

$$\begin{cases} R_g = R_{g0} + R_g ' \cdot s \\ L_g = L_{g0} + L_g ' \cdot s \end{cases} \tag{3}$$

where s is the position of armature, R_g' is the resistance gradient of railgun.

When capacitors finished discharge, diodes will be conducting. And the circuit equation can be written as

$$\frac{\mathrm{d}i}{\mathrm{d}t} = -\frac{R_s + R_c + n \cdot R_g + n \cdot v \cdot L_g '}{L_s + L_c + n \cdot L_g} i \tag{4}$$

When the initial capacitor voltage U_{c0}, railgun length l, initial armature velocity v and position s are given, the breech current and armature position can be obtained by solving Eqs. (1) and (4).

2.2 Theory of field model.

In order to analyze the transient electromagnetic field of railgun, MAXWELL's equations are employed:

$$\begin{cases} \nabla \cdot B = 0 \\ \nabla \times E = -\dfrac{\partial B}{\partial t} \\ \nabla \times H = \sigma E \end{cases} \tag{5}$$

where B is magnetic induction intensity, E is induced electrical field intensity, H is magnetic field intensity, and σ is conductivity of material. Thus, a simple relation is given by

$$\nabla \times \frac{1}{\sigma} \nabla \times H + \frac{\partial B}{\partial t} = 0 \tag{6}$$

According to the T-Ω formulation, and applying Galerkin's scheme [8], the electromagnetic field equations can be expressed in matrix form as

$$\left([S]+[Q]\frac{\mathrm{d}}{\mathrm{d}t}\right)\{T\}+\frac{\mathrm{d}}{\mathrm{d}t}[B]\{\Omega\}+\frac{\mathrm{d}}{\mathrm{d}t}[H]\{i_{\omega}\}=\{g\} \tag{7}$$

where $\{T\}$ and citation $\{\Omega\}$ are the column vector of the edge values of current vector potential and the column vector of the nodal values of magnetic scalar potential, respectively. $\{i_{\omega}\}$ is the column matrix of current in railgun. The matrices $[S]$, $[Q]$, $[B]$, $[H]$, and column matrix citation $\{g\}$ are from the standard Galerkin's method [9].

Based on the breech current and armature position solved above, eddy-current loss can be calculated by the following equations with FE analysis code (Ansys Maxwell 3D):

$$P=\frac{1}{\sigma}\int_{V} J_{e}^{2}\mathrm{d}V \tag{8}$$

$$Q=\int_{0}^{t} P\mathrm{d}t \tag{9}$$

where J_e is eddy-current density, V is containment volume, P is eddy-current loss power, and Q is the eddy-current loss of whole containment.

3 Simulation and analysis

3.1 Launch efficiency of EMG system.

Table 1 shows the parameters of full circuit model. R_g' and L_g' depend on the material and size of containment. A 10-MJ pulse power system consisting of 20 independently triggered modules is employed. The armature mass is 583.2g and rail length is 6 m. Armature initial velocity and position are both zero in our calculations.

The breech current and armature velocity curves under two conditions (with and without stainless steel containment) are shown in Figure 2. It is shown that breech peak currents are 2.73mA and 2.71mA, respectively. And reach the peak current at the same time. The armature muzzle velocities are 1393m/s

Table 1: Parameters of the full circuit model.

	R_g' (μΩ/m)	L_g' (μH/m)	U_{c0} (kV)	C (mf)	R_s (mΩ)	L_s (μH)	R_c (mΩ)	L_c (μH)
Without containment	20.4	0.347	10	10	2	20	10	2
With stainless steel containment	305	0.273						

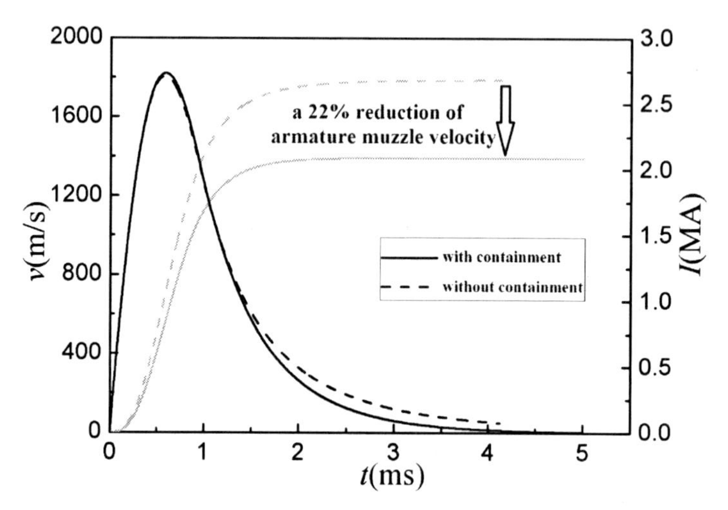

Figure 2: Breech current and armature velocity curves.

and 1787m/s, respectively. Thus, the containment has no effect on breech current almost, but a great influence on armature muzzle velocity.

The eddy current in containment has an opposite electromagnetic field with rails'. And it weakened magnetic field around armature, decreased propulsive force and velocity.

Defined the efficiency of EMG system as [10]

$$\eta = \frac{\text{Kinetic energy}}{\text{Electrical energy used}} = \frac{\frac{1}{2} m \cdot v^2}{\frac{1}{2} C \cdot U_{c0}^2} \tag{10}$$

Calculation results show that the efficiency of EMG system with stainless steel containment is 5.66%, much less than without containment, which is 9.31%.

3.2 Eddy-current loss.

For discussing the eddy-current loss, the transient electromagnetic field of railgun is simulated and analyzed. The parameters of railgun's structure and materials in our simulations are shown in Table 2.

Figure 3 shows the breech current and eddy-current loss power. There are two peaks in the curve of eddy-current loss power. When pulse current increasing, eddy current is also increased. When pulse current decreasing, reverse eddy current is induced, result in eddy-current density and loss power reduction. With the movement of armature, reverse eddy-current density and induction area of containment are both increased, that's why the second peak higher than the first one.

Table 2: Parameters of railgun's structure and material.

	Length (mm)	Width (mm)	Height (mm)	Material	σ (S/m)
Armature	60	60	60	aluminum	3.8×10^7
Rail	6000	90	45	copper	5.8×10^7
Containment	6000	Distance from rails	Thickness	Stainless steel	1.1×10^6
		10	20		

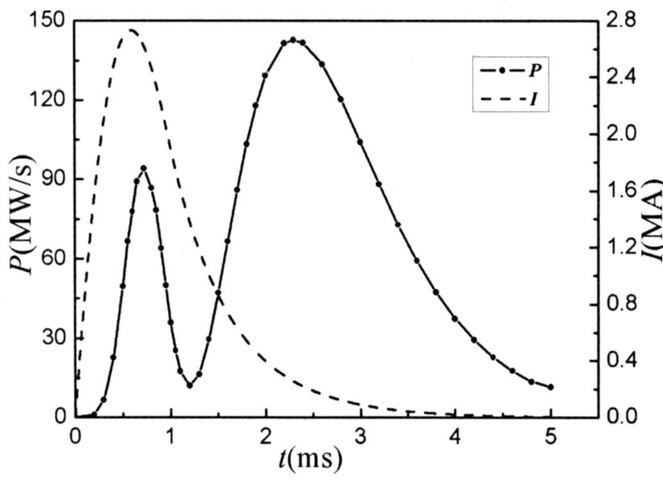

Figure 3: The breech current and eddy-current loss power.

Through Eq. (9), the eddy-current loss is obtained, which is 0.321MJ, 3.21% of the total energy. Obviously it is more than half of the muzzle kinetic energy (E_k= 0.566MJ). Therefore, the eddy-current loss in containment must be considered in EMG system simulation.

4 Influence of different containments

Research shows that using laminated containment will cut off the eddy current loop and restrain the eddy-current loss. And magnetic material containment can compress the magnetic field of railgun, leading to improve the magnetic induction intensity around the armature [4]. The armature velocity versus different containments are shown in Figure 4. The armature muzzle velocity of laminated stainless steel containment is 1764m/s, slightly lower than without containment (1787m/s), while much higher than stainless steel containment (1393m/s). And the armature muzzle velocity of laminated silicon steel containment is 2790m/s.

The efficiency versus different containments is shown in Table 3. Obviously, the efficiency of laminated silicon steel containment is almost four times of the value with stainless steel containment. And the efficiency of laminated stainless steel containment is close to without containment.

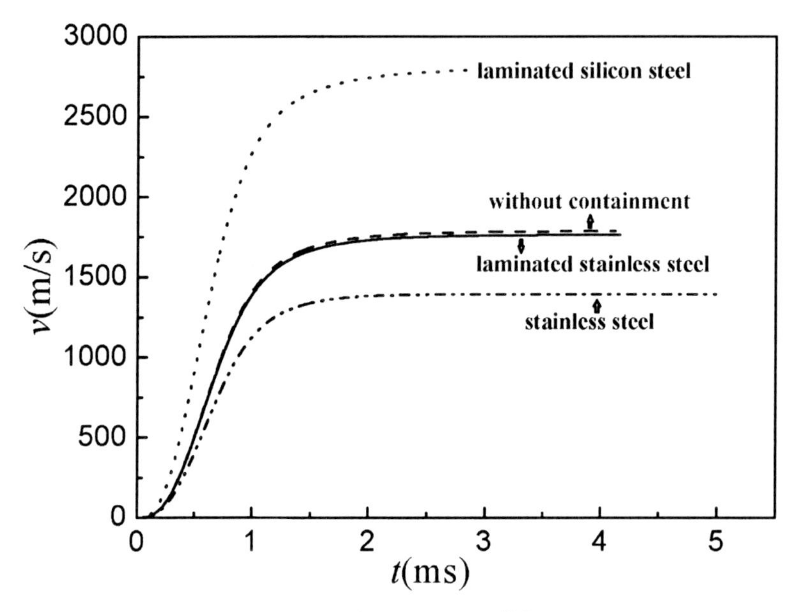

Figure 4: Armature velocity versus different containments.

Table 3: Efficiency versus different containments.

Containment	Without	Stainless Steel	Laminated Stainless Steel	Laminated Silicon Steel
η (%)	9.31	5.66	9.07	22.7

5 Conclusions

The efficiency of EMG system is weakened by eddy current of containment. For calculating and analyzing the efficiency and eddy-current loss, a full circuit model in combination with finite element method was employed in this paper. And the influence of efficiency with different containments was discussed. Results show that:

1) The efficiency of EMG system with stainless steel containment is much less than without containment.
2) Eddy-current loss is 3.21 percent of the total energy, more than half of the muzzle kinetic energy of armature. If ignoring the eddy-current loss in EMG system simulation, calculation and analysis will be inaccurate.
3) The efficiency of EMG system with laminated silicon steel containment is four times more than stainless steel containment. And the efficiency of laminated stainless steel containment is close to without containment.

References

[1] Pascale, L., Minh, D.V., Walter, W., Comparative study of railgun housings made of modern fiber wound materials, ceramic, or insulated steel plates. *IEEE Transactions on Magnetics*, **41(1)**, pp. 200–205, 2005.

[2] Elham, S.M., Asghar, K., S. Mahmoud, N.Z., Effects of shielding on railgun inductance gradient. *3rd International Conference on Computational Electromagnetics and Its Applications*, pp. 44–47, 2004.

[3] Galanin, M.P., Popov, Y.P., Khramtsovsky, S.S., The use of currents induced within a conducting shield for railgun performance control. *IEEE Transactions on Magnetics*, **33(1)**, pp. 544–548, 1997.

[4] Jerald, V.P., Scott, L., Loss of propulsive force in railguns with laminated containment. *IEEE Transactions on Magnetics*, **35(1)**, pp. 442–446, 1999.

[5] Dwight, D., Sikhanda, S., Eddy current effects in the laminated containment structure of railguns. *IEEE Transactions on Magnetics*, **43(1)**, pp. 150–156, 2007.

[6] He Yong, Guan Yong-Chao, Gao Gui-shan, et al., Efficiency analysis of an electromagnetic railgun with a full circuit model. *IEEE Transactions on Plasma Science*, **38(12)**, pp. 3425–3428, 2010.

[7] Deadrick, F.J., Hawke, R.S., Scudder, J.D., Magrac–a railgun simulation program. *IEEE Transactions on Magnetics*, **18(1)**, pp. 94–104, 1982.

[8] Zhou, P., Fu, W.N., Cendes, Z.J., Numerical modeling of magnetic devices. *IEEE Transactions on Magnetics*, **40(4)**, pp. 1803–1809, 2004.

[9] Du Zhi-ye, Zhan Ting, Ruan Jiang-jun, et al., Research on electromagnetic performance affected by shielding enclosure of a coil launcher. *IEEE Transactions on Plasma Science*, **41(5)**, pp. 1077–1083, 2013.

[10] Thomas, G.E., Jesse, M.N., Michael, J.V., The velocity and efficiency limiting effects of magnetic diffusion in railgun sliding contacts. *14th Symposium on Electromagnetic Launch Technology*, Victoria, BC, pp. 1–5, 2008.

Study on the compression sensing method in magneto-acousto-electrical tomography with magnetic induction

Liang Guo[1–3], Guoqiang Liu[1], Hui Xia[1]
[1]*Institution of Electrical Engineering, Chinese Academy of Science, Beijing, China*
[2]*University of Chinese Academy of Sciences, Beijing, China*
[3]*College of Control Theory and Engineering, China University of Petroleum, Qingdao, China*

Abstract

Magneto-acousto-electrical tomography with magnetic induction (MAET-MI) is an imaging modality proposed for noninvasive conductivity imaging of high spatial resolution. In our work of study, a conductivity reconstruction algorithm based on the compression sensing method in the MAET-MI is presented with the reciprocal theorem. Firstly, the reciprocal theory was introduced to account for the forward and inverse problem of the MAET-MI. And then the current vector in reciprocal process was reconstructed by the CS method. The vector, which was reconstructed by the CS method above, was used for reconstructing the conductivity. At last, the result of the computer simulation showed the capability and reliability of the CS method in conductivity reconstruction of MAET-MI.
Keywords: CS method, magneto-acousto-electrical tomography, magnetic induction, reciprocal theory.

1 Introduction

Because the change of the conductivity is a typical character of the most tumor in early stage, the conductivity imaging becomes an important assessment tool for the early diagnosis of the cancer. Lots of imaging methods are developed for the conductivity reconstructing, such as electrical impedance tomography (EIT) [1], magnetic induction tomography (MIT) [2] and magnetic resonance electrical

WIT Transactions on Engineering Sciences, Vol. 107, © 2015 WIT Press
www.witpress.com, ISSN 1743-3533 (on-line)
doi:10.2495/SEEEII140131

impedance tomography (MREIT) [3]. However, these methods are mainly based on the electromagnetic excitation and electromagnetic measurement, which has lower spatial resolution and contrast than the medical ultrasonic imaging. In order to overcome this shortcomings, the Magneto acoustic tomography (MAT) [4] and magneto-acousto-electrical tomography (MAET) [5] are proposed recently.

In MAET, different from MAT-MI [6], the biological tissue is under a static magnetic field and a beam of high intensity narrow pulse ultrasound is applied to the tissue. The ultrasonic vibration leads to the separation of the charges due to the Lorenz force and produces the current distributed in the tissue. A voltage signal can be detected by the electrodes or the coils arbitrarily arranged out of the tissue.

The forward and inverse problems of the MAET are well studied substantially by Haider et al. [5]. They obtained the MAET signals of a slab by electrodes or coils and imaged the conductivity by ultrasonic scan on a plane[7]. Leonid [8] has proposed a set of methods for MAET in electrode detection mode.

In our work of study, we focus on the inverse problem of the conductivity reconstruction. Firstly, we discuss the reciprocal theory of the MAET-MI and the formula for conductivity reconstruction by the use of current vector. Secondly, the CS method is discussed in detail for reconstructing the current vector. At last, the proposed algorithms are applied to the numerical simulations.

2 Discussed problems

In order to formulize the MAET-MI, both the actual physical process and the reciprocal process should be discussed. The actual physical process of MAET-MI can be described as a partial differential equation (PDE) of the electromagnetic fields that is excited by the ultrasound waves. The coupling of the ultrasound and electromagnetic field is described as follows [9]:

$$\boldsymbol{J}_{e1} = \sigma \boldsymbol{V}_1 \times \boldsymbol{B}_0 \tag{1}$$

where $\boldsymbol{J}_{e1}$ is the current density stimulated by ultrasound waves with the vibration velocity $\boldsymbol{V}_1$. $\boldsymbol{B}_0$ is the uniform static magnetic field and σ is the conductivity of the medium. Under the excitation of the source current density $\boldsymbol{J}_{e1}$, the corresponding electrical field $\boldsymbol{E}_1$ is generated in the conductive medium. By ignoring the displacement current, the magnetic field $\boldsymbol{H}_1$ that is stimulated by the total current density can be represented as follows:

$$\nabla \times \boldsymbol{H}_1 = \boldsymbol{J}_{e1} + \sigma \boldsymbol{E}_1 \tag{2}$$

At the same time, according to the Faraday's law of electromagnetic induction, the magnetic field and electrical field also satisfy the following equation:

$$\nabla \times \boldsymbol{E}_1 = -\mu \frac{\partial \boldsymbol{H}_1}{\partial t} \tag{3}$$

where μ is the permeability of the medium. Because the electrical field $\boldsymbol{E}_1$ is rotational, an induced voltage can be detected by a coil. The induced voltage of the coil can be described as a line integration of the induced electric field along the coil:

$$u(t) = \int_{r \in l_{cir}} \boldsymbol{E}_1(t) \cdot \mathrm{d}\boldsymbol{l}_{cir} \tag{4}$$

where $\mathrm{d}\boldsymbol{l}_{\mathrm{cir}}$ represents the tiny line element along the coil of wire.

The external current density $\boldsymbol{J}_{\mathrm{e}2}$ in the reciprocal process is represented as follows:

$$\boldsymbol{J}_{e2}(\boldsymbol{r}) = \delta(\boldsymbol{r} - \boldsymbol{r}_{cir}) s(t) \boldsymbol{e}_{e2}(\boldsymbol{r}) \tag{5}$$

where $\boldsymbol{r}$ denotes an arbitrary field point of the medium, $\boldsymbol{r}_{\mathrm{cir}}$ denotes the field point of the coil, $\delta(\cdot)$ represents the Dirac function in the area of three dimensions and $s(t)$ is the time component of $\boldsymbol{J}_{e2}$. The unit vector of $\boldsymbol{J}_{e2}(\boldsymbol{r})$ is represented by $\boldsymbol{e}_{\mathrm{e}2}(\boldsymbol{r})$.

According to the Ampere's law and the Faraday's law of electromagnetic induction, the electrical field and magnetic field induced by the external current in reciprocal process can be described as follows:

$$\nabla \times \boldsymbol{H}_2 = \boldsymbol{J}_{e2} + \sigma \boldsymbol{E}_2 \tag{6}$$

$$\nabla \times \boldsymbol{E}_2 = -\mu \frac{\partial \boldsymbol{H}_2}{\partial t} \tag{7}$$

where $\boldsymbol{E}_2$ and $\boldsymbol{H}_2$ are the eddy electrical field and magnetic field of the reciprocal process, respectively. The subscript 2 denotes the variable in the reciprocal process and the subscript 1 denotes the variable in the actual physical process.

According to the reciprocal theory, the relationship between the actual process and the reciprocal process can be described as follows:

$$u(t) = \iiint_{r \in \Omega} \boldsymbol{J}_2^o(r) \cdot \boldsymbol{V}_1'(r,t) \times \boldsymbol{B}_0 \mathrm{d}\Omega \tag{8}$$

where $\boldsymbol{V}_1'(r,t)$ represents the derivative of $\boldsymbol{V}_1(r,t)$ with respect to time t. $\boldsymbol{J}_2^o(r)$ is the spatial component of the current density in reciprocal process. Considering the 2D models in *xoy* plane, the Eq. (8) can be rewrite as follows:

$$u(t) = \iiint_{r \in \Omega} [J_{2x}^o(r) V_{1y}'(r,t) B_{0z} - J_{2y}^o(r) V_{1x}'(r,t) B_{0z}] \mathrm{d}\Omega \tag{9}$$

where $V_{1x}'(r,t)$ and $V_{1y}'(r,t)$ represent the x and y component of the $\boldsymbol{V}_1'(r,t)$, respectively, $J_{2x}^o{}'(r)$ and $J_{2y}^o{}'(r)$ represent the x and y component of the $\boldsymbol{J}_2^o(r)$ respectively, B_{0z} is the z component of the static magnetic field.

3 CS method in MAET-MI

The inverse problem of MAET-MI can be stated in two steps. The first step is to acquire the spatial component of the current density in reciprocal process $\boldsymbol{J}_2^o(r)$ and the second is to acquire conductivity from $\boldsymbol{J}_2^o(r)$.

The forward problem of Eq. (9) can be rewritten as a matrix equation:

$$\boldsymbol{U} = \boldsymbol{\Phi D} \tag{10}$$

where $\boldsymbol{D}$ is the column vector of the current vector satisfying:

$$\boldsymbol{D}(\boldsymbol{r}_j) = \begin{bmatrix} \boldsymbol{J}_{2x}^o(\boldsymbol{r}_j) \\ -\boldsymbol{J}_{2x}^o(\boldsymbol{r}_j) \end{bmatrix} \tag{11}$$

$\boldsymbol{U}$ is a column vector of the MAET-MI signal u_i. The $\boldsymbol{\Phi}$ in Eq. (10) is the known forward operator that we discretize as Eq. (12), where $\boldsymbol{V}_y'(r_j,t_i)$ and $\boldsymbol{V}_x'(r_j,t_i)$ are component matrixes of discretized vector $\boldsymbol{V}_1'(r,t)$. Here we discretize the vector by its component in different part of rows.

$$\boldsymbol{\Phi}_{i,j} = \begin{bmatrix} \boldsymbol{V}_y'(r_j,t_i)B_{0z} \\ \boldsymbol{V}_x'(r_j,t_i)B_{0z} \end{bmatrix}^T = \begin{bmatrix} \dfrac{\partial}{\partial y}\dfrac{\delta'(t_i - |\boldsymbol{r}_j - \boldsymbol{r}_d|/c)}{4\pi|\boldsymbol{r}_j - \boldsymbol{r}_d|}B_{0z} \\ \dfrac{\partial}{\partial x}\dfrac{\delta'(t_i - |\boldsymbol{r}_j - \boldsymbol{r}_d|/c)}{4\pi|\boldsymbol{r}_j - \boldsymbol{r}_d|}B_{0z} \end{bmatrix}^T \tag{12}$$

Vector $\boldsymbol{D}$ can be represented in this basis as: $\boldsymbol{D} = \boldsymbol{\Psi\theta}$, where $\boldsymbol{\theta}$ is a column vector. According to the CS theory, if $\boldsymbol{D}$ is sparse in the basis of $\boldsymbol{\Psi}$, $\boldsymbol{\theta}$ is the solution to the l_0 norm minimization [10]:

$$\min\|\boldsymbol{\theta}\|_{l_0} \text{ s.t.} \boldsymbol{U} = \boldsymbol{\Phi\Psi\theta} \tag{13}$$

with the overwhelming probability if $\boldsymbol{\Phi\Psi}$ is a CS matrix.

Donoho et al. have shown that the l_0 minimization problem above is equivalent to the l_1 problem for most $\boldsymbol{\Phi\Psi}$ [11]:

$$\min \|\boldsymbol{\theta}\|_{l_1} \ \text{s.t.} \ \boldsymbol{U} = \boldsymbol{\Phi\Psi\theta} \tag{14}$$

In order to satisfy the CS matrix condition, $\boldsymbol{\Phi\Psi}$ should be particularly designed according to the CS theory. In previous work of Candès et al. [12], if $\boldsymbol{\Psi}$ is the Fourier basis and $\boldsymbol{\Phi}$ is a random matrix satisfying the Gaussian distribution, the $\boldsymbol{\Phi\Psi}$ can be regarded as a CS-matrix. In the previous work of Jean [13], the kernel of Radon transform sampling the Fourier domain is used in the PAT. In MAET-MI, if the velocity potential function can be regarded as a linear independent function, the matrix that is composed of the product between the forward operator in the Fourier domain and the derivative of the velocity potential will be a CS matrix. In order to solve Eq. (14), the orthogonal matching pursuit (OMP) algorithm [14] is employed in our study.

4 Method

The schematic of the MAET-MI imaging setup is shown in Figure 1. The ultrasound wave is transmitted from a wide band transducer with center frequency of 500 KHz. A pair of coils with 150 turns winded tightly is arranged with a separation of 6 cm in the z direction to detect the induced voltage. The static magnetic field is assumed to be uniform in the imaging area and pointing to the z direction with the flux density set to 0.3 T. The FEM simulation can then be implemented at the object slice with a thickness of 1 mm at $z = 0$ [15, 16].

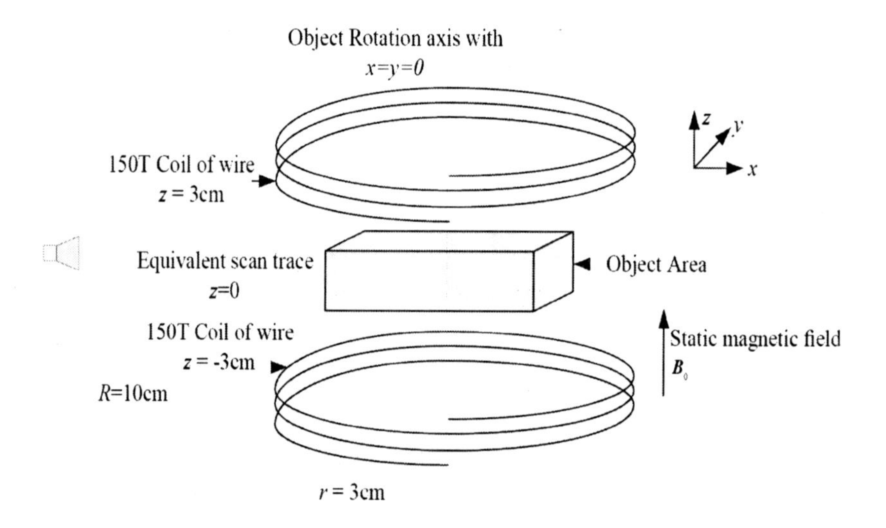

Figure 1: Schematic diagram of MAET-MI setup used in this study.

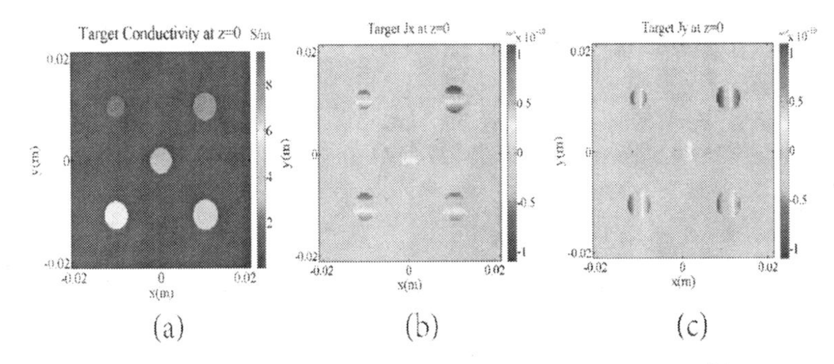

Figure 2: (a) Original piecewise uniform conductivity distribution, (b) x component of the induced current density vector, (c) y component of the induced current density vector.

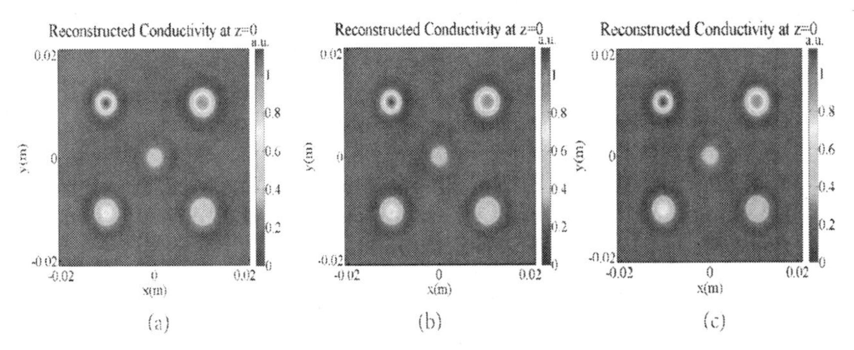

Figure 3 (a)–(c) shows the result of the reconstructed conductivities for SNR of 100, 20, 2, respectively.

5 Result

Figure 2(a) shows the original piecewise uniform conductivity distribution and Figure 2(b) and (c) show the x and y components of the induced current density vector, respectively.

The reconstructed conductivity by CS-method in reciprocal process is seen in Figure 3 for various noise levels. Figure 3(a)–(c) shows the result of the reconstructed conductivities for SNR of 100, 20, 2, respectively. We can see from Figure 3 that the conductivity can be estimated approximately by these reconstructed images using our proposed methods.

6 Conclusion

In conclusion, we have studied on the mechanism of the MAET-MI in coil detection mode and have implemented the CS method in the current vector reconstruction. Computer simulation has been conducted to demonstrate the validity and the reliability of the proposed algorithms.

Acknowledgement

The research work was supported by National Natural Science Foundation of China under Grant Nos. 61401514, 61271424 and 51137004.

References

[1] M. Cheney, D. Isaacson, J.C. Newell, Electrical impedance tomography. *SIAM Rev.*, **41(1)**, pp. 85–101, 1999.
[2] H. Griffiths, Magnetic induction tomography. *Meas. Sci. Technol.*, **12**, pp. 1126–1131, 2001.
[3] S. Lee, J. Seo, C. Park, B. Lee, J. Woo, S. Lee, O. Kwon, J. Hahn, Conductivity image reconstruction from defective data in MREIT: numerical simulation and animal experiment. *IEEE Trans. Med. Imag.*, **25(2)**, pp. 168–176, 2006.
[4] X. Li, Y. Xu, B. He, Magnetoacoustic tomography with magnetic induction for imaging electrical impedance of biological tissue. *J. Appl. Phys.*, **99**, pp. 066112–066112, 2006.
[5] S. Haider, A. Hrbek, Y. Xu, Magneto-acousto-electrical tomography: a potential method for imaging current density and electrical impedance. *Phys. Meas.*, **29(6)**, pp. S41–S50, 2008.
[6] Yuan Xu, Bin He, Magnetoacoustic tomography with Magnetic Induction (MAT-MI). *Phys. Med. Biol.*, **50(21)**, pp. 5175–5187, 2005.
[7] Guoqiang Liu, Xin Huang, Hui Xia, Magnetoacoustic tomography with applied current. *Chin. Sci. Bull.*, **58(30)**, pp. 3600–3606, 2013.
[8] Leonid Kunyansky, A mathematical model and inversion procedure for magneto-acousto-electric tomography. *Inv. Probl.*, **28(3)**, p. 35002, 2012.
[9] Guo Liang, Liu Guo-qiang, Xia Hui, Chenjing, Forward procedure of magneto-acousto-electric signal in radial stratified medium of conductivity for logging models. *Chin. Phys. Lett.*, **30(12)**, p. 124303, 2013.
[10] D.L. Donoho, Compressed sensing. *IEEE Trans. Inf. Theory*, **52(4)**, pp. 1289–1306, 2006.
[11] D.L. Donoho, For most large underdetermined systems of equations, the minimal l(1)-norm near-solution approximates the sparsest near solution. *Commun. Pure Appl. Math.*, **59(7)**, pp. 907–934, 2006.
[12] E.J. Candès, J. Romberg, T. Tao, Robust uncertainty principles: exact signal reconstruction from highly incomplete frequency information. *IEEE Trans. Inf. Theory*, **52(2)**, pp. 489–509, 2006.
[13] Jean Provost, Frédéric Lesage, The application of compressed sensing for photo-acoustic tomography. *IEEE Trans. Med. Imag.*, **28(4)**, pp. 585–94, 2009.
[14] G. Davis, S. Mallat, M. Avellaneda, Adaptive greedy approximations. *Constr. Approx.*, **13(1)**, pp. 57–98, 1997.

[15] Liang Guo, Guoqiang Liu, Hui Xia, Conductivity reconstruction algorithms and numerical simulations for magneto-acousto-electrical tomography with piston transducer in scan mode. *Chin. Phys. B*, **23(10)**, p. 104303, 2014.
[16] Xin Huang, Guoqiang Liu, Hui Xia, Study of pulsed magnetic field used in magnetioacoustic tomography with magnetic induction. *Trans. China Electrotech. Soc.*, **28(2)**, pp. 67–72, 2013.

The transient grounding impedance calculation for corroded tower grounding body based on the non-uniform FDTD algorithm

Zhu Ling-feng, Zhou Li-xing, Wang pan, Cao Zhi-ping
Changsha University of Science and Technology
Institute of Electrical and Information, Changsha, China

Abstract

This paper adopts the Finite Difference Time Domain method (FDTD) to analyze the influence on the transient grounding impedance of the corrosion of the grounding electrode, establishes the FDTD calculation model of localized corrosion of the grounding electrode. By using the non-uniform grid FDTD method, this paper gets the transient grounding impedance when part of the localized corrosion of grounding impedance happens, and the conclusion that the calculation of transient grounding impedance based on the non-uniform grid FDTD algorithm is more timesaving than the original FDTD algorithm and can improve the computational accuracy when there are fine draws inside the grounding electrode. Finally, verify the accuracy and feasibility of the method through Matlab simulation.

Keywords: FDTD, localized corrosion, non-uniform grid, transient grounding resistance.

1 Introduction

Currently, most of the researches on the impulse characteristics of tower grounding body are based on the assumption of quasi-steady field [1], which will artificially introduce errors, and ignored the time-varying characteristics during lightning current discharge. According to the long-term operational experience of power system, transmission line accidents are mainly caused by lightning strike on transmission line or trips happening on the tower [2]. The statistics of grid failure shows that: Accidents caused by lightning strike on the transmission line account for a large proportion in the total grid accidents, especially in the areas of

www.witpress.com, ISSN 1743-3533 (on-line)
doi:10.2495/SEEEII140141

strong lightning activity and high soil resistivity [3]. Therefore, to reduce the transmission line lightning accident rate and protect safe and stable operation of the power system has a big significance.

The main function of transmission line tower grounding body is to rapidly release the lightning current into the earth, and its transient impulse characteristics include the transient surge impedance and transient ground potential rise of the grounding body after the lightning current flows through the tower, which is an important factor of the insulation coordination design of grounding system [4]. Because there is a big calculation error in the impulse grounding resistance of the impact coefficient and that the grounding body is perennial buried underground and reacts with the surrounding environment while subjected to the mechanical damage, soil moisture changes and other factors, the grounding body is easy to corrode, suffer surface slit even fracture failure [5], and the impulse grounding resistance of grounding body may exceed a safe value and cause a lightning accident. Therefore, establishing a reliable calculation model of the transmission line grounding transient impulse characteristics and accurately simulating the discharge process of the lightning current when the grounding body suffer a corrosion is the basis for the correct calculation of the transient impedance when the tower grounding body corrodes.

2 The FDTD model of the transient grounding impedance when the tower grounding body corrodes

When the lightning current flows through the grounding, the diffuser underground is very complex. The impedance of grounding body not only have relationship with its own structure size, the lightning current parameters, the depth, the soil resistivity and other factors, but also changes with time [6]. Therefore, using the analytical method to study its transient process is relatively difficult. Based on the transient performance analysis experimental architecture of the grounding system [7], establish the FDTD computational model of the vertical grounding body as shown in Figure 1.

Grounding electrode is a rectangular steel vertical grounding body, and its cross-sectional area is with its length is vertically buried underground, with its upper end parallel with the ground surface. The current loop pole of the left discharge circuit is deeply buried 1.5 m, which is consist of a 0.135m-diameter circumference with three copper bars of a 0.01m diameter and 120° angle. The height of the connection line from the ground is 1 m, with its length 20m. The diameter of connecting line and leading up wire is 0.135m. Soil conductivity $e_g = 0.02$ s/m, relative dielectric constant $e_{rg} = 0.8$.

3 Non-uniform grid FDTD algorithm

FDTD method is the corresponding time domain propulsion formula after the difference divergence of the differential form Maxwell's equations. In space, the electromagnetic field components are configured in accordance with the grid

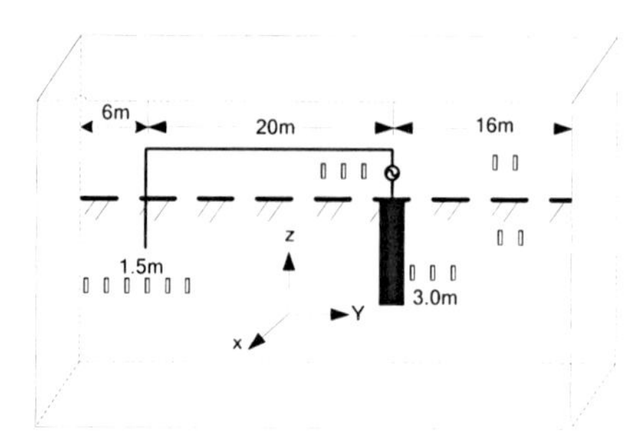

Figure 1: FDTD computational model of the vertical grounding body.

consist of Yee cellular; in time, the electric field components and magnetic components alternating intervals sampled in every half time step, so that Maxwell curl equations become more suitable for difference calculation [8, 9]. Therefore, when the initial value and boundary conditions of the electromagnetic problems are given, we can gradually get the electromagnetic field distribution in each time and space by using FDTD method. Generally speaking, the entire computation space adopts the uniform mesh generation. However, in the special areas with large changes in the electromagnetic field strength or complex structure(e.g., the object region of the structures, such as sheet, the thin aperture and thin coating, etc.), it adopts the even coarser uniform grid, which is difficult to truly reflect the electromagnetic properties of the structure that is less than the mesh size; if the entire computing space still uses the uniform mesh refinement, it will inevitably result in a surge in the number of grid and the calculated amount. In order to achieve a more accurate simulation of the geometrical shape or the fine structure in special areas and truly reflect the electromagnetic properties of the object of study while saving the calculating memory and computing time, we need to adopt the fine mesh generation in areas that need to be refine, and use the coarse mesh generation in the remaining areas. This meshing method is the FDTD algorithm of the non-uniform grid.

The principle of non-uniform mesh includes mutation mesh and expanded mesh [8]. Mutation mesh refers to divide the mesh into coarse and fine mesh according to the different structure of objects, which has no specific rule, there is no obvious transition region between the coarse mesh and fine mesh, directly from coarse to fine mesh. Extended mesh refers to the one that the adjacent meshes are the same dimension scale and the mesh is divided according to the exponential increases of the stretch factor and a certain rule.

In non-uniform grid, the formula is as follows:

$$\left\{h_{xi} = (\Delta x_i + \Delta x_{i-1})/2; i = 2,...,n_x\right\} \tag{1}$$

$$\left\{h_{yi} = (\Delta y_i + \Delta y_{i-1})/2; i = 2,...,n_y\right\} \tag{2}$$

$$\left\{h_{zi} = (\Delta z_i + \Delta z_{i-1})/2; i = 2,...,n_z\right\} \tag{3}$$

So, the symbols for electric-field component and magnetic-field component in non-uniform grid can be shown:

$$E_x^n(i+\frac{1}{2},j,k)=E_x(x_{i+\frac{1}{2}},y_j,z_k,n\Delta t) \tag{4}$$

$$H_x^{n+\frac{1}{2}}(i,j+\frac{1}{2},k+\frac{1}{2})=H_x(x_i,y_{j+\frac{1}{2}},z_{k+\frac{1}{2}},(n+\frac{1}{2})\Delta t) \tag{5}$$

The algorithm of non-uniform grid FDTD is obtained by discrete of Maxwell equation in integral form, especially faraday's law and ampere's law.

In the formula of faraday's law, the right surface integral is the surface integral of unit grid, and the left loop integral is the integral of loop circuit around, seen in Figure 2. Correspondingly, in the formula of ampere's law, the right surface integral is the surface integral of complex grid, seen in Figure 3.

By conducting the field discretization for faraday's law and ampere's law and getting central difference from derivative of time, the formulas (6) (7) can be gained.

$$H_z^{n+\frac{1}{2}}(i+\frac{1}{2},j+\frac{1}{2},k)=H_z^{n-\frac{1}{2}}(i+\frac{1}{2},j+\frac{1}{2},k)+\frac{\Delta t}{\mu_{(i+\frac{1}{2},j+\frac{1}{2},k)}}\cdot\begin{bmatrix}\frac{1}{\Delta y_j}\left(E_x^n(i+\frac{1}{2},j+1,k)-E_x^n(i+\frac{1}{2},j,k)\right)\\-\frac{1}{\Delta x_i}\left(E_y^n(i+1,j+\frac{1}{2},k)-E_y^n(i,j+\frac{1}{2},k)\right)\\-J_{mz}^{n+\frac{1}{2}}(i+\frac{1}{2},j+\frac{1}{2},k)\end{bmatrix} \tag{6}$$

$$E_z^{n+1}(i,j,k+\frac{1}{2})=\frac{2\varepsilon_{(i,j,k+\frac{1}{2})}-\Delta t\sigma_{(i,j,k+\frac{1}{2})}}{2\varepsilon_{(i,j,k+\frac{1}{2})}+\Delta t\sigma_{(i,j,k+\frac{1}{2})}}\cdot E_z^n(i,j,k+\frac{1}{2})-\frac{2\Delta t}{2\varepsilon_{(i,j,k+\frac{1}{2})}+\Delta t\sigma_{(i,j,k+\frac{1}{2})}}\cdot\begin{bmatrix}\frac{1}{h_{yj}}\left(H_x^{n+\frac{1}{2}}(i,j+\frac{1}{2},k+\frac{1}{2})-H_x^{n+\frac{1}{2}}(i,j-\frac{1}{2},k+\frac{1}{2})\right)\\-\frac{1}{h_{xi}}\left(H_y^{n+\frac{1}{2}}(i+\frac{1}{2},j,k+\frac{1}{2})-H_y^{n+\frac{1}{2}}(i-\frac{1}{2},j,k+\frac{1}{2})\right)\\+J_z^{n+1}(i,j,k+\frac{1}{2})\end{bmatrix} \tag{7}$$

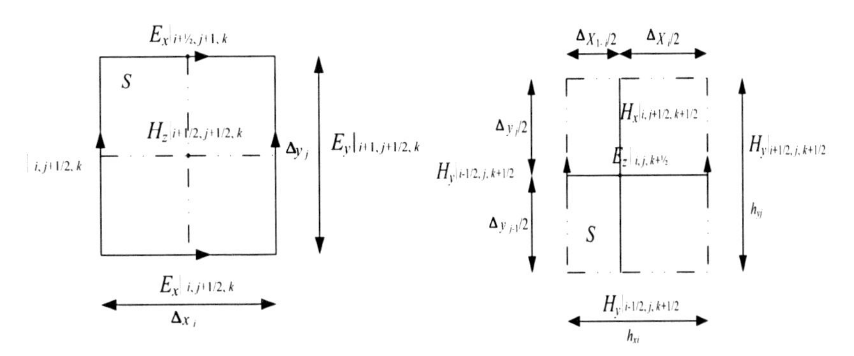

Figure 2: Around the contour of the Hz.

Figure 3: Around the contour of the Ex.

Similarly, according to right hand rule, the calculation formula of time advances of other electric-field components can be deduced.

This paper uses the principle of grdriving instructorent mesh to divide the whole computation space. FDTD difference equation can only apply the above formula on the interface of thick and thin grids. Obviously, when the electric-field component on the interface is updated, FDTD difference equation can not be used, and only the average value of the size of thick and thin grids can replace the size of single grid. On the other hand, due to the relative position relationship of the distribution of magnetic-field component and electric-field component, FDTD difference equation of uniform grid can be adapted for solving.

4 Simulating calculation of impedance ground of transient state upon partial corrosion of grounding body

FDTD model image of grounding body is seen in Figure 1. The size of the grounding body's corrosion area is set as: $t = 2.0\,\mu s$. The non-uniform arriving instructorent mesh is used to divide computational domain: the total number of grids in the whole computation space is: $58 \times 176 \times 298$, the total time step is $t_{num} = 200$. The formula for concrete grid division is as follows. The thin grid is divided on the nearby area of grounding body and its *x*, *y*, *z* direction, with the size of $\Delta s' = \Delta s / 5 = 0.05\,m$ (Δs is the size of thick grid), and in other areas, the grid with the size of Δs is uniformly divided.

Figure 4 shows the results of calculations by adapting uniform grid division and non-uniform division. The simulation result shows that the costing time is 1916.224 seconds when the non-uniform grid FDTD technique is applied, with 709.251 seconds that is shorter than the time calculated by uniform grid. It evidently increases the calculation speed.

In the process of calculation, when $t = 0.6\,\mu s$, $t = 1.3\,\mu s$, $t = 2.0\,\mu s$ in non-uniform grid, the potential distribution of different time steps is [10–11] seen in Figure 5(a)–(c).

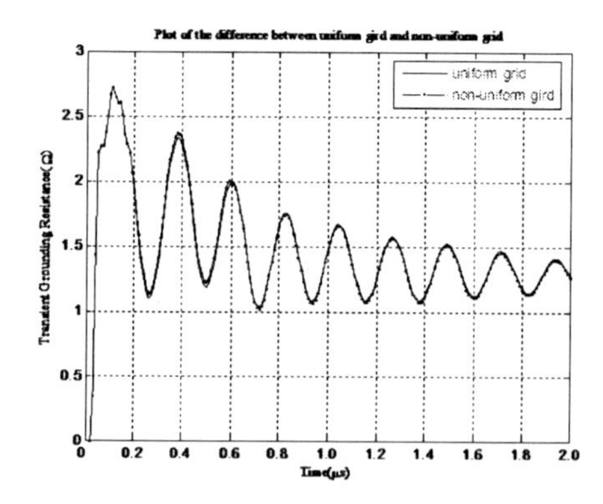

Figure 4: Non-uniform grid and uniform meshing calculation results.

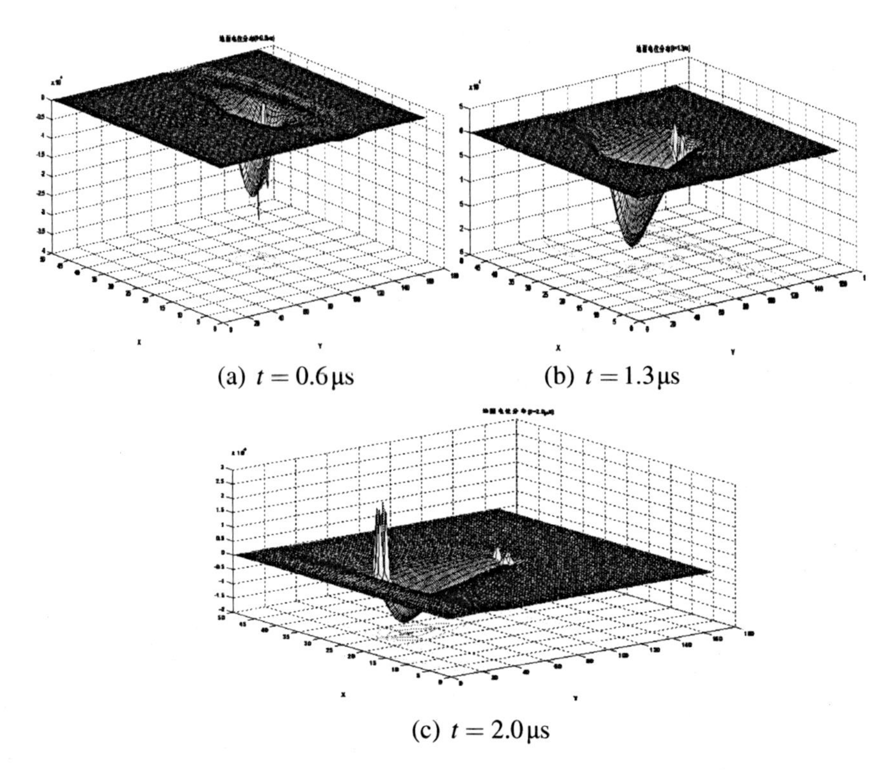

(a) $t = 0.6\,\mu s$ (b) $t = 1.3\,\mu s$

(c) $t = 2.0\,\mu s$

Figure 5 (a)–(c) are the distribution of electric field intensity of the ground at different time by Matlab programming calculation in the process of lightning current flowing through the corrosion grounding body into soil under the condition of partial corrosion of vertical grounding body.

From Figure 5(a), it can be seen that, after lightning current expressed by the standard lightning current waveform of $2.6/5.0\,\mu s$ gets into the ground, the electric potential suddenly increases, especially the electric field intensity around vertical grounding body reaches tens of volts per meter; Figure 5(b) shows: the electric field intensity around current loop and grounding body throughout current of $0.7\,\mu s$ evidently decreases; Figure 5(c) shows that lightening current of $2.0\,\mu s$ almost relieves to the ground.

Figure 6 shows the calculation results by respectively adapting uniform grid division and non-uniform grid division. Upon non-uniform grid division, the finedraw area takes up 3 and 5 thin grid size on the levels of x and y, which describes the grid structure in more detailed way than uniform grid division, and improves the calculation accuracy. In the same condition, the time for calculation is 1242.204000 seconds, with 674.02 seconds that is shorter than the time for uniform grid division calculation, which evidently increases the calculation efficiency.

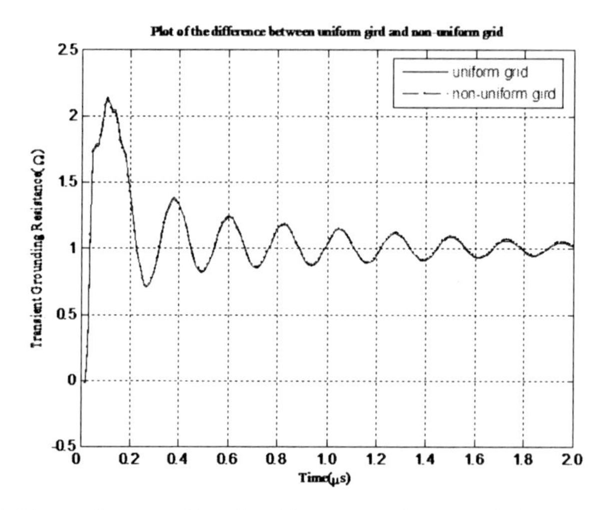

Figure 6: Non-uniform grid and uniform meshing calculation results.

5 Conclusion

This paper uses empirical formula to deduce the non-uniform grid FDTD promotion formula on all levels with dissipative medium, shows the division method of grdriving instructorent mesh in detail, and applies this method to the simulating calculation of impedance ground of transient state upon partial corrosion of grounding body. Finally, through Matlab programming calculation and analysis to get that non-uniform FDTD algorithm can not only save the calculation time but also achieve higher accuracy, which is obviously better than uniform FDTD algorithm.

References

[1] Yu Hao, *Modern Lightning Protection Technology Basis*. Tsinghua University Press, Beijing, 2005.

[2] Xie Guang-Run, *Power System Grounding Technology*. Water Conservancy and Electric Power Press, Beijing, 1991.

[3] He Jing-Liang, *Power System Grounding Technology*. Science Press, Beijing, 2007.

[4] Zhang Min, Cao Xiao-Bing, Li Rui-Fang, Transmission line tower grounding impact analysis of grounding resistance characteristics. *Insulators and Surge Arresters*, (4), pp. 5–9, 2012.

[5] Yang Tao, *Corrosion of Grounding Network State Test and Life Prediction*. Hunan University, Hunan, 2011.

[6] Deng Chang-Zheng, Yang Ying-Jian, Tong Xue-Fang, Grounding device impact characteristic analysis. *High Voltage Technology*, **38(9)**, pp. 2447–2545, 2012.

[7] Ge De-Biao, Yan Yu-Bo, *The Finite Difference Time Domain Method for Electromagnetic Wave*, 3rd ed.. Xi'an Electronic and Science University Press, Xi'an, 2011.
[8] Kazuo Tanabe, Computer analysis of transient performance of grounding electrodes based on FDTD/FI method. *2011 7th Asia-Pacific International Conference on Lightning*, Chengdu, China, 2008.
[9] Allen Taflove, Susan C. Hagness, *Co-Mutational Electrodynamics, The Finite-Difference Time-Domain Method*, 3rd ed. Artech House, Norwood, MA, 2005.
[10] Zhang Bo, Xue Hui-Zhong, Jin Zu-Shan, Transient potential distribution of transmission tower and its grouding device under lighting. *High Voltage Technology*, **39(2)**, pp. 393–398, 2013.
[11] Zeng Fan-Chunchen, Jia-Qing, Liang Miao-Yuan, FDTD analysis of the distribution of potential conductor discharging grouding impulse current. *Safety and Electromagnetic Compatibility*, 2010.

Analysis of reducing electron density in reentry plasma using electric and magnetic fields method

Li Wei, Suo Ying, Feng Qiang
School of Electronics and Information Engineering,
Harbin Institute of Technology, Harbin, China

Abstract

A joint method of electric and magnetic fields is able to reduce electron density in plasma. In this paper, the method of reducing the electron density by electric field and magnetic field in reentry plasma is presented. The ratio of reducing electron density in plasma with different conditions is analyzed. The current density and magnetic induction intensity can influence the plasma electron density ratio. For higher electron density plasma, the increasing of the current density, the magnetic induction intensity and the electrode length can reduce the plasma electron density ratio evidently.
Keywords: electric and magnetic fields, electron density, plasma.

1 Introduction

Electromagnetic wave attenuation increase is able to occur in reentry stage. The phenomenon is closely related to the plasma generated by the interaction between spacecraft and atmosphere. The electron density is very high in the reentry plasma, and the electron can interact with the electromagnetic wave emitted by aircraft antenna. More additional electromagnetic wave energy losses will produce so that communication signal is attenuated seriously [1].

Static magnetic field is able to bound random motion of free electrons. The free electrons hardly collide with positive ions or neutral molecules in the direction of the field. The magnetic field can reduce electron density, so the electromagnetic wave attenuation can be decreased.

A joint application of electric and magnetic fields have been proposed, in which the electric field is also able to reduce the electron density. But the method is still in the preliminary study stage [2].

WIT Transactions on Engineering Sciences, Vol. 107, © 2015 WIT Press
www.witpress.com, ISSN 1743-3533 (on-line)
doi:10.2495/SEEEI1140151

The method of reducing the electron density in reentry plasma by electric field and magnetic field is presented in this paper. The ratio of reducing electron density in plasma with different conditions is analyzed.

2 The method of reducing the electron density

Based on the theory of magnetic fluid mechanics, if magnetic field with a certain strength is applied near the antenna window on a fight vehicle, the motion of free electron in the plasma region can be controlled. When the plasma cross the magnetic field and electric field region, the electron is able to change its direction of motion under the action of electric field and magnetic field [3].

The electron density of plasma can be decreased near the aircraft antenna, thus a called magnetic window effect is formed. The theoretical and experimental results show that the establishment of strong constant magnetic field around the antenna, the propagation characteristics of electromagnetic wave in plasma can be improved [4].

In order to produce magnetic window effect better and reduce the weight of the equipment, the joint application of electric and magnetic fields have been proposed, in which the electric field is in order to eliminate the free electron magnetic window. Plasma has shielding effect to the electric field when the electrode does not contact with the plasma or both insulation[5]. Once the electrode contact with the plasma, a closed loop in the plasma will be formed by connecting two electrodes. Charge accumulation is not able to occur on the plasma surface, and shielding effect to the external electric field disappears. Through the discussion of one and two-dimensional results of plasma distribution of the interaction of electric and magnetic field, electric field and magnetic field that when vertical position, can significantly reduce the electron density region is shown in Figure 1.

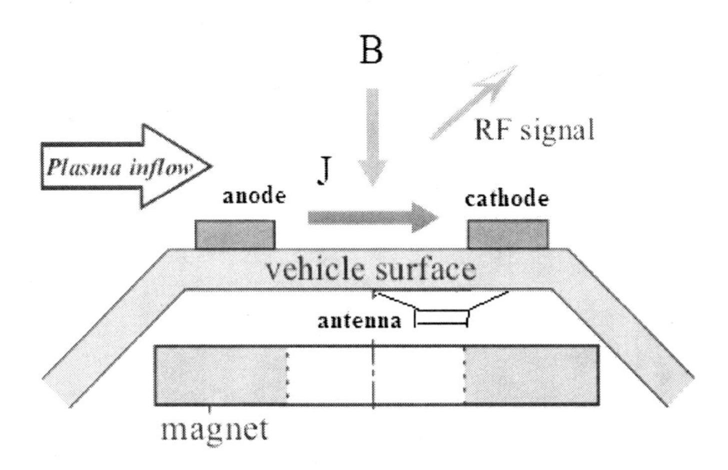

Figure 1: The electric field and magnetic field on a vehicle.

3 Analysis of the ratio of reducing electron density of plasma

The reducing ratio of plasma electron density N is able to be calculated by the following formula:

$$N = \frac{n}{n_{0\,i}}$$

$$= \frac{2V_{i,0}}{(\frac{k(T_i+T_e)}{m_i V_{i,0}} - V_{i,0} + \frac{JB}{m_i}\frac{L}{n_0 V_{i,0}}) + \sqrt{(-\frac{k(T_i+T_e)}{m_i V_{i,0}} + V_{i,0} - (\frac{JB}{m_i}\frac{L}{n_0 V_{i,0}}))^2 - 4\frac{k(T_i+T_e)}{m_i}}}$$

Where N is reducing ratio of plasma electron density, n_0 is initial electron density, n is electron density in plasma, B is magnetic induction intensity, k is Boltzmann constant, J is current density, L is length of electrode, $V_{i,0}$ is velocity of ions, T_i is temperature of ions, T_e is temperature of electrons, m_i is mass of ions.

By changing the initial electron density, plasma layer thickness and other parameters, the change of characteristic parameters of plasma can be calculated under different conditions. At the same time, change the current density of J, the magnetic induction intensity B and the electrode length parameters such as L, characteristic parameters of plasma can be obtained with the different conditions of electric field and magnetic field conditions.

Applied current density J, a certain degree of magnetic induction B and electrode length L, the ratio N of reducing electron density of plasma can be got corresponding to different initial plasma electron density n_0. Figure 2 shows the reducing ratio of plasma electron density with different n_0. Where B is 0.5T, $J = 500\text{A/m}^2$, $L = 0.04\text{m}$.

The results show that the emission spectra are enhanced by the applied electrical and magnetic field, while their frequency components and bandwidths are not changed. For the same current density and magnetic induction intensity, the plasma electron density ratio N is increased with the increasing of plasma electron density.

For the same plasma electron density, with increasing of the current density and the magnetic induction intensity, the plasma electron density ratio N is decreased.

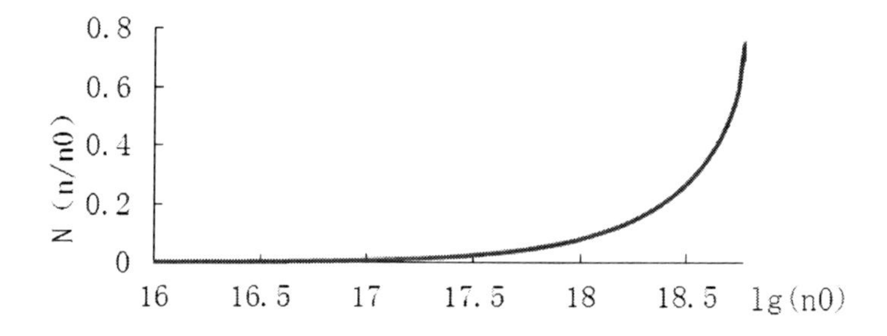

Figure 2: The reducing ratio with different initial plasma electron density.

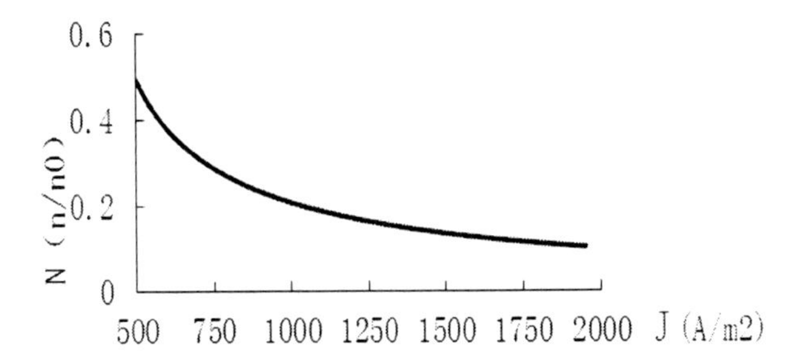

Figure 3: The reducing ratio with different current density.

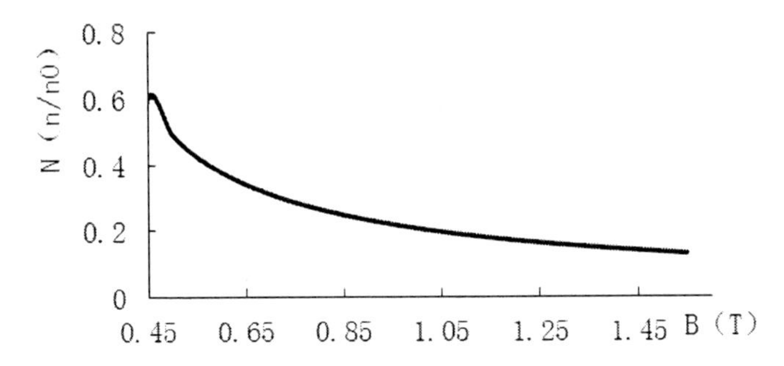

Figure 4: The reducing ratio with different magnetic induction intensity.

Figure 3 shows the reducing ratio of plasma electron density with different n_0. Where B is 0.5T, $n_0 = 5\times10^{18}$A/m^3, $L = 0.04$m.

In the analysis of current density on the plasma electron density, the first fixed magnetic induction B and electrode parameters of L size, to observe the plasma electron density than N by changing the size of the applied current density J. Said the length of the electrode in the magnetic induction intensity, reduced, when different current density ratio.

Reduce the electron density of the plasma current density increased with the ratio decreases. The more current density is applied, the more reduce the electron density in the plasma. For higher electron density plasma, the increasing of the current density can reduce the plasma electron density ratio N evidently.

Figure 4 shows the reducing ratio of plasma electron density with different n_0. Where $J = 500$A/m^2, $n_0 = 5\times10^{18}$A/m^3, $L = 0.04$m.

In the analysis of the change of the magnetic induction intensity of the plasma electron density, the first fixed current density J and the electrode parameters of L size, the magnetic induction intensity B change and size of plasma electron density decreases N. Indicates the length in the current density, electrode, reduction of different magnetic induction intensity B ratio. Reduced plasma electron density ratio increases with the applied of the magnetic induction intensity decreases. The more magnetic induction intensity is applied, the more the electron density in the plasma reduce. For higher electron density and plasma, the increasing of the magnetic induction intensity can reduce the plasma electron density ratio N.

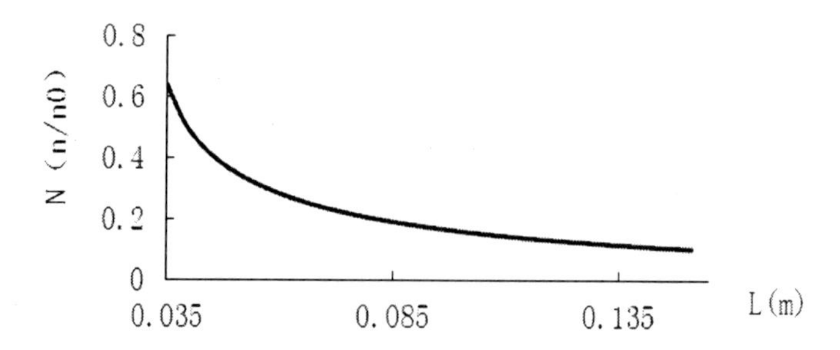

Figure 5: The reducing ratio with different electrode length.

Figure 5 shows the reducing ratio of plasma electron density with different n_0. Where B is 0.5T, $n_0 = 5\times10^{18}$A/m^3, J = 500A/m^2.

In the analysis of the electrode length changes of plasma electron density, the first fixed current density J and the magnetic induction B size, the electrode length L change and size of plasma electron density decreases N. The magnetic induction intensity, reduce the ratio of different electrode length L.

Reduced plasma electron density ratio increases with the external electrode length decreases, the electrode length and larger, decrease the electron density in the plasma more. For higher electron density of plasma, the increasing of the additional electrode length can reduce the plasma electron density ratio N.

4 Conclusions

The structure of reducing the electron density in reentry plasma by electric field and magnetic field is presented on a fight vehicle. The ratio of reducing electron density in plasma with different conditions is analyzed. When external electric field and magnetic field is selected properly, this method can reduce the electron number density of plasma significantly. The current density and magnetic induction intensity can influence the plasma electron density ratio. For higher electron density plasma, the increasing of the current density, the magnetic induction intensity and the electrode length can reduce the plasma electron density ratio N evidently.

Acknowledgement

The authors would like to express their sincere gratitude to funds supported by "the National Natural Science Funds" (Grant No. 61201014), and "the Fundamental Research Funds for the Central Universities" (Grant No. HIT.NSRIF. 2012025).

References

[1] D.D. Morabito, The spacecraft communications blackout problem encountered during passage or entry of planetary atmospheres. *IPN Progress Report*, pp. 1–23, 2002.

[2] Melvik J. Kofoui, Permanently magnetized ferrite antenna windows for improving electromagnetic wave transmission trough a plasma. *IEEE Transactions on Antennas and Propagation*, pp. 251–252, 1966.

[3] M. Kim, M. Keidar, I.D. Boyd, Two-dimensional model of an electromagnetic layer for the mitigation of communications blackout. *47th AIAA Aerospace Sciences Meeting Including The New Horizons Forum and Aerospace Exposition*, Florida, USA, pp. 10–17, 2009.

[4] M. Kim, M. Keidar, I.D. Boyd, Electrostatic manipulation of a hypersonic plasma layer: images of the two-dimensional sheath. *IEEE Transactions on Plasma Science*, **36(4)**, pp. 1198–1199, 2008.

[5] Minkwan Kim, Iain D. Boyd, Michael Keidar, Modeling of electromagnetic manipulation of plasma for communication during reentry flight. *Journal of Spacecraft and Rockets*, **47(1)**, pp. 29–35, 2010.

Design and analysis of rectangular waveguide edge slot array

Ying Suo[1,2], Wei Li[1,2], Weibo Deng[1], Shuangbin Yin[1]
[1]*School of Electronics and Information Engineering, Harbin Institute of Technology, Harbin, China*
[2]*Electronic Science and Technology Postdoctoral Station, Harbin Institute of Technology, Harbin, China*

Abstract

In order to achieve a narrow beam character, a rectangular waveguide edge slot array is designed. In this paper, it is discussed the design theories of waveguide edge slot array, puts forward some optimization methods. Firstly, FEKO software is used in the design and simulation of a slot structure element, and then the parameters of model is selected and optimized, the theoretical model was modified, and lastly one-dimensional slot array antenna model is formed by the slot element, which can actual the requirements.
Keywords: waveguide edge slot antenna, array, narrow beamwidth.

1 Introduction

The slot antenna is designed in a waveguide, coaxial line, metal plate or resonant cavity, wherein, planar slot antenna has a low profile, integrated, easy to form an array and other characteristics, which has been widely applied. The common used planar slot antennas are two important forms, which are microstrip slot antenna and waveguide slot antenna. Microstrip slot antenna has the advantages of easy conformal, convenient processing, low cost, but it has the disadvantage of high loss, low efficiency. While the waveguide slot array antenna has the advantages of low loss, large power capacity, high radiation efficiency, easy to achieve low side lobe and many other advantages. This type of antenna is especially suit for use in radar, satellite communication system using [1].

WIT Transactions on Engineering Sciences, Vol. 107, © 2015 WIT Press
www.witpress.com, ISSN 1743-3533 (on-line)
doi:10.2495/SEEEIl140161

For waveguide slot antenna array, there are two common used forms, which are longitudinal slot in wide wall waveguide and the inclined slot in the narrow wall waveguide. The resonant length is very strict in wide wall slot waveguide, and phase changes acutely. While the length change in the inclined edge slot has an inconspicuous effect on the phase change. Therefore, edge slot in narrow wall is used for radiation element [2, 3].

2 Antenna design

In the actual antenna design, there are two main methods to determine the antenna parameters: software simulation and actual measurement. With the rapid development of commercial software and the greatly improvement of computer operating speed, software simulation becomes a fast, accurate method of antenna design. Firstly determine the initial parameters of the antenna through the theoretical calculation, secondly, establish and simulate model using EMSS Company's FEKO software, lastly, according to the design target, modify the model parameters, to obtain the antenna with optimized parameters. This method can meet the design requirements of the antenna, avoid a lot of repeated work, and improve the efficiency of design.

For transmitting antenna, to reduce the energy loss, use the same gaps with same size and angle to form array. The conductance in resonant frequency of each slot is same. The rectangular waveguide slot antenna is used. The model is shown in Figure 1. L is the wide side length of waveguide, w is the edge side length of waveguide. Only TE_{10} mode wave is permitted to transmit in the waveguide. The gap is cut in the edge side. θ is the inclined tangle of slot, δ is cutting depth in the wide wall, the width of the slot is a, s is the slot length, λ is the work wave length and λ_g is waveguide wavelength. For the inclined slot in the waveguide edge, as the gap cutting the lateral current in waveguide edge, the model can be equivalent to the parallel with equivalent transmission line.

According to the analysis method of Stevenson theory, when the following conditions are met:

1) $s/a >> 1$
2) $s = \lambda/2$
3) only the TE_{10} mode is allowed
4) t is approximately equal to 0, waveguide can be seen as a good conductor.

At the same time, by using the equivalent transmission line theory and the waveguide Green Function, normalized conductance expressions of the gap can be calculated as following,

$$g=g_2'[\sin\theta\cos(\pi\lambda\sin\theta/2\lambda_g)/(1-(\lambda/\lambda_g)^2\sin^2\theta))]^2 \qquad (1)$$

where $g_2' = (30/73\pi)(\lambda_g/\lambda)(\lambda^4/a^3b)$.

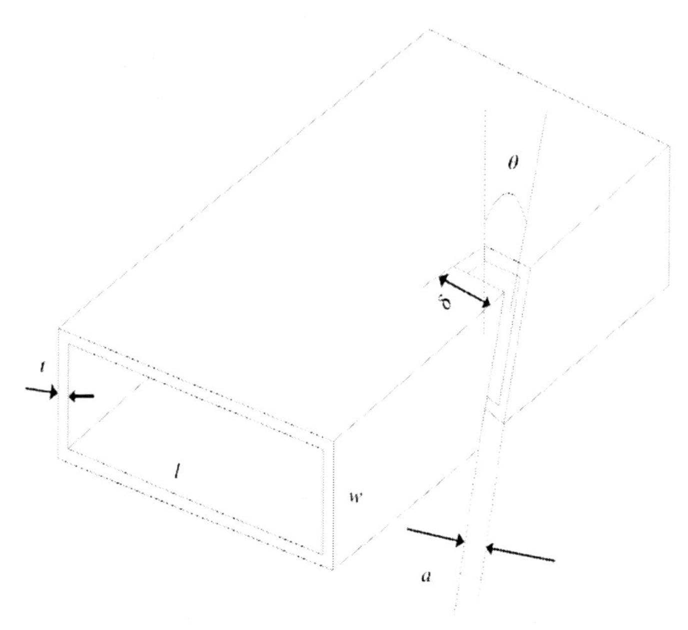

Figure 1: Model of rectangular waveguide edge slot element.

3 Element simulation

Geometric model of the single slot structure is firstly established by using Feko software. As shown in Figure 2, the slot is cutting in the edge waveguide. The radiation intensity is determined by the slot inclined angle, and radiation intensity is related to radiation conductance. So the parameter conductance g is needed to calculate to decide the inclined angle of the gap.

With the increase of the inclined angle θ, the conductivity increases, which means that the radiation increases, but does not infinitely increase. By calculation, determine θ is equal to 15°, slot length s is 50mm, that is $\lambda/2$. The selection of slot width also have great affects to antenna's radiation. The max value of conductance and 0 point of susceptance will drift with the increase of the gap width. The width of gap need to avoid the gap medium from breakdown at the same time, and therefore the parameter a should be chosen according to the actual need. The model is meshed by the geometric structure. The methods of moment (MOM) is an exact numerical method based on integral equation, the calculation is accuracy. So the simulation of electromagnetic properties of geometric model in this study using the methods of moment.

The waveguide slot antenna can be seen as a two port network, as given in Figure 2. The simulation result of the model's S parameter S_{21} is shown in Figure 3, which is the imaginary part varies with frequency curve. As can be seen from the chart, the frequency point whose imaginary part equal to 0 is about 3.01GHz, namely the susceptance is 0, which is the resonant point, approximate travelling wave propagation.

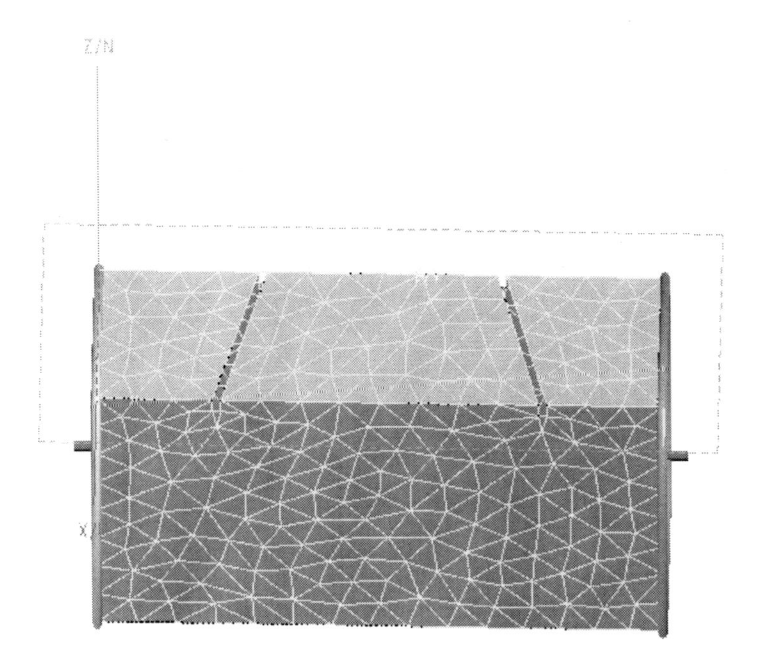

Figure 2: Simulation model of the antenna element.

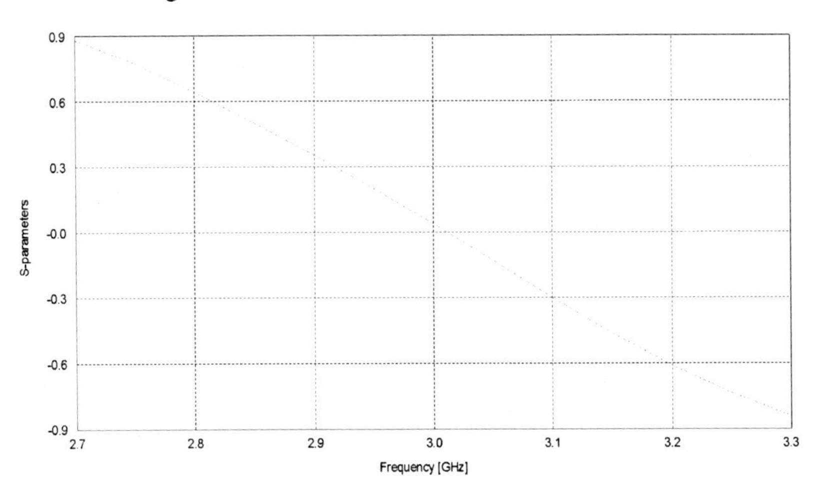

Figure 3: S_{21} curve versus frequency.

The far-field radiation pattern of slot antenna element is calculated. According to the direction of the electric field, we can judge the waveguide slot antenna's E-plane is yoz plane, H-plane is xoz plane, the two planes of the far field radiation pattern are shown in Figures 4 and 5. It can be seen in the two figures that, the 3dB beam width of E-plane is about 40°, while the 3dB beam width of H-plane is about 100°. The directivity is more than 8.2dB.

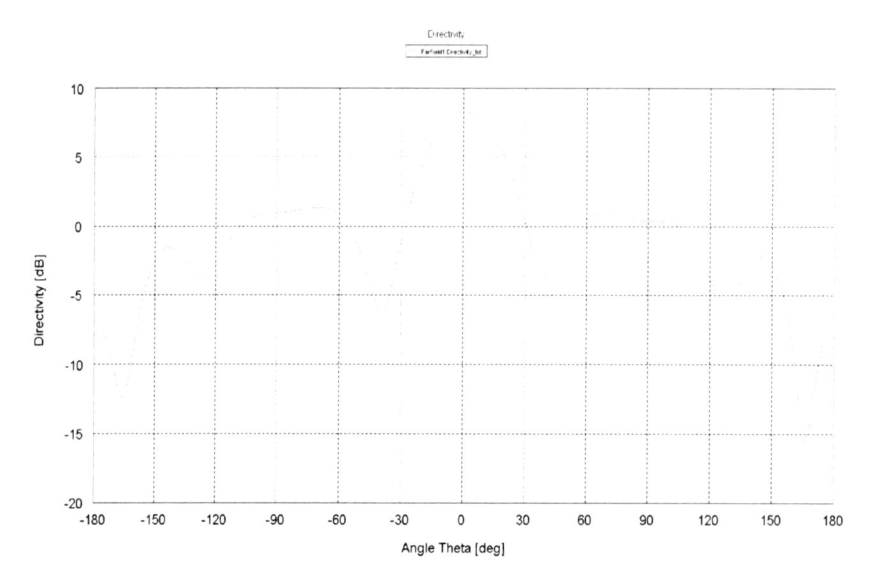

Figure 4: E-plane radiation pattern.

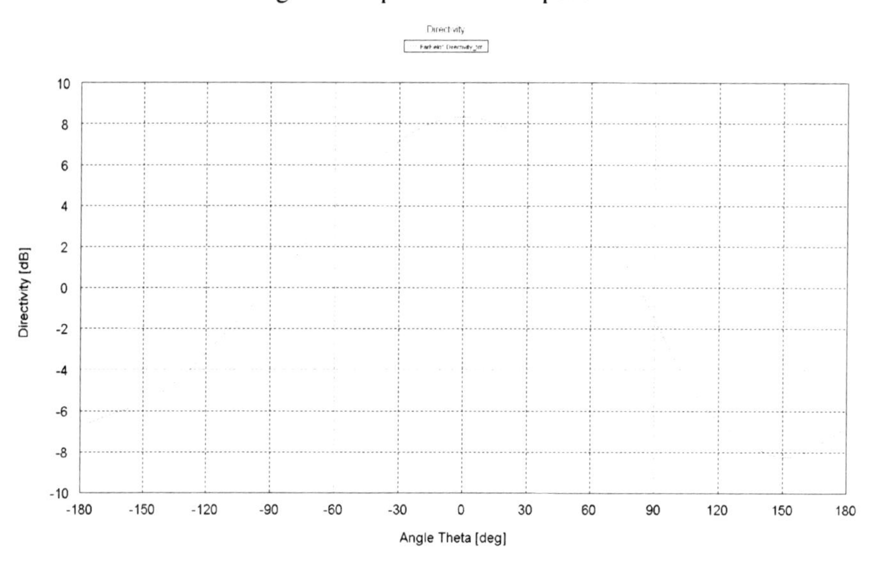

Figure 5: H-plane radiation pattern.

4 Waveguide slot array antenna

Designed (two elements antenna) waveguide slot elements are spaced along the longitudinal direction of waveguide. Each element structure is consistent, means the array has same feeding amplitude and phase. Calculation of the reflection coefficient as shown in Figure 6, as can be seen from the graph, at the resonant point 3.01GHz, the reflection coefficient is about −12.5dB, which means a good matching, can meet the needs of engineering. The far field radiation pattern of the antenna array of 10 elements is given in Figures 7 and 8. It is shown in the two figures, the 3dB beam width of E-plane is about 4°, which can realize good performance in narrow beam. While the 3dB beam width of H-plane is about 96°, which means the beam width is not be compressed. The directivity is nearly 19dB. The ratio of main lobe and side lobe is 13dB, which is consistent with the general law of equal magnitude and phase array.

Discussed above are the basic model of rectangular waveguide slot antenna, for how to broaden the bandwidth, how to change the array amplitude distribution and realize the low side lobe, research and analysis in-depth can be based on the present article discussed.

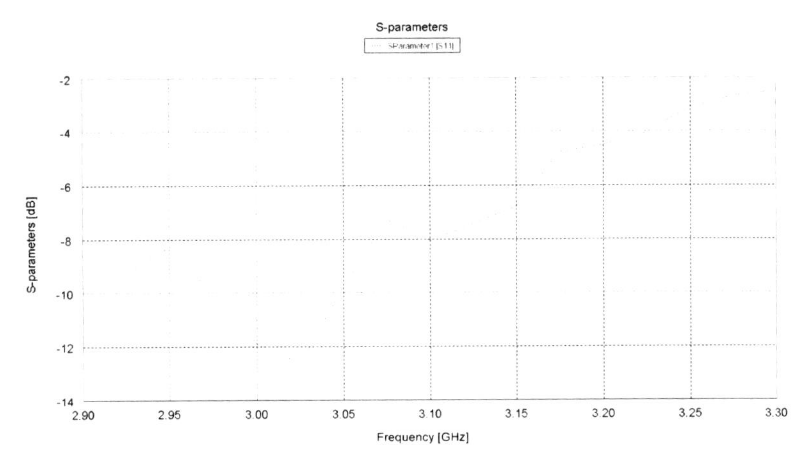

Figure 6: Return loss versus frequency curve.

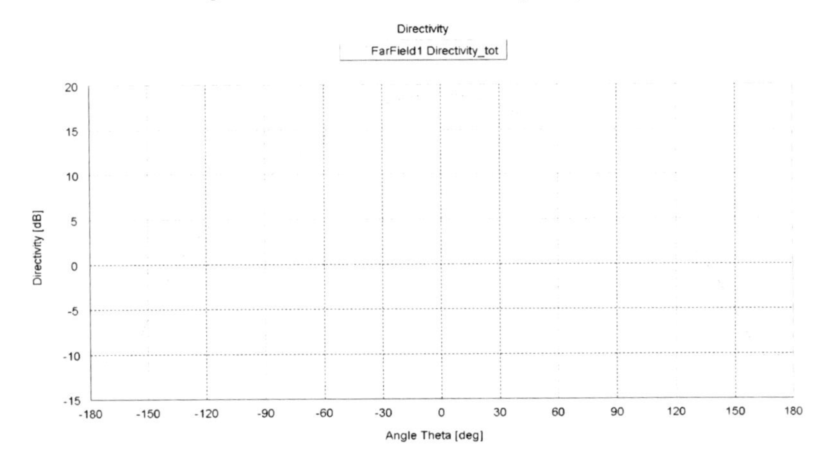

Figure 7: E-plane radiation pattern.

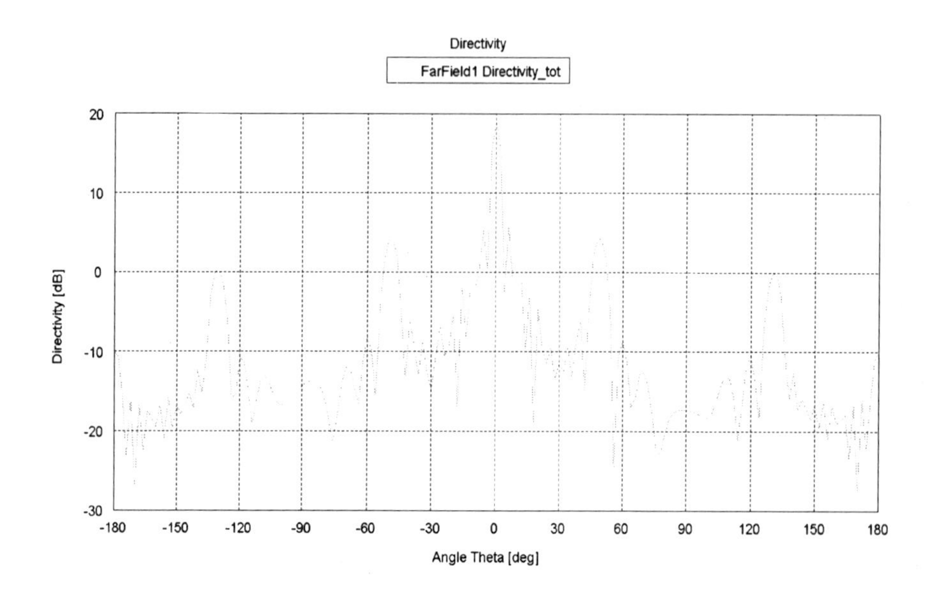

Figure 8: H-plane radiation pattern.

5 Conclusion

Rectangular waveguide edge slot array working at S band is simulated with the application of Feko software. By the merits of edge slot array, this designed antenna can realize high directivity, low reflection and other good character. Future plan is to widen the working frequency band and high efficient design.

Acknowledgment

The authors would like to express their sincere gratitude to funds supported by "the National Natural Science Funds" (Grant No. 61201014), "the Fundamental Research Funds for the Central Universities" (Grant No. HIT.NSRIF. 2014023), also, the authors would like to express their sincere gratitude to FEKO software.

References

[1] C.H. Wei, G. Li, Research and design of edge slotted-waveguide antenna. *Radar Science and Technology*, **11(5)**, pp. 557–560, 2013.
[2] J.Y. Li, G.Y. Zhang, Theory analysis of rectangular waveguide longitudinal slot arrays. *Modern Radar*, **21(6)**, pp. 77–81, 1999.
[3] J.C. Coetzee, J. Joubert, M. Derek, A off-center-frequency analysis of a complete planar slotted-waveguide array consisting of sub-arrays. *IEEE Transactions on Antennas and Propagation*, **48(11)**, pp. 1746–1755, 2000.

Transmission line analysis for the electromagnetic wave propagation characteristics in tunnels

Feng De-wang[1], Lan Jian-rong[2], Sun Lei[1], Lin Yu-Qing[1]
[1]*School of Computer and Information Science, Fujian Agriculture and Forestry University, Fuzhou, Fujian, China*
[2]*School of Mechanical and Electrical Engineering, Fujian Agriculture and Forestry University, Fuzhou, Fujian, China*

Abstract

The tunnel environment is so complex that the study on the characteristics of electromagnetic waves propagation in a tunnel is very difficult. Based on the equivalent transmission line method, which combines electromagnetic field and circuit theory, the wave-guiding characteristics in a tunnel are easily analyzed. As examples, the transmission line model of horizontal polarization and vertical polarization electromagnetic waves in rectangular tunnel are built. Furthermore, numerical analysis is performed. The results indicate that a tunnel radio signal consists of a summation of many propagation modes. The electric field strength decreases with the increase of the propagation distance, and the higher frequency, the lower loss. The receiving signal of the center tunnel is the strongest. In a rectangular tunnel, the height has a great effect upon the propagation characteristics of the vertical polarization wave modes, and the width has the same effect upon the horizontal polarization wave modes. Moreover, the relative dielectric constant of wall rock exerts little influence.

Keywords: electromagnetic wave, transmission line theory, polarization, attenuation, rectangular tunnel.

1 Introduction

The tunnel is a special restricted space, the propagation characteristics of electromagnetic waves in which are affected by many factors, such as the size

www.witpress.com, ISSN 1743-3533 (on-line)
doi:10.2495/SEEEII140171

of cross section and shape of the tunnel, the frequency of the electromagnetic waves, the electric parameters of surrounding rocks in tunnels, coal dust, water vapor [1–5]. In recent years, scholars at home and abroad study and analyze the propagation characteristics of electromagnetic wave in tunnels based on the model which takes tunnels as a hollow dielectric wage-guide [6, 7]. The theories and computations in these dissertations which use electromagnetic fields theory to analyze the propagation characteristics of electromagnetic waves in tunnels are more complex. However, the transmission line theory which converts the analyzing of the electric and magnetic fields into the calculating of the voltage and current has widely used in microwave systems, so that the combination of the circuit theory and the network analyze can be well used in microwave systems [8–14]. In this paper, it used the equivalent transmission line method which combines fields and circuits to analysis the characteristics of wave guiding in tunnels. The basic idea is that the tunnel is taken as a symmetrical lossy system, and the corresponding RLCG transmission line model is built by using the electromagnetic field theory. Then, propagation characteristics of electromagnetic waves for various modes is analyzed based on the circuit theory.

2 Theoretical model

In general orthogonal coordinate systems, the horizontal electromagnetic field of any wave guiding system can be expressed as [15]

$$\begin{cases} \mathbf{E}_t(x,y,z) = \sum \mathbf{e}_{tk}(x,y)\mathbf{U}_k(z) \\ \mathbf{H}_t(x,y,z) = \sum \mathbf{h}_{tk}(x,y)\mathbf{I}_k(z) \end{cases} \tag{1}$$

where $\boldsymbol{e}_{tk}(x,y)$ and $\boldsymbol{h}_{tk}(x,y)$ are vector mode functions which indicate the distribution of electromagnetic field in cross-section under one propagation mode; $\boldsymbol{U}_k(z)$ and $\boldsymbol{I}_k(z)$ are the equivalent voltage and the equivalent current which indicate the distribution of the electromagnetic field on this propagation direction.

If the equality (2) is given

$$\int_{\mathbf{S}} \mathbf{e}_{tk}(x,y) \times \mathbf{h}_{tk}(x,y) \cdot d\mathbf{S} = 1 \tag{2}$$

the wave guiding system can be equivalently seen as the transmission line. The tunnel underground coal mine has characteristics of guiding waves, so it can be equivalent to a transmission line. Take the rectangular tunnel as an example, building the corresponding equivalent transmission line RLCG model.

The distribution parameter of horizontal polarization $\boldsymbol{E}_h(m, n)$ wave model of the transmission line mode can be obtained as

$$\begin{cases} R = \dfrac{\lambda^2 a\sqrt{\mu_0/\varepsilon_0}}{2b}\left(\dfrac{m\varepsilon_1}{a^3\sqrt{\varepsilon_1 - 1}} + \dfrac{n}{b^3\sqrt{\varepsilon_2 - 1}}\right) \\ L = \dfrac{a\mu_0}{b} \\ C = \dfrac{b\varepsilon_0}{a} \\ G = \dfrac{\lambda^2 b}{2a\sqrt{\mu_0/\varepsilon_0}}\left(\dfrac{m\varepsilon_1}{a^3\sqrt{\varepsilon_1 - 1}} + \dfrac{n}{b^3\sqrt{\varepsilon_2 - 1}}\right) \end{cases} \tag{3}$$

where the width of the rectangular tunnel is a, the height is b, the dielectric constant of the tunnel is ε_0, the relative dielectric constant of the two side walls is ε_1, the relative dielectric constant of the top and the bottom of the tunnel is ε_2, the permeability outside and inside the tunnel is μ_0.

Similarly, distribution parameter of vertical polarization $\boldsymbol{E}_v(m, n)$ wave model of the transmission line mode can be obtained as

$$\begin{cases} R = \dfrac{\lambda^2 b\sqrt{\mu_0/\varepsilon_0}}{2a}\left(\dfrac{m}{a^3\sqrt{\varepsilon_1 - 1}} + \dfrac{n\varepsilon_2}{b^3\sqrt{\varepsilon_2 - 1}}\right) \\ L = \dfrac{b\mu_0}{a} \\ C = \dfrac{a\varepsilon_0}{b} \\ G = \dfrac{\lambda^2 a}{2b\sqrt{\mu_0/\varepsilon_0}}\left(\dfrac{m}{a^3\sqrt{\varepsilon_1 - 1}} + \dfrac{n\varepsilon_2}{b^3\sqrt{\varepsilon_2 - 1}}\right) \end{cases} \tag{4}$$

When take dB/m as the measure, the attenuation constant of horizontal polarization $\boldsymbol{E}_h(m, n)$ wave model and vertical polarization $\boldsymbol{E}_v(m, n)$ wave model in the rectangular tunnel are

$$\alpha_{E_h(m,n)} = 4.343\lambda^2\left(\frac{m^2\varepsilon_1}{a^3\sqrt{\varepsilon_1 - 1}} + \frac{n^2}{b^3\sqrt{\varepsilon_2 - 1}}\right) \tag{5}$$

This conclusion is exactly same as the rectangular tunnel electromagnetic wave attenuation formula deduced by Emslie using the mode matching method, so the effectiveness of the equivalent transmission line theory in the rectangular tunnel can be proved.

$$\alpha_{E_v(m,n)} = 4.343\lambda^2 \left(\frac{m^2}{a^3\sqrt{\varepsilon_1 - 1}} + \frac{n^2\varepsilon_2}{b^3\sqrt{\varepsilon_2 - 1}} \right) \tag{6}$$

3 Numerical analysis

The basic parameters of the rectangular tunnel is a = 4m, b = 2.5m, ε_0 = 8.854 × 10^{-12}F/m, $\varepsilon_1 = \varepsilon_2$ = 5, μ_0 = 1.257 × 10^{-6}H/m, f = 600MHz, m = 11, n = 11.

3.1 The effect of the propagation distance

Various modes of electromagnetic waves can be propagated in tunnels in the UHF band, therefore, the field strength distribution of one point is the superposition result of all modes of electric fields on this point. Each mode has the transmission line model, according those models, the electric field components E_x, E_y, E_z of all modes can be obtained firstly, then by using superposition principle, the size of composited electric field in the tunnel can be obtained as

$$|\mathbf{E}| = \sqrt{\left(\sum_m \sum_n E_x\right)^2 + \left(\sum_m \sum_n E_y\right)^2 + \left(\sum_m \sum_n E_z\right)^2} \tag{7}$$

Figure 1 shows the law that field strength changes along the distance. Where (a) simulates the situation of horizontal polarization wave mode $\boldsymbol{E}_h(m, n)$, (b) simulates the situation of vertical polarization wave mode $\boldsymbol{E}_v(m, n)$. From Figure 1, there exists many high-order modes in the near field region at the same time which make the synthetic field strength decrease rapidly with the distance increase.

3.2 The effects of communication frequency

Figure 2 shows the law that field strength changes along the frequency. From Figure 2, the attenuation rate of the synthetic field strength decreases with the frequency increases, while the transmission loss of the high frequency electromagnetic waves is more low.

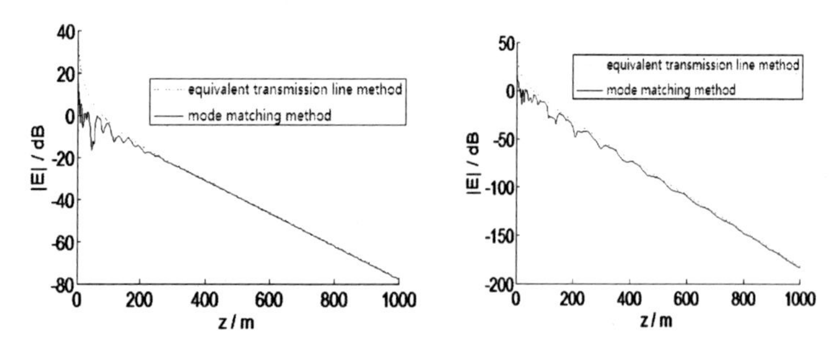

Figure 1: The relationship between the field strength and distance. (a) horizontal polarization. (b) vertical polarization.

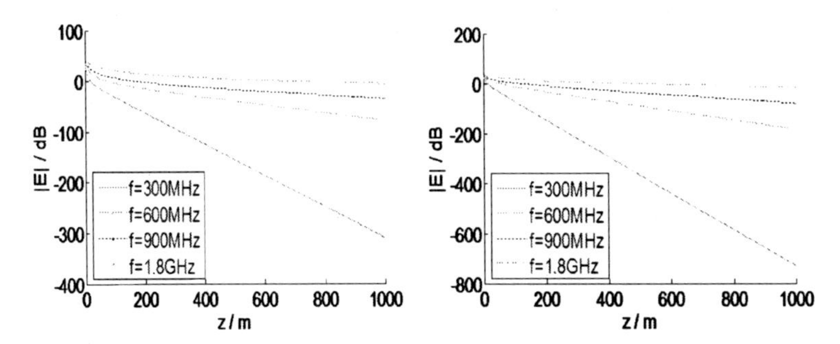

Figure 2: The relationship between the field strength and frequency. (a) horizontal polarization. (b) vertical polarization.

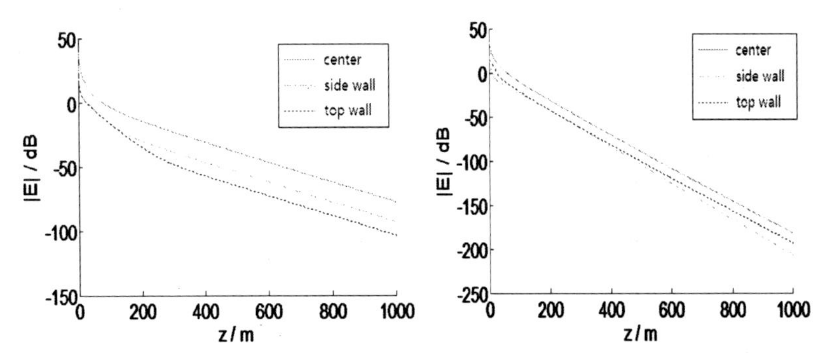

Figure 3: The relationship between the field strength and receive location. (a) horizontal polarization. (b) vertical polarization.

3.3 The effects of receive location

Figure 3 simulates the signals received situation of different locations. From Figure 3, the signal in the center position of the tunnel is strongest and the signal near to the tunnel wall arises a greater attenuation. Compared Figure 3(a) and 3(b), the differences of horizontal polarization electromagnetic waves in the three locations are greater than that of the vertical polarization electromagnetic waves in the same tunnel.

3.4 The effects of the size of section

3.4.1 Variation of height

Figure 4 shows distributions of field strength in rectangular tunnels with different heights. From Figure 4, when widths of tunnels keep the same, with the height increases, the attenuation of the field strength decreases obviously. Compared Figure 4(a) and 4(b), the changes of height have a great effect on vertical polarization electromagnetic waves but relatively little on horizontal polarization electromagnetic wave.

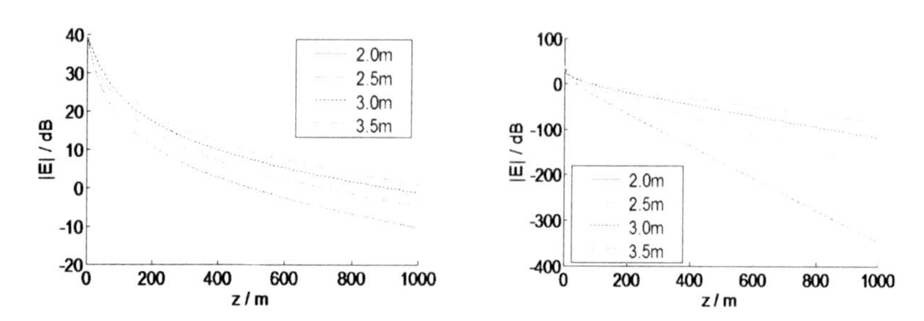

Figure 4: The relationship between the field strength and height. (a) horizontal polarization. (b) vertical polarization.

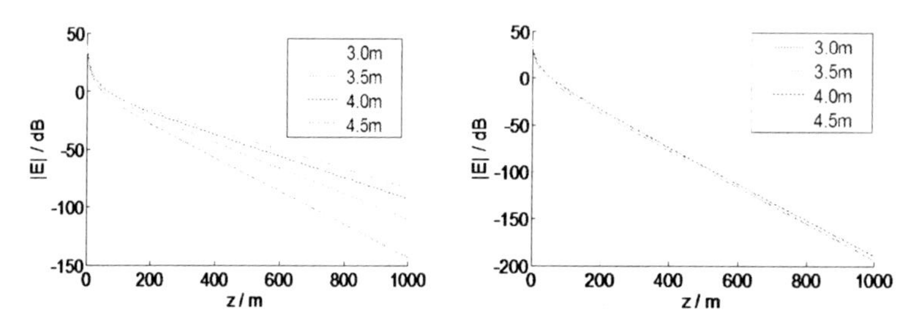

Figure 5: The relationship between the field strength and width. (a) horizontal polarization. (b) vertical polarization.

3.4.2 Variation of width

Figure 5 shows distributions of field strength in rectangular tunnels with different widths. From Figure 5, when heights of tunnels keep the same, with the width increases, the attenuation of the horizontal polarization electromagnetic wave field decreases obviously, while the attenuation of the vertical polarization electromagnetic wave field changes little.

3.5 The effects of the relative dielectric constant of surrounding rocks.

3.5.1 Surrounding rocks of side walls

Figure 6 shows the effects on the field strength by the relative dielectric constant of side wall surrounding rocks. From Figure 6(a), with the increase of the relative dielectric constant of side wall surrounding rocks, the attenuation of the horizontal polarization electromagnetic wave field just increases a little. From Figure 6(b), the changes of the relative dielectric constant of side wall surrounding rocks will have little effect on the vertical polarization electromagnetic waves.

3.5.2 Surrounding rocks of the roof of the tunnel

Figure 7 shows the effects on the field strength by the relative dielectric constants of surrounding rocks of the tunnel roof. From Figure 7,with the increase of the relative dielectric constant of tunnel roof surrounding rocks, the attenuation of the horizontal polarization electromagnetic wave field will decreases a little, while the attenuation of the vertical polarization electromagnetic wave field will increases a little.

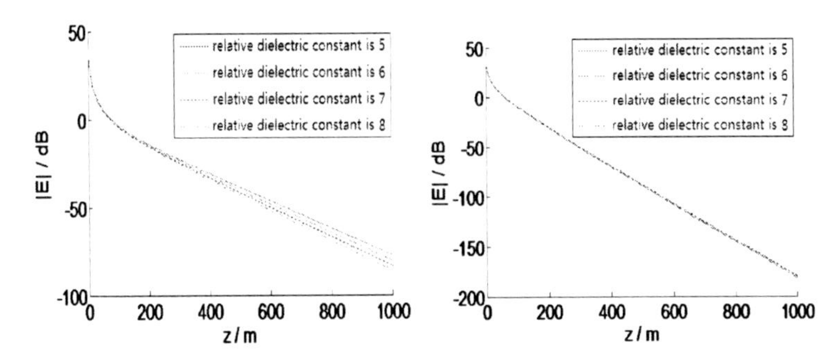

Figure 6: The relationship between the field strength and the relative dielectric constant of side wall surrounding rocks. (a) horizontal polarization. (b) vertical polarization.

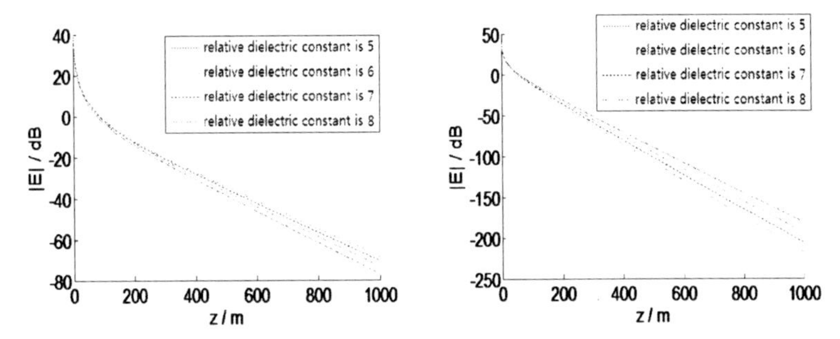

Figure 7: The relationship between the field strength and the relative dielectric constant of side wall surrounding rocks. (a) horizontal polarization. (b) vertical polarization.

4 Conclusions

1. In the near field region, because of the existence of many high-order modes, the electromagnetic wave field will decrease rapidly along the increase of distance. In the far field region, the low-order modes specially fundamental mode play the main role, the fluctuation of the field strength is more gentle. With the increase of the frequency, the attenuation rate of the synthetic field will decrease, which indicates that high-frequency electromagnetic waves have less transmission loss and more easy to spread. The attenuation rate of horizontal polarization electromagnetic waves are less than that of vertical polarization electromagnetic waves obviously.
2. The received signals in the center position of tunnels are strongest and the ones near to the tunnel walls will arise a greater attenuation. The horizontal polarization electromagnetic waves arise a greater attenuation on the top walls than the side walls, while the vertical polarization electromagnetic waves have the opposite characteristics.

3. With the increase of the height, the attenuation of the field strength will decrease obviously. And the height has a big effect on horizontal polarization electromagnetic waves but a little effect on vertical polarization electromagnetic waves. With the increase of the width, the attenuation of the horizontal polarization electromagnetic wave field will decrease obviously, while the attenuation of the vertical polarization electromagnetic wave field just decreases a little.
4. With the increase of the relative dielectric constant of side wall surrounding rocks, the attenuation of the horizontal polarization electromagnetic wave field will increase a little, while the changes of the relative dielectric constant of side wall surrounding rocks only have a little effect on vertical polarization electromagnetic wave field. With the increase of the relative dielectric constant of tunnel roof surrounding rocks, the attenuation of the horizontal polarization electromagnetic wave field will decrease a little, while the attenuation of the vertical polarization electromagnetic wave field will increase a little.

Acknowledgement

The research work was supported by Provincial Natural Science Foundation of Fujian under Grant No. 2012J01190 and the Project of Fujian Provincial Education Department under Grant No. JA11098.

References

[1] Wei Zhan-yong, Sun Ji-ping, Lu Jian-guo, Effects of cross sections of tunnel on propagation characteristics of electromagnetic waves. *Journal of UEST of China*, **32(6)**, pp. 620–623, 2003.
[2] Sun Ji-ping, Cheng Ling-fei, Zang Chang-sen, Influence of transverse dimensions on electromagnetic waves propagation in rectangular tunnels. *Journal of China University of Mining & Technology*, **34(5)**, pp. 596–599, 2005.
[3] Sun Ji-ping, Wei Zhan-yong, Propagation characteristic of low-frequency guided electromagnetic waves in tunnel. *Chinese Journal of Radio Science*, **18(2)**, pp. 203–206, 2003.
[4] Sun Ji-ping, Cheng Ling-fei, Zang Chang-sen, Influence of conductivity on radio waves propagation in tunnels. *Journal of Liaoning Technical University*, **26(1)**, pp. 96–98, 2007.
[5] Wei Zhan-yong, Moisture in tunnel influenced to transmission features of electromagnetic wave. *Coal Science and Technology*, **31(2)**, pp. 39–41, 2003.
[6] Sun Ji-ping, Zang Chang-sen, The propagation characteristic of electromagnetic wave in a round tunnel. *Chinese Journal of Radio Science*, **18(4)**, pp. 408–412, 2003.
[7] Sun Ji-ping, Cheng Ling-fei, Analysis of electromagnetic wave propagation modes in rectangular tunnel. *Chinese Journal of Radio Science*, **20(4)**, pp. 522–525, 2005.

[8] Feng Xi-yao, Xiao Gao-biao, Mao Jun-fa, Design of broadband couplers with non-uniform coupled transmission lines. *Journal of Applied Sciences*, **24(5)**, pp. 453–457, 2006.
[9] Wang An-guo, Wu Yong-shi, Ding Run-tao, Derivation of S-parameters in general condition for asymmetric coupled lines in inhomogeneous Medium. *Journal of Microwaves*, **20(1)**, pp. 10–14.
[10] Pan Yong-mei, Xu Shan-jia, Equivalent transmission line analysis for the loss characteristics of nonideal waveguiding structure. *Journal of Microwaves*, **20(2)**, pp. 40–42.
[11] Xing Feng, Liu Yao-wu, Song Wen-miao, Equivalent circuit of rectangular waveguide based on the generalized transmission line equation. *Chinese Journal of Radio Science*, **19(1)**, pp. 32–35.
[12] Bai Xue Xu Lei-jun, Research on the effects of substrates on the crosstalk in microstrips. *Journal of Electronics & Information Technology*, **32(11)**, pp. 2768–2771, 2010.
[13] Yan Xu, Li Yu-shan, Gao Song, et al., A passive transmission-line macro-model based on the method of characteristics. *Journal of Electronics and Information Technology*, **33(4)**, pp. 927–931, 2011.
[14] Liu Yao-wu, Hong Jing-song, Kenneth KM, Analysis of a double step microstrip discontinuity using generalized transmission line equations. *IEEE Transactions on Advanced Packaging*, **26(4)**, pp. 368–374, 2003.
[15] Hang Yu-lan, *Magnetic Field and Microwave Technology*. People's Posts and Telecommunications Press, Beijing, 2007.

Design and optimize a compensated linear induction launcher

Liang Gao, Zhenxiao Li, Baoming Li
National Key Lab of Transient Physics, Nanjing University of Science, Nanjing, China

Abstract

In recent decades, with advances in power supply technology, electromagnetic (EM) launch has received great attention because of their potential applications. Putting micro satellites into orbit could be the use of electromagnetic launch technology to replace traditional chemical boosters, while attractively increasing launch safety and reliability. The linear induction launcher (LIL) is one kind of electromagnetic launcher, this paper proposes a high-efficiency compensated LIL (CLIL) which can accelerate a projectile of 1kg in weight from 4 km/s to 8km/s in 4.69m/s by a group of high-frequency generators. A finite element method of modeling a LIL was developed in this paper which based on the electromagnetic field theory. After we obtained the reference value of driving force, velocity and displacement, a compensator was settled to improve the distribution of magnetic field. The comparison about the compensator in different thickness and material was done by this paper, also including the best initial position. Compared with the traditional LIL, the proposed solution has higher efficiency. The permeability of the compensator is an important factor to improve the performance, the research process mentioned in this paper could help to optimize the design of LIL in future applications.
Keywords: electromagnetic field, linear induction launcher, compensated, finite element.

1 Introduction

There are three main categories of EM launch techniques, namely, linear motors, coil guns, and rail guns. With the development of power supply, we can obtain a higher muzzle energy than ever before, so it seems possible to launch payloads by

www.witpress.com, ISSN 1743-3533 (on-line)
doi:10.2495/SEEEI140181

EM launch techniques [1, 2], such as micro satellites. The linear induction launcher (LIL) belongs to the electromagnetic launcher, it is essentially a group of linear electric motors, and the velocities are greater than those achievable with conventional cannons. Excitation by a group of capacitors or poly-phase generators is design to create an electromagnetic wave packet, which can travel with increasing velocity of the projectile [3–6]. The LIL has a wide application prospect in military field because it can be used to accelerate heavy projectile or electromagnetic armor [7, 8].

In recent studies, great progress has been made in structure optimization and high speed multi-stage cylinder linear induction motor (HSMCLIM) [9–12]. The efficiency of LIL is in direct ratio to the velocity of projectile, so a primary difficult and the focus of a great amount of attention is how to improve the efficiency in recent years.

Several geometries and designs about compensated pulsed alternator (CPA) evolved in the past, of these, three main configurations are presented since they are most suitable for applications: active CPA, passive CPA, and selective passive CPA [13]. Passive CPA is the simplest type of compulsator in which an armature winding spins relative to the excitation field and the compensation for the armature winding is provided by a continuous conductive shield, CPA provides some reference for this paper, and the design and verification about CLIL is presented and compared with an example in previous published case [9], after optimizing the magnetic field distribution by setting the compensator, the launcher can obtain better performance than that of the traditional LIL at the same conditions.

Ref. [9] assumed that using a gas gun to obtain the initial 4 km/s velocity at the input to the coil launcher, and accelerate a projectile of 1kg in weight from 4km/s to 8 km/s in at = 5.76 ms, the total length of coil launcher is 34.38m. Compared with the example in Ref. [9], the CLIL significantly improved the performance: the acceleration time only need 4.69 ms, and the total length of coil launcher can be shorten to 28.17m.

The intent of this paper is to provide a new method for EM launch system for reducing cost, and find out how to achieve a maximum muzzle velocity. Further and more details are presented as follows.

2 Working principle and model

2.1 The basic principle of LIL

The LIL usually consist of power supply, coil, projectile and control circuit. The drive coil of LIL is fixed, and when the drive coils are fed in sequence by a group of five-phase generators, projectile currents are induced in an attempt to exclude magnetic flux from the projectile, the interaction of the net radial magnetic field with the azimuthal projectile current results in an axial force that accelerates the projectile.

The drive coils of this LIL consist of 20 segments, each segment of the array which has 10 drive coils is fed at a constant frequency, but each succeeding section, traversed by projectile at higher and higher velocity, is energized at an increasingly higher frequency.

This paper takes the first 10 drive coils for an example. The drive coils are fed by a five-phase generator. The parameter of the generator is as follows: peak phase voltage is 16kV, average peak phase current is 40kA, and the frequency is 21.3kHz, configuration of each phase is shown in Figure 1(a).

This paper assumes that the projectile does not get in touch with any other objects in which moment frictional force is conventionally neglected. The equivalent circuit diagram of system is shown in Figure 1(b).

2.2 Mathematical model

The drive coil can be equivalent to m pieces of copper ring while projectile can be equivalent to n pieces of mutual insulating aluminum ring. The force on the projectile can be expressed

$$F_z = \sum_{i=1}^{m} \frac{\mathrm{d}M_{0i}(t)}{\mathrm{d}x} I_i(t) I_0(t) \tag{1}$$

2.3 3D FEM model

The structure of before improved is shown in Figure 2. The adjacent coils keep insulation. This model has been simulated in Ansoft and the results of force, velocity and displacement are shown in Figure 3.

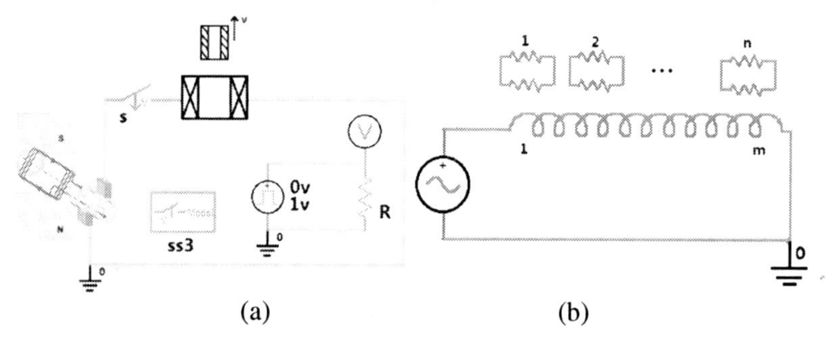

(a) (b)

Figure 1: The equivalent circuit diagram of system.

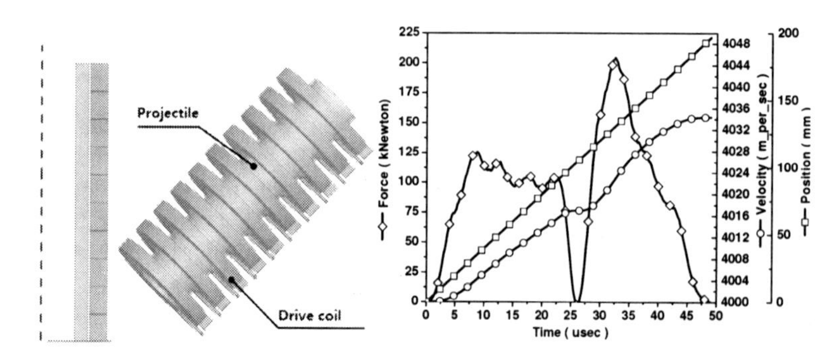

Figure 2: The 3D FEM model. Figure 3: The simulation results of FEM.

After the acceleration during the first 10 coils, the velocity of the projectile is increased to 4034.38m/s, and the acceleration, am, of the projectile is 80kg.

3 Design and optimize

3.1 Effect of LIL with compensator

In the same condition, we set a compensator with outer radius less than the drive coil. The air gap between the compensator and drive coil is 1mm, we assume that the compensator does not get in touch with any other objects, structure of the model is shown in Figure 4.

As shown in Figure 4, the green areas indicate the compensator, in order to study the effect of the compensator, simulation and analysis on different material and thickness of the compensator are compared with traditional LIL. The results in Figure 5 show the motion situation of the projectile with and without the compensator.

3.2 Effect on the performance of different materials

The speed of projectile has been improved after setting the compensator, then we study the effect about different material of the compensator. The compensator is loaded with three types of materials, including ferrite, steel and copper. Properties of these materials are shown in Table 1 and Figure 6.

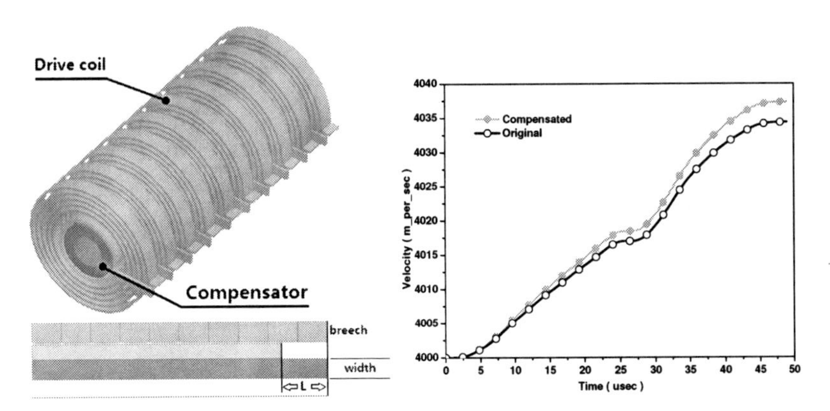

Figure 4: Structure of LIL.

Figure 5: Plots of velocity for 10 coils.

Table 1: Material property.

No.	Material	Relative Permeability	Bulk Conductivity (m/s)
1	Ferrite	1000	0.01
2	Steel	B-H in Figure 7	2e+6
3	Copper	0.999991	5.8e+7

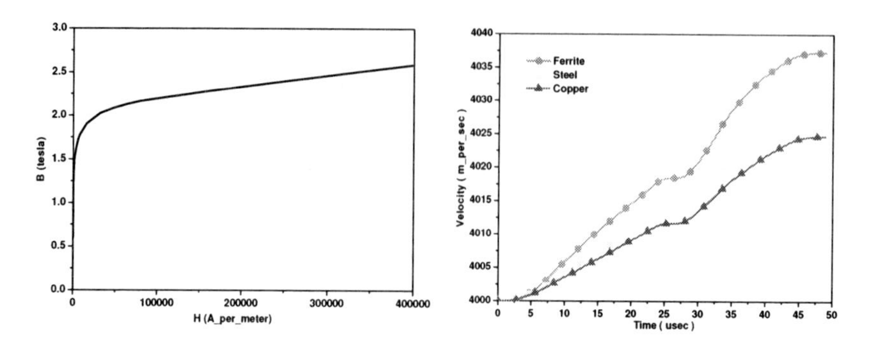

Figure 6: Nonlinear B-H curve of steel.

Figure 7: Comparison of different material.

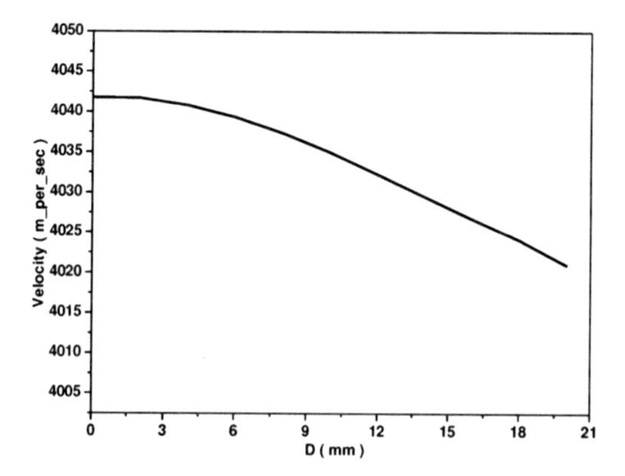

Figure 8: Transient simulation results at different width of compensator.

The simulation results are shown in Figure 7. It is shown that compensator with a magnetic nonconductive material such as ferrite can greatly improve the velocity of the projectile in the same time, while a conductive material would reduce the axial force, this is because the induced eddy current can exclude magnetic lines from projectile. From this conclusion, the magnetic material with low conductivity should be the best choice.

3.3 Effect on the performance of different geometry

Study on the effect caused by different thickness of compensator. The width marked in Figure 4 indicates the radius difference of the compensator. Magnetic material was selected to make the study about the relationship between velocity and thickness of the compensator, results are shown in Figure 8.

Simulation shows that the muzzle velocity is in direct ratio to the thickness of the compensator. The flux lines can be constrained to the center and easily pass through by setting a magnetic compensator. Through the above analyses, the best structure of compensator is a magnetic cylinder.

3.4 Effect on the performance at different initial position

Study on the relationship between initial position and the velocity. The L marked in Figure 4 indicates the distance between projectile and the bottom of the drive coils. Effect caused by different thickness of compensator is discussed as follows. The compensator is a ferrite cylinder with radius of 24 mm. Study the velocity characteristics by changing d from 0 to 20 mm every 1 mm. After accelerated by the first 10 coils, the relationships between velocity and initial position are shown as Figure 9.

The curve in Figure 9 shows that, with the increase of the distance between projectile and drive coils, there is always an optimal initial position that can obtain the maximum velocity of the armature. By this configuration, the velocity can be increased to 4041.75 m/s, the acceleration, a_m, of the projectile can be increased to 98.2 m/s^2.

4 Analysis and discussion

CPA provides some reference for the design of CLIL, magnetic distribution can be improved by setting a compensated device, and the velocity is in direct ratio to the relative permeability of material, and in inverse ratio to conductivity.

The comparison of axial force after final optimization is shown as Figure 10. The average acceleration force is 105.8kN.

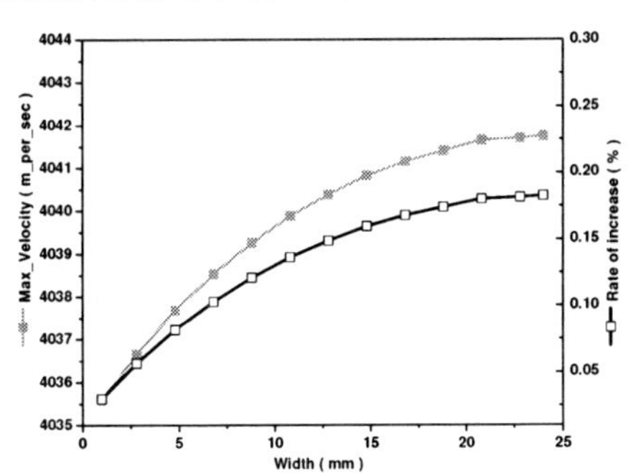

Figure 9: Transient simulation results at different initial position of projectile.

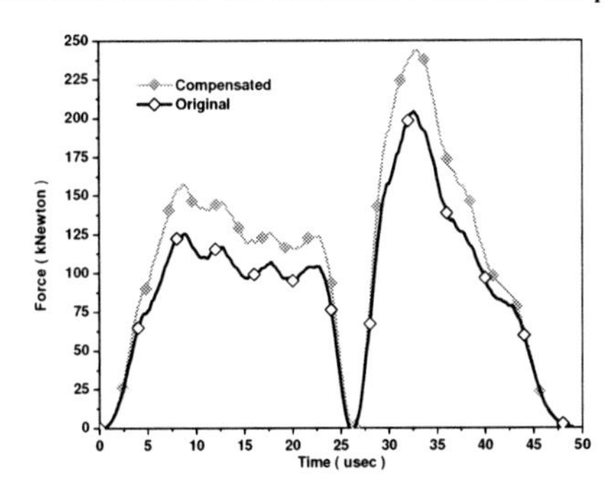

Figure 10: Plots of force for 10 coils with or without compensator.

Table 2: Comparison of performance.

	Average Force (kN)	Time (ms)	Length (m)
Reference	88	5.76	34.38
Compensated	105.8	4.69	28.17
Rate of increase %	+20.23	−18.58	−18.06

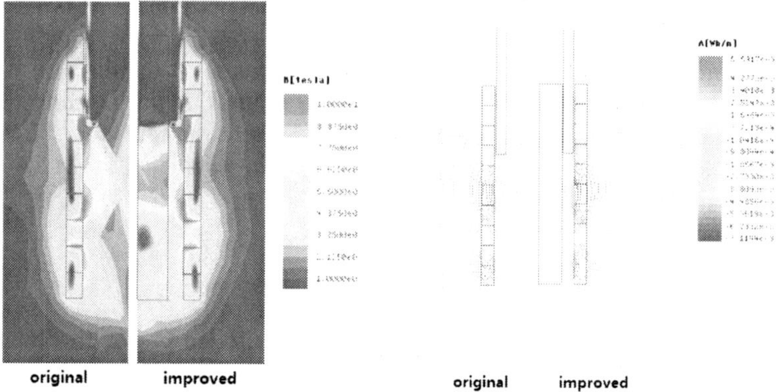

Figure 11: The magnetic distribution. Figure 12: The magnetic flux lines.

Comparison of the magnetic distribution is shown in Figures 11 and 12 when the force is maximum.

Comparison shows that the flux density is enhanced, that is the purpose of compensation. Comparison of the performance is shown in Table 2.

Based on the calculation above, accelerating a projectile of 1kg in weight from 4km/s to 8km/s only need 4.69ms, and the length of the drive coils can be shorten to 28.17m, Points should be noted that, the frequency of generator should be modified to adapt to the projectile when the velocity is higher.

5 Summary and conclusion

The LIL is studied in this paper, in order to improve the magnetic distribution, a compensated device is considered in the design, and this paper have made optimization by FEM. Analysis shows that by the method of setting a compensator, the acceleration time and the total length of the section can be respectively reduced by 18.58% and 18.06%. The result of analysis possesses certain guiding significance on parametric selection and design of CLIL.

References

[1] I.R. McNab, A research program to study airborne launch to space. *IEEE Transactions on Magnetics*, **43(1)**, pp. 486–490, 2007.

[2] I.R. McNab, Electromagnetic augmentation can reduce space launch costs. *IEEE Transactions on Plasma Science*, **41(5)**, pp. 1047–1054, 2013.

[3] A. Martin, R. Eskridge, Electrical coupling efficiency of inductive plasma accelerators. *Journal of Physics. D: Applied Physics*, **38(23)**, pp. 4168–4179, 2005.
[4] B.M. Novac, I.R. Smith, Studies of a very high efficiency electromagnetic launcher. *Journal of Physics. D: Applied Physics*, **35(12)**, pp. 1447–1457, 2002.
[5] M.D. Driga, W.F. Weldon, H.H. Woodson, Electro-magnetic induction launchers. *IEEE Transactions on Magnetics*, **22(6)**, pp. 1453–1458, 1986.
[6] W.R. Cravey, G.L. Devlin, E.L. Loree, et al., Design and testing of a 25 stage electromagnetic coil gun. *IEEE Transactions on Magnetics*, **31(1)**, pp. 1323–1328, 1995.
[7] L. Zhiyuan, M. Sun, F. Ming, The finite element analysis for the magnetic of the active EM armor projectile interceptor. *12th IEEE Electromagnetic Launch Symposium*, pp. 441–443, 2004.
[8] Cao Yanjie, Wang Huijin, Wang Chengxue, et al., Acceleration process of the interception projectile in active electromagnetic armor. *IEEE Transactions on Magnetics*, **45(1)**, pp. 631–634, 2009.
[9] A. Balikci, Z. Zabar, L. Birenbaum, et al., On the design of coilguns for super-velocity launchers. *IEEE Transactions on Magnetics*, **43(1)**, pp. 107–110, 2007.
[10] L. Gherman, M. Pearsica, C. Strimbu, et al., Induction coilgun based on "E-Shaped" design. *IEEE Transactions on Plasma Science*, **39(1)**, pp. 725–729, 2011.
[11] Zou Bengui, Cao Yanjie, Wu Jie, et al., Magnetic structural coupling analysis of armature in induction coilgun. *IEEE Transactions on Plasma Science*, **39(1)**, pp. 65–70, 2011.
[12] Tao Zhang, Wei Guo, Fuchang Lin, et al., Design and testing of 15 stage synchronous induction coilgun. *IEEE Transactions on Plasma Science*, **41(5)**, pp. 1089–1093, 2013.
[13] M.L. Spann, S.B. Pratap, M.D. Werst, et al., Compulsator research at the University of TEXAS at AUSTIN. *IEEE Transactions on Magnetics*, **25(1)**, pp. 529–537, 1989.

The recoil force distribution of electromagnetic railgun

Jiangbo Shi, Baoming Li
National Key Laboratory of Transient Physics,
Nanjing University of Science and Technology, Nanjing, China

Abstract

Over the past few years, there have been many researches on the recoil force of electromagnetic railgun, but these studies didn't focus the recoil force distribution. This paper analyzed the electromagnetic railgun recoil force mechanism on the bus-bar. Based on Maxwell's equations, the current density and magnetic flux distribution on the bus-bar were calculated first, and then the recoil force distribution. Through finite element method, we calculated the current density, magnetic flux and recoil force on the bus-bar of the two different cable arrangement models. The distribution of current density, magnetic flux and recoil force of the bus-bar were analyzed. We also found the concentrate phenomenon in the areas where bus-bars and rails contacted. To improve the launching safety of railgun, the bus-bar plates which contact with the rails should be strengthened according to the recoil force concentrate.

Keywords: electromagnetic railgun, current density, magnetic flux, recoil force distribution.

1 Introduction

Electromagnetic railgun is an important part of electromagnetic emission [1, 2]. The electromagnetic railgun using the force created by the strong electromagnetic field to push the armature moving along the railgun barrel, and the armature can be accelerated to very high speed in a short period. Compared with the conventional emission, electromagnetic emission needs simpler energy, lower cost, and better working stability, but muzzle velocity has a great improve [2, 3].

The ISL research center and the Texas at Austin electromechanical research center found the railgun recoil phenomenon in electromagnetic emission

www.witpress.com, ISSN 1743-3533 (on-line)
doi:10.2495/SEEEII140191

experiments, the scientists have done a lot of research on it [4–6]. Wm. F. Weldon and other researches point out that the recoil force mainly concentrated in the feeding parts, and found the recoil force can be existed in the transmission structure, power switching structure and power components through theoretical analysis [7]. R. A. Marshall and Jia Qiang also found this law, and they thought the recoil force in power switching structure and power components can be ignored [8, 9]. Usually bus-bar was the transmission structure, so when studying the railgun recoil force, it can only take the force of bus-bar into account only.

Two different finite element analysis models were founded in this paper to study the force distribution law when the bus-bar and electric cables were arranged along the barrel or perpendicular to the rail. Calculated the current density and magnetic flux of the bus-bar based on Maxwell's equations, the recoil force distribution law can be obtained. The differences between the two models were discussed, and analyzed the advantages and disadvantages of the models. Finally, there are some suggestions for the bus-bar structure to improve the safety of railgun.

2 Theory and computational model

In this paper, based on Maxwell's equations, we calculated the electric field and magnetic field of the railgun [10]. We assume the armature is fixed with the rail and no deformation. Let H, J, B and E be the magnetic intensity, current density, magnetic flux and the electric field intensity. The governing Maxwell's equations can be expressed as follow:

$$\begin{cases} \nabla \times H = J \\ \nabla \times E = -\dfrac{\partial B}{\partial t} \\ \nabla \bullet B = 0 \\ \nabla \bullet J = 0 \end{cases} \tag{1}$$

In Eq. (1), we have neglected the displacement current as has been discussed in previous work [10], this approximation is reasonable for electromagnetic railgun calculations.

The associated constitutive relations are expressed as follows:

$$B = \mu H \tag{2}$$

$$J = \sigma E \tag{3}$$

where μ is magnetic permeability of free space and σ is electrical conductivity.

For solutions of problems, boundary conditions and electrical excitations must be prescribed. Magnetic vector potentials at far field position are set to zeroes. Two natural boundary conditions can be defined as the tangential component of magnetic intensity and the normal component of current density as follows:

$$n \times H = 0 \tag{4}$$

$$n\ J = 0 \tag{5}$$

where n is the unit vector of the boundary.

After current density and magnetic flux were calculated, the recoil force can be described as the following:

$$F_r = \iiint_{\Omega} J \times B dv \tag{6}$$

where F_r is the recoil force, Ω is the volume of the recoil parts, and dv is the volume integral infinitesimal.

To make the calculation efficiency, the models were simplified as followed: the isolation components in railgun barrel were expurgated, both for the isolation tiers of the electric cable. Each cable was assigned a 70kA current excitation, and there are 24 cables arranged in different ways. The finite element models are shown in Figure 1.

As shown in Figure 1, the electric cables and the bus-bar plates were numbered with certain regularity. In transverse model, the cables were arranged in three rows perpendicular to the rails, and numbered as 1–1 to 1–8 and 3–1 to 3–8. Bus-bar plates were numbered as 1–1 to 1–3 and 2–1 to 2–3. In vertical model, the cables were arranged in 4 columns parallel to the rails, numbered as 1–1 to 1–6 and 4–1 to 4–6. Bus-bar plates were numbered as 1–1 to 1–6 and 2–1 to 2–6. There were extraction electrode plates to connect bus-bar and rail.

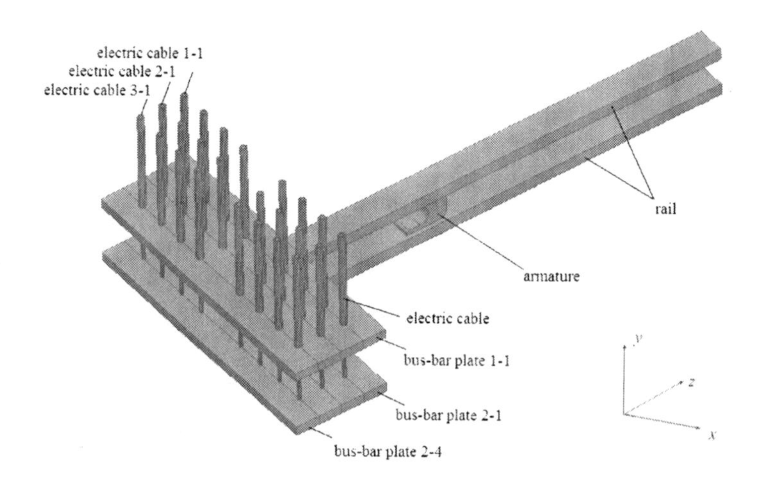

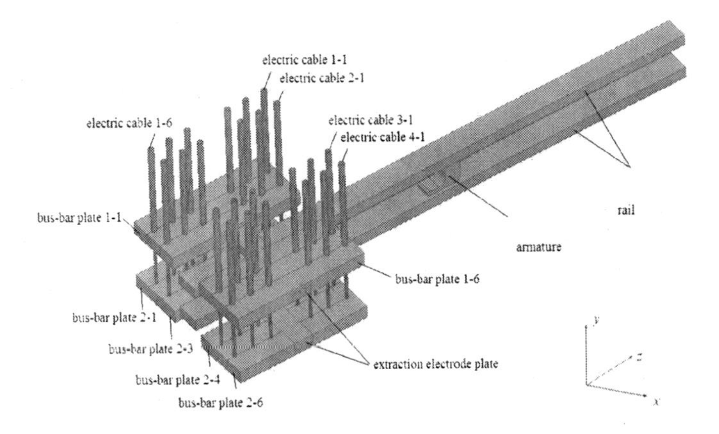

Figure 1: The finite element analysis models. (a) transverse model: electric cables arranged perpendicular to the barrel. (b) vertical model: electric cables arranged along the barrel.

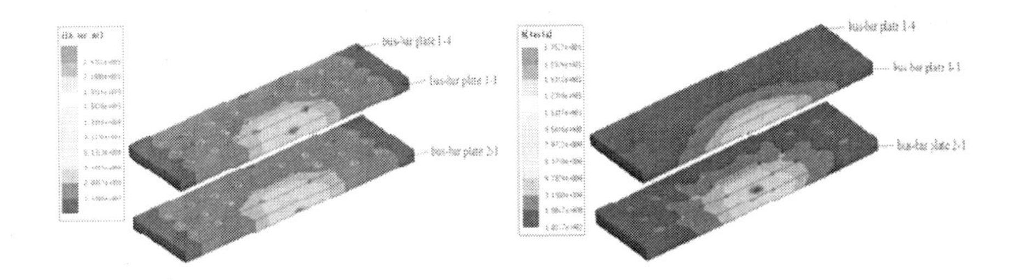

Figure 2: The bus-bar's current density and magnetic flux distribution of transverse model. (a): the bus-bar's current density distribution of transverse model. (b): the bus-bar's magnetic flux distribution of transverse model.

3 Simulation and result

3.1 Current density and magnetic flux distribution

Using the finite element solution, the current density distribution and magnetic flux distribution of the two models were obtained and shown in Figures 2 and 3.

As shown in Figure 2(a), the current concentrated in the middle area of the bus-bar, and as the distance between bus-bar plates and armature getting bigger, the current density getting smaller. The maximum current density is 2435MA/m^2 in middle of bus-bar plate 1–1 and 2–1 where the bus-bar contacted with the rail, and minimum is about 2.159kA/m^2 in plate 1–4 and 2–4. The bus-bar's current density distributions are symmetrical with the *x*-*z* plane, and also symmetrical with *y*-*z* plane.

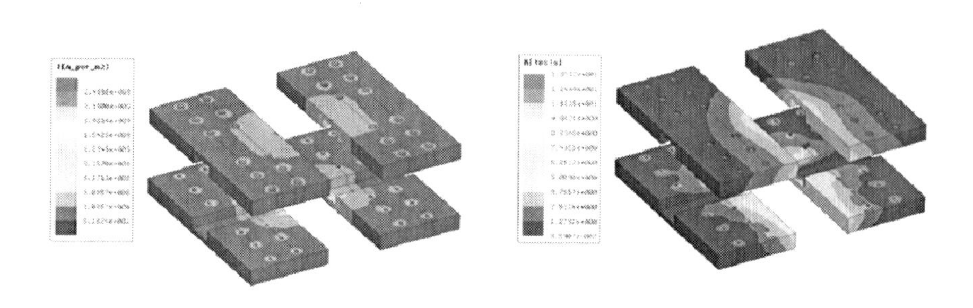

Figure 3: The bus-bar's current density and magnetic flux distribution of vertical model. (a): the bus-bar's current density distribution of vertical model. (b): the bus-bar's magnetic flux distribution of vertical model.

The magnetic flux has the same distribution law with the current density as shown in Figure 2(b). The maximum magnetic flux is 17.52T in bus-bar plate 1–1 and 2–1, almost the same position where the maximum current density appears.

In this model, there are extraction electrode plates to connect bus-bars and rails. And current density and magnetic flux in the electrode plates were much higher than in the other structures. In Figure 3, we ignored the electrode plates for a batter observation of bus-bar's condition.

As the structure has symmetrical with the *x-z* and *y-z* plane, the current density distribution also have the symmetrical characteristic with *x-z* plane and *y-z* plane. In the middle areas of the bus-bar plates where they contacted with the electrode plates have higher current density. The maximum current density is 2468.5MA/m^2 in plate 1–3, 1–4 and 2–3, 2–4, concentrated in the contacted areas. The minimum current density is 518.2A/m^2 in plate 1–1, 1–6 and 2–1, 2–6.

The magnetic flux distributions were much different with the current density. Magnetic flux didn't distribute in the region of the current density distribution. In bus-bar plates which were nearest to the rails almost half volume had high magnetic flux, and it appeared on the side near the armature. The maximum magnetic flux is 13.712T. In the other plates, the magnetic flux is much small.

3.2 Recoil force distribution

After calculated the current density and magnetic flux of the bus-bar, use Eq. (6) to calculate the recoil force. Figure 4 shows the recoil force vector in bus-bar of the two models. And the recoil force distributions were obtained.

As shown in Figure 4, recoil force has obvious similar relation with current density and magnetic flux. The recoil force is much larger where has higher current density and magnetic flux. In transverse model, recoil force mainly concentrated in the middle area of bus-bar plate 1–1 and 2–1 and in the vertical model recoil force concentrated in bus-bar plate 1–3, 1–4 and 2–3, 2–4. Calculated the force of each plate using Eq. (6), the recoil force changing law of each plate was shown in Figure 5.

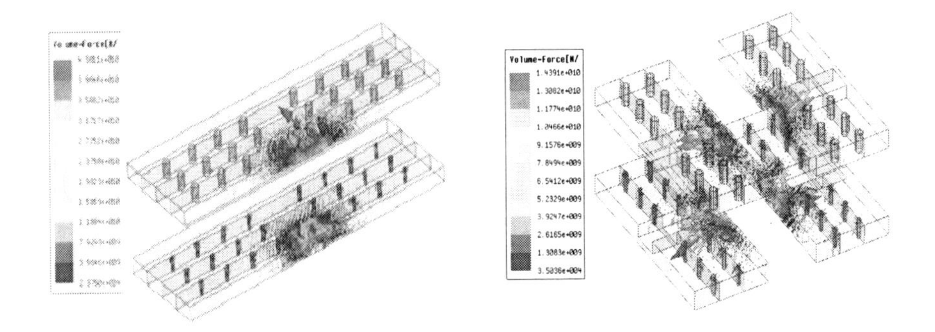

Figure 4: The bus-bar's recoil force distribution of the two models. (a): the bus-bar's recoil force distribution of transverse model. (b): the bus-bar's recoil force distribution of vertical model.

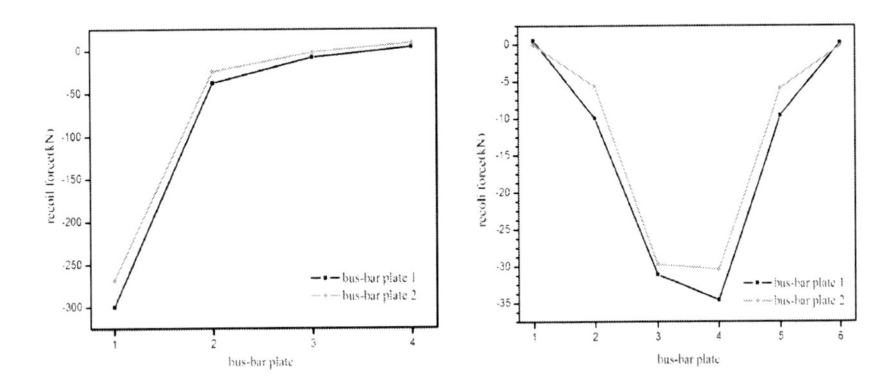

Figure 5: The recoil force of bus-bar plate for the two models. (a) the recoil force of every bus-bar plate of transverse model. (b) the recoil force of every bus-bar plate of vertical model.

As there have extraction electrode plates in vertical model, the largest recoil force comes to 299.26kN of transverse model and 34.58kN of vertical model. The bus-bar plate's recoil force of vertical model is much smaller than the recoil force of transverse model. The recoil force is mainly concentrated on the extraction electrode plates. The edge plates of bus-bar's recoil force are approximate to zero of the two models. The recoil force of bus-bar plate decreased rapidly as the distance to contact area increased.

In the point of bus-bar's recoil force, the structure of the vertical model is much better than transverse model. As the recoil force distribution have been clear, there should be some structure to strengthen the bus-bar plate. For transverse model, the bus-bar plate 1–1 and 2–1 should be strengthened. And for vertical model, the extraction electrode plates should be designed strengthen enough to make it safe.

4 Conclusions

Based on Maxwell's equations, two different finite element analysis models were founded by different electric cable arrangement. And current density and magnetic flux on electromagnetic railgun bus-bar were calculated in this paper. The result shows that: (1) current density and magnetic flux have concentrate phenomenon, for the transverse model, it concentrated in the area where the bus-bar contacted with the rail, and for the vertical model concentrated in the area where the bus-bar plate and extraction electrode plate contacted; (2) the current density and magnetic flux getting smaller as the distance to the contact area increase; (3) the recoil force has the obvious similar relation with current density and magnetic flux, and as the distance to the contact area increased, the recoil force getting smaller rapidly; the vertical model's bus-bar plate recoil force is much smaller than the transverse model's; (4) the vertical model's recoil force mainly concentrated in the extraction electrode plates. As the recoil force distribution is clear, there should be strengthen structure to make the bus-bar safe. For the transverse model, the bus-bar plates which contacted to the rails should be strengthened. And there should have some structure improvement to make the extraction electrode plates safe for the vertical model.

References

[1] J.E. Allen, T.V. Jones, Relativistic recoil and the railgun. *Journal of Applied Physics*, **67(1)**, pp. 18–21, 1990.

[2] Peter Graneau, Railgun recoil and relativity. *Journal of Physics D: Applied Physics*, **20**, pp. 391–393, 1987.

[3] Ching-Chuan Su, Mechanisms for the longitudinal recoil force in railguns based on Lorentz force law. *IEEE Transactions on Magnetics*, **42(9)**, pp. 2193–2195, 2006.

[4] Michael Putnam, *An Experimental Study of Electromagnetic Lorentz Force and Rail Recoil*. Naval Postgraduate School, Monterey, CA, 2009.

[5] Hilmar Peter, Francis Jamet, Volker Wwgner, Technical aspects of railguns. *IEEE Transactions on Magnetics*, **31(1)**, pp. 348–353, 1995.

[6] J.H. Price, E.P. Fahrenthold, W.G. Fulcher, Design and testing of large-bore, ultra-stiff railguns. *IEEE Transactions on Magnetics*, **25(1)**, pp. 460–466, 1989.

[7] Wm.F. Weldon, M.D. Driga, H.H. Woodson, Recoil in electromagnetic railguns. *IEEE Transactions on Magnetics*, **22(6)**, pp. 1808–1811, 1986.

[8] R.A. Marshall, Comment: origin, location, magnitude and consequences of recoil in the plasma armature railgun. *IEEE Proceedings – Science, Measurement Technology*, **144(1)**, pp. 49–52, 1997.

[9] Jia Qiang, Gao Yuefei, Analysis of recoil force in the electromagnetic railgun. *Academic Conference Proceedings of China Electrotechnical Society*, **1**, pp. 1–4, 2011.

[10] Kuo-Ta Hsieh, Bok-ki Kim, Implementing tri-potential approach in EMAP3D. *IEEE Transactions on Magnetics*, **35(1)**, pp. 166–169, 1999.

Design of non-lethal weapons platform for 8×8 wheeled armored anti-riot vehicles

Song Wang, Renjun Zhan
Equipment Engineering College, Engineering University of Armed Police Force of China, Xi'an Shaanxi, China

Abstract

To effectively deal with the increasingly serious situation of anti-antiterrorism and urgency treatment, the advanced weapons platform of 8×8 wheeled armored anti-riot vehicles should be developed. In order to avoid the limitation of precision strikes and covert shooting function of existing armored anti-riot vehicles, the overhead remote control weapon stations concept was proposed, as well as to improve the new armored anti-riot vehicles non-lethal strike capability, offered the loading scheme of non-lethal weapons, which can cover the strike range of 0–300 m. The non-lethal weapons platform scheme provides a better reference for development of new 8×8 wheeled armored anti-riot vehicles.
Keywords: wheeled armored vehicles, armored anti-riot vehicles, vehicle-mounted weapons, remote weapon station, non-lethal weapons.

1 Introduction

Recently, mass incidents and violent terrorist activities within China showed a high incidence and frequent situation of occurrence, the number and size of group incidents are rising year by year, which severely affect the country's social stability and economic development. China has entered a "universal terror" era, which fully demonstrates China's current seriousness and urgency situation and the reality of the fight against terrorism. What's more, the armed forces urgently need the advanced weaponry to carry out anti-terrorism task or other multifarious tasks.

Armored anti-riot vehicles is the ideal equipment to enforce urgency treatment and counter-terrorism mission, which has a high mobility, good protection and a certain aggressiveness, can safely and quickly transport personnel and supplies to the destination, and own the ability of providing fire support and effective protection

WIT Transactions on Engineering Sciences, Vol. 107, © 2015 WIT Press
www.witpress.com, ISSN 1743-3533 (on-line)
doi:10.2495/SEEEI140201

for front-line combat soldiers [1], which is important to ensure that suddenly anti-terrorism task is triumphantly completed by armed Police Force.

2 Limitation of current armored anti-riot vehicles

Compared with tracked armored vehicles, wheeled armored vehicle has advantages such as, a light weight, maneuverability, simple structure, ride comfort, short development cycle, cost-effective, versatile and low life-cycle cost and so on, which has become a national focus on the development of equipment [2]. At present, the development of wheeled armored vehicles are mainly 4×4, 6×6, 8×8 models, considering that 8×8 wheeled armored vehicles have advantage performance of mobility, armor protection, internal volume, load capacity, which lead to widely development and equipment in and around. Wheeled armored vehicles use high-tech and new technology to continuously increase its firepower, mobility, protection performance, information warfare capabilities.

The development of wheeled armored anti-riot vehicles, although in recent years has been rapid degree of upgrading, but because of a late start compared with foreign advanced models there are still large gaps. Mainly reflected in:

For fire systems, vehicle-mounted lethal weapons cannot meet anti-terrorism, defense and combat firepower requirements; non-lethal weapons onboard configuration features a single, technical content is clearly insufficient;

For protective performance, with the escalation of terrorist activities, mines, improvised explosive devices, car bombs, suicide bombers and rockets and other threats increased threats faced by armored anti-riot vehicles in the future operational environment will come from any direction, existing equipment Key protection is no longer a positive way to adapt to operational requirements;

For situational awareness, the existing anti-riot armored vehicle design prevalence of congenital defects. One window is narrow, limited observation. The basic trend of the existing equipment perception artificial optical periscope mode is mainly to obtain information accuracy is low, inefficient and there is a narrow field of view observation, comparative perspective fixed the problem. Second, which lack the technique of air observation.

3 Overhead remote weapon station design

3.1 Overview of remote weapon station

Since the development of armored fighting vehicles in firepower, protection and mobility have reached a very high level, but there are still two shortcomings: Firstly, observation of vehicle-mounted automatic weapon, aim and shoot are using the original manual operation, longer reaction time, lower shooting precision, higher requirements for occupant; secondly, operator in the operation of these weapons must be exposed, completely abandoned the armor protection, greatly reducing the battlefield survivability.

To solve these problems, competing abroad in recent years launched a remote weapon station – equipped with multiple types of weapons, with the goal of the search, identification, tracking, aiming, shooting and other remote control function, can be installed on a variety of military platforms relatively independent modular Weapon System [3]. A remote weapon station, also known as a remote weapon system (RWS), is a remotely operated weaponries system for light and medium calibres weapons which can be installed on ground combat vehicle or sea and air-based combat platforms. Such equipment is used on modern military vehicles, as it allows a gunner to remain in the relative protection of the vehicle.

Remote weapon station usually consists of fire systems, fire control systems and integrated suite. Fire systems include weapons subsystems, holder subsystems, which can be configured to a variety of weapons subsystems caliber machine guns, automatic grenade launchers, machine guns and missiles. Weapon fire control system is mainly used for remote control operation, day and night observation targets, ballistic solver, stability control, which includes sighting subsystem, servo control subsystem, fire control computer, control terminal, wherein sighting subsystem Configuring optical sight, CCD camera, thermal imager, laser range finder. Integration Kit includes power supply subsystem, rotary connector, connectors and cables, etc.

3.2 Typical automotive overhead remote weapon station

(1) "Heroes" non-lethal remote weapon stations (Israel)

2012, Israeli Rafael company launched "Heroes" non-lethal remote weapon station (NL-RWS). The weapon station is equipped with a 140dB acoustic transmitter, a non-coherent xenon light blinding device, a six remote grenade launcher (which can be fired tear gas, flash – shock bombs and smoke bombs, etc.), as well as the company developed launch non-lethal ammunition 10–40 mm grenade launcher, and comes with a 5.56 mm rifle at close range precision shooting. Some of the weapons station with a full array of photoelectric sensors, with surveillance, tracking and engagement capabilities, to prevent stones and firebombs attacks.

"Heroes" remote weapon station has six kinds of sea floor models and two kinds of models, and supports a variety of variations, compatible with a variety of weapons and sensors, including the 5.56/7.62/12.7 mm machine guns, 40 mm automatic grenade launcher as well as non-lethal weapons.

Figure 1: Raphael's "Heroes" remote weapon station.

(2) Full Spectrum Effects Platform (United States)
"Full Spectrum Effects Platform" is mounted on the United States, "Stryker" armored vehicle, carrying a variety of non-lethal weapons and different functions. The platform includes a 12.7mm machine gun, a GBD-IIIC laser glare control, a strong white beam emitter and 12 EROS hinged front emitter a remote weapon station, and is equipped with a set of remote acoustic emission device (LRAD) and one on the 12th automatic shotgun rear remote weapon station.

In addition to the installation of a laser rangefinder independent weapon system (pre-installed on the remote weapon station) and acoustic sniper detection sensor, the platform also includes a number of other sensors and an improved fire control unit (FCU), for control hinged launchers smokescreen or other non-lethal grenades, and delay fuze programming.

Currently, with a "full-spectrum effects platform" 2 "Stryker" prototype has been put into the battlefield for the military police forces.

Features combination of foreign and Chinese anti-terrorism weapon station, the actual task at the sudden, new armored anti-riot vehicles should be designed to include the following fire systems remote weapon station:

(1) a deadly weapon: 30mm small-caliber gun, 12.7mm caliber machine gun, 5.8mm close range precision shooting rifles, 40mm automatic grenade launcher;
(2) Non-lethal weapons: laser dazzling, a strong white beam emitter (bright lights);
(3) auxiliary devices: laser rangefinder, the whole array of photoelectric sensors.

4 Vehicle-mounted non-lethal weapons platform design

4.1 Existing vehicle-mounted non-lethal weapons platforms

The Active Denial System (ADS) is a weapon that uses millimeter wave energy to heat up water molecules in the subcutaneous layers of the skin, causing a painful burning sensation [4]. The ADS projects a focused beam of millimeter waves to induce an intolerable heating sensation on an adversary's skin, repelling the individual without causing permanent injury [5].

Multi-Frequency Radio-Frequency Vehicle Stopper (shown in Figure 2(a)) launched and mounted on vehicles, the system is designed to provide a flash-bang effect for sensor overload with the effect scalable and determined by the operational at the time of use. The systems would be used support of missions such as urban patrolling, convoy operations and crowd control [6].

Strong sound to disperse anti-riot vehicles use multifunction strong maneuverability "warriors" military off-road vehicles as a platform to integrate science blinding bright lights, tear gas launchers, mobile multimedia communication systems and non-fatal strong voice has international advanced level, etc. equipment, has formed around the clock at the sudden, riot against terrorism combat maneuvering systems for Chinese domestic initiative.

Figure 2: Full spectrum effects platform installed on the "Stryker" armored vehicle.

4.2 Design of new type non-lethal weapons platform

The existing non-lethal weapons which mounted on armored anti-riot vehicles, mostly are multiple tube launchers, sound drive spreader, bright lights, vehicle-mounted riot disperse, etc. [7, 8], just covering the chemical incapacitating, bulk and other non-lethal acoustic drive technology, but effective distance is relatively short, although the technical content gradually improve, but for the purposes of disposal of increasingly complex anti-terrorism and anti-riot situation, still existing a bit thin. In addition, non-lethal weapons vehicle-mounted platform will focus on the application of laser abroad, acoustic and other advanced non-lethal technologies, while increasing the conventional multi-tube launcher range, making tactical and technical indicators have been further enhanced. Therefore, in addition to new armored anti-riot vehicles should be loaded overhead remote weapon stations, we must also equipped with multi-functional non-lethal weapons, in particular:

(1) The location of each vehicle before and after the installation of two with six launchers and two 64 mm anti-riot equipment 938 mm riot transmitter for tear gas to disperse achieve 60–300 m range.
(2) the installation of overhead remote weapon station centered in the right body position, the center position to the left of the vehicle, the installation of remote strong voice to disperse the means used to achieve the riots elements within 0–300 m range warnings, and strong voice within 150m range to disperse rioting activists [9].
(3) Installation of a bright lights in both sides of the front portion, 360° rotation, combined with laser blinding device overhead remote weapon stations on the surrounding elements to achieve blinding disperse riots.

(4) The front part of the center position of the installation of an anti-riot gun [10] can be achieved within 20–60 m range precision momentum strike for implementing a single riot elements within close range of kinetic energy to disperse.
(5) The installation of the vehicle around 8–12 tear powder spray device for riots elements within 0–20 m range to achieve the chemical to disperse.
(6) The body surrounding the installation of electric shock device for violent extremists trying new armored anti-riot vehicles overturned implement high-voltage low-current electrical shock.

5 Conclusions

Armored anti-riot vehicles is important equipment for urgency treatment and anti-terrorism, the current severe urgency treatment and anti-terrorism situation put forward higher requirements for overall performance of armored anti-riot vehicles. By analyzing the limitations of existing armored anti-riot vehicles, proposed design concept of loading overhead remote weapon station, in order to improve the accuracy and observation capability under covert conditions, through the use of sound, light, electricity, kinetic energy, chemical and other means to achieve non-lethal disperse within the range of 0–300m. The scheme has significant instructional sense for the designing of new onboard weapons platform mounted on armored anti-riot vehicles.

Acknowledgement

The research work was supported by National Natural Science Foundation of China under Grant No. 71401179 and Basic Research Foundation of Engineering University of Armed Police Force of China under Grant No. WJY201410.

References

[1] Meng Qinghua, Gao Wei, Guo Xi, et al., A study on dynamic collision detection of armored vehicle in visual simulation. In: *AASRI Conference on Computational Intelligence and Bioinformatics*, **1**, pp. 505–511, 2012.
[2] P.P. Papados, Simulation of projectile impact on the composite armored vehicle. In: *Proceedings Second MIT Conference on Computational Fluid and Solid Mechanics,* June 17–20, pp. 544–547, 2003.
[3] Erwin, I.M., Prihatmanto, A.S., Motor driver program on heavy machine gun model for remote controlled weapon station (RCWS). In: *2013 International Conference on ICT for Smart Society (ICISS)*, pp. 1–6, 2013.
[4] M.R. Murphy et al., Bio-effects research in support of the Active Denial System (ADS). In: *Proceedings of the 2nd European Symposium on Non-Lethal Weapons*, pp. 13–14, 2003.
[5] Chen Yu, Meng Cui, Numerical Study of deposition of energy of Active Denial Weapon in human skin. In: *2012 Asia-Pacific Symposium on Electromagnetic Compatibility (APEMC)*, pp. 361–364, 2012.

[6] Robert H. Stark, Franke Sonnemann, Jürgen Urban, Dieter Weixelbaum, Threat and potential of high power electromagnetic technology (HPEM). *Future Security*, CCIS 318, pp. 363–365, 2012.
[7] Nick Lewer, Neil Davison, Non-lethal technologies – an overview. *Science, Technology and the CBW Regimes*, pp. 37–51, 2005.
[8] Adam Baddeley, The future of non-lethal weapons and technologies. *Military Technology-MILTECH*, pp. 54–55, 2012.
[9] Strode, C., Integrating non-lethal response measures within existing surface platform security instructions. *2010 International Waterside Security Conference (WSS)*, pp. 1–7, 2010.
[10] Widder, J., Rascoe, J., Perhala, C., Non-lethal ballistic system with point accuracy for facility and border security. *2011 IEEE International Conference on Technologies for Homeland Security (HST)*, pp. 123–128, 2011.

A study on antenna element with dual-mode features

Xiaoyan Zhang[1,2], Guohao Wang[1], Zhiwei Liu[1,2]
[1]*School of Information Engineering, East China Jiaotong University, Nanchang, China*
[2]*The State Key Laboratory of Millimeter Wave, Nanjing, China*

Abstract

In this letter, a dual-mode microstrip antenna with two shorting pins is designed. To make the antenna element has a dual-mode features, the method of adding different shorting pins to different resonator is presented. With a metal patch and a rectangular ground, the sizes of the proposed antenna is 66mm×20mm, which is designed on a FR-4 substrate. Two shorting pins are used to make the superposition of two different resonant frequencies and avoid the mutual interference, which can extend the bandwidth. It covers the LTE low frequency and has linear polarization. By the results, the proposed antenna has a good performance on the return loss.
Keywords: Dual-mode, LTE antenna, two shorting pins.

1 Introduction

With the development of the wireless communication, more and more efforts have been devoted on the long-term evolution (LTE). As a core part of the communication system, LTE antenna also attracts the attention of designers [1–5]. Yonghun Cheon and Jungyub Lee presented a magneto-dielectric (MD) material, which has allowably low magnetic loss and moderate permeability for the wireless communication band. It is demonstrated that the MD material is advantageous for an antenna miniaturization and bandwidth expansion for the LTE 700 MHz [6]. And then, Shih-Hsun Chang and Wen-Jiao Liao designed a broadband LTE/WWAN antenna for Tablet PC. The antenna provides an extensive coverage for existing and upcoming mobile communication bands [7]. PIFA will produce

WIT Transactions on Engineering Sciences, Vol. 107, © 2015 WIT Press
www.witpress.com, ISSN 1743-3533 (on-line)
doi:10.2495/SEEEI140211

an additional mode by having the two resonant elements share a common shorting pin to extend the bandwidth [8]. But an interaction is existed when different modes are close, which will prevent the return loss lower. And PIFA usually have a strict demand on its height.

By the conclusion of structures above, we can see that a great number of LTE antennas have a large ground plane which increase the size of the design. What's more, many antennas select –6dB as a standard of S parameter for it is difficult to extend the bandwidth or reduce the return loss between different modes at such a lower frequency. In Ref. [9], two different resonators will produce two different modes by a common shorting pin, but there is an obvious interference between them. If we can eliminate the influence, it will be easy to make the antenna covers 6 or 7 bands.

In this letter, two shorting pins are used to combine different modes and reduce the return loss between them. By this way, we want to study the influence which is produced by different resonators and using different short pins to reduce the interference.

2 Element design

Figure 1 shows the structure of the proposed antenna. The overall size of the proposed antenna is 66mm×20mm, and it is designed on a low-cost FR-4 double sided board with a thickness of h=1.6mm. It consists of four parts: a radiating

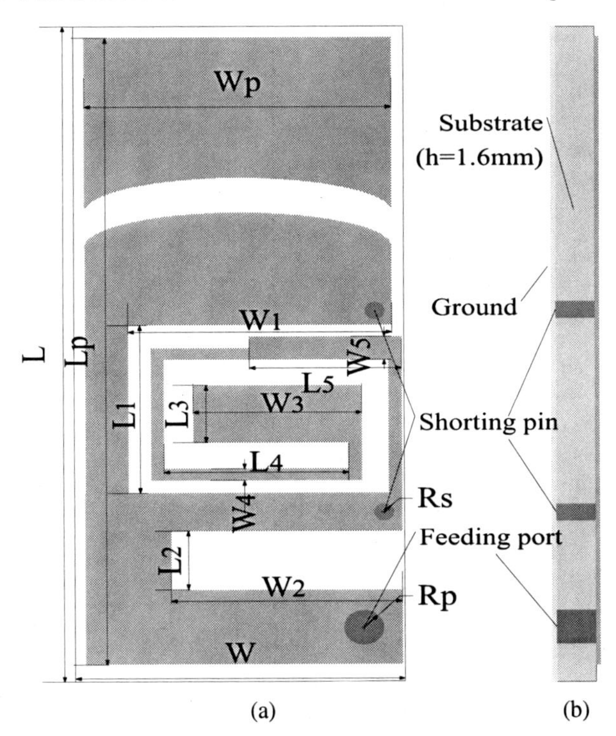

Figure 1: Structure of the proposed LTE antenna. (a) Top view (b) side view.

patch, a small hole for coaxial feeding, a rectangular ground plane, and two shorting pins located at different parts of the patch. In addition, a standard SMA connector in practical dimensions is also modeled to produce more reliable simulation results. The width of the substrate is 20mm (W) while the length is 66mm (L). The radiation patch (Wp×Lp) is consisted of two resonators to produce the different frequency. W5 is 1.5mm width while other parts (W4) of the microwave is 1mm to improve the impedance matching (All the parameters will be seen in the Table 1). What'more, positions of shorting pins are also optimized by Ansoft HFSS to have a good performance on the S11 parameter. The fabric is shown in Figure 2 and the different sides of the antenna is linked by a copper line. Moreover, a circular hole is made through the substrate to install the SMA connector.

Table 1: The dimensions of the designed antennas.

Parameter	Value (mm)	Parameter	Value (mm)
W	20	L2	3
L	66	W3	7.6
Wp	19	L3	2.8
Lp	60.1	W4	0.6
W1	17	L4	10.4
L1	13	W5	1.5
W2	15	L5	9

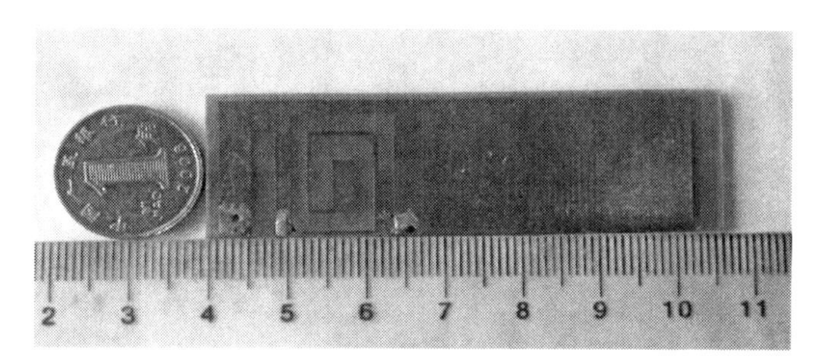

(1) Radiation patch

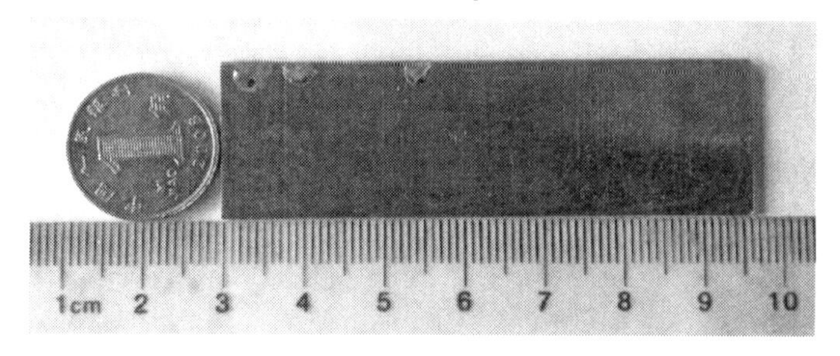

(2) Ground

Figure 2: The fabric of the proposed antenna.

The geometric design of the radiating patch consists of a rectangle and a bending line with different width. Both of them act as two resonators for different frequency. Figure 3 shows the current distributions on the radiant patch. As we can see from Figure 3(a), the current is collected at the bending line when the frequency is at 710MHz, which proves the bending line is in resonant modes. The bending line is in different width for impedance matching. And then, when the solution frequency is set up at 732MHz, the current will mostly locate on the rectangular patch, which can be seen from Figure 3(b).

3 Results and analysis

Figure 4 shows the influence about the bending line position on the return loss. For comparison, the bending line structure is connected to the different shorting pin. Common shorting pin can be seen in most design and in this paper, we also give the S parameter in different cases. From the return loss we can see there will be an interval between two modes. Although parameter optimization has been done, the S parameter between two modes failed to lower when the bending line share a same shorting pin [10]. When two resonators using different shorting pin, the interval will be eliminated and the S parameter between the different modes is successful to be lower.

When in our simulation, position of two short pins also affect the S11 and the final location is decided by parameter sweep. To get a more reliable result, a PML box is created in the project. Figure 5 shows the measured S parameter for the proposed antenna which is compared to the simulated one. We can see the measured result is closed to the simulated one and from the cooperation, we can see that the propose antenna has a good performance on the S parameter, which is lower than −10dB, especially in the middle of the different resonant frequency.

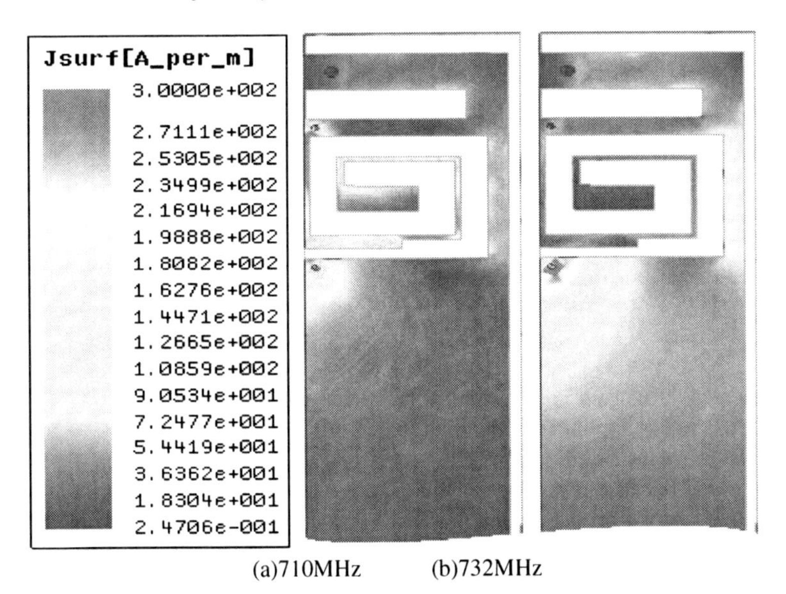

(a)710MHz (b)732MHz

Figure 3: The simulated surface current on the radiating patch.

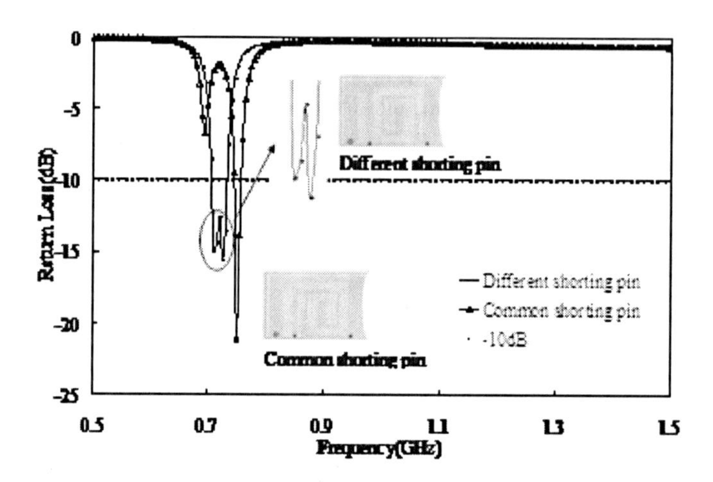

Figure 4: The simulated s-parameter for using different shorting pin.

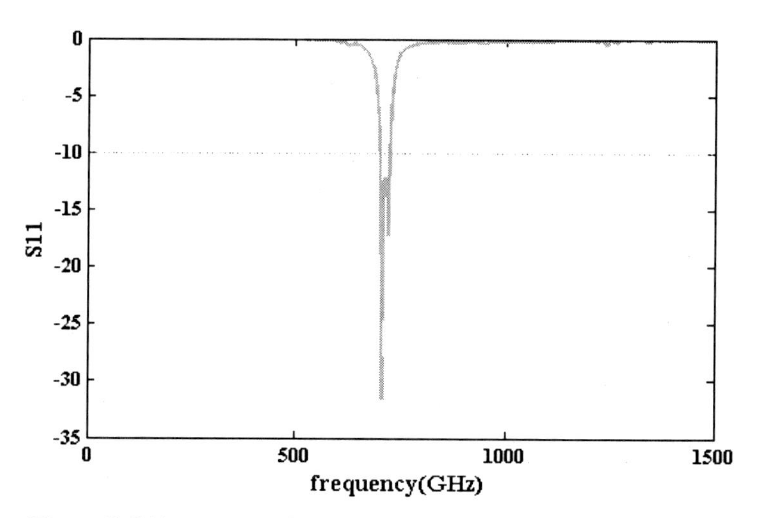

Figure 5: S11 parameter for the proposed antenna in measurement.

Vertical polarization is the ideal polarization for it will not produce a current parallel to the earth and cause the heat loss. So it is also used in this structure [11]. Figure 6 shows the polarization of the proposed antenna. The antenna is placed in the YZ plane because it is more easily to judge the polarization. From the picture we can see that the proposed antenna has a linear polarization for its directivity is about 1.43dB on the phi direction, which is approach to the total directivity, while the directivity on the theta direction is blew zero, which indicates the antenna is almost no radiation in this direction.

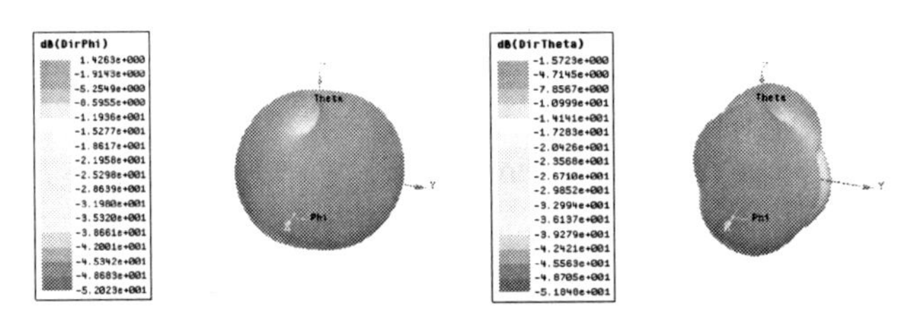

(a) Directivity on the phi direction (b) Directivity on the theta direction

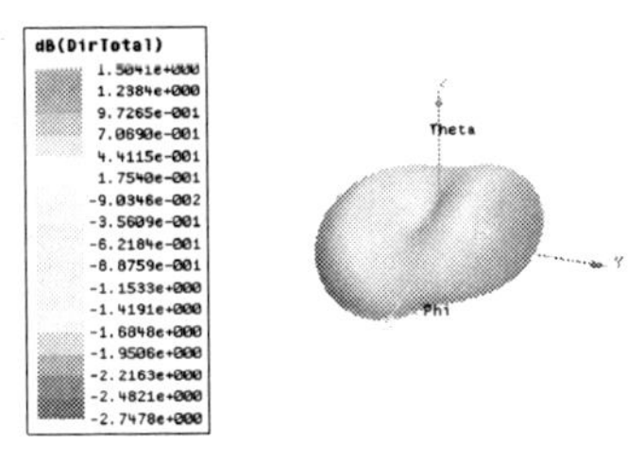

(c) Total directivity

Figure 6: The Directivity of the proposed antenna at 732MHz.

4 Conclusion

In this paper, a compact LTE antenna with two short pins s proposed. By the measurement and simulation, we can see the antenna has a good performance on lower the return loss. As we knew that mostly antenna for Smartphone chose –6 dB as the standard of S11 for it is different to cover 6 or 7 bands by the standard of –10 dB. Method of adding two short pins can be used in the future design on smartphone antenna design. If the antenna prefers to cover several frequencies, we can use a lot of resonators with respective shorting pins to reduce the interaction. This structure will give designer method to add different bands to the antenna and much effort need to be done in future work.

Acknowledgement

This work was supported in part by the National Natural Science Foundation of China (No. 61061002, 61261005), the Graduate Special Fund Innovative Projects in Jiangxi Province (No. YC2013-S158), Jiangxi Provincial Department of Education Project funded (No. GJJ13352) and the Open Project of State Key Laboratory of Millimeter Wave (No. K201325).

References

[1] Yu-Jiun Ren, Ceramic based small LTE MIMO handset antenna. *IEEE Transactions on Antennas and Propagation*, **61(2)**, pp. 934–938, 2013.

[2] Chan-Woo Yang, Young-Bae Jung, Chang Won Jung, Octaband internal antenna for 4G mobile handset. *IEEE Antennas and Propagation Letters*, **11**, pp. 551–554, 2012.

[3] Risto Valkonen, Mikko Kaltiokallio, Clemens Icheln, Capacitive coupling element antennas for multi-standard mobile handset. *IEEE Transactions on Antennas and Propagation*, **61(5)**, pp. 2783–2791, 2013.

[4] Yi Bo-nian, Turbo code design and implementation of high-speed parallel decoder. *TELKOMNIKA*, **11(4)**, pp. 2116–2123, 2013.

[5] Liu Jinhu, Ma Jianting, Improved algorithm of LTE random access preamble detection under the high speed condition. *TELKOMNIKA*, **11(8)**, pp. 4646–4650, 2013.

[6] Yonghun Cheon, Jungyub Lee, Joonghee Lee, Quad-band monopole antenna including LTE700MHz with magneto-dielectric material. *IEEE Antennas and Wireless Propagation Letters*, **11**, pp. 137–140, 2012.

[7] Shih-Hsun Chang, Wen-Jiao Liao, A broadband LTE/WWAN antenna design for tablet PC. *IEEE Transactions on Antennas and Propagation*, **60(9)**, pp. 4354–4359, 2012.

[8] Kin-Lu Wong, Ming-Fang Tu, Tzuenn-Yih Wu, Wei-Yu Li, Small-size couple-fed printed PIFA for internal eight-band LTE/GSM/UMTS mobile phone antenna. *Microwave and Optical Technology Letters*, **52(9)**, pp. 2123–2128, 2010.

[9] Hanyang Wang, Ming Zheng, A multi-band internal antenna. In: *Loughborough Antennas & Propagation Conference*, 14–15 November, pp. 1–4, 2011.

[10] Kin-Lu Wong, *Planar Antennas for Wireless Communications*. John Willey & Sons Ltd., pp. 34–37, 2003.

[11] Bo Yin, Feng Yang, Ping Wang, Zu-Fan Zhang, Radiation characteristics of a plasma column antenna. *TELKOMNIKA Indonesian Journal of Electrical Engineering*, **12(1)**, pp. 113–121, 2014.

Multilayer microstrip array antenna with box-type reflecting plate

Guoqi Ni, Lina Wang, Baiping Yu
School of Information and Communication Engineering, Guilin University of Electronic Technology, Guangxi, China
The 2nd Department, Air Force Airborne Force Academy, Guangxi, China

Abstract

At present, most of the researches on electromagnetic and microwave technology are focused on antenna design. In this paper, a broadband and high gain 16-element microstrip array antenna is presented. It is operating at 12.4GHz. The array antenna consists of double-layered substrates which are separated by a layer of air. The parasitic patches and pins are adopted to achieve broadband. The box-type reflecting plate is used to improve the gain. The measured results indicate that the impedance bandwidth is as good as 24.91% and the gain is 19.31dB. The design provides reference for developing broadband and high gain microstrip array antenna.

Keywords: antenna structure, antenna element, microstrip array antenna.

1 Introduction

The concept of microstrip antenna was proposed by Deschamps in 1953. It is an attractive type of antenna due to its conformability, low cost, and ease of manufacture and is widely used in radar, communication, aerospace, electronic countermeasures and other fields [1, 2]. However, the primary barrier to implementing these antennas in many applications is their limited bandwidth and gain. Because of these facts, much work has been devoted to improving the bandwidth and gain of microstrip antennas. D.M. Pozar proposed aperture-coupled microstrip antenna in 1985. This design involves two substrates separated by a small apertured ground plane. One substrate contains the radiating patch, with the other substrate containing the feed network. This method can avoid the interfered radiation from

www.witpress.com, ISSN 1743-3533 (on-line)
doi:10.2495/SEEEI140221

the feed network and can increase the bandwidth [3, 4]. In 1991, D.M. Pozar designed a stacked patches microstrip antenna which is applied to phased array antenna [5]. Siyang Sun proposed another bandwidth enhancement technique in 2009. In his design, the bandwidth is expanded by cutting down a square at the corner of the ordinary rectangular patch antenna, three times of bandwidth can be achieved [6]. A technique that has been used extensively is forming array to improve gain [7].

In this paper, we present a new broadband and high gain 16-element microstrip array antenna. Through analyzing the simulation and experimental data in detail, the results indicate that the antenna has certain practical value.

2 Analysis and design of the microstrip antenna element

Figure 1 shows the structure of the microstrip antenna element. For our design, parasitic patches and short pins are adopted, through these measures the resonant mode and current distribution are changed. Therefore, the bandwidth of the microstrip antenna is increased. The box-type reflecting plate is added at the back of the antenna array, it is used to reduce the back radiation of the aperture coupled microstrip antenna and improve the gain. Multilayer structure is adopted in our design, which is composed of two layers of medium plates and an air layer.

The top layer: a DiClad880 substrate with thickness 1mm and a relative dielectric constant $\varepsilon_r = 2.2$ is utilized. The bottom layer: a RO4003C substrate with thickness 0.813mm and a relative dielectric constant $\varepsilon_r = 3.38$ is used.

The rectangle patch with corner and edge cutting and the parasitic patches [8, 9] are placed in the positive side of the upper layer medium plate. The short pin [10] is inserted into the upper layer medium plate, which is not coppered at its back side. The ground plane with H-shape coupling slot is placed in the positive side of the bottom layer medium plate, the 50-Ω transmission line is on the other side.

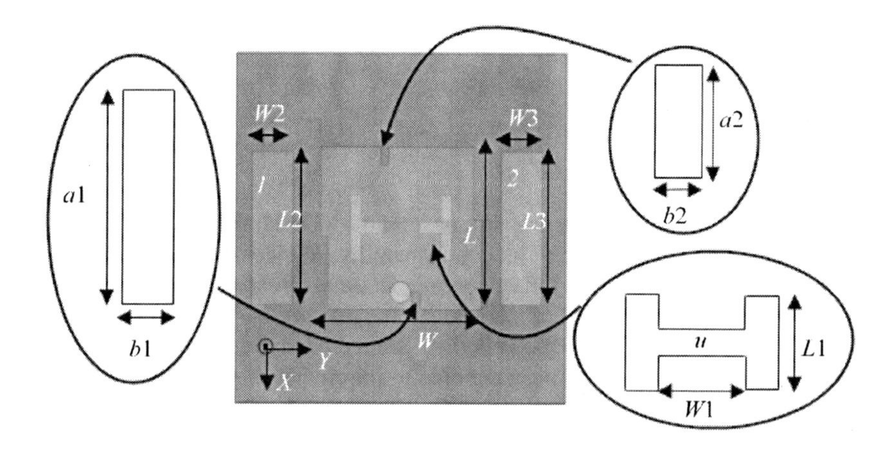

Figure 1: Main geometries and dimensions of the microstrip antenna element.

The element dimensional parameters, shown in Figure 1 are: $L = 7.6$ mm, $W = 8.0$ mm, $L1 = 3$ mm, $W1 = 4$ mm, $L2 = 7.2$ mm, $W2 = 2$ mm, $L3 = 7.2$ mm, $W3 = 2$ mm, $a1 = 3.4$ mm, $b1 = 0.4$ mm, $a2 = 0.8$ mm, $b2 = 0.4$ mm, $u = 0.5$ mm. The length of the feed line is 9.8 mm and the width is 1.8mm. The distance between parasitic patch 1 and radiation patch is 1.4 mm, and the distance between parasitic patch 2 and radiation patch is 1 mm. The dimension of diameter and height of the circular cylinder is 1mm.

3 Analysis and design of the microstrip array antenna

Figure 2(a) depicts the model of the array antenna which is arrayed in 4×4 elements. The antenna is designed to resonate at 12.4GHz. Figure 2(b) shows the photograph of prototype 4×4 elements array antenna. The whole antenna array with the total area is $(94.88 \times 76.38 \times 14.11)\,\text{mm}^3$.

The anti-phase feeding and quarter-wave transformer techniques are applied in our design, to decrease the cross-polarization and realize good matching. A 180° phase shift is introduced by increasing the length of the feed line, then using anti-phase feeding technique to achieve the same phase. The parameters are: y_dis1 = 19.48 mm, y_dis2 = 37 mm, x_dis1 = 19.48 mm, x_dis2 = 38.45 mm. As the antenna is arrayed by elements, the distance between elements has great effect on the gain. The microstrip antenna theory has noted that $d / \lambda_0 = 0.7 \quad 0.9$ is perfect. In our project, we designed the array antenna by two steps: constructing the 1×4 array, madding the 4×4 array based on the former structure. The simulation result shows that $d_1 = 19.48\,\text{mm}$ ($0.805\lambda_0$), $d_2 = 18.98\,\text{mm}$ ($0.784\lambda_0$), $d_3 = 17.52\,\text{mm}$ ($0.724\lambda_0$).

In our design, short pins are applied to aperture-coupled microstrip antenna as shown in Figure 1. One side of the probe is connected to the radiation patch, the other side is suspended. The width of 50Ω, 70.7Ω, 100Ω transmission line are 1.8mm, 1mm, 0.5mm.

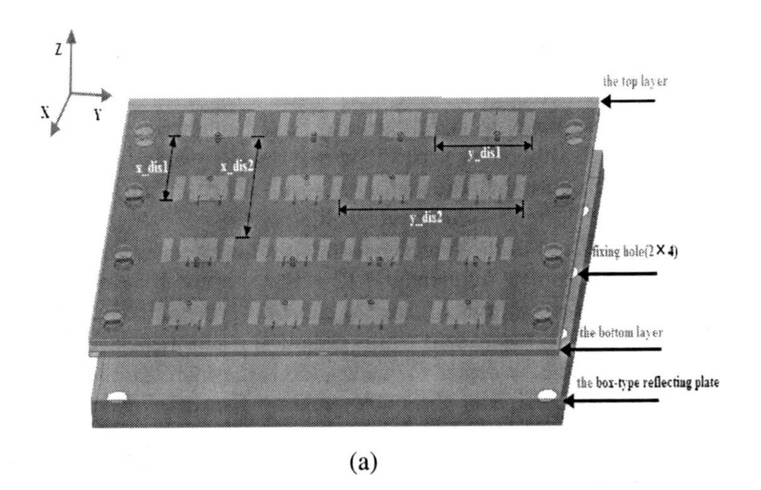

(a)

(b)

Figure 2: (a) Geometries of the proposed antenna array using 4×4 elements. (b) photograph of prototype antenna array.

4 Simulated and measured results

The proposed 4×4 array antenna has been designed and simulated with HFSS based on the finite-element method (FEM). With the multilayered configuration, the fabrication process of the array is complex. Figure 2(b) shows a photograph of prototype 4×4 elements array antenna with internal anti-phase feeding network.

S-parameter of the prototype array has been measured by Agilent N5230A. The comparison of simulated and measured S-parameter of the antenna array is shown in Figure 3. The simulated bandwidth of VSWR ≤ 2 is 21.50%, from 10.67 GHz to 13.24 GHz, and the measured bandwidth of VSWR ≤ 2 reaches 24.91%, from 11.21 GHz to 14.40 GHz. The measured VSWR of the prototype array is slightly shifted higher compared with the simulated one. It should be noted that the discrepancy is mainly caused by the inhomogeneous of dielectric layers and air layer, although a series of measures have been taken into account. It is obvious in the figure that, the measured result is better than the simulated one, which can meet the design requirements well.

The radiation patterns of the prototype array have been measured in an anechoic chamber, and the selective frequency is 12.4GHz. Figure 4 shows the simulated and measured radiation patterns of the 4×4 elements array antenna.

Figure 4 shows the comparisons of the radiation performances of this antenna. The results show that at the center frequency, the highest simulated gain of E-plane and H-plane is 20.42dBi, and the corresponding measured gain of E-plane is 19.28dBi, H-plane is 19.31dBi. It is obvious that the measured result is good agreement with the simulated one.

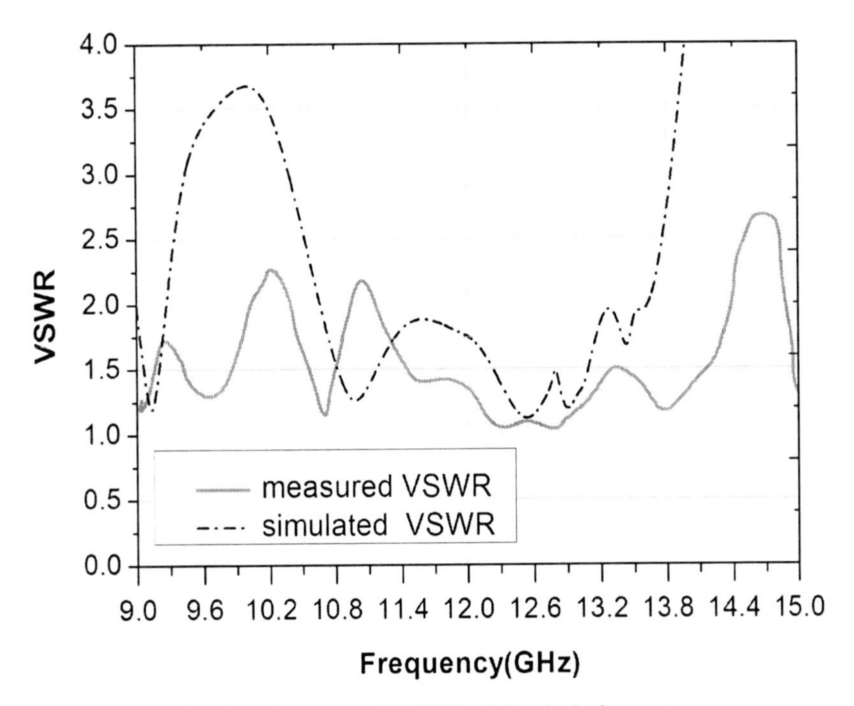

Figure 3: Simulated and measured VSWR of the 4×4 elements array antenna.

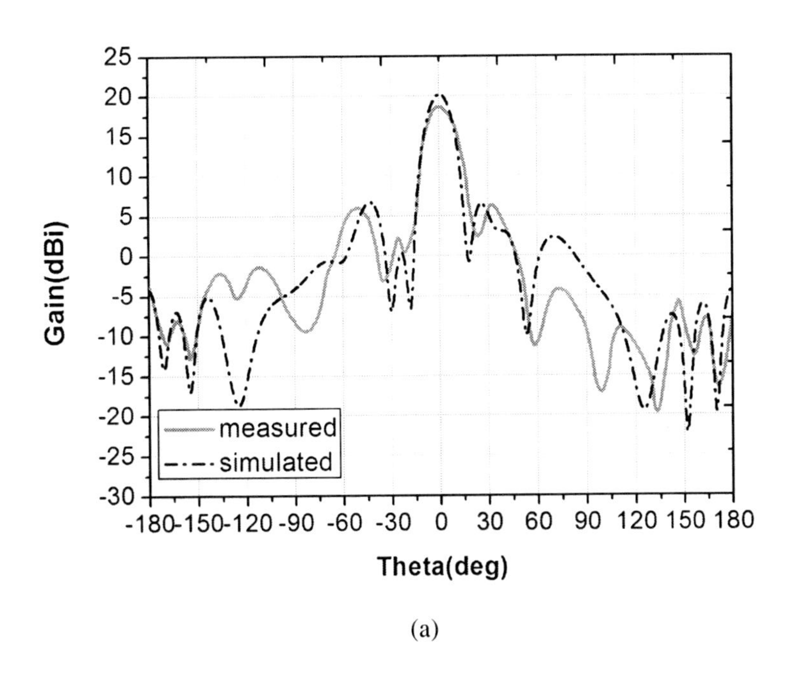

(a)

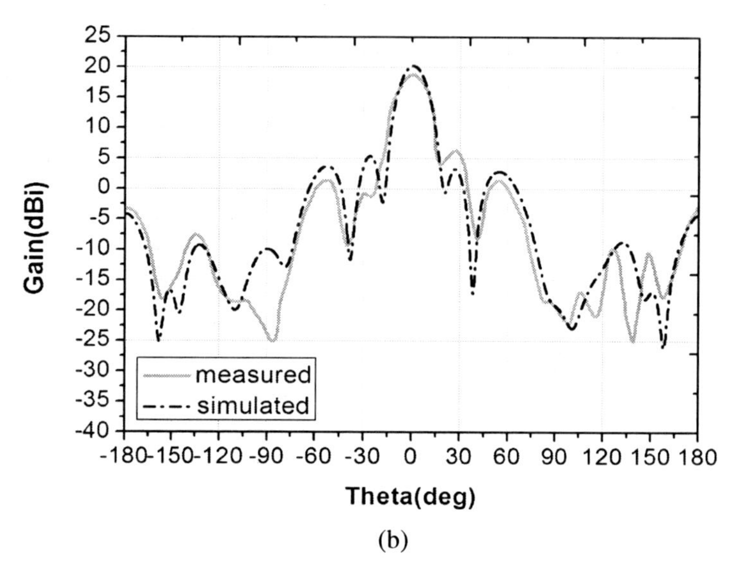

(b)

Figure 4: Comparison between the measured and simulated radiation patterns in the (a) E-plane and (b) H-plane, for the 4×4 elements array antenna at 12.4 GHz.

5 Conclusions

The design of a broadband and high gain 4×4 elements microstrip array antenna is proposed. Adding pins to aperture-coupled feeding technique and adopting parasitic patches have been shown to be effective measures to expand the bandwidth of multilayer microstrip antenna. The box-type reflecting plate can improve the gain. The simulated and measured S-parameter, as well as the radiation pattern have been presented. Good agreement between simulated and measured results has been achieved for the array antenna, except for some shifts in the return loss. The experimental results have verified the practicability of the broadband and high gain multilayer microstrip array antenna, which will be an attractive reference for radar, communication, aerospace and electronic countermeasures.

Acknowledgement

This work was supported by the Scientific Research Foundation of Guangxi Province for Returned Chinese Scholars under Grant No. 0991021.

References

[1] Targonski, S.D., Waterhouse, R.B., Pozar, D.M., Design of wide-band aperture-stacked patch microstrip antenna. *IEEE Transactions on Antennas and Propagation*, **46(9)**, pp. 1245–1251, 1998.

[2] Yang, W.C., Che, W.Q., Wang, H., High-gain design of a patch antenna using stub-loaded artificial magnetic conductor. *IEEE Antennas and Wireless Propagation Letters*, **12**, pp. 1172–1175, 2013.
[3] Pozar, D.M., Targonski, S.D., A shared-aperture dual-band dual-polarized microstrip array. *IEEE Transactions on Antennas and Propagation*, **49(2)**, pp. 150–157, 2001.
[4] Ghassemi, N., Rashed-Mohassel, J., Sh. Mohanna, et al., A wideband aperture-coupled microstrip antenna for S and C bands. *Microwave and Optical Technology Letters*, **51(8)**, pp. 1807–1809, 2009.
[5] Frederic Crop, David, Millimeter-wave design of wide-band aperture-coupled stacked microstrip antenna. *IEEE Transactions on Antennas and Propagation*, **39(12)**, pp. 1770–1776, 1991.
[6] Sun Si-Yang, Lin Xin, Gao You-Gang, et al., Design of a wideband microstrip patch antenna. *Chinese Journal of Radio Science*, **24(2)**, pp. 307–310, 2009.
[7] Ni Guo-Qi, Liang Jun, Yu Bai-Ping, Design of a16-element broadband and high gain microstrip array antenna. *Telecommunication Engineering*, **53(6)**, pp. 786–790, 2013.
[8] Kumar, G., Gupta, K. C., Nonaddicting edges and four edges gap-coupled multiple resonator broadband microstrip antennas. *IEEE Transactions on Antennas and Propagation*, **33(2)**, pp. 173–178, 1985.
[9] Hong, J.S., Lancaster, M.J., Couplings of microstrip square open-loop resonators for cross-coupled planar microwave filters. *IEEE Transactions on Microwave Theory and Techniques*, **44(12)**, pp. 2099–2109, 1996.
[10] Wong, S.W., Huang, T.G., Mao, C.X., et al., Planar filtering ultra-wideband (UWB) antenna with shorting pins. *IEEE Transactions on Antennas and Propagation*, **61(2)**, pp. 948–953, 2013.

A design of broadband 90° coplanar waveguide phase shifter based on composite right/left hand transmission lines for circularly polarized RFID reader antenna application in UHF band

Bo Xu, Qingchong Liu, Jun Hu
Centre for Optical and Electromagnetic Research,
Zhejiang University, Hangzhou, China

Abstract

At present, 3-dB axial-ration bandwidth of most researches on circularly polarized (CP) UHF RFID reader antennas is very narrow. This paper presents a novel design of a broadband 90° phase shifter on coplanar waveguide for CP UHF RFID application. In order to reduce the large dispersion of the traditional 90° phase delay line, we introduced two different sets of composite right/left hand (CRLH) transmission lines (TLs) with lumped elements to produce wideband 90° phase delay. By adjusting the values of lumped elements and introducing an extra TL, a broadband 90° phase shifter with almost zero-dispersion can be acquired. The simulation results show that, in UHF band, $|S_{11}|$ of the phase shifter is below −15 dB. $|S_{12}|$ and $|S_{13}|$ is nearly −3 dB. $|S_{23}|$ is below −25 dB. The phase difference is 90° ± 2° from 675 MHz to 1093 MHz.
Keywords: circular polarization, RFID, phase shifter, composite right/left hand, zero-dispersion.

1 Introduction

Radio frequency identification (RFID) system in the ultra-high frequency (UHF) band has attracted a lot of attentions in numerous fields, such as inventory management, logistics, cold chain management, and bio-security. In China, UHF RFID systems have two separate bands, 840.5–844.5 MHz and 920.5–924.5 MHz. In North America, they operate between 920 and 928 MHz. In Europe, they work

WIT Transactions on Engineering Sciences, Vol. 107, © 2015 WIT Press
www.witpress.com, ISSN 1743-3533 (on-line)
doi:10.2495/SEEEI140231

in 866–869 MHz, and 952–955 MHz in Japan. A typical RFID system includes a reader and tags. The reader transmits an electromagnetic signal to a tag. The reader then receives a backscattered signal from the tag with modulated information. Since most commercial tag antennas are linearly polarized, a circularly polarized (CP) antenna is preferred for readers so as to avoid polarization mismatch and reduce multi-path effects between the tag antenna and the reader antenna.

There are two basic approaches to generate CP waves. The first approach is to use two separate and spatially orthogonal feeds with a 90° phase difference by utilizing a 90° hybrid or 90° power divider. The second approach is to excite two amplitude degenerate orthogonal modes by introducing asymmetry in the structure. One mode will increase in frequency by a 45° and the other mode will decrease in frequency by 45°. Thus a CP radiation is produced [1]. This paper adopts the first approach as a basic feeding network to obtain circular polarization.

In these years, narrow band CP antennas for RFID reader application have been widely researched [2–5], but few work has been done on broadening the CP operation band of reader antenna to cover all the UHF RFID bands mentioned above, which is 840.5–955 MHz. [6] proposed a planar broadband antenna with square slot by insetting an arc-shaped strip into square slot. The proposed antenna of [6] achieves CP radiation with impedance bandwidth (return loss $\geq$ 10 dB) of 142 MHz (15.3% at 931 MHz), 3 dB axial-ratio (AR) bandwidth of 166 MHz (17.7% at 940 MHz), 6.8dBic peak gain and 98% radiation efficiency. [7] proposed a CP antenna including a square-loop radiating element with two feed ports and a broadband feeding network in the same plane, which is composed of a Wilkinson power divider, a high-pass filter and a low-pass filter, connected to the radiating element, producing broadband right-hand circular polarization. The measured broad impedance bandwidth ($S_{11} < -10$ dB) of [7] is 710–1150 MHz. The measured 3-dB AR bandwidth of [7] is 800–940 MHz. The peak gain of [7] is 1.3 dBic at 910 MHz.

This paper proposes a broadband 90° coplanar waveguide phase shifter based on composite right/left hand (CRLH) transmission lines (TLs) [8], which can be used for broadband circularly polarized RFID reader antenna application to cover all the UHF-RFID bands mentioned above.

The paper is organized as follows. In the next section, we propose a novel phase shifter design, and some definitions and values are given. In Section 3, the results simulated by Ansoft HFSS v13.0 [9] are presented. Finally, we conclude our paper in Section 4.

2 Phase shifter design

The configuration of the proposed structure is shown in Figure 1, in which a Wilkinson power divider and two sets of CRLH TLs proposed on [8] are etched on the top of Rogers RT/duroid 6006 (tm) substrate, with thickness of 1.27 mm and relative permittivity of 6.15. The phase shifter is fed through Port1. One set of CRLH TL with lumped inductance L_1 and lumped capacitance C_1 connects one port of the Wilkinson power divider to Port 2. The other set of CRLH TL with

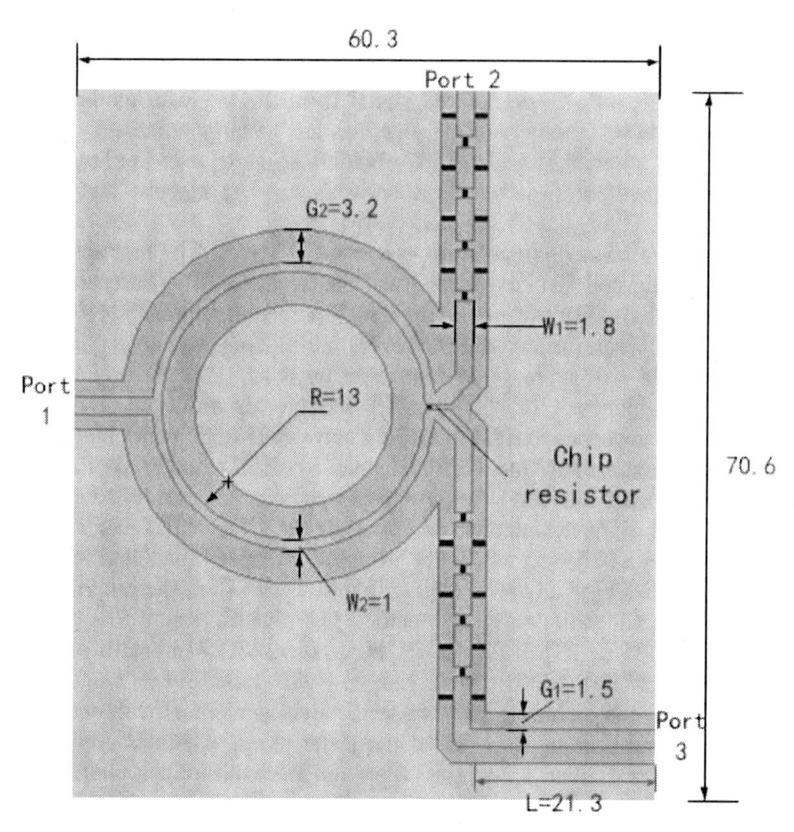

Figure 1: Configuration of proposed phase shifter. The phase shifter is fed through Port 1. Port 2 and Port 3 are two outputs connecting to radiation elements.

lumped inductance L_2 and lumped capacitance C_2 connects the other port of the Wilkinson power divider to Port 3 through an extra right hand TL. Each set of the CRLH TLs is composed of 4 unit cells with the same structure but different lumped element values.

The circuit model of one unit cell of CRLH TLs is shown in Figure 2 [8]. The matching condition for CRLH TLs was originally derived in [9]:

$$Z_0 = \sqrt{\frac{L_1}{C_1}} = \sqrt{\frac{L_2}{C_2}} = \sqrt{\frac{L}{C}} \tag{1}$$

where Z_0 is the input impedance of the phase shifter, as well as the impedance of right hand part of CRLH TL, or host TLs. For a unit cell of CRLH TLs with periodicity d and phase shift of the host TLs $\varphi_{TL} = \beta_{TL} d$, the total

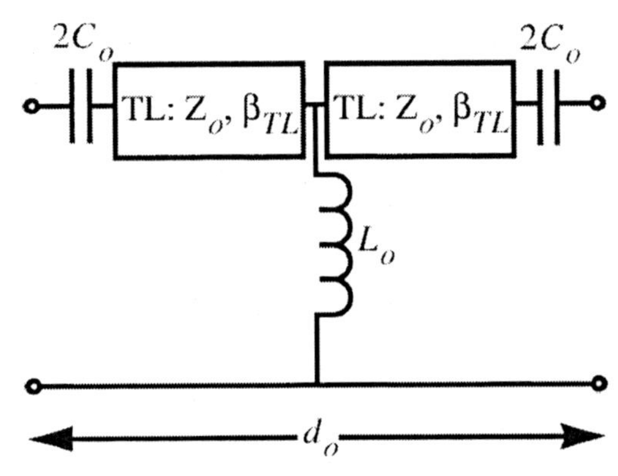

Figure 2: Circuit model of unit cell [8].

phase shift per unit cell under the matching condition of (1) can be written by [8]:

$$|\varphi_1| = \varphi_{\text{TL}} + \frac{-1}{\omega\sqrt{L_1C_1}} \tag{2}$$

$$|\varphi_2| = \varphi_{\text{TL}} + \frac{-1}{\omega\sqrt{L_2C_2}} \tag{3}$$

Because the length of host TLs in each set is the same, the phase difference of between two ports is caused merely by the lumped elements:

$$\Delta\varphi = |\varphi_2|\text{-}|\varphi_1| = \frac{1}{\omega\sqrt{L_1C_1}} \text{-} \frac{1}{\omega\sqrt{L_2C_2}} \tag{4}$$

In this case, $C_2 = 9\text{pF}$, $2L_1 = 45\text{nH}$, $C_2 = 18\text{pF}$, $C_1 = 90\text{nH}$. But the dispersion of two sets of CRLH TLs is different, which will cause obvious dispersion in UHF band, thus an extra RH TL is needed to compensate it. By inserting an extra RH TL with L = 21.3 mm in the branch corresponding to Port 3, a broadband 90° phase shift is acquired.

3 Results and discussions

The proposed phase shifter is designed to operate at the center frequency of 900 MHz in UHF band for RFID readers. The simulated $|S_{11}|$, $|S_{12}|$, $|S_{13}|$, and $|S_{23}|$ are presented in Figure 3. It shows that, between 800MHz and 960 MHz, $|S_{11}| < -15\text{dB}$; $|S_{12}|$ and $|S_{13}|$ is almost the same about −3 dB; $|S_{23}| < -25\text{dB}$. Therefore, the proposed design can equally split power with little loss. Figure 4

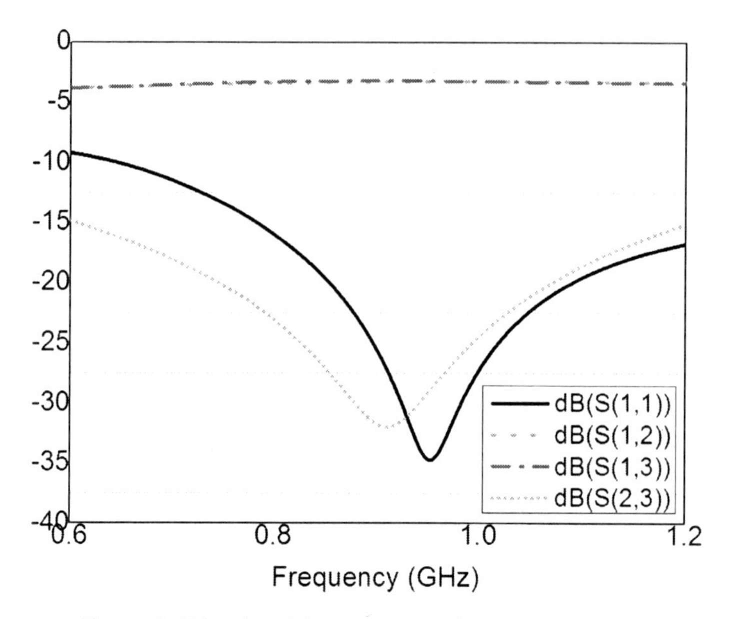

Figure 3: Stimulated S parameters of proposed design.

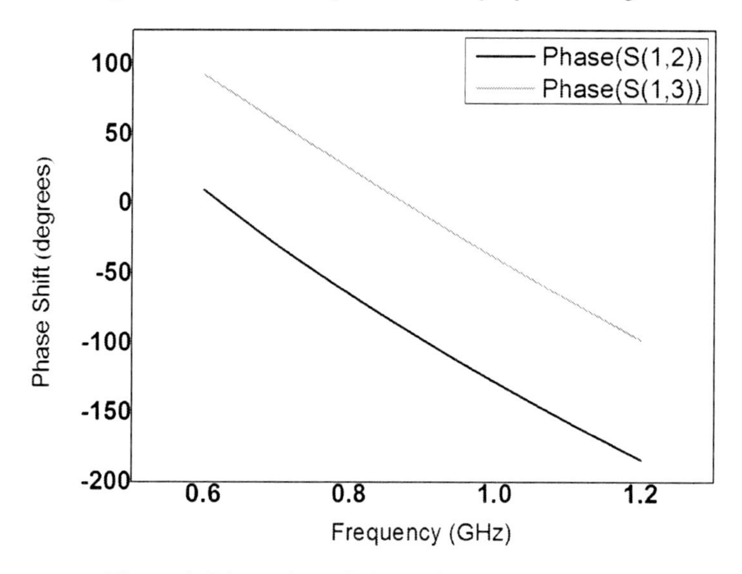

Figure 4: Dispersion relations of Port 2 and Port 3.

shows the dispersion relations of Port 2 and Port 3. The dispersion of the CRLH TL with L_1 and C_2 is 323° per GHz. The dispersion of the CRLH TL and the additional RH TL together, with L_2 and C_2 is 317° per GHz. The dispersion difference is no larger than 6° per GHz. The phase shift versus frequency of the design is presented in Figure 5. As we can see, the phase difference between Port 2

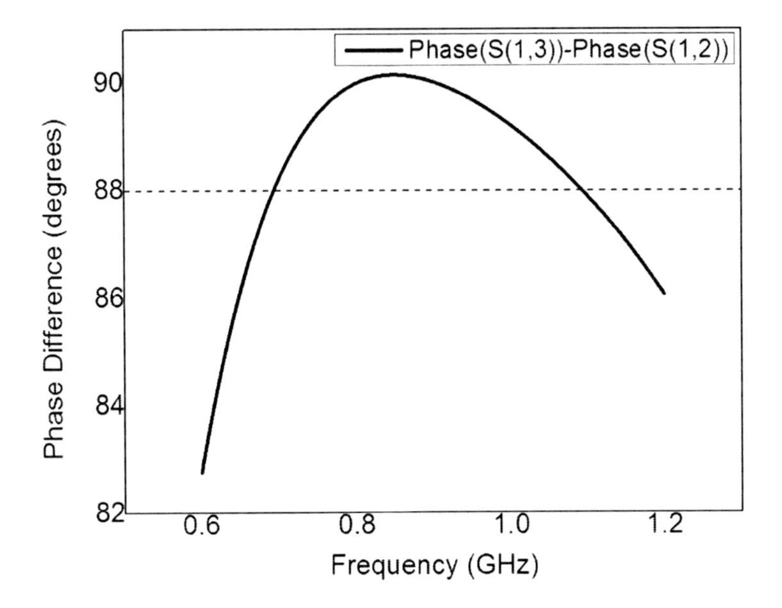

Figure 5: Phase difference between Port 2 and Port 3.

and Port 3 is between 88° and 91° from 695 MHz to 1093 MHz. It indicates that the CP antenna using the proposed design can achieve purely circular polarization and very low axial ratio in UHF band.

4 Conclusion

A CRLH TLs based broadband phase shifter has first been presented in UHF band. Based on selecting appropriate values of lumped inductance and capacitance, by inserting an additional RH TL to compensate, the two CRLH TLs can produce low-dispersion 90° phase difference with low loss and almost no phase variation. The proposed structure is designed on CPW, on which low profile radiating elements can be easily loaded, such as printed monopole, loop and PIFA. It is a meaningful work to implement a broadband CP RFID reader antenna for universal application. Because in overseas trade, RFID tags in one country may need to be read by another country's RFID reader.

Acknowledgement

This work was supported by a grant from the National High Technology Research and Development Program of China (863 Program) (No. 2012AA030402).

References

[1] K.R. Carver, J.W. Mink, Microstrip antenna technology. *IEEE Transactions on Antenna and Propagation*, **AP-29(1)**, pp. 2–24, 1981.

[2] J.-H. Bang, C. Bat-Ochir, H.-S. Koh, E.-J. Cha, B.-C. Ahn, A small and lightweight antenna for handheld RFID reader applications. *IEEE Antennas and Wireless Propagation Letters*, **11**, pp. 1076–1079, 2012.

[3] R. Caso, A. Michel, M. Rodriguez-Pino, P. Nepa, Dual-band UHF-RFID/WLAN circularly polarized antenna for portable RFID readers. *IEEE Transactions on Antenna and Propagation*, **62(5)**, pp. 2822–2826, 2014.

[4] Y.F. Lin, H.M. Chen, C.H. Chen, C.H. Lee, Compact shorted inverted-L antenna with circular polarization for RFID handheld reader. *Electronics Letters*, **49(7)**, 2013.

[5] Y.-F. Lin, Y.-K. Wang, H.-M. Chen, Z.-Z. Yang, Circularly polarized crossed dipole antenna with phase delay lines for RFID handheld reader. *IEEE Transactions on Antenna and Propagation*, pp. 1221–1227, **60(3)**, 2012.

[6] J.-H. Li, S.-F. Wang, Planar broadband circularly polarized antenna with square slot for UHF RFID reader. *IEEE Transactions on Antenna and Propagation*, **61(1)**, pp. 45–53, 2013.

[7] B. Xu, Q. Liu, Y. Liu, Broadband circularly polarized loop antenna based on high-pass and low-pass filters for handheld RFID reader applications. *PIERS Proceedings 2014*, pp. 500–503, Guangzhou, 2014.

[8] M.A. Antoniades, G.V. Eleftheriades, Compact linear lead/lag metamaterial phase shifters for broadband applications. *IEEE Antennas and Wireless Propagation Letters*, **2**, pp. 103–106, 2003.

[9] http://www.ansys.com/Products/Simulation+Technology/Electronics/Signal+Integrity/ANSYS+HFSS.

[10] G.V. Eleftheriades, A.K. Iyer, P.C. Kremer, Planar negative refractive index media using periodically L–C loaded transmission lines. *IEEE Transactions on Microwave Theory and Techniques*, **50**, pp. 2702–2712, 2002.

Ship shaft-rate electric field signal measurement method based on DUFFING oscillator

Yi Liu, Dou Ji, Xiangjun Wang
College of Electrical and Information Engineering, Naval University of Engineering, Wuhan, China

Abstract

Ship shaft-rate electric field signal is difficult to detected because it is very weak. In this paper, according to the characteristics of ship shaft-rate electric field, propose a chaotic detection method based on DUFFING oscillator. Determine the amplitude of the signal by observing the phase trajectory, and using the oscillator array to determine the frequency of signal. Experiments show that this method is of high accuracy of measurement and fast speed of computation time.
Keywords: shaft-rate, DUFFING oscillator, oscillator array.

1 Characteristics of Ship shaft-rate electric field

Ship material in the marine environment, will inevitably occur electrochemical corrosion. There is an electric current flowing in seawater corrosion process, and the stability of the electric current can establish a steady state of the electric field. When electric current is periodic alternating, it can establish an alternating electromagnetic field. In order to prevent corrosion occurred in the shell, propeller, and other underwater parts, ships are equipped with passive sacrificial anode cathodic protection system (PCP) and impressed current cathodic protection system (ICCP) [1]. Whatever the anti-corrosion methods, there will be part of the electrochemical corrosion and cathodic protection current through the water flow to the propeller, and then through the bearing and mechanical loop returns to the hull. The total resistance of circuit will be cyclical change because of the rotation of propeller, thus modulating the electrochemical corrosion and cathodic protection current through the bearing, as shown in Figure 1. This creates a shaft-rate electromagnetic fields, which has far transmission distance, having a very high military value.

WIT Transactions on Engineering Sciences, Vol. 107, © 2015 WIT Press
www.witpress.com, ISSN 1743-3533 (on-line)
doi:10.2495/SEEEI140241

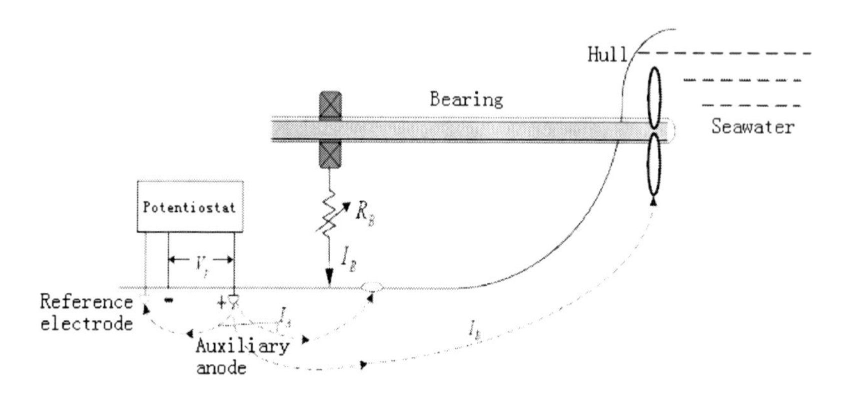

Figure 1: Corrosion current circuit diagram using ICCP system.

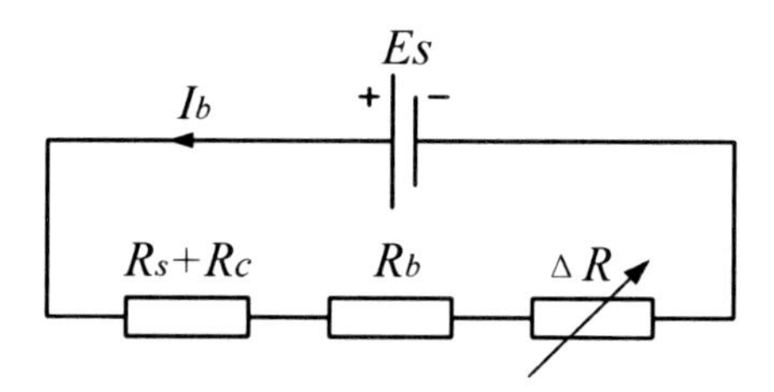

Figure 2: The low frequency current equivalent circuit.

The equivalent circuit is shown in Figure 2.

In diagram 2, ΔR is shaft bearing contact equivalent alternating resistance, its value is $\Delta R = R_b(5+5*\sin 2\pi ft)$.

In the formula, f is the rotation frequency of propeller.

Then, shaft current is:

$$I_b = \frac{Es}{R_s + R_c + R_b + \Delta R} \tag{1}$$

Alternating voltage of shaft's alternating resistance is u_{rf} : $u_{rf} = I_b \Delta R$

The parameters are as follows:

$$Es = 0.45\text{V},\ R_s + R_c = 0.1\Omega,\ R_b = 0.000507\Omega.$$

But in the marine environment, shaft-rate electric field signal is very weak, usually in the order of magnitude of μV/m. At the same time, the ocean there are many other electric signals will be the interference sources of shaft-rate electric field, such as: 1. electric field of the water movement; 2. electric field of the earth's magnetic field change; 3. electric field of the ionospheric current; 4. electric field of submarine, waterfront rock and metal ore body; 5. natural electric field of seismic activity; 6. electric field of marine biological activity.

2 DUFFING oscillator detection system

Chaotic system is dynamic system determined by a special class of deterministic nonlinear mathematical model. When the parameter of the system change, this dynamic system generates a transition from simple motion to chaotic motion state. So chaotic system model is a relationship established between the simple motion and chaotic motion [2]. Common chaos parameter mathematical models are: chaotic system dynamics model is set up by the DUFFING equation, chaotic system dynamics model is set up by the LOGISTIC equation, chaotic system dynamics model is set up by the LORENZ equation.

Nonlinear dynamics system DUFFING system represents the performance of the nonlinear dynamics of the rich, can exhibit complex dynamic of oscillation, bifurcation, chaos, has become one of the most commonly used model to study the nonlinear damped oscillation, bifurcation, chaos [3].

One of the characteristics of the DUFFING equation is the plus forcing term $f\cos(\omega t)$ on the right-hand side of DUFFING equation. It is due to the intrinsic frequency of the system and the interactions of periodic forced frequency, makes this equation contains extremely rich content: period-doubling bifurcation, strange attractor, attractor coexist with the popularity of complex structure and so on. The specific form of equation is:

$$x(t)+kx(t)-x(t)+x^3(t)=f\cos(\omega t) \tag{2}$$

k is damping coefficient, $f\cos(\omega t)$ is system driving signal.

When k is a fixed value (usually is 0.5) and f gradually increases from zero, the system state appears regular change: homoclinic trajectory, the bifurcation trajectory, chaotic trajectory, the large-scale periodic state [4].

Analysis of the above system time domain waveform and phase plane trajectory changes to be seen:

(1) When $f=0$, the saddle point of system phase plane is (0, 0), and system focus is (±1, 0). Point will eventually stay in one of two focus, as shown in Figure 3.

(2) When $f>0$, System is divided into stages exhibit complex dynamics shape, which can be divided into the following several cases. When F is small, the phase trajectory for the performance of attractors under the meaning of Poincare mapping, and phase points around the focus as a periodic oscillation. f gradually increase to a critical value f_c (the value of f_c can be obtained by the Melnikov method), along with the increase of f, the system undergoes homoclinic, periodic bifurcation until it reaches the state of chaos. This process as the change of f is very quick, and in a long time, the system will be in a state of chaos movement. Further increases beyond a threshold f_d, the system with the frequency of periodic force large-scale cycle oscillation. At this point the phase trajectory will be surrounded by focus, saddle point, the corresponding Poincare mapping is also a fixed point [5].

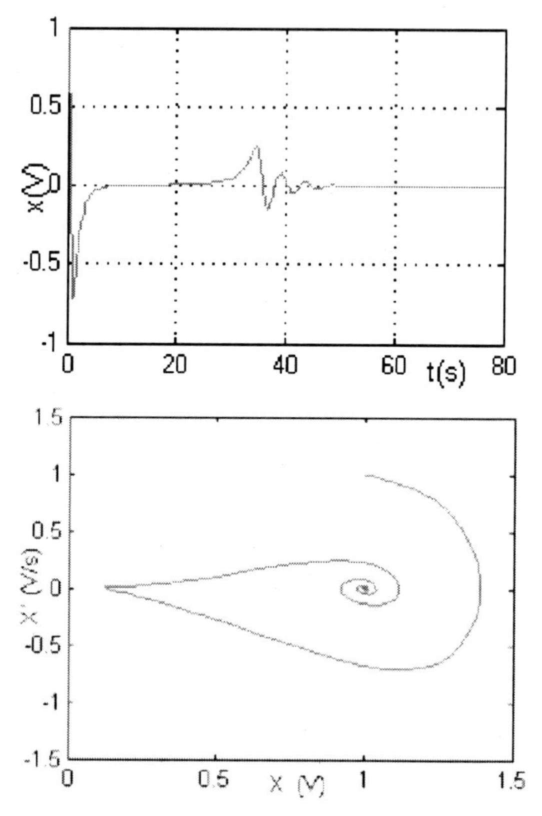

Figure 3: $f = 0$, initial state of $[x(0), x(0)]$.

3 The shaft-rate electric field signal chaos detection principle

The mathematical model is set up based on DUFFING oscillator equation, and it can detect frequency and amplitude of ship shaft-rate electric field. The model is shown in Figure 4.

(1) The principle of amplitude detection

First regulate system of the driving amplitude f, making $f = f_d$, and the system is in a critical state that a chaotic state transiting to a large-scale periodic state. When a sinusoidal signal having small amplitude, frequency similar with periodic driving force and a white noise are on DUFFING chaotic oscillator for perturbation, the system will enter the large-scale periodic state from chaotic state. Through the observation on computer chaos system phase track changes, known sine signal to be whether the detection signal contains to detection. At this time, continue to adjust the driving signal amplitude f, and make the system again to the critical state from chaos to large scale period. By getting the driving signal amplitude f_d' of this time, the amplitude of measured sine signal is $a = f_d' - f_d$.

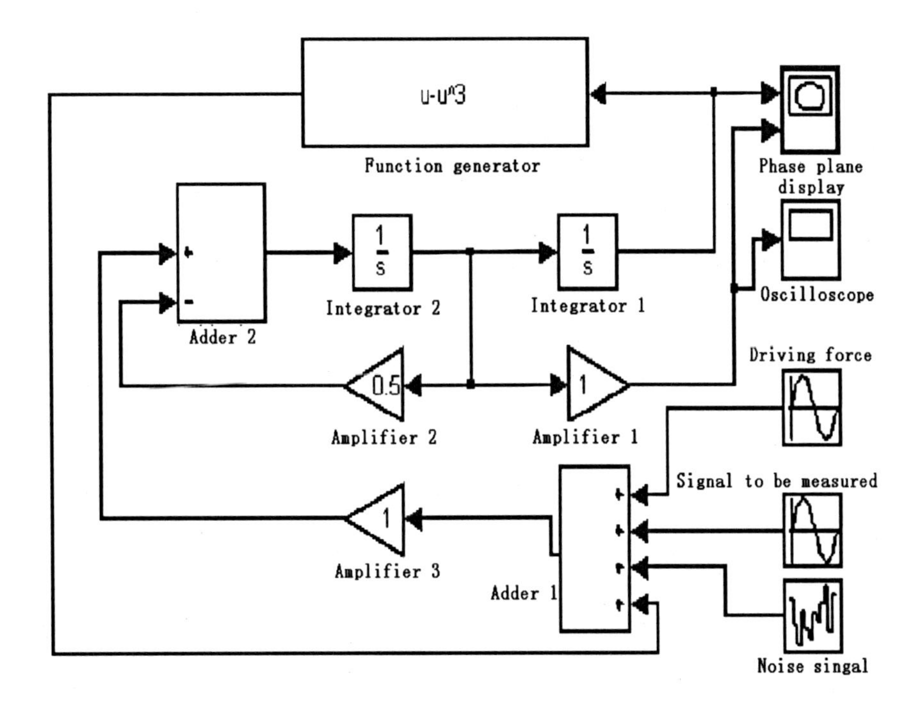

Figure 4: Detection model diagram.

(2) The principle of frequency detection

Considering the frequency of shaft-rate electric field is determined by the ship propeller speed, which could transform only within a certain range, an oscillator array method can be used for corresponding signal frequency detection. Imagine using a limited set of array, whose frequency are limited in 1–10 Hz, and establish a geometric array with common ratio of 1.03. The dipole array has 78 element composition, which are $\omega_1 = 1,\ \omega_2 = 1.03,\ \omega_3 = (1.03)^2,\quad ,\ \omega_{78} = (1.03)^{77} = 9.738$.

When the measured shaft-rate signal is be put into the array, there will be stable intermittent chaos in and only in two adjacent oscillators. If besides the k and k+1 oscillator, the other oscillators still are in chaotic state, the frequency of measured shaft-rate signal must meet: $\omega_k \le \omega \le \omega_{k+1}$, so the frequency of measured shaft-rate signal can be confirmed [6].

4 Experimental verification

In order to verify the ability of DUFFING chaos oscillator to detect weak ship shaft-rate electric field signal, we design a sinusoidal signal $0.008\cos(3.6t)$ which be used to simulate ship shaft-rate electric field signal and a strong noise signal 0.04randn, and the signal to noise ratio is $\mathrm{SNR} = 20\log(0.008/0.4) = -34\mathrm{db}$. By observing the change of DUFFING system phase plane and oscillator array detection, determine the amplitude and frequency of the signal.

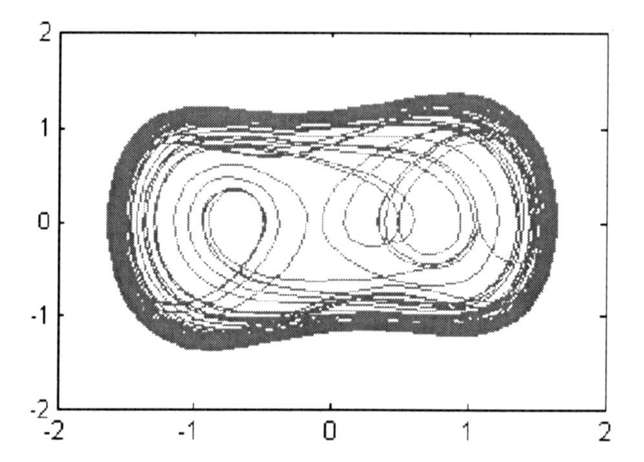

Figure 5: Critical state diagram of DUFFING oscillator chaotic to cycle.

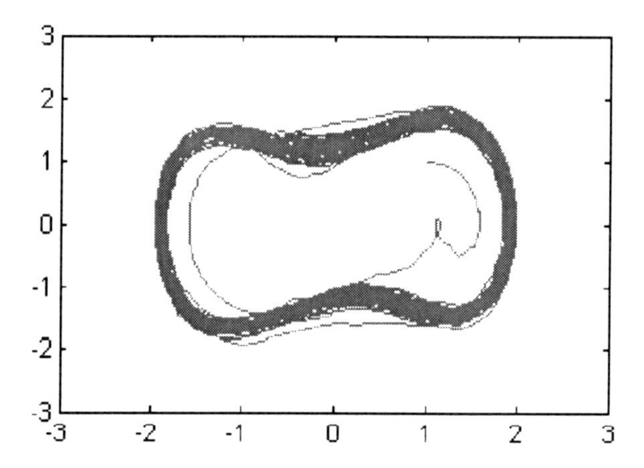

Figure 6: Phase diagram of DUFFING oscillator cyclical movement.

For DUFFING detection system, select the parameter $k = 0.5$, adjust driving amplitude $f_d = 0.7258$, and make the system in the critical state from chaos to large scale period as shown in Figure 5.

When we put mixed signal perturbation to DUFFING oscillator, the system is transient state transition from chaos state to cycle. The system trajectories are shown in Figure 6. You can see that the system phase trajectories change into limit cycle, due to the existence of noise, which makes the limit cycle bounds a bit rough, and system has actually made chaos state transition to a stable periodic motion state.

In order to detect the periodic signal amplitude, the computer in the process of calculation, each time through the loop, it will automatically decreased 0.0001 of the driving amplitude f, until the system is composed of a periodic state to chaotic state. At this time, f is reduced to 0.718. Then, can calculate the signal amplitude is $a = 0.7258 - 0.718 = 0.0078$, there are some errors compared with the given amplitude, that is because it is difficult to determine the precise value of f_d.

Then use the oscillator array to detect signal frequency. Through the generation of value calculation, system is stable intermittent chaos in 43 and 44 oscillator. According to the theoretical basis of previous, the frequency of ship shaft-rate electric field signal ω must meet: $(1.03)^{43} \leq \omega \leq (1.03)^{44}$, namely $3.5645 \leq \omega \leq 3.6714$, so it can determine the frequency of the measured signal.

The experimental results show that the method for weak signal, especially the signal in the submerged in strong noise background, is better than the traditional time domain method.

Acknowledgement

The research work was supported by National Natural Science Foundation of China under Grant No. 41476153.

References

[1] He Zhaoyou, The communication technology based on chaos theory. *Information Security*, **7**, pp. 23–26, 2006.

[2] Kyriazis M, Practical applications of chaos theory to the modulation of human ageinganture prefers chaos to regularity. *Biogerontology*, **4(2)**, pp. 75–90, 2003.

[3] Guanyu W, Dajun C, Jianya L, Xing C, The application of chaotic oscillators to weak signal detection. *IEEE Transactions on Industrial Electronics*, **46(2)**, pp. 440–444, 1999.

[4] C. Piccard, Controlling chaotic oscillations in delay-differential systems via peak-to-peak maps. *IEEE Transactions on Circuits and Systems I: Fundamental Theory and Applications*, **48(8)**, pp. 1032–1037, 2001.

[5] Guanyu Wang, Sailing He, A quantitative study on detection and estimation of weak signals by using chaotic duffing oscillators. *IEEE Transactions on Circuits and Systems-I: Fundamental Theory and Applications*, **50(7)**, pp. 945–953, 2003.

[6] Zhao Xiangyang, Liu Junhua, Shi Baizhou, A new method of weak frequency variation detection in silicon microresonator. *International Conference on Sensor Technology, Proceedings of SPIE*, 4414, pp. 445–448, 2001.

An improved real-time gesture recognition algorithm based on image sequences detecting old people

Chuncai Wang[1], Yuanyuan Sun[2], Xiaoqiang Liu[3]
[1]*Department of Science and Technology, Changchun University of Science and Technology, Changchun, China*
[2]*Engineering Research Department, Changchun WHY-E Science and Technology Co. Ltd., Changchun, China*
[3]*Research Department, Jilin Zhencai Information Technology Co. Ltd., Changchun, China*

Abstract

Gesture recognition based on the video is the basic task of computer vision research. At present there exists lots of research fruit about gesture recognition which limits on the research of mild daily action. For instance walking, jogging, sitting down, bent down, lying down, and so on, but the effect of real-time is not effect and now process the video sequence too much. It is not effect on the process image process. And research on the action recognition little. This paper presents an improved real-time gesture recognition algorithm based on image sequence detecting old people. At first connect the da hua camera, which of the resolution of 1300 thousand real-time gather monitoring data, collect 20 frame data per second; second video image data compress the resolution of 320*240, which transfer an improved Gaussian mixture model algorithm, morphology operation, extract outline, according to the stature, calculate the ratio of width and height, main Fourier descriptors recognize the gesture. And make use of the template match to classify gesture classification.

Keywords: real-time monitor, image sequences, an improved Gaussian Mixture Model, gesture recognition.

WIT Transactions on Engineering Sciences, Vol. 107, © 2015 WIT Press
www.witpress.com, ISSN 1743-3533 (on-line)
doi:10.2495/SEEEI140251

1 Introduction

Automatic detection and identification of the abnormal behavior of the old people is important and difficult in the field of computer vision, and is a key content for computer intelligent video surveillance, especially our country has entered the aging society in 1999, was an early into the aging society of one of the developing countries. At the end of 2008, China has reached 159.9 million old people aged 60 and above, 12.0% of the population; First 20 years of this century is rapidly aging stage, predicts 2020, the aging population will reach 248 million people, ages level will reach 17.17%; In 2050, the elderly population will be over 400 million people, ages level above 30%. Anomaly detection in the field of home endowment faced by multi-level and multimodal sensor combination perception to behaviour [1], a large number of scholars at home and abroad investment behavior recognition in the field of research and exploration.

On the human behavior of video feature extraction is the first step of behavior recognition system, gesture recognition classification at present basically has the following several kinds of methods, method one of using human body contour parameters to establish human posture model, by matching profile code book, hierarchical identification method or nearest neighbor classifier achieve the cognition of human body posture; Two by identifying the space-time interest points of human body movement characteristic, a new space-time interest point detector; To detect the interest of space and time as the center, based on polyhedron model of space and time gradient descriptor the visual characteristics of human body movement in time and space [2]; Three state space method, based on the analysis of video human behavior interaction analysis modeling method, the hidden Markov model based on the observation vector decomposition describes people interaction. In this model, people interaction behavior will analyze from two levels: the individual layer and interaction. Among them, each individual layer is used to describe the behavior of the individual characteristics of the movement, multiple target layer is used to describe the interaction behavior of the behavior characteristics of the share [3, 4]. The two levels complement each other, common to describe people interaction; Method four word bag model extracted from video sequences space–time interest points, according to the point of interest extension is interested area or interest in space, to extract the feature vector[5], through the study of the clustering of feature vector to construct thesaurus, to solve the frequency matching identification[6,7].

2 Discussed problems

Algorithm including data collection, video frame image preprocessing, feature vector extraction, feature recognizer, training and recognition is established.

Database collection, avi format of dahua shot in hd camera fixed near field data, the background static, single stand, bend, squat down, lay in wait for form different attitude data acquisition.

Preprocess using an improved gaussian mixture model to extract the human posture movement areas quickly, using morphological operator for opening and

closing operation, eliminate jitter noise points. Using sobel operator, canny implementation to extract the contour model, feature vectors extracted by some mathematical methods statistical outline on the perimeter, area, area ratio, main Fourier fusion feature vector extracted some of the key into a feature vector.

Build character recognizer using the statistical feature vector for storage. Classification by entering a different posture action compared with the code in the template library book, recognition is standing, lying down, sitting, bend over, or squat movement, if the movement characteristic isn't in the template library, the new recognition of dynamic characteristics of the rich in the template library.

3 The lower solution based on interpolation

3.1 The preprocess of the algorithm

$$P(x_t) = \sum_{i=1}^{K} \omega_{i,t} * \eta(x_t, \mu_{i,t}, \textstyle\sum_{i,t}) \tag{1}$$

Using an extraction is an improved adaptive gaussian mixture model from foreground area and background region, using *p.* KadewTraKuPong and R. Bowden improved adaptive gaussian mixture model of modeling, the model with K gaussian Distribution to the statistical characteristic of the same pixel in each frame. Set as a pixel value x_t at the t moment, the probability density function can be expressed as: among them, at the t moment of the weight value $\omega_{i,t}$ of the ith a gaussian distribution and the averages of value μ, and covariance matrix of the value $\sum_{i,t}$ is the first weight at the t moment a gaussian distribution [8].

$$\eta(x_t, \mu_t, \textstyle\sum_t) = \frac{1}{(2\Pi)^{n/2} |\sum_t|^{1/2}} \exp[-\frac{1}{2}(x_t - \mu_t)^T \textstyle\sum_t^{-1} (x_t - \mu_t)] \tag{2}$$

Gaussian distribution parameters value $\omega_{i,t}$ by sort, take out before K gaussian distribution as the background, the remaining for the foreground.

$$B = \arg\min(\textstyle\sum_{i=1}^{K} \omega_{i,t} > T) \tag{3}$$

Weights updated to gaussian distribution

$$\omega_{i,t} = (1-\beta)\omega_{i,t-1} + \beta M_{i,t}, i = 1, 2, \ldots K \tag{4}$$

Updating for the parameters of the gaussian distribution

$$M_{i,t} \begin{cases} 1, i = j \\ 0, i \neq j \end{cases} \tag{5}$$

$$\mu_{j,t} = (1-\alpha)\mu_{j,t-1} + \alpha x_t \quad (6)$$

$$\sigma^2_{j,t} = (1-\alpha)\sigma^2_{j,t-1} + \alpha(x_t - \mu_{j,t})^T (x_t - \mu_{j,t})^2 \quad (7)$$

3.2 The extraction of feature vector

Statistical contour boundary rectangle of the height, width, aspect ratio of the external rectangle, outline on the edge of the perimeter, area, Fourier descriptor. Using Fourier descriptor, as one of the important characteristics of posture model, draw the extracted contour in the Fourier spectrum, spectrum diagram can be analyzed by Fourier descriptor contains more information, the Fourier descriptor (Main Fourier Descriptors, MFD) for storage, reduce the storage space.

3.3 Establish character identifier

Using Fourier transform can be contour features into a one-dimensional feature vector, and at the same time due to the frequency component of Fourier spectrum most concentrated in the low frequency part, extract Fourier coefficients of low frequency parts of the Fourier descriptor, can greatly reduce the dimension of characteristic vector and computational complexity. MFD before 10 Fourier descriptor as the main description.

Establish a classification recognizer

To identify according to the method described in the above algorithm to extract image 12 character data: N, R and MFD.

Image in the target body posture to stand, the feature vector and the standard code positive vector feature vector, the side of the book compared with standard code book, determine what kind of attitude.

To extract human body contour and its external rectangle around the edge of the distance, to calculate the normalized after image of human body contour of the rectangular box on the left side of the distance, the distance on the left side of the boundary of group H data, and the same outline on the right side of the border. The left and right sides of the distance value form the description of the human body outline a set of H*2 dimension vector. Because of the normalized after image width is not the same, need to the extraction of the each group of data normalization, take them with the corresponding external rectangle width ratio W as for training Vector data. Set after the normalization of left and right side distance [x_k, y_k], respectively, k = 1, 2, it is expressed as the plural form:

$$s(k) = x(k) + jy(k), \quad k = 1,2...H \quad (8)$$

the fourier transform on the set of data

$$S(\omega) = \frac{1}{H}\sum_{k=1}^{H} s(k)\exp[-j2\pi\omega k / H], \quad \omega = 1,2...H \quad (9)$$

Have H Fourier coefficient, and according to descending order

Extract before 10 normalized modulus value as the main vector Fourier description

$\mathrm{MFD} = [d_1, d_2, ..., d_{10}]$

$S = [S_1, S_2, ..., S_H]$

Among them

$$d_i = \|S_i\| / \| S_l \|, \quad i = 1,2,...10 \tag{10}$$

Human body posture in the model has a total of 12 characters vector-valued: N, R and 10 main Fourier descriptor.

4 Genetic algorithm

5 Computational examples and analysis

Avi format of dahua shot in HD camera real-time collect database fixed near field data, make use of an improved adaptive Gaussian mixture model and morphology operator partly results as follows. For instance the background image, stand, bend, squat down, lay in wait for form different attitude data acquisition.

The area is extracted based on the morphological operator results for peripheral contour of the width of the external rectangle high ratio N values to initial stance, this article selects each posture 10 kind of image data, selected four kinds of gesture. As you can see by the picture, standing and lying posture is easy to distinguish, squat down and bent down to two kinds of attitude is not obvious, can easily distinguish with R values. Four features of Fourier descriptor energy which squat down, bent down to distinguish not obvious by multiplication can be divided with the R value. Flowchart of the algorithm by Figure 1, Use the method [9], Different gesture detecting results can be experimented by Figure 2 to Figure 13. Different gesture recognization results can be experimented under Table 1.

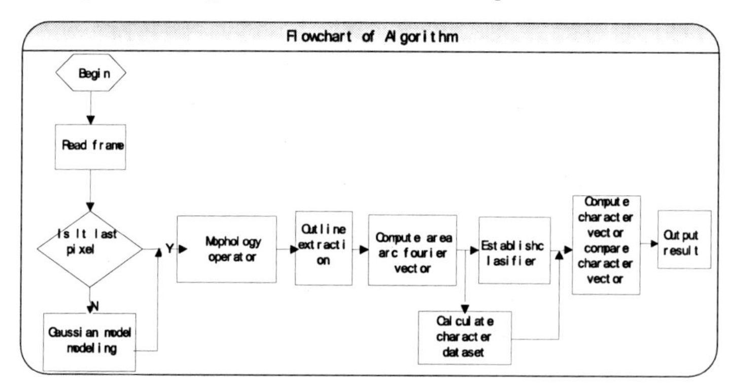

Figure 1: An algorithm of improve real-time old people's behavior analyses based on image sequence.

Figure 2: Original background video image

Figure 3: Standing.

Figure 4: A half lying posture.

Figure 5: Lying posture.

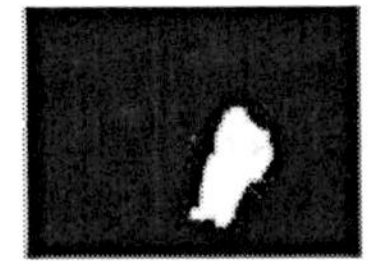

Figure 6: Sit up.

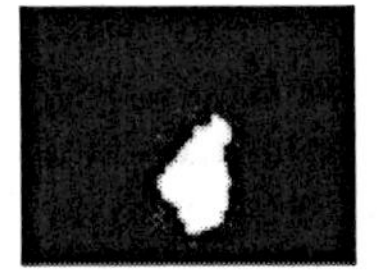

Figure 7: Squat.

Figure 8: Bow down.

Figure 9: Sit up.

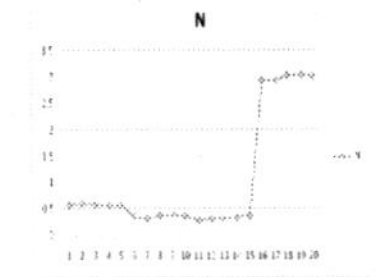

Figure 10: External rectangular wide high ratio *N*.

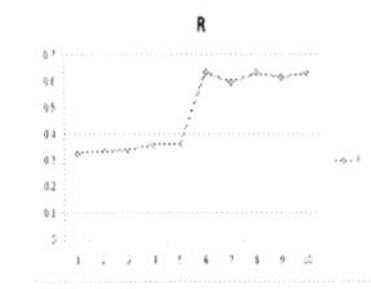

Figure 11: Regional area and external rectangular area ratio *R*.

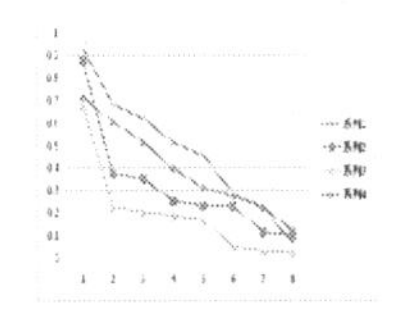

Figure 12: The attitude of the four main fourier descriptor.

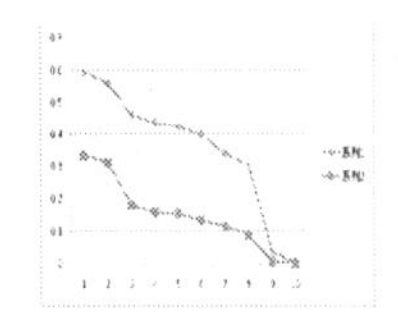

Figure 13: Squat down and bent down to two kinds of attitude R main fourier descriptor.

Table 1: Four characteristics of gesture recognition rate.

Gesture Result	Stand	Bend Down	Lie Down	Squat down	Image (images)	Accuracy
Stand	970	30	0	0	1000	97%
Bend down	5	950	0	45	1000	95%
Lie down	0	5	980	15	1000	98%
Squat down	0	50	0	940	1000	94%

6 Conclusions

Current human gesture detection and recognition is conducted in the primary stage of study, with few papers in elderly gesture recognition, USES in this paper, an improved image sequence based on home endowment behavior analysis algorithm of real-time, could improve the accuracy of the algorithm, basically meet the requirements of the real-time, real-time detection and identification of actions of the old man's attitude, meet unexpected situation in a timely manner to report to the police, when surveillance cameras detected the old man suddenly appear unexpected gesture of the sitting room, sitting, lying posture, a posture for a long time did not move in a timely manner alarm, notify the family or community. Rich profile template library in the future we will gradually, and detect the characteristics of each match, falls in a timely manner to all kinds of gesture features used to identify the fall of the detected.

Acknowledgements

The research work was supported Science and Technology development plan of JiLin in China in 2014 key Scientific Research Project under Grant No. 20140204059 SF and science and technology of Jilin province small and medium enterprises technique innovates funds under Grant No. SC201201015.

References

[1] Yang Lizhao, Behavior Recognition Algorithm Based on Feature Fusion Research. Master degree theses of University of Electronic Science and Technology, pp. 1–30, 2010.

[2] Yao Lixiu, Wang Xiaonian, Yang Jie, Liu Jie, Human action recognition based on polyhedron space and time gradient descriptor. *Journal of South China University of Technology* (natural science edition), **40(6)**, pp. 56–62, 2012.

[3] Jiang ZhaoLin, Human Motion Detection and Identification Method Based on Computer Vision Research. PhD thesis of South China University of Technology, pp. 1–25, 2010.

[4] Li Diping, An Movement to Identify Human Gesture Detection Method Based on Video. PhD thesis. Central South University, pp. 1–15, 2012.

[5] Wang Chuncai, Sun Yuanyuan, LiZhe, A histogram equalization pentagon fitting the elderly abnormal behavior detection algorithm research. *Journal of Changchun University of Science* (natural science edition) **5(2)**, pp. 128–132, 2014.

[6] Du Youtian, ChenFeng, Based on the visual recognition of motion review. *Journal of Electronic Journal*, **35(1)**, pp. 84–89, 2007.

[7] Hu Qiong, Qin Lei, Huang Qing Ming, Human action recognition based on visual review. *Journal of Computer*, **4(12)**, pp. 2512–2524, 2013.

[8] P. KaewTraKulPong, R. Bowden, An improved adaptive background mixture model for real-time tracking with shadow detection. *Video Based Surveillance Systems: Computer Vision and Distributed Processing*, pp. 1–5, 2001.
[9] Deng Tiantian, Based on the hierarchical model of human behavior recognition research. Master degree of university of science and technology of China, pp. 1–68, 2010.

Beam spread upon specular reflection of the GSM beam on slant turbulent atmosphere

Ningjing Xiang[1,2], Xinfang Wang[1], Qiufen Guo[1], Zhensen Wu[2], Mingjun Wang[1], Xuanni Zhang[1]
[1]*School of Physics and Electronic Engineering, Xian Yang Normal College, Xianyang, China*
[2]*School of Science, Xidian University, Xi'an, China*

Abstract

Based on ABCD-ray matrices and the refractive-index structure constant model in atmospheric turbulence by ITU-R, the expression for the mean-squared beam width of reflected Gaussian Schell-model (GSM) on slant turbulence is derived. It is found that the spread of GSM beam reflected by mirror through turbulence is depended on wavelength, object altitude and mirror size.
Keywords: turbulent atmosphere, slant path, double passage, beam spread.

1 Introduction

In recent years, the propagation of optical beams through atmospheric turbulence is an important subject for many applications such as remote sensing, tracking, and a free space optics communications link, because it determines the loss of power at the receiver. Much work has been carried out focused on the spreading of laser beams in atmospheric turbulence, and it was demonstrated that partially coherent beams are less sensitive to the effects of turbulence than fully coherent ones [1–4]. So, the theories of all kinds of the partially coherent beam propagation in the atmospheric turbulence have been studied by many scholars [5–8]. As yet, a fewer papers have dealt with the influence of turbulence on target return beam spread. The reflected wave may experience considerable beam spreading along the path back to the receiver, particularly for smaller targets [9].

The aim of this paper is to study the spread of partially coherent beams coming to a smooth reflector and its return wave through the slant turbulence.

www.witpress.com, ISSN 1743-3533 (on-line)
doi:10.2495/SEEEII140261

2 ABCD theory parameter

In this analysis, a lowest-order Gaussian beam wave is assumed to be incidence upon a mirror at distance L from the transmitter and received in the source plane ($L = 0$) after reflection.

Using the method of ABCD ray matrices, transmitted lowest order Gaussian beam wave is characterized by beam parameters [10]

$$\Theta_0 = 1 - \frac{L}{F_0}, \qquad \Lambda_0 = \frac{2L}{kW_0^2} \tag{1}$$

where k is the optical wave number, is a vector in the transverse direction, F_0 and W_0 denote the phase front radius of curvature and beam radius, respectively. In the absence of the atmosphere, the Gaussian beam wave incident upon the mirror at $z = L$ is described by the related beam parameters

$$\Theta_{11} = \frac{\Theta_0}{\Theta_0^2 + \Lambda_0^{\ 2}} = 1 + \frac{L}{F_{11}}, \ \bar{\Theta}_{11} = 1 - \Theta_{11}, \ \Lambda_{11} = \frac{\Lambda_0}{\Theta_0^2 + \Lambda_0^2} = \frac{2L}{kW_{11}^2} \tag{2}$$

In what follows, in term of our former paper, we have replaced the receiver beam parameters for a coherent beam in free space with the equivalent free-space receiver parameters for a partially coherent beam

$$\Theta_1 = \frac{\Theta_0}{\Theta_0^2 + \varsigma_s \Lambda_0^{\ 2}} = 1 + \frac{L}{F_1}, \ \bar{\Theta}_1 = 1 - \Theta_1, \ \Lambda_1 = \frac{\varsigma_s \Lambda_0}{\Theta_0^2 + \varsigma_s \Lambda_0^2} = \frac{2L}{kW_1^2} \tag{3}$$

where $\varsigma_s = 1 + W_0^2 / \sigma_g^2$, F_1 and W_1 denote the radius of curvature and beam radius of the incident beam, σ_g^2 is the variance of the Gaussian process determined by the characteristics of diffuser.

Neglecting effects of atmospheric turbulence, if the target is a smooth circular mirror of radius W_R, we see that the reflected wave back in the source–receiver plane is described by beam parameters

$$\Theta_2 = \frac{2 - \Theta_1}{(2 - \Theta_1)^2 + (\Lambda_1 + \Omega_R)^2} = 1 + \frac{L}{F}, \ \bar{\Theta}_2 = 1 - \Theta_2 \tag{4}$$

$$\Lambda_2 = \frac{\Lambda_1 + \Omega_R}{(2 - \Theta_1)^2 + (\Lambda_1 + \Omega_R)^2} = \frac{2L}{kW^2} \tag{5}$$

F and W denote the radius of curvature and beam radius at the receiver, and Fresnel ratio $\Omega_R = 2L/kW_R^2$, characterizes the finite size of the retroreflector. The complex amplitude of the received wave is $P = \Theta - i\Lambda = (\Theta_1 - i\Lambda_1)(\Theta_2 - i\Lambda_2)$, where $\Theta = \Theta_1\Theta_2 - \Lambda_1\Lambda_2$, $\Lambda = \Lambda_1\Theta_2 + \Theta_1\Lambda_2$.

3 Mutual coherence function

The MCF associated with the reflected Gaussian beam wave in the plane of the receiver is defined by the ensemble average [10].

$$\Gamma_2(r_1, r_2, 2L) = < U(r_1, 2L)U^*(r_2, 2L) > = \Gamma_2^0(r_1, r_2, 2L)M_2(r_1, r_2) \quad (6)$$

where

$$\Gamma_2^0(r_1, r_2, 2L) = < U_0(r_1, 2L)U_0^*(r_2, 2L) > = \frac{W_0^2}{W^2(1+\Omega_R/\Lambda_1)} \exp\left[-\frac{(r_1^2 + r_2^2)}{W^2} - \frac{ik(r_1^2 - r_2^2)}{2F}\right] \quad (7)$$

is the MCF in the absence of atmospheric turbulence

$$M_2(r_1, r_2) = P(r_1, r_2)N(r_1, r_2) \quad (8)$$

In the absence of mutual correlations between incident and reflected waves, the factor $P(r_1, r_2)$ describes the angular spread and loss of transverse spatial coherence of the received wave due to turbulence. The factor $N(r_1, r_2)$ describes the redistribution of energy in the reflected beam caused by correlations between the complex phase perturbations of the incident and reflected waves.

In the case of a biostatic system, the factor $N(r_1, r_2)$ becomes unity. Let $r_1 = r_2 = r$, the mean irradiance

$$< I(\vec{r}, 2L) > = \Gamma_2(\vec{r}, \vec{r}, 2L) = \frac{W_0^2 P(\vec{r})}{W^2(1+\Omega_R/\Lambda_1)} \exp(-2r^2/W^2) \quad (9)$$

For $r \neq 0$, a useful approximation due to the diffractive beam edge is [9].

$$P(\vec{r}) \cong \frac{1}{1+H_i+H_R} \exp\left[-\frac{2r^2}{W^2}(H_i + H_R)\right], \quad r < W \quad (10)$$

Based on small arguments in the exponential function, quantities H_i and H_R are defined explicitly by

$$H_i = 4\pi^2 k^2 L \int_0^1 \int_0^\infty \kappa \Phi_n(\kappa) \left\{ 1 - \exp\left[-\frac{\Lambda_1 L \kappa^2}{k} \xi(\Theta_2 + \bar{\Theta}_2 \xi + \Theta \xi) \right] \right.$$
$$\left. \times \exp\left[-\frac{L\kappa^2}{k} (1 - \bar{\Theta}_1 \xi)(\Lambda_2 - \Lambda_2 \xi + \Lambda \xi) \right] \right\} \mathrm{d}\kappa \mathrm{d}\xi \qquad (11)$$

$$H_R = 4\pi^2 k^2 L \int_0^1 \int_0^\infty \kappa \Phi_n(\kappa) \left[1 - \exp\left(-\frac{\Lambda_2 L \kappa^2 \xi^2}{k} \right) \right] \mathrm{d}\kappa \mathrm{d}\xi \qquad (12)$$

where $\Phi_n = 0.033 C_n^2(h) \kappa^{-11/2}$, $C_n^2(h)$ is the model for the refractive-index structure constant in the atmosphere obtained by the International Telecommunication Union [11].

The mean-squared beam width is defined as

$$W_s^2(2L) = 4 \frac{\iint r^2 I(\vec{r}, 2L) \mathrm{d}\vec{r}}{\iint I(\vec{r}, 2L) \mathrm{d}\vec{r}} \qquad (13)$$

Using Eqs. (9), (10) and (11), after some lengthy mathematical manipulations, due to small argument approximately $H_i << 1, H_R << 1$, Eq. (13) is given by

$$W_s(2L) = W\sqrt{1 + H_i + H_R} \qquad (14)$$

Eq. (14) is the main analytical result obtained in this paper. W represents beam size of free-space.

Under the assumption of a Kolmogorov power-law spectrum for the atmosphere, we obtain

$$H_i = 3.5\sigma_0^2 \int_0^1 \left[\Lambda_1 \xi (\Theta_2 + \bar{\Theta}_2 \xi + \Theta \xi) + (1 - \bar{\Theta}_1 \xi)(\Lambda_2 - \Lambda_2 \xi + \Lambda \xi) \right]^{5/6} \mathrm{d}\xi \qquad (15)$$

where $\sigma_0^2 = 1.23 C_n^2 k^{7/6} L^{11/6}$ is the Rytov variance. For $C_n^2(\xi H) = C_{n0}^2$, Eq.(15) can reduce to the horizontal path.

Using similar method, Eq. (12) is written as

$$H_R = 3.5\sigma_0^2 \int_0^1 \frac{C_n^2(\xi H)}{C_{n0}^2} (\Lambda_2 \xi^2)^{5/6} \mathrm{d}\xi \qquad (16)$$

4 Numerical analysis

Using Eq. (14) together with Eqs. (15) and (16), relative beam width numerical examples are given for collimated beam in Figures 1–3. We can see that with distance increase, beam spread becomes more and more large. Figures 1–3 show the normalized beam size which is the ratio of beam width in turbulence to beam width in free space as a propagation distance. Figure 1 compares spread of reflected GSM with different wavelength. As Figure 1 illustrates, we can see that beam size increases with decreasing wavelength. It is apparent that, for a fixed value L=1000m, $\frac{W_s(2L)}{W(2L)}$ of wavelength $\lambda = 0.85\mu\text{m}$ is larger than $\lambda = 1.55\mu\text{m}$ where $W_s(2L)$ and $W(2L)$ are beam spread in turbulence and in free space respectively. In Figure 2, with propagation distance increases, beam spread affected by turbulence is more apparent. When mirror size becomes large, the difference between $\frac{W_s(2L)}{W_0}$ in turbulence and in free space decreases. Figure 3 shows relative width $\frac{W_s(2L)}{W(2L)}$ of the GSM beam versus propagation L for different H. As can be seen in Figure 3, we choose $H = 700\,\text{m}$, the relative beam width after propagation of 4 km through atmospheric turbulence will be 1.10, and for $H = 1000\,\text{m}$ the relative beam width will be 1.06. As object altitude increases, relative width becomes small.

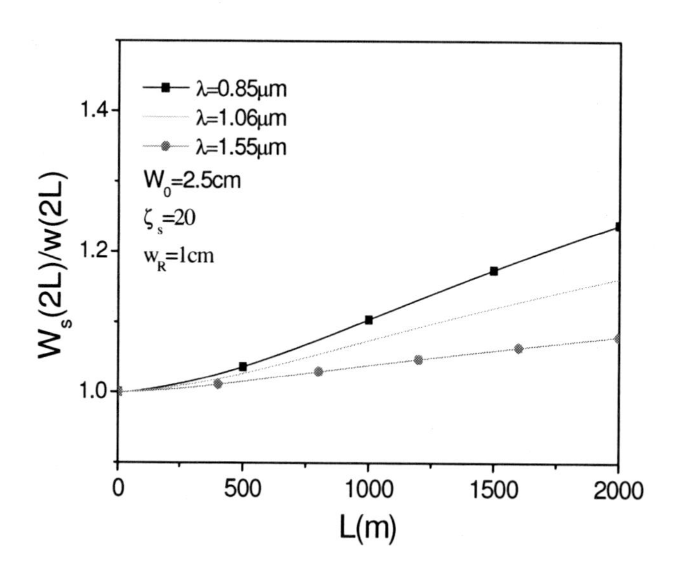

Figure 1: Relative width $\frac{W_s(2L)}{W(2L)}$ of GSM beam versus propagation L for different λ.

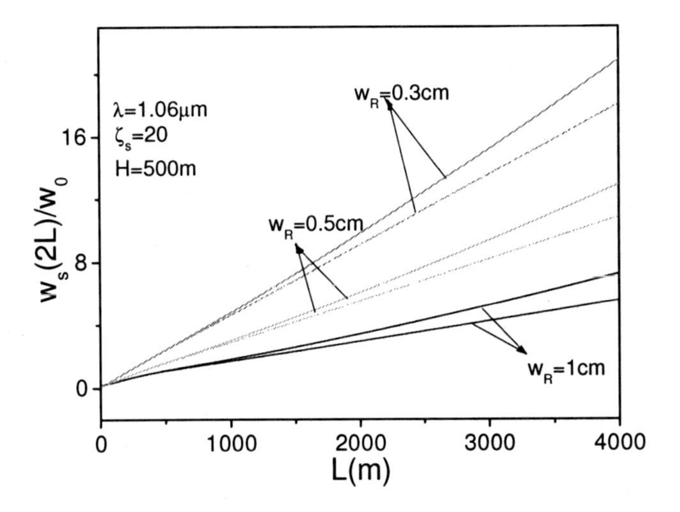

Figure 2: Relative width $\frac{W_s(2L)}{W(2L)}$ of GSM beam versus propagation L.

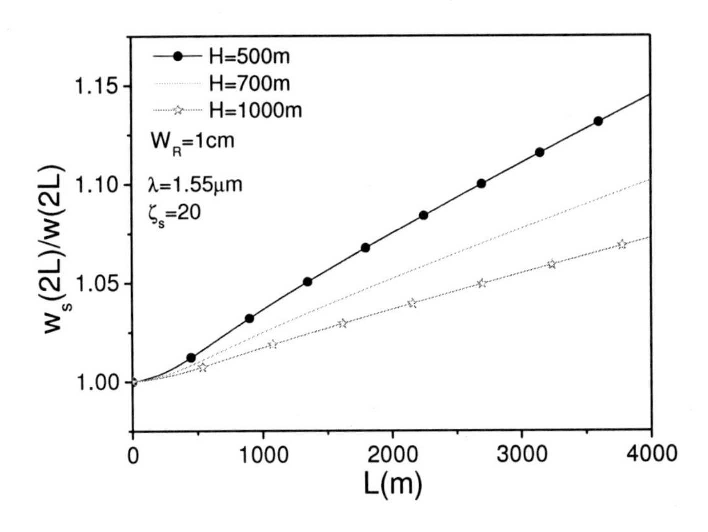

Figure 3: Relative width $\frac{W_s(2L)}{W(2L)}$ of GSM beam versus propagation L for different H.

5 Concluding remarks

In this paper, the reflected GSM beam in turbulence has been studied in detail. In comparison with previous work our results are more general, because spread of the reflected GSM beam can be reduced to the spread of GSM beam in single passage horizontal turbulence. When the GSM beam is reflected by mirror, mirror size plays an important role in beam width. With the mirror size increases, beam width becomes more and more small.

Acknowledgement

This work is partly supported by the National Natural Science Foundation of China (Grant Nos. 61271110, 61307002), New Scientific and the Natural Science Foundation of Shaanxi Province education office, China (Grant No. 14JK1798).

References

[1] G. Gbur, E. Wolf, Spreading of partially coherent beams in random media. *J. Opt. Soc. Am. A*, **19**, pp. 1592–1598, 2002.
[2] A. Dogariu, S. Amarande, Propagation of partially coherent beam: turbulence-induced degradation. *Opt. Lett.*, **28**, pp. 10–12, 2003.
[3] T. Shirai, A. Dogariu, E. Wolf, Mode analysis of spreading of partially coherent beams propagating through atmospheric turbulence. *J. Opt. Soc. Am. A*, **20**, pp. 1094–1102, 2003.
[4] M. Salem, T. Shirai, A. Dogariu, E. Wolf, Long distance propagation of partially coherent beams through atmospheric turbulence. *Opt. Commun.*, **216**, pp. 261–265, 2003.
[5] X. Ji, X. Chen, B. Lü, Spreading and directionality of partially coherent Hermite-Gaussian beams propagating through atmospheric turbulence. *J. Opt. Soc. Am. A*, **25**, pp. 21–28, 2008.
[6] H.T. Eyyuboğlu, Propagation of Hermite-cosh-Gaussian laser beams in turbulent atmosphere. *Opt. Commun.*, **245**, pp. 37–47, 2005.
[7] L.W. Casperson, A.A. Tovar, Hermite–sinusoidal-Gaussian beams in complex optical systems. *J. Opt. Soc. Am. A*, **15**, pp. 954–961, 1998.
[8] G. Wu, H. Guo, et al., Spreading and direction of Gaussian-Schell model beam through a non-Kolmogorov turbulence. *Opt. Lett.*, **35**, pp. 715–717, 2010.
[9] L.C. Andrews, R.L. Phillips, *Laser Beam Propagation through Random Media*. SPIE Press, Bellingham, Wash, 1998.
[10] O. Korotkova, A model for a partially coherent Gaussian beam in atmospheric turbulence with application for lasercom and LIDAR systems, 2003.
[11] ITU-R, Document 3J/31-E, On propagation data and prediction methods required for the design of space-to-earth and earth to-space optical communication systems. *Radio Communication Study Group Meeting*, Budapest, 2001.

Inversion for atmospheric duct from radar clutters by Metropolis factor particle swarm optimization

Rongxu Hu[1], Zhensen Wu[1], Jinpeng Zhang[2]
[1]*School of Physics and Optoelectronic Engineering, Xidian University, Xi'an, China*
[2]*China Research Institute of Radiowave Propagation, Qingdao, China*

Abstract

At present, most of the researches on refractivity from clutter (RFC) were focused on how to improve its performance. This paper proposed to apply Metropolis factor particle swarm optimization (M-PSO) to RFC in order to obtain more accurate estimate of RFC. This method possessed of self-adjusting acceleration coefficients compared with standard particle swarm optimization (S-PSO). Inversions on a set of synthetic clutter power for vertical spatial distribution of refractivity (M-profile) of surface-based ducts were performed by M-PSO. Results show that M-PSO exhibits a slightly slower convergence speed, similar computational time and slight improvements in accuracy compared with S-PSO in RFC.
Keywords: refractivity from clutter, Metropolis factor particle swarm optimization, Inversion.

1 Introduction

Lower atmospheric ducts were commonly observed in oceanic environments. Anomalous propagation events occur when the radar operates in atmospheric ducts. Events include strong variations in the radar detection range and creation of radar holes in which the radar is blind. They significantly affect the performance of shipboard radars. Using ducts effectively can be performed when a detailed refractivity profile against altitude is available for large marine regions. Measures for refractivity are classified into three categories as follows: (1)

www.witpress.com, ISSN 1743-3533 (on-line)
doi:10.2495/SEEEI140271

conventionally direct sensing techniques (e.g., radiosonde, microwave refractometer, and rocketsonde); (2) remote sensing techniques [1] (e.g., radar clutter and GPS signals based on); and (3) a data fusion approach [2] that syncretizes information from direction sensing, remote sensing, and numerical weather prediction.

Refractivity from clutter (RFC) [3, 4] is an active research field, in which M-profile is estimated based on radar sea-surface returns in maritime environments. Some of the advantages of RFC include high temporal and spatial resolutions, low operating costs, and operation convenience. RFC belongs to (2) remote sensing techniques and is developed during the 1990s.

When using RFC to estimate M-profile, an accurate and speedy search approach is necessary to avoid temporal errors [5]. GA [3], particle swarm optimization (PSO) [4], and LS-SVM [5] have been proposed as rapid and nonlinear optimization methods. However, GA had complex program and LS-SVM had poor accuracy [5]. PSO was simple and recommended method [4].

In order to further improve the performance of RFC including PSO, metropolis learning factor particle swarm optimization (M-PSO) [6] with self-adjusting acceleration coefficients is proposed. M-PSO is improved combination of PSO and simulated annealing (SA). This paper focuses on testing the performance of RFC employing M-PSO, through comparison of the estimates by M-PSO and S-PSO inversions based on synthetic clutter powers with noiseless signals. Test results show that M-PSO can upgrades precision and thus improves performance of RFC system to a certain extent.

The paper is organized as follows. In the second section, we set forth briefly tropospheric ducts and theory about radar clutters used in RFC system. Section 3 presents the inversion method named as M-PSO based on S-PSO. In Section 4, the simulative inversion is made to illustrate the efficiency of the algorithm. Finally, we conclude our paper in Section 5.

2 Theory and model

2.1 Tropospheric duct

Low-altitude atmospheric refractivity varies with meteorological conditions in time and space. Ducts are formed under several particular meteorological conditions and comprise three duct types: evaporation duct (Figure 1(a)), Surface-based duct (SBD) (Figure 1(b)) and Elevated duct (Figure 1(c)). Evaporation duct permanently exist in oceanic environments, with a worldwide average height of approximately 13 m. SBD persist and extend over oceanic environments for up to several hundreds of meters. SBD and elevated duct are less common than evaporation duct, but their effects on radar sea surface returns are highly marked.

The duct vertical refractivity structure is often expressed by M-profile parameterized. SBD and elevated duct are characterized by a parameterized trilinear M-profile represented by vector $\mathbf{m} = (s_1, s_2, h_1, h_2)$ [7],

$$M(z) = M_0 + \begin{cases} s_1 z, & 0 < z \le h_1 \\ s_1 h_1 + s_2(z - h_1), & h_1 < z \le h_1 + h_2 \\ s_1 h_1 + s_2 h_2 + 0.118(z - h_1 - h_2), & z > h_1 + h_2 \end{cases} \tag{1}$$

where (s_1, h_1) and (s_2, h_2) respectively represent the slope and thickness of the base and inversion layer. The top layer slope is assumed to be a standard atmosphere with a vertical gradient of 0.118 M-units/m. This parameterized trilinear SBD model was selected in this study. M-profiles are hypothesized to remain constant along the entire propagation path in open sea to simplify the RFC inversion. In addition, it is worth reminding that elevated duct is not obtained by RFC because of their low influence on sea clutter at sea level [5].

2.2 Sea clutter power

Given a refractive maritime environment that depend on range and height, the sea clutter signal power $P_c(x,\mathbf{m})$ received by the radar in the absence of receiver noise is expressed as follows [3]:

$$P_c(x,\mathbf{m}) = -2L(x,\mathbf{m}) + 10\log_{10}(x) + \sigma^0(x,\theta) + C \tag{2}$$

where $L(x,\mathbf{m})$ is the one-way propagation loss in a ducted medium represented by parameter vector $\mathbf{m}$, $\sigma^0(x,\theta)$ is the normalized radar cross section (NRCS) of the sea surface at range x, which is determined from empirical model of sea clutter for simplify. θ is grazing angle which can be determined by geometric optics [8]. And C accounts for constant terms induced from radar parameters [3].

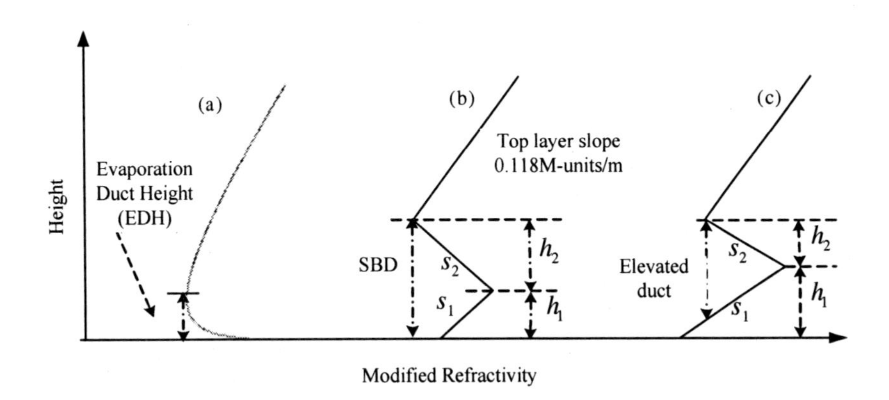

Figure 1: M-profiles parameterized model for atmospheric duct. (a) Evaporation Duct, (b) Surface-Based Duct (SBD), (c) Elevated Duct.

The propagation loss $L(x,\mathbf{m})$ is given by the following equation:

$$L(x,\mathbf{m}) = 20\log_{10}(4\pi x/\lambda) - 20\log_{10} F \tag{3}$$

where λ is wavelength, F is propagation factor, which is expressed as follows:

$$F = \sqrt{x}\left|u(x,z)\right| \tag{4}$$

where $u(x,z)$ is electric or magnetic field at range x and height z for horizontal or vertical polarization respectively, and can be obtained by following recursive split-step FFT numerical solution [9, 10] to parabolic wave equation (PWE) [11, 12]:

$$u(x+\Delta x,z) = \exp\left[ik\Delta x M(x,z)\times 10^{-6}\right]\Im^{-1}\left\{\exp(-ip^2\Delta x/2k)\Im\left[u(x,z)\right]\right\} \tag{5}$$

where $u(x+\Delta x,z)$ is field at range $x+\Delta x$ and height z, Δx is the horizontal range increment, $\Im$ and $\Im^{-1}$ respectively denote the discrete Fourier transform and inverse discrete Fourier transforms along the height z, k is the free space radar wave number, M is the modified refractivity, $p = k\sin\alpha$ is the spatial frequency, and α is the propagation angle from the horizontal direction. z and p are associated by Nyquist's criteria $z_{\max} p_{\max} = N\pi$, N is the discrete Fourier transform size [11]. For calculations of spatial field $u(x,z)$ by PWE, the initial field $u(x=0,z)$ being the antenna aperture distribution is essential and can be obtained by taking the Fourier transform of the far-field antenna pattern [11].

It should be noted that $P_c(x,\mathbf{m})$ is calculated at the effective scattering height which is usually approximated to 1 m or 0.6 times the mean wave height [13].

3 Inversion methods: Metropolis factors-PSO

PSO is an evolutionary computation inspired by the social behavior of animal groups and developed by Kennedy and Eberhart [14]. The method has been successfully applied in several fields [15]. Detailed procedures and steps of the application of PSO to RFC are found in Ref. [4].

M-PSO is an improvement of the combination of PSO and SA, and different from S-PSO in that it used Metropolis factor acceleration coefficients $c_{1id}^{\mathrm{m}}(t)$ and $c_{2id}^{\mathrm{m}}(t)$ replacing acceleration coefficients c_1 and c_2 in S-PSO. The updated equations are given as follows:

$$\begin{aligned} v_{id}(t+1) = {} & w_i(t)v_{id}(t) + c_{1id}^{\mathrm{m}}(t)r_{1d}(t)\left[p_{id}(t) - x_{id}(t)\right] \\ & + c_{2id}^{\mathrm{m}}(t)r_{2d}(t)\left[p_{gd}(t) - x_{id}(t)\right] \end{aligned} \tag{6}$$

$$x_{id}(t+1)=x_{id}(t)+v_{id}(t+1) \tag{7}$$

Eqs. (9) and (10) constitute M-PSO algorithm. Where metropolis factor acceleration coefficients are given as follows [6]:

$$c_{1id}^{m}(t)=c_1\exp(\left|p_{id}(t)-x_{id}(t)\right|/t) \tag{8}$$

$$c_{2id}^{m}(t)=c_2\exp(\left|p_{gd}(t)-x_{id}(t)\right|/t) \tag{9}$$

$c_{1id}^{m}(t)$ and $c_{2id}^{m}(t)$ vary according to the distances between particle's current optimum $p_{id}(t)$ and global optimal positions $p_{gd}(t)$, t represents update generation. The subscripts i and d denote the ith particle and the dth dimension state vector. For example, $v_{id}(t)$ and $x_{id}(t)$ represents the local velocity and position of the dth dimensional state of ith particle at tth update generations. $w_i(t)=w_{max}-t(w_{max}-w_{min})/ST_{max}$ is the inertia weight for the ith particle, where $w_{max}=0.95$, $w_{min}=0.4$, and ST_{max} is the maximum generation. c_1 and c_2 are acceleration coefficients, usually $c_1=c_2=2$. $r_{1d}(t)$ and $r_{2d}(t)$ are random variables uniformly distributed in the interval $[0,1)$.

4 Inversion experiment

4.1 Inversion layout

RFCs using M-PSO and S-PSO were respectively performed on synthetic clutter power. Inversions were carried out from 10–60 km with a range point placed for every 500 m. 20 inversions for true M-profile were executed and each executes 500 generations. Random numbers used by each update step in M-PSO were same as that in S-PSO. A least-squares objective function [3] was selected and the population size was set at 20. Moreover, as noted, the forward propagation PWE runs in each calculation of the objective function for a possible M-profile expressing true SBD in the following implementations.

4.2 Inversion results

A set of clutter power (solid and green -) in Figure 3(b) is synthetic, which is obtained from true M-profile (solid and green -) parameterized by vector $\mathbf{m}=(0.62,\ -1.34,\ 32,\ 66)$. Space Range Radar (SPANDAR) parameters for the Wallops'98 observation are adopted in addition to Gaussian antenna pattern. Lower and upper search limits are set as $(0,\ -2,\ 0,\ 50)$ and $(1.5,\ -0.5,\ 50,\ 100)$.

Table 1 shows the mean, standard deviation (STD), and relative error (RE) of 20 inversions at 300 generations. From Table 1, the inversions of s_1 and h_1 by

M-PSO was significantly accurate, h_2 was improved slightly, and s_2 was nearly unchanged compared with those by S-PSO.

Figure 2 illustrates the M-profiles based on mean in Table 1 and corresponding clutters. The plots (-. and --) describe the inversion by S-PSO and M-PSO, respectively. Here, clutter plot (--) was achieved by substitution of M-profile (--) to PWE (5) and application of Eq. (2). Clutter plot (-.) ibid. From the M-profiles and clutter plots (Figure 2), the results by M-PSO were better match true plots than those by S-PSO. Conclusions As a result, it is drawn that M-PSO yielded more accurate estimates for RFC than S-PSO, especially s_1 and h_1.

Figure 3 shows the convergence characteristics of both PSO methods. The convergence speed of M-PSO was slightly slower than that of S-PSO. Both M-PSO and S-PSO converged at about 300 generations.

Table 1: Synthetic data: S-PSO and M-PSO estimated results.

Parameters	Units	True Parameters	S-PSO			M-PSO		
			Mean	STD	RE	Mean	STD	RE
s_1	M-units/m	0.62	0.556	0.078	0.115	0.602	0.057	0.031
s_2	M-units/m	−1.34	−1.343	0.021	0.002	−1.345	0.021	0.003
h_1	m	32	32.620	0.884	0.019	32.224	0.664	0.007
h_2	m	66	76.021	10.875	0.132	74.739	10.914	0.117

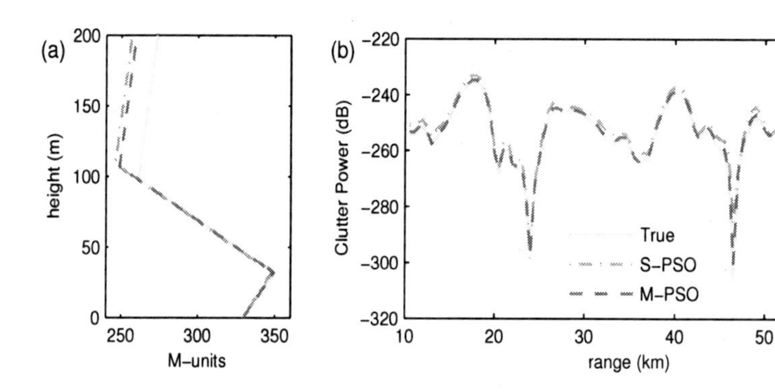

Figure 2: Inversion results for synthetic data: (a) estimated (-. and ö) and true (-) M-profiles, and (b) clutters computed with the true parameters and estimated parameters by the S-PSO and M-PSO.

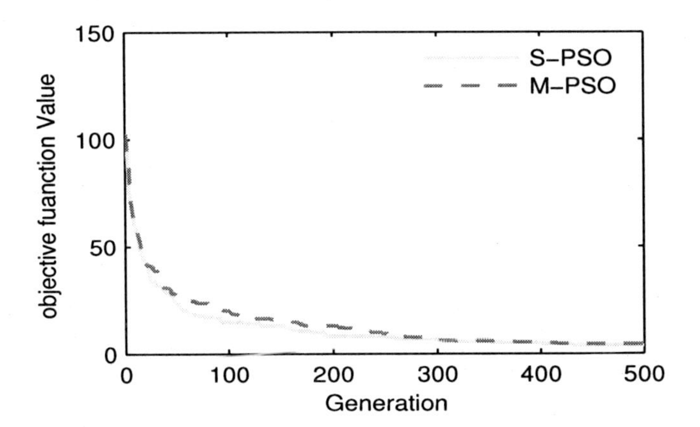

Figure 3: Convergence characteristics for synthetic data.

5 Summary and conclusion

An M-PSO method was proposed to apply to RFC. The algorithm has self-adjusting acceleration coefficients and requires no more time compared with S-PSO. Results show that M-PSO brings slight improvements in precision of the estimate for RFC and it exhibits slightly slow convergence characteristics.

References

[1] L.T. Rogers, C.P. Hattan, J.L. Krolik, Using radar sea echo to estimate surface layer refractivity profiles. *Proceedings of IEEE International Geoscience and Remote Sensing Symposium*, **1**, pp. 658–662, 1999.

[2] A. Karimian, C. Yardim, T. Haack, P. Gerstoft, W.S. Hodgkiss, L.T. Rogers, Toward the assimilation of the atmospheric surface layer using numerical weather prediction and radar clutter observations. *J. Appl. Meteor. Climatol.*, **52**, pp. 2345–2355, 2013.

[3] P. Gerstoft, L.T. Rogers, J.L. Krolik, W.S. Hodgkiss, Inversion for refractivity parameters from radar sea clutter. *Radio Sci.*, **38(3)**, p. 8053, 2003.

[4] B. Wang, Z.-S. Wu, Z.-W. Zhao, H.-G. Wang, Retrieving evaporation duct heights from radar sea clutter using particle swarm optimization (PSO). *Prog. Electromagn. Res.*, **9**, pp. 79–91, 2009.

[5] R. Douvenot, V. Fabbro, P. Gerstoft, C. Bourlier, J. Saillard, A duct mapping method using least squares support vector machines. *Radio Sci.*, **43**, RS6005, 2008.

[6] J. Xing, D.Y. Xiao, New metropolis coefficients of Particle Swarm Optimization. *Control and Decision Conference*, pp. 3518–3521, 2008.

[7] C. Yardim, P. Gerstoft, W.S. Hodgkiss, Estimation of radio refractivity from radar clutter using Bayesian Monte Carlo analysis. *IEEE Trans. Antennas Propag.*, **54(4)**, pp. 1318–1327, 2006.

[8] A.E. Barrios, Considerations in the development of the advanced propagation model (APM) for U.S. navy applications. *Radar Conference, 2003 Proceedings of the International*, pp. 77–82, 2003.
[9] J.R. Kuttler, R. Janaswamy, Improved Fourier transform methods for solving the parabolic wave equation. *Radio Sci.*, **37(2)**, pp. 5-1–5-11, 2002.
[10] G.D. Dockery, J.R. Kuttler, An improved impedance-boundary algorithm for fourier split-step solutions of the parabolic wave equation. *IEEE Trans. Antennas Propag.*, **44(12)**, pp. 1592–1599, 1996.
[11] A.E. Barrios, A terrain parabolic equation model for propagation in the troposphere. *IEEE Trans. Antennas Propag.*, **42(1)**, pp. 90–98, 1994.
[12] D.J. Donohue, J.R. Kuttler, Propagation modeling over terrain using the parabolic wave equation. *IEEE Trans. Antennas Propag.*, **48(2)**, pp. 260–277, 2000.
[13] J.P. Reilly, G.D. Dockery, Influence of evaporation ducts on radar sea return. *IEE Proc. Radar Signal Process.*, **137(F-2)**, pp. 80–88, 1990.
[14] J. Kennedy, R. Eberhart, Particle swarm optimization. *Proceedings of IEEE International Conference on Neural Networks*, **4**, pp. 1942–1948, 1995.
[15] R. Poli, Review article: analysis of publications on the applications of particle swarm optimization. *Journal of Artificial Evolution and Applications*, 2008.

Section 2
Electrical engineering

Application of a transformer type FCL for mitigating the effect of DC line fault in VSC-HVDC system

Yuwei Dai[1], Jian Fang[2], Wei AI[1], Lei Chen[3]
[1]*Jilin Electric Power Survey and Design Institute, Changchun, China*
[2]*Fujian Electric Power Survey and Design Institute, Fuzhou, China*
[3]*School of Electrical Engineering, Wuhan University, Wuhan, China*

Abstract

Aiming at the DC fault issue occurred in a voltage source converter (VSC)-based HVDC system, this paper introduces a transformer type FCL as the technical means for mitigating the fault current and protecting the system. The suggested FCL is placed in series with the head of a DC transmission line, and under fault condition, its current-limiting impedance will be inserted into the main circuit for affecting the system's transient behaviors. Related theoretical analysis is conducted, and the detailed simulation model of a 110 kV class VSC-HVDC system integrated with the FCL is built in PSCAD/EMTDC. From the demonstrated simulation results, employing the transformer type FCL can obviously suppress the DC fault current's rate of rise and improve the DC voltage sag. As a result, the VSC-HVDC system's robustness against the short-circuit fault can be well enhanced.

Keywords: DC line fault, transformer type FCL, transient simulation analysis, VSC-HVDC transmission system.

1 Introduction

Currently, high-voltage direct-current (HVDC) power transmission systems and technologies associated with the flexible ac transmission system (FACTS) continue to advance as they make their way to commercial applications [1]. In regards to a voltage source converter-based HVDC (VSC-HVDC) system, it can carry out rapid and independent control of the active and reactive power, and may

www.witpress.com, ISSN 1743-3533 (on-line)
doi:10.2495/SEEEI140281

provide greater operational flexibility which suits for renewable energy integration and power grid upgrading [2, 3].

Nevertheless, when a DC-line fault appears in a VSC-HVDC system, although the voltage source inverter's (VSI's) power switches are turned-off during a short time, the fault current is not interrupted since the anti-parallel diodes conduct as a rectifier bridge, and introducing an additional limiting device is needed to guarantee the system's operation performance.

In this paper, based on a transformer type fault current limiter (FCL), the application study of the FCL in a VSC-HVDC transmission system is performed. This paper is organized in the following manner. Section 2 presents the FCL's structural principle, and discusses its effects on the DC fault current at different times. In Section 3, the model of a 110 kV class VSC-HVDC system integrated with the FCL is created in PSCAD/EMTDC, and simulation analyses are carried out to access the FCL's performance. In Section 4, conclusions are summarized and next steps are suggested.

2 Theoretical analysis

2.1 Structural principle of the transformer type fault current limiter

The schematic configuration of the transformer type fault current limiter is shown in Figure 1 [4–7]. The FCL consists of a series transformer and a superconducting current limiting device. Herein, the suffixes 1 and 2 of current I, voltage V, and self-inductance L indicate that the values are the primary and secondary side values, respectively. V_S is the voltage of power source, and I_S is the main current. M is the mutual inductance between the primary and secondary winding. R_2 is recorded as the resistance of the superconducting current limiting device.

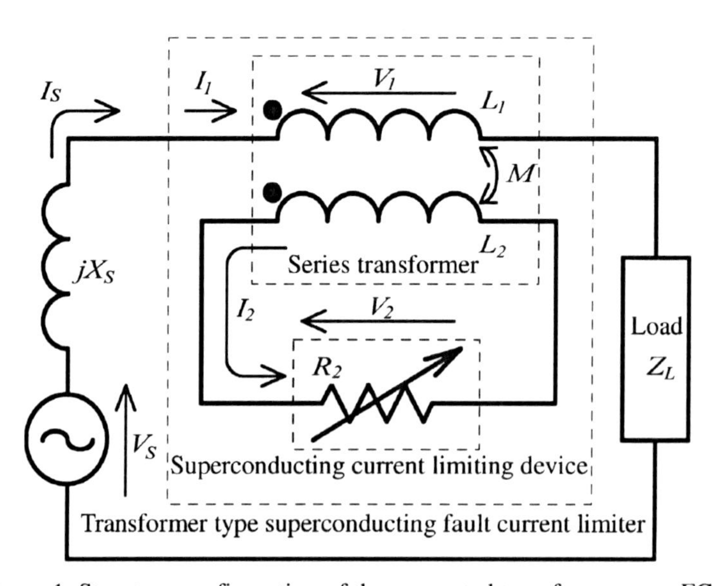

Figure 1: Structure configuration of the suggested transformer-type FCL.

In normal (no fault) condition, the current limiting device is maintained in the superconducting state, R_2=0 and the transformer's secondary side is "short-circuited." Consequently, the FCL shows a very low impedance, and it will not affect the main circuit.

After the fault happens, the S-N transition (transition from superconducting state to normal conducting state) of the current limiting device occurs. Due to this normal resistance, the FCL will show a high impedance (current limiting mode), and the fault current can be reduced to a limited value. According to Figure 1, the current-limiting impedance can be expressed as:

$$Z_{FCL} = [R_2\, j\omega L_1 + \omega^2 (M^2 - L_1 L_2)] / (R_2 + j\omega L_2) \tag{1}$$

The transformer type FCL has many advantages such as design flexibility of the current limiting device, and it may be suitable for a high-voltage electrical grid and can theoretically enhance the transient performance more efficiently.

2.2 Application of the FCL into a VSC-HVDC system

The schematic diagram of a VSC-HVDC system integrated with the FCL is shown in Figure 2, where the FCL is placed in series with the head of the DC transmission line. $E_{HVDC\text{-}R}$ and $E_{HVDC\text{-}I}$ are recorded as the two equivalent AC sources, and Z_{sR}/Z_{sI} is expressed as a model of Thevenin impedance of the AC rectifier/inverter power network.

After the fault appears, the VSI's power switches are immediately gated off, but the DC current is not interrupted, and the entire dynamic process is mainly consisting of two stages [8], [9]. For stage 1, it occurs as the dc-link capacitor discharging, and this stage is actually an RLC circuit until the positive dc voltage drops to below any grid phase voltage. In regards to stage 2, the main grid will contribute to feed the fault current. Once the fault is detected, the controlled switch will be timely triggered, and the limiting inductance L_2 will be inserted into the DC line for affecting the fault current during different stages. Detailed mathematical relationships are expressed as follows:

$$i_{\text{f-stage1}} = C\frac{\mathrm{d}u_{\mathrm{dc1}}}{\mathrm{d}t} = -\frac{I_0\omega_0}{\omega}e^{-\delta t}\sin(\omega t - \alpha) + \frac{U_0}{\omega L}e^{-\delta t}\sin(\omega t) \tag{2}$$

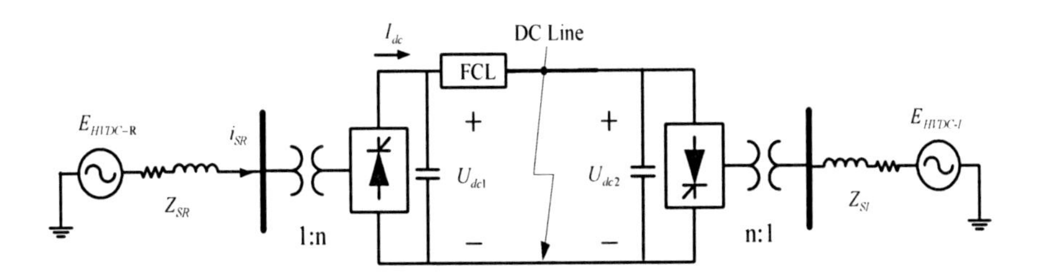

Figure 2: Schematic diagram of a VSC-HVDC system integrated with the FCL.

where $\delta = (R + R_g)/2L$, $\omega_0 = \sqrt{\delta^2 + \omega^2}$, $\alpha = \arctan(\omega/\delta)$, $\omega^2 = 1/LC - [(R + R_g)/2L]^2$. ($I_0/U_0$ is the given initial DC voltage/current. R, L denote the DC-line resistance and inductance, and R_g is the fault ground resistance),

$$\begin{pmatrix} \dot{u}'_{dc1} \\ \dot{i}_{f\text{-stage}2} \\ \dot{i}_{choke} \end{pmatrix} = \begin{pmatrix} 0 & -1/C & 1/C \\ 1/L & -(R_g + R)/L & 0 \\ -1/L_{choke} & 0 & 0 \end{pmatrix} \begin{pmatrix} u'_{dc1} \\ i_{f\text{-stage}2} \\ i_{choke} \end{pmatrix} + \begin{pmatrix} 0 \\ 0 \\ 1/L_{choke} \end{pmatrix} u_{g\,a,b,c} \quad (3)$$

where u'_{dc1}, $i'_{f\text{-stage}2}$, and i_{choke} are the state variables (L_{choke} denotes the equivalent choke inductance of the AC main grid), and a theoretical method of determining the short-circuit fault response is provided in (3). For the calculation of the limited DC-line current, "R_2," and "L_2" should be taken into account.

3 Simulation analysis

For purpose of quantitatively evaluating the transformer type FCL's performance behaviors in a VSC-HVDC transmission system, the simulation model corresponding to Figure 2 is created in PSCAD/EMTDC, and parts of simulation parameters are indicated in Table 1. During the simulation, different cases are imitated and stated.

3.1 Different current-limiting inductances

Simulation conditions are described as that, a grounded fault occurs in the DC line's middle segment at t=2 s, and the fault grounded resistance is R_g = 1 Ω (duration of the fault is set as 50 ms). Figure 3 shows the response characteristics of the DC-line fault current (near the VSC system's rectifier side). The DC-line current's normal level is about 0.48 kA, and after the fault, its first peak value will increase to 2.51 kA under the condition without any limiting devices.

Table 1: Main parameters of the VSC-HVDC system integrated with the FCL.

VSC-HVDC Transmission System	
V_{DC}/P_{DC}	110 kV/75 MW
$Z_{S\text{-}R}/Z_{S\text{-}I}$	(4.58+j26.05) Ω/(5.15+j25.81) Ω
$Q_{f\text{-}R}/Q_{f\text{-}I}$	15 Mvar/15 Mvar
DC line length	200 km
Transformer Type FCL	
Coil ratio (L_1/L_2)	4
Coupling coefficient	0.99
Secondary inductance	40 mH/80 mH/160 mH
superconducting device	3 Ω

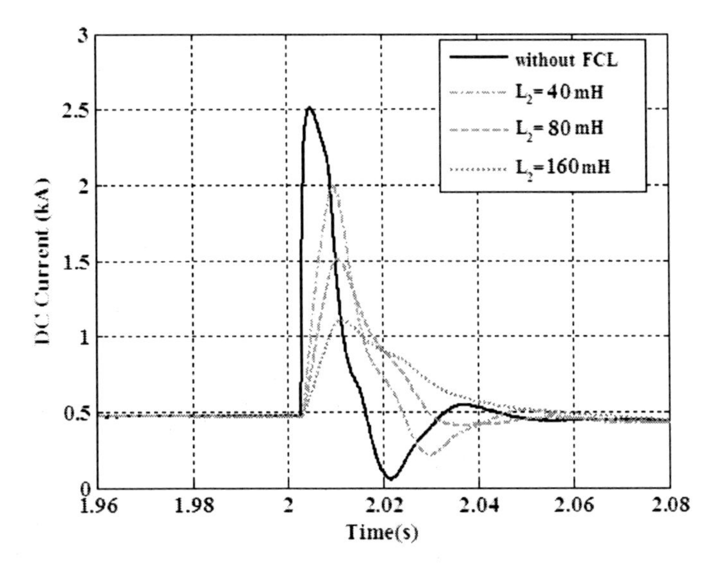

Figure 3: Dynamic response characteristics of the DC-line fault current.

When the FCL is employed, the fault current's rate of rise can be suppressed effectively. For that the inductance L_2 is, respectively, set as 40, 80 and 160 mH, the fault current's first peak value will be limited to 1.94, 1.54 and 1.15 kA. Compared with the value of 2.51 kA without FCL, the inhibition rate is, respectively, calculated as 22.7%, 38.6% and 54.2%.

Figure 4 signifies the waveform of the voltage across the FCL's two terminals. This terminal-voltage under normal condition is approximate to zero, and it will rise with the appearance of the short-circuit current. For that the inductance L_2 is respectively set as 40, 80 and 160 mH, the terminal-voltage's first peak amplitude can be up to 38.2, 43.4 and 47.8 kV, respectively. Moreover, since the FCL is placed in series with the DC line, it can also provide a compensation effect on the DC-bus voltage sag.

3.2 Different fault locations

On this occasion, it is assumed that the distance between the fault location and the rectifier side's outgoing interface changes step by step, and the other fault conditions remain unchanged (L_2=80 mH is adopted). Figure 5 shows the current-limiting characteristics of the suggested FCL under different fault locations, and Table 2 expresses the detailed data.

When the fault distance is set as 40, 80, 120 and 160 km, respectively, the expected inhibition rate will be calculated as 49.7%, 44.3%, 41.5%, and 40.1%, respectively. Note that, although the FCL's positive effects are in the tendency of decline along with the rise of the fault distance, the decline is acceptable.

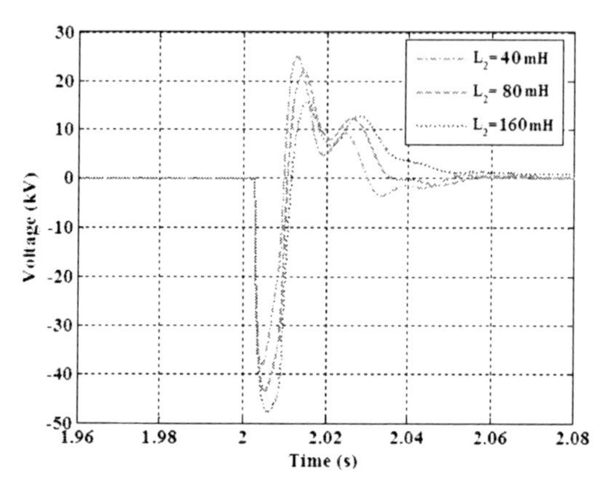

Figure 4: Main voltage across the FCL's two terminals under different current-limiting inductances.

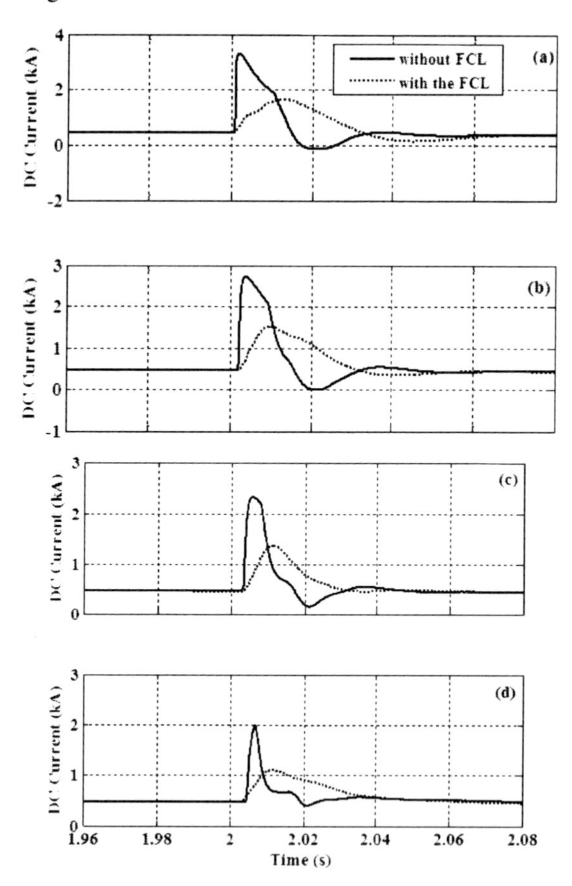

Figure 5: Current-limiting characteristics of the FCL under different fault distances: (a) 40 km, (b) 80 km, (c) 120 km and (d) 160 km.

Table 2: Influence of the fault distance on the FCL's performance behavior.

Fault Distance (km)	First Peak Value of the DC Fault Current (kA)	
	Without FCL	With the FCL
40	3.31	1.68
80	2.73	1.56
120	2.34	1.37
160	2.02	1.22

4 Conclusions

In this paper, directing at the DC fault issue occurred in a VSC-HVDC power transmission system, the transformer type FCL is suggested as a technical means for mitigating the DC short-circuit current. Theoretical discussion and simulation analysis are carried out. The results demonstrate that, employing the FCL can effectively suppress the DC fault current's rate of rise and improve the DC voltage sag. Consequently, the VSC-HVDC system's robustness against the short-circuit fault can be well enhanced.

In the near future, regarding multi-terminal VSC-HVDC systems, the fault analysis will be carried out, and the FCL's engineering feasibility will be systematically accessed. The results will be reported in later articles.

References

[1] N. Flourentzou, V.G. Agelidis, G.D. Demetriades, VSC-based HVDC power transmission systems: An overview. *IEEE Transactions on Power Electronics*, **24(3)**, pp. 592–602, 2009.

[2] Lidong Zhang, Lennart Harnefors, H.-P. Nee, Modeling and control of VSC-HVDC links connected to island systems. *IEEE Transactions on Power Systems*, **26(2)**, pp. 783–793, 2011.

[3] A. Egea-Alvarez, F. Bianchi, A. Junyent-Ferre, et al., Voltage control of multiterminal VSC-HVDC transmission systems for offshore wind power plants: Design and implementation in a scaled platform. *IEEE Transactions on Industrial Electronics*, **60(6)**, pp. 2381–2391, 2013.

[4] Hiroshi Yamaguchi, Kazushi Yoshikawa, Masahiro Nakamura, et al., Current limiting characteristics of transformer type superconducting fault current limiter. *IEEE Transactions on Applied Superconductivity*, **15(2)**, pp. 2106–2109, 2005.

[5] H. Yamaguchi, T. Kataoka, An experimental investigation of magnetic saturation of a transformer type superconducting fault current limiter. *IEEE Transactions on Applied Superconductivity*, **19(3)**, pp. 1876–1879, 2009.

[6] Sayaka Oda, Sho Noda, Hideyoshi Nishioka, et al., Current limiting experiment of transformer type superconducting fault current limiter with rewound

structure using BSCCO wire in small model power system. *IEEE Transactions on Applied Superconductivity*, **21(3)**, pp. 1307–1310, 2011.

[7] G. Wojtasiewicz, T. Janowski, S. Kozak, et al., Experimental investigation of a model of a transformer-type superconducting fault current limiter with a superconducting coil made of a 2G HTS tape. *IEEE Transactions on Applied Superconductivity*, **24(3)**, pp. 560–1005, 2014.

[8] Jin Yang, John E. Fletcher, John O'Reilly, Short-circuit and ground fault analyses and location in VSC-based DC network cables. *IEEE Transactions on Industrial Electronics*, **59(10)**, pp. 3827–3837, 2012.

[9] D.M. Larruskain, I. Zamora, O. Abarrategui, et al., A solid-state fault current limiting device for VSC-HVDC systems. *International Journal of Emerging Electric Power Systems*, **14(5)**, pp. 375–384, 2013.

Research and application of the power supply reliability evaluation models based on the status of China's distribution network

Wan Lingyun[1], Wu Gaolin[1], Li Qiuhua[2], Song Wei[1], YaoQiang[1], Yue Xingui[1]
[1]*State Grid Chongqing Electric Power Research Institute, Chongqin, China*
[2]*State Grid Chongqing Electric Power Company, Yuzhong District, Chongqing, China*

Abstract

Recently, there do not exist a universal model, method and index system in the reliability evaluation of the distribution system, which brought a challenge of the analysis of the evaluation. The challenge lied in the different assumption and reduction techniques when evaluated the reliability. This paper proposed a model and method which not only enhanced the computing performance but also considered different factors that can sensitively influence the assessment. The result of the reliability evaluation of cases demonstrated the validity and engineering value of the proposed model.
Keywords: mid-voltage distribution network, distribution network model, facility outage model, reliability evaluation method.

1 Introduction

In recent years, with the development of China's economy and society, the users' demand for power supply reliability is higher and higher. The evaluation and analysis through only the statistics of the outage accidents is difficult to adapt to the high demand for power supply reliability. Since power supply reliability evaluation can effectively guide the power system planning, design, construction, reconstruction, operation and management, improve the power supply reliability of the system fundamentally and increase the investment profits of the power grids, more and more domestic power supply enterprises are carrying out or planning to carry out this work.

WIT Transactions on Engineering Sciences, Vol. 107, © 2015 WIT Press
www.witpress.com, ISSN 1743-3533 (on-line)
doi:10.2495/SEEEII140291

At present, the domestic mid-voltage distribution network reliability evaluation has the following problems [1–7]: (1) the influence of scheduled outage cannot be considered, or the model calculating the influence of scheduled outage greatly contradicts reality; (2) the common switching devices are not subdivided according to the functions and the established switch device model cannot accurately simulate different influences of circuit breaker, load switch and fuse on the outage process; (3) the excessive simplification of the outage and restoration processes cannot accurately reflect the actual outage process; (4) it fails to consider all factors affecting the index calculation results or the assumption simplification is not reasonable; (5) The reliability evaluation divorces from the power supply system reliability evaluation to a certain extent, and the achievements of the statistical evaluation on the power supply reliability are not fully utilized.

Thus, based on the reality of mid-voltage distribution network and power supply reliability management in China, this paper puts forward a set of power supply reliability evaluation models and methods of mid-voltage distribution network systematically, which can more accurately and flexibly calculate the power supply reliability indexes of mid-voltage network, take into account and analyze the influence of various factors on the power supply reliability, so as to exert an active role in the promotion and application of the power supply system reliability evaluation work in China.

2 Analysis of the outage process

China's mid-voltage distribution network can be divided into cable network, overhead wire network and hybrid network of cables and overhead wires. The commonly used distribution devices include bus, overhead line, cable line, ring main unit, cable connection box, breaker, disconnecting switch, car switch, load switch, fuse and distribution transformer.

2.1 Relevant definitions

Fault location-isolation time: the time from fault occurrence and fault isolation, unit: h.

Fault repairing time: the time from the outage caused by the device fault to the power restoration through device repairing or change, unit: h.

Fault outage-interconnection switch switching time: the time from the isolation of the fault point to the completion of the load transfer, unit: h.

Fault outage-load transfer time: the time from the occurrence of fault outage to the completion of load transfer, including the fault location and isolation time and the fault outage-interconnection switch switching time, unit: h.

Power restoration operation time at the upstream of the fault point: the time from the isolation of the fault point to the restoration of power supply after the re-connection of the switching devices on the upstream, unit: h.

Power restoration time on the upstream of the fault point: the time from fault occurrence to power restoration on the upstream of the fault point, including fault location and isolation time and the time of power restoration on the upstream of the fault point, unit: h.

Scheduled outage- isolation time: the time from the occurrence of the scheduled outage to the isolation of the scheduled feeder, unit: h.

Scheduled outage-interconnection switch switching time: the time from the isolation of the scheduled outage feeder to the completion of load transfer, unit: h.

Scheduled outage-load transfer time: from the occurrence of scheduled outage to the completion of load transfer, including the isolation of the scheduled outage to the scheduled outage-interconnection switch switching time, unit: h.

Power restoration operation time on the upstream of the scheduled outage feeder section: from the isolation of scheduled outage feeder to power restoration on the upstream after the re-connection of the switch devices on the upstream of the section, unit: h.

Power restoration time on the upstream of the scheduled outage feeder: from the occurrence of scheduled outage to power restoration on the upstream of the scheduled outage wire section, including scheduled outage isolation time and power restoration operation time on the upstream of the scheduled outage wire section, unit: h.

2.2 Analysis of the fault outage process

The feeder lines and fault shown in Figure 1 are taken as an example to analyze the process from fault outage to restoration of power supply. After the occurrence of the fault, the breaker on the upstream of the fault location is disconnected. Then, the fault location is identified and the isolation switches on both sides of the fault location are disconnected to isolate the fault. The power on the downstream of the line where the fault is located is restored through switching on the interconnection switch. After the repair of the fault, the original operation way is restored through the switch operation, so as to realize the power restoration of the line where the fault is located. The outage times of different parts of the line and the relationship among different times are shown in Figure 1. In this figure, the power restoration time of the line where the fault is located is the fault repairing time; the power restoration time of the upstream of the line where the fault is located is the power restoration time of the upstream of the fault; the power restoration time of the downstream is the fault outage load transfer time.

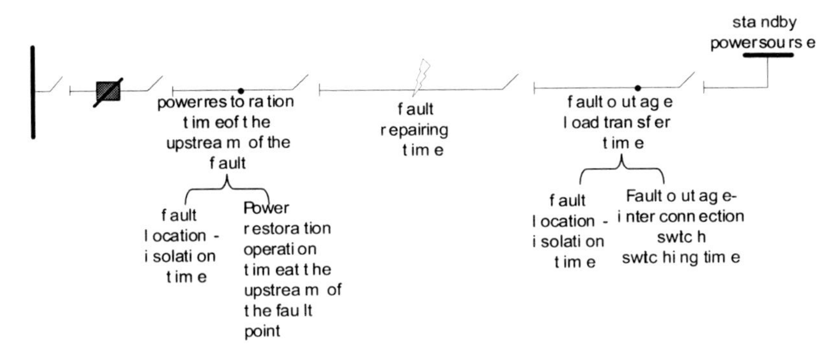

Figure 1: The fault process of transmission line.

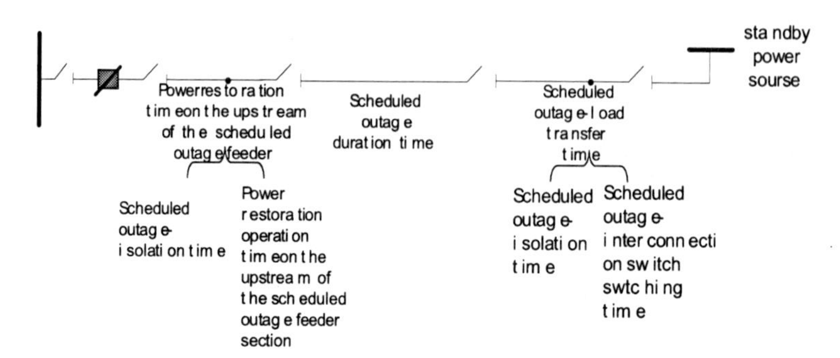

Figure 2: The scheduled outage process of transmission line.

2.3 Analysis of the scheduled outage process

Figure 2 shows the scheduled outage and power restoration process with the feeder line and scheduled outage work as an example. First of all, the breaker on the upstream of the outage line is disconnected. Then, the isolation switches on both sides of the scheduled outage lines are disconnected to isolate the line section. The power on the downstream of the scheduled outage line is restored through switching on the interconnection switch. After the end of the scheduled outage work, the original operation way is restored through the switch operation, so as to realize the power restoration of the scheduled outage line. The outage times of different parts of the line and the relationship among different times are shown in Figure 2. In the figure, the power restoration time of the scheduled outage line is the scheduled outage duration time; the power restoration time of the upstream of the scheduled outage line is the power restoration time of the upstream of the scheduled outage line; the power restoration time of the downstream is the scheduled outage load transfer time.

2.4 Explanation of the outage process

At present, most domestic and foreign reliability evaluation models ignore the power restoration time of the upstream of the fault point (the power restoration time of the upstream of the scheduled outage line). Such simplification directly causes the too ideal evaluation result of the power supply reliability indexes, especially for the power supply enterprises with poor management. The power restoration time of the upstream of the fault point may be so short for the enterprises with good management that it has little influence on the evaluation result if being ignored. However, the poorly managed enterprises may need quite a long time to restore the power supply of the upstream of the fault point, which has great influence on the evaluation result if being ignored. Thus, the power restoration time of the upstream of the fault point shall not be ignored (the power restoration time of the upstream of the scheduled outage line).

3 Reliability evaluation index system

The reliability evaluation index system can be divided into two categories, namely, load point indexes and system indexes. The commonly used load point indexes [6] include load point outage rate (λ_{LP}, time per year), load point outage time (u_{LP}, h per year), average service availability index-load point (ASAI-LP, %) and energy not supplied-load point (ENS-LP, kWh per year). System indices include system average interruption frequency index (SAIFI, time per household • year), system average interruption duration index (SAIDI, h per household • year), Average Service Availability Index (ASAI, %), energy not supplied (ENS, kWh) and average energy not supplied (AENS, kWh per household • year).

4 Calculation model

4.1 Distribution network model

In order to make full use of the historical data of the reliability statistical evaluation and establish an organic connection between reliability evaluation and statistical evaluation, the establishment of the distribution network model should be combined with the research results of the reliability statistical evaluation. According to Electric Reliability Management Code issued by the reliability center of China Federation of Electric Power Enterprises, China's power distribution devices for reliability of customer service in power supply system can be divided into nine categories, including overhead line, cable line, devices on the column, outdoor distribution transformer platform, box-type sub-station, civil sub-station, switching station (ring main unit), user equipment and unknown equipment. Each category is subdivided into several small classes [8]. Based on the Influence Table of the Distribution Devices on the Fault Outage Rate of the Equipment in Work Guidance of Customer Service Reliability in Power Supply System [9], the models of the main distribution devices can be established, so as to construct the distribution network model. The methods of establishing the models of different devices are as follows:

- The faults of the conductor of the overhead line and its accessories all affect the calculation of the fault outage rate of the overhead line. Thus, the conductor of the overhead line and its accessories are combined into the overhead line model. The cable line is treated similarly.
- Arrester, anti-bird devices, high voltage capacitor, high voltage metering box and voltage transformer on columns do not belong to the main electric connecting devices and have no influence on the topology of the network, so they can be neglected. Breaker, load switch, high voltage fuse and isolation switch on columns are connected in the network to directly affect the topology of the network and the outage, so they can be retained in the network model.
- The faults of transform platform, high voltage lead, low voltage distribution facilities and arrester in the outdoor distribution transformer platform do not affect the calculation of the fault outage rate of the transformer, so they are neglected. Only the transformer model is kept.

The following can be obtained after the similar analysis and treatment:

- Breaker, load switch, fuse and transformer in the box-type distribution substation are retained and common devices, box (wall), basic and transformer low voltage distribution facilities in the sub-station are ignored.

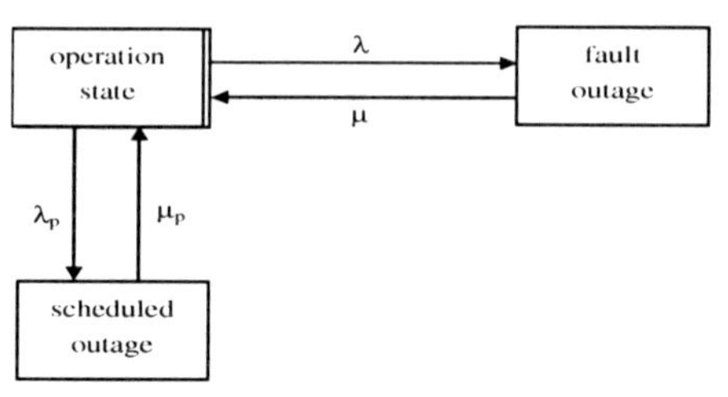

Figure 3: Three state model.

- Breaker, load switch, fuse, isolation switch, transformer in the civil distribution sub-station are retained and common devices, box (wall), basic and transformer low voltage distribution facilities in the sub-station are ignored.
- Breaker, load switch, fuse and isolation switch in the switching station are retained and common devices in the container, box (wall), basic and transformer low voltage distribution facilities in the station are ignored. The ring main unit is treated similarly.
- User devices and unknown devices are neglected.

Due to the different influences of breaker, load switch, isolation switch and fuse on the outage process, the models of those four switching devices are established, respectively.

Given all that, the distribution network model generally includes substation 10(6, 20) kV bus, overhead line, cable line, distribution transformer, breaker, load switch, isolation switch and fuse models and the connection relations.

4.2 Device outage model

The facility outage model is mainly divided into independent outage model and related outage model. Considering the complex calculation of the reliability evaluation of the mid-voltage distribution network as well as the difficulty in obtaining the reliability parameters, the related outage model is not adopted generally. At the same time, due to the large proportion of the scheduled outage in China at present, the influence of the scheduled outage must be taken into accounted. Thus, the three-state model should be adopted for the reliability evaluation.

The three-state model mainly considers the operation state, fault outage state and scheduled outage state of the devices. It can be simulated through the transition graph of the stable operation state-fault outage-scheduled outage state.

As shown in Figure 3, λ is the fault outage rate; λ_p is the scheduled outage rate, unit: time per year; μ is the repairing rate of the fault outage, namely, the inverse of the average time to repair the fault; μ_p is the repairing rate of scheduled outage, namely, the inverse of the average scheduled outage duration time, unit: time per year.

In addition, due to the small probability of concurrence of two and above independent outage incidents (i.e. the second-order and above outage incidents), only the first-order outage incident is considered in the reliability evaluation.

5 Reliability evaluation methods

5.1 Complex distribution network reliability evaluation methods

Mid-voltage distribution network reliability evaluation methods can be divided into two categories, namely, the simulation method and the analytical method. The former includes sequential Monte Carlo simulation method and non-sequential Monte Carlo simulation method. The latter includes failure mode and effects analysis method, minimal path method, Bayesian network method and Markov method [2]. Due to the slow convergence rate of the simulation method in calculation, Bayesian network method and Markov method are not suitable for the complex distribution network. Failure mode and effects analysis method and minimal path method enjoy fast calculation with high accuracy. In addition, they are suitable for the complex distribution network, so the complex distribution network reliability evaluation calculation should adopt failure mode and effects analysis method or minimal path method. In the actual application, in order to further increase the calculation efficiency, those two methods can be improved properly.

6 Theoretical example

The reliability evaluation is carried out on the distribution network shown in Figure 4. In the example, the capacity of the standby power is 1,500 kW; the type of the overhead line is LGJ-240; the type of the cable line is YJV-300; the type of the distribution transformer is S9; the load points are ranked to be e, d, c, d, a according to the importance degree. The rest parameters are shown in Tables 1 and 2. When doing reliability evaluation calculation, the followings are simplified: not considering the influence of the higher level power grid and the isolation switch faults on both sides of breaker (load switch). The distribution transformers connected by load points in Figure 4 are expressed by Ta, Tb, Tc, Td and Te, and the fuses are expressed by Fa, Fb, Fc, Fd, Fe and F5, respectively.

The reliability evaluation index system can be divided into two categories, namely, load point indexes and system indexes. The commonly used load point indexes [6] include outage time-load point (λ_{LP}, time per year), outage time-load point (u_{LP}, h per year), average service availability index-load point (ASAI-LP, %) and energy not supplied-load point (ENS-LP, kWh per year). System indices include system average interruption frequency index (SAIFI, time per household • year), system average interruption duration index (SAIDI, h per household • year), average service availability index (ASAI, %), energy not supplied (ENS, kWh) and average energy not supplied (AENS, kWh per household • year).

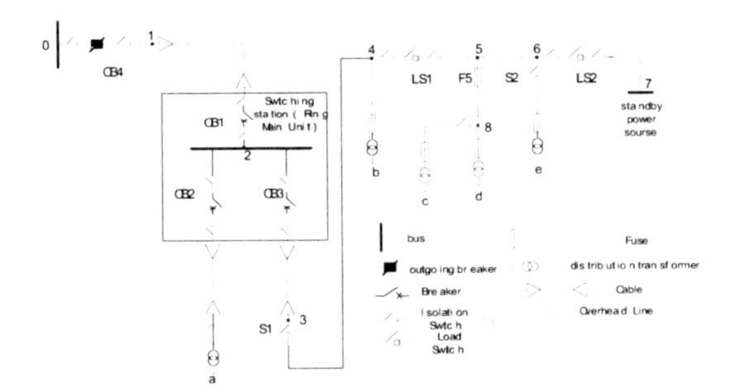

Figure 4: Network diagram of an example.

Table 1: Network data 1 of the example.

Device		L	λ	T	N	W
Supply Main	0-1	0.5	0.3	2		
	1-2	2	0.1	5		
	2-3	3	0.1	5		
	3-4	3	0.3	2		
	4-5	1	0.3	2		
	5-6	2	0.3	2		
	6-7	2	0.3	2		
	5-8	1	0.3	2		
Branch	2-a	1	0.1	5		
	4-b	1.5	0.3	2		
	8-c	1.5	0.3	2		
	8-d	0.5	0.3	2		
	6-e	1	0.3	2		
Breaker			0.25	3		
Load Switch			0.2	2.5		
Isolation Switch			0.25	2.5		
Fuse			0.2	2		
Transformer			0.35	4		
Load Point a					1	800
Load Point b					1	200
Load Point c					1	50
Load Point d					1	100
Load Point e					1	315

Table 2: Network data 2 of the example.

Device Category	Average Fault Location and Isolation Time (h)	Average Interconnection Switch Switching Time after Fault Outage (h)	Average Power Restoration Time of the Upstream of the Fault Section (h)
Switch	1	0.5	0.3
Device category	Average Scheduled Outage Isolation Time (h)	Average Interconnection Switch Switching Time after Scheduled Outage (h)	Average Power Restoration Time of the Upstream of the Scheduled Outage Section (h)
Switch	1	0.1	0.1
Device category	Scheduled Outage Rate (time per hundred km per year)	Average Scheduled Outage Duration Time (h)	–
Overhead (cable) line	6	7	–

Table 3: Reliability index of the load points.

Index	Load Points				
	a	*b*	*C*	*d*	*e*
Outage rate λ_{LP}	0.23	0.6585	1.094	1.094	0.8985
Outage time u_{LP}	0.4965	2.05815	2.78725	2.1707	1.9065
ASAI-LP (%)	99.994	99.977	99.968	99.975	99.978
ENS-LP	397.2	411.63	139.3625	217.07	600.55

Table 4: Reliability index of the system.

System Index	Value
SAIDI (h per household per year)	1.88382
ASAI	99.978%
SAIFI (time per household per year)	0.795
ENS (kWh)	1765.81
AENS (kWh per household per year)	353.162

Table 5: The reliability evaluation of the pilot areas.

Object	ASAI (%)	Improvement Measures	ASAI after Improvement
Xiamen A+ class area	99.9974	Introducing the second power source; increasing the cable operation and maintenance level etc.	99.9990
Dezhou A- class area	99.9885	Adding segmental switches; adding line interconnection	99.9910
Jiangbei A- class area	99.9750	Automation transformation; line cutting and change	99.9934
Tongchuan B-class area	99.9295	Ring network transformation; insulation transformation; strengthening the comprehensive outage management	99.9397
Tongliang C- class area	99.8890	Optimizing the network structure, changing the old equipments and strengthening the comprehensive outage management	99.9198

The reliability evaluation calculation is carried out with the evaluation models and methods proposed in this paper to obtain the reliability values of different load points and the system, as shown in Tables 3 and 4.

7 Actual engineering application

To verify the scientificity, adaptability and operability of the models and methods, the center of the city (A+ class: load density $\sigma \geq 30$ MW/km^2, realizing distribution automation), urban area (A class: $15 \leq \sigma < 30$ MW/km^2, realizing distribution automation, B class: $6 \leq \sigma < 15$ MW/km^2) and town (C class: $1 \leq \sigma < 6$ MW/km^2), are selected for application. The evaluation results of five pilot areas are shown in Table 5.

The application of the pilots shows that the theoretical power supply reliability levels and differences of different areas are basically consistent with the

actual statistics and the weak links identified by the software are consistent with the judgments of the experts, which verifies the scientificity, adaptability and operability of the models and methods proposed in the paper.

8 Conclusion

Based on the distribution network and power supply reliability management reality in China, considering the common problems existing in the mid-voltage distribution network reliability evaluation at present, this paper systematically puts forward a set of reliability evaluation models and methods. Its innovations mainly reflect in the following aspects:

- Based on the achievements of the customer power supply reliability statistical evaluation, it proposes new ideas and methods to establish the distribution device models and establishes the organic connection between reliability evaluation and reliability statistical evaluation.
- Based on the realistic mid-voltage distribution network operation and management in China, it proposes a model to calculate the influence of the scheduled outage according to the segmental scheduled outage mode. The calculation models can accurately reflect the influence of scheduled outage and shorten the distance between the theoretical models and the reality.
- It subdivides the functions of the commonly used switches in China and establishes different switch calculation models to accurately simulate the influence of breaker, load switch, breaker and fuse on the outage process.

The theoretical calculation and engineering pilot application verify the science, adaptability and operability of the models proposed in the paper. They have good engineering application and promotion value.

References

[1] Wenyuan Li, Risk *Assessment of Power Systems: Models, Methods and Applications*, 2006.
[2] Xie Kaigui, Yang Qunying, Wan Lingyun (compilers), *Theoretical Foundation of Power Reliability. China Electrical Power Press*, 2012.
[3] Chen Wengao, *Fundaments of Distribution System Reliability. China Electrical Power Press*, 1998.
[4] Guo Yongji, *Analysis of the Reliability of the Power System*. Tsinghua University Press, 2003.
[5] Fan Mingtian, Liu Jian (translaters), *Power Distribution Planning Reference*. China Electrical Power Press, 2013.
[6] Wang Chengshan, Luo Fengzhang, *Comprehensive Evaluation Theory and Method of Distribution System*. Science Press, 2011.
[7] Lan Yujun, Li, Wang Honggang, *Applied Techniques of Power Supply Reliability Management*. China Electrical Power Press, 2008.
[8] Power Reliability *Management Center of China Electricity Council, Power Reliability Management Code*. China Electrical Power Press, 2013.
[9] State Grid Corporation of China, *Working Guide for Power Supply Reliability for Customer Service in Power Supply System*. China Electrical Power Press, 2012.

Analysis of optimal allocation of FACTS for large AC and DC hybrid power system

Junyong Wu[1], Kaijun Lin[1], Yanmei Hu[1], Hongjun Fu[2], Fang Li[3], Xiaoxiao Yu[4]
[1]*School of Electrical Engineering, Beijing Jiao Tong University, Beijing, PR China*
[2]*State Grid Henan Electric Power Company, Zhengzhou, Henan, PR China*
[3]*China Electric Power Research Institute (EPRI), Beijing, PR China*
[4]*Beijing Electric Power and Economic Research Institute, Beijing, PR China*

Abstract

The multi-objective evolutionary algorithm based on grouping selection (MOEA-GS) is proposed in this paper, to solve the multi-type FACTS devices multi-objective optimal placement problem considering the system loadability and the cost of FACTS installations. In order to shrink the search space and speedup the optimizing process, the modal analysis approach is proposed to select the candidate buses for SVC installations, and the sensitivity analysis approach is utilized to select the candidate lines for TCSC installations. Application results on China Henan AC and DC hybrid power system show that MOEA-GS outperforms NSGA-II in terms of diversity-preservation, avoiding prematurity, and achieving the Pareto frontier with more complete distribution and better uniformity. The research achievements obtained in this paper provide the theoretical foundations for optimizing the power flow patterns and the enhancement of system loadability in China Henan Power Grid.

Keywords: FACTS, multi-objective evolutionary algorithm based on grouping selection (MOEA-GS), system loadability, Pareto frontier.

WIT Transactions on Engineering Sciences, Vol. 107, © 2015 WIT Press
www.witpress.com, ISSN 1743-3533 (on-line)
doi:10.2495/SEEEI140301

1 Introduction

As an important alternative approach, FACTS devices can improve the efficiency of the existing networks by re-dispatching line flow patterns in such a way that the thermal limits are not exceeded, while fulfilling contractual requirements between grid stakeholders and the increasing system loadability [1]. However, the benefits of these FACTS devices are severely dependent on their type, location, size and number in the transmission system [2]. Some researches focus on a single type of FACTS device to be sited at a given number of optimally chosen locations, such as: SVC [3], STATCOM [4], TCSC [5] and UPFC [6], while the others adopt a mix of different types of FACTS devices, such as SVC, TCSC, TCPST, TCVR and UPFC, to allow the benefits of each single type to be included. From a series of methods on the literature of multi-type FACTS devices multi-objective optimal placement problem, multi-objective evolutionary algorithms (MOEAs) such as NSGA [7] and NSGA-II [8] have the distinctive advantages to obtain the Pareto-optimal set, called Pareto frontier.

This paper proposes a novel multi-objective evolutionary algorithm, the MOEA-GS, to solve the multi-type FACTS devices multi-objective optimal placement problem considering the system loadability and the cost of installation. This approach is improved from NSGA-II. Considering the present FACTS application situations in China power systems and under the requirements of the utility, SVC and TCSC are comparatively sophisticated and widely installed, so only these two types of FACTS devices are included in this paper, to make the optimizing results more meaningful and practical to the utility managers.

2 Problem formulation

The goal of this paper is to maximize the system loadability and minimize the cost of installation of the selected FACTS devices simultaneously. These two objectives are contradictory to each other. This is a typical multi-objective optimization problem (MOP) under constraints.

Generally speaking, a MOP with k objectives, n decision variables and m constraints can be formulated as follows:

$$\begin{cases} \min\ y = f(x) = \left(f_1(x), f_2(x), f_3(x), \ldots, f_k(x)\right) \\ \text{s.t.}\ \ g(x) = \left(g_1(x), g_2(x), g_3(x), \ldots, g_m(x)\right) \le 0 \end{cases} \tag{1}$$

where $x \in R^n$ is the decision variables vector in the decision space X, $y \in R^k$ is the objectives vector in the objective space and $g(x)$ is the inequation constraints vector, which decides the available ranges of the decision variables.

2.1 Margin of system loadability

In order to assess quantitatively the effects of FACTS devices improving the system loadability and voltage stability, the margin of system loadability index is proposed as follows:

$$\begin{cases} \lambda = \dfrac{P_{L-\mathrm{limit}} - P_{L0}}{P_{L0}} \times 100\% \\ \max f_1 = \max(\lambda) \end{cases} \tag{2}$$

where P_{L0} and $P_{L-\mathrm{limit}}$ are the system total active power loads under the initial status and the static voltage stability limit status, respectively. This index is straight and simple, can take the system nonlinearity and various constraints under consideration, the larger the index the more powerful loadability of the system.

2.2 Cost of installation

For the electricity utilities, their economic benefits should be considered in the optimal placement problem. Under the requirements of the utility and considering the present FACTS application situations in China electricity utilities, SVC and TCSC are widely equipped already, as the second optimizing objective in this paper, the cost of these two types of FACTS devices can be formulated as follows:

$$\min f_2 = \min\left(\sum_{i=1}^{I} u_{isvc} Q_{isvc} + \sum_{k=1}^{K} u_{kTCSC} Q_{kTCSC}\right) \tag{3}$$

where f_2 is the cost of installation, i and I are the index and the total number of SVC, and k and K are the index and the total number of TCSC, respectively. Q_{isvc} and u_{isvc} are the capacity and unit capacity price of the ith SVC, Q_{kTCSC} and u_{kTCSC} are the capacity and unit capacity price of the kth TCSC, respectively.

3 Multi-objective evolutionary algorithm based on grouping selection (MOEA-GS)

3.2 Pareto operators

In the selection procedure of MOEA-GS, the Pareto operators are used, and they are necessary operators in MOEA-GS. The Pareto operators are defined as below:

1) *Non-dominated sorting*

In the non-dominated sorting, each individual is compared with every other individual in the same population according to their multiple fitness. If nobody else has higher margin of system loadability and lower cost of installation, this individual is so-called Pareto optimal individual and belongs to the first non-dominated frontier.

2) *Individual crowded distance*
The individual crowded distance is defined as the perimeter of the quadrangle with the two nearest individuals besides the selected individual. We prefer the lower one for compared individuals with the different ranks. For the individuals with same rank, their crowded distances are calculated, and sorted from larger to smaller, then all individuals in this population are sorted at last.

3) *Grouping selection strategy*
The Pareto non-dominated sorting is conducted at first, and then we get the average rank. Put the individuals whose ranks are below the average into the dominant group (*N*1), and the others go into the inferior group (*N*2). *N*1+*N*2=*N*, *N* is the population scale.

4) *Crossover ratio and mutation ratio*
The crossover ratio and the mutation ratio are self-adaptive genetic operators which are based on the ranks, and they help to perform deep global searching. They are defined as follows:

$$P_c = \begin{cases} P_{c1} - (P_{c1} - P_{c2})\dfrac{(\mathrm{R}_{better} - \mathrm{R}_{avg})}{(\mathrm{R}_{best} - \mathrm{R}_{avg})} & \mathrm{R}_{better} <= \mathrm{R}_{avg} \\ P_{c1} & \mathrm{R}_{better} > \mathrm{R}_{avg} \end{cases} \tag{4}$$

$$P_m = \begin{cases} P_{m1} - (P_{m1} - P_{m2})\dfrac{(\mathrm{R}_{best} - \mathrm{R}_{ind})}{(\mathrm{R}_{best} - \mathrm{R}_{avg})} & \mathrm{R}_{ind} <= \mathrm{R}_{avg} \\ P_{m1} & \mathrm{R}_{ind} > \mathrm{R}_{avg} \end{cases} \tag{5}$$

P_c is the crossover ratio, P_m is the mutation ratio, R_{avg} is the rank average, R_{best} is the minimum rank, R_{better} is the dominant rank between the two crossover individuals, R_{ind} is the rank of the individual to perform mutation, $P_{c1} = 0.9$, $P_{c2} = 0.6$, $P_{m1} = 0.1$ and $P_{m2} = 0.01$.

3.2 Procedure of MOEA-GS

To summarize the above descriptions, here is the flow chart of MOEA-GS. (Figure 1).

4 Application on China Henan Grid

4.1 Introduction of China Henan Grid

The case study in this paper is based on 2014 planning data of China Henan Grid, with 2,153 buses, 3,235 lines and 5.5×10^4 MW loads. The operation of the Ha-Zheng HVDC line will bring obvious changes to the power flow patterns of Henan Grid,

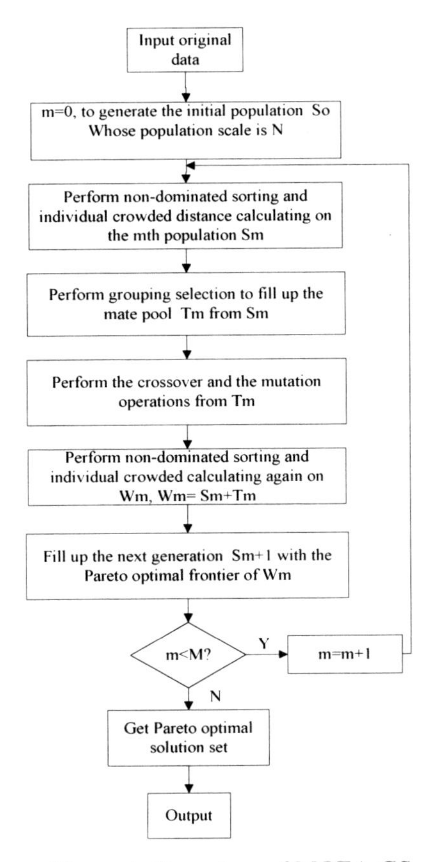

Figure 1: Procedure of MOEA-GS.

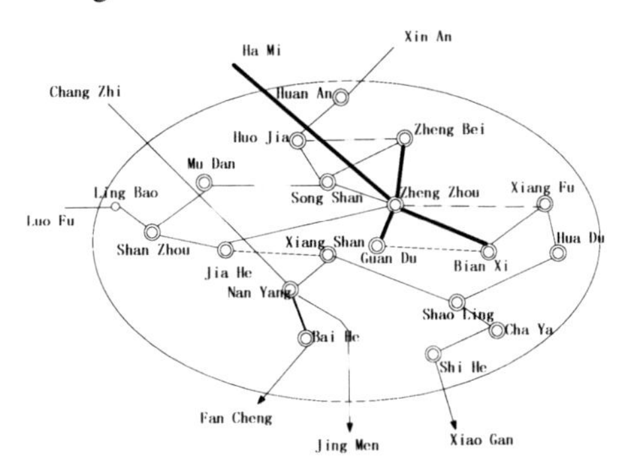

Figure 2: Diagram of henan power grid.

and there arises the necessity to optimize the power flow patterns with FACTS devices to enhance the loadability of Henan Grid within the security constraints. The diagram of Henan Grid is shown in Figure 2, in which the border black lines represent the Ha-Zheng $\pm$ 800 kV HVDC line and the corresponding 500 kV dispatching lines.

4.2 FACTS devices candidate buses and lines selection

4.2.1 Candidate bus selection for SVC installations

Based on the modal analysis approach presented in [9], the *Bus Participating Factors* of Henan Grid buses after Ha-Zheng HVDC put into operation are calculated, the first 10 buses with larger factors are listed in descending order in Table 1, which are considered to be suitable for SVC installation.

4.2.2 Candidate line selection for TCSC installations

There are so many transmission lines in the complicated Henan Grid, to simplify the problem, the 500 kV backbone network are considered for TCSC installations. This overhead transmission lines are LGJQ-300×4 , the thermal limit is 2,450 MW. Based on the sensitivity analysis approach presented in [10], the sensitivity coefficients of these 500 kV backbone lines are calculated, the first 10 lines with larger coefficients are listed in descending order in Table 2, which are considered to be suitable for TCSC installations.

Table 1: Bus participation factors.

Order	Bus Index	Bus Name	Bus Participating Factor
1	1541	Hehua 35-1	0.00545
2	1490	Hehua 110	0.00505
3	1591	Tongqiu 10-1	0.00494
4	1584	Shuizhai 10-2	0.00465
5	1583	Shuizhai 10-1	0.00463
6	1572	Mingzhong 35-1	0.00454
7	1501	Tongqiu 110	0.00453
8	1611	Zhaocun 10	0.00447
9	1499	Shuizai 110	0.00444
10	1551	Huaiyang 10-2	0.00441

Table 2: Branch sensitivity coefficients.

Order	Line Index	From Bus	To Bus	Coefficients
1	391320/391321	Huojia 500	Jiaozuo 500	3.0540
2	391119/391120	Baihe 500	Nanyang 500	2.2492
3	391318/391319	Xinbei 500	Jiaozuo 500	1.9329
4	300922	Shaolin 500	Xiangshan 500	1.9145
5	312239/312240	Huojia 500	Hebi 500	0.5147
6	392289	Shaolin 500	Zhanhe 500	0.4165
7	300845/391225	Tapu 500	Xiangfu 500	0.3794
8	391113/392178	Xiangfu 500	Huadu 500	0.3406
9	391223/319224	Shaolin 500	Huadu 500	0.2801
10	392202/392203	Xinbei 500	Boai 500	0.2201

4.3 Results of FATCS multi-objective optimal placement

4.3.1 Optimizing process of MOEA-GS

Based on the MOEA-GS proposed in Section 3, the multi-type FACTS devices multi-objective optimizations are conducted on the AC and DC hybrid Henan Grid considering the margin of system loadability and the cost of installation within the security constraints. The population scale is 200, the optimizing generation is 300, the SVC capacity is from −150 to 150 MVar, the compensating degree of TCSC is from −0.2 to 0.8. The obtained chaotic initial population is shown in Figure 3.

During the optimizing process of MOEA-GS, the evolutionary process of the Pareto frontiers of initial generation, 100, 200 and 300 generation are shown in Figure 4.

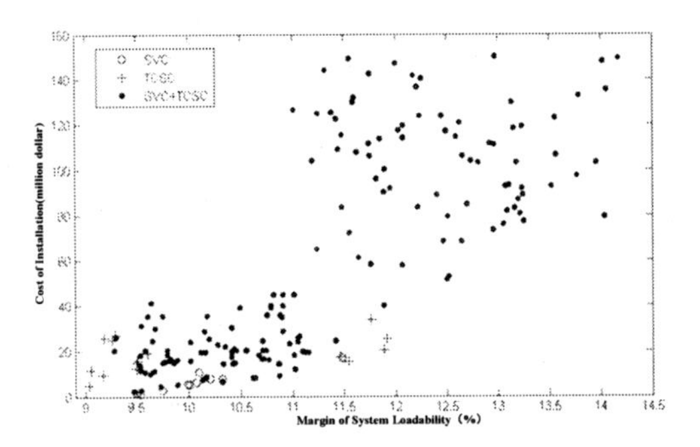

Figure 3: Chaotic initial population.

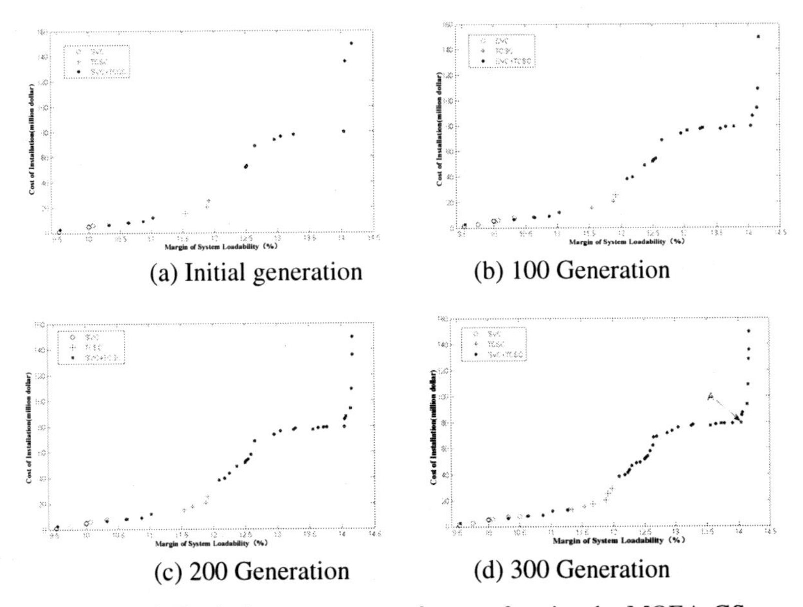

(a) Initial generation (b) 100 Generation

(c) 200 Generation (d) 300 Generation

Figure 4: Evolutionary process of pareto frontiers by MOEA-GS.

4.3.2 Comparison between MOEA-GS and NSGA-II

In order to compare the effects of the proposed MOEA-GS with the popular NSGA-II, NSGA-II is also conducted on the same problem, and the last optimizing results are compared and shown in Figure 5. It can be seen that the optimizing result of NSGA-II on this problem is not ideal, the distribution of the Pareto frontier obtained by NSGA-II is not completed, with worse diversity and uniformity. Comparatively, the results of MOEA-GS are with more complete distribution in the Pareto frontier, better diversity and uniformity, showing the better global searching ability and effectiveness.

4.3.3 Determination of the optimal placement scheme of FACTS

From Figure 5 we can see that, with the increase of the FACTS devices investment, the system loadability increases gradually, but has some limits, for example 14.3% for the maximum margin of system loadability in the current power flow pattern in Figure 5(a). Considering comprehensively both of the system loadability and the cost, solution A in Figure 5(a) maybe the best choice for the utility managers. Decoding the solution A and the details for this optimal placement scheme are shown in Table 3, in which zero means no need for FACTS devices installation.

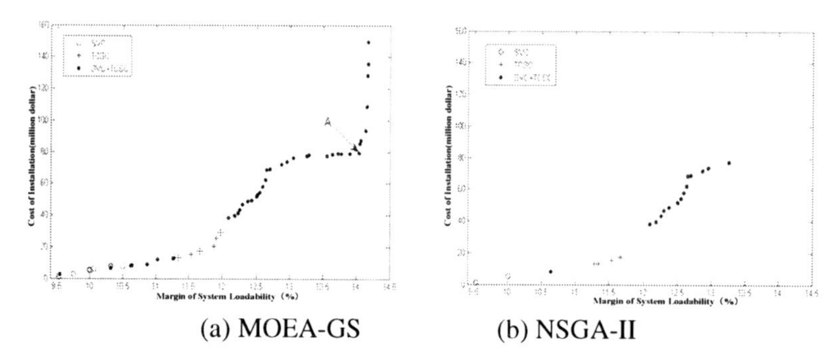

(a) MOEA-GS (b) NSGA-II

Figure 5: Comparison between MOEA-GS and NSGA-II.

Table 3: Optimal placement scheme of FACTS devices in solution A.

SVC		TCSC	
Bus Index	Capacity (MVar)	Line Index	Compensating Degree
1541	10	391320/391321	0.2
1490	80	391119/391120	0.2
1591	110	391318/391319	0
1584	50	300922	0.1

1583	10	312239/312240	0.1
1572	70	392289	0.5
1501	0	300845/391225	0.5
1611	20	391113/392178	0.5
1499	0	391223/319224	0
1551	0	392202/392203	0.7
Note: 1. The cost of this scheme is 79.23 million dollars; 2. The margin of system loadability increases from 9.26% to 14.04%, upgrading 4.78%, the maximum system loadability increases from 60,093 to 62,722 MW.			

5 Conclusions

A novel multi-objective evolutionary algorithm, the multi-objective evolutionary algorithm based on grouping selection (MOEA-GS) is proposed in this paper, to solve the multi-type FACTS devices multi-objective optimal placement problem considering the system loadability and the cost of FACTS installation. In order to shrink the search space and speedup the optimizing process, the modal analysis approach is proposed to select the candidate buses for SVC installations, and the sensitivity analysis approach is utilized to select the candidate lines for TCSC installations. MOEA-GS is improved from NSGA-II. Application results on China Henan AC and DC hybrid power system show that MOEA-GS outperforms NSGA-II in terms of diversity -preservation, avoiding prematurity, and achieving the Pareto frontier with more complete distribution and better uniformity. It has also been found that, with the increase of the FACTS devices investment, the system loadability increases gradually, but has some limits. The best FACTS placement scheme should balance both the system loadability and the cost of installations.

Acknowledgments

The authors thank the State Grid Henan Electric Power Company and China Electric Power Research Institute for their financial supports and their granted access to the data required for this study.

References

[1] N. Hingorani, Flexible AC transmission. *IEEE Spectrum*, **30(4)**, pp. 40–45, 1993.
[2] S. Rahimzadeh, M. Tavakoli Bina, A. Viki, Simultaneous application of multi-type FACTS devices to the restructured environment: Achieving both optimal number and location. *IET Generation, Transmission and Distribution*, **4(3)**, pp. 349–362, 2009.
[3] S.R. Najafi, M. Abedi, S.H. Hosseinian, A novel approach to optimal allocation of SVC using genetic algorithms and continuation power flow. In: *Proceedings of the 2006 IEEE International Power and Energy Conference*, pp. 202–206, November 28–29, 2006.
[4] E.N. Azadani, S.H. Hosseinian, M. Janati, et al., Optimal placement of multiple STATCOM. In: *Proceedings of the 2008 IEEE International Middle-East Conference on Power Systems (MEPCON'08)*, pp. 523–528, March 12–15, 2008.
[5] Y.-C. Chang, Multi-objective optimal thyristor controlled series compensator installation strategy for transmission system loadability enhancement. *IET Generation, Transmission and Distribution*, **8**, pp. 552–562, 2014.
[6] A. Kazemi, D. Arabkhabori, M. Yari, et al., Optimal location of UPFC in power systems for increasing loadability by genetic algorithm. In: *Proceedings of the 2006 IEEE Univ. Power Engineering Conference* (Vol. 2), pp. 774–779, September 6–8, 2006.
[7] N. Srinivas, K. Deb, Multi-objective function optimization using non-dominated sorting genetic algorithms. *IEEE Transactions on Evolutionary Computation*, **2(3)**, pp. 221–248, 1995.
[8] K. Deb, A. Pratap, S. Agarwal, et al., A fast and elitist multiobjective genetic algorithm: NSGA-II. *IEEE Transactions on Evolutionary Computation*, **6(2)**, 2002.
[9] Z. Li, R. Wang, W. Xing, Voltage stability analysis based on continuation power flow and modal analysis. *Electric Power Automation Equipment*, **29(9)**, pp. 81–84, 2009.
[10] T. Kamel, G. Tawfik, H.A. Hsan, et al., Optimal location and parameter setting of TCSC based on sensitivity analysis. In: *2012 First International Conference on Renewable Energies and Vehicular Technology*, pp. 420–424, 2012.

Grid energy efficiency evaluation system based on Beidou leading multimode systems

Xiaoming Li, Rui Song, Lingjun Yang, Peng Zhao
School of Electrical Engineering, Wuhan University, Wuhan, China

Abstract

To achieve sustainable energy development, energy saving became a serious problem in today's society waiting to be solved. We have to make a scientific evaluation of the grid first to act appropriately to reduce the energy loss for grid, only by establishing the correct evaluation of the power grid, we can take the right measures to improve the level of power grid. Considering the dynamic characteristics of the grid operation, this paper uses the Beidou timing system to accurate time and acquire the high-precision synchronous sampled data, evaluating the energy efficiency of the grid through the both sides of line losses and power quality effectively.
Keywords: energy efficiency, evaluation, Beidou timing system, energy loss, power grid, line loss.

1 Introduction

The premise work of network loss reduction and energy saving is to determine the efficiency level of the power grid, but the lack of energy efficiency evaluation system effectively is one of the current energy problems. Energy efficiency assessment system proposed in this paper considers the effect of both line loss and the quality of electric energy, through the analysis of the two, considering quality of the energy efficiency level of power grid [1].

To truly master power grid energy efficiency level, real-time dynamic evaluation of energy efficiency is necessary. The emergence of GPS is very good to improve this status. Through the whole network unified timing, the synchronous error of data will maintain in a microsecond or less. However, using a navigation system based on other countries' control is always risky. Beidou replaces GPS and becomes the main timing system of the power system will be good to avoid the threat and ensure the safe and reliable operation of the power grid.

www.witpress.com, ISSN 1743-3533 (on-line)
doi:10.2495/SEEEI140311

2 Index system of power grid real-time energy efficiency assessment

2.1 Concept and advantages

Beidou system has technical defects in user capacity, survival ability, data sharing and performance price ratio, therefore the single Beidou navigation system will be unable to meet the needs of the application, and it need to combine with other navigation systems to better play its advantages. Beidou multimode system is led by the Beidou navigation system, and provides all-weather, all-time positioning information of satellite navigation, with collaboration of GPS, GLONASS and Galileo satellite navigation systems. It makes the multi system compatible navigation becomes the development trend of future satellite navigation [2].

2.2 Demand analysis

The sources of errors are mostly data acquisition being not synchronous, as Beidou multimode system is developing, its application to SCADA system, can realize the high precision real-time synchronous acquisition of raw data. The index that affects the power quality level most is harmonic. Because the amplitude and frequency of the harmonic electric quantity are unchanged, the monitoring of harmonic is monitoring of phases of each electrical quantity. The main method of phase measuring is the zero crossing detection method, through comparison of the zero crossing time with 1PPS moment, we can achieve absolute phases of the each node voltage based on UTC absolute time.

3 Index system of power grid real-time energy efficiency assessment

3.1 Establishment of index system

For the real-time monitoring and analysis system of the power grid efficiency proposed in this paper, it is mainly used for real-time monitoring of running state of power network. Power grid efficiency real-time evaluation system established in this paper evaluates the energy efficiency level, mainly based on the loss and the power quality [3].

3.2 Real-time quantization of evaluation index

3.2.1 Loss index

(1) *The line loss rate*

According to the rating criteria, the proposal in this paper: "good" is $\eta_1 \leq 2\%$, "middle" is $2\% < \eta_1 \leq 5\%$ and "bad" is $\eta_1 > 5\%$.

(2) *Power factor qualification rate*

According to the rating criteria, the proposal in this paper: "good" is $\eta_2 \geq 80\%$, "middle" is $60\% \leq \eta_2 < 80\%$ and "bad" is $\eta_2 < 60\%$.

(3) *The load qualification rate*

According to the rating criteria, the proposal in this paper: "good" is $\eta_3 \geq 80\%$, "middle" is $60\% \leq \eta_3 < 80\%$, "bad" is $\eta_3 < 60\%$.

3.2.2 Power quality index

(1) Voltage deviation

According to the rating criteria, the proposal is shown in Table 1:

Table 1: Voltage deviation grade evaluation.

Grade	220 V	20 kV	More than 35 kV
Bad	$\Delta U\% < -10\%$ $\Delta U\% > 7\%$	$\Delta U\% < -7\%$ $\Delta U\% > 7\%$	$\Delta U\% < -10\%$ $\Delta U\% > 10\%$
Middle	$-10\% < \Delta U\% < -2\%$ $2\% < \Delta U\% < 7\%$	$-7\% < \Delta U\% < -2\%$ $2\% < \Delta U\% < 7\%$	$-10\% < \Delta U\% < -3\%$ $3\% < \Delta U\% < 10\%$
Good	$-2\% < \Delta U\% < 2\%$	$-2\% < \Delta U\% < 2\%$	$-3\% < \Delta U\% < 3\%$

(2) *Harmonic analysis*

According to the rating criteria, the proposal in this paper is shown in Table 2:

Table 2: Harmonic ratio grade evaluation table.

Grade	0.38 kV	6, 10 kV	35, 66 kV	110 kV
Bad	$THD_u > 5\%$	$THD_u > 4\%$	$THD_u > 3\%$	$THD_u > 2\%$
Middle	$2\% < THD_u < 5\%$	$2\% < THD_u < 4\%$	$1\% < THD_u < 3\%$	$1\% < THD_u < 2\%$
Good	$0 < THD_u < 2\%$	$0 < THD_u < 2\%$	$0 < THD_u < 1\%$	$0 < THD_u < 1\%$

(3) *Three-phase unbalance degree*

According to the rating criteria, the proposal in this paper: "good" is $\varepsilon_U > 2\%$, "middle" is $1\% < \varepsilon_U < 2\%$ and "bad" is $0 < \varepsilon_U < 1\%$.

4 Evaluation system for real-time energy grid based on fuzzy evaluation method

4.1 Implementation steps of the method of fuzzy comprehensive evaluation

Aimed at the power grid efficiency evaluation system proposed in this paper, the steps in the fuzzy comprehensive evaluation can be divided into the following steps. Firstly, formulate the ladder models of evaluation index; secondly, determinate single index membership degree of measures layers; thirdly, determine the single index scores of measures layers; fourthly, determine total ranking each of single factor of measures layers and finally, calculate the score and rating of comprehensive evaluation [4].

4.2 Determination of the membership model

The steps of determination of the single index fuzzy evaluation model are as follows: firstly, according to the property of single index, determine the index applicable model from three kinds of membership function model; secondly, according to the index value limit, divide the specific ranges of "good," "middle" and "bad" of membership function; thirdly, make the corresponding grade score F_1, F_2, F_3, and according to the index memberships degrees belonging to "good," "middle" and "bad" grade: μ_1, μ_2, μ_3 [5]. Each single index score formula:

$$F_{kj} = \frac{\mu_1(x)\times F_1 + \mu_2(x)\times F_2 + \mu_3(x)\times F_3}{\sum_{i=1}^{3}\mu_i(x)} \tag{1}$$

among them F_1, F_2, F_3 are, respectively, the indexes belonging entirely to "good," "middle" and "bad" grades and they are, respectively, 90, 75 and 50.

5 Computational examples and analysis

5.1 Raw data

In this paper, through MATLAB we built IEEE14 model to simulate the real empower grid operation and use real-time data that simulation model gathered to assess the power grid efficiency (Figure 1).

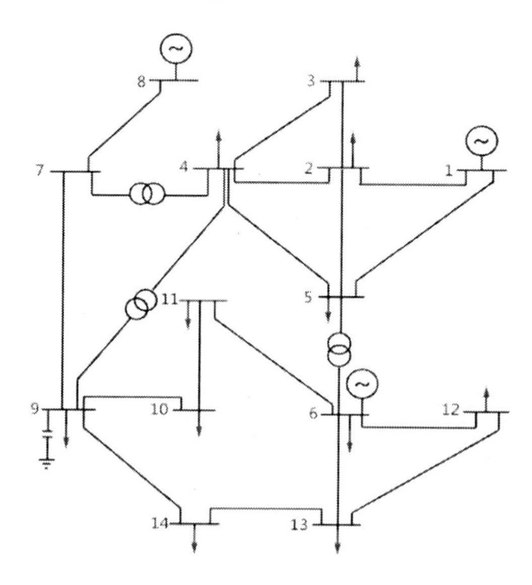

Figure 1: Grid structure diagram.

Table 3: Related data to net loss index.

Generator Data								
Generator G1			Generator G6			Generator G8		
P	Q	U	P	Q	U	P	Q	U
11	82	220	121	31	110	38	20	110
Load Data								
Load	P	Q	U	$\cos\varphi$				
Load 2	23.2	13.58	227.5	0.863				
Load 3	88.72	21.02	231.4	0.9731				
Load 4	51.63	−4.213	228.7	0.9958				
Load 5	6.407	1.709	227.4	0.9662				
Load 6	11.2	7.5	110	0.8309				
Line loss rate	$\eta=\frac{\Sigma P_{in}-\Sigma P_{out}}{\Sigma P_{in}}\times 100\%=3.87\%$							
Power factor qualification rate	$\eta=\frac{k_p}{k_\Sigma}\times 100\%=63.64\%$							
Qualified rate of load	$\eta=\frac{P_p}{P_{max}}\times 100\%=17.43\%$							

Table 4: Related data to power quality indicators.

Voltage Deviation, Harmonic Voltage and Three-Phase Unbalance Degree Data						
Points	Point 1	Point 2	Point 3	Point 4	Point 5	Point 6
ΔU (%)	3.41	5.19	3.95	3.37	0	2.55
THD_U (%)	0.9	0.3	2.93	0.6	0	2.378
ε_U (%)	0	0	0	0	0	0

Because the data are too much, Tables 3 and 4 list only the relevant data to energy efficiency evaluation index.

5.2 Assessment of the overall efficiency level of power grid

According to the scoring method, we can calculate the final score of each energy efficiency evaluation index, then according to the formula: $F=\sum_{k=1}^{n} w_k \times F_k$, we can calculate the final score of the electrical energy efficiency. The results of the power grid real-time energy efficiency assessment are shown in Table 5:

Table 5: Results of the power grid real-time energy efficiency assessment.

Energy Efficiency Index	Line Loss Rate	Power Factor Qualification Rate	The Qualified Rate of Load	Voltage Deviation	Total Harmonic Distortion	Three-Phase Unbalance Degree
w_k	0.236	0.097	0.083	0.104	0.135	0.149
F_k	68.825	56.075	50	86.30	65.334	50
Score	$F = \sum_{k=1}^{n} w_k \times F_k$=64.89					

So, real-time energy efficiency evaluation system proposed in this paper can reflect the real-time energy efficiency level of power grid operation and it is reliable to the evaluation of actual power network.

6 Conclusions

This paper is from the perspective of mastering grid dynamic efficiency and establishes the system of power grid real-time energy efficiency mainly based on energy quality index and loss index based on the principle that Beidou multimode system timing get high precision synchronous data. Real-time energy efficiency evaluation system proposed in this paper provides a theoretical basis for the dynamic analysis of power grid operation and it has the important practical significance in ensuring the economic operation of the power grid and enhancing power enterprise economic benefit and social benefit.

Acknowledgments

This research work was supported by the Hubei Province Key New Product New Technology Research and Development (Project No. 2012BAA2005), Jiangsu Province Natural Science Fund (Project No. BK2011347) and National Natural Science Fund (Project No. 51277134).

References

[1] Liu Xiaolong, Research on the Operation of Economic Evaluation of Distribution Network. North China Electric Power University, 2011.
[2] Yu Xiaofen, Fu Dai, A review of the comprehensive multi index evaluation method. *Statistics and Decision*, **(11)**, pp. 119–121, 2005.
[3] An Xiaobo, Research on Navigation Support Beidou/GPS Dual Mode Navigation Terminal. Chongqing University, 2012.
[4] Xiao Jun, Wang Chengshan, Zhou Min, Comprehensive evaluation of city power grid planning decision based on interval analytic hierarchy process. *Chinese CSEE*, **24(4)**, pp. 50–57, 2004.

[5] Li L., Shen L., An improved multilevel fuzzy comprehensive evaluation algorithm for security performance. *The Journal of China Universities of Posts and Telecommunications*, **13(4)**, pp. 48–53, 2006.

Power fault transient information extraction based on improved atomic decomposition

Xiaoming Li, Xianyong Yu, Xiaodong Deng, Peng Zhao
School of Electrical Engineering, Wuhan University, Wuhan, China

Abstract

The power fault transient information is the significant basis of fault identification and system restoration in a power system. In view of the problems of current power fault transient analysis, a new method of power fault transient information extraction based on improved atomic decomposition is proposed. The transient signal atomic functions library is built up according to the characteristic of power system transient signal. Signal parameters are estimated roughly with fast Fourier transform and wavelet transform. Then this method applies particle swarm optimization algorithm to find out the best matched atoms, extract the accurate parameters of the transient signal, and realizes power fault transient information extraction. Simulative results demonstrate the accuracy, adaptively and anti-noise capability of the proposed method.
Keywords: transient information, improved atomic decomposition, particle swarm optimization, parameters estimation.

1 Introduction

Analyzing and processing the power fault transient signal can provide theoretical basis for the power grid dispatching center to know the real situation of the failure [1, 2], to determine the cause of the problem, and to exclude the fault state after failure. At present, the improved Fourier iteration algorithm, wavelet analysis method, least square method and Kalman filter algorithm are commonly used for fault signal analysis. The improved Fourier iteration algorithm is restricted by signal synchronous sampling [3].

This paper proposes a method based on the improved atom-decomposition to extract fault transient information of electric power system. This method makes use of fast Fourier transform and wavelet analysis to estimate signal parameters

WIT Transactions on Engineering Sciences, Vol. 107, © 2015 WIT Press
www.witpress.com, ISSN 1743-3533 (on-line)
doi:10.2495/SEEEII140321

roughly and adopts the improved particle swarm optimization (PSO) algorithm to optimize the atomic decomposition process at the same time. The simulation results show that the method can effectively separated the fundamental component, the harmonic components and the fault transient components from the fault signal.

2 Improved atom-decomposition algorithm

Mallat and Zhang proposed the thought of atom-decomposition based on over-complete dictionary of atoms in 1993. It can choose basis function adaptively to realize the sparse decomposition of the signal [4].

The logistic chaotic iteration is adopted to create the initial values of other parameters (such as phase). Logistic map expression is as follows [5]:

$$Z_{n+1} = uZ_n(1 - Z_n) \tag{1}$$

where $0 < Z < 1$; $0 < u \leq 4$, Z, u are the logistic parameters. When the value of u is more close to 4, the value of Z is more average distribution between 0 and 1, the argotic is better. This paper sets $u = 4$.

Assume that the size of particle swarm is N, then randomly select Z_0, $Z_0 \in (0, 1)$ to carry on the logistic map $Zn+1=4Zn(1-Zn)$ (n=0,1,2,...). Select $Z_i, Z_{i+1}, \ldots, Z_{N+i-1}$ (i>30), a total of N values after many iterations. Map them to (a, b), the range of atomic parameter values $X=a+(b-a)Z$. For example, if the phase parameters φ scope for $(-\pi, \pi)$, the initial value selection of the phase is: $\varphi = (2Z - 1)\pi$.

Then the inertia factor ω in the PSO arithmetic should be corrected as follows:

$$\omega^k = \omega_{\max} - \frac{\omega_{\max} - \omega_{\min}}{iter_{\max}} \bullet k \tag{2}$$

where ω^k is the inertial factor after iterating k times. $\omega_{\max}$ is the maximum value of the inertia factor, $\omega_{\min}$ is the minimum value of the inertia factor. $iter_{max}$ represents the maximal number of iterations.

3 The fault transient information of electric power system extraction method based on the improved atom-decomposition

3.1 Characteristics of power fault transient signal

Fault transient signal model is set up as follows:

$$x = \sum_{m=1}^{p} A_m \sin(\omega_{1m}t + \phi_{1m}) + B_n e^{-\lambda_n t} \sin(\omega_{2n}t + \phi_{2n}) + N(t) \tag{3}$$

where x represents the fault transient signal, $N(t)$ represents for the noise. A_m, ω_{1m}, ϕ_{1m}, respectively, represent the amplitude, angular frequency and phase angle of the fault steady components, P represents for the number of steady-state components, B_n is on behalf of the maximum amplitude of transient component, $\lambda_n > 0$ is the attenuation coefficient, ω_{2n} is the oscillation frequency and ϕ_{2n} is the phase angle.

3.2 Function model of matching atom

(1) The function model of matching atomic for steady component is as follows:

$$g_{\gamma 0} = K_0 \sin\left(\omega_0 t + \varphi_0\right)(u(t - t_1) - u(t - t_2)) \tag{4}$$

where $\gamma_0 = [\omega_0, \varphi_0, t_1, t_2]$, respectively, represents the angular frequency, phase angle and start–stop time of the fault steady-state component. $u(t)$ is a step function, K_0 makes $\|g_{\gamma_0}\| = 1$, ensuring its unitization. Calculation of the amplitude of the matching signal is as follows:

$$A_0 \left|\left\langle x', g_{\gamma_0} \right\rangle\right| / K_0 \tag{5}$$

where x' represents the original signal this matching.
(2) The transient component matching function model is as follows:

$$g_{\gamma_1} = K_1 e^{-\lambda(t - t_3)} \sin(\omega_1 t + \varphi_1)(u(t - t_3) - u(t - t_4)) \tag{6}$$

where $\gamma_1 = [\lambda, \omega_1, \varphi_1, t_3, t_4]$, respectively, represents the attenuation coefficient angular frequency, phase angle and start–stop time of the transient component. K_1 makes $\|g_{\gamma_1}\| = 1$. Calculation of the amplitude of the matching signal is as follows:

$$A_1 = \left|\left\langle x', g_{\gamma_1} \right\rangle\right| / K_1 \tag{7}$$

4 The simulation analysis

To test whether the proposed method for power fault transient information extraction is effective in the form of (9), this paper chooses Gaussian white noise with signal-to-noise of 40 db. The sampling frequency is 10 kHz. The sampling data length is 1,024 points. The parameters corresponding to each component are shown in Table 1.

Table 1: The simulation signal parameters.

Components	Amplitude (A)	Frequency (Hz)	Phase Angle (rad)	Attenuation Coefficient	Begin Time (s)	Suspension Time (s)
1	1.000	50.00	0.000			
2	0.120	150.00	0.349			
3	0.100	250.00	0.524			
4	0.090	183.00	0.698			
5	0.800	500.00	0.524	100.0	0.0200	0.0500
6	1.250			50	0.0600	0.1000

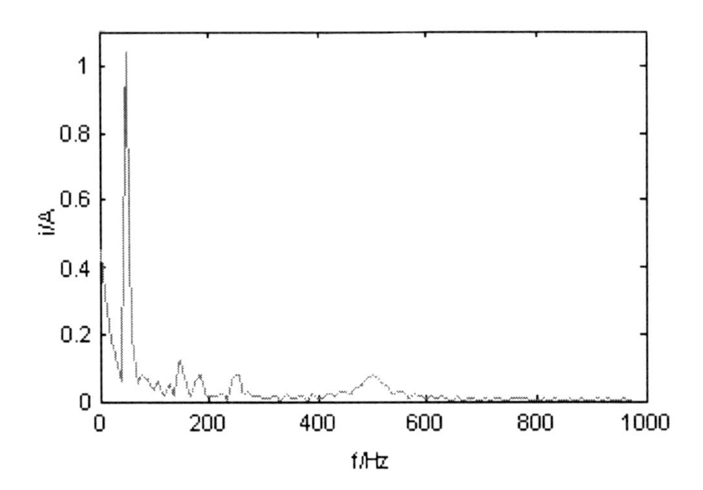

Figure 1: FFT spectrum diagram of the simulation signal.

Table 2: Signal parameter estimation based on the fourier transform.

Compo-nents	Amplitude (A)	Amplitude Error (%)	Frequeny (Hz)	Frequency Error (%)
1	1.038	3.8	48.83	2.3
2	0.1261	5.0	146.50	2.3
3	0.0819	18.1	253.90	1.6
4	0.0844	6.2	185.50	1.4
5	0.0808	89.8	498.00	0
6	0.4492	64.1	0	

The spectrum diagram of the signal after making the fast Fourier change is as shown in Figure 1.

The signal can be analyzed with the help of fast Fourier transform. The estimate parameters and the error analysis of the signal are shown in Table 2.

Table 2 shows that the signal component of each frequency can be identified through Fourier transform. However, because the sampling frequencies are not synchronized, the errors of the Fourier analysis for the amplitude and frequency of signal components are big. With the use of the wavelet base of "db4," the signal can be decomposed into three-layers wavelet. The results are shown in Figure 2.

In Figure 2, *a*3 represents for the low-frequency component of Layer 3. d3, d2 and d1 are on behalf of the high-frequency part of signal decomposition, respectively. The transient moment of the signal can be known through the layer 1 of high frequency, about 0.02 and 0.06 s.

According to the coarse estimation of signal parameters, the corresponding component parameters and error analysis can be obtained by using the method of sampling signals atomic decomposition proposed before. The results are as follows.

The extraction results of the component parameters are shown in Table 3. The method proposed in this paper can decompose all kinds of signal components including the inter-harmonic component and it can locate the start time of transient signal component more accurately without being limited by the sampling frequency. Table 4 shows that the proposed method can extract the parameters of each component in the complex synthetic signal more comprehensively and accurately and the error is small. The algorithm also has good noise resistance.

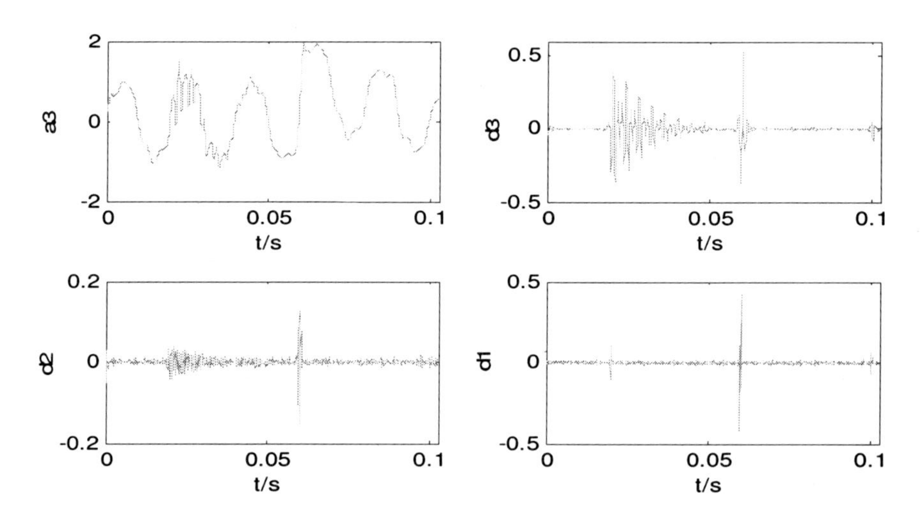

Figure 2: Signal decomposition based on the db3 wavelet.

Table 3: Atom-decomposition results of signal.

Compo-nents	Amplitude (A)	Frequency (Hz)	Phase Angle (rad)	Attenua-tion Coefficient	Begin Time (s)	Suspen-sion Time (s)
1	0.9992	49.948	0.0323			
2	0.1199	149.842	0.6869			
3	0.0989	250.294	0.4241			
4	0.0894	183.273	0.6561			
5	0.7869	500.385	0.2084	98.89	0.0200	0.0488
6	1.2613			50.58	0.0599	0.0999

Table 4: Error analysis of the atom-decomposition results.

Compo-nents	Amplitude (A)	Amplitude Error (%)	Frequency (Hz)	Frequency Error (%)	Attenua-tion Coefficient	Attenuation Coefficient error (%)
1	0.9992	0.08	49.948	0.10		
2	0.1199	0.083	149.842	0.11		
3	0.0989	1.1	250.294	0.12		
4	0.0894	0.67	183.273	0.15		
5	0.7869	1.64	500.385	0.07	98.89	1.11
6	1.2613	0.91			50.58	1.16

5 Conclusions

This paper proposed a kind of method based on improved atom-decomposition to extract electric power fault transient information. The method is proposed after studying the characteristics of electric power fault transient signal, and it is combined with fast Fourier transform and wavelet analysis to make the estimation of signal parameter. The simulation example shows that the improved atomic decomposition algorithm has the advantages of high accuracy, adaptive, and resistance to noise when extracting information of power fault transient. Whether the transient signal model established before is suitable for more complex power failure remains to be further researched.

Acknowledgments

This research work was supported by the Hubei Province Key New Product New Technology Research and Development (Project No. 2012BAA2005), Jiangsu Province Natural Science Fund (Project No. BK2011347) and National Natural Science Fund (Project No. 51277134).

References

[1] Liu Zhzchao, Huang Jun, Cheng Wenxin, Implementation of management information system for protective relaying and fault recorder. *Automation of Electric Power Systems*, **(1)**, pp. 72–75, 2003.

[2] Yang Guangliang, Yue Quanming, Yu Weiyong, et al., A fault classification method based on wavelet neural networks and fault record data. *Proceedings of the Chinese Society for Electrical Engineering*, **26(10)**, pp. 99–103, 2006.

[3] S.G. Mallat, Z. Zhang, Matching pursuits with time-frequency dictionaries. *IEEE Transactions on Signal Processing*, **41(12)**, pp. 3377–3415, 1993.

[4] J. Kennedy, R.C. Eberhart, Particle swarm optimization. In: *Proceedings of the IEEE International Conference on Neural Networks*, Piscataway, NJ, USA (Vol. 4). IEEE Press, pp. 1942–1948, 1995.

[5] Wang Ling, Zheng Dazhong, Li Qingsheng, Survey on chaotic optimization methods.*Computing Technology and Automation*, **20(1)**, pp. 1–5, 2001.

Study on optimization and control method to the voltage quality and reactive power of rural low voltage area

Li Xiaoming[1], Zhao Peng[1], Deng Xiaodong[1], Tian Zhen[2], Wangzhu Tao[3]
[1]*School of Electrical Engineering, Wuhan University, Wuhan, China*
[2]*School of Electronic Information and Electrical Engineering, Shanghai JiaoTong University, Shanghai, China*
[3]*Power Supply Company of Suizhou, Suizhou, China*

Abstract

The aim of this paper is to increase the power factor and reduce the loss of power grid. After studying the rural low voltage area power line intimately, we propose an optimization and control method to improve the voltage eligibility rate and reduce active power lose. The technology is about to use low-voltage reactive power dynamic compensation technology with zero transition in the low-pressure side of the transformer, whose rated capacity is larger than a specific one, and use the timing services of BeiDou Navigation Satellite System to coordinate the control process. In the high pressure side, make use of sensitivity analysis techniques and follow the principles of zonal compensation to compensate in the high sensitivity with the static compensation technology. The example shows that the proposed optimization and control technique can effectively improve the quality of the voltage along the rural low voltage area power line and reduce its total active power losses.
Keywords: rural area, optimal technology, reactive power compensation, voltage quality, line loss.

1 Introduction

Due to the aging power grid, problem of client's voltage quality is increasingly prominent while the power supply reliability cannot meet the requirements, and it has become one of the principal contradictions of the current power quality [1, 2].

www.witpress.com, ISSN 1743-3533 (on-line)
doi:10.2495/SEEEI140331

This paper is based on the main characteristics of rural low voltage areas. We propose an optimized technology to improve the voltage eligibility rate and reduce active power lose of rural low voltage areas power line. The technology is to use low-voltage reactive power dynamic compensation technology with zero transition in the low-pressure side of the transformer, whose rated capacity is larger than a specific one. In the high pressure side, taking use of sensitivity analysis techniques and follow the principles of zonal compensation to compensate in the high sensitivity with the static compensation technology. We make full use of complementary advantages of static and dynamic compensation.

2 Mathematical model

The objective functions of voltage quality of rural low voltage areas are set as follows:

$$\min\ P_L = f(Q) \tag{1}$$

where P_L is active network loss and Q is reactive power injected in the reactive power source node.

2.1 The sensitivity analysis model

Sensitivity analysis model [3] of network loss each node can be shown as follows:

$$\frac{\partial P_L}{\partial Q_i} = \sum_{k=1}^{n} \{ (2\sum_{j\in k} U_j G_{kj} \cos\theta_{kj}) \frac{1}{U_i G_{ik} \sin\theta_{ik}} \} \tag{2}$$

$$s.t. \begin{cases} Q_G - Q_L - Q_D - Q = 0 \\ U_i^{\min} \le U_i \le U_i^{\max} \left(i = 1,2,\ldots\ldots,n\right) \\ I_j \le I_j^{\max} \left(j = 1,2,\ldots\ldots,m\right) \end{cases} \tag{3}$$

In Eq. (3), Q_G is the total input reactive power from generator, Q_L is the line reactive power loss and Q_D is the total reactive load.

2.2 The dynamic reactive power compensation model

We firstly use dynamic compensation technology with zero transition [4].

As shown in Figure 1, the dynamic reactive power compensation can be applied to the low voltage side of the multiple-user and non-uniform power lines. If the nominal capacity of the *i*th transformer is higher than a specific capacity, we install low-voltage dynamic reactive power compensation device with zero transition on the low voltage side of it.

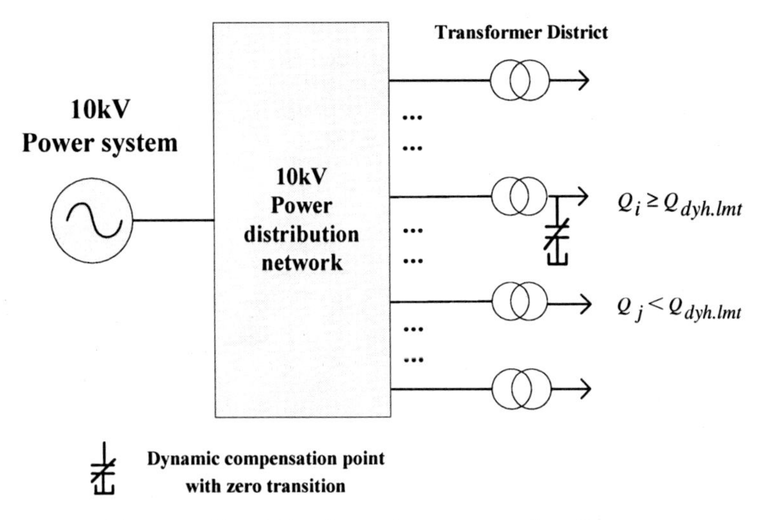

Figure 1: Simplified dynamic compensation diagram.

Introducing the 0–1 variable λ_i, and setting up the total capacity calculation model about dynamic reactive power compensation on the low voltage side as follows:

$$Q_d = \sum_{i=1}^{n} Q_i \lambda_i \tag{4}$$

$$\lambda_i = \begin{cases} 1, Q_i \geq Q_{dyh.lmt} \\ 0, Q_i < Q_{dyh.lmt} \end{cases} \quad (i = 1,2,\ldots\ldots,n) \tag{5}$$

Among them, Q_d is the total capacity of dynamic reactive power compensation, n is the total number of nodes on the low voltage side of the transformer, Q_i is the capacity of dynamic reactive power compensation on the low voltage side of the ith transformer and $Q_{dyh.lmt}$ is the minimum capacity limits for dynamic compensation optimization.

2.3 The static reactive power compensation partition model

This paper secondly proposed a method of high voltage side partition compensation [5], which not only satisfies the requirement of economy, but also can make the optimization effect of the voltage quality and line loss best (Figure 2).

Supposing $j \in [1,M]$ and M is the total number of the areas, Q_j is the capacity of reactive power flowing into area j. So the capacity of the static compensating device installed in the installed node of area j can be calculated as follows:

$$Q_{Sj} = Q_j - Q_{dj} \tag{6}$$

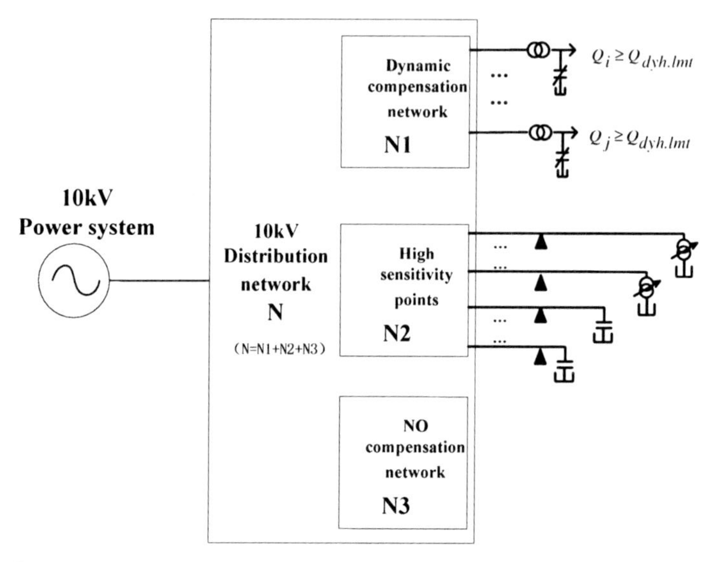

Figure 2: Simplified partition diagram of multiple-user power lines.

Introducing the 0–1 variable λ_j, and setting up the total capacity calculation model about static reactive power compensation on the high voltage side as follows:

$$Q_S = \sum_{j=1}^{M} Q_{sj} \lambda_j \tag{7}$$

$$\lambda_j = \begin{cases} 1, S_j \geq S_{Jyh.lmt} \\ 0, S_j < S_{Jyh.lmt} \end{cases} \tag{8}$$

Among them, Q_s is the total capacity calculation model about static reactive power compensation on the high voltage side. S_j is the total capacity of all transformers contained area j. $S_{Jyh.lmt}$ is the minimum capacity limits for static compensation optimization.

3 Optimization techniques for voltage quality and line loss of multiple-user line

This paper proposes a method that combines two compensation techniques—the optimization method for voltage quality and line losses.

As shown in Figure 3, we choose to install dynamic reactive compensation device at the low-pressure side of some large-capacity transformer (e.g. capacity

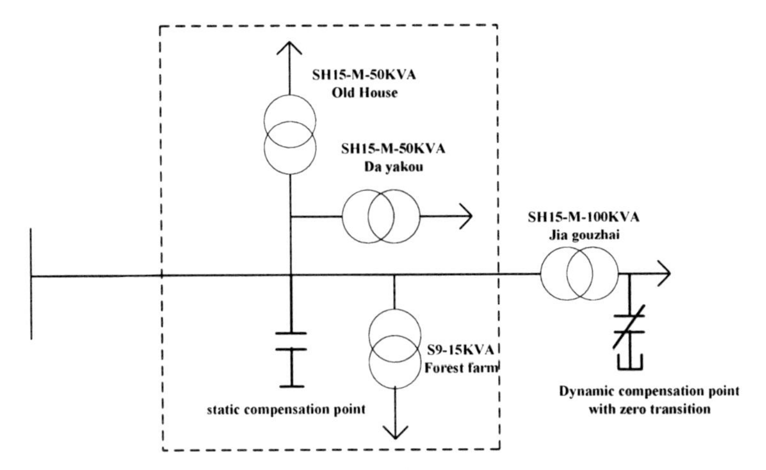

Figure 3: Simplified installation diagram of the optimization technology.

Table 1: Comparison before and after optimization under different load conditions.

Load	Voltage Qualification Rate (%)	Losses Ratio (%)	Maximum Voltage Deviation (%)	Total Capacity (kVar)
1610 kW before compensation	30.27	37.69	14.4	0
1610 kW after compensation	100	10.78	5	1751.17
0.6 SN before compensation	15.68	30.31	30.8	0
0.6 SN after compensation	100	18.08	5	4010.02
0.7 SN before compensation	3.72	44.17	34.8	0
0.7 SN after compensation	100	22.08	5	4912.97

above 100 kVA), the remaining reactive power vacancy can be compensated by static compensation devices, based on sensitivity analysis and zoning principles.

4 Computational examples and analysis

The 053 lines of 10 kV distribution network in a city of Guangdong province belong to typically multiple-user and non-uniform power lines. Considering the operating conditions separately of maximum load 1610 kW, 0.6 SN (SN is the total load of all the transformers while they are in rated operation, and the unit is kW) and 0.7 SN, this paper uses the optimization method and the result is shown in Table 1.

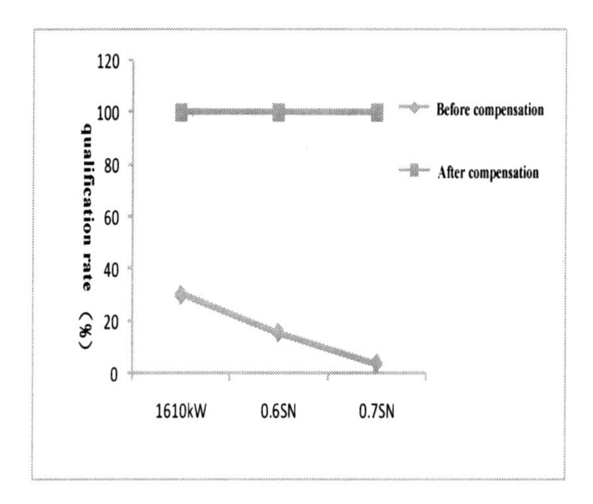

Figure 4: Comparison of voltage qualification rate before and after optimization.

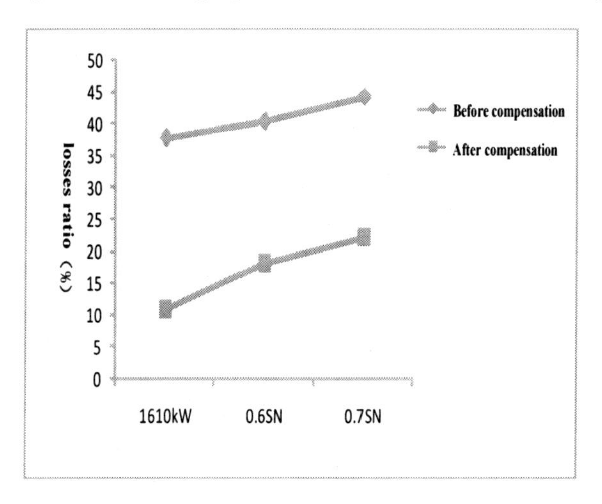

Figure 5: Comparison of line loss rate before and after optimization.

Under different load conditions, the voltage qualified rate and line loss whether the technology is taken or not are shown separately in Figures 4 and 5.

The voltage quality is raised from previously severely substandard to a pass rate of 100%, and the line loss is also obviously reduced.

5 Conclusions

In this paper, aiming at the high line loss and low voltage qualified rate of rural low voltage areas 10 kV power lines, the method to optimizing voltage quality and reactive power is proposed. This method combines low-voltage reactive power dynamic compensation technology with zero transition with static reactive power compensation technology to complement each other's advantage. This method can be applied to that power line which has large users, small capacity, long distance and low load rate. In order to reduce line loss and improve voltage quality, it has broad application prospects.

Acknowledgments

This research work was supported by Hubei Province Key New Product New Technology Research and Development (Project No. 2012BAA2005), Jiangsu Province Natural Science Fund (Project No. BK2011347) and National Natural Science Fund (Project No. 51277134).

References

[1] Sun Hongbin, Zhang Boming, Guo Qinlai, et al., Design for global optimal voltage control system based on soft identification of secondary control zones. *Automation of Electric Power Systems*, **27(8)**, pp. 16–20, 2003.
[2] Tan Dongming, Studies on Intelligent Technology of Reactive Power Optimization in Rural Distribution Power Network. Information and Electrical Engineering College of Shenyang Agricultural University, Liaoning, pp. 1–126, 2011.
[3] Yu Jianming, Du Gang, Yao Lixiao, Application of genetic algorithm combining sensitivity analysis to optimized planning of reactive power compensation for distribution networks. *Power System Technology*, **26(7)**, pp. 46–49, 2006.
[4] Standardization Administration of the People's Republic of China, GB/T 25839-2010. *Low Voltage Reactive Power Dynamic Compensation Equipment with Zero Transition*. China Standard Press, Beijing, 2010.
[5] Sun Yuanzhang, Wang Zhifang, Lu Qiang, The effect of SVC on the voltage stability. *Proceeding of the CSEE*, **17(6)**, pp. 373–376, 1997.

Electrical fast transient/burst study of the operation of relays

Wang Yu-feng, Guo Ren-zhao, Sun Peng
University of Science and Technology LiaoNing, AnShan, China

Abstract

For relay-protection of power system, EFT/B (electrical fast transient/burst) often disturbs the microprocessor protection devices. By the analysis of switch operations, EFT/B was measured when control relay operated. Disturbance observed in output relay coil interfered badly microprocessor protection devices. The results indicate that EFT/B can be controlled effectively from interference source and coupling path.
Keywords: EFT/B, microprocessor protection device, relay.

1 Introduction

Microprocessor protection devices based on technology of computer and micro-electronics are applied widely in power systems with improvement of power system autoimmunization and its reliability threatens power system security. Electrical fast transient (EFT/B) with relay operations often caused malfunction of microprocessor protection devices. Electromagnetic interference such as high-voltage, surge and electrostatic discharge can be controlled effectively now. But EFT/B is still a problem for microprocessor protection devices because it has such features as short rise-time, short duration, high voltage and high repetition frequency [1–7].

Countermeasure was used to protect microprocessor protection devices from EFT/B through analyzing EFT/B's forming process caused by relay operations.

2 EFT/B's forming process

Equivalent circuit model for EFT/B caused by switch operations is shown in Figure 1 [8]. L_0, R_0 and C_0 are stray inductance, stray resistance and parasitic

WIT Transactions on Engineering Sciences, Vol. 107, © 2015 WIT Press
www.witpress.com, ISSN 1743-3533 (on-line)
doi:10.2495/SEEEI140341

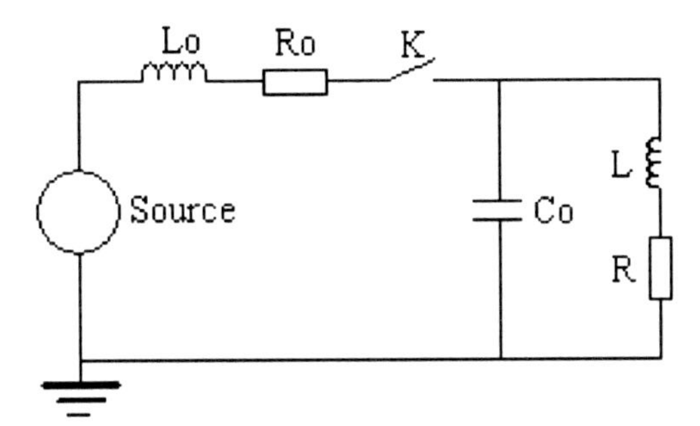

Figure 1: Equivalent circuit model for EFT/B.

capacitance, respectively. Parasitic capacitance is charged because current cannot chop in inductance while switching off inductive load. And over-voltage occurs between switch contacts. Switch are restrict while voltage between switch contacts is higher than the medium recovery voltage. Parasitic capacitance discharge and switch are extinct. Over-voltage occurs between switch contacts again. The above process occurs again and again until parasitic capacitance voltage cannot break down switch contacts.

3 Study EFT/B in secondary circuit

Electromagnetic relays were applied widely in secondary circuit and its operating voltage is low. But, EFT/B caused by electromagnetic relay operation interfered with microprocessor protection devices badly because relays are close to microprocessor protection devices [9].

Generally, microprocessor protection devices perform tripping operation, alarm and junction closure through controlling auxiliary relay and contactor with output relay [10]. EFT/B caused by tripping operation of DC output relay was analyzed.

SRD-24VDC-SL-C and JQX-13F were used as output relay and auxiliary relay, respectively. Voltage in output relay coil was measured when output relay tripped. Figures 2(a) and (b) show output relay coil voltage when output relay carry no load and trip auxiliary relay, respectively. It can be seen that transient over-voltage occurs while disconnecting output relay coil because coil is inductive load. Comparing Figure 2(a) with 2(b), there is more transient over-voltage, while output relay tripping auxiliary relay than carrying no load. And the redundant transient over-voltage is named RTOV.

Voltage waveforms shown in Figure 3 reflected the relationship between the DC power supply and the output relay coil voltage of microprocessor protection devices. It can be seen that DC power supply was interfered mainly by RTOV.

The relationship between coil voltage and contact current of output relay is shown in Figure 4. Contact current was measured with TEK A621 current probe. Measuring range of 1,000 A was adopted here and conversion relationship between voltage and current is 1 mV/A. It can be seen that RTOV is induced by contact current.

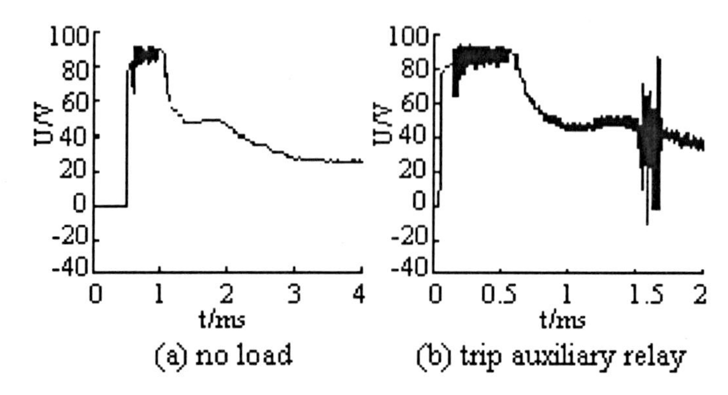

Figure 2: Voltage waveforms of output relay coil.

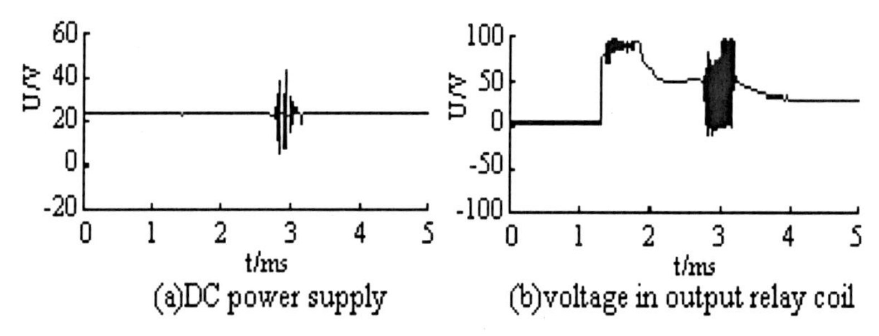

Figure 3: DC power supply and output relay coil voltage.

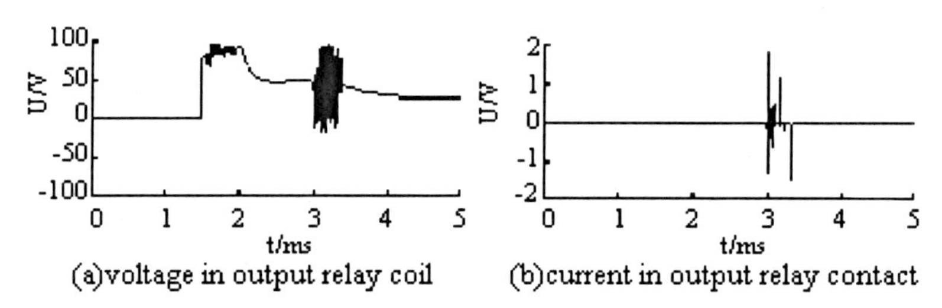

Figure 4: Coil voltage and contact current of output relay.

Output relay consists of coil and contact, and its model is shown in Figure 5. i:contact current, dl: contact unit, $P(x,y,z)$: solution point, $P_1(x_1,y_1,z_1)$: source of current unit, r: distance from source of current unit to solution point. Time change rate of magnetic flux coupling inside coil is big and RTOV is induced when transient current flow through contacts.

Vector magnetic potential at $P(x,y,z)$ when transient current of i flow through contacts:

$$\vec{A} = \frac{\mu_0}{4\pi} \int \frac{i\,\mathrm{d}\vec{l}}{r} \tag{1}$$

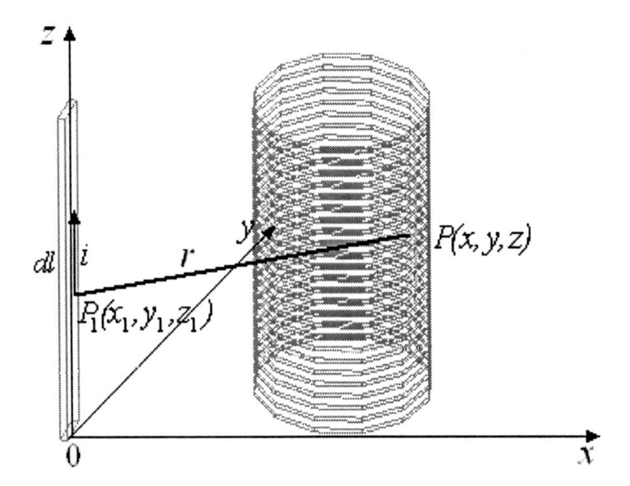

Figure 5: Output relay model.

μ_0: vacuum permeability and l: integral route of contact. Each component of vector magnetic potential:

$$\begin{cases} A_x = \dfrac{\mu_0}{4\pi}\displaystyle\int \dfrac{i\,\mathrm{d}x}{r} = \dfrac{\mu_0 i}{4\pi}\displaystyle\int \dfrac{\mathrm{d}x}{\sqrt{(x-x_1)^2+(y-y_1)^2+(z-z_1)^2}} \\ A_y = \dfrac{\mu_0}{4\pi}\displaystyle\int \dfrac{i\,\mathrm{d}y}{r} = \dfrac{\mu_0 i}{4\pi}\displaystyle\int \dfrac{\mathrm{d}y}{\sqrt{(x-x_1)^2+(y-y_1)^2+(z-z_1)^2}} \\ A_z = \dfrac{\mu_0}{4\pi}\displaystyle\int \dfrac{i\,\mathrm{d}z}{r} = \dfrac{\mu_0 i}{4\pi}\displaystyle\int \dfrac{\mathrm{d}z}{\sqrt{(x-x_1)^2+(y-y_1)^2+(z-z_1)^2}} \end{cases} \tag{2}$$

Because i only has Z direction component, vector magnetic potential at $P(x,y,z)$:

$$\vec{A} = \frac{\mu_0 i}{4\pi}\int \frac{\mathrm{d}z}{\sqrt{(x-x_1)^2+(y-y_1)^2+(z-z_1)^2}}\,\vec{a_z} \tag{3}$$

Magnetic flux density at $P(x,y,z)$:

$$\vec{B} = \nabla \times \vec{A} = \frac{\partial A_z}{\partial y}\vec{a_x} - \frac{\partial A_z}{\partial x}\vec{a_y} \tag{4}$$

Induction field in coil:

$$\nabla \times \vec{E} = -\frac{\partial \vec{B}}{\partial t} \tag{5}$$

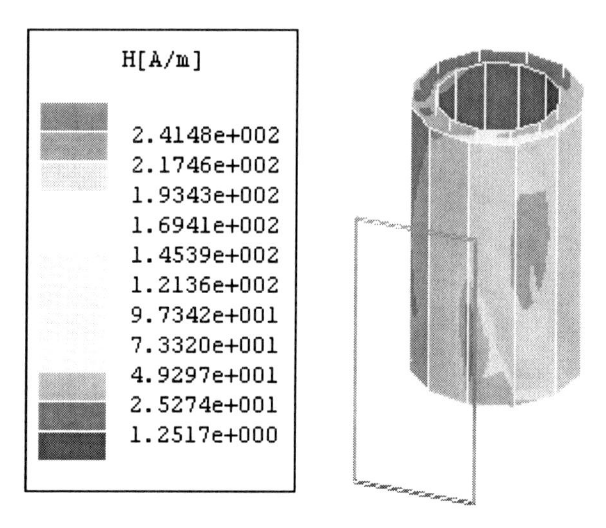

Figure 6: Magnetic field distribution in output relay coil.

Therefore, induced electromotive force in coil:

$$U = \int \vec{E} \cdot \mathrm{d}\vec{l} = -\frac{\mathrm{d}}{\mathrm{d}t}\int_{s} \vec{B} \cdot \mathrm{d}\vec{S} \tag{6}$$

Three-dimension eddy-current solver of ANSOFT was used to simulate magnetic field induced by transient current through contact to couple inside coil [11]. Firstly, output relay model of Figure 5 was built in geometric modeling solver. Materials of coil and contact were assigned as copper in material management solver. Transient current flowing through contact was set 1,000 A and 70 MHz and eddy current effect was concerned in boundary management solver. Secondly, compute according to specific rules in solver. Lastly, simulation results in post processor that magnetic density distributed in coil are shown in Figure 6. It can be seen that the maximum of magnetic density reached 241 A/m.

4 Genetic algorithm

EFT/B while output relay tripping auxiliary relay can be controlled through following methods: on one side, transient over-voltage is restrained while disconnecting output relay coil [12]. On the other hand, coupling route inside coil is cut.

By the analysis of switch operations, EFT/B occurred because current in inductive load charge parasitic capacitance when switch operated. Therefore, fly-wheel diodes were used to provide path for DC output relay coil current, shown in Figure 7. A fly-wheel diode paralleled with output relay coil. Then, coil current flowed through fly-wheel diode when switch operated. Energy consumed in coil resistance and transient over-voltage did not occur in coil.

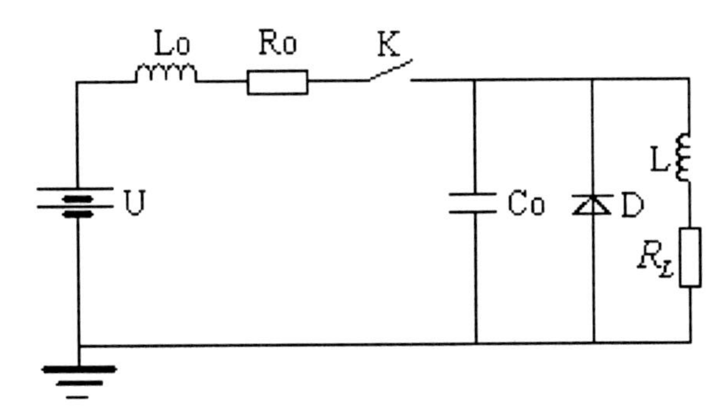

Figure 7: Parallel circuit of fly-wheel diode and output relay coil.

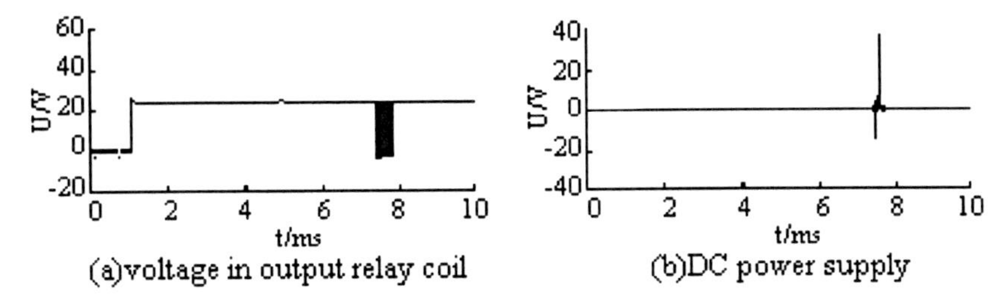

Figure 8: DC power supply and output relay coil voltage with fly-wheel diode.

Voltage waveforms of DC power system and output relay coil are shown in Figure 8 after fly-wheel diode was used. Compared with Figure 3, no-load over-voltage can be controlled effectively while disconnecting output relay coil. But RTOV still interfered with DC power supply.

5 Conclusions

In order to protect microprocessor protection devices from EFT/B, not only improve microprocessor protection devices immunity but also study EFT/B's forming, coupling and working mechanism. EFT/B caused by relay operations in secondary circuit was analyzed. Such initiative countermeasures as fly-wheel diode and electromagnetic shielding were used to restrain interference source and cut coupling path.

References

[1] Liang Zhi-cheng, Fu Jing-bo, Li Fu-tong et al., Electrical fast transient burst immunity of protection equipment. *Automation of Electric Power Systems*, **27(11)**, pp. 65–68, 2003.

[2] Xiao Bao-ming, Wang Ze-zhong, Lu bin-xian, et al., Research on the susceptivity8 on microprocessor based protection equipment. *Automation of Electric Power Systems*, **29(3)**, pp. 61–64, 2005.

[3] Feng Li-min, Chen Ping-ping, Chen Wei et al., Study on improving the immunity of SMPS to EFT/B test. *Automation of Electric Power systems*, **30(5)**, pp. 78–82, 2006.
[4] Cheng Li-jun, Deng Hui-qiong, Research on electrical fast transient/burst immunity for numerical protection. *Electric Power Automation Equipment*, **22(6)**, pp. 5–8, 2002.
[5] Li Qing-quan, Li Yan-ming, Niu Ya-min, The transient electromagnetic fields caused by the operation of disconnector. *High Voltage Engineering*, **27(4)**, pp. 35–37, 2005.
[6] Wiggins C.M., Wright S.E., Switching transient fields in substations. *IEEE Transactions on Power Delivery*, **6(2)**, pp. 591–600, 1991.
[7] Wang Yu-feng, Li Li-wei, Zou Ji-yan et al., Nanosecond-risetime high-voltage electrical fast transient/burst generator. *Automation of Electric Power Systems*, **30(22)**, pp. 96–99, 2006.
[8] Wang Mei-yi, *Applications of Relay Protection in Power Network*. China Electric Power Press, Beijing, 1999.
[9] Cai Min, Cheng Li-jun, Research on transient saturation of opto-couplers for electrical fast transient/burst. *Automation of Electric Power Systems*, **28(8)**, pp. 72–75, 2004.
[10] Zhang Bao-hui, Yin Xiang-gen, *Relay Protection of Power System*. China Electric Power Press, Beijing, 2005.
[11] Liu Guo-qiang, Zhao Ling-zhi, Jiang Ji-ya, *Ansoft Finite Element Analysis of Engineering Electromagnetic Field*. Publishing House of Electronics Industry, Beijing, 2005.
[12] Qin Xiao-hui, Huang Shao-feng, Model and MATLAB simulation of the electrical fast transient/burst disturbance in protection and control secondary circuit. *Relay*, **34(4)**, pp. 17–21, 2006.

Modeling and simulation on TRV characteristics of the circuit breaker for highly compensated UHV transmission lines

Yonggang Guan[1], Bing Fang[1], Peiqi Guo[2], Ya'nan Han[3], Qiyan Ma[3], Bin Zheng[3]
[1]The State Key Laboratory of Control and Simulation of Power Systems and Generation Equipments, Department of Electrical Engineering, Tsinghua University, Beijing, China
[2]Guangzhou Power Supply Co. Ltd., Guangzhou, China
[3]China Electric Power Research Institute, Beijing, China

Abstract

The use of series compensation in UHV transmission lines can improve the transmission capacity of system, increase the transmission distance and enhance the transient stability. However, after installation of series capacitors, the transient conditions of circuit breakers when clearing line faults will change. In this paper, PSCAD/EMTDC was used to establish a simulation platform of series compensated UHV line, and to simulate the transient recovery voltage of the circuit breaker during the breaking of typical faults. Various conditions were considered in the simulation, the impact of series compensation degree from 0% to 70% on the maximum of TRV was analyzed. The simulation results show that higher degree of series compensation will lead to higher TRV level, and the maximum of TRV peak value and RRRV on 40% or higher compensated transmission lines have already exceeded the current TRV standards for circuit breakers used in UHV lines when interrupting 3LG faults.

Keywords: UHV, series compensation, circuit breaker, TRV, PSCAD.

1 Introduction

In order to improve the power grid's transmission capacity, to increase the transmission distance and to improve the transient stability of the system, series

WIT Transactions on Engineering Sciences, Vol. 107, © 2015 WIT Press
www.witpress.com, ISSN 1743-3533 (on-line)
doi:10.2495/SEEEII140351

Table 1: Chinese standard TRV values for circuit breakers rated 1,100 kV.

Type of Experiment	k_{pp}	k_{af}	TRV peak (kV)	Rate of Rise of TRV ($kV\cdot\mu s^{-1}$)
T100	1.3	1.40	1635	2.0
T60	1.3	1.50	1751	3.0
T30	1.3	1.53	1786	5.0
T10	1.3	1.53	1786	7.0
OP1-OP2	2.0	1.25	2245	1.54

capacitors (SC) are required to be installed in UHV transmission lines. However, the installation of series capacitors changes the switching conditions and phenomena. The trapped charge on SC often leads to markedly higher transient recovery voltage (TRV) to circuit breakers than the ones used in non-compensation lines, and sometimes even can result in interrupting failure.

Accompanying with the development of UHV power grid, it is imperative that compensation of higher degree should be employed, which may lead to higher TRV level and considerable difficulties in manufacturing of circuit breakers. Therefore, the TRV of compensated UHV lines is drawing increasing academic attention.

A set of TRV standards on 1,100 kV circuit breakers have been proposed as an expansion of IEC standards based on researches and experiments, which are shown in Table 1 [1].

A number of studies on the impact of series capacitors on interrupting conditions on HV/EHV systems have been carried out [2–4]. Recently, it is found through the researches on series compensated UHV lines that are already put into operation that under certain circumstances TRV peak on compensated lines can exceed the present standards for circuit breakers [5]. However, studies on interrupting conditions of series compensated UHV lines are currently comparatively scarce, especially when the compensation degree is higher than 40%. Thus, in this paper, a typical radial network has been used to simulate the TRV characteristics on highly compensated UHV lines and the impact of compensation degree.

2 System conditions and modeling

In this paper, the transient simulation software PSCAD/EMTDC has been employed to build a simulation platform of compensated UHV transmission lines based on the real data of UHV systems that are already put into operation in China.

2.1 System conditions

The voltage of UHV system in China is rated 1,000 kV, and normally ranges from 900 to 1,100 kV. The rated frequency of the system is 50 Hz and normally ranges between 50±0.2 Hz.

Typical data of UHV transmission system such as voltage and power flow used in simulation are shown in Table 2.

Table 2: Typical voltage and power flow in 1,000 kV transmission system.

Bus-bar Voltage on UHV Lines (kV)		Transmission Power (GW)	
Sending End	Receiving End	Double Circuit	Single Circuit
1065	1075	3.5	3.0
		4.0	3.5
		4.5	3.0

Table 3: Equivalent capacitance of common electric devices used on typical 1,000 kV UHV power grid.

Name of Devices	Value (pF)
High impedance	5,000
Lighting arrester	20
Ground switch	150
Isolator switch	300
Bus capacitor	8,980

The lower system of 500 kV rated voltage associated with UHV network is simplified as a power source with equivalent impedance. The modeling of the main components in UHV transmission system are described as below:

(1) *Towers and lines*

The configuration of UHV transmission tower with multi sub-conductor bundles employs vertical double circuit drum-type tower arrangement. The circuit was transposed at 1/3 and 2/3 lengths of transmission line.

The frequency-dependent (phase) model was chosen to simulate the transmission line, which is the most common used and accurate model of transmission line for time domain analyze at present.

(2) *Transformers*

Three-winding transformers used on UHV system in China were modeled by transformer model in PSCAD with saturation characteristic, with an entrance capacitance of 5,000 pF and inductance of 20 mH [6].

(3) *Circuit breakers*

Circuit breakers were modeled as ideal switches, which can be easily opened at first current zero crossing after receiving operating signal, without taking account of arc. Capacitance between the gaps of the breaker after opening is 540 pF.

(4) *Other stray parameters*

Other components such as bus-bar, lighting arresters, ground switches and isolator switches are all simplified as capacitors in concentrated parameters, the values of which are shown in Table 3.

2.2 Compensation devices used in UHV system

A typical module of series compensation device is shown in Figure 1(a), which comprises a capacitor in parallel with a metal oxide varistor (MOV), a bypass switch, a spark gap and a damping part.

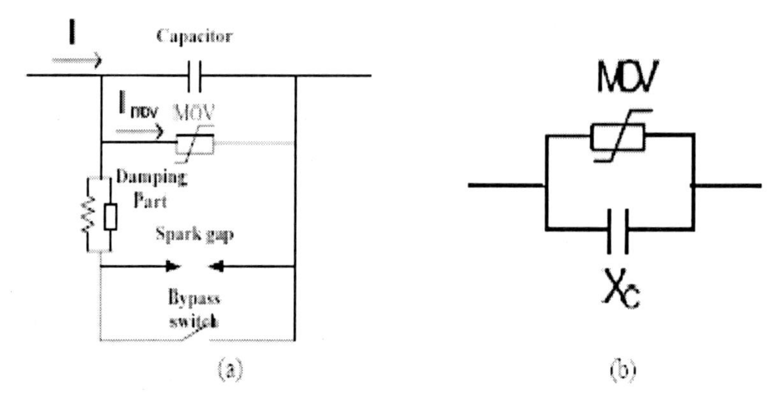

Figure 1: Series capacitor arrangement.

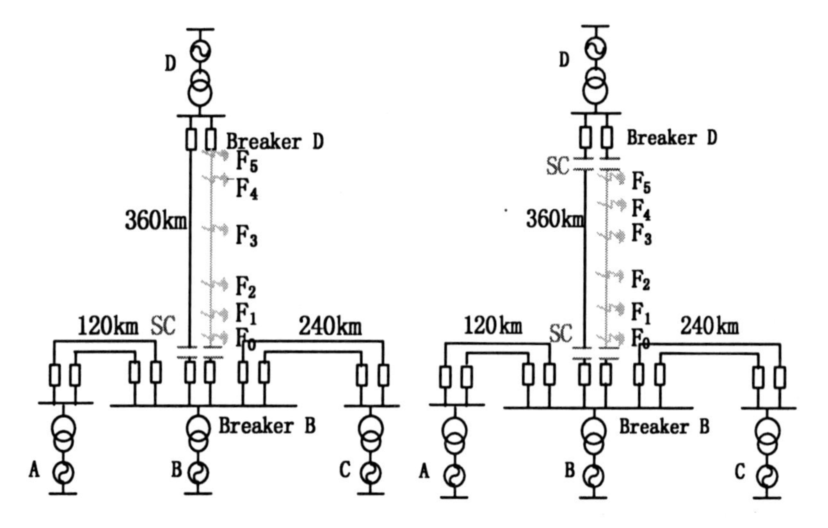

(a) SCs located on terminal B (b) SCs located on both ends of line

Figure 2: Radial network of UHV compensated lines.

The simulation model of SC devices can be seen in Figure 1(b), which is simplified as a capacitor paralleled with a MOV. The capacitance of X_c is an equivalence of the whole SC devices, which can be calculated by the compensation degree and transmission line parameters. The *V–I* characteristics of MOV were valued according to the real parameters of ASEA XAP-A lighting arresters, and the rated voltage can be computed by the compensation degree.

2.3 Simulation condition

Figure 2 illustrates the typical radial network with four power sources recommended by CIGRÉ for simulation of TRV, which is applicable for most transmission lines of voltage level higher than 800 kV [7].

Compensation devices are located in the line B–D. Due to the series capacitor's withstanding voltage, when the compensation degree is relative high, for example 70%, series capacitors have to be separated and installed at both ends of the transmission line [8]. Normally, the compensation capacitors are separated into two parts with equal impedance (Figure 2(b)). When the compensation degree is relative low, for example, 30%, the compensation capacitors are totally installed at side B (Figure 2(a)).

Fault points F_0–F_5 are located along 360 km transmission line between B and D, separately 0.1, 30, 60, 150, and 270 km away from the circuit breaker, which provides TRVs for the common fault types such as bus terminal fault (BTF), short-line fault (SLF) and long-line fault (LLF) breaking conditions. Circuit breaker operates 70 ms after the fault occurs. The step of simulation is 2 ns.

The employment of lighting arresters and opening resistors can reduce the TRV level across circuit breakers. Thus, to study the most severe TRVs in this network, neither of them are employed in the simulation model.

3 TRV characteristics of highly series compensated situation

3.1 TRV characteristics of 40% compensation

Figure 3 illustrates the statics of TRVs across circuit breaker B and D when interrupting the most severe 3LG fault line with 40% compensation and the corresponding short circuit value. SCs are located in the vicinity of terminal B.

As can be seen, TRVs across circuit breaker B, which is located on the central terminal of the radial network, are usually higher compared with breaker D, can greatly exceed the standards under certain interrupting circumstances. When clearing LLF fault, the severest TRV peak is as high as 2771.8 kV, which is far beyond the standard value for OP1–OP2 of 2245 kV.

The RRRVs in this case are mostly within the standard except when the fault is of SLF type. The highest RRRV reaches 8.083 kV/μs, much higher than the T60 standard of 3.0 kV/μs.

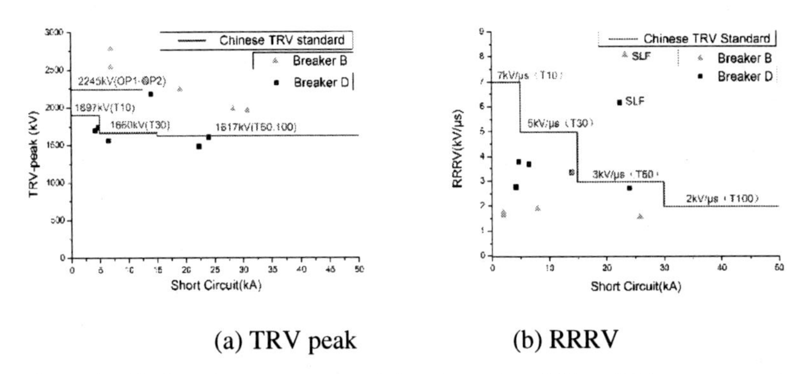

(a) TRV peak (b) RRRV

Figure 3: TRV statics across circuit breakers B and D at 40% compensation.

3.2 Influence of compensation degree to TRV

To analyze the influence of compensation degree on TRV characteristics, TRVs across circuit breakers interrupting 3LG fault lines at various compensation degrees were simulated. Fault points are F_0–F_5 as shown above.

Figure 4 illustrates the TRV characteristics on circuit breaker B and D of compensation degree from 0% to 40% and SCs are totally installed near terminal B. When the compensation degree is higher than 40%, SCs are separated and located on both ends of the transmission line, the simulation outcomes on which conditions are shown in Figure 5.

As can be seen, TRV level across both circuit breakers B and D rises greatly accompanying with the increasing of compensation degree no matter if the SC devices are installed on one or both ends of the line. For 70% compensation, the greatest TRV across circuit breaker B and D, respectively, reaches 2778.1 and 2640.9 kV, both of which are far beyond the standards given above, and significantly higher than the ones in non-compensated lines, which are 2164.9 and 1858.1 kV. The highest RRRV on 40% separately compensated lines is 7.265 kV/μs, while on 70% compensated line this value reach 7.927 kV/μs.

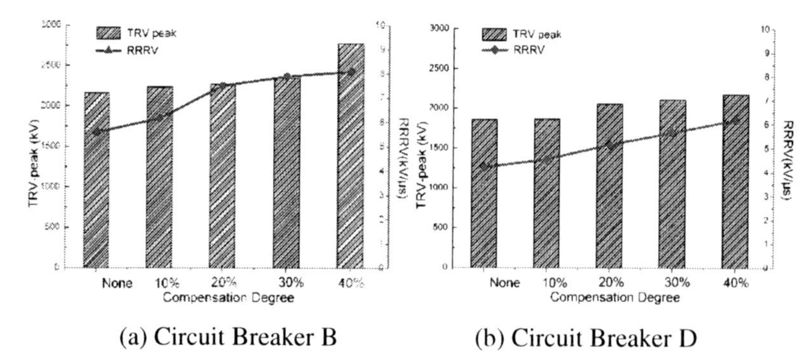

(a) Circuit Breaker B (b) Circuit Breaker D

Figure 4: Maximum TRV peak and RRRV when interrupting 3LG fault line of compensation degree from 0% to 40%. SCs are totally installed on end B.

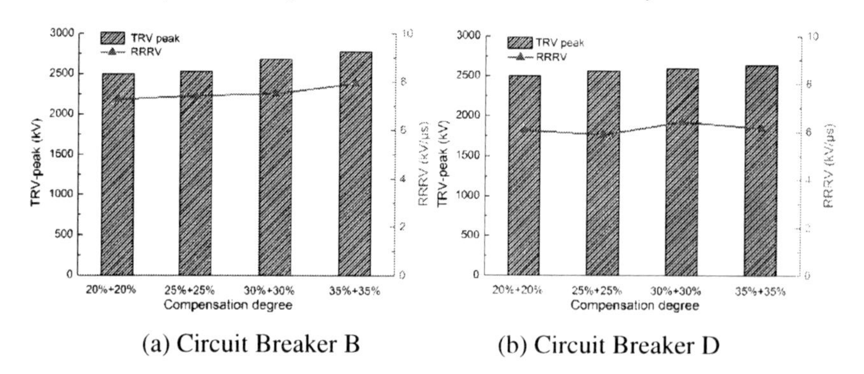

(a) Circuit Breaker B (b) Circuit Breaker D

Figure 5: Maximum TRV peak and RRRV when interrupting 3LG fault line of compensation degree from 40% to 70%. SCs are installed on both ends.

It also can be seen that TRVs across circuit breaker B, which is located on the central terminal of the radial network, are usually higher compared with breaker D, especially when the SC is totally installed on terminal B.

4 Conclusions

A simulation platform based on the radial network recommended by CIGRÉ conference and the typical parameters of the UHV power grid in China has been established. Based on this platform, TRV characteristics on UHV lines at 0–70% compensation during different types and points of faults are simulated in this paper. The conclusions are as below:

(1) The peak value and RRRV of TRV when interrupting highly compensated lines can greatly exceed the present standards of UHV circuit breakers, and the circuit breaker located in central terminal of the radial network often suffers more severe TRV than the other ones. For 40% compensation, the maximum of TRV reaches 2771.8 kV when clearing LLF fault, and the severest RRRV is 8.083 kV/μs for SLF fault.
(2) Higher compensation degree will rise greater TRV level. The maximum of TRV increases from 2494.5 to 2778.1 kV when the compensation level changes from 40% to 70%, while the severest TRV peak is 2164.9 kV for non-compensated lines. The increment of RRRV is not so obvious.
(3) Measures should be taken to resolve the TRV problem in highly compensated UHV transmission lines. Either UHV circuit breaker with higher TRV endurance or effective TRV suppressing methods should be developed.

Acknowledgments

This research work was supported by the project of State Grid Corporation of China (XT17201200037).

References

[1] X. Guozheng, Z. Jierong, Q. Jiali, et al., *Theory and Application of High Voltage Circuit Breaker*. Tsinghua University Press, Beijing, 2000.
[2] D.D. Wilson, Series compensated lines voltages across circuit breakers and terminals caused by switching. *Power Apparatus and Systems*, pp. 1050–1056, 1973.
[3] C.L. Wagner, H.M. Smith, Analysis of transient recovery voltage (TRV) rating concepts. *IEEE Transactions on Power Apparatus and Systems*, pp. 3353–3363, 1984.
[4] C.L. Wagner, D. Dufournet, G.F. Montillet, Revision of the application guide for transient recovery voltage for AC high-voltage circuit breakers of IEEE C37.011: A working group paper of the High Voltage Circuit Breaker Subcommittee. *Power Delivery*, **22**, pp. 161–166, 2007.

[5] B. Han, J. Lin, L. Ban, et al., Study on transient recovery voltage of circuit breakers in UHV AC double circuit system. In: *Proceedings of the 2010 International Conference on Power System Technology* (*POWERCON*), pp. 1–6, 2010.
[6] C.W.G, Guidelines for representation of network elements when calculating transients. *CIGRE Technical Brochure,* 1990.
[7] CIGRÉ Working Group, TRV Evaluations in Radial and Meshed Network Models Using System/Equipment Parameters or Different Projects (Version C3), CIGRE Technical Brochure, 2013.
[8] X. Qin, H. Shen, Q. Zhou, et al., Voltage distribution along the line and disposition scheme of series capacitors for UHV transmission lines with series capacitors. *Zhongguo Dianji Gongcheng Xuebao* (*Proceedings of the Chinese Society of Electrical Engineering*), **31(25)**, pp. 43–49, 2011.

Analysis and determination of auxiliary capacitance for self-excited induction generators

Yang Zhang, Xinzhen Wu
Department of Electrical Engineering,
Qingdao University, Qingdao, China

Abstract

Based on the equivalent circuit of an induction generator, its steady state is predicted by setting both the real part and imaginary part of loop impedance equal to zero. For the convenience of computation, the pattern search method as a kind of optimization method is used to solve the equations for calculating the capacitances, which is simpler and more effective than Newton–Raphson method. The external characteristics of the auxiliary capacitors' connecting to the stator port are evaluated, which show that the secondary capacitors not only modify the inherent external characteristics of an induction generator but also increase its ability to carry load. In view of the voltage regulation, the importance of selecting suitable auxiliary capacitors is emphasized.
Keywords: induction generators, steady-state operation, auxiliary capacitors.

1 Introduction

With people's improving awareness of environment and the decreasing reserves of fossil fuels, renewable energy is supposed to be widely utilized for electricity generation in the world. As an operating mode of low-capacity isolated power system, the induction generator is an ideal energy conversion device to generate power with clean energy such as hydro and wind energy due to its low cost, reliable operation, simple structure and control strategy, which is significant for environmental protection [1]. However, there are still some shortages limiting their application. For example, when the load changes, both output voltage and frequency will be affected in company [2–4]. Therefore, the self-excited capacitance must be adjusted with the help of auxiliary capacitor to improve the stabil-

WIT Transactions on Engineering Sciences, Vol. 107, © 2015 WIT Press
www.witpress.com, ISSN 1743-3533 (on-line)
doi:10.2495/SEEEI140361

ity of isolated power system [5, 6]. This paper concentrates on the twice-input auxiliary capacitors and the alterations of voltage and frequency in different steady states when the load increases continuously.

Since the auxiliary capacitor in fact works as a part of self-excited capacitor, the general analysis method for steady-state operation of induction generators can still be adopted [7]. The expression of loop-impedance or node-admittance derived from the equivalent circuit should be equal to zero. A practical direct search method is adopted in this paper, which simplified the solving procedure by changing the binary high-degree equation group into an optimizing problem with two unknown quantities based on the loop-impedance of equivalent circuit. In this way, the programming is simpler with a faster convergence rate and better stability. The curves of external characteristics and frequency when the auxiliary capacitor is input in two times are given with resistive load and resistive-inductive load (lag).

2 The loop-impedance of equivalent circuit

When the self-excited induction generator is operating in steady state, the machine parameters, self-excited capacitance, load, output voltage, current and frequency will have to satisfy some balance relationship [8]. The analysis of steady-state operation is based on the equivalent circuit in reduced frequency with the real part and imaginary part of loop-impedance equal to zero [9].

C_0 and C_1 are self-excited capacitor and auxiliary capacitor, respectively. The total capacitor is

$$C = C_0 + C_1 \tag{1}$$

Ignoring iron loss and magnetizing resistance, all the parameters are constant in above equivalent circuit except the magnetizing reactance, which is affected by magnetic saturation and operating state.

The loop voltage equation can be derived from Figure 1:

$$\dot{I}_1 \left\{ Z_{ab} + \frac{\mathrm{j} X_m [R_2 + \mathrm{j}(f_1 - n) X_2]}{R_2 + \mathrm{j}(f_1 - n)(X_2 + X_m)} \right\} = 0 \tag{2}$$

where Z_{ab} is the equivalent impedance look from port ab, which can be derived as

$$Z_{ab} = R_{ab} + \mathrm{j} X_{ab} \tag{3}$$

$$R_{ab} = \frac{R_L X_c^2}{f_1 [R_L^2 f_1^2 + (X_L f_1^2 - X_c)^2]} + \frac{R_1}{f_1} \tag{4}$$

$$X_{ab} = -\frac{R_L^2 X_c + (X_L f_1^2 - X_c)}{R_L^2 f_1^2 + (X_L f_1^2 - X_c)^2} + X_1 \tag{5}$$

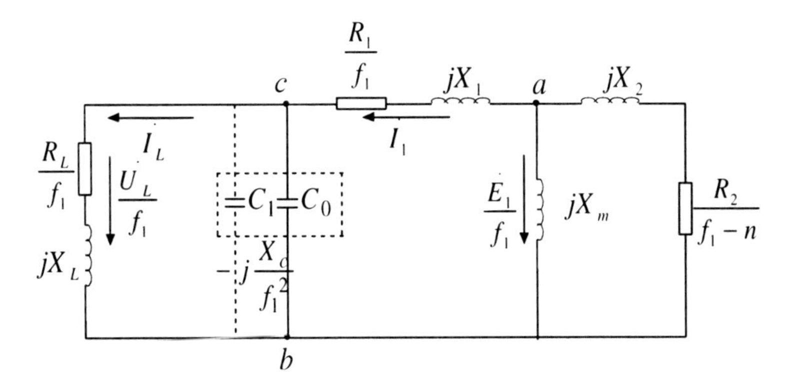

Figure 1: The equivalent circuit of induction generator in steady-state operation under reduced frequency.

3 The optimization mathematical model

Since the loop current I_1 is unequal to zero in steady-state operation, the loop-impedance must satisfy the equation that

$$Z_s = Z_{ab} + \frac{\mathrm{j}X_m[R_2 + \mathrm{j}(f_1 - n)X_2]}{R_2 + \mathrm{j}(f_1 - n)(X_2 + X_m)} = 0 \tag{6}$$

Traditionally, for a given induction generator system this high-degree equation group with stator frequency f_1 and magnetizing reactance X_m as unknown variables can be derived from Eq. (6) by setting both the real and imaginary parts of loop impedance to be zero. However, as the total capacitance C is also an unknown variable, the output voltage U_L is selected as an extra condition for calculating due to its constant value for carrying loads. The optimization objective function can be expressed as

$$f(X_c, X_m, f_1) = \mathrm{Re}[Z_s] + \mathrm{Im}[Z_s] \tag{7}$$

The goal of optimization is to calculate the minimum value of this objective function. And the optimization method adopted in this paper is pattern search method, which is a kind of direct search method with good stability, fast convergence rate and simple programming. The X_c and f_1 are selected as independent variables, which means that in every iteration step the values of these two variables are known. So the induced electromotive E_1 can be derived from the equivalent circuit with the constraint condition of output voltage U_L, and the expression can be expressed as

$$E_1 = \left|\frac{Z_{ab}}{Z_{bc}}\right| U_L \tag{8}$$

where Z_{bc} can be expressed as

$$Z_{bc} = \frac{-\mathrm{j}\frac{X_C}{f_1^2}\left(\frac{R_L}{f_1} + \mathrm{j}X_L\right)}{\frac{R_L}{f_1} + \mathrm{j}\left(X_L - \frac{X_c}{f_1^2}\right)} \tag{9}$$

With calculated E_1 in every iteration step, the magnetizing reactance X_m can be derived from the magnetizing curve $X_m = g(E_1)$. The value of X_m has to be substituted into the loop-impedance equation to calculate the objective function in every iteration step. And the pattern search method is used to optimize the objective function.

4 External characteristics of the prototype with calculated capacitance

The rated data of prototype are as follows: rated active power P_N = 2.2 kW, Y-connection, rated voltage U_N = 380 V, rated current I_N = 5 A, rated frequency (base value) f_b = 50 Hz, rated speed corresponding to base value frequency n_N = 1,500 r/min. And the parameters in p.u. are: stator resistance R_1 = 0.0636, stator reactance X_1 = 0.1356, reduced rotor resistance $R_2^{'} = 0.0727$, reduced rotor reactance in base frequency $X_2^{'} = 0.1356$.

The relationship between E_1 and X_m is obtained from the synchronous experiment, which can be piecewise linearized as

$$\frac{\dot{E}_1}{f_1} = \begin{cases} 1.6154 - 0.3449X_m, & X_m < 1.5838 \\ 1.6804 - 0.3860X_m, & 1.5838 \le X_m < 2.2672 \\ 1.9209 - 0.4920X_m, & 2.2672 \le X_m < 2.4424 \\ 2.3773 - 0.6789X_m, & 2.4424 \le X_m < 3.5018 \\ 0, & X_m \ge 3.5018 \end{cases} \tag{10}$$

When the induction generator is going to carry load, it should build up its no-load output voltage with the help of self-excited capacitance C_0 first. The self-excited capacitance is calculated to be 39.6 μF with the load in Figure 1 as infinite. And the external characteristic curves of induction generator with no-load self-excited capacitance C_0 carrying resistive-inductive loads are shown in Figure 2. However, the output voltage begins to collapse when the current equals 0.15. Therefore, if the induction generator is going to reach its rated state, there must be an extra capacitance paralleled to the load, which is called the auxiliary capacitance C_1. As the impedance module value of rated load is 44 Ω, the total capacitance is calculated to be 92.2 μF when the power factor of load is 0.89 lag. In this way the auxiliary capacitance C_1 should be 52.6 μF. And the external characteristic curve of induction generator with auxiliary capacitance carrying resistive-inductive load is shown in Figure 3, which improves a lot with the help of auxiliary capacitance.

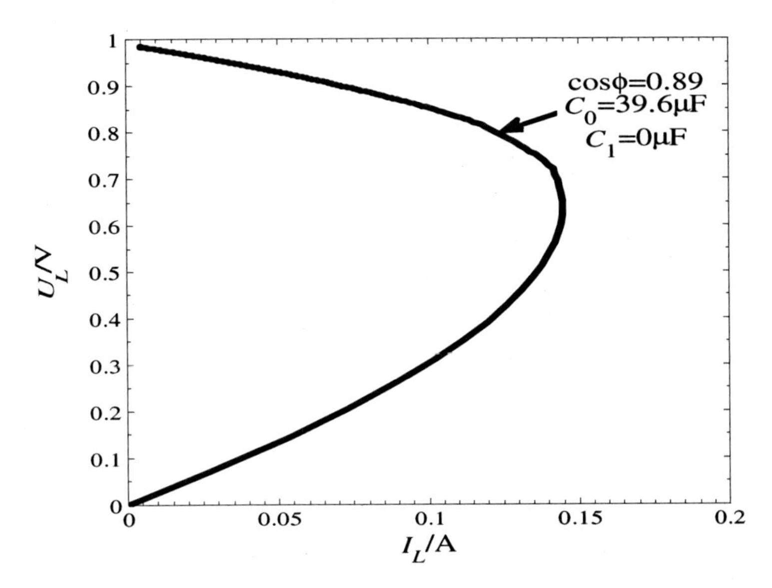

Figure 2: External characteristics of induction generator without auxiliary capacitance.

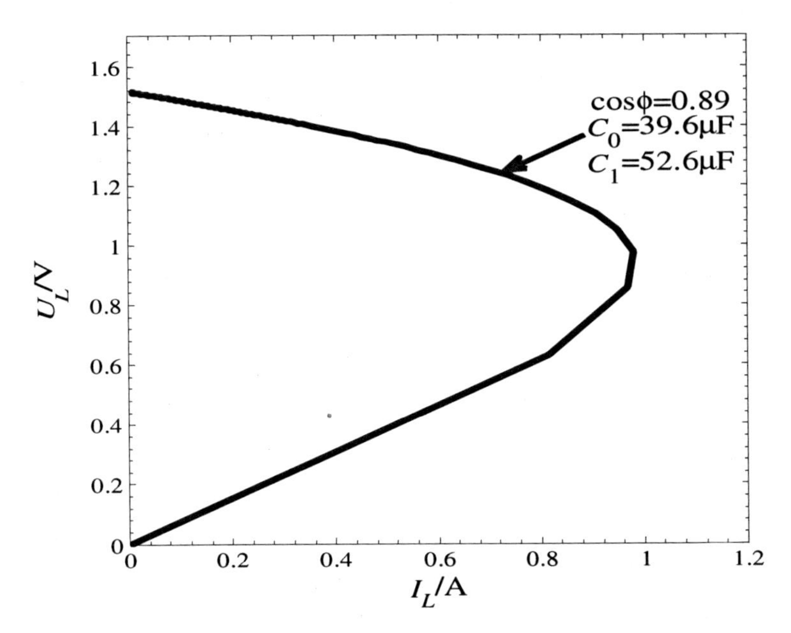

Figure 3: External characteristics of induction generator with auxiliary capacitance as $C_1 = 52.6$ μF.

5 Conclusion

This paper proposed an optimization method in steady-state analysis of induction generators, which avoids the differentiation and matrix operation for a high rapidity of convergence and better stabilization. And the self-excited capacitance for no-load voltage buildup is calculated with this method. When the induction generator connects the load, the output voltage would drop and even collapse. Therefore, the auxiliary capacitance for rated operation is necessary, which is also calculated in this paper. The characteristics of induction generator with only self-excited capacitance and that with both self-excited capacitance and auxiliary capacitance are analyzed.

Acknowledgment

This research was financially supported by the National Science Foundation 51377086.

References

[1] S. Rahman, Going green-the growth of renewable energy. *IEEE Power and Energy Magazine*, **1(6)**, pp. 16–18, 2003.
[2] S.S. Murthy, O.P. Malik, A.K. Tandon, Analysis of self-excited induction generators. *IEE Proceedings, Pt. C.*, **129(6)**, pp. 260–265, 1982.
[3] T.F. Chan, Steady-state analysis of self-excited induction generators. *IEEE Transactions on EC*, **9(2)**, pp. 288–296, 1994.
[4] S. Kumari, G. Bhuvaneswari, Voltage regulation of a stand-alone three-phase SEIG feeding single-phase loads. In: *2014 IEEE Students' Conference on Electrical, Electronics and Computer Science* (*SCEECSI*), pp. 1–6, 2014.
[5] N.H. Malik, A.H. Al-Bahrani, Influence of the terminal capacitor on the performance characteristic of a self-excited induction generator. *IEE Proceedings, Pt. C*, **137(2)**, pp. 168–173, 1990.
[6] Wang Li, Su Jian-Yi, Effects of long-shunt and short-shunt connections on voltage variations of a self-excited induction generator. *IEEE Transactions on EC*, **12(4)**, pp. 368–374, 1997.
[7] Wang Long-peng, Wu Xin-zhen, Yuan Bo-qiang, Determination of short-shunt connected capacitance for self-excited induction generator, *Proceedings of the CSEE*, **30(18)**, pp. 85–90, 2010.
[8] Y.K. Chauhan, S.K. Jain, Bhim Singh, Operating performance of static series compensated three-phase self-excited induction generator. *Electrical Power and Energy Systems*, **49**, pp. 137–148, 2013.
[9] R.R. Chilipi, B. Singh, S.S. Murthy, Performance of a self-excited induction generator with DSTATCOM-DTC drive-based voltage and frequency controller. *IEEE Transactions on Energy Conversion*, **29(3)**, pp. 545–557, 2014.

Analytical determination of optimal winding-power-splitting-ratio for the automatic MPPT generator system

Yinru Bai, Baoquan Kou, C.C. Chan
School of Electrical Engineering and Automation, Harbin Institute of Technology, Harbin, China

Abstract

A novel generator system was proposed in 2013. This generator system, which is called automatic maximum power points tracking generator system, is able to track maximum power from wind automatically without using converters and control circuits, thus it has advantages of higher reliability and lower loss. The new system contains two sets of three-phase winding, so it is very necessary to study the relation between the amounts of one winding's output power and the other's. This paper presents an analytical determination on the optimal winding-power-splitting-ratio of the new generator system, and the simulation result validates the analysis.
Keywords: automatic, MPPT, off-grid wind power generator.

1 Introduction

With the advantages of flexible, low cost, and clean, stand-alone wind power generations (SWPGs) have the largest utilizations. Maximum power points tracking (MPPT) is a research hot spot for SWPGs. Many MPPT methods have been proposed. Many MPPT methods have been proposed, and these methods can be classified into two categories [1–4]. One is called optimum relationship based control method which depends on a pre-known relationship. References [5–7] proposed various relationships for MPPT. The other is well known as hill climb searching (HCS) control method that does not require any prior knowledge of the system and is absolutely independent of the turbine, generator, and wind characteristics [8, 9]. Refs. [10–13] studied several improved HSC algorithms.

WIT Transactions on Engineering Sciences, Vol. 107, © 2015 WIT Press
www.witpress.com, ISSN 1743-3533 (on-line)
doi:10.2495/SEEEI140371

The MPPT methods mentioned above can also be classified into the same category—electric control method, because all of them depend on control algorithms and electrical control devices. SWPGs are used outdoors, and thus high reliability, high efficiency, and low cost are the necessary features. However, electrical control devices are prejudicial to these aims.

In [14], we proposed a novel MPPT method: automatic MPPT method that is independent of any electrical devices and control algorithms. The new generator system has a characteristic of P_{in} vs. n curve which fits to the MPPs curve of a wind turbine. This method is passive and, therefore, it can absolutely eliminate the disadvantages of the automatic MPPT methods, such as low reliability and inefficiency.

This study presents a study on the optimal power distribution ratio of the new generator system.

2 Winding-power-splitting-ratio

Figure 1 shows the schematic of the new system and Figure 2 shows the characteristics.

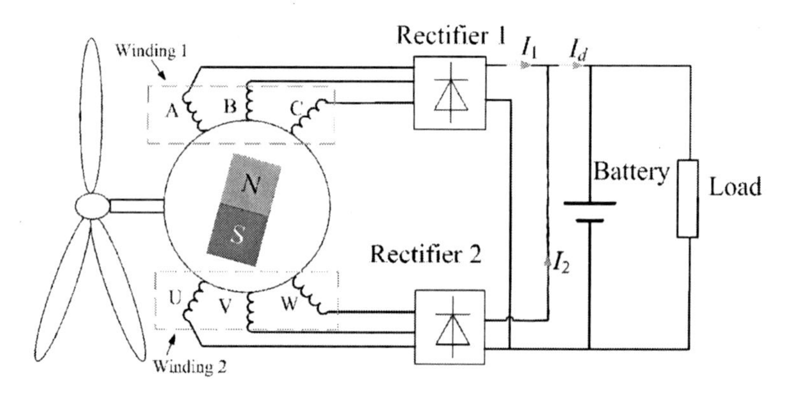

Figure 1: Schematic of the new system.

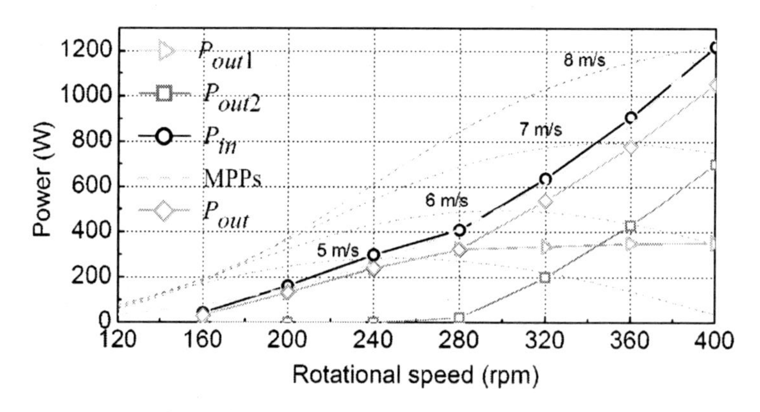

Figure 2: Characteristics of the new system.

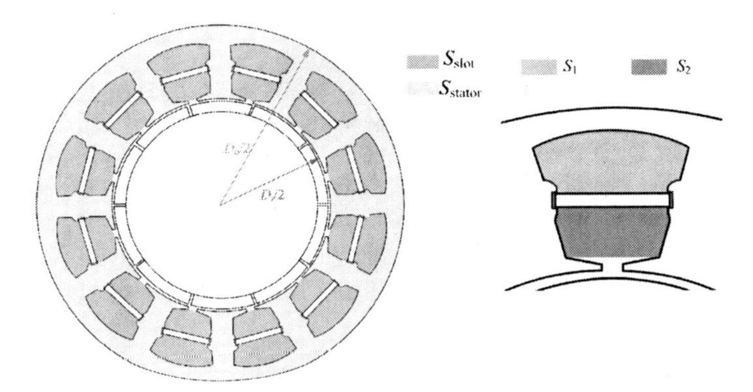

Figure 3: Structure of the new generator: (a) stator; (b) slot.

In this paper, the power distribution ratio is defined as

$$k_i = \frac{E \cdot I_{1r}}{E \cdot I_{dr}} = \frac{I_{1r}}{I_{dr}} \tag{1}$$

Define calculated capability of the system as

$$P_i = E_{1r} I_{1r} + E_{2r} I_{2r} \tag{2}$$

For simplicity, per phase EMF is considered as sinusoidal wave, so the following equations can be obtained according to the rectification theory:

$$E_{1r} = 2.34 E_{ph1r} \tag{3}$$

$$E_{2r} = 2.34 E_{ph2r} \tag{4}$$

Per phase EMF of winding 1 at rated speed can be formulated as:

$$E_{ph1r} = 4.44 \cdot \frac{n_r}{60} p K_w N_1 \Phi_0 K_\Phi \tag{5}$$

Per phase EMF of winding 2 at rated speed can be formulated as:

$$E_{ph2r} = 4.44 \cdot \frac{n_r}{60} p K_w N_2 \Phi_0 K_\Phi \tag{6}$$

According to (5)–(8) and (19), it is easy to know that

$$\frac{E_{2r} I_{2r}}{E_{1r} I_{1r}} = \frac{N_{2r} I_{2r}}{N_{1r} I_{1r}} = \frac{J_2}{J_1 k_s} \approx \frac{1}{k_s} \tag{7}$$

where I_{1r} is the DC side average current output by winding 1 at rated speed; I_{2r} is the DC side average current output by winding 2 at rated speed; I_{1r} is the DC side average current output by winding 1 at rated speed; I_{dr} is the DC side total current at rated speed; E_{1r} is the rectified voltage of winding 1 at no load state at rated speed; E_{2r} is the rectified voltage of winding 2 at no load state at rated speed; N_1 is the per phase number of turns of winding 1; N_2 is the per phase number of turns of winding 2; n_r is the rated rotational speed; Φ_0 is the per phase flux of a generator at no load state; and n_0 is the voltage buildup speed. The following equation can be derived from (2) (the details of the derivation are omitted).

$$\frac{k_d k_1 n_r \left(\frac{4V}{k_1 \pi}\right)^{\frac{4}{3}} \left(1-k_d^2\right)}{P_i} = \frac{72.11}{J \frac{k_{st}}{1+k_{st}} B_\delta \alpha_i K_w K_\Phi} \tag{8}$$

where k_f is the fill factor; K_w is the winding coefficient; K_Φ is the wave form factor of the air gap flux density; k_s is the ratio of S_1:S_2 (see Figure 3); k_{st} is the ratio of the area of slot to the area of tooth and yoke; k_d is the split ratio; k_l is the length-width ratio; B_δ is the air gap flux density at no load state; L_{ef} is the core thickness; J is the current area density; and $V = D_2^2 \pi L_1/4$ is the volume of the new generator. E_{1r} can be expressed by voltage build up speed n_0, shown as follows:

$$E_{1r} = \frac{n_r}{n_0} E \tag{9}$$

Compared to the conventional generator parameters, k_i is a unique parameter of this novel generator and also a very key parameter. For given k_d, k_l, and k_{st}, obviously, decreasing P_i helps to reduce V. P_i has a high correlation with k_i, thus it is necessary to analyze the optimal value of k_i. According to Figure 2, when double windings is in operation, the output power of winding 1 increases very little with the rise of speed, thus for simplicity, the output power of winding 1 can be roughly considered as a constant at the speed higher than n_2, and therefore the calculated capability of the generator is

$$P_i = E_{1r} I_{1r} + E_{2r} I_{2r} = E_{1r}(I_{dr} - I_{2r}) + E_{2r} I_{2r} \tag{10}$$

According to (7), (9), and (10), P_i can be expressed as

$$P_i = \frac{n_r}{n_0} E I_{dr} - n_r E I_{2r} \left(\frac{1}{n_0} - \frac{1}{n_2}\right) \tag{11}$$

To track the maximum power curve of the wind turbine, the output power of winding 1 must approach the maximum power of the wind turbine, so according to Figure 2, the following equation can be obtained:

$$EI_{1r} = E(I_{dr} - I_{2r}) \approx C_1 n_2^{\ 3} \tag{12}$$

where $C_1 = \frac{1}{2}\eta \rho A C_{pmax} (\frac{2\pi R}{60\lambda_{op}})^3$, thus substituting (12) into (11), we obtain

$$P_i \approx \frac{n_r}{n_0} EI_{dr} - n_r E \cdot \left(I_{dr} - \frac{C_1 n_2^{\ 3}}{E} \right) \cdot \left(\frac{1}{n_0} - \frac{1}{n_2} \right) \tag{13}$$

and P_i also can be expressed by k_i

$$P_i \approx n_r EI_{dr} \cdot \left[\frac{1}{n_0}(1-k_i) + \frac{k_i}{\sqrt[3]{\frac{EI_{dr}(1-k_i)}{C_1}}} \right] \tag{14}$$

P_i vs. k_i curve has a minimal value. To calculate the minimal value point, the derivative of (13) with respect to n_2 is

$$\frac{dP_i}{dn_2} \approx -n_r E \cdot \left(-\frac{3C_1 n_2^{\ 2}}{En_0} + \frac{2C_1 n_2}{E} - \frac{I_{dr}}{n_2^{\ 2}} \right) \tag{15}$$

When (15) equals 0, P_i has extremisms, so an equivalent equation can be obtained as follows:

$$-\frac{3C_1 n_2^{\ 4}}{En_0} + \frac{2C_1 n_2^{\ 3}}{E} - I_{dr} = 0 \tag{16}$$

and because speed is greater than 0, its root is

$$n_{2op} \approx -\frac{b}{4a} + \frac{1}{2}\sqrt{\frac{b^2}{4a^2} + \Delta} + \frac{1}{2}\sqrt{\frac{b^2}{2a^2} - \Delta + \frac{b^3}{a^3} \Bigg/ \left(4\sqrt{\frac{b^2}{4a^2} + \Delta} \right)} \tag{17}$$

where

$$a = -\frac{3C_1}{En_0}\,;\, b = -\frac{2C_1}{E}\,;\, \Delta_1 = 12ae\,;\, \Delta_2 = 27b^2e$$

$$\Delta = \frac{\sqrt[3]{2}\Delta_1}{3a\sqrt[3]{\Delta_2 + \sqrt{-4\Delta_1{}^3 + \Delta_2{}^2}}} + \frac{\sqrt[3]{\Delta_2 + \sqrt{-4\Delta_1{}^3 + \Delta_2{}^2}}}{3\sqrt[3]{2}a}$$

Therefore, the optimal power split ratio can be calculated as

$$k_{\text{iop}} = \frac{I_{\text{1op}}}{I_{\text{dr}}} \approx \frac{C_1 n_{\text{2op}}{}^3}{EI_{\text{dr}}} \tag{18}$$

Table 1 shows the designs of different generator models having different values of k_i, and these models are tested by 2D finite element software in order to ensure they meet the same design requirements. Figure 4 shows the correlation between the volume of the generator and k_i.

Table 1: Different generator models.

	k_i	k_d	k_l	D_o/mm	n_0/rpm	P_out/kW
Model A	0.24	0.61	0.54	170	150	1,000
Model B	0.34	0.61	0.54	165	150	1,000
Model C	0.45	0.61	0.54	168	150	1,000
Model D	0.55	0.61	0.54	175	150	1,000

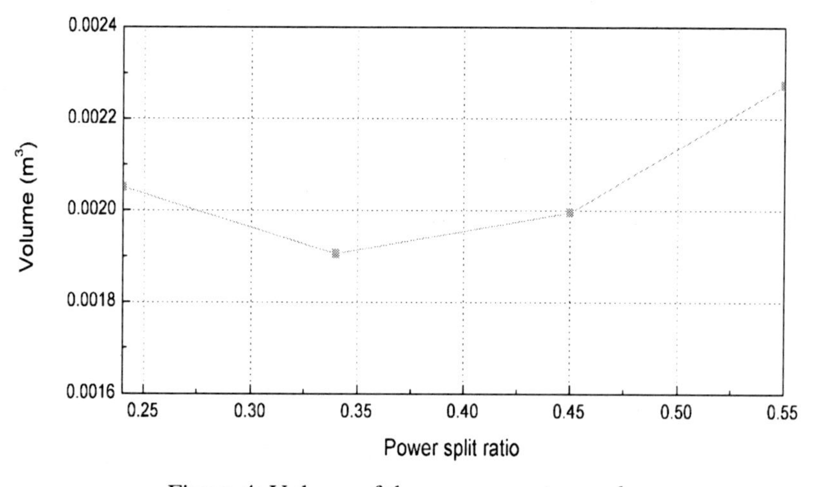

Figure 4: Volume of the new generator vs. k_i.

3 Conclusions

In [1], we proposed the automatic MPPT method. By specially designing the structure of the generator system, the new system is able to complete MPPT automatically. Therefore, the electrical control devices are eliminated, which makes the system more reliable. The new system contains two sets of three-phase winding, so the optimal winding-power-splitting-ratio of the new generator system must be analyzed. This paper gives a calculation method for obtaining the optimal power distribution ratio. This method is validated by the FEA results.

References

[1] S.M. Barakati, M. Kazerani, J.D. Aplevich, Maximum tracking control for a wind turbine system including a matrix converter. *IEEE Transactions on Energy Conversion*, **24(99)**, pp. 705–713, 2009.

[2] C. Pan, Y. Juan, A novel sensor less MPPT controller for a high-efficiency microscope wind power generation system. *IEEE Transactions on Energy Conversion*, **25(1)**, pp. 207–216, 2010.

[3] R. Pena, R. Cardenas, J. Proboste, et al., Sensor less control of doubly-fed induction generators using a rotor-current-based MRAS observer. *IEEE Transactions on Industrial Electronics*, **55(1)**, pp. 330–339, 2008.

[4] S. Ichikawa, M. Tomita, S. Doki, et al., Sensor less control of permanent-magnet synchronous motors using online parameter identification based on system identification theory. *IEEE Transactions on Industrial Electronics*, **53(2)**, pp. 363–372, 2006.

[5] K. Tan, S. Islam, Optimum control strategies in energy conversion of PMSG wind turbine system without mechanical sensors. *IEEE Transactions on Energy Conversion,* **19(2)**, pp. 392–399, 2004.

[6] Z. Chen, E. Spooner, Grid power quality with variable speed wind turbines. *IEEE Transactions on Energy Conversion,* **16(2)**, pp. 148–154, 2001.

[7] Y. Xia, K. Ahmed, B.W. Williams, Wind turbine power coefficient analysis of a new maximum power point tracking technique. *IEEE Transactions on Industrial Electronics*, **60(3)**, pp. 1122–1132, 2013.

[8] R. Datta, V.T. Ranganathan, A method of tracking the peak power points for a variable speed wind energy conversion system, *IEEE Transactions on Energy Conversion*, **18(1)**, pp. 163–168, 2003.

[9] E. Koutroulis, K. Kalaitzakis, Design of a maximum power tracking system for wind-energy-conversion applications. *IEEE Transactions on Industrial Electronics*, **53(2)**, pp. 486–494, 2006.

[10] Q. Wang, L. Chang, An intelligent maximum power extraction algorithm for inverter-based variable speed wind turbine systems. *IEEE Transactions on Power Electronics*, **19(5)**, pp. 1242–1249, 2004.

[11] S.M.R. Kazmi, H. Goto, H. Guo, et al., A novel algorithm for fast and efficient speed-sensor less maximum power point tracking in wind energy conversion systems. *IEEE Transactions on Industrial Electronics*, **58(1)**, pp. 29–36, 2011.

[12] D. Sera, R. Teodorescu, J. Hantschel, et al., Optimized maximum power point tracker for fast-changing environmental conditions. *IEEE Transactions on Industrial Electronics*, **55(7)**, pp. 2629–2637, 2008.
[13] V. Agarwal, R.K. Aggarwal, P. Patidar, et al., A novel scheme for rapid tracking of maximum power point in wind energy generation systems. *IEEE Transactions on Energy Conversion*, **25(1)**, pp. 228–236, 2010.
[14] B. Kou, Y. Bai, L. Li, A novel wind power generator system with automatic maximum power tracking capability. *IEEE Transactions on Energy Conversion*, **28(3)**, pp. 632–642, 2013.

The study of buffering methods of the fast mechanical switch

Cheng Lin[1], Yulong Huang[1], Weijie Wen[1], Tiehan Cheng[2], Shutong Gao[2], Keke Sun[2], Zhihua Ma[2]
[1]*State Key Laboratory of Power System, Department of Electrical Engineering, Tsinghua University, Beijing, China*
[2]*Pinggao Group Co. Ltd., Pingdingshan, Henan Province, China*

Abstract

As a key component of the hybrid HVDC circuit breaker, the fast mechanical switch conducts its opening and closing operation by electromagnetic repulsion, and should open to a position that can withstand the system voltage within several milliseconds. The moving parts of the switch can move at a velocity of 10 m/s, even faster, so it brings a problem of buffering inevitably. At present, the study of the fast mechanical switch focuses mainly on electromagnetic operating mechanism, but ignores the buffering problem. This paper studies the buffering problem of the fast mechanical switch with moving parts 4.75 kg, moving velocity 8 m/s and full stroke 45 mm. To solve this problem, this paper analyses three buffering methods: polyurethane (PU) buffers, electromagnetic buffering, and hydraulic buffering, and also shows the experimental results to confirm their buffering effects. As to hydraulic buffering, this paper builds the motion model of buffering process and conducts simulation in software MSC EASY5. Through comparing results of experiment and simulation, it proves that the model is correct and the hydraulic buffer fulfills the buffering requirements pretty well.
Keywords: fast mechanical switch, electromagnetic repulsion, buffering.

1 Introduction

In MTDC system, in order to avoid overall collapse in case a short-circuit fault occurs, it's absolutely essential that the DC circuit breaker can cut off and isolate the short-circuit fault within several milliseconds to keep the system

WIT Transactions on Engineering Sciences, Vol. 107, © 2015 WIT Press
www.witpress.com, ISSN 1743-3533 (on-line)
doi:10.2495/SEEEI140381

working normally and stably [1, 2]. The proposed solution is hybrid HVDC circuit breaker [3], as shown in Figure 1. When a short-circuit fault occurs, the power electronic devices turn off to transfer the fault current to the arrestors once the fast mechanical switch opens to a position that can withstand the voltage across the arrestors, so the breaking time depends on the operating speed of the fast mechanical switch. Traditional spring, permanent magnet and hydraulic operating mechanism can't meet this requirement, but the electromagnetic repulsion operating mechanism this paper refers to can meet this requirement well [4 ,5], as shown in Figure 2.

To reduce the breaking time, the fast mechanical switch should improve its opening speed. But if there's no buffering or not enough buffering, the moving parts will impact heavily on the fixed parts, with serious results as follows [6]:

1) Failure of the opening operation. In this situation, the moving contact will experience such a big bounce that returns back to the close position.
2) The opening operation will be a success, but the moving parts will bounce back and forth, resulting in continuous impact, which may damage the metal plate, the coils, the contacts, and bring a mechanical looseness. So it has greatly negative effects on the stability of the switch, and reduces the switch's mechanical endurance significantly.

According to the above analysis, the key to ensure the switch works reliably and stably is to find an effective buffering way. This paper chooses PU, electromagnetic and hydraulic buffering methods as its object of study [7-9], and conducts experiments to compare their buffering effects. Besides it builds the motion model of hydraulic buffering process and runs simulation in software MSC EASY5 to calculate the results. Through experiments, it proves that the simulation model is correct and the hydraulic buffer can be applied to the fast mechanical switch for buffering.

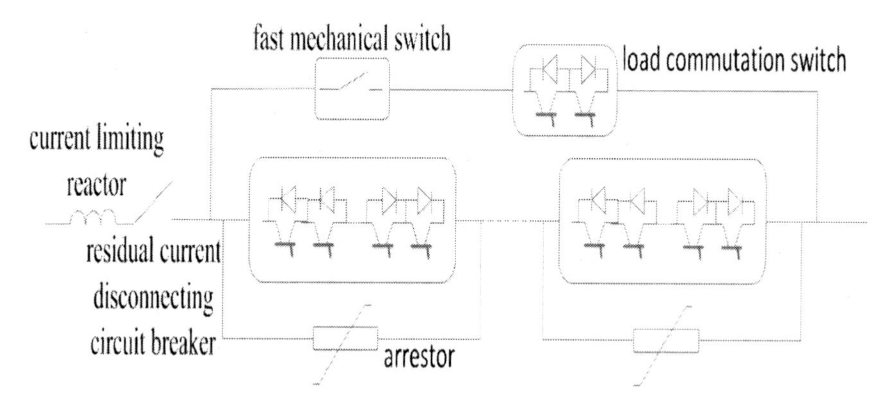

Figure 1: Hybrid HVDC circuit breaker.

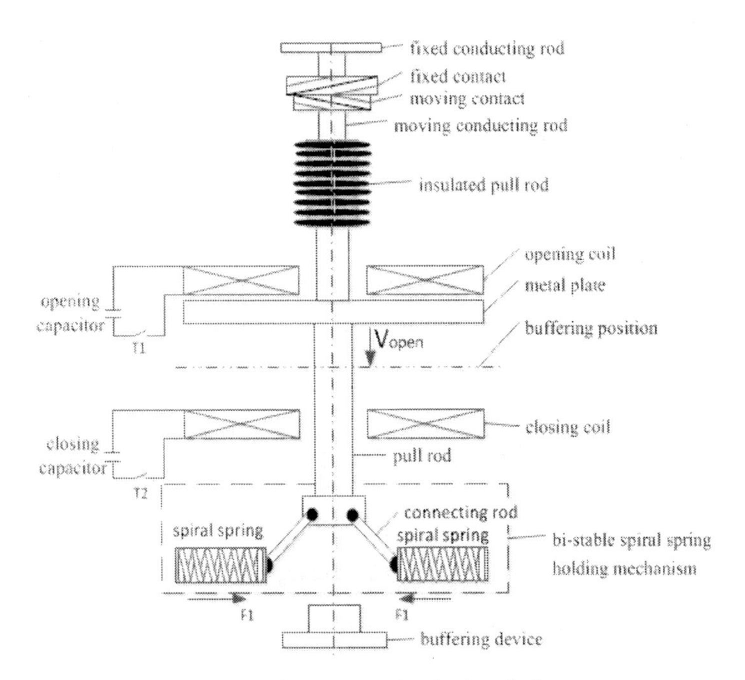

Figure 2: Fast mechanical switch.

Figure 3: Experimental platform.

2 Experimental platform

An experimental platform related to this research is set up as shown in Figure 3. The fast mechanical switch's full stroke is 45 mm, with bi-stable spiral spring as its holding mechanism and dumbbells replacing its moving parts, total mass 4.75kg. The opening and closing capacitor is impulse capacitor, charged by DC voltage-stabilized power supply with a maximum voltage 3000V, so we can control the charging voltage to change the opening speed, and the opening moment is controlled by triggering signal sent by single-chip. To measure the displacement, high-speed camera is adopt while linear displacement transducer is not used because of its large error. The buffering device is installed on a thick steel plate.

3 Buffering experiments

3.1 PU buffers

PU buffers absorb kinetic energy by the special microcellular structure of PU material, equivalent to many springs with air damping when buffering, so obvious nonlinear deformation. When the impact load is heavier, the more elastic potential energy is accumulated. After cutting process, 4 PU buffers are installed in parallel, with a pre-compression of 2 mm, as shown in Figure 4. With the opening capacitor charged to 800V, 900V, 1000V, the switch with PU buffers installed conducts its opening operation, and the experimental results are shown in Figure 5. As we can see, the higher the opening velocity, the longer the buffering stroke, also the more serious bounce at the end of stroke. When the voltage is 800V, 900V, although there's some bounce, the switch can still hold steadily at the open position with the force of holding mechanism, while if the voltage rises to 1000V, the switch bounces back to the close position. So the PU buffers can't meet the buffering requirements of the switch with a high opening speed.

3.2 Electromagnetic buffering

At the moment when the switch opens into the buffering stroke, the closing capacitor starts to discharge, so an electromagnetic repulsion is produced between the closing coil and the metal plate, and the moving parts begin to decelerate. In this experiment, the parameters are as follows: the total mass of moving parts 4.9 kg, full stroke 26 mm, both the opening and closing capacitor charged to

Figure 4: PU buffers.

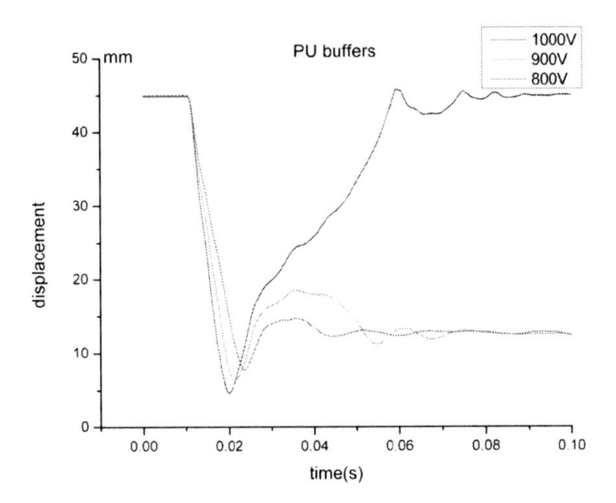

Figure 5: The experimental results of PU buffers.

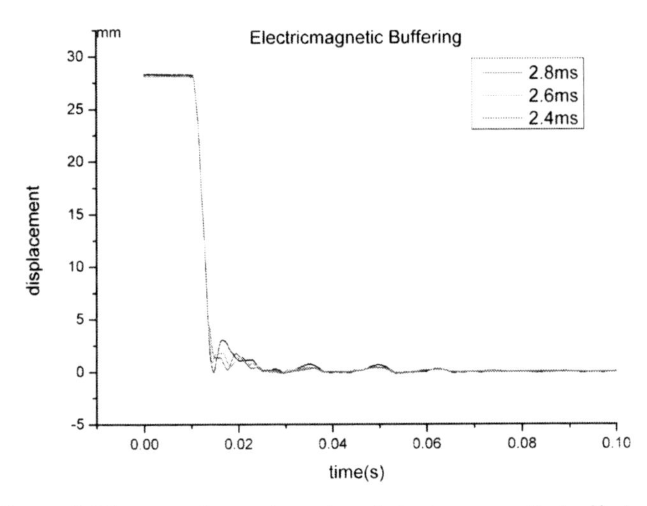

Figure 6: The experimental results of electromagnetic buffering.

1400V. We can control the discharging moment of closing capacitor to regulate the electromagnetic buffering effects. In this experiment, the closing capacitor discharges 2.4 ms, 2.6 ms, 2.8 ms after the opening capacitor discharges, and the results are shown in Figure 6. As we can see, its buffering effect is comparatively ideal with a little bounce at the end of stroke, and the switch can hold steadily at the open position eventually. The later the closing capacitor discharges, the less buffering effect it produces. So if the charging voltage and discharging moment of the opening and closing capacitor can match reasonably, electromagnetic buffering can meet the buffering requirements well. But the disadvantage is that if the switch needs to re-close, it takes a long time for capacitor to be charged, so there should be extra buffering capacitors, which is uneconomical.

4 Hydraulic buffering

4.1 Working principle and theoretical model

Hydraulic buffers absorb the kinetic energy of the moving parts by converting it to the internal energy of oil. Based on different structures, they can be divided into different groups: mandrel-type, and orifice-type. This paper chooses orifice-type, as shown in Figure 7.

After the moving parts impact on rod head 1, they move together with piston 4 and piston rod 3. Pressure of oil in chamber A will increase with extrusion, resulting in oil flowing to chamber B. The oil storage chamber C is connected with chamber B by oil drain holes 8 of large cross-sectional area, so pressure in chamber B and C can be seen equal. The volume occupied by piston rod moving into chamber A can be compensated by the compressed volume of the air in chamber C. With different pressure in chamber A and B, a resistance force will act on the moving parts and lead to its deceleration. Over the stroke length, the number of orifices will decrease and the resistance force remains unchanged approximately. When it stops, the buffering process is over. The return spring can push the piston and piston rod to their initial position, waiting for a next buffering process.

Now make a theoretical analysis of hydraulic buffering process during opening process. The parameters of the hydraulic buffer are shown in Figure 7. Set the

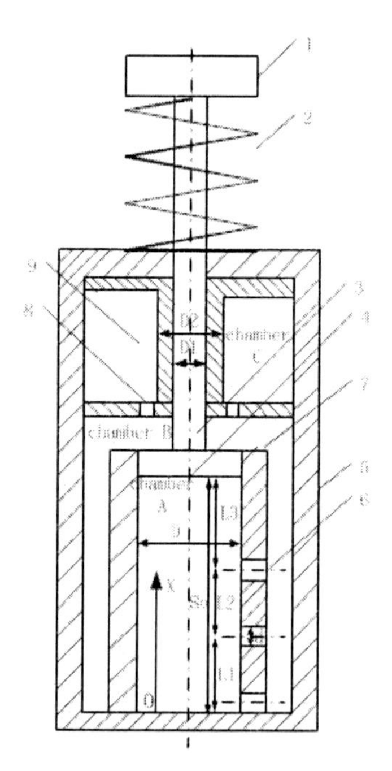

Figure 7: Hydraulic buffer. 1 – rod head, 2 – return spring, 3 – piston rod, 4 – piston, 5 – outer cylinder, 6 – orifice, 7 – inner cylinder, 8 – oil drain hole, 9 – oil storage chamber.

ending point of the buffering stroke as the origin of x coordinate axis. At the initial time, the piston locates at $x = s_0$ because of the return spring's force. The total mass of the switch's moving parts is M_1, and its velocity is V_0 at the moment when arrives at the buffering position. The total mass of rod head, piston and piston rod is M_2. The initial collision is seen as completely inelastic collision, so the velocity becomes $\frac{M_1 V_0}{M_1+M_2}$. The hydraulic buffering process is too complex that in order to establish the model we should simplify some factors as follows:

1) Ignore the friction force during buffering process;
2) As the piston with annular spring and the inner cylinder fit closely, we can assume that all the oil in chamber A flows to chamber B through orifices; compared with kinetic energy and pressure energy, gravitational potential energy of oil is so little that can be ignored; ignore the effect of increased temperature on density and viscosity of oil; ignore the air dissolved in oil;
3) Ignore the deformation of chamber A because of the increased pressure.
   When the piston moves to the position x at time t, its velocity becomes v, the pressure in chamber A is P_A and the pressure in chamber B and C is P_B. As the velocity of oil flowing through orifices is high, Re is bigger than 2000 during buffering process that oil flow can be seen as turbulent flow. So the volume of oil flowing from chamber A to chamber B is:

$$q = C_d A(x)\sqrt{\frac{2(P_A - P_B)}{\rho(P_A, T)}}$$

C_d is discharge efficient, usually 0.6 – 0.65, depending on the diameter of orifices, the thickness of inner cylinder and the sharpness of the margins; $A(x)$ is the total outflow area of orifices between 0 and x; ρ is the density of oil, a function of P_A and temperature T.

The oil in chamber A is compressed with P_A increasing, and the flow continuity equation is:

$$q\Delta t + \frac{V_1 \Delta P_A}{K} = A_1 v \Delta t$$

$V_1 = A_1 x$ is the volume of oil in chamber A, K is the bulk elastic modulus of oil. Then the motion equation is:

$$\frac{\mathrm{d}v}{\mathrm{d}t} = \frac{1}{M_1 + M_2}[-p_A A_1 + p_B A_2 + F_s + F_M - F_R - F_f + (M_1 + M_2)g]$$

$$v = -\frac{\mathrm{d}x}{\mathrm{d}t}$$

F_M is the electromagnetic repulsion, decreasing to a small value so as to be ignored;

F_R is the force of return spring; F_f is the total friction force between piston rod and seal ring, between piston and inner cylinder, and the dynamical friction exerted on metal plate and pull rod. F_s is the force of bi-stable helical spring holding mechanism with a compression x_0, expressed as:

$$F_s = 2\text{k}\left(\sqrt{L^2 - (\frac{S_0}{2} - x)^2} - \sqrt{L^2 - (\frac{S_0}{2})^2} + x_0\right)\frac{(\frac{S_0}{2} - x)}{\sqrt{L^2 - (\frac{S_0}{2} - x)^2}}$$

So run a simulation in software MSC EASY5 to calculate the above model, we can get the hydraulic buffering characteristics curves.

4.2 Results of simulation and experiments

Parameters are defined as follows: $M_1 = 4.75$ kg; $M_2 = 0.2$ kg; $s_0 = 23$ mm . $k = 63$ kN/m; $L = 50$ mm; $x_0 = 10$ mm return spring: $k_1 = 1.338$ kN/m , pre-compression $x_1 = 52$ mm . To meet the requirement of opening time, the max moving velocity reaches 8 m/s. Run a simulation in software MSC EASY5, and the results of displacement and velocity curves are shown in Figure 8. It indicates that the buffering effect of this hydraulic buffer is pretty good, even no bounce at the end of stroke.

To verify the results of simulation, conduct an experiment that the opening capacitor is charged to 1300V to ensure the moving velocity of 8 m/s and get the displacement and velocity curves shown in Figure 8. The results in Figure 8 show the excellent agreement between simulation and experiment.

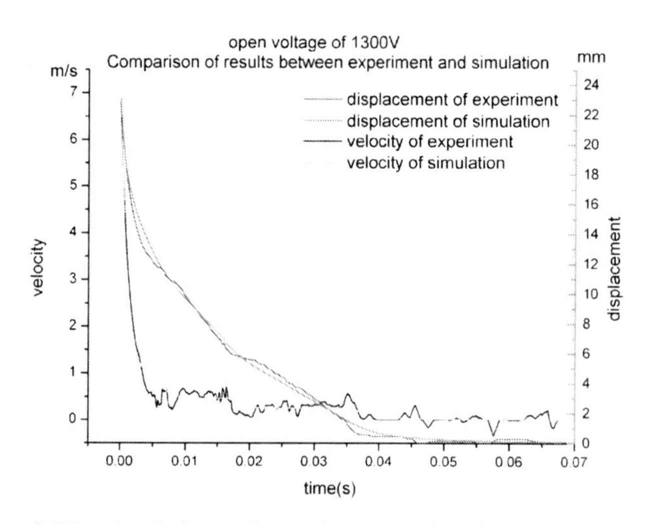

Figure 8: The simulation and experiment results of the hydraulic buffer.

5 Conclusion

1) PU buffers have some bounce at the end of buffering stroke if applied to the fast mechanical switch, and the higher the opening speed, the more serious the bounce. In extreme situations the switch returns back to its close position after opening, so PU buffers can't meet the buffering requirements.
2) The effect of electromagnetic buffering is comparatively ideal with a little bounce at the end of buffering stroke. If the charging voltage and discharging moment of the opening and closing capacitor match well, electromagnetic buffering can meet the buffering requirements fairly well, but the disadvantage is that it's so costly.
3) The hydraulic buffer has the best buffering effect with no bounce at the end of buffering stroke compared with above two buffering methods, so it will be the first choice to solve the buffering problem. Based on the working principle of hydraulic buffer, the motion model of buffering process is built and calculated in software MSC EASY5. The simulation and experimental results fit pretty well, so it proves the validity of the model.

Acknowledgements

This work is financially supported by Science and Technology Project of SGCC (Key Technology Study and Prototype Development on HVDC Circuit Breaker), the National Natural Science Foundation of China under the reference number 51377090, the Science and Technology Project of SGCC (KJ[2012]227), and the Excellent Young scientists Fund of Pinggao Group Ltd. (PGKQ2013-002).

References

[1] Zhu Jie-Bei, Future multi-terminal HVDC transmission systems using voltage source converters. *Universities Power Engineering Conference (UPEC)*, pp. 1–6, 2010.
[2] Zhang Wen-Liang, Multi-terminal HVDC transmission technologies and its application prospects in China. *Journal of Power System Technology*, **34(9)**, pp. 1–6, 2010.
[3] Wei Xiao-Guang, A novel design of high-voltage DC circuit breaker in HVDC flexible transmission grid. *Journal of Automation of Electric Power System*, **37(15)**, pp. 95–102, 2013.
[4] Mao Hai-Tao, Finite element analysis of fast electromagnetic repulsion mechanism. *Journal of High Voltage Engineering*, **35(6)**, pp. 1420–1425, 2009.
[5] Liu Jia-Yu, *Simulation and Optimization of Vacuum Switch Electromagnetic Repulsion Mechanism*. Tsinghua University, Beijing, 2010.
[6] Jiang Nan, *Research on the Design Method of the Electromagnetic Repulsion Mechanisms*. China Ship Research and Development Academy, Beijing, 2013.

[7] Li Yun-Peng, *Research on Design and Optimal Matching of the Buffer Made in Polyurethane*. Huazhong University of Science and Technology, Wuhan, 2013.

[8] Sun Shuang, *Simulation and Optimum Design to the Multi-Orifice Hydraulic Buffer*. Dalian University of Technology, Dalian, 2006.

[9] Zou Gao-Peng, The simulation model and optimum of hydraulic buffers in spring operating mechanism. *Science and Technology Innovation Herald*, **(25)**, pp. 21–22, 2011.

Switched nonlinear optimal excitation control of power system

Yalu Li, Baohua Wang
Department of Automation, Nanjing University of Science and Technology, Nanjing, China

Abstract

The research of generator excitation control method has theory significance and engineering application value. In the paper, we designed a switched nonlinear optimal excitation which co-ordinately improved the transient stability and voltage regulation. Considering the power system is nonlinear and dynamic, we realized the feedback linearization of the system with the inverse system and optimum control method. Then, by building Lyapunov energy function, we found nonlinear compensation for the control rule with zero dynamics. So the designed excitation can suppress the nonlinear part of the system better under big disturbance. Simulation results on MATLAB prove that the proposed switched nonlinear optimal excitation can effectively improve the transient stability and voltage regulation performance.
Keywords: excitation control, optimum control, switched system.

1 Introduction

Synchronous generator's magnetic field is established and controlled by excitation controller [1]. Excitation controller can improve system damping, control terminal voltage and adjust reactive power allocation, which guarantee the power system stability and good dynamic response performance [2].

Most of these controllers are based on single theory [3]. The linear optimal excitation finds optimal control strategy to meet certain condition, so that the performance of the system reaches the optimum [4, 5]. Nonlinear control rules provide improved performance compared with linear ones, which take the changing of operating point into account. Like the exact feedback linearization control method [6] and control rules based on the hierarchical control principle in order

www.witpress.com, ISSN 1743-3533 (on-line)
doi:10.2495/SEEEI140391

to co-ordinately enhance the voltage regulation accuracy and transient stability of power systems [7].

It's common knowledge that the application of variety theories has been the key to solve problems. And controllers based on multivariable feedback are more tally with the actual situation [8]. In this paper, based on the theory of feedback linearization, we modify the linear optimal excitation. The feedback linearization of the system is realized via the inverse system method. It takes the local linearization parameters' changing into account when the running mode changes. Then, we partially linearism the system which contains zero dynamics. We use the Lyapunov function method to evaluate the stability of the system and design an additional part to suppress the effect of the nonlinear part. Finally, a switched excitation controller is designed for the power system.

The paper is organized as follows. In the next section, we propose the model that we research in this paper, and some parameters are given. A switched optimal excitation has been designed in Section 3. In Section 4, computational simulation based on MATLAB is made to illustrate the efficiency of the controller, compared with previous design. Finally, we conclude our paper in Section 5.

2 Discussed model

The discussed model is given in Figure 1.

The parameters are set as shown in Table 1.

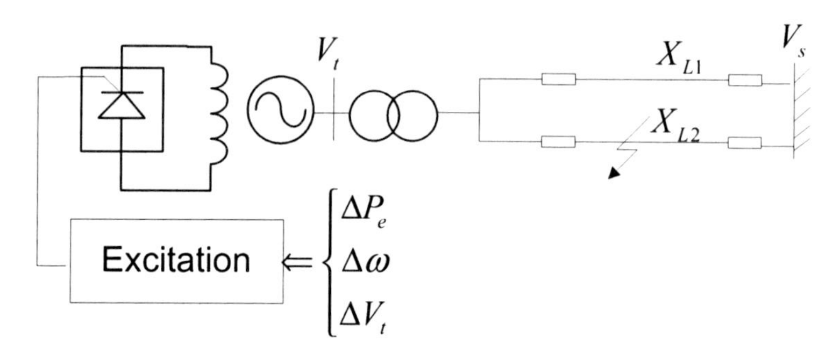

Figure 1: Infinite bus power system model.

Table 1: The parameters of the model.

Generator		Transformer	Transmission	Initial Steady State
$P_n = 200\,\text{MVA}$	$X_d' = 0.6$	$X_T = 0.01$	$X_L = 1.46$	$\delta_0 = 70°$
$V_n = 13.8\,\text{kV}$	$T_{d0} = 10\,\text{s}$	13.8kV/230kV		$P_n = 0.5\,\text{p.u}$
$f_n = 50\,\text{Hz}$	$H = 8\,\text{s}$	$\Delta - \text{Yg}$		$V_s = 1\,\text{p.u}$
$X_d = 2.543$	$D = 5\,\text{s}$			

3 Designs of the excitation

3.1 Linear optimal excitation

The transition process of synchronous generator is quite complex. The general state space equation of linear system can be written as follow:

$$\dot{\mathbf{X}}(t) = \boldsymbol{A}(t)\mathbf{X}(t) + \boldsymbol{B}(t)\mathbf{U}(t) \tag{1}$$

where $\mathbf{A}(t) \in R^{n\times n}$ is state matrix, $\mathbf{X}(t) \in R^{n}$ is state variables, $\mathbf{B}(t) \in R^{n\times r}$ is control matrix, $\mathrm{U}(t) \in R^{r}$ is control vector. The optimal control vector will be $\mathbf{U}^{*}(t) = -\mathbf{K}^{*}(t)\mathbf{X}(t)$, where $\mathbf{K}^{*}(t)$ can be solved by lqr function.

In order to establish the equation of the power system, we need to linearize the above equation based on general three order model of generator.
Set $\mathbf{X}(t) = [\Delta P_e \quad \Delta\omega \quad \Delta V_t]^T$, the equation can be changed into:

$$\begin{bmatrix} \Delta\dot{P}_e \\ \Delta\dot{\omega} \\ \Delta\dot{V}_t \end{bmatrix} = \begin{bmatrix} \dfrac{S_E - S_v}{T_d' S_v} & S_E & -\dfrac{R_v S_E}{T_d' S_v} \\ -\dfrac{\omega_0}{H} & -\dfrac{D}{H} & 0 \\ \dfrac{S_E - S_v}{T_d' R_v S_v} & \dfrac{S_{E'} - S_v}{R_v} & -\dfrac{S_E}{T_d' S_v} \end{bmatrix} \begin{bmatrix} \Delta P_e \\ \Delta\omega \\ \Delta V_t \end{bmatrix} + \begin{bmatrix} \dfrac{R_{E'}}{T_{d0}} \\ 0 \\ \dfrac{R_{E'}}{T_{d0} R_v} \end{bmatrix} \Delta E_f \tag{2}$$

where $\Delta P_e = S_E \Delta\delta + R_E \Delta E_q$, $\Delta P_e = S_{E'} \Delta\delta + R_{E'} \Delta E_q'$, $\Delta P_e = S_v \Delta\delta + R_v \Delta V_t$. We can get

$$\Delta E_f = U = -\mathbf{KX} \tag{3}$$

3.2 Nonlinear optimal excitation

The method of linearization can meet general engineering needs for small disturbance. However, when the system has big disturbance, the operating equilibrium point will change. Thus, the linear optimal excitation may be insufficient or even wrong.

Considering the complexity of power system and diversity of linear equations, we use the direct feedback linearization method.

The terminal voltage of generator can be expressed as:

$$\begin{cases} V_t = \sqrt{V_{td}^2 + V_{tq}^2} \\ V_d = x_q V_s \sin\delta / x_{q\Sigma} \\ V_q = [(x_{d\Sigma}' - x_d')E_q' + x_d' V_s \cos\delta] / x_{d\Sigma}' \end{cases} \tag{4}$$

Hence, we can get

$$\Delta\dot{V}_t = C_1 + C_2 E_f \tag{5}$$

Set the linear feedback control rule as $v_1 = \Delta\dot{V}_t$,

$$E_f = \frac{v_1 - C_1}{C_2} \tag{6}$$

According to Section 3.1,

$$\begin{bmatrix}\Delta\dot{P}_e \\ \Delta\dot{\omega} \\ \Delta\dot{V}_t\end{bmatrix} = \mathbf{AX} + \mathbf{B}U = (\mathbf{A} - \mathbf{BK})\mathbf{X} = \begin{bmatrix} a_{11} & a_{12} & a_{13} \\ a_{21} & a_{22} & a_{23} \\ a_{31} & a_{32} & a_{33}\end{bmatrix}\begin{bmatrix}\Delta P_e \\ \Delta\omega \\ \Delta V_t\end{bmatrix} - \begin{bmatrix} b_1 \\ b_2 \\ b_3\end{bmatrix}\begin{bmatrix}k_p & k_\omega & k_v\end{bmatrix}\begin{bmatrix}\Delta P_e \\ \Delta\omega \\ \Delta V_t\end{bmatrix}$$

Hence,

$$v_1 = \Delta V_t = a_{31}\Delta P_e + a_{32}\Delta\omega + a_{33}\Delta V_t - b_3(k_p\Delta P_e + k_\omega\Delta\omega + k_v\Delta V_t) = -\mathbf{K}'\mathbf{X}$$

3.3 Switched nonlinear optimal excitation

A partial state feedback linearization is realized by employing inverse system method. Zero dynamics will appear when the relative degree is less than system degree. Zero dynamic system can be generally expressed as $\dot{\eta} = p(\xi, \gamma, \eta)$, which can be transformed into Eq. (6)

$$\dot{\eta} = A_1\xi + B_1\gamma + C_1\eta + \tilde{g} \tag{7}$$

where $A_1 = \left.\frac{\partial p}{\partial \xi}\right|_{(\xi_0,\gamma_0,\eta_0)}$, $B_1 = \left.\frac{\partial p}{\partial \gamma}\right|_{(\xi_0,\gamma_0,\eta_0)}$, $C_1 = \left.\frac{\partial p}{\partial \eta}\right|_{(\xi_0,\gamma_0,\eta_0)}$.

Then the mathematical function can be expressed as:

$$\dot{\mathbf{X}} = \mathbf{A}'\mathbf{X} + \mathbf{B}'\varphi + \mathbf{g} \tag{8}$$

Set $\mathbf{X} = \begin{bmatrix}\Delta P_e & \Delta\omega & \Delta V_t\end{bmatrix}^T$ like Section 3.1. Choose the output $y = \Delta V_t$, then the linearized system contains two-order zero dynamic:

$$\begin{cases} \Delta\dot{\omega} = -\dfrac{\omega_0}{H}\Delta P_e - \dfrac{D}{H}\Delta\omega \\ \Delta\dot{P}_e = c_1\Delta P_e + c_2\Delta\omega + c_3\Delta V_t + g(\Delta P_e, \Delta\omega, \Delta V_t) \end{cases} \tag{9}$$

Set the feedback control rule $\varphi = v_1 + v = \dot{y}$, v_1 is same to Section 3.2.

$$\dot{\mathbf{X}} = \mathbf{A}'\mathbf{X} + \mathbf{B}'\varphi + \mathbf{g} = \mathbf{A}'\mathbf{X} + \mathbf{B}'(-\mathbf{K}'\mathbf{X} + v) + \mathbf{g} = \mathbf{A}_c\mathbf{X} + \mathbf{B}'v + \mathbf{g} \tag{10}$$

We construct the energy function $V = \mathbf{X}^{\mathrm{T}}\mathbf{P}\mathbf{X}/2$ for nonlinear system where $\mathbf{P} = \mathbf{I}$, and select it as Lyapunov function. To make the closed-loop system stable, we need $\dot{V} < 0$. So we select the nonlinear compensation for the control rule as:

$$v = \begin{cases} -\dfrac{1}{\left|\mathbf{X}^{\mathrm{T}} p_3\right|}\left|\mathbf{X}^{\mathrm{T}} p_1 g\right|, \left|\mathbf{X}^{\mathrm{T}} p_3\right| \geq \varepsilon \\ -\dfrac{\mathbf{X}^{\mathrm{T}} p_3}{\varepsilon^2}\left|\mathbf{X}^{\mathrm{T}} p_1 g\right|, \left|\mathbf{X}^{\mathrm{T}} p_3\right| < \varepsilon \end{cases} \tag{11}$$

where ε is a proper positive value.
So,

$$E_f = \frac{(v_1 + v) - C_1}{C_2} = \begin{cases} \dfrac{-\mathbf{K}'\mathbf{X} - \dfrac{1}{\left|\mathbf{X}^{\mathrm{T}} p_3\right|}\left|\mathbf{X}^{\mathrm{T}} p_1 g\right| - C_1}{C_2}, \left|\mathbf{X}^{\mathrm{T}} p_3\right| \geq \varepsilon \\ \dfrac{-\mathbf{K}'\mathbf{X} - \dfrac{\mathbf{X}^{\mathrm{T}} p_3}{\varepsilon^2}\left|\mathbf{X}^{\mathrm{T}} p_1 g\right|}{C_2}, \left|\mathbf{X}^{\mathrm{T}} p_3\right| < \varepsilon \end{cases} \tag{12}$$

4 Computational analysis

In order to illustrate the feasibility and effectiveness of the excitations, we construct a model according to Section 2. We test two different approaches. The first is the linear optimal excitation, and the second is the switched excitation proposed in this paper. We use the MATLAB to simulate. The terminal voltage and rotor angle are simulated when the fault occurred at t_f and disappeared at t_c. The performance of the excitations is shown as follows:

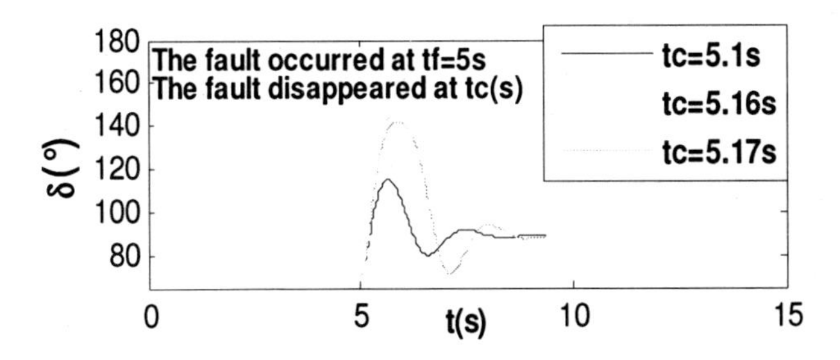

Figure 2: Performance of approach 1.

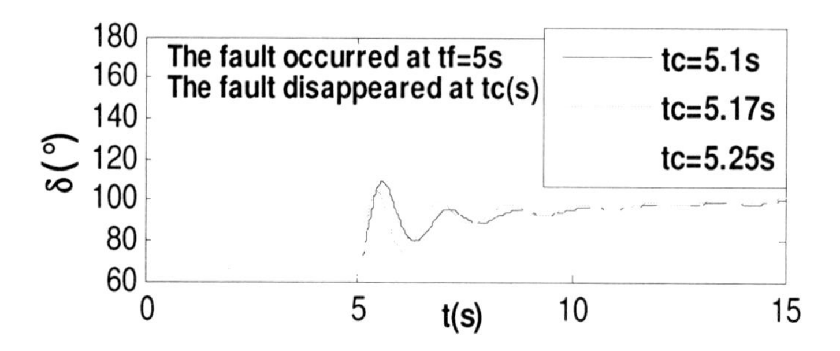

Figure 3: Performance of approach 2.

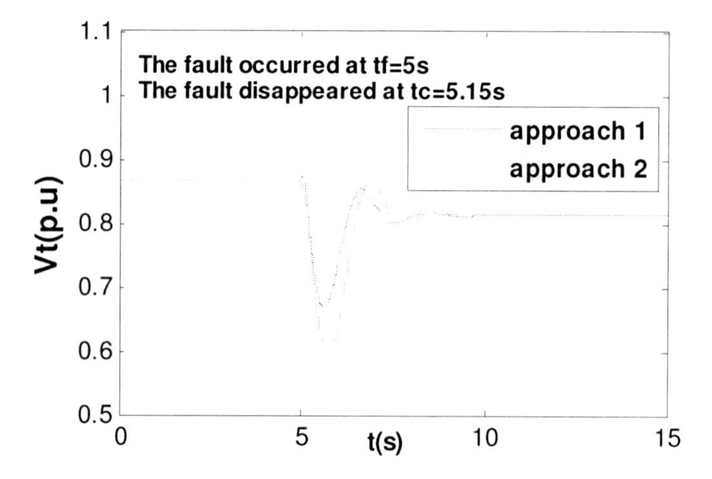

Figure 4: Terminal voltage comparison between two approaches.

As can be seen from Figures 2 and 3, both of the excitations can improve the transient stability and voltage regulation. But the t_c of approach 2 (t_c = 5.25s) is longer than that of approach 1 (t_c = 5.16s) when the stability is guaranteed, which means approach 2 can allow more time to remove the fault when the power system meets big disturbance.

From Figure 4, we can see the terminal voltage comparison between two approaches. When we choose approach 2, the terminal voltage recovers more rapidly and the oscillation amplitude is smaller in contrast with approach 1. It can be seen that the proposed excitation is effective.

5 Conclusions

The proposed switched nonlinear optimal excitation has different control rules when the system's structure changes and can effectively restrain the effect of nonlinear part, so the terminal voltage can recover better. All in all, the designed

excitation can effectively improve the transient stability and has better voltage regulation performance.

References

[1] Lu Qiang, Mei Sheng-Wei, Sun Yuan-Zhang, *Nonlinear control of power system*. Tsinghua University Press, Beijing, 1984.

[2] Liu Qu, *Power System Stability and Generator Excitation Control*. China Electric Power Press, Beijing, 2007.

[3] Cheng Qi-Ming, Cheng Yi-Man, Xue Yang, Hu Xiao-Qing, Development and prospects of excitation control methods. *Chinese Journal of Electric Power Automation Equipment*, **32(5)**, pp. 108–117, 2012.

[4] Liu Ming-Jin, Design and simulation of generator excitation system based on linear optimal control theory. *Chinese Journal of Electric Switcher*, **(1)**, pp. 52–55, 2010.

[5] An Qiao-Jing, Zhu Chang-Qing, GuZhi-Feng, Liu Feng, Zhao Yuan-Ping, Research of the optimal excitation control technology for synchronous generator based on AC tracking. *Chinese Journal of Industry and Mine Automation*, **39(2)**, pp. 58–61, 2013.

[6] Liu Hui, Li Xiao-Cong, Wei Hua, Nonlinear excitation control for generator unit based on NCOHF. *Journal of Proceedings of the CSEE*, **27(1)**, pp. 14–18, 2007.

[7] Chen Tie, ShuNai-Qiu, Study of nonlinear excitation based on direct feedback linearization. *Chinese Journal of Electric Power Science and Engineering*, **(1)**, pp. 30–31, 42, 2005.

[8] O. Akhrif, F.A., Okou, Application of a multivariable feedback nonlinearization scheme for rotor scheme for rotor angle stability and voltage regulation of power systems. *IEEE Transactions on Power Systems*, **14(2)**, pp. 620–628, 1999.

Non-contact power system voltage sensor based on the theory of electric field coupling

Songnong Li[1], Xingzhe Hou[1], Kongjun Zhou[1], Ruixi Luo[2], Qiang Zhou[2], Wei He[2]
[1]*State Grid Chongqing Electric Power Co. Electric Power Research Institute, Chongqing, China*
[2]*State Key Laboratory of Power Transmission Equipment & System Security and New Technology, Chongqing University, Chongqing, China*

Abstract

Traditional voltage transformer has shown more and more inadaptability in the develop trend of smart grid due to its size, weight, insulating cost, steady-state accuracy and transient response. A voltage sensor is proposed based on the theory of electric field coupling to realize non-contact voltage measurement with power system device. A new difference output structure is introduced to improve measurement accuracy. The sensor is manufactured into PCB to measure voltage of a bus in a high voltage transformer's insulating bush without keeping electrical connection with it. Test of the sensor demonstrates the possibility and advantage of this technique.
Keywords: voltage measurement, electric field, non-contact.

1 Introduction

Voltage sensor, which obtains voltage signal from power system devices, holds a very important position in areas of power metering [1], relay protection [2], automation device control [3] and on-line monitoring [4]. The accuracy and reliability of voltage sensor is meaningful in maintaining fair trade settlement of power and ensuring safe operation of the grid [5, 6].

Inductive voltage transformer (IVT) and capacitive voltage transformer (CVT), which are usually known as traditional voltage transformer, are used most widely in today's voltage measurement application of power system [7].

WIT Transactions on Engineering Sciences, Vol. 107, © 2015 WIT Press
www.witpress.com, ISSN 1743-3533 (on-line)
doi:10.2495/SEEEII140401

As IVTs and CVTs need to keep electrical connection to the high-voltage conductors to be measured, then these traditional voltage transformer have to get a complicated insulation structure which results in their cumbersome size and expensive costing. Additionally, IVTs are also limited in narrow bandwidth, small dynamic range, and ferromagnetic resonance due to iron core saturation [8]. On the other hand, CVTs can achieve a larger dynamic range when compared with IVTs as the high-voltage is divided linearly with capacitive voltage divider firstly. However CVTs get a poor performance in stability over time and temperature which is caused by the drifting in the dielectric properties of their paper-oil insulation. CVTs are also restricted by its transient response of which secondary voltage may not catch up with the change of primary voltage.

Since the electric field intensity around a high-voltage conductor is proportion to the potential of the conductor to be measured. It is possible to measure conductor's voltage by measuring electric field, which means an electric-field-coupling-based voltage sensor can work in a non-contact way and avoid the used of insulation structure. So, the sensor can be manufactured into a very small size without worrying about insulation breaking down. In addition, the dielectric of the electric field coupling voltage sensor and the medium between conductor and sensor are linear, which make the sensor to have large dynamic range and well-performed transient response [9].

In this paper, the theory of electric field coupling is explained. Following this a difference output structure is proposed to improve the performance of the sensor and finite element calculation for optimal design is carried out based on the error analysis of the sensor. Finally, the designed sensor is tested within a 10 kV experimental system.

2 Operating principle of voltage sensor

When an electrode is placed near the tested conductor, it can be regarded as a current source by using Gauss Theorem. A load resistance is connected to this electrode and the voltage-drop on the resistance is proportional to the first time derivative of the incident electric fields, which is also proportional to the high voltage to be measured. After integrating the signal from the load resistance, the attenuated waveform of the high voltage can be obtained. By using Charge Simulation Method (CSM) and Gauss Theorem, the output of the sensor is given in Eq. (1) when it is used for a cylindrical conductor measurement.

$$V_0 = \varepsilon_0 A_{eq} R_m \frac{\mathrm{d}}{\mathrm{d}t} E(t) = \varepsilon_0 A_{eq} R_m \arctan \frac{2r_0 + R}{2r_0 - R} \frac{\mathrm{d}\varphi(t)}{\mathrm{d}t} \quad (1)$$

where A_{eq} is the equivalent area of the sensor which is determined by positions of sensor and conductor, R_m is the value of the load resistance, R_0 is the distance between the sensor and the conductor, r_0 is the radius of the conductor, V_0 is the voltage-drop on the load resistance, $E(t)$ is the electric field strength where the sensor is mounted, $\varphi(t)$ is the voltage of the conductor. Since the output of the sensor is proportional to the first derivative of the conductor's voltage. In some cases, it needs an integrator to reconstruct voltage signal.

Considering the influence of frequency, the sensor can serve as a capacitive divider to analyze its transfer function with inductance of the connections neglected [10]. The equivalent circuit of the sensor and passive integrator is shown in Figure 1.

Then the transfer function of the sensor is given by Eq. (2).

$$H(s) = \frac{V_i}{V_1} = \frac{C_m R_m s}{(C_m + C_s) R_m s + 1} \tag{2}$$

where V_i is the test high voltage, V_1 is the output voltage of the sensor, V_o is integrated voltage of the sensor, C_m is the equivalent mutual capacitance between test conductor and the probe, C_s is the stray capacitance of the probe.

When $(C_m + C_s) R_m \ll 1$, the output of the sensor V_1 is proportional to the derivative of V_i, then the output of integrator V_o is proportional to V_i and it works in Differential Mode. While, when $(C_m + C_s) R_m \gg 1$, V_1 is proportional to V_i without being integrated and it works in Self-Integrating Mode.

The bandwidth of the measurement system (including sensor and integrator) is limited by the integrator when the sensor works in Differential Mode, wave distortion may occur with the influence due to spurious inductance of integrator [11]. Since the frequency in power system ranges from tens Hertz to millions Hertz, it is more suitable for choosing a self-integrating sensor to have a broader frequency range.

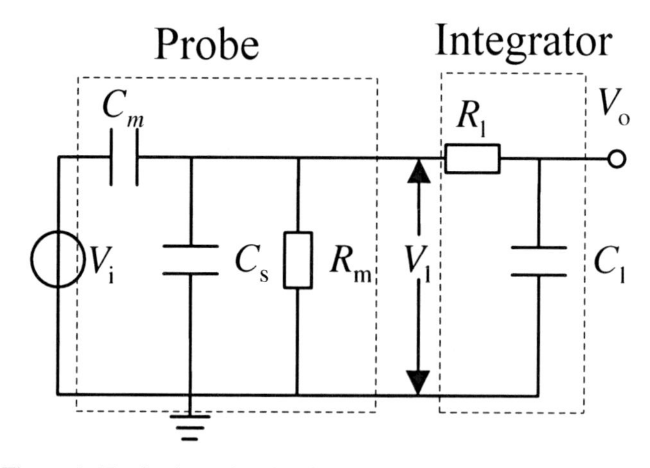

Figure 1: Equivalent circuit of the sensor and passive integrator.

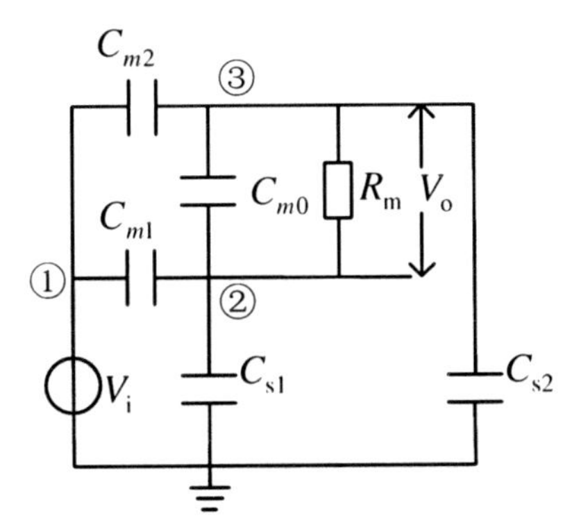

Figure 2: Difference output circuit structure.

3 Design of the sensor

3.1 The difference output circuit structure

If the sensor works in self-integrating mode at the frequency of 50 Hz or lower, it can be seen that there must be a large value for C_m and C_s through Eq. (2). However the equivalent capacitance between two conductors (which is always in the level of tens pF) cannot achieve such a large number. A difference output circuit structure in Figure 2 is introduced and the load resistance is replaced by a differential amplifier to ensure that the sensor works in self-integrating mode at the low frequency.

In Figure 2, C_{m1} and C_{m2} are equivalent mutual capacitance between test conductor and electrodes, C_{s1} and C_{s2} are stray capacitance of the electrodes, C_{m0} is mutual capacitance between electrodes. Node ② and node ③ are potential of two unearthed and disparate electrodes with different A_{eq}. The potential of electrodes is floating, so the maximum electric field intensity in and around the probe would be smaller. Then the insulation is strengthened and the probe would be easily installed with smaller size. The transfer function of the probe is given in Eq. (3).

$$H(s)=\frac{V_o}{V_i}=\frac{C_1R_m s}{C_2R_m s+1}$$
$$C_1=\frac{C_{m1}C_{s2}-C_{m2}C_{s1}}{C_{m1}+C_{m2}+C_{s1}+C_{s2}},C_2=\frac{1}{\frac{1}{C_{m1}+C_{s1}}+\frac{1}{C_{m2}+C_{s2}}}+C_{m0} \tag{3}$$

It can be seen that the difference output circuit structure did not change the transfer function of sensor but to introduce a new parameter C_{m0} into the polo of transfer function. It is possible for C_{m0} to achieve a large value as the electrodes could be placed much closed to each other. Capacitance between different electrodes can also be added up by getting the electrodes in parallel. The amplitude-frequency function and phase-frequency function can be deduced according to Eq. (3), which are shown in Eqs. (4) and (5).

$$|H(\omega)| = \frac{C_1 R_m}{\sqrt{(C_2 R_m)^2 + \frac{1}{\omega^2}}} \tag{4}$$

$$\angle H(\omega) = \arctan\frac{1}{C_2 R_m \omega} \tag{5}$$

It can be referred from Eqs. (4) and (5) that the sensor could achieve a higher divide ratio and a smaller phase error by increasing C_{m0}. It is better to get a large mutual capacitance value C_{m0} to have the sensor performs better when it comes to the optimization of the sensor.

3.2 Design of the electrode

The design of the probe is shown in Figure 3.

In purpose of insulation and achieving a higher sensitivity, semi-toroid electrodes are applied in the design of the sensor. The sensor is manufactured into PCBs for high precision. A multiple-electrode structure is used for increasing value of C_{m0} as shown in Figure 2. There are tens of toroid electrodes with different radius placed on the top layer and bottom layer of a PCB. Mutual capacitance exists between each electrode. These mutual capacitance of electrodes are added up by connecting them in parallel. To obtain a larger mutual capacitance, the top layer electrodes of several PCBs are paralleled and the same configuration is applied on the bottom layer electrodes. Electrodes on the same layer are equipotential and output voltage of the probe is the potential difference between top and bottom layer electrodes of PCBs. In this way, a large mutual capacitance value can be achieved in order to minimize sensor's phase error in low frequency.

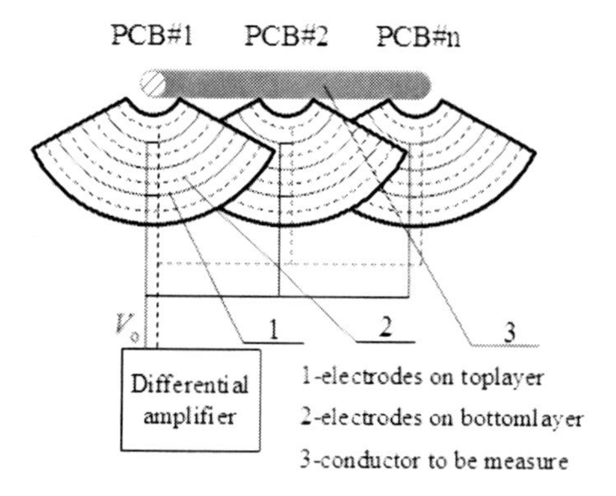

Figure 3: Structure design of sensor.

As the surface of semi-toroid electrodes are approximate to the equipotential surface of electric field around the conductor, the electric field can be distributed evenly at the surface of electrodes of the sensor, which makes the local maximum electric field intensity be in a low level and the possibility of insulation breakdown could be reduced.

4 Test of the sensor

4.1 Test setup for sensor

The designed sensor for testing comprises 5 PCBs in parallel and the parameter of each PCB is shown in Table 1.

The sensor is installed on the high voltage insulating bush of a boost transformer which is shown in Figure 4.

The test setup is shown in Figure 5.

Table 1: The design parameters of the PCB sensor.

	Number of Electrodes	Electrode Radius (mm)	Electrode Width (mil)
Top layer	25	44	6
Bottom layer	30	49.08	10

Figure 4: The installation of the voltage sensor.

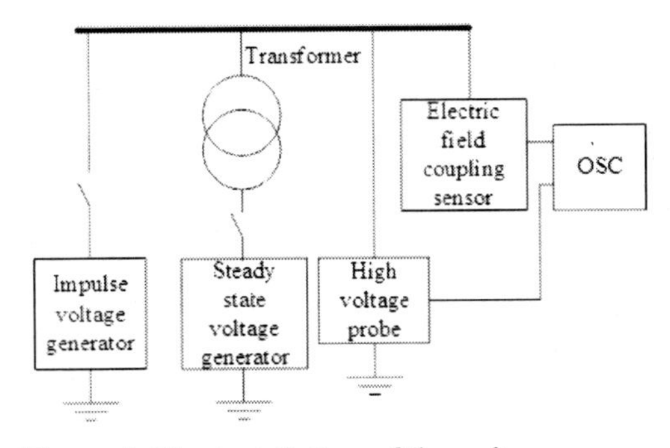

Figure 5: The installation of the voltage sensor.

In Figure 5, the steady state voltage generator is used to generate a sinusoidal voltage signal of which frequency ranges from 10 Hz to thousands Hz. The impulse voltage generator is used to generate 1.2/50μs lightening waveform for sensor's transient performance testing. The sensor is compared with a Tektronix produced P6015A capacitance high voltage probe which needs to connect to high voltage conductor (1000:1, 75MHz, 3pF, 100 MΩ, 40 kV peak pulse, 100 ms maximum duration, 20 kA direct current). The voltage level is up to 10kV and the generated voltage is measured by the capacitance high voltage probe and electric field coupling sensor simultaneously. The outputs are sent to a 3Gs/s Agilent oscilloscope (OSC) for observing.

4.2 Steady state test

The steady state test is carried out to examine the dynamic range of the sensor. The voltage generates a 50 Hz voltage of which voltage changes from $5\%U_n$ to $120\%U_n$. The output of electric field coupling sensor and high voltage probe are compared and error results are shown in Table 2.
where $\varepsilon\%$ is the ratio error of sensor and φ_u is the phase error of sensor. Ratio error and phase error are defined in Eqs. (6) and (7) according to standard IEC60044–7.

$$\varepsilon\% = \frac{K_n U_s - U_p}{U_p} \times 100\% \tag{6}$$

$$\varphi_u = \varphi_s - \varphi_p \tag{7}$$

where K_n is divide ratio of sensor, U_s is the output voltage of sensor, U_p is the voltage measured by high voltage probe. φ_s and φ_p are the initial phase angle of two sensor's output voltage waveform respectively.

The results from Table 2 shows that the sensor achieve an accuracy by 0.5 class voltage transformer level ($\varepsilon\%$ <0.5%, φ_u<20′) and holds a wide dynamic range.

Table 2: The design parameters of the PCB sensor.

	U_p/kV	U_s/V	$\varepsilon\%$	φ_u/(′)
2%Un	0.214	0.213	–0.35	19
5%Un	0.512	0.511	–0.29	19
10%Un	1.018	1.016	–0.23	18
20%Un	2.000	1.998	–0.11	17
40%Un	3.987	3.986	–0.03	16
60%Un	6.120	6.123	0.05	16
80%Un	8.033	8.043	0.13	17
100%Un	10.050	10.069	0.19	18
120%Un	12.138	12.171	0.27	19

4.3 Harmonic test

The electric field coupling sensor is also test with odd harmonic voltage which ranges from 3rd harmonic to 25th harmonic in order to test its frequency range in low frequency. The gain of the sensor along frequency is shown in Figure 6.

In Figure 6 the fundamental frequency in test is 50 Hz. The result shows that the sensor keep its gain at about −196 stably along with the frequency changing. The sensor performs well with harmonic voltage measurement in low frequency.

4.4 Lightening wave test

In order to determine the transient characteristics of the sensor in high frequency, a test was carried out with lightning wave. The impulse generator is used to generate 1.2/50μs lightning waveform with 10kV peak voltage for testing. The result obtained by the electric field coupling sensor and capacitance high voltage probe is shown in Figure 7.

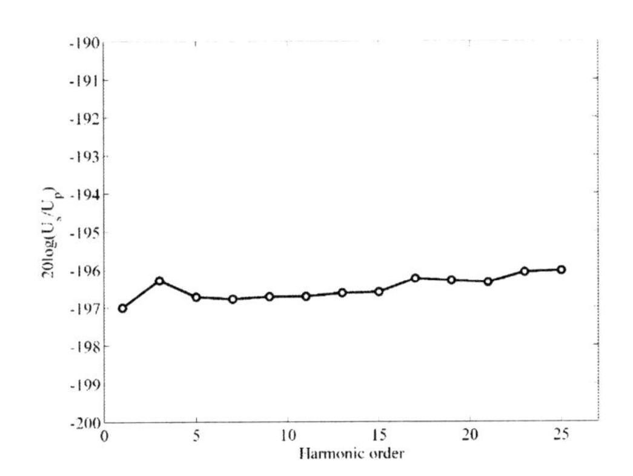

Figure 6: Harmonic test result.

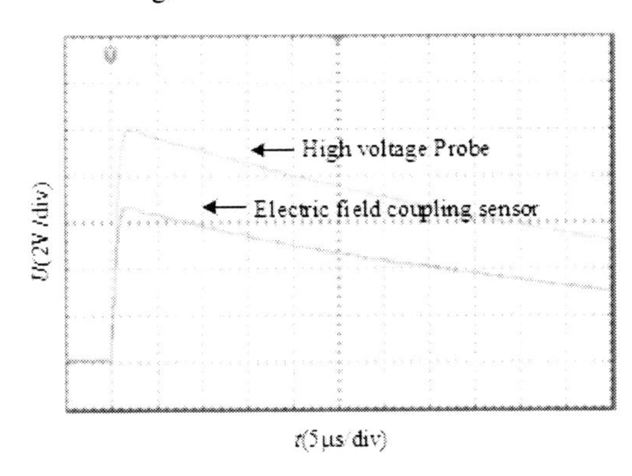

Figure 7: Result obtained with a 1.2/50μs lightning wave measurement by the sensors (5μs/div; 2V/div).

The result from Figure 7 shows that the electric coupling sensor transmits the measured signal without distortion or noise. The designed sensor has fast response and good agreement with the high voltage probe.

5 Conclusion

A concept for measuring high voltage through electric field coupling has been introduced. A new difference output structure was proposed based on the principle of electric field coupling sensor. The new structure sensor was manufactured into PCB style. The sensor has been tested on experimental platform with its steady-state accuracy, harmonic voltage response, and transient response. The results shown that the steady-state accuracy of the sensor has reached a 0.5 class voltage transformer level and a wide frequency range. It was shown that voltage measurement can be accomplished by electric field measurement which keeps voltage measurement from the large, heavy, and expensive insulating structure. It is looking forward to form a new method in power system voltage measurement.

References

[1] Duncan, B.K., Bailey, B.G., Protection, metering, monitoring, and control of medium-voltage power systems. *IEEE Transactions on Industry Applications*, **40(1)**, pp. 33–40, 2004.
[2] Girgis, A.A., Hart, D.G., Peterson, W.L., A new fault location technique for two-and three-terminal lines. *IEEE Transactions on Power Delivery*, **7(1)**, pp. 98–107, 1992.
[3] Ye, H., Xu, P., Zong, H., et al., The configuration and switch-over of voltage transformer in digital substation. *Automation of Electric Power Systems*, **24**, pp. 93–95, 2008.
[4] Sadeghkhani, I., Ketabi, A., Feuillet, R., Artificial-intelligence-based techniques to evaluate switching overvoltages during power system restoration. *Advances in Artificial Intelligence*, p. 1, 2013.
[5] Barker, P.P., Mancao, R.T., Kvaltine, D.J., et al., Characteristics of lightning surges measured at metal oxide distribution arresters. *IEEE Transactions on Power Delivery*, **8(1)**, pp. 301–310, 1993.
[6] Liu, G.Y., Kang, L., Peng, W.N., On-line monitoring system for transformer partial discharge. *Applied Mechanics and Materials*, **303**, pp. 464–467, 2013.
[7] Chavez, P.P., Rahmatian, F., Jaeger, N.A.F., Accurate voltage measurement with electric field sampling using permittivity-shielding. *IEEE Transactions on Power Delivery*, **17(2)**, pp. 362–368, 2002.
[8] Xu, D., Zhao, J., Zhang, A., et al., Application of electronic transformers in digital substation. *High Voltage Engineering*, **1**, p. 17, 2007.
[9] Chavez, P.P., Jaeger, N.A.F., Rahmatian, F., Accurate voltage measurement by the quadrature method. *IEEE Transactions on Power Delivery*, **18(1)**, pp. 14–19, 2003.

[10] Gockenbach, E., Aro, M., Chagas, F., et al., Measurements of very fast front transients. *Electra*, **181**, pp. 71–91, 1998.
[11] Wei, B., Fu, Z., Wang, Y., et al., Frequency response analysis of passive RC integrator. *High Voltage Engineering*, **1**, p. 14, 2008.

Discussion on the current practice of the integration of operation and maintenance of substation and carrying out measures

Yitao Jiang[1], Li Mu[1], Yanfei Ma[2], Yang Jiang[1]
[1]*State Grid of Technology College, Jinan, China*
[2]*State Grid Jinan Power Supply Company, Jinan, China*

Abstract

It's very important of conducting the integration of operation and maintenance of substation to improve the work efficiency of production departments during the process of "SANJIWUDA" system construction. Based on the implementation goal of integration of operation and maintenance, the operation personnel are required to do part of maintenance personnel's work. However, the long-term independent model of operation and maintenance is a bottleneck of developing the integration. The implementation effects of integration of operation and maintenance in some provinces are summarized in this paper. After that, effective and reasonable trainings are put forward to push the integration of operation and maintenance. At the same time, the development of integration of operation and maintenance is prospected in the end.

Keywords: integration of operation and maintenance of substation, "SANJIWUDA", training.

1 Introduction

In the process of "SANJIWUDA" system construction, in order to improve the efficiency of power production sector and the professional quality of the substation operation personnel, the purpose, principle, mode, main task, supporting measures and schedule requirements of the integration of operation and maintenance are proposed.

The purpose of integration of operation and maintenance is restructuring and integrating equipment inspection, operation, maintenance, operation and maintenance

WIT Transactions on Engineering Sciences, Vol. 107,
www.witpress.com, ISSN 1743-3533 (on-line)
doi:10.2495/SEEEII140411

personnel, which differ from the traditional manner of power production, such as equipment inspection and operation are carried out by operation personnel and maintenance work is carried out by maintenance personnel. The integration of operation and maintenance will be carried out according to short, medium and long term plans. [1]

The traditional manner of power production is based on definite division of work, employees who fulfill their duties are divided into operation personnel, maintenance personnel, dispatchers and so on. [2]The use of this manner ensures the safety of power production significantly, however it has some drawbacks such as low efficiency, the integration of operation and maintenance will solve these problems.

2 The issues and challenges faced by integration of operation and maintenance

Dividing work and duty definitely is an important feature of power production units, which means different preparations have to be done before operation work and maintenance work, this is one of the bottlenecks to carry out integration of operation and maintenance.

2.1 The status of independent professional

Operation and maintenance work of substation has direct impact on safe operation of substation, meanwhile substation connects power plant and client, play an important role of transforming and dispatching electric power, so safe and economical operation of substation has significant impact on power system, industrial and agricultural production and people's lives. [3]

Operation personnel's daily work includes substation inspection, switch operation, exception handling and so on, the places where daily work is carried out are generally close to charged equipment.

Maintenance personnel carry out their work when danger points are told to them. [4]Maintenance work has to be checked by operation personnel before the whole maintenance work end.

The two posts described previously have little in common, so they belong to different departments.

2.2 Current distribution of human resources

Operation personnel's comprehensive quality has been improved greatly, which is resulted from the join of graduates recently, but their operational capacity can hardly meet the demand of rapidly changing power system. [6]Even if the original academic of the substation staff is undergraduate and above all who is new to join the power system, China's traditional educational content and mode still difficult to adapt to the power system needs of job skills, due to the continuous improvement of the degree of substation automation, extensive application of new technologies in substation equipment and the rapid development of smart grid. Meanwhile, the expansion of substation makes the number of substation staff generally tight. [7]

2.3 Current development of substation equipment

With the rapid development of the power system, the number of substation in China is also rising. With the level of substation automation improved, substation basically become unattended. And with the advancement of "SANJIWUDA" construction in State Grid Corporation of China, the number of unattended substation is increasing. The significantly improvement of manufacturing and operation level in substation equipment, the average service life of the main substation equipment less than 10 years from the past is raised to 20–30 years, and availability factor of it is above 99.9%. [8]A large number of intelligent substations, without excessive personnel are on duty for intelligent substation. Operation personnel need to improve the comprehensive quality and constantly adjust to the development of new technology with the addition of new equipment.

3 The developmental situation of provincial integration of operation and maintenance

3.1 State grid Zhejiang electric power company

Operation and maintenance professional of substation has been carried out the integration of operation and maintenance training. Currently Operation personnel have been carried out maintenance overhaul work related live detection, including replacement of arrester line monitor, protection inverter, PD monitoring of switchboard, etc. [9]

Through training, the smoothly development of the integration of operation and maintenance makes the operation and maintenance of substation team more professional, and qualification accreditation for the qualified operation personnel.

3.2 State grid Fujian electric power company

During the process of integration of operation and maintenance, State Grid Fujian Electric Power Company not only assigns the operation personnel some work that belongs to maintenance personnel, but also makes one department be responsible for some easy and inefficient work that needed the cooperation of operation personnel and maintenance personnel originally, which promotes the work efficiency of operation work group and maintenance work group. [10]

3.3 State grid Liaoning electric power company

State Grid Liaoning Electric Power Company adopts the way of classified training and classified handover and the maintenance department has handed over some C-type and D-type maintenance work to operation department, which makes the operation personnel have more understanding of the principle and structure of different kinds of equipment, increase the quality of daily supervision and acceptance and avoid the possible misunderstanding and omission during the process. [11]This significantly reduces the cooperation time with maintenance department and makes the duty interface cleaner between holistic operation and professional maintenance. As a result, there are more definite condition maintenance process, complete administration system, better work quality and benefits, which achieves the results of optimizing equipment operation and maintenance and makes the equipment operation and maintenance can be controlled and be under control.

3.4 State grid Xinjiang electric power company

State Grid Xinjiang Electric Power Company establishes the specific training leading group based on the principle of "training first, proper motivation, evaluation mount guard and training the operation and maintenance talents with more ability." Then Xinjiang Electric Power Company mainly develops the training of integration of operation and maintenance on different stages and the implementation work.

4 Effective measures to develop the integration of substation operation and maintenance

4.1 Developing the professional training of the integration of operation and maintenance

Due to the profession independence, it's hard for the operation personnel to master some relative maintenance task. So it's an effective way to solve the problem by developing the professional training of the integration of operation and maintenance. The well-skilled teachers can be chosen to conduct the professional training from both the theory and operation. [12] After passing the evaluation, the operation personnel are ensured to grasp relative skills and conduct the integration programs smoothly and make sure the safety during the implementation.

4.2 Conducting the technology certification for the integration of operation

In order to increase the enthusiasm of operation personnel to ensure the integration of operation and maintenance going smoothly, the technology certification can be used for the operation personnel. The certificate is awarded to those who pass the technology certification and only people with the certificate can do the maintenance operation. The technology certification can be classified into different levels that cover different maintenance programs, which can promote the enthusiasm of operation personnel and increase the comprehensive quality of operation personnel.

4.3 Holding the exchange meeting between professional maintenance personnel and operation personnel

According to "Integration of substation operation and maintenance programs directory" which issued by state grid corporation, the operation personnel will carry out the 100 operation and maintenance projects. Since the practical ability of the operation and maintenance staff is limited, once difficulties, professionals are needed to improve the whole capacity of the staff.

By holding the exchange meeting, we can analyze the problems after integration of operation and maintenance and reach an improvement opinion to solve the problems as soon as possible.

4.4 Standardization working instruction book

In order to improve the management infrastructure, site requirements and quality of enterprises, Standardization Working Instruction Book is needed. Also the book is required to guide the staff through the work of projection and maintenance. The book should include objection, basis, application scope, preparation, procedure, defection and remarks (Table 1).

Table 1: Substation maintenance projects standardization working instruction book template.

XXX Maintenance Projects Standardization Working Instruction Book

1 Objection

2 Basis

3 Application Scope

4 Preparation

4.1 Personnel division of labor

√	serial number	content	remarks

4.2 Maintenance instrument

√	serial number	name	specifi-cations	unit	quantity	remarks

4.3 Dangerous point analysis and control counter measures

serial number	dangerous analysis	precaution

5 Procedure

6 Defection

serial number	device name	defec-tions	recorder	reporter	results	handler

7 Remarks

5 Conclusions and prospects

Integration of substation operation and maintenance is meant to let the operation personnel conduct some maintenance work that can be done without cutting off the power basically. After that, it promotes the operation personnel's work efficiency, shortens the time of handling the abnormal situations and avoids the problems during the handover, which increases the reliability of the power grid equipment. At the same time, the maintenance personnel can focus on the more professional maintenance skills. Therefore, the integration of substation of operation and maintenance is good at the development of the power grid and increases the comprehensive quality of operation personnel.

Acknowledgement

The research work was supported by State Grid of Technology College under training new employees and the short-term training. And thanks for Zhejiang, Fujian, Liaoning, and Xinjiang Electric Power Company providing lots of information to support this research.

References

[1] Xu Yanyang, The scheme discussion on the Integration of operation and maintenance of substation. *China Electric Power Education*, **(33)**, 2012.

[2] Luo Yanjuan, Research on training, practice and development of the integration of operation and maintenance of substation. *China Electric Power Education*, **(20)**, 2013.

[3] Zhou Zhuojun, Cai Hongjuan, Wang Huifeng, Discussion on operation and maintenance work of substation. *Urban Construction Theory Research*, **(32)**, 2013.

[4] Xiong Yanbin, Peng Yanchun, The analysis on problems during the implementation of the integration of operation and maintenance of substation. *Power Technology*, **(10)**, 2013.

[5] Wu Kexin, The discussion on problems during the implementation and strategy on the integration of operation and maintenance of substation. *Chinese E-Commerce*, **(6)**, 2014.

[6] Luan Fengkui, Zeng Ming, Liu Baohua, et al., Survey of equilibrium model in electricity market modeling research. *Proceedings of the CSU-EPSA*, **(1)**, 2008.

[7] Hu Ming, Chen Heng, Survey of power quality and its analysis methods. *Power System Technology*, **24(2)**, 2000.

[8] Lin Haixue, On improving electric power quality. *Power System Technology*, **18(4)**, 1994.

[9] Chen Shuyong, Song Shufang, Li Lanxin, et al., Survey on smart grid technology. *Power System Technology*, **33(9)**, 2009.

[10] Li Qingquan, Li Yanming, Niu Yamin, The transient electromagnetic fields caused by the operation of disconnector in substation and protection. *High Voltage Engineering*, **27(4)**, 2001.
[11] Wang Mingjun, Some highlights in relation to smart grid. *Power System Technology*, **33(18)**, 2009.
[12] Gu Dingxie, Xiu Muhong, Dai Min, et al., Study on VFTO of 1000kV GIS substation. *High Voltage Engineering*, **33(11)**, 2007.

Design of power grid smart operation and maintenance platform considering power grid operation information and equipment state monitoring

Chen Xiangyu[1], Zhao Jianning[2], Wang Qi[2]
[1]*EHV Power Transmission Company,*
China Southern Power Grid, Guangzhou, China
[2]*State Key Lab of Control and Simulation of Power Systems and Generation Equipment, Department of Electrical Engineering, Tsinghua University, Beijing, China*

Abstract

In the paper, many demands in many different levels such as basic information, functional module, key indicator, application service, business department, etc. are analyzed. It is proposed that construction of power grid operation and equipment management smart platform becomes optimum solution under the background of information-based power grid. Construction plan and design of the platform are introduced in the aspects of basic platform, network operation monitoring (control) analysis subsystem, monitoring and analysis subsystem of substation equipment, transmission equipment monitoring and analysis subsystem, emergency command subsystem and power station information access. Finally, active influence of China Southern Power Grid EHV Power Transmission Company on power grid operation and management is expected after formal operation of the platform.
Keywords: operation and maintenance, smart management, equipment management, information-based power grid, equipment condition monitoring.

1 Introduction

Information is gradually developed and matured currently. Advanced information integration pattern, advanced intelligent and integrated information technology are introduced into equipment operation and maintenance of power systems, which

www.witpress.com, ISSN 1743-3533 (on-line)
doi:10.2495/SEEEI140421

has become an inevitable trend of smart power grid with information-based features [1–3]. Construction of power grid operation and equipment management smart platform is produced under the major trend of information integration.

Construction scope of power grid operation and equipment management smart platform includes the follows: Operation maintenance automatic system foundation system of Southern Power Grid EHV Company, power grid monitoring (control) analysis subsystem operated on foundation system, monitoring and analysis subsystem of transformer equipment, monitoring and analysis subsystem of transmission equipment and emergency command subsystem. Power grid model, parameter information and other production operation information are centralized in a centralized mode and uniformly maintained in the basic platform. Meanwhile, various public services are mutually shared. All subsystems are mainly used for achieving power grid monitoring and analysis, monitoring and analysis of power transmission and transformation equipment, emergency command and professional applications and algorithms in other aspects. Analysis results are called for human-machine service exhibition provided by basic platform.

2 Analysis of function demand

General construction goals and key elements, as shown in Fig. 1.
Basic information level: The system should be fully integrated Equipment account, operation conditions, test data, monitoring data, environmental information, defect information, maintenance cost and other comprehensive data information about operation and maintenance of EHV power transmission and transformation equipment are sufficiently integrated in the system.

Functional module level: the system should be provided with basic functional modules of monitoring and early warning, fault diagnosis, condition assessment, risk warning, supporting decision, indicator analysis, etc.

Key indicator level: the system should be used for comprehensively analyzing and sufficiently showing key indicator parameters such as reliability, economy, environment influence, etc. in many dimensions at different levels of equipment, power grid, etc.

Application service level: various application services directly faced to production business practice can be constructed rapidly and flexibly aiming at practical business demands of different business departments by basic functional modules based on the system, including condition monitoring, maintenance decision, asset optimization, performance management, emergency command, scheduling assistance, etc.

Business department level: actual business demand of different business departments is basic guidance of system construction. System construction effect is reflected in service effect and efficiency on actual business demand of business department. The system should serve business demands of Production Technical Department, Security Supervision Department, Inspection Test Center and other business department of South Power Grid EHV Company as well as South Power Grid General Dispatching, South Power Grid Research Institute and other cooperative units.

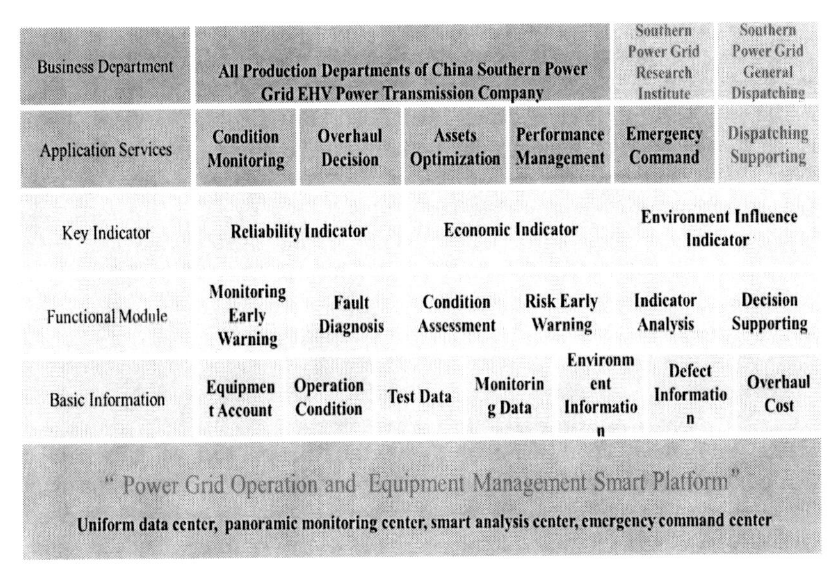

Figure 1: General construction goals and key elements.

3 Overall structure of solution

China Southern Power Grid EHV Power Transmission Company intends to construct the overall solution of 'power grid operation and equipment management smart platform' aiming at the above demand. The overall structure consists of three levels: equipment condition monitoring, wide-area information integration and application platform, advanced and intelligent analysis application, as shown in Fig. 2.

3.1 Equipment condition monitoring level

Equipment condition monitoring level is mainly used for realizing perception and recording of various conditions of electrical equipment. Monitoring objects include various substation equipment and transmission equipment. Monitoring means include online monitoring, test detection, inspection record, etc. [4–6].

Therefore, the following contents should be mainly researched in equipment condition monitoring level: more comprehensive monitoring items are adopted for realizing equipment condition as sufficiently as possible, thereby discovering other hidden danger of faults thereof; more effective condition parameters can be studied for assessing equipment condition and diagnosing equipment failure more accurately; intelligence of existing electrical equipment and intelligent power transmission equipment can be researched, thereby enhancing self-reliability of monitoring unit.

3.2 Wide area information integration and application basic platform level

Online monitoring, test detection, line inspection records and other data of various power transmission equipment distributed in wide area are gathered according to hierarchical classification system on wide area information integration and application basic platform level, data in SCADA, MIS, GIS (geographic information system) and other power business information systems are fully integrated together, thereby providing advanced and intelligent analysis application level with basic data support, general system management, workflow, public services, human-machine interface, development operation environment and other basic function frames [7, 8].

The following contents should be researched at wide-area information integration level: building high-performance communications networks and systems based on uniform standards; building globally unified electrical equipment model cross business system platform; achieving organic integration of information in different business systems..

3.3 Advanced and smart analysis application level

Electrical equipment condition monitoring information can be presented and processed at advanced and smart analysis application level, thereby providing decision suggestion for electrical equipment condition maintenance and other electrical services, including monitoring and early warning, fault diagnosis, condition assessment, risk early warning, indicator analysis, decision supporting and other functional modules.

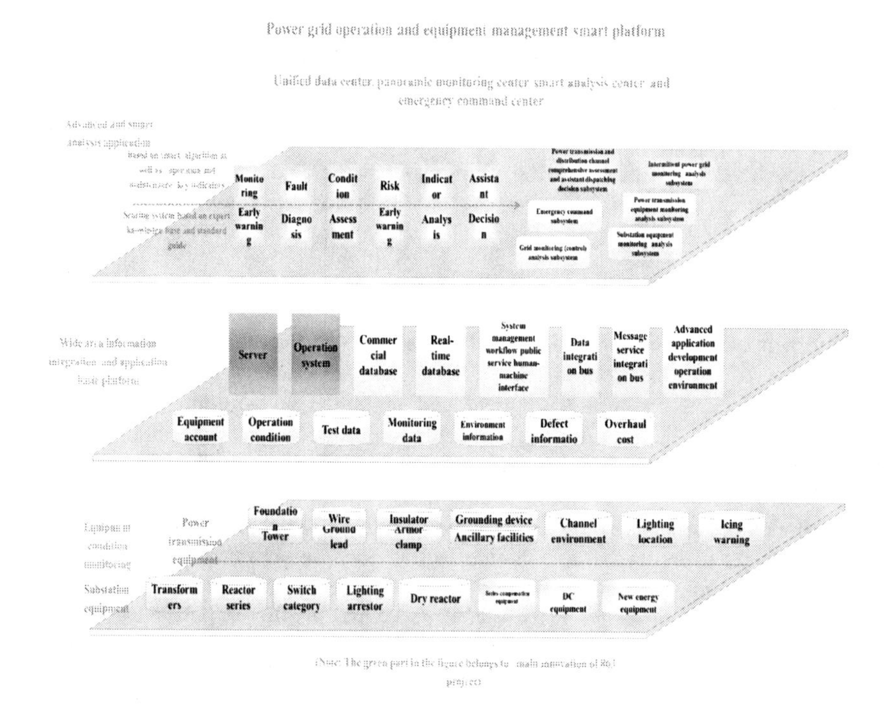

Figure 2: Overall structure of smart platform.

The following contents should be mainly researched at advanced and smart analysis application level: more comprehensive monitoring items based on power equipment, more efficient condition parameters, history fault record provided by integrating other power business systems, network topology structure, geography meteorological information and other more comprehensive information. Smart analysis algorithm is adopted for achieving more comprehensive and accurate assessment and prediction on condition of electrical equipment, and providing power business assistant decision with more guidance, operability and economic benefits.

4 Function development plan of basic platform and subsystems

(1) Basic platform

Basic platform is basically built. Integration bus, public services, human-machine interface, system management, data acquisition and processing, and other modules are configured. Perfect operating environment is built for upper application. Only two-dimensional view can be realized in the stage at the aspect of visual display aspect of basic platform. Three-dimensional view is not implemented temporarily due to higher workload and investment. Icing monitoring, lightning location and GIS three-dimensional transmission line management systems are connected in a linkage mode. Dedicated interface is developed for communicating with Nanning area control, production management system, electric energy metering system, video surveillance system, emergency command system, South Network icing early warning and monitoring centre, equipment condition monitoring centre, lightning monitoring centre and technical supervision centre.

Basic platform is further perfected on the technology, thereby providing operation environment for developing advanced application functions.

(2) Power grid monitoring (control) analysis subsystem

Power grid operation monitoring (control) analysis subsystem is basically completed. Power grid operation monitoring analysis (including trend monitoring, substation centralized monitoring, substation control, DC system monitoring and other modules), protection and fault information management functions are configured. Fault location functions are developed, and substation fault distance measurement devices are accessed.

(3) Substation equipment monitoring and analysis subsystem

Substation equipment monitoring analysis subsystem is basically built. Oil chromatography and casing pipe monitoring devices are mostly installed in EHV substation. Online monitoring functions of power transformer and capacitor-type equipment are mainly constructed under the background. Live data are accessed. Meanwhile, frames of other function modules of substation equipment monitoring analysis subsystem are constructed. The frames can be provided with ability of debugging once the live information is accessed. Other online monitoring information can be gradually accessed according to company demand. Substation equipment condition dynamic assessment, early warning and condition decision supporting functions are constructed.

(4) Transmission equipment monitoring and analysis subsystem

The GIS three-dimensional transmission line management system is linked and integrated into operation and maintenance automatic system. Login information is synchronized, and unified entrance can be realized. Other functions can be directly accessed after operation and maintenance automatic systems are logged. Related data of icing online monitoring and warning system constructed by EHV Company as well as China Southern Power Grid lighting location system can be accessed through standard interfaces (webservice). The data is integrated and uniformly exhibited in the system. Basic data is provided for advanced application. Other transmission equipment online monitoring functional modules are constructed according to company demand, and power transmission equipment condition dynamic assessment, early warning and condition detection decision supporting functions are constructed.

(5) Emergency command subsystem

Emergency command system is connected for daily information management, contingency plan management, emergency resource management, emergency training exercises, emergency duty management, emergency early warning management, information comprehensive statistics and other functional modules can be shown in the system.

(6) Plant station information access

Oil chromatography and casing pipe monitoring information are accessed in three pilot stations of Nanning, Mawo and Xingren in integrated data network mode. Substation monitoring information and converter station remote information are accessed from Nanning Area Control. Protection and fault recorder information substation are accessed in the mode of schedule data network. When the substation is provided with conditions, substation monitoring system is connected in the mode of two-way special line, and fault distance measurement device information is connected in the mode of one-way special line.

Monitoring system in the station, protection and fault recorder information management system and fault distance measurement device are transformed on the basis. Therefore, it can be provided with direct communication ability with operation and maintenance automatic system. It is recommended that the functions can be realized through data network channel. Comprehensive treatment unit is configured on substation, thereby accessing substation oil chromatogram and casing pipe online monitoring information.

5 Conclusion

After the platform is constructed, it will act as panoramic monitoring and analysis center of power grid operation condition and equipment condition of EHV Company. It becomes information collection and dissemination platform of own power grid and standby system of regional centralized control center, and provides complete technical support for emergency command of EHV Company. It becomes a part of Southern Power Grid icing early warning monitoring center, equipment condition monitoring center, lightning location monitoring center and technical supervision center in EHV Company. Core business needs based on

West-East electricity transmission project for EHV Power Transmission Company are eventually established. Various demands and problems of power transmission and distribution equipment condition monitoring under smart power grid development new trend are mainly solved. Various business related information and application platforms of EHV Power Transmission Company are constructed uniformly and orderly after coordinated planning and integration. On the basis, production business practice of EHV Power Transmission Company is closely followed for making all efforts to construct power grid operation and equipment management intelligent platform integrating SCADA, power transmission equipment operating condition collection and analysis, power transmission condition information (including new energy access) wide area surveillance and intelligent analysis early warning, emergency command and other functions. It can become unified data center, panoramic monitoring center, smart analysis center and emergency command center demanded for production business of EHV Power Transmission Company, thereby realizing long-term vision of constructing EHV overhaul testing center information platform.

Acknowledgement

The research work was supported by National High Technology Research and Development Program of China (863 Program) under Grant No. 2012AA050209.

References

[1] Yang Yang, Information Modeling and Security Strategy of Condition Monitoring. Hunan University, 2009.
[2] Zhao Xilin, Research on Fault Diagnostic Technology for Power System Based on Information Fusion Theory. Huazhong University of Science & Technology, 2009.
[3] Zhang Yagang, Research on Fault Component Localization of Electric Power Systems Based on Wide Area Measurement Information. North China Electric Power University, 2011.
[4] Wang Dewen, Wang Yan, Di Jian, Design scheme of condition monitoring system for smart substation. *Automation of Electric Power Systems*, **35(18)**, pp. 51–56, 2011.
[5] Liu Ji, Huang Guo-fang, Xu Shi-ming, Development of smart grid condition monitoring. *Electric Power Construction*, **30(7)**, pp. 1–3, 2009.
[6] Li Peng, Huang Xinbo, Zhao Long, Zhu Yongcan, Development of smart grid condition monitoring. *Proceedings of the CSEE*, **33(16)**, pp. 153–161 + 6, 2013.
[7] Wei Zu-kuan, Jiang Nan, Kim Jae-hong, Fast research on key technology of three-dimensional GIS in electric power information system. *Computer and Modernization*, **5**, pp. 83–88, 2010.
[8] Wang Baoyi, Research on Key Techniques of Information Security in Electric Power Information System. North China Electric Power University, 2009.

Harmonic analysis and prediction of power grid with electric vehicle charging station

Yanhua Ma[1], Chen Dong[1], Xuan Zhang[1], Zhiming Li[1], Jie Zhou[2]
[1]*State Grid Electric Power Research Institute, Nanjing, China*
[2]*School of Automation, Nanjing University of Science and Technology, Nanjing, China*

Abstract

Taking Matlab/Simulink as the main tool for the study. Discuss the influence of harmonic current at the point of common coupling (PCC) with the number of chargers, the capacity of transformer, the capacity of system, length of lines and charging methods, when the electric vehicle charging station access the power grid. To solve the problem of large workload of Matlab simulation and long computing time, change the power system and equipment parameters, and do the prediction of the change of harmonic current by using BP neural network. Provide the necessary theoretical basis and technical support for the construction of the charging station.
Keywords: electric vehicle, harmonic current, neural network.

1 Introduction

In the power system, electric vehicle chargers are non-linear loads, which can produce harmonic current as a kind of pollution [1–3]. So before the construction of the charging station, it is necessary to do the calculation and evaluation of the harmonic current. We must ensure that the harmonic current of power grid can meet the requirement of power quality standards. In view of the influence of harmonic current at the PCC, this article analyses the changes of the 5th harmonic by changing the number of chargers, the capacity of transformer, the capacity of system, length of lines and charging methods. At the same time, do the prediction of the harmonic current by using BP neural network. And provide the necessary theoretical basis and technical support for the construction of the charging station.

www.witpress.com, ISSN 1743-3533 (on-line)
doi:10.2495/SEEEII140431

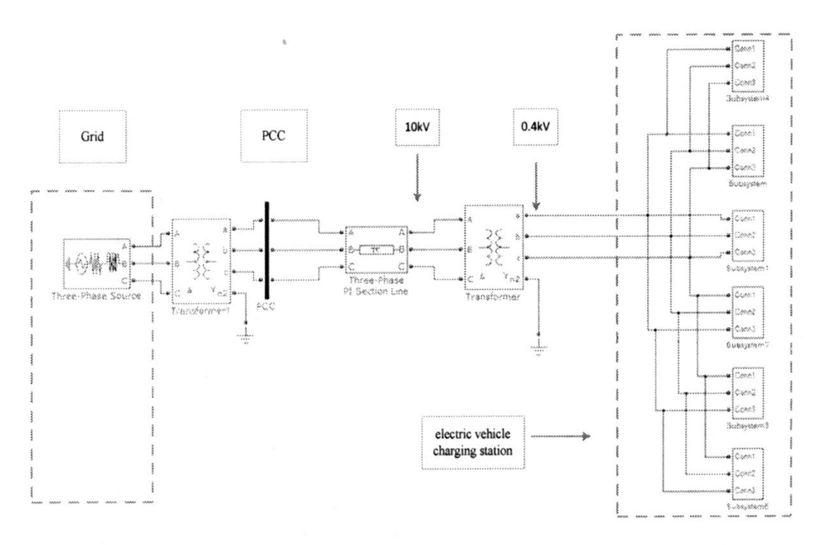

Figure 1: The model of electric vehicle charging station with six chargers.

2 The establishment of the model of electric vehicle charging station

The electric vehicle charging station consists of multiple charging machines in parallel when accessing the power grid. We should inspect the changes of harmonic current at the PPC.

Set up the model of power grid with electric vehicle charging station by using Matlab/Simulink. Choose the parameters of type I charger and set up a 6-pulse diode uncontrolled rectifier charger. Adopt 10kV power supply. The transmission line is LGJ-120/25, whose per kilometer parameters include: $r = 0.223$, $x = 0.348$. The voltage of transformer is 10kv/0.4kV and Dyn11 connection. The model of charging station with six chargers is established in Figure 1.

3 The analysis of the influencing factors of harmonic current

In view of the analysis of the power quality of electric vehicle charging station, the impact of electric vehicles on the power quality is mainly reflected in the effect of harmonic current [4–6]. Because the 5th current is relatively large, this article only discussed the 5th.

3.1 The influence of the number of chargers

Table 1 shows the simulation results by changing the number of chargers when the electric vehicle charging station accesses the power grid. Notice that with an increase in the number of chargers, the value of the 5th current is getting larger. It's worth noting that the increase of harmonic current is not linear. It is offset with each other.

3.2 The influence of the capacity of transformer

Considering the influence of distribution transformers, from Table 1 the value of the 5h harmonic current grows gradually with the transformer capacity increasing. At the same time, it is noticed that the harmonic limits are also increasing and the rate of which is higher than the rate of the growth of harmonic current value. Therefore, under the same conditions, enlarging the capacity of transformer can make more chargers access the power grid without exceeding the limit.

3.3 The influence of the capacity of system

Table 2 shows that the value of the 5th harmonic current hardly changes when the capacity of system increases. However, during this process, the harmonic limits increase. It can be concluded that changes of the power grid will not have an effect upon the harmonic current at the point of PPC, but the increasing of the capacity of system can ensure adding more chargers without exceeding the limit.

3.4 The influence of the length of lines

If the distribution transformer of the charging station is far away from higher power, the impact of the transmission lines which are at the high-voltage side of the transformer should be considered. Changing the length of the transmission line, we can see the simulation result in Table 3. As shown in Table 3, the longer the transformer line is, the less the value of harmonic current is. Since the power line itself has impedance, the harmonic current is inverse proportion to the impedance when the voltage of higher power is unchanged. On the other hand, power line itself has a voltage drop, so the voltage distortion at long-term line points of PPC is worse than which at short-term line points of PPC.

Table 1: The influence of the capacity of transformer on harmonic current.

The Capacity of Transformer (MVA)	Harmonic Number	Harmonic Limit (A)	Number			
			1	2	5	6
0.8	5	0.43	**0.77**	**1.53**	**3.6**	**4.20**
1.6	5	0.76	0.76	**1.54**	**3.81**	**4.53**
4	5	1.63	0.75	1.51	**3.84**	**4.62**

Table 2: The influence of the capacity of system on harmonic current.

The Capacity of System (MVA)	Harmonic Number	Harmonic Limit (A)	Number			
			1	2	3	4
150	5	0.68	**0.76**	**1.54**	**2.31**	**3.07**
289.3	5	0.76	0.76	**1.54**	**2.31**	**3.07**
400	5	0.80	0.76	**1.54**	**2.31**	**3.07**

3.5 Charging methods

Change the charging methods of the electric vehicle charging station [7, 8]. The detailed changes are shown in Figures 2 and 3. A set of tests are 6 chargers charging simultaneously, another are chargers charging in stages, which means that let a charger access the power grid every 50 minutes. In Figure 2, 5th harmonic is 4.8A with 6 Chargers Charging Simultaneously. In Figure 3, 5th harmonic is 3.25A while 6 Chargers Charging In Stages. It is generally accepted that the harmonic current reduced significantly when 6 chargers charges in stages.

Table 3: The influence of length of lines on harmonic current.

Length (kM)	Harmonic Number	Harmonic Limit (A)	Number		
			1	3	6
0	5	0.76	0.76	**2.31**	**4.53**
5	5	0.76	**0.76**	**2.30**	**4.46**

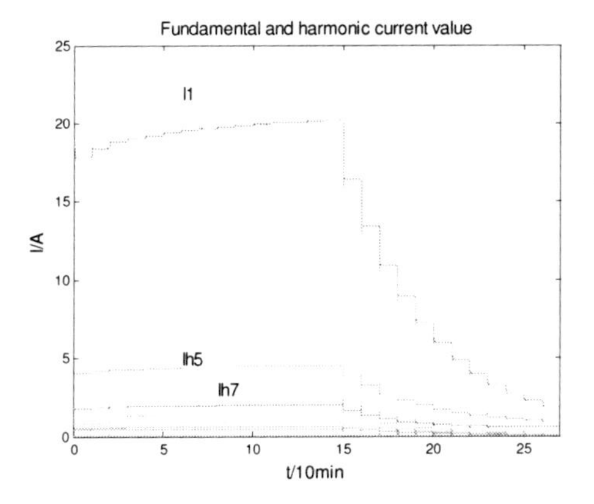

Figure 2: Charging simultaneously.

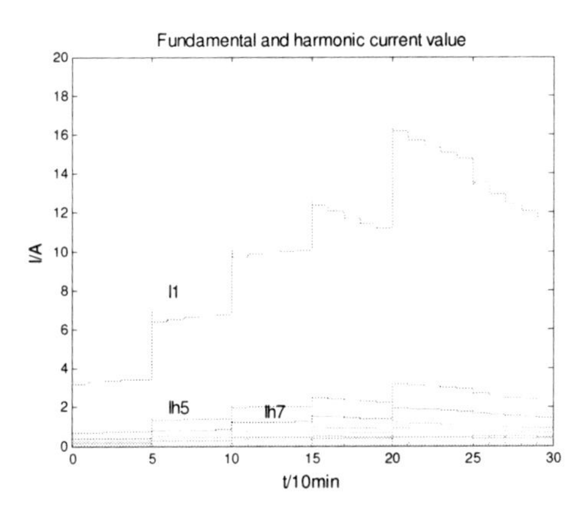

Figure 3: Charging in stages.

3.6 Suggestions for the construction of charging station

According to the analysis of the factors influencing the harmonic current, we draw the suggestions for the construction of charging station as follows:

i) To reduce the voltage distortion, the charging station ought to adopt the higher power supply which is near the station.
ii) We consider that designers can cut down the capacity of harmonic suppression device when the system need to carry a large number of chargers.
iii) In consideration of economic conditions, it is advantageous to choose the transformers with larger capacity.
iv) To cut down the harmonic current flowing into the grid, it's advantageous to choose the charging method that chargers charge in stages.

4 The prediction of harmonic current by using BP network

Since the power system and equipment parameters are complex, to solve the problem of large workload of Matlab simulation and long computing time, change the power system and equipment parameters and do the prediction of the harmonic current by using BP neural network. BP network is a multi-layer neural network having three or more layers, as shown in Figure 4 [9].

We simulate five hours of the electric vehicle charging station. Change the number of chargers per 50 minutes. The capacity of transformer, the capacity of system, length of lines and the number of chargers during the six time periods are the inputs. And the maximum of the 5th harmonic current is output. Select 103 groups of data obtained by simulation. The 93 groups of data are training data, and the other 10 groups of data are test data. The inputs and the outputs should be normalized to make them the data between [0, 1]. The prediction results are shown in Figures 5 and 6. The prediction errors are shown in Table 4. It can be

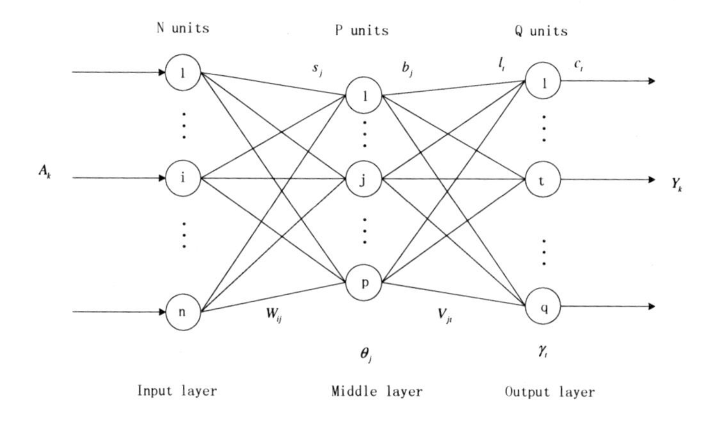

Figure 4: BP network with three layers.

concluded that BP neural network shows its better predictive ability, the prediction error of which is about 1% and will fall below 5%. However, during the process of prediction, the errors of some sample points are relatively large, so we can improve the weight and the threshold of the network by taking genetic algorithm or particle swarm algorithm to optimize the BP neural network [10].

Table 4: The prediction result of the fifth harmonic current.

S_T (MVΛ)	S_N (MVA)	L (kM)	T1	T2	T3	T4	T5	T6	误差 (%)
0.8	289.3	0	1	1	1	1	1	1	3.77
1.25	289.3	0	3	3	3	3	3	3	0.17
1.6	289.3	0	4	4	4	4	4	4	0.37
2.5	289.3	0	6	6	6	6	6	6	1.79
1.6	150	0	2	2	2	2	2	2	0.08
1.6	200	0	3	3	3	3	3	3	0.21
1.6	350	0	4	4	4	4	4	4	0.62
1.6	500	0	5	5	5	5	5	5	4.39
1.6	289.3	5	1	2	3	4	5	6	0.91
1.6	289.3	0	6	5	4	3	2	1	1.75

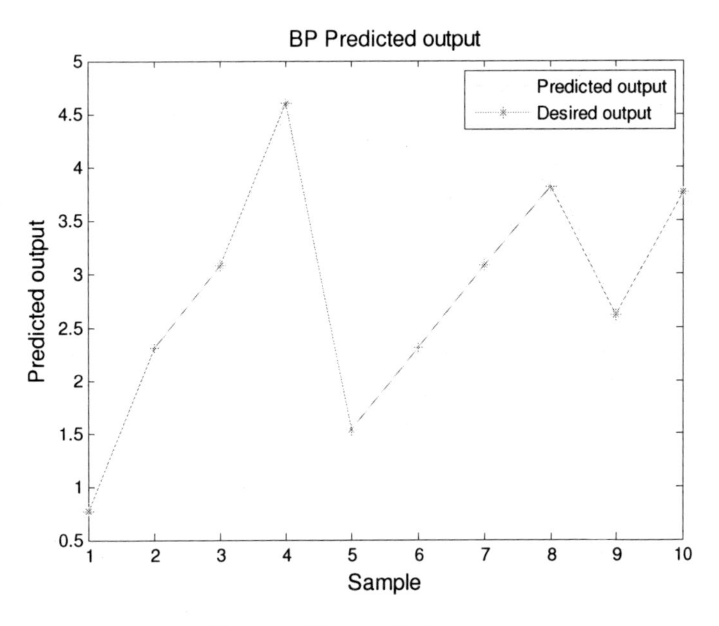

Figure 5: The prediction output.

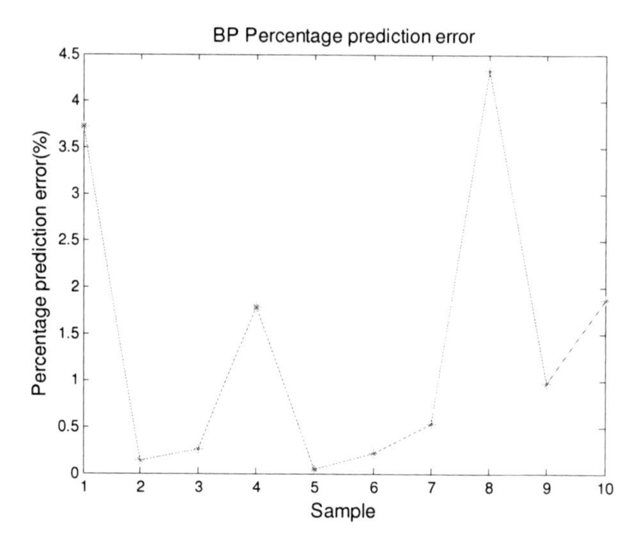

Figure 6: The prediction error percentage.

5 Conclusion

In the power system, electric vehicle chargers are non-linear loads which will produce harmonic current injected into the grid and pollute the power grid.

We establish the Matlab simulation model for electric vehicle charging station to analyze the harmonic current when the electric vehicle charging station accesses the power grid and the influencing factors of harmonic current are also received.

At the same time, to solve the problem of complicated parameter change and long computing time, we put forward a prediction model based on BP neural network which can better predict the value of harmonic current in each case and provide data basis for the construction of charging station.

Acknowledgement

This work was financially supported by the National High Technology Research and Development Program (863Program) (2011AA05A110), National Energy Board technology projects (NY20110702-1) and State Grid Corporation Headquarters in IT Projects (The research of the electric vehicle's charging-swapping experiment and the detection technique).

References

[1] Cui Yufeng, Yang Qing, Zhang Linshan, Research on the domestic and foreign development status of EV and the charging technology. *Yunnan Electric Power*, **38(2)**, pp. 9–12, 2010.

[2] Zhen Pan, Zhou Yusheng, Zeng Long, Harmonic analysis of high-frequency electric vehicle charger. *Journal of Electric Power*, **28(2)**, pp. 100–113, 2013.
[3] Huang Mei, Huang Shaofang, Jiang Jiuchun, Harmonic analysis of power grid with electric vehicle charging station. *Journal of Beijing Jiaotong University*, **32(5)**, pp. 85–88, 2008.
[4] State Bureau of Technical Supervision. GB/T 14549-1993 The Electric Energy Quality of Public Power Grid Harmonic. State Bureau of Technical Supervision, 1993.
[5] Wu Jiang, Li Weiguo, Ma Jixian, Research on the influence of the Olympic vehicle charger on the electrical energy quality of the distribution network. *Power Electronics*, **(12)**, pp. 41–48, 2009.
[6] Ma Linlin, Yang Jun, Fu Cong, Research review of the influence of electric vehicle charging and discharging on the power grid. *Relay*, **41(3)**, pp. 140–148, 2013.
[7] Tian Liting, Zhang Mingxia, Wang Huanling, Assessment and solutions on the impact of electric vehicles on the power grid. In: *Proceedings of the CSEE*, **32(31)**, pp. 43–49, 2012.
[8] Hu Zechun, Song Yonghua, Xu Zhiwei, Influence and use of electric vehicles accessing the power grid. In: *Proceedings of the CSEE*, **32(4)**, pp. 1–10, 2012.
[9] Zhou Ying, Ying Bangde, Ren Lin, Research on the short-term power load forecasting model by using BP neural network. *Electrical Measurement and Instrumentation*, **48(542)**, pp. 68–71, 2011.
[10] Zhuang Yuanyuan, Short-term load forecasting of power system based on particle swarm. *Control and Management*, **3(3)**, pp. 9–11, 2007.

Design of high voltage capacitor charging power supply for ETCG applications

Yazhou Zhang[1], Zhenxiao Li[1], Wangsheng Li[2], Baoming Li[1]
[1]*National Key Laboratory of Transient Physics, Nanjing University of Science and Technology, Nanjing, China*
[2]*Nanjing Electronic Technology Research Institute, Nanjing, China*

Abstract

A compact and high power capacitor charging power supply (CCPS) has been designed for electrothermal-chemical gun (ETCG). Through the comparison and analysis, the full bridge zero current switching series resonance circuit has better performance, and was selected for CCPS. Calculated the devices parameters and simulated the circuit current. Multiple practical protecting measures are designed and adopted in the CCPS, which can prevent malfunctions from occurring. At last, the CCPS has been assembled and experimented. The experimental results have proved that the CCPS operates reliable and satisfy the requirements of ETCG.
Keywords: ETCG, CCPS, series resonance circuit, zero current switching.

1 Introduction

ETCG is one of the adopted standard solid propellant gun hardware technology. It adopts electrical energy input to conduct electric explosion of the metal wires in the plasma generator to control propellant combustion. It is a new concept of transmission techniques which can improve the ballistic performances [1].

The experimental system of ETCG includes pulse power supply (PPS), launcher, ammunition which are equipped with plasma generator, control and test system. The primary energy storage of ETCG usually uses PPS which is based on capacitor energy storage. This is because the PPS can produce enough pulse energy in a very short time, and one of the important ways to get the pulse energy is through the fast charge and discharge of pulse capacitor. The capacitor provides energy by CCPS charging. With the further development

WIT Transactions on Engineering Sciences, Vol. 107, © 2015 WIT Press
www.witpress.com, ISSN 1743-3533 (on-line)
doi:10.2495/SEEEII140441

of pulse power technology and extensive application, the production of CCPS which has high power density, efficiency and security performance has become an urgent need.

A powerful high frequency and high voltage CCPS for ETCG has been designed in this paper. The CCPS can charge 10 1400μF/12kV capacitors in a short time. When the charging voltage reaches 12kV, the PPS can provide 1MJ of energy storage. Basically it has same energy storage supply with for the other ETCGs in the world [2]. Through the analysis and simulation, the charging circuit of CCPS uses full bridge zero current switching series resonance circuit. Multiple practical protecting measures are designed and adopted in the CCPS which can prevent malfunctions from occurring. The CCPS performances. It can satisfy the requirements of ETCG.

2 Analysis of full bridge series resonant charging circuit

In the field of pulse power technology, the primary energy storage capacitor has two ways of charging: constant voltage charge mode and constant current charge mode. For constant voltage charge mode, the efficiency and power is relatively lower and the supply volume is higher. And for constant current charge mode, it has the characteristics of constant current charging, small volume, high efficiency, high power density and suitable for wide range load, which is better for CCPS than the constant voltage charge mode [4, 5].

The constant current charge mode for high frequency and high voltage usually uses the resonant inverter structure. It has three topological structures: series resonance, parallel resonance, and series-parallel resonance. The series resonance can realize constant current charging and raise voltage step by step, resulting advantages such as low switching loss, easy control, high precision, and strong anti-interference ability. The series resonance structure is better for the CCPS [6–9].

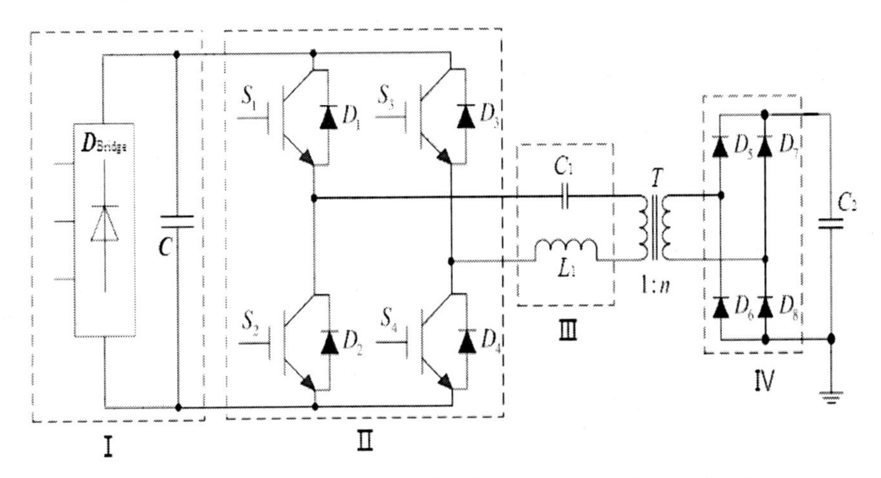

Figure 1: Principle diagram of series resonance charge circuit.

Figure 1 is a principle diagram of series resonance charge circuit. Part I is a rectifier bridge which rectifies the 380V/AC to constant voltage DC source; part II is a full bridge circuit. It has four switching units, including four switches $S_1 - S_4$ and four direction parallel diodes $D_1 - D_4$; part III is a series resonance circuit, it composed by a resonance inductor L_1 and resonance capacitor C_1. Part IV is an output rectifier silicon consisting of four equivalent diodes $D_5 - D_6$; T is a high voltage transformer; ratio of T is $1{:}n$. C_2 is a charging capacitor. The characteristic of the series resonant circuit are as follows:

$$\begin{cases} Z = \sqrt{L_r / C_r} \\ f_r = 1 / (2\pi\sqrt{\mathrm{L}_r\mathrm{C}_r}) \\ T_r = 2\pi\sqrt{L_r C_r} \end{cases} \tag{1}$$

where Z is the characteristic impedance of the resonant circuit, f_r is the resonant frequency, T_r is the resonant period. According to the relation of switching frequency f_r and resonant frequency f_s, three operating modes of series resonant circuit can be defined, which are $f_s < f_r/2$, $f_r/2 < f_s < f_r$, and $f_s > f_r$. Using the full bridge series resonant charging circuit at operating mode $f_s < f_r/2$ can satisfy the charging power supply of ETCG.

At operating mode $f_s < f_r/2$, the resonant current is interrupted, which is the discontinuous current mode. During the whole switching period, the switching units can obtain both zero-current-on and zero-current-off. This is the soft switch state and has the highest efficient. The resonance current of the positive half-cycle can return to the power supply at the negative half-cycle when the load is short-circuited. At this moment, the average charging current is small, and the electromagnetic interference is also small because the wave of circuit current is sine. The leakage inductance of the high power transform can be used as the part of the resonant inductor to participate in circuit resonance. This mode can automatically resist the short circuit of the load, and repeatedly charges and discharges in a short time. It is suitable for ETCG continuous emission.

It is the continuous current mode when $f_r/2 < f_s < f_r$. The switching units can achieve hard switching state and zero-current off. It has high switch-on loss and strong electromagnetic interference and needs to bear a certain degree of voltage and current conduction; the switch devices have to sustain high current stress. The circuit still work at continuous current mode when $f_s > f_r$. The switching units zero voltage and zero current are switch on and hard is switch off. It has high switch-on loss and strong electromagnetic interference. Due to the circuit inductance, the peak voltage is high when the switching units are turn off. When the load is short-circuited, the circuit current is high. Since the circuit itself has no ability to resist short circuit of load, it can be easily damaged.

3 Designs of the CCPS

3.1 Parameter and simulation of the CCPS

The primary performance indicators of CCPS are summarized in Table 1.

The switching frequency is 18kHz and resonant frequency is 40kHz in the circuit. It satisfies $f_s < f_r/2$ and ensures the charging circuit to work at discontinuous current mode. The resonant capacitance C_1 is 1.98μF, and by calculation of Eq. (1), the resonant inductor L_1 is about 8μH, including the transformer leakage inductance. Matlab/Simulink is used to simulate the circuit; the outcomes are shown in Figure 2.

3.2 Choose main components of CCPS

3.2.1 Switching units

For charging power supply that works below 50 kHz, the use of the IGBT as switching units is the most typical choice; because IGBT can work at high voltage and current and can be easily controlled [10]. Compared with traditional PWM, the IGBT has no switching loss but only has conduction loss. The IGBT in part II Figure 1 uses FF200R12KE3 (1200V–200A) manufactured by Infineon technologies AG.

Table 1: Performance indicators of CCPS.

Parameter of CCPS	
Input voltage	380V/AC±10%
Output voltage	12kV
Average charging rate	25kJ/s
Capacitance	14mF
Error of charging voltage	<2%
Volume	600×800×1200mm

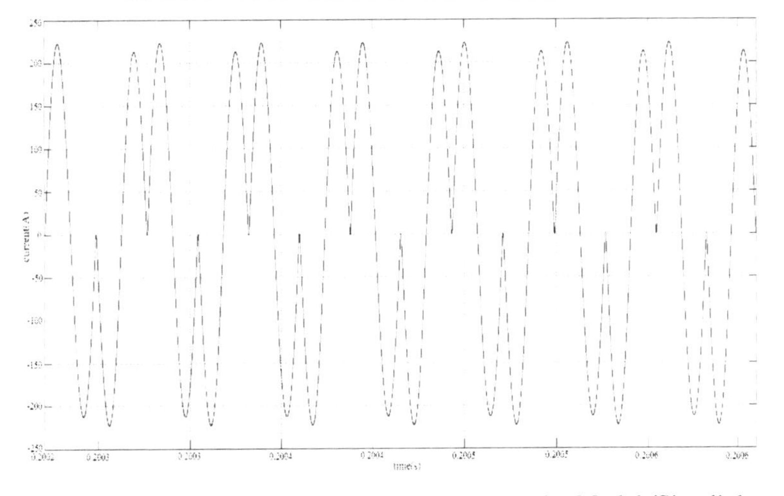

Figure 2: The waveform of series resonant current by Matlab/Simulink.

3.2.2 High voltage transformer

In the series resonant circuit, as the stray inductance of the high voltage transformer is part of the resonant inductance, its value is far less than the resonant inductance. The turn ratio of the transformer is 1:27. The core of the transformer is two ferrite cores (Type UF110); the secondary winding uses subsection design composed by 8 windings. By this way, stray inductance and capacitance of transformer can be effectively reduced. The transformer adopts insulating frameworks to prevent high voltage breakdown, and an oil bank with a wind-cooling heat exchanger is installed in to avoid overheating damage. Figure 3 shows the high voltage transformer of the CCPS.

3.2.3 High voltage rectifier

The high voltage rectifier composed of 64 fast recovery diodes DSEI60–12A (1200V-52A) manufactured by IXYS Corp. Each bridge consists of two series diodes, and four bridges are connected with one winding of the transformer secondary winding. The principle diagram of rectifier circuit is shown in Figure 4.

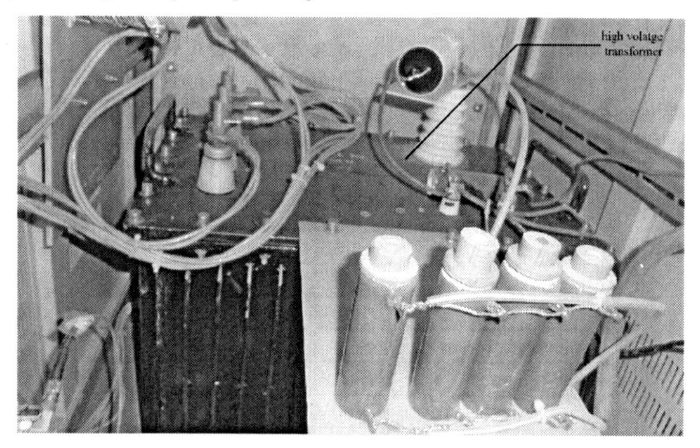

Figure 3: The high voltage transformer of the CCPS.

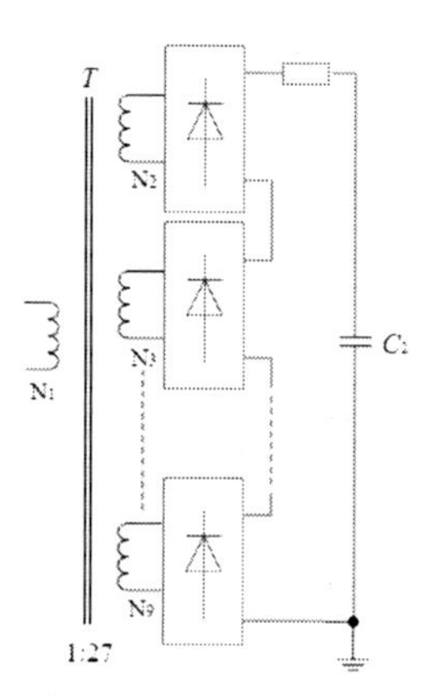

Figure 4: Principle diagram of rectifier circuit.

3.2.4 Control unit

The control unit of CCPS needs high reliability, strong anti-jamming capability, easy programming and quick signal processing. The control unit uses Programmable Logic Controller (Type FP0-C16CT) manufactured by Panasonic Corp. The Programmable Logic Controller has a RS-232 port which can through optical fiber connect with the photoelectricity isolation device and communicate with the master computer. Safety of the control unit can be ensured by photoelectricity isolation.

3.3 Protections of CCPS

When charging, the circuit can cause high reverse voltage and high current if the charging line is disconnected. When the capacitor is discharged, the discharge current can inflow into the supply internal and damage the supply if the ground potential floats. An isolation resistance and a 15kV small capacitor added between the CCPS and the capacitor, which can avoid these problems and make sure the supply will not be damaged. The CCPS also has overvoltage and overheating protections. It will stop working when detects that the voltage and temperature exceed the safe value. Photograph of the CCPS is shown in Figure 5.

Figure 5: Photograph of the CCPS.

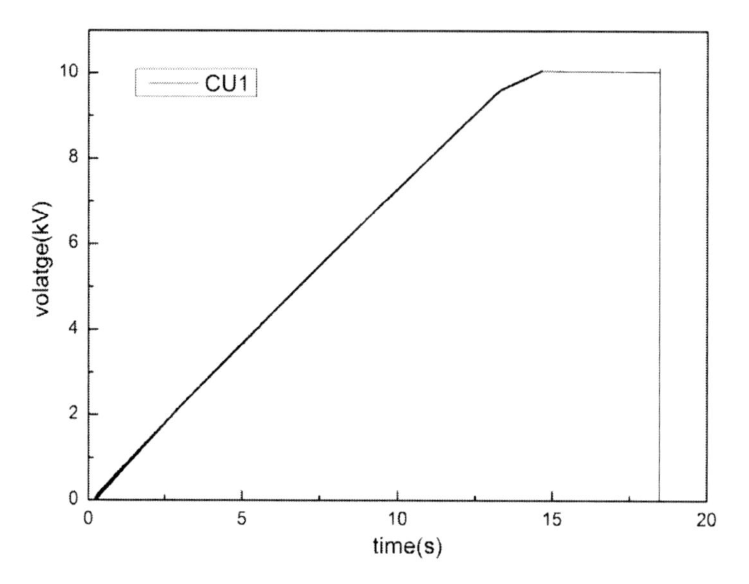

Figure 6: The voltage of two CCPS charging for ten capacitors.

4 Experimental result

Figure 6 shows the charging voltage of two CCPS in parallel charge for ten capacitors. When charging stops, the measured peak voltage is 10.16kV and minimum voltage is 9.82kV. Those voltages are within the error range of charging voltage, which is <2%. The figure shows the voltage of capacitor is rising steadily with constant current charging and voltage is raised step by step. The charging time is within 15s with properly worked of the charger. It is also proven that CCPS can use multiple parallel charging to shorten charging time.

5 Conclusion

This paper designs a CCPS for 1MJ PPS of ETCG. The output voltage is 12kV and average charging current is about 5A. It utilizes full bridge series resonance zero-current off technology. Higher reliability can be assured by using multiple protecting measures. Further research will focus on the improvement of the efficiency and the power density.

References

[1] Dong Jian-Nian, Chui Dong-Min, Zhang Jun, Huang Xin-Hua, Design of PFN in the ETC. *Chinese Journal of Ballistics*, **7(4)**, pp. 83–87, 1995.

[2] Xiang Yang, Gu Gang, Zhang Jian-Ge, The overview of the electrothermal-chemical launch technology abroad. *Chinese Journal of Ship Science and Technology*, **29**, pp. 159–162, 2007.
[3] Liu Xi-San, *High Pulsed Power Technology*. National Defense Industry Press, Beijing, 2005.
[4] Su Jian-Cang, Wang Li-Min, Ding Yong-Zhong, Song Xiao-Xin, Analysis and design of series resonant charging power supply. *Chinese Journal of High Power Laser and Particle Beams*, **12(12)**, pp. 1611–1614, 2004.
[5] Ma Xun, Li Hong-Tao, Design of high power full bridge series resonant charging power supply. *Chinese Journal of Electronic Design Engineering*, **21(8)**, pp. 116–118, 2013.
[6] Yang Jing-He, Zhang Li-Feng, Wang Guo-Bao, Design and application of series resonant charging power supply for electron linear accelerator. *Chinese Journal of Atomic Energy Science and Technology*, **48(7)**, pp. 1296–1299. 2014.
[7] G.H. Rim, I.W. Jeong, G.I. Gusev, Y.W. Choi, H.J. Ryoo, J.S. Kim, A constant current high voltage capacitor charging power supply for pulsed power applications. *Journal of Pulsed Power Plasma Science*, **(2)**, pp. 1284–1286, 2001.
[8] H.J. Ryoo, S.R. Jang, Y.S. Jin, J.S. Kim, Y.B. Kim, Design of high voltage capacitor charger with improved efficiency, power density and reliability. *Transactions on Dielectrics and Electrical Insulation IEEE*, **20(4)**, pp. 1076–1084, 2013.
[9] Gao Ying-Hui, Sun Yao-Hong, Yan Ping, Shi Yi, High power capacitor charging power supply for EML applications. *14th Symposium on Electromagnetic Launch Technology IEEE*, pp. 1–4, 2008.
[10] Yu Yue-Hui, Liang Lin, *Pulse Power Device and Application*. China Machine Press, Beijing, 2010.

A novel duty ratio control to reduce torque ripple for DTC of IM drives with constant switching frequency

Zhengxue Li[1,3], Zhengxi Li[2], Xiaojuan Ban[1]
[1]*College of Computer and Communication Engineering, University of Science and Technology, Beijing, China*
[2]*North China University of Technology, Beijing, China*
[3]*Yanbian University, Yanji, China*

Abstract

The conventional switching-table-based direct torque controlled (DTC) induction motor drives are usually afflicted by large torque ripple as well as inconstant switching frequency in a low speed range. A novel DTC strategy of induction motor with duty ratio control is proposed based on the detailed analysis of effect factors of flux linkage and torque ripples. In this method the pulse duration of active vector is calculated according to equalizing the mean torque with its desired value over one control cycle. The proposed scheme is able to reduce the torque ripple significantly while maintaining the simplicity and robustness of conventional DTC at the most. Simulations and presented experimental results validate the effectiveness of the proposed scheme as compared with the conventional method.
Keywords: direct torque control, duty ratio control, ripple reduction.

1 Introduction

Since its introduction by Takahashi [1] and Depenbrock [2] for induction machine drives, direct torque control (DTC) has become a powerful control scheme for AC machine drives. The conventional switching-based DTC scheme uses hysteresis regulators and a switching table for the control of both stator flux magnitude and electromagnetic torque [3, 4]. Compared to the FOC scheme, DTC does not require current regulators, nor coordinate transformations or pulse width modulation (PWM) block. Hence, the transient torque control

www.witpress.com, ISSN 1743-3533 (on-line)
doi:10.2495/SEEEII140451

performance could be significantly improved. However, the DTC strategy using a switching table has some drawbacks. First, switching frequency varies according to the motor speed and the hysteresis bands of torque and flux [5]. Second, large torque ripple is generated, in particular, in a low speed range because of small back EMF of an induction machine, and third, high control sampling time is required to achieve good performance [6]. Constant-switching-frequency strategy seems to have been achieved to some extent by several studies [7, 8]. However, these literature have mainly focused on the constant switching frequency regulation, thus, these methods have not shown a large improvement in reducing instantaneous torque ripple.

To solve the aforementioned problems, various modified DTC-based schemes have been proposed. Among them, a commonly used way is the DTC incorporated of space vector modulation (SVM-DTC) [9, 10], which can adjust the torque and flux more accurately, hence the torque and flux ripples were reduced while obtaining fixed switching frequency. In the SVM-based DTC schemes, rotary coordinate transformation is often needed [10], which is more computationally intensive than the conventional DTC [11].

In this paper, a new DTC technique is proposed which reduces torque ripple by keeping constant switching frequency. Output voltage vector $\boldsymbol{V}_s$ is selected using the conventional DTC switching table, but the pulse duration of $\boldsymbol{V}_s$ is calculated according to equalizing the mean torque with its desired value over one control cycle. The instantaneous torque variation of the motor can be expressed as a function of the applied voltage vector $\boldsymbol{V}_s$. To further improve the performance of DTC system, a full order observer with constant gain matrix [12] is adopted to acquire accurate stator flux and torque estimation over a wide speed range. The proposed scheme is able to reduce the torque ripple significantly while maintaining the simplicity and robustness of conventional DTC at the most. Simulation and experimental results prove that the proposed DTC strategy has the advantages of conventional DTC and effectively reduces torque ripple, which improve the performance of conventional DTC.

2 Steady-state performance in DTC

2.1 Direct torque control principle

According to the principle of conventional DTC, the inverter switching signals are generated by using a switching table. Stator flux and torque can be controlled directly and independently by properly selecting the inverter voltage vectors, which can be represented, in the three-phase systems, by six active voltage vectors and two zero voltage vectors, Figure 1.

By way of example, operating principle of conventional DTC is presented in Figure 1 when the stator flux vector is located at sector 1 and rotates anticlockwise. Torque increases if voltage vectors with positive tangential component are applied ($\boldsymbol{V}_2$ or $\boldsymbol{V}_3$), and decreases if vectors with negative tangential component are applied ($\boldsymbol{V}_5$ or $\boldsymbol{V}_6$). At the same time, the stator flux magnitude increases if vectors with positive radial component are applied ($\boldsymbol{V}_2$ or $\boldsymbol{V}_6$), and decreases if

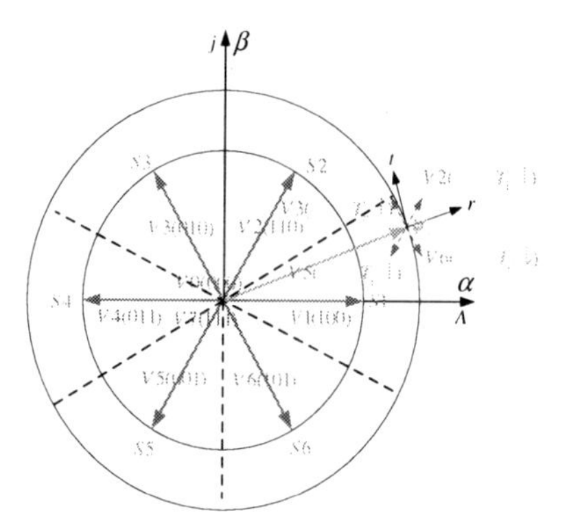

Figure 1: Conventional DTC operating principle.

Table 1: Voltage-vector table for the proposed DTC algorithm.

ψ_s is lying in sector m		ε_T		
		1	0	−1
ε_ψ	1	$\boldsymbol{V}_{m+1}$	$\boldsymbol{V}_0$, $\boldsymbol{V}_7$	$\boldsymbol{V}_{m-1}$
	0	$\boldsymbol{V}_{m+2}$	$\boldsymbol{V}_0$, $\boldsymbol{V}_7$	$\boldsymbol{V}_{m-2}$

vectors with negative radial component are applied (V_3 or V_5). A switching table can be deduced as shown in Table 1.

2.2 Effects of voltage on torque

The dynamic behavior of induction motor can be described in terms of space vectors in the α–β stationary reference frame as

$$\begin{cases} \dfrac{d\boldsymbol{\psi}_s}{dt} = -\dfrac{R_s}{\sigma L_s}\boldsymbol{\psi}_s + \dfrac{R_s L_m}{\sigma L_s L_r}\boldsymbol{\psi}_r + \boldsymbol{V}_s \\ \dfrac{d\boldsymbol{\psi}_r}{dt} = \dfrac{R_r L_m}{\sigma L_s L_r}\boldsymbol{\psi}_s + (j\omega_r - \dfrac{R_r}{\sigma L_r})\boldsymbol{\psi}_r \end{cases} \tag{1}$$

where $\boldsymbol{\psi}_s$ and $\boldsymbol{\psi}_r$ are stator and rotor flux complex vector, $\boldsymbol{V}_s$ is stator voltage complex vector, R_s and R_r are stator and rotor resistance, L_s and L_r are stator and rotor self-inductance, L_m is mutual inductance, $\sigma = 1 - L_m^2 / L_s L_r$ is leakage coefficient, and ω_r is motor angular velocity.

The electromagnetic torque can be written in terms of stator flux and stator current as

$$T_e = \frac{3}{2} N_p \boldsymbol{\psi}_s \otimes \boldsymbol{i}_s \tag{2}$$

where N_p is the number of pole pairs.

For voltage source inverter-fed DTC, the voltage vector is the sole controllable input variable, so it is desirable to analytically derive the relationship between the torque and voltage vector. From Eq. 2), the torque differentiation with respect to time t is

$$\frac{\mathrm{d}T_e}{\mathrm{d}t} = \frac{3}{2} N_p \left(\frac{\mathrm{d}\boldsymbol{\psi}_s}{\mathrm{d}t} \otimes \boldsymbol{i}_s + \boldsymbol{\psi}_s \otimes \frac{\mathrm{d}\boldsymbol{i}_s}{\mathrm{d}t} \right) \tag{3}$$

Substitute Eq. (1) into Eq. (3) and omit the tedious derivation process, the final torque differentiation is

$$f_1 = \frac{\mathrm{d}T_e}{\mathrm{d}t} = -\left(\frac{R_s}{\sigma L_s} + \frac{R_r}{\sigma L_r}\right) T_e - \frac{3}{2} N_p \frac{L_m}{\sigma L_s L_r} \omega_r \boldsymbol{\psi}_r \odot \boldsymbol{\psi}_s + \frac{3}{2} N_p \frac{L_m}{\sigma L_s L_r} \boldsymbol{\psi}_r \otimes \boldsymbol{V}_s \tag{4}$$

It is seen from Eq. (4) that the torque differentiation is composed of three parts. The first part is always negative with respect to T_e; the second part is also negative and proportional to rotor speed; the last term is the positive one and it reflects the effect of stator voltage on T_e. When zero voltage vector is selected, the torque differentiation becomes

$$f_2 = \frac{\mathrm{d}T_e}{\mathrm{d}t} = -\left(\frac{R_s}{\sigma L_s} + \frac{R_r}{\sigma L_r}\right) T_e - \frac{3}{2} N_p \frac{L_m}{\sigma L_s L_r} \omega_r \boldsymbol{\psi}_r \odot \boldsymbol{\psi}_s \tag{5}$$

so the zero voltage vector always decreases the torque.

3 Proposed DTC with duty ratio control

3.1 New DTC to improve the torque performance

With a small t_{sp}, slopes f_1 and f_2 can be considered to be constant during the duration of t_{sp} because the dynamics of flux and speed are not so fast. Thus, as can be seen from Figure 2, the increase and decrease of torque are approximated as straight lines. Here, for the convenience of description, time is reset to 0 when a new control period begins. Applying active voltage vector first, the torque increases at the beginning. Then, applying zero vector, the torque decreases. In Figure 2, active and zero voltage vectors are consecutively applied during one control period.

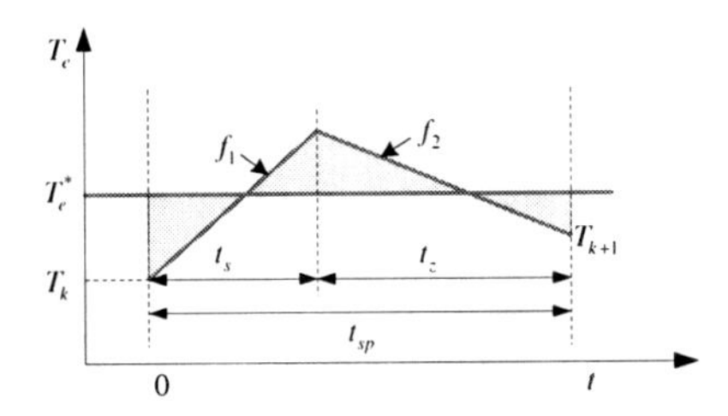

Figure 2: The typical steady-state torque wave form by DTC algorithm.

Then, the switching events are directly scheduled in a way that the mean torque over the cycle is equal to the desired value, can be expressed as

$$T_e^* = \frac{1}{t_{sp}} \int_0^{t_s} (T_k + f_1 t) \mathrm{d}t + \frac{1}{t_{sp}} \int_{t_s}^{t_{sp}} (T_k + f_2 t - f_2 t_s + f_1 t_s) \mathrm{d}t \tag{6}$$

where t_s is the switching instant at which the voltage vector is changed from active to zero vector, T_k is a torque initial value, and T_e^* is a torque command.

By solving Eq. (6) with respect to t_s, the optimal switching instant t_s is obtained as

$$t_s = t_{sp} - \sqrt{|t_z|} \tag{7}$$

where t_z is $(2T_k - 2T_e^* + f_1 t_{sp}) t_{sp} / (f_1 - f_2)$, ascending and descending torque slopes f_1 and f_2 are calculated from Eqs. (4) and (5), so duty ratio d is t_s/t_{sp}.

With the results above, the proposed strategy can be briefly described as follows.

1. The stator voltage vector V_s is selected using Table 1, where signs of torque and flux error, ε_T and ε_ψ, are determined with a zero hysteresis band.
2. In normal cases, the active vector will be first applied over the duration, and then switched to the appropriate zero vector for the rest time of the period. The appropriate zero vector means that vectors $V_1(100)$, $V_3(010)$, and $V_5(001)$ will be followed by $V_0(000)$ while other vectors followed by $V_7(111)$. If $d>1$ or $d<0$ (i.e., torque is not in the steady state), selected voltage vector V_s is fully turned on during the whole sampling period t_{sp}.
3. At each new switching cycle, time is reset to zero and the above sequences are repeated.

3.2 Flux and torque estimation

Accurate estimation of flux and torque is essential to ensure good performance of DTC. In this paper, a full order observer is adopted due to its accuracy over a wide speed range. By introducing the error feedback of stator current, the accuracy of estimation is increased and the observer is more robust against the motor parameter variations. The mathematical model of the observer is based on the IM model in Eq. (1), which is expressed as

$$\frac{\mathrm{d}\hat{\boldsymbol{i}}_s}{\mathrm{d}t} = -[(\frac{1}{\sigma T_s} + \frac{1}{\sigma T_r}) - j\omega_r]\hat{\boldsymbol{i}}_s + \frac{1}{\sigma L_s}(\frac{1}{T_r} - j\omega_r)\hat{\boldsymbol{\psi}}_s + \frac{1}{\sigma L_s}\boldsymbol{V}_s + G_1(\boldsymbol{i}_s - \hat{\boldsymbol{i}}_s) \tag{8}$$

$$\frac{\mathrm{d}\hat{\boldsymbol{\psi}}_s}{\mathrm{d}t} = -R_s \hat{\boldsymbol{i}}_s + \boldsymbol{V}_s + G_2(\boldsymbol{i}_s - \hat{\boldsymbol{i}}_s) \tag{9}$$

In classical observer for IM drives, the poles of observer are designed to be proportional to the poles of IM (factor $k>1$), which produces high imaginary part at high speed and is harmful to the system stability. To address this issue, it is suggested to shift the real part of observer poles to the left in the complex plane compared to the poles of IM, and the imaginary part of observer poles are not changed. However, this leads to complicated expressions of observer gains. In this paper, a very simple constant gain matrix $\boldsymbol{G}$ is employed to improve stability of the observer, which is expressed as [12]

$$\boldsymbol{G} = [-2b \quad -\sigma b L_s] \tag{10}$$

where b is a negative constant gain.

4 Simulation results

In order to validate the effectiveness of proposed DTC with duty ratio control, simulations were done in the environment of MATLAB/Simulink. The results obtained from the conventional DTC and the proposed DTC with duty ratio control are presented for the aim of comparison. The sampling frequency is 10kHz for the two methods, unless explicitly indicated. The parameters of motor and control system are listed in Table 2. The overall control diagram for the duty-based DTC is illustrated in Figure 3. The stator flux reference is kept at 0.94Wb, slightly lower than the rated value to avoid magnetic saturation.

Table 2: Simulated and experimental parameters.

U_{dc}	540V
$P_n/U_n/f_n/N_p/\psi_{sn}$	2.2kW/380V/50Hz/2/0.9876Wb
R_s	2.99Ω
R_r	1.468Ω
L_m	0.221H
L_s	0.230H
L_r	0.230H

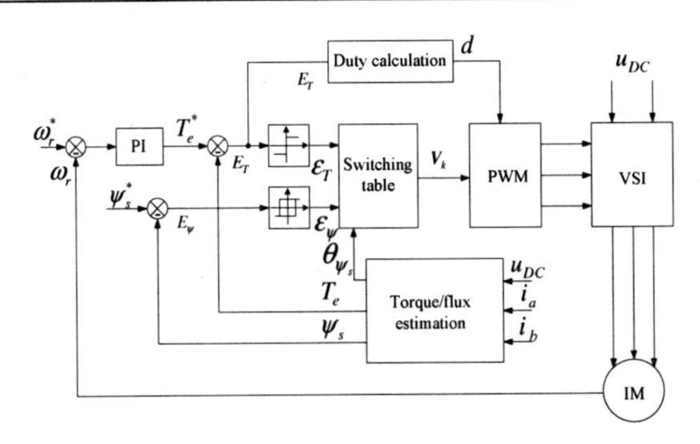

Figure 3: Control diagram of the proposed DTC.

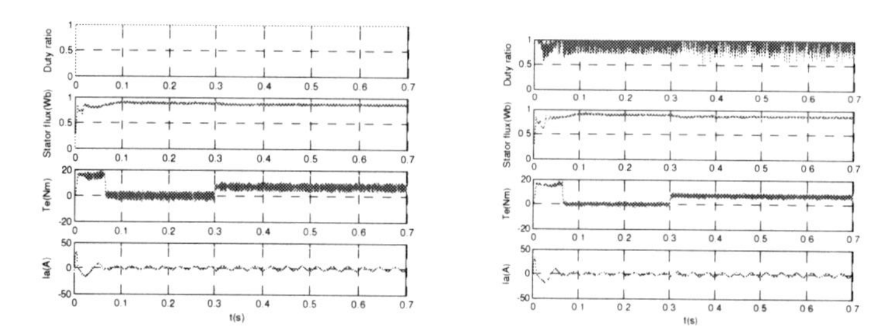

Figure 4: Simulated responses at 1050r/min with 50% rated load. (a) conventional DTC. (b) proposed DTC.

Figure 4 presents the simulated staring response from standstill to 1050r/min without load, and steady-state responses at 0.3s with 50% rated load for conventional DTC and the proposed DTC. From top to bottom, the curves shown in Figure 4 are duty ratio, stator flux amplitude, electromagnetic torque, and one-phase stator current. It is seen that the conventional DTC presents relatively large torque ripple about 7Nm. The torque ripple is much reduced in proposed DTC about 4Nm. The proposed DTC exhibits the better steady-state performance in terms of torque ripple, which confirms the effectiveness of incorporating the duty ratio control.

5 Experimental results

Apart from simulation study, the proposed DTC is experimentally tested on a two-level inverter-fed IM drives platform. The motor and control system parameters are the same as those listed in Table 2. The results obtained from conventional DTC and the proposed DTC are also presented for the aim of comparison. The sampling frequencies of conventional DTC and the proposed DTC are 20kHz and 10kHz, respectively. In the following tests, all variables are displayed on digital oscilloscope via onboard digital-to-analogue (DA) inverter except the stator current, which is measured directly by a current probe.

The steady-state response with rated load at low and high speeds are presented in Figures 5 and 6, respectively. From top to bottom, the curves shown in Figures 5 and 6 are rotor speed, electromagnetic torque, stator flux, and stator current. It can be seen that, at low-speed operation, the proposed DTC exhibits the lower torque ripple, followed by conventional DTC. There are much less harmonics in the stator current of the proposed DTC, compared to conventional DTC. Similar results can be observed at high-speed operation. Again, the better performance in terms of current harmonics and torque ripple is observed in the proposed DTC than that in conventional DTC.

The dynamic responses of conventional DTC and the proposed DTC are also compared. Figure 7 presents the staring response from standstill to 1500r/min without load. From top to bottom, the curves shown in Figure 7 are rotor speed, electromagnetic torque, stator flux, and stator current. The stator flux is first established before starting the machine. It is seen that the motor accelerates

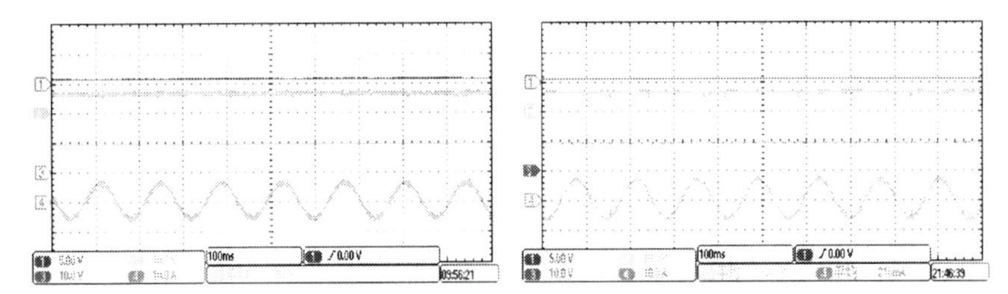

Figure 5: Low-speed operation at 150r/min with rated torque. (a) conventional DTC. (b) proposed DTC.

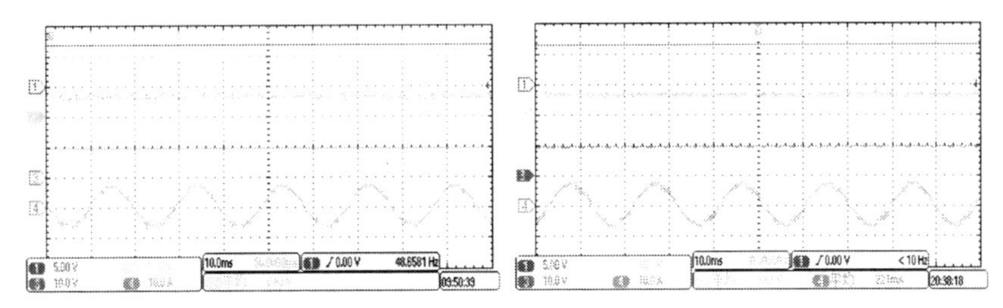

Figure 6: High-speed operation at 1500r/min with rated torque. (a) conventional DTC. (b) proposed DTC.

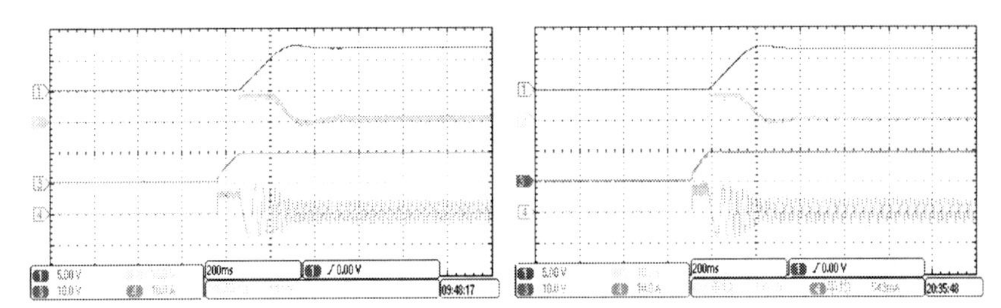

Figure 7: Starting from standstill to 1500r/min. (a) conventional DTC. (b) proposed DTC.

quickly to 1500r/min without large starting current. The dynamic responses of two kinds of methods are very similar, but the proposed DTC presents much lower torque ripple. The results prove that proposed DTC performs well with much lower sampling frequency during the dynamic process.

6 Conclusions

This paper proposes an improved DTC by taking the duty ratio control into account. Better torque performance with much lower sampling frequency can be observed in the proposed DTC at various speeds, especially in the low-speed range. Meanwhile, the quick dynamic response of conventional DTC is maintained. A comparative study with conventional DTC and the proposed DTC with duty ratio control is carried out. Both simulation and experimental results confirm the effectiveness of the proposed method.

References

[1] I. Takahashi, T. Noguchi, A new quick-response and high-efficiency control method of an induction-motor. *IEEE Trans. Ind. Appl.*, **22(5)**, pp. 820–827, 1986.

[2] M. Depenbrock, Direct self-control (dsc) of inverter-fed induction machine. *IEEE Trans. Power Electron.*, **3(4)**, pp. 420–429, 1988.

[3] S. Mathapati, J. Boecker, Analytical and offline approach to select optimal hysteresis bands of dtc for pmsm. *IEEE Trans. Ind. Electron.*, **60(3)**, pp. 885–895, 2013.

[4] A. Jidin, N.R.N. Idris, A.H.M. Yatim, T. Sutikno, M.E. Elbuluk, An optimized switching method for quick dynamic torque control in dtc-hysteresis-based induction machines. *IEEE Trans. Ind. Electron.*, **58(8)**, pp. 3391–3400, 2011.

[5] M.P. Kazmierkowski, A.B. Kasprowicz, Improved direct torque and flux vector control of PWM inverter-fed induction motor drives. *IEEE Trans. Ind. Electron.*, **42**, pp. 344–350, 1995.

[6] P. Tiitinen, The next motor control method – DTC direct torque control. In: *Proc. Int. Conf. Power Electronics, Drives and Energy System for Industrial Growth*, Delhi, India, 1996, pp. 37–43.

[7] X. Roboam, A special approach for the direct torque control of an induction motor. In: *1st European Technical Scientific Report*, Ansaldo-CRIS, Naples, Italy, 1996, pp. 216–225.

[8] Y. Li, J. Shao, B. Si, Direct torque control of induction motors for low speed drives considering discrete effects of control and dead-time of inverters. In: *Conf. Rec. IEEE-IAS Annual Meeting*, pp. 781–788, 1997.

[9] Y. Zhang, J. Zhu, W. Xu, Y. Guo, A simple method to reduce torque ripple in direct torque-controlled permanent-magnet synchronous motor by using vectors with variable amplitude and angle. *IEEE Trans. Ind. Electron.*, **58(7)**, pp. 2848–2859, 2011.

[10] G. Foo, M.F. Rahman, Sensorless direct torque and flux-controlled IPM synchronous motor drive at very low speed without signal injection. *IEEE Trans. Ind. Electron.*, **57(1)**, pp. 395–403, 2010.

[11] J. Beerten, J. Verveckken, J. Driesen, Predictive direct torque control for flux and torque ripple reduction. *IEEE Trans. Ind. Electron.*, **57(1)**, pp. 404–412, 2010.

[12] Y. Zhang, J. Zhu, Z. Zhao, W. Xu, D. Dorrell, An improved direct torque control for three-level inverter-fed induction motor sensorless drive. *IEEE Trans. Power Electron.*, **27(3)**, pp. 1502–1513, 2012.

Bi-directional prediction edge-preserving algorithm for the removal of salt-and-pepper noise from images

Ming Yang1, Beichen Chen[1], Wei Liu[1]
[1]*College of Information and Control Engineering, Jilin Institute of Chemical Technology, Jilin, China*

Abstract

The shortage of improved bi-directional prediction algorithm for the removal of salt-and-pepper noise from images is lack of edge-preserving for the processed images. So, a new edge-preserving algorithm based on it is proposed. First, noise pixels are detected, and then removed with improved bi-directional prediction algorithm. Next, the pixel to be processed in our algorithm is estimated to be an edge pixel or not, so that we can preserve the edge pixels. Experimental results prove that the edge-preserving algorithm proposed in this paper performances a strong de-noising ability and edge-preserving effect.

Keywords: image de-noising, salt-and-pepper noise, edge-preserving, bi-directional prediction.

1 Introduction

In the process of acquisition and transmission of images, the sensor device and the transmission channel is often affected by various factors, from the self and the outside world, to produce a series of impulse noise, and the noise is often black and white points in the image. We vividly called it as salt-and-pepper noise. Practice shows that in terms of salt-and-pepper noise filtering, nonlinear filtering is more effective than linear filtering. Median filtering is current widely used in filtering of it, and it is the first suitable nonlinear filtering algorithm [1]. But its biggest drawback is the rapid decline in de-noising performance while the noise density is increasing. Therefore, in recent years, there have been a variety of improved algorithms, such as tri-state median filtering algorithm (TSM) [2], center weighted median filtering

WIT Transactions on Engineering Sciences, Vol. 107, © 2015 WIT Press
www.witpress.com, ISSN 1743-3533 (on-line)
doi:10.2495/SEEEI140461

method (CWM) [3], soft-switching median filter method (SWM) [4] as well as some improved median filter.

Prediction algorithm in speech enhancement, image coding, and other fields has been applied widely [5]. With this thought, bidirectional prediction algorithm and its improved method is presented for the removal of image salt-and-pepper noise [6]. But, due to the limitations of prediction algorithm, bi-prediction algorithms and its improved method have common drawbacks: the judgment of noise pixel and the protection of image edges. For these two drawbacks, this paper presents a bi-directional prediction edge-preserving algorithm based on noise pixels detection. This algorithm divides image pixels into two categories firstly: suspicious noise pixels and true image pixels. Next, detect suspicious noise pixels again to determine whether they are noise pixels. If it is not the noise pixel, keep it value. Else, it is noise pixel, to ascertain it is edge pixel or not based on the magnitude between it and other pixels. And then, edge pixel is protected with new algorithm.

2 Improved bi-directional prediction algorithm

Xiaofeng Meng et al. presented bi-directional prediction algorithm and its improved method for the removal of salt-and-pepper noise from images [6]. In this algorithm, pixels of images are divided into categories: noise pixels and true pixels. The gray value of salt-and-pepper noise usually is 0 or 255, and the more general form is distributed within a certain range [7]. In other words, noise pixels distribute in $[0,\delta]\cup[255-\delta,255]$, and δ is the range of noise. If the pixel gray vale is fall in $[0,\delta]\cup[255-\delta,255]$, the pixel is a noise pixel. Otherwise, it is a true image pixel.

$$f(i,j)\in\begin{cases}S, & \delta < f(i,j) < 255-\delta \\ N, & 0 \le f(i,j) \le \delta \text{ or } 255-\delta \le f(i,j) \le 255\end{cases} \tag{1}$$

In Eq. (1), S are true image pixels, N are noise pixels.

In the subsequent processing, noise pixels are filtered with prediction algorithm, true image pixels are remained. Compared with bidirectional prediction algorithm, the prediction coefficients of improved bidirectional prediction algorithm consists of two parts: adaptive weights (Figure 1) and fixed coefficients (Figure 2). Adaptive weights $a_{s,t}$ $(-1 \le s,t \le 1)$ is determined by the nature of the pixel $f(i+s,j+t)$. If $f(i+s,j+t)$ is a noise pixel, $a_{s,t}=0$. If $f(i+s,j+t)$ is a filtered noise pixel, $a_{s,t}=1$. If $f(i+s,j+t)$ is a true image pixel, $a_{s,t}=2$. In the prediction algorithm, you can adjust the prediction coefficients according to the nature of pixel to get accurate result. Fixed coefficients $b_{s,t}(-1 \le s,t \le 1)$ is determined by the correlation between pixels. The pixels is closer, the correlation is higher. The value of $b_{s,t}$ is shown in the brackets of Figure 2. The initial weight $w_{s,t}=a_{s,t}\times b_{s,t}$.

$b_{-1,-1}\left(\frac{1}{4}\right)$	$b_{-1,0}\left(\frac{1}{2}\right)$	$b_{-1,1}\left(\frac{1}{4}\right)$
$b_{0,-1}\left(\frac{1}{2}\right)$	$b_{0,0}(0)$	$b_{0,1}\left(\frac{1}{2}\right)$
$b_{1,-1}\left(\frac{1}{4}\right)$	$b_{1,0}\left(\frac{1}{2}\right)$	$b_{1,1}\left(\frac{1}{4}\right)$

Figure 1: Adaptive weights.

$a_{-1,-1}$	$a_{-1,0}$	$a_{-1,1}$
$a_{0,-1}$	$a_{0,0}$	$a_{0,1}$
$a_{1,-1}$	$a_{1,0}$	$a_{1,1}$

Figure 2: Fixed coefficients.

$$d_{s,t} = \frac{w_{s,t}}{\sum_{k=-1}^{1}\sum_{r=-1}^{1} w_{k,r}} \tag{2}$$

$f^{'}(i,j)$ is the improved bi-directional prediction result of noise pixel $f(i,j)$:

$$f^{'}(i,j) = \left[\sum_{s=-1}^{1}\sum_{t=-1}^{1} f(i+s, j+t)\times d_{s,t}\right] \tag{3}$$

3 Bi-directional prediction edge-preserving algorithm

3.1 Noise pixels detection

Although the distribution of salt-and-pepper noise is usually in $[0,\delta]\cup[255-\delta,255]$, but not to say that all the pixels in that range are necessarily noise point. So, in this paper, if the pixel gray value falls in this range, the pixel will be suspicious noise pixel. And then, the suspicious noise pixel is determined again with the method proposed in ref. [8]. After that, more accurate noise pixels are obtained.

3.2 Bi-directional prediction edge-preserving algorithm

3.2.1 Basic idea of algorithm

The first step, determine the pixel to be processed is a noise pixel or not:

If it is a true image pixel, maintain its value, don't carry out any operation. Else, this pixel is filtered with improved bi-directional prediction algorithm in 3×3

neighborhood. And then, update the image database. At the same time, count the number of non-noise pixels P in 3×3 neighborhood.

The second step, detect the edge of updated image:

(1) Calculate the average value of non-edge pixels in 3×3 neighborhood;
(2) Calculate the change amplitudes of pixels in 3×3 neighborhood with the average value.
(3) Take a given threshold T, and calculate the number of pixels whose change amplitude is larger than T, denote it by Q.
(4) If the number of pixels exceeding the threshold value is smaller than a half of the pixel number of the non-noise pixel in 3×3 neighborhood, that is $Q < P/2$. This indicates that the number of mutational pixels in 3×3 neighborhood is small, and then this can be considered the center pixel of 3×3 neighborhood is not the original edge pixel; otherwise, the center pixel is an edge pixel of the original image.
(5) If the pixel is not the edge pixel, the filter result of improved bi-directional prediction is final result; otherwise, filter the edge pixel with bi-directional prediction edge-preserving algorithm.

3.2.2 Processing of algorithm

Compared with bi-directional prediction algorithm, prediction coefficients of new algorithm consists of three parts, adaptive weights (Figure 1), fixed coefficients (Figure 2) and edge weights (Figure 3).

And the prediction coefficient $d_{s,t}$ is changed correspondingly:

$$d_{s,t} = \frac{w_{s,t}}{\sum_{k=-1}^{1}\sum_{r=-1}^{1} w_{k,r}} \tag{4}$$

In Eq. (4), $w_{s,t} = a_{s,t} \times b_{s,t} \times c_{s,t}$, $\left(i-1 \le s \le i+1, j-i \le t \le j+1\right)$.

Bi-directional prediction value of the edge pixel:

$$f'(i,j) = \left[\sum_{s=-1}^{1}\sum_{t=-1}^{1} f(i+s, j+t) \times d_{s,t}\right] \tag{5}$$

$c_{-1,-1}$	$c_{-1,0}$	$c_{-1,1}$
$c_{0,-1}$	$c_{0,0}$	$c_{0,1}$
$c_{1,-1}$	$c_{1,0}$	$c_{1,1}$

Figure 3: Edge weights.

Compared with improved bi-directional prediction methods, this algorithm has more weights (edge weights). Using the previously described method of image edge pixels detection, if $f(i,j)$ is the edge pixel, the edge weight $c_{i,j}=4$; otherwise, $c_{i,j}=1$.

4 Experiment

For comparison, δ is 3 in the new algorithm and prediction algorithms, salt-and-pepper noise is distributed in [0, 3] and [252, 255]. When edge detection, threshold T is one fourth of the maximum and minimum difference in 3×3 neighborhood. Table 1 shows the results of different algorithms for each test image. It can be seen from Table 1, compared with improved bi-directional prediction algorithm, noise density at 0.1, the PSNR of new algorithm is increased by 0.58db, 0.60db, and 0.49db respectively; noise density at 0.7, the PSNR of new algorithm is increased by 1.21db, 1.37db, and 1.04db respectively. Comparing simulation results in Figures 4 and 5, it can be found that the proposed bi-directional prediction edge-preserving algorithm is better in removing salt-and-pepper noise, reduces the risk of noise misjudgement, protects the edge pixels of image.

Table 1: The PSNR of filtered images with different algorithms (unit: db).

Source Image	Noise Density	Noise Image	Bi-directional Algorithm	Improved Bi-Direction	New Algorithm
Lena	0.1 0.7	15.41 6.95	41.62 29.21	43.11 30.63	43.69 31.84
Peppers	0.1 0.7	15.31 6.85	39.33 27.49	41.12 29.21	41.72 30.58
Barbara	0.1 0.7	15.23 6.84	33.71 23.89	34.59 24.38	35.08 25.42

(a)

(b)

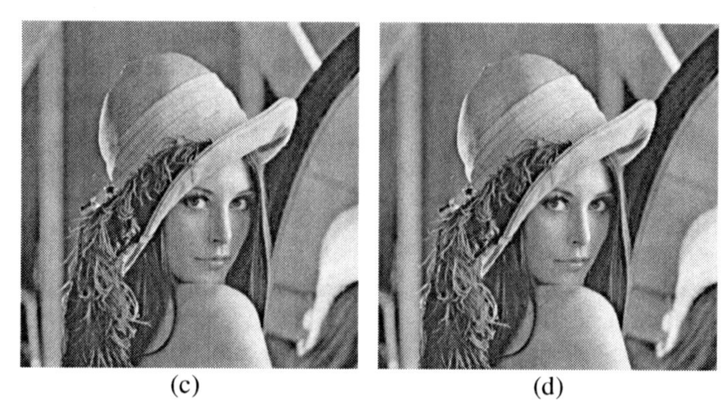

(c) (d)

Figure 4: Filtered images with different algorithms (lena, noise density = 0.1). (a) noise image. (b) bi-directional. (c) improved bi-direction. (d) new algorithm.

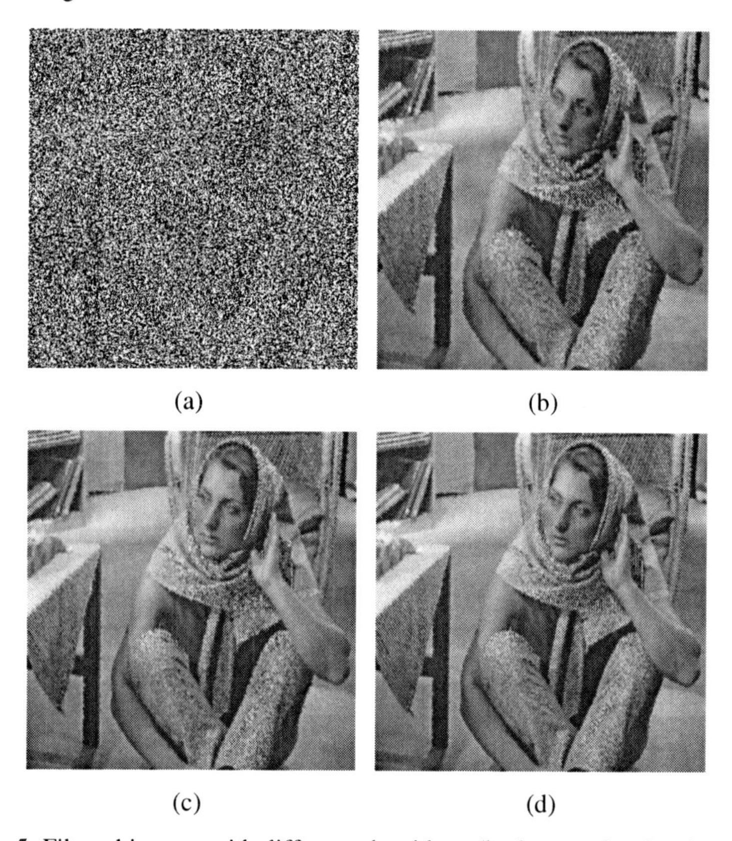

(a) (b)

(c) (d)

Figure 5: Filtered images with different algorithms (barbara, noise density = 0.7). (a) noise image. (b) bi-directional. (c) improved bi-direction. (d) new algorithm.

5 Conclusion

This paper analyzes the shortcoming of bi-directional prediction algorithm and its improved method, and gives a new algorithm based on improved bi-directional prediction algorithm. Compared with foregoing algorithm, the new algorithm detects noise pixels more accurately, and maintains more edge information in filtered image.

Acknowledgement

The research work was supported by "Twelfth Five-year" Jilin province education department scientific research program (No. [2013] 325).

References

[1] Kun Qiao, Chaoyong Guo, Dong Mao, An adaptive switch median filtering algorithm for salt and pepper noise. *Computer Applications and Software*, **28(10)**, pp. 254–256, 2011.
[2] Tao Chen, Kai-Kuang Ma, Li-Hui Chen, Tri-state median filter for image denoising. *IEEE Trans. on Image Processing*, **8(12)**, pp. 1834–1838, 1999.
[3] S.J. Ko, Y.H. Lee, Center weighted median filters and their application to image enhancement. *IEEE Trans. on Circuits Systems*, **38(9)**, pp. 984–993, 1991.
[4] H.L. Eng, K.K. Ma, Noise adaptive soft-switching median filter. *IEEE Trans. on Image Processing*, **10(2)**, pp. 242–252, 2001.
[5] Zhigang Zhu, *Digital Image Processing*. Publishing House of Electronics Industry, Beijing, pp. 50–53, 2002.
[6] Xiaofeng Meng, Pei Liu, Zhongke Yin, Jianying Wang, Bi-directional prediction for the removal of salt-and-pepper noise from images and its improvements. *Journal of Circuits and Systems*, **12(6)**, pp. 47–51, 2008.
[7] Vladimir Crnojevic, Vojin senk, Zeljen Trpovski, Advanced impulse detection based on pixel-wise MAD. *IEEE Signal Processing Letter*, **11(7)**, pp. 589–592, 2004.
[8] Lianghai Jin, Dehua Li, An image denoising algorithm based on noise detection. *Pattern Recognition and Artificial Intelligence*, **21(3)**, pp. 298–302, 2008.

Research on predictive method of target characteristics based on electrostatic detection technology

Li Yanxu, Bu Dingxin
Department of Electrical Engineering,
JiangSu University, ZhenJiang, China

Abstract

The character research of electrostatic target has played an important role in the electrostatic detection. The obtained characteristic curve was judged and identified by detecting the static electric field of the moving object with high velocity in the air. This paper designs BP neural networks and linear neural networks to predict the characteristic signal of electrostatic target. Simulation results show that the BP neural networks is superior to predict the effectiveness and timeliness of the target.
Keyword: electrostatic target characteristics, electrostatic detection, linear neural network, BP neural network.

1 Introduction

The static electricity will be produced for the various charge processes in the usage of engine or the moving objects. The electrostatic detection technology is a new detective technology, which obtains the target electrostatic field as an informational source [1]. For an air flying object, static electricity can't be removed for the friction or electrostatic induction. So the static charge of aircraft is very big. For example, the static electricity of helicopters and cruise missiles can be up to 10^{-6}C–10^{-4}C and the static electricity of jet aircraft can be up to 10^{-7}C–10^{-3}C [2]. The static electricity field can be detected by air target on the fly in its surrounding space. However, the larger range error is produced by the signal processing circuit delay when the target moving with high speed. Therefore, the method of signal prediction research is very important [3].

WIT Transactions on Engineering Sciences, Vol. 107, © 2015 WIT Press
www.witpress.com, ISSN 1743-3533 (on-line)
doi:10.2495/SEEEI140471

2 Analysis on the characteristics of electrostatic target

The characteristics of electrostatic target is mainly determined by the quantity of target charge, the shape, the volume and other factors. The relation between the changed rules of target field and detector output current [4]:

$$i = 2\varepsilon_0 S \frac{\mathrm{d}E_n}{\mathrm{d}t} \tag{1}$$

The ε_0 as the vacuum dielectric constant; S as the surface area of detected electrode. E_n as the normal vector component of electric field intensity on the target electrode. As shown in formula (1), $\frac{\mathrm{d}E_n}{\mathrm{d}t}$ (the changed rate of target field) decided the characteristics of the response signal. Target characteristics of electrostatic detection expression [4]:

$$\frac{\mathrm{d}E_n}{\mathrm{d}t} = \frac{Qv}{4\pi\varepsilon} \frac{\sin\theta(y^2 - 2x^2) - 3xy\cos\theta}{(x^2 + y^2)^{\frac{5}{2}}} \tag{2}$$

The ε as the dielectric constant of dielectric between the electrodes; Q as the goal of charged amount; y as the detector get the shortest distance perpendicular to the relative motion trajectory; x as the detected distance of the relative motion trajectory; θ as angle between detected electrode surface and the relative motion trajectory. When $\theta = 90°$, the detector is a short axial type of electrostatic detector [4]. The expression of target characteristics can be simplified as formula (3), the response curve is shown in Figure 1.

$$\frac{\mathrm{d}E_n}{\mathrm{d}t} = \frac{Qv}{4\pi\varepsilon_0} \frac{(y^2 - 2x^2)}{(x^2 + y^2)^{\frac{5}{2}}} \tag{3}$$

The target will encounter a variety of natural and man-made static in the air, so it not only to locate the target, but also can be identified according to the characteristics of the target. The system from the input signal to the final solution exists delay time, so put forward a method to predict the performance of electrostatic will quickly and effectively improve greatly the detection system.

3 The predictive method of electrostatic target

3.1 Linear neural network prediction

Linear neural network uses linearized transfer function. Based on the weights and bias values to comply with the given input to produce a desire output according to the minimum mean square error rules. The expression of the linear neural network as follows [5]:

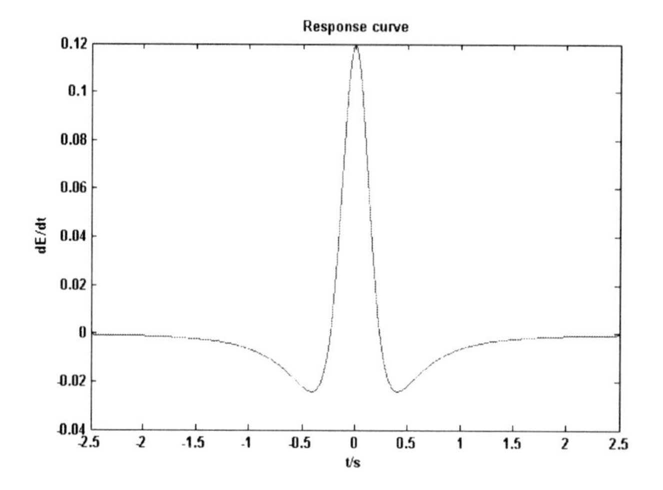

Figure 1: The characteristics of response curve.

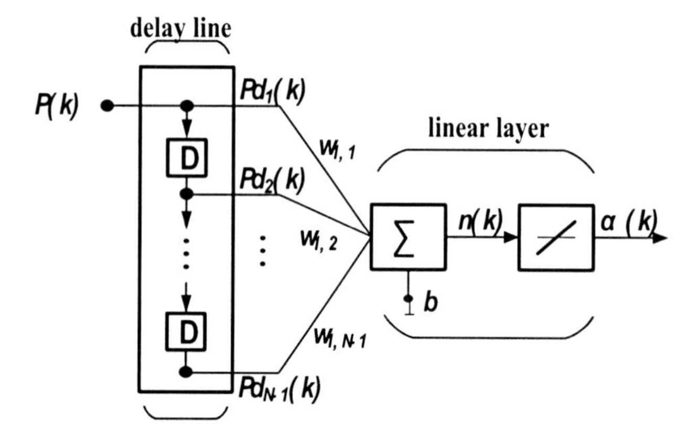

Figure 2: The construction of adaptive linear filter.

$$\alpha = purelin(Wp + b) = Wp + b \tag{4}$$

The α as the actual output of the network; W as the weights of the network; P as the N input vector; B as the threshold. Linear neural network combined with the delay line can constitute an adaptive linear filter, which can automatically adjust the present time filter parameters by obtaining the result of the previous time. Therefore, it can automatically adapt to unknown or the time varying signal [6]. As shown in Figure2, the input signal output N–1 dimensional vector as the input signal of the neuron layer in current time. It will get the current filter output after linear layer.

The miss distance y = 200m, the target charge $Q = 1\times10^{-4}$C, the target and detector relative velocity v = 600 m/s, the distance between the detector and the target x = 3000m, required time t = 5s. In the configuration of neural network, it is important to select a suitable learning rate. The higher learning rate will lead to instability of the learning process and too low will cause the train spend more time [7]. Selected 0.4 as the training of the learning rate and the training goal is 1×10^{-5}.

According to the simulation condition, the emulational results of the linear neural network predictive filter is shown in Figure 3:

The solid line represents the actual target output, the dotted line represents the line neural network prediction. The error between them as shown in Figure 4:

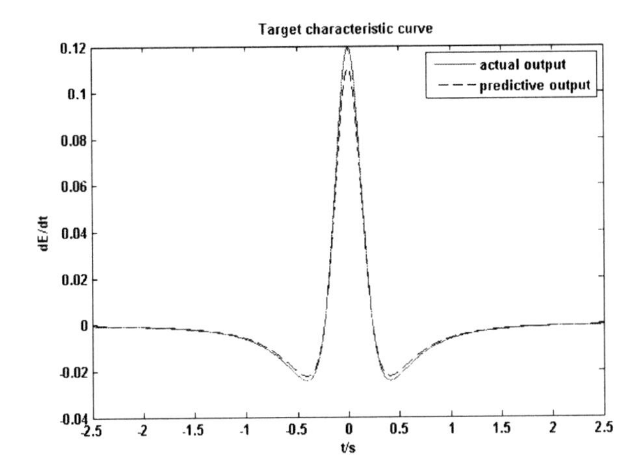

Figure 3: The simulation results of the linear neural network.

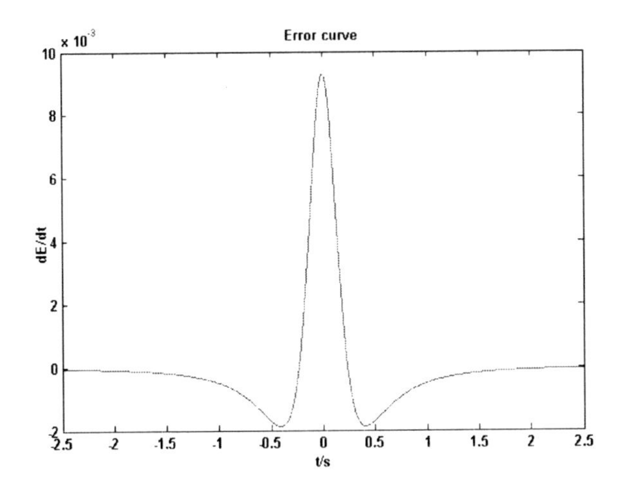

Figure 4: The error of the linear neural network prediction.

As can be seen from Figure 4, the maximum predictive error of the linear neural network prediction is 9.4×10^{-3}, which appeared in the rendezvous time of detector and target. The linear transfer function of the neural network is a linear relationship, which caused the error present linear relationship [7]. Therefore, the trend of the predictive error corresponds with the trend of the output signal. As shown in Figure 5, the predictive time needs 5 seconds. The prediction of the linear neural network needs for a longer time and will create a larger range error. It is difficult to meet the requirements of electrostatic detection system.

BP is also called the back propagation neural network. Its each layer of the neuron state only affects the next layer of neurons state by learning the rules of the steepest descent method. If the output layer of BP neural network cannot get the expected output, it will turn the back propagation. Predicted output of BP neural network can become closer and closer to the desired output by predicting error to adjust the weights and thresholds [8]. As shown in Figure 6, the construction of the BP neural network. The expression of the BP neural network as follows:

$$\alpha = f(Wp + b) \tag{5}$$

The f as the nonlinear transfer function of neurons, this paper chose trainlm function as transfer function. There are many network transfer function in BP neural network, the commonly used are the following functions [8]:

The traingdm function adjust the weights at the current moment in the direction of the negative gradient; the traincgf function can adjust the weight and bias along the fastest decline of error function; but the two function will lead to slow convergence problem [9].

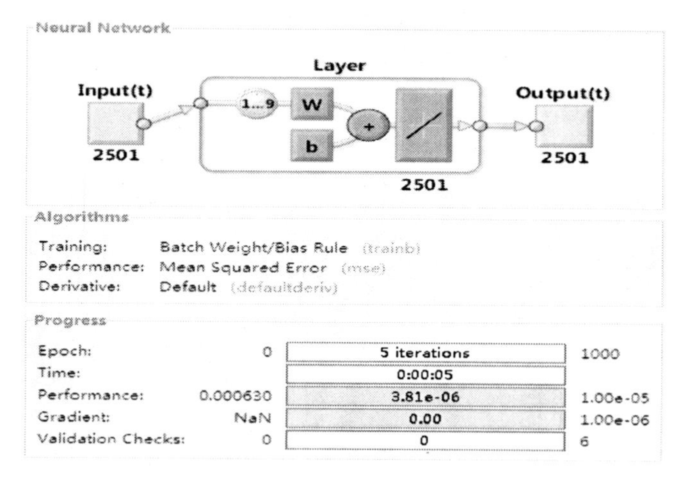

Figure 5: The predictive time of the linear neural network. 3.2 BP neural network prediction.

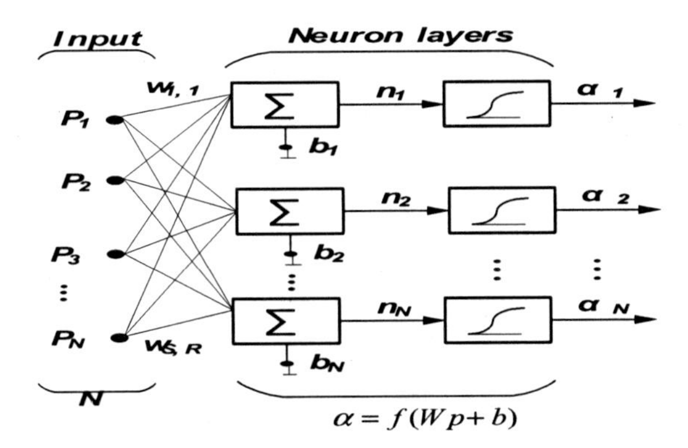

Figure 6: The construction of the BP neural network.

Table 1: The maximum error and the predictive time of different functions.

No.	Transfer Function	Maximum Error	Predictive Time (s)
1	Traingdm	0.0615	5.08
2	Trainlm	0.0032	0.73
3	Trainbfg	0.0029	5.67
4	Trainrp	0.0328	5.51
5	Traincgb	0.0935	5.17
6	Traincgf	0.0163	3.42
7	Trainoss	0.0132	12.2

The trainbfg function has a faster convergence rate; the traincgb function can improve training efficiency by optimizing gradient direction; the trainrp function can reduce the input range. But the three types of functions have very large calculation [9]. The trainoss function can reduce the storage and computation resources but relatively increase algorithm and is suitable for large-scale network.

The trainlm function has fast convergence speed and low memory for small and medium-sized network [10]. The maximum error and the predictive time of different functions were listed in Table 1. The trainlm function can be seen best in predictive accuracy and real-time performance in Table 1.

In order to compare the two kind of neural networks, the emulational conditions as same as linear neural network.

In addition, BP neural network need to set an extra iterations and the number of iterations set 500. The emulational results of the BP neural network is shown in Figure 7.

The transfer function of the BP neural network belongs to nonlinear function, which caused error produce a plurality of local pole [10]. The trend of the predictive error completely inconsistent with the trend of the output signal and the maximum error is 3.6×10^{-3}. The error of the BP neural network prediction is shown in Figure 8.

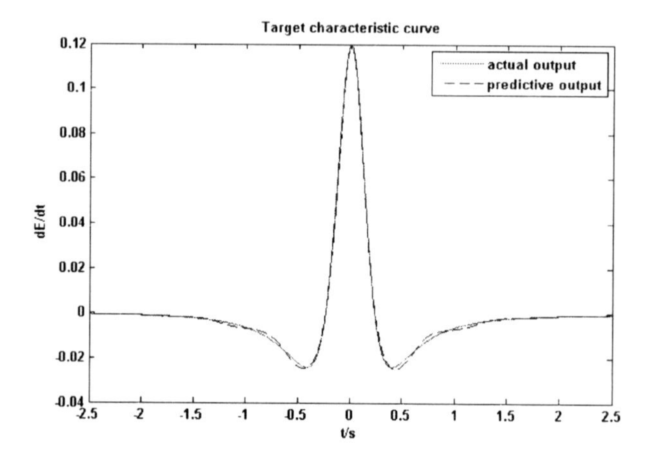

Figure 7: The simulation results of the BP neural network.

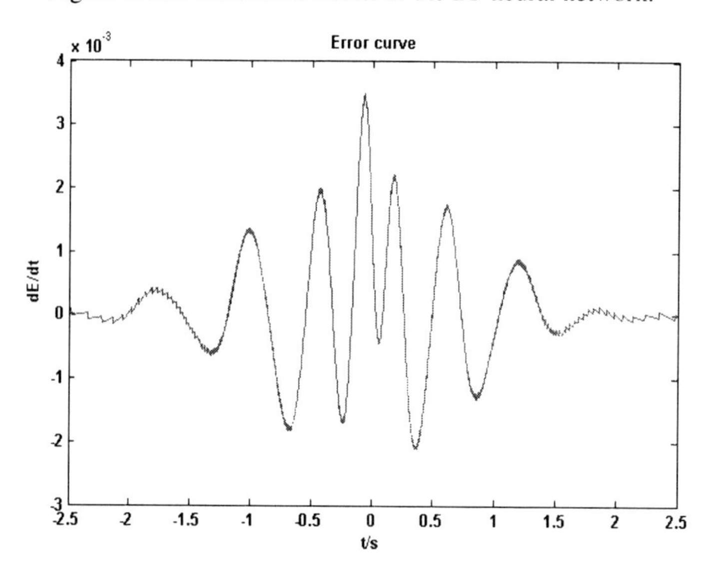

Figure 8: The error of the BP neural network prediction.

4 Conclusion

In this paper, there are two kinds of network to predict electrostatic target signal. Comparison of Figure 4 and Figure 8 shows that the predictive error of BP neural network is smaller. The peak of linear neural network appeared in the rendezvous time of detector and target. But BP neural network in the intersection of the error did not reach the peak. Thus the prediction precision of BP neural network has more advantages. For the detection of air targets with fast movement speed, the calculations of positioning prediction required to complete the work in a very short time. The predictive time of BP neural network needs 0.73 seconds and the predictive time of the linear neural network needs 5.76 seconds. But the effective time of the whole electrostatic target detection needs 5 seconds. So from signal real-time consideration, the BP neural network is superior to the linear neural network. In summary, the BP neural network is more appropriate to meet real-time and effective prediction of electrostatic target characteristic requirements.

References

[1] Gerald KleiTomasBohlen, Inversion of dispersive seismic waves in shallow marine environments. *Marine Geophysical Researches*, 26(3), pp. 287–315, 2005.

[2] Liu Shi, Research on the predictive method of electrostatic detector signal processing and target characteristics. Beijing: *Journal of Detection & Control*, 10(2), pp.138-144. 2007.

[3] Li Yanxu, *The Retical Research on the Influence of Plume Flow on Aero Stat's Charging*. School of Mechatronics Engineering, Beijing:Beijing Institute of Technology, 2006.

[4] Li YinLing, *Technology and Information Processing of Detection of Passive Electrostatic Fuze*. Beijing Institute of Technology, Beijing, 2000.

[5] Zhu CaoKai, *Proficient in MATLAB Neural Network*. Beijing:Electronic Industry Press, 29(2),pp.132–138, 2009.

[6] Vatazhin AB, Starik AM, Kholshchenikova EK, *Electric Charging of Soot Particles in Aircraft Engine Exhaust Plumes*. Beijing:China Machine Press, 2003.

[7] Cao Ge, *MATLAB Tutorial and Training*. Beijing:Mechanical Industry Press, 2010.

[8] Reddy SN, Performance evaluation and error analysis of monopulse radar comparator. *IEEE Instrument and Measurement Technology Conference*, 1996.

[9] Cong Shuang, *Neural Network Theory and Application for MATLAB Toolbox*. University of Science and Technology of China Press, 2009.

[10] Shi Feng, *Thirty Cases of MATAB Neural Network Analysis*. Beijing University of Aeronautics and Astronautics Press, 2010.

An improved artificial bee colony algorithm based on expanding foraging

Ye Tian, Ming Fang
School of Computer Science and Technology,
Changchun University of Science and Technology, Changchun, China

Abstract

Artificial bee colony (ABC) algorithm invented recently is a biological-inspired optimization algorithm, which simulates the foraging behaviors of honey bee swarm. As one of the global optimization algorithms, ABC is good at exploration but poor at exploitation. A modified artificial bee colony (MABC) algorithm based on expanding foraging is proposed for slow convergence of basic ABC. MABC enhances local search ability and increases the convergence speed by means of optimizing the candidate source again. The proposed algorithm is tested on five different scale problems and compared with basic ABC. The comparison results show that MABC is better than basic ABC in not only the convergence speed, the solution quality, but also the robustness.
Keywords: Artificial intelligence, global optimization, artificial bee colony, expand foraging.

1 Introduction

Artificial bee colony (ABC) algorithm is one of the swarm intelligent techniques, which originated from observation and simulation of bee foraging [1]. For its features of being easily implemented, having fewer parameters, ABC has been successfully applied to functions optimization, scheduling problem, data mining, etc. [2–7]. However, like many other swarm intelligent algorithms, ABC is also good at exploration but poor at exploitation. Therefore, an improved artificial bee colony (MABC) algorithm, which adopted an expanding foraging strategy, is proposed to improve the performance of standard ABC.

The paper is organized as follows. In the next section, we describe the main aspects of PSO algorithm. The details of IMPSO are described in Section 3.

doi:10.2495/SEEEII140481

Section 4 illustrates efficiency and accuracy of the proposed approach through computational comparisons. Finally, major results of the paper are summarized in Section 5 along with some remarks of areas for future.

2 ABC algorithm

The artificial bee colony (ABC) algorithm, which simulating the foraging behavior of honey bee swarm, was initially introduced by Karaboga [1] in 2005. In ABC algorithm, each food source is a possible solution for the problem under consideration and the nectar amount of a food source represents the quality of the solution represented by the fitness value. The number of food sources is the same as the number of employed bees and there is exactly one employed bee for every food source. When the employed bee found a specific food source, it carried information about this source and shared it with the onlookers. The onlooker will select a food source according to the nectar amount. The more the nectar amount (fitness) of the food source is, the greater probability of selection happens. The scout searches around the nest randomly to find a better food source.

3 Improved ABC

3.1 Interpolation function

In ABC, the candidate solution (new food source) is generated through a difference disturbance with other selected solution randomly. So, the algorithm has good global search ability, but a less weak ability to exploit, which results in a slow convergence speed [2]. Therefore, an expanding foraging strategy is adopted to guide the search process of bees. Specifically, after the employed bee choose a new food source (candidate solution), this new food source is likely to be more close to the optimal solution if it is better than the old one. So, if the employed bee continue along this direction, it is possible to find or near the optimal solution faster. As can be seen in Figure 1. Left picture shows the initial state, in where the red solid dot denotes the potential global optima and the blue solid dots X_1, X_2 denote two food

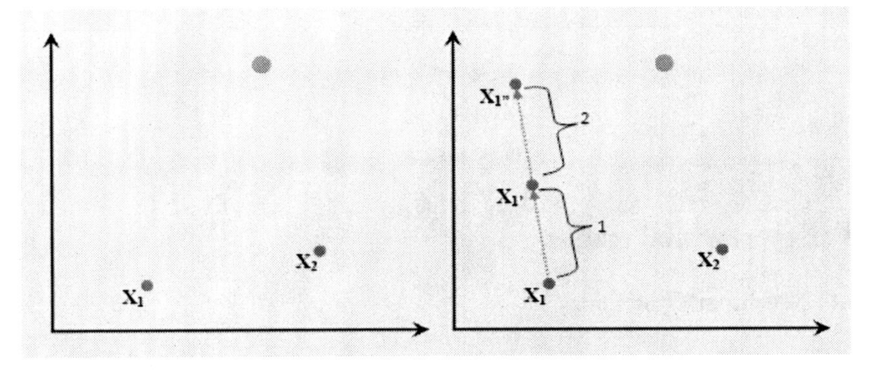

Figure 1: Expand foraging.

MABC algorithm is presented below:

MABC algorithm	
Step 1	Initialize the swarm (population size *NP* and the dimension *D*);
Step 2	Evaluate the swarm, and compute the fitness of the food source;
Step 3	Move employed bee to candidate food source, select new food source through greedy selection and expanding foraging strategy;
Step 4	Compute the fitness and the selection probability value;
Step 5	Move onlooker to candidate food source based on the probability value. Select a new food source again by expanding foraging;
Step 6	Scout executes a random selection and replace the abandoned one;
Step 7	Record the best food source so far;
Step 8	If the stop criterion is satisfied, output the best position and fitness, otherwise, go to Step2.

source position. Right picture depicts the moving state and X_1' is the candidate solution through perturbation on X_1 randomly. If the fitness of X_1' is better than that of X_1, it means X_1 can move a new position than the old position along a dotted line (line 1). So, it is possible that we can find a position X_1'' better than X_1' if we continue moving X_1' along the dotted line (line 2). In this way, MABC can effectively improve the convergence speed, and balance of exploration and exploitation.

The process of expanding foraging can be obtained by using the two following equations: where x_{ij} represent *j*th-dimension of food source *i*. x_{kj} a selected *k*th food source randomly, *r* is a random number uniformly distributed in [−1, 1]. v_{ij} is the new candidate food source. If the fitness value of the new one is better than that of the previous one, the bee would memorize the new position and move in the same direction to find the better food source $v_{ij}^{'}$. Otherwise, it keeps the position of the previous one in its memory on the standard ABC.

$$v_{ij} = x_{ij} + r * (x_{ij} - x_{kj}) \quad (1)$$

$$v_{ij}^{'} = v_{ij} + r * (x_{ij} - x_{kj}) \quad (2)$$

4 Experimental results

4.1 Benchmark functions

In this section, the proposed MABC was compared with the standard ABC 0 on some benchmark functions, which includes unimodal and multimodal functions.

These benchmark functions were Griewank ($f_1(x)$), Rastrigin ($f_1(x)$), Rosenbrock $f_3(x)$, Ackley ($f_4(x)$), and Schwefel ($f_5(x)$), and listed as follows:

$$f_1(x) = \frac{1}{4000}\sum_{i=1}^{D} x_i^2 - \prod_{i=1}^{D} \cos(\frac{x_i}{\sqrt{i}}) + 1 \qquad (-600 \le x_i \le 600)$$

$$f_2(x) = \sum_{i=1}^{D} (x_i^2 - 10\cos(2\pi x_i) + 10) \qquad (-15 \le x_i \le 15)$$

$$f_3(x) = \sum_{i=1}^{D-1} (100(x_{i+1} - x_i^2)^2 + (x_i - 1)^2) \qquad (-15 \le x_i \le 15)$$

$$f_4(x) = -20\exp\left(-0.2\sqrt{\frac{1}{D}\sum_{i=1}^{D} x_i^2}\right) - \exp\left(\frac{1}{n}\sum_{i=1}^{D} \cos 2\pi x_i\right) + 20 + e \qquad (-32.768 \le x_i \le 32.768)$$

$$f_5(x) = D * 418.9829 + \sum_{i=1}^{D} -x_i \sin(\sqrt{|x_i|}) \qquad (-500 \le x_i \le 500)$$

4.2 Parameters setting

There are two major control parameters used in the MABC and standard ABC: The number of the food source (*NP*), the value of limit, which are fixed according to literature [1]. Table 1 lists all the parameters' value in this paper.

4.3 Results and discussion

In this part, we compared MABC with standard ABC. According to literature [1], for each function three dimensions sizes 10, 20, and 30 were tested, and corresponding to the maximum numbers of iterations were set to 500, 750, and 1000. Each algorithm was simulated with 30 independent runs for each test.

Table 1: Parameters setting of MABC.

NP = 125			Limit
Employed Bee	Onlooker	Scout	
62	62	1	*NP**D*0.5

Table 2: Comparison of ABC and MABC.

Fun	D	Gen	ABC		MABC	
			Mean	SD	Mean	SD
f_1	10	500	1.730E–3	0.004	**2.030E–10**	7.510E–10
	20	750	8.080E–8	2.650E–7	**2.520E–12**	6.274E–12
	30	1000	1.950E–8	7.360E–8	**4.990E–13**	1.268E–12
f_2	10	500	1.150E–14	1.950E–14	**0**	0
	20	750	8.300E–10	1.999E–9	**4.790E–13**	1.213E–12
	30	1000	1.310E–5	5.013E–5	**1.650E–11**	2.908E–11
f_3	10	500	5.980E–2	0.064	**1.920E–2**	0.015
	20	750	2.060E–1	0.220	**8.540E–2**	0.078
	30	1000	5.980E–1	0.687	**2.080E–1**	0.193
f_4	10	500	2.180E–10	1.360E–10	**1.110E–13**	4.066E–14
	20	750	5.520E–7	1.868E–7	**1.810E–9**	1.239E–9
	30	1000	7.760E–6	2.788E–6	**6.960E–8**	3.291E–8
f_5	10	500	**1.270E–4**	2.589E–12	**1.270E–4**	4.458E–13
	20	750	1.600E+1	40.892	**2.550E–4**	1.770E–9
	30	1000	1.030E+2	110.678	**5.600E+1**	68.134

As can be seen from Table 2, MABC achieves the smaller fitness value and standard deviation than ABC algorithm for the all test problems. For Schwefel problem, MABC and ABC achieve the same mean fitness value 1.27E–4 when the dimension is 10 and max iteration is 500. However, the standard deviation of MABC is smaller than that of ABC slightly. Hence, in view of the algorithm's effectiveness and robustness on the whole, MABC remains quite competitive against the standard ABC algorithm.

In order to compare the performance of these two algorithms conveniently, we listed the two evolution curve in Figures 2 and 3. Figure 2 shows the evolution of ABC and MABC for Griewangk problem. When the iteration is less than 150, MABC is close to ABC algorithm in the convergence speed. With the increase of the iteration, the convergence speed of MABC is faster than that of ABC algorithm, and MAB provides a better result at the end of evolution.

Similar results can be achieved in Figure 3. We can see that the convergence rates of the two algorithms are very close at the beginning stage, and become difference during the later stage of the evolution, and MABC outperforms NPSO in the total solution quality thoroughly.

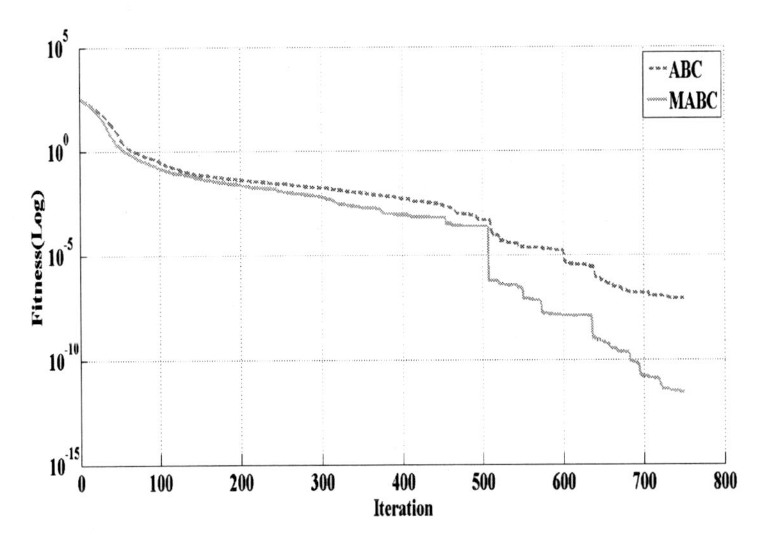

Figure 2: Evolution of ABC and MABC for griewangk problem.

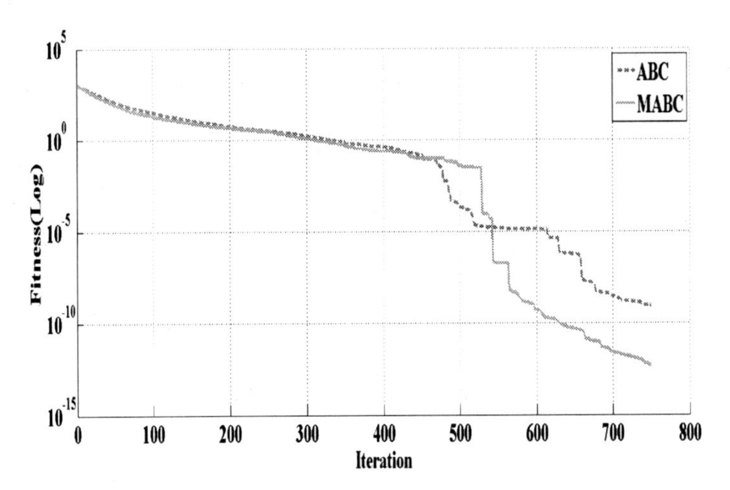

Figure 3: Evolution of ABC and MABC for rastrigin problem.

5 Conclusions

In this paper, to overcome the disadvantage of the standard ABC on local searching ability, an improved artificial bee colony algorithm based on expanding foraging strategy is presented for locating the global optima of continuous unconstrained optimization problems. The proposed ABC algorithm can enhance local searching ability and increase the convergence speed through expanding foraging on the better candidate food source found during the searching stage.

To validate the effectiveness of MABC, we compared MABC with the standard ABC on some different scale benchmark functions. The experimental results show that MABC algorithm has an obvious improvement in whatever the solution quality, the convergence speed as well as the stability. Our further research is to test the performance of the MABC algorithm for more benchmark functions with high dimension, multi-objective.

Acknowledgement

The research work was supported by Jilin Scientific and Technological Development Program, Jilin public computing platform under Grant No. 20130101179JC-11, Natural Science Foundation of Jilin Province of China under Grant No. 20130101054JC.

References

[1] Karaboga Dervis, Basturk Bahriye, A powerful and efficient algorithm for numerical function optimization: artificial bee colony (ABC) algorithm. *Journal of Global Optimization*, **39(3)**, pp. 459–471, 2007.
[2] Zhu Guopu, Kwong Sam, Gbest-guided artificial bee colony algorithm for numerical function optimization. *Applied Mathematics and Computation*, **217(7)**, pp. 3166–3173, 2010.
[3] Karaboga Dervis, Akay Bahriye, A comparative study of Artificial Bee Colony algorithm. *Applied Mathematics and Computation*, **214(1)**, pp. 108–132, 2009.
[4] Zhang Changsheng, Ouyang Dantong, Ning Jiaxu, An artificial bee colony approach for clustering. *Expert Systems with Applications*, **37(7)**, pp. 4761–4767, 2010.
[5] Yan Xiaohui, Zhu Yunlong, Zou Wenping, Wang Liang, A new approach for data clustering using hybrid artificial bee colony algorithm. *Neurocomputing*, **97**, pp. 241–250, 2012.
[6] Tasgetiren M. Fatih, Pan QuanKe, Suganthan P.N., Chen Angela H-L, A discrete artificial bee colony algorithm for the total flowtime minimization in permutation flow shops. *Information Sciences*, **181(16)**, pp. 3459–3475, 2011.
[7] Zhang Rui, Song Shiji, Wu Cheng, A hybrid artificial bee colony algorithm for the job shop scheduling problem. *International Journal of Production Economics*, **141(1)**, pp. 167–178, 2013.

Research on intelligent analysis method of sneak circuits based on learning mechanism

He Hui-ying[1,2], Li Zhi-gang[1]
[1]*College of Electrical Engineering, Hebei University of Technology, Tianjin, China*
[2]*Fundamental Department, Academy of Military Transportation, Tianjin, China*

Abstract

Sneak circuits are common in a circuit system, this paper has summarized the current various analysis methods of sneak circuits and introduced the general flow of intelligent sneak circuits analysis method based on learning mechanism. And this method is applied to analyze the sneak circuits that exists in the basic step-down resonant switched capacitor converter. It reduces the larger workload of system input, also avoids the difficult work of finding the clues table in traditional sneak circuits analysis method. And the results verify the accuracy and practicability of this method.

Keywords: sneak circuit analysis, learning mechanism, reliability, power electronic converter.

1 Introduction

More domestic and foreign research results show that there are generally sneak circuits in large and complex system inevitably, and sneak circuit is one of the main expression forms of sneak circuits [1]. The so-called "sneak circuit" (sneak circuit, SC) refers to a state that exists in electrical and electronic circuits. Under certain conditions, it can lead to undesired circuit system functions or inhibit the desired functions, seriously sometimes it still can cause serious system faults [2, 3].

In the world at present, sneak circuit analysis technique is a more effective way to detect these sneak questions [1]. At present, there are a variety of analysis methods of sneak questions. Taking the analysis process of sneak circuits of some power electronic circuits as an example, this paper has introduced the analysis

www.witpress.com, ISSN 1743-3533 (on-line)
doi:10.2495/SEEEI140491

method of sneak circuits based on learning mechanism. Through this method, it can avoid the difficult problem to obtain clues table [4], and can reduce the larger workload of system input, etc.

2 Review of analysis method of sneak circuits

It is worth noticing that the causing reason of sneak circuits is different from the common failure. They are not caused by components, equipments or system failures, but the designer brings them into the design process inadvertently in order to achieve the design expectation [3]. Sneak circuits hide in the system under normal working state, only in some specific conditions it will be activated.

At present there are mainly three analysis methods of sneak circuits, such as manual analysis method, semiautomatic analysis method, automatic and intelligent analysis method [1].

(i) Manual analysis method of sneak circuits
Based on the characteristics of designers' control ability of accurate design, the complex system can be divided into some simple subsystems with specific patterns that designers can grasp their behaviors and functions more easily and accurately through visual analysis. Through contrasting design experience to a simple subsystems and discriminant analysis, it can confirm whether there are any sneak circuits in the system. Feasible corrective actions should be taken for specific problems if there are some sneak circuits.

(ii) Semiautomatic (or computer-assisted) analysis method of sneak circuits
CapFast/SCAT, a sneak circuit analysis system that developed by American is more internationally representative of the computer aided analysis system of the sneak circuit. It can automatically track all the paths from the specified source points to the designated earthing points, so some work of manual analysis of sneak circuits is assisted by computer to artificial sneak circuit analysis, thus it makes the analysis of sneak circuits have great efficiency improvement.

Semiautomatic analysis method of sneak circuits works mainly with the combining of topological pattern recognition and clues table method to analyze sneak circuits. During current system design process, sneak circuits analysis must be completed with existing technology by very experienced experts according to the long-term accumulation of experience, and then they make analysis and judgment for network topology model that comes from computer assistant identification. But this analysis process has a large workload and a long term, thus causing some missed judgment and some incompletement, so affects the system integrated design personnel to use the analysis system [5].

(iii) Automatic and intelligent analysis method of sneak circuits
There are some automatic and intelligent analysis methods of sneak circuits such as the analysis method based on qualitative simulation and quantitative simulation, analysis method based on artificial neural network learning mechanism etc. But there is not a set of relatively complete intelligent analysis method of sneak circuits that has general design criteria or circuit solution to overcome the exist-

ing of sneak circuits [6]. So the related research should be strengthened, but the completely automatic and intelligent development direction of sneak circuits is indisputable.

3 Analysis process of sneak circuits based on learning mechanism in general

Analysis steps of sneak circuits based on learning mechanism are shown in Figure 1 [1].

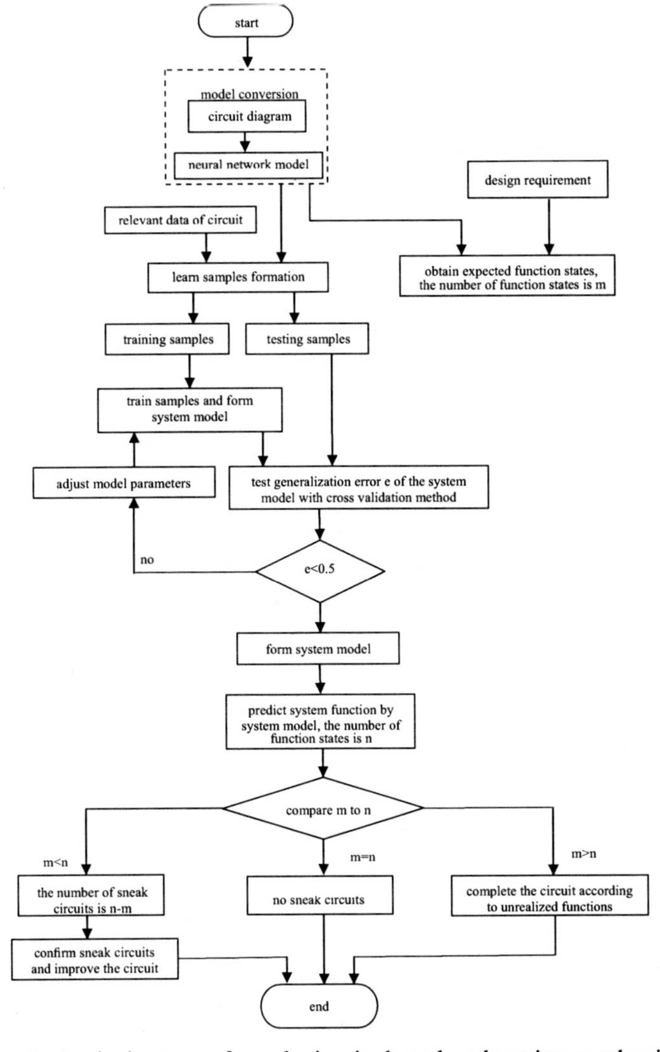

Figure 1: Analysis steps of sneak circuits based on learning mechanism.

As the output of neural network training model Y is the state of system function, namely function realized or not. "1" represents the function realized, and "0" represents the function unrealized. But during the training process, the output of neural network training model Y may be a value between 0 and 1, at this point, the threshold value filtering method is used to deal with the calculation results, that is, the threshold parameter is set with value of α. When, we think system functions realized, namely, otherwise unrealized and so the identification results of sneak circuits is directly affected by the parameter value of α, and α is generally set by 0.5 or higher.

4 Sneak circuits analysis of a basic step-down RSC converter

4.1 Basic working principle

The basic working principle of a basic step-down RSC converter [7] is shown in Figure 2.

4.2 Analysis process of sneak circuits

If the electric circuit can realize resonance function, "1" represents the function state, otherwise "0". According to the design function of the circuit, the expected function state is obtained, as shown in Table 1.

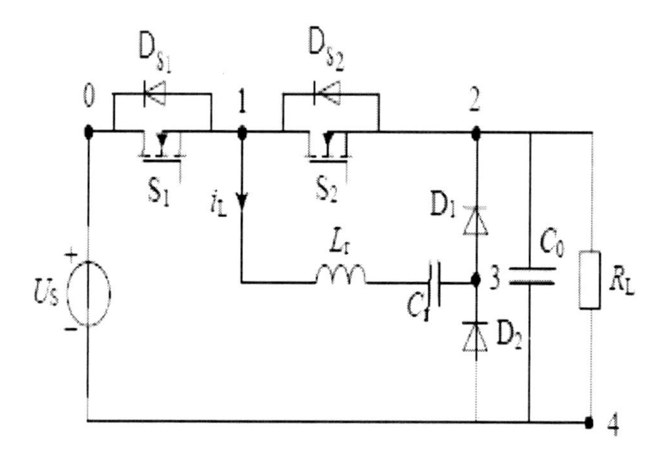

Figure 2: A basic step-down RSC converter.

Table 1: Expected function state.

No.	Switch State						Function State	Function Description
1	0	0	0	0	0	0	1	Discharge
2	0	0	1	0	0	1	1	Discharge
3	1	0	0	0	1	0	1	Charge

Step 1: Samples formation

According to the actual working principle of basic step-down RSC converter, whole control devices cannot conduce with parasitic diodes themselves or directly parallel diodes with them, namely and, and cannot conduce at the same time. The two switches S_1 and S_2 as shown in Figure 2 should conduct in turn according to the requirement of resonant circuit, that is, and cannot conduce at the same time. So 32 invalid circuit paths can be removed from all the paths from the ergodic work [8]. Then the training sample space is obtained, as shown in Table 2.

Step 2: Training model formation

(i) Parameter setting

The population size of clonal selection algorithm is 10; the total number of clone is 500; the learning algebra is 200; the error less than 0.1 as the termination condition; the threshold value of affinity degree between antibodies is 0.1; the threshold value of affinity degree between antibodies and antigens is 0.1; the output threshold value is 0.5.

(ii) Sample classification and network training

Set the last five samples as test samples, and the rest of the samples as the training samples. The network is trained with six times cross validation method.

Step 3: Analysis of generalization error and robustness of the model

Using the average error on the training model in the validation set as the network generalization error estimation.

Table 2: Training sample space.

No.	Switch State						Function State
1	0	0	0	0	0	0	1
2	0	0	0	0	0	1	0
3	0	0	0	0	1	0	0
4	0	0	0	0	1	1	0
5	0	0	0	1	0	0	0
6	0	0	0	1	0	1	0
7	0	0	0	1	1	1	0
8	0	0	1	0	0	0	0
9	0	0	1	0	0	1	1
10	0	0	1	0	1	0	0
11	0	0	1	0	1	1	0
12	0	1	0	0	0	0	0
13	0	1	0	0	1	0	0
⋮	⋮	⋮	⋮	⋮	⋮	⋮	⋮
25	1	0	0	0	1	0	1
26	1	0	0	0	1	1	0
⋮	⋮	⋮	⋮	⋮	⋮	⋮	⋮
30	1	0	0	1	1	1	0

Table 3: Switch and function states according to sneak circuit paths.

No.	Switch State						Function State	Function Description
1	0	0	0	1	1	0	0	Reverse continued
2	0	1	0	0	0	1	0	Reverse continued

Step 4: Sneak circuit confirmation

The function states of the circuit, as shown in Table 3, are not consistent with the expected ones through the prediction work of training network. The circuit paths corresponding to those states are confirmed sneak circuits. And this analysis result corresponds with that in paper [8].

Step 5: Circuit improvement, sneak circuit elimination.

5 Conclusions

The sneak circuit analysis method based on intelligent learning mechanism is verified accurate and practical through sneak circuit analysis process of a basic step-down resonant switched capacitor converter in power electronic circuit. The advantages embodied during the course are incomparable to the traditional sneak circuits analysis method. But the sneak circuit analysis method based on intelligent learning mechanism has its own shortcomings, such as parameter setting problem, relativity problem between network model parameters and circuit parameters, reliability analysis of the predicted results, etc. Those problems are still key projects in the later, so can realize the goal of short-term analysis process and extensive use of sneak circuit analysis method [5].

Acknowledgement

The research work was supported by National Natural Science Foundation of China under Grant No. 51377044.

References

[1] Hu Changhua, Chen Binwen, Liu Bingjie, *Theory and Application of Sneak Circuit Analysis for Complex System*. Science Press, Beijing, 2008.

[2] Rankin JP, Sneak circuit analysis. *Nuclear Safety*, **14(5)**, pp. 461–468, 1973.

[3] Ren Liming, *Analysis Technology and Application of Sneak Circuit*. National Defense Industry Press, Beijing, 2011.

[4] Hu Changhua, Liu Bingjie, Sneak circuit analysis based on novel coadjacent neural network model for reliability control of complex system. *Acta Automatica Sinica*, **34(2)**, pp. 179–194, 2008.

[5] Zhou Tao, Hu Changhua, Xia Qibing, Sneak circuit analysis in reliability engineering. *Network Security and Reliability of the System*, **1(1)**, pp. 24–28, 2003.

[6] Yin Shenyan, Liu Min, Li Liang, Sneak Circuit Analysis (SCA) and isolation methods under power – off low impedance path. *Telecommunication Engineering*, **8(50)**, pp. 159–161, 2010.
[7] Qiu Dongyuan, Zhang Bo, Study of sneak circuit in resonant switched capacitor converters. In: *Proceeding of the CEE*, **25(21)**, pp. 35–37, 2005.
[8] He Huiying, Li Zhigang, Zhang Feifei, Fu Lanfang, Excluding method of invalid paths in sneak circuits for power electronic converters. *Application of Electronic Technique*, **12(39)**, pp. 72–74, 2013.

Design and realization of switch capacitive readout circuit

Xiangliang Jin, Mengliang Liu, Liang Xie
Faculty of Material, Optoelectronics and Physics,
Xiangtan University, China,
Hunan Engineering Laboratory for Microelectronics,
Optoelectronics and System on a Chip, Xiangtan, China

Abstract

This paper presents a fully differential low noise parasitic-insensitive switch-capacitor readout circuit that is used for capacitive sensors. The proposed capacitor readout circuit has been fabricated in a three-metal two-poly 0.35μm CMOS process. Measured results show that the readout circuit can sense difference and single capacitor with 78.1dB dynamic.
Keywords: capacitive readout, Chopper Switch.

1 Introduction

Over the past years, low cost and high resolution MEMS acceleration are used in many applications in automotive stability control and GPS inertial navigation [1, 2]. It is costly and rarely available to integrate a sensing device and its readout circuit in bulk micromachining process [4]. The readout circuit should be designed to suppress these effects. The switched capacitor (SC) readout circuit is the best way to tradeoff between complexity and performance. The readout circuit should be designed to suppress these parasitic effects. There are four kinds of capacitance-to-voltage structures: the switched capacitor (SC), the AC-bridge with voltage amplifier, the trans-impedance amplifier and the trans-capacitance circuits. Among the four kinds of readout circuits, the switched capacitor readout circuit is the best way to tradeoff between complexity and performance.

In this paper, a full differential SC readout circuit is designed. Measured results show that the readout circuit can sense capacitor with 78.1dB dynamic.

WIT Transactions on Engineering Sciences, Vol. 107, © 2015 WIT Press
www.witpress.com, ISSN 1743-3533 (on-line)
doi:10.2495/SEEEII140501

In the next section, the design of the SC readout circuit is introduced in Section 2. Test results are given in Section 3 and some conclusions are presented in Section 4.

2 Circuit design

The SC charge integration manner has been widely used in the MEMS accelerometer sensor interface readout circuit [3]. The SC interface circuit which is conveniently fabricated in CMOS technology is designed in the MEMS accelerometer to convert capacitance-to-voltage [6].

A schematic of the interface readout circuit is given in Figure 1. The full differential charge transfer operation is achieved with switches S1–S4. In this method, the common node A is biased at a stable voltage, so the driving force influence is eliminated arising from voltage fluctuation on the common node.

Charge transferring technique and basic working principle of the whole interface circuit are presented in this part. As shown in Figure 1, during the phase PE, the integration capacitors CF are discharged, and the sensed capacitors CS+ and CS– are charged with VDD and VSS (gnd). The charges stored in the sensed capacitors are transferred to the CF when the phase PS is high. The opposite signs are used to charge the sensed capacitors, and the charge arrived at CF is proportional to the difference between the sensed capacitors. During phase ch1 and ch2, the chopper switch begins to work. The $1/f$ noise and offset are translated to high frequency by the chopping square wave [5], then the noise and offset can be removed by a low pass filter without affecting the signal.

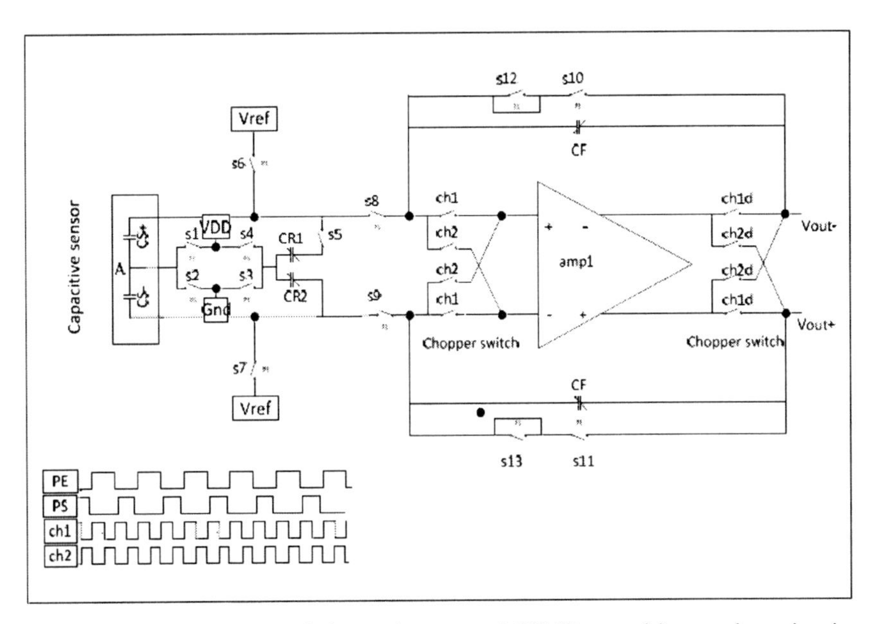

Figure 1: A commonly used charge integrator MEMS capacitive readout circuit.

The readout circuit is designed to work under 100 kHz clock. The output of the readout circuit is given by Eq. (1) under differential capacitance detection when the switch S5 is high as shown in Figure 1. And when the S5 is disconnected with single capacitance detection, the output of the readout circuit is Eq. (2):

$$V_{\mathrm{out}+} - V_{\mathrm{out}-} = \frac{V_{DD}\left(CS_{+} - CS_{-}\right)}{C_F} * \frac{R_F}{R} \tag{1}$$

$$V_{\mathrm{out}+} - V_{\mathrm{out}-} = \frac{\mathrm{Vref} * CS}{C_F} * \frac{R_F}{R} \tag{2}$$

where $V_{\mathrm{out+}}$ and $V_{\mathrm{out-}}$ are the output of the voltage buffer. The capacitance-to-voltage gain of the readout circuit is between 0.56 and 13.3 mV/fF through controlling the integration capacitor CF in the front end charge transfer.

In order to model the transient output of the MEMS acceleration, a simplified model is established for the readout circuit simulation. The model is composed of two timing-varying capacitors whose capacitances are resonating at 100 Hz.

Because there is no time-varying capacitor in traditional SPICE simulator tool, a voltage-controlling-capacitor is proposed which consists of a fixed capacitor, two multipliers and a voltage-controlled voltage source. The working principle of the voltage-controlling-capacitor is explained in Figure 4. The capacitance between the nodes M and O is

$$C_{\mathrm{MO}}(t) = \left[1 - G_{\mathrm{mult}} V_a(t)\right] C_0 \tag{3}$$

where G_{mult} and V_{a} are the gain of the multiplier and the voltage that controls the capacitance, respectively. It is shown in Eq. (3) that the voltage $V_{\mathrm{a}}(t)$ can linearly control the equivalent capacitance between node M and O with the help of the multipliers and the voltage-controlled voltage source.

The output result from the readout circuit is shown in Figure 2 when the model is used to represent the MEMS acceleration. The control signal V_{a} is a sine wave with 100 Hz frequency and the capacitance-voltage coefficient of the SC readout circuit is set to 5mV/fF.

3 Experiment result

The SC readout circuit chip has been fabricated in a 3-metal/2-poly 0.35-μm CMOS process, which measures 2.0×2.5mm^2. Its Micrograph is shown in Figure 3. A 5-V supply provided power for the readout circuit and an Agilent E5061B network analyzer was used. Figure 3 indicates test equipment of the experiments. Its power consumption is about 17mW under a single 5V supply, which the four amplifiers in the SC readout circuit expand about eighty percent of the whole power consumption.

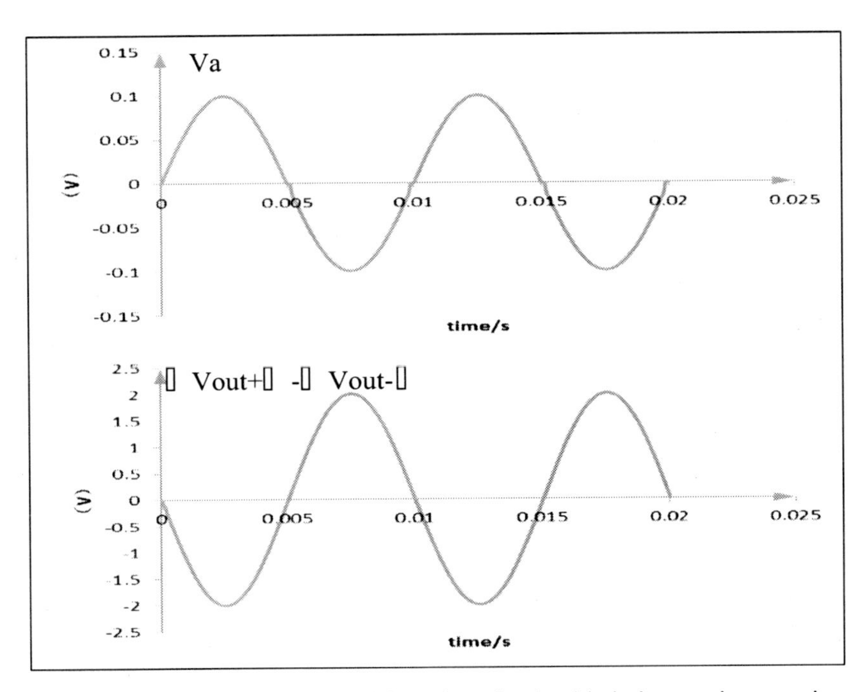

Figure 2: Simulation output of the SC readout circuit with timing varying capacitors for MEMS acceleration.

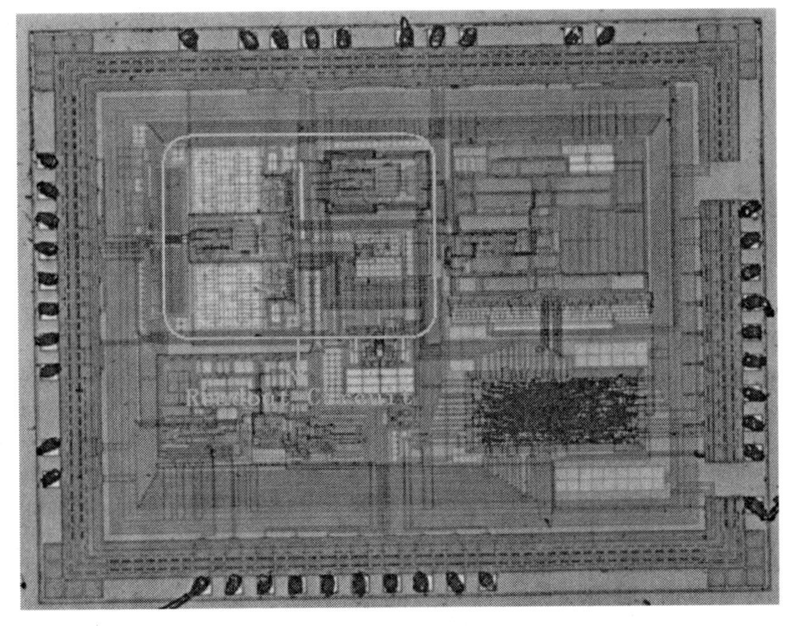

Figure 3: Micrograph of the SC readout circuit.

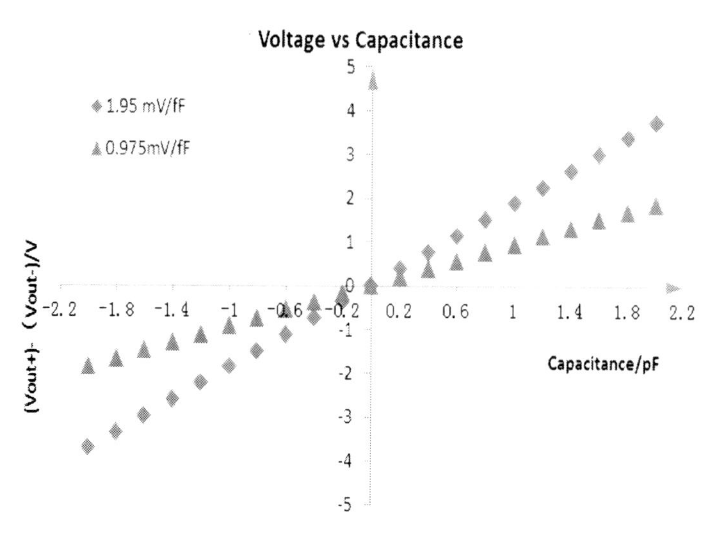

Figure 4: Output of the readout circuit as a function of input differential capacitance.

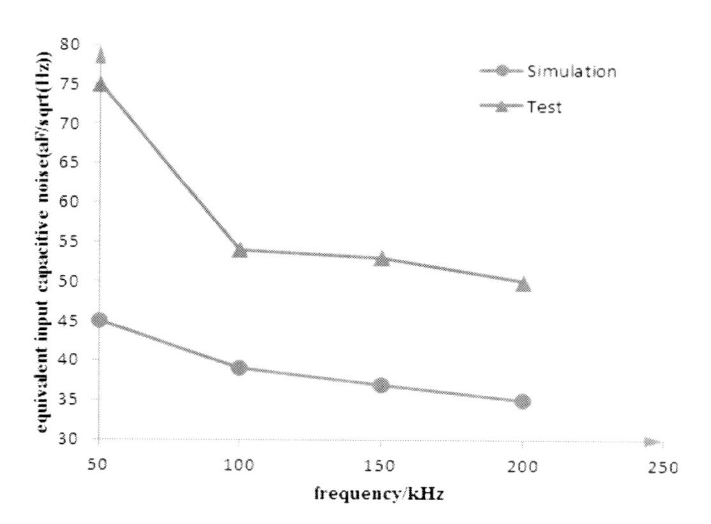

Figure 5: Test result and simulated equivalent input capacitance noise density at 1 kHz as a function of the chopper frequency with 1mV/fF gain setting.

As shown in Figure 4, the measured result of the SC readout circuit is plotted at different gain settings when fixing one of the input capacitance and changing the other in 0.2pF steps. The result shows that the output voltage of the SC readout circuit is linear with measured capacitance at different capacitance voltage coefficient. And all output linearity is better than one percent in the range of –4V to 4V.

Figure 5 shows that the measured output noise of the readout circuit is high than the simulation result with the same frequency dependence. Because the noise

Table 1: Performance summary comparison.

	This Work*	[2]*	[6]*	[7]*
Sensitivity (mV/fF)	0.5	9980	7880	155
Resolution (aF)	4.0	0.5	2.8	N/A
Clock (kHz)	100	125	250	4000
Technology (μm)	0.35	0.8	0.8	0.35
Supply (V)	5	5	5	3.3
Power (mW)	3.4	9.38	N/A	3.78

bandwidth of the output is far wider than chopper frequency, so the high chopper frequency could help to reduce the noise component. The output noise voltage at 1 kHz is 0.5mV in 200 Hz resolution bandwidth, corresponding to 500aF input equivalent capacitive noise with gain setting as 1mV/fF in 100 Hz resolution bandwidth. So the input equivalent capacitance noise power spectral density of the readout circuit is about 50aF/sqrt (Hz) and the dynamic range could be as large as 78.1dB.

A performance summary of the design is provided in Table 1, where a comparison with other paper is also presented. It should be noticed that the results are based on measurements from fabricated chips.

4 Conclusion

This paper introduces a fully differential SC readout circuit for a MEMS capacitive acceleration. The readout circuit has been fabricated in a 0.35-μm CMOS process. The SC readout circuit includes a fully differential charge transfer front end. Test results on the circuit have been performed and the feasibility of the capacitive readout circuit structure has been proved. Several performance parameters are measured and the test results show that the SC readout circuit can resolve input capacitance variation as low as 50aF and the readout circuit achieves a measured result with a dynamic as large as 78.1dB. Such result is promising because it could meet the specification required by the micro acceleration.

Acknowledgment

Project supported by the National Natural Science Foundation of China (Grant No. 61274043), Program for New Century Excellent Talents in University of Ministry of Education of China (NCET-11-0975) and Hunan Provincial Natural Science Foundation of China (Grant No. 12JJ4064).

References

[1] J. Bryzek, S. Roundy, B. Bircumshaw, et al., Marvelous MEMS. *IEEE on Circuits and Devices Magazine*, **22(2)**, pp. 8–28, 2006.

[2] J. Shiah, H. Rashtian, S. Mirabbasi, A low-noise parasitic-insensitive switched-capacitor CMOS interface circuit for MEMS capacitive sensors. *Proceedings of IEEE NEWCAS*, pp. 470–473, 2011.

[3] N. Yazdi, H. Kulah, K. Najafi, Precision readout circuits for capacitive micro-accelerometers. In: *Proceedings of IEEE Conference on Sensor*, pp. 28–31, 2004.

[4] H.D. Roh, J. Roh, Q.Z.D. Duanquanzhen, All MOS transistors bandgap reference using chopper stabilization technique, ISOCC, pp. 353–357, 2010.

[5] C.C. Enz, G.C. Temes, Circuit techniques for reducing the effects of op-amp imperfections: autozeroing, correlated double sampling, and chopper stabilization. *Proceedings of the IEEE*, **84(11)**, pp. 1584–1614, 1996.

[6] J. Shiah, H. Rashtian, S. Mirabbasi, A low-noise high-sensitivity readout circuit for MEMS capacitive sensors, ISCAS, pp. 3280–3283, 2010.

[7] H. Rodjegard, A. Loof, A differential charge transfer readout circuit for multiple output capacitive sensors. *Sensors and Actuators A*, **119(2)**, pp. 309–315, 2005.

Complex linearized Bregman iteration reconstructed algorithm for compressed sensing

Wenfeng Chen, Bin Xia, Jun Yang
Air Force Early Warning Academy, Wuhan, China

Abstract

Compressed sensing (CS) theory has attracted much interest since it can effectively alleviate the data transmission, storage and processing pressure. Meanwhile, the CS theory has been successfully applied to real-valued problem. However, there are many complex-valued problems in practice. In order to solve complex-valued problem in CS effectively and efficiently, we propose a complex linearized Bregman iteration (CLBI) algorithm to apply to signal recovery. The CLBI algorithm is implemented by the complex mathematical framework and using the Bregman distance and Taylor expansion. Simulation results show that the novel algorithm is effective, convergent, and faster than the conventional linearized Bregman iteration algorithm.

Keywords: compressed sensing, complex-valued reconstruction, linearized Bregman iteration.

1 Introduction

Since the CS theory can reconstructed signal effectively even lower than the Nyquist Sampling theory, it can alleviate the data transmission, storage and processing pressure, has shown great appeal and broad prospect in large-scale sparse and compressible signal processing [1–3]. Many researchers only discussed the CS reconstruction of real valued problem. However, in most of practical situations, data or signal are complex-valued, such as synthetic aperture radar (SAR), inverse synthetic aperture radar (ISAR) and magnetic resonance imaging (MRI), etc. Therefore, it is necessary to expand the CS reconstructed algorithms to complex-valued domain. In fact, Ender presented a method to

WIT Transactions on Engineering Sciences, Vol. 107, © 2015 WIT Press
www.witpress.com, ISSN 1743-3533 (on-line)
doi:10.2495/SEEEII140511

solve this problem by arranging the real and imaginary parts of the data in single real-valued array [4]. Also, the problem can be solved through original linearized Bregman iteration (real-valued LBI). But this method has two disadvantages, one is the dimension of signal is doubled than that the real-valued case, it will need more computational burden. The other is any prior phase information is not exploited, which may lead to rough or noisy after acquisition, or discontinuities in both amplitude and phase. Another approach for solving complex-valued problem is estimating the phase and the magnitude of the complex-valued data at the same time, thus the quality of reconstruction results can be improved [5].

In this paper, a novel CLBI algorithm in CS, which has effectiveness and computational efficiency, is proposed to apply to signal recovery. The paper is organized as follows: in the second section, the complex linearized Bregman iteration is presented with complex-valued in CS. In the third section, experimental results are implemented to show convergence, effectiveness and advantage of the proposed signal recovery algorithm. The last section is some conclusions about the paper.

2 The CLBI algorithm

The mathematical model of complex sparse signal recovery in CS theory can be expressed as follows:

$$\boldsymbol{y} = \boldsymbol{\Phi s} = \boldsymbol{\Phi \Psi x} = \boldsymbol{\Theta x} \tag{1}$$

where $\boldsymbol{s}$ is a complex signal, it can be represented as $\boldsymbol{s} = \boldsymbol{\Psi x}$; $\boldsymbol{x}$ is the coefficient vector; $\boldsymbol{y}$ is the measured signal; $\boldsymbol{\Psi} \in C^{N \times N}$ is the sparse basis matrix; $\boldsymbol{\Phi} \in C^{M \times N}$ is the measurement matrix; and $\boldsymbol{\Theta} = \boldsymbol{\Phi \Psi}$ is the sensing matrix. Note: only the real part of measurement matrix is used for the measurement matrix.

It has been proved that signal can be precisely recovered, when the sensing matrix satisfies the restricted isometry property (RIP) condition, thus $\boldsymbol{s}$ can be recovered by solving the following basis pursuit problem [2]:

$$\min_{x \in C^N} \| \boldsymbol{x} \|_1 \ \text{s.t.} \ \ \boldsymbol{y} = \boldsymbol{\Theta x} \tag{2}$$

In the case of measurement noise, the basis pursuit problem (2) can be relaxed to the regularization model as Eq. (3).

$$\hat{\boldsymbol{x}} = \arg \min_{x \in C^N} \mu \|\boldsymbol{x}\|_1 + \frac{1}{2} (\boldsymbol{\Theta x} - \boldsymbol{y})^H (\boldsymbol{\Theta x} - \boldsymbol{y}) \tag{3}$$

where $\mu > 0$ is the regularization parameter.

We also apply the Bregman iterative regularization to solve the problem (3) by using the Bregman distance of regularization term instead of regularization term [6], thus there exists the following iteration:

$$\boldsymbol{x}^{k+1}=\underset{\boldsymbol{x}\in C^N}{\operatorname{argmin}}\, D_J^{p^k}\left(\boldsymbol{x},\boldsymbol{x}^k\right)+\frac{1}{2}\left(\boldsymbol{\Theta x}-\boldsymbol{y}\right)^H\left(\boldsymbol{\Theta x}-\boldsymbol{y}\right) \tag{4}$$

where $J(\boldsymbol{x})=\|\boldsymbol{x}\|_1$, $\boldsymbol{x}^0=\boldsymbol{0}$, $\boldsymbol{p}^0=\boldsymbol{0}$, $k=0$, and $D_J^{p^k}\left(\boldsymbol{x},\boldsymbol{x}^k\right)$ is the Bregman distance. The Bregman distance of function $J(\boldsymbol{x})$ between points $\boldsymbol{x}$ and $\boldsymbol{x}^k$ is defined as [7]:

$$D_J^{p^k}\left(\boldsymbol{x},\boldsymbol{x}^k\right)=J(\boldsymbol{x})-J\left(\boldsymbol{x}^k\right)-\left\langle \boldsymbol{p}^k,\boldsymbol{x}-\boldsymbol{x}^k\right\rangle \tag{5}$$

where $\langle\cdot,\cdot\rangle$ is the inner product, $\boldsymbol{p}^k\in\partial J\left(\boldsymbol{x}^k\right)$ is a sub-gradient in the sub-differential of $J(\boldsymbol{x})$ at point $\boldsymbol{x}^k$. The sub-differential of convex function $J(\boldsymbol{x})$ with domain S at point $\boldsymbol{x}$ is defined as in Ref. [8], that is

$$\partial J(\boldsymbol{x})=\left\{\boldsymbol{p}\,\middle|\,J(\boldsymbol{v})\geq J(\boldsymbol{x})+\langle\boldsymbol{p},\boldsymbol{v}-\boldsymbol{x}\rangle,\forall\boldsymbol{v}\in S\right\} \tag{6}$$

According to the stationary point sub-differential condition,

$$\boldsymbol{0}\in\left(\partial J(\boldsymbol{x})-\boldsymbol{p}^k+\boldsymbol{\Theta}^H\left(\boldsymbol{\Theta x}-\boldsymbol{y}\right)\right) \tag{7}$$

Since $\boldsymbol{p}\in\partial J(\boldsymbol{v})$, and then $\boldsymbol{p}^{k+1}\in\partial J\left(\boldsymbol{x}^{k+1}\right)$ at point $\boldsymbol{x}^{k+1}$. Therefore,

$$\boldsymbol{p}^{k+1}=\boldsymbol{p}^k-\boldsymbol{\Theta}^H\left(\boldsymbol{\Theta x}^{k+1}-\boldsymbol{y}\right) \tag{8}$$

Then we have complex Bregman iterative regularization as Eqs. (4) and (8). The linearized Bregman iteration was proposed to improve computational efficiency [6]. By using the linearization method of linearized Bregman iteration, we obtained linearized form of complex linearized Bregman iterative regularization. First, approximating the term $\|\boldsymbol{\Theta x}-\boldsymbol{y}\|^2$ by its Taylor expansion around $\boldsymbol{x}^k$ and ignoring the constant term, then we have the following equation.

$$\boldsymbol{x}^{k+1}=\underset{\boldsymbol{x}\in C^N}{\operatorname{argmin}}\, D_J^{p^k}\left(\boldsymbol{x},\boldsymbol{x}^k\right)+\left\langle\boldsymbol{x},\boldsymbol{\Theta}^H\left(\boldsymbol{\Theta x}^k-\boldsymbol{y}\right)\right\rangle+\frac{1}{2\delta}\left(\boldsymbol{x}-\boldsymbol{x}^k\right)^H\left(\boldsymbol{x}-\boldsymbol{x}^k\right) \tag{9}$$

According to the stationary point sub-differential condition, then we have

$$\boldsymbol{0}\in\left(\partial J(\boldsymbol{x})-\boldsymbol{p}^k+\frac{1}{\delta}\left(\boldsymbol{x}-\left(\boldsymbol{x}^k-\delta\boldsymbol{\Theta}^H\left(\boldsymbol{\Theta x}^k-\boldsymbol{y}\right)\right)\right)\right) \tag{10}$$

Since $\boldsymbol{p} \in \partial J(\boldsymbol{v})$, and then $\boldsymbol{p}^{k+1} \in \partial J\left(\boldsymbol{x}^{k+1}\right)$ at point $\boldsymbol{x}^{k+1}$. Therefore,

$$\boldsymbol{p}^{k+1} = \boldsymbol{p}^{k} - \frac{1}{\delta}\left(\boldsymbol{x}^{k+1} - \boldsymbol{x}^{k}\right) - \boldsymbol{\Theta}^{H}\left(\boldsymbol{\Theta}\boldsymbol{x}^{k} - \boldsymbol{y}\right) \tag{11}$$

By utilizing the recurrence formula, Eq. (11) can be written as:

$$\boldsymbol{p}^{k+1} = \boldsymbol{p}^{k} - \frac{1}{\delta}\left(\boldsymbol{x}^{k+1} - \boldsymbol{x}^{k}\right) - \boldsymbol{\Theta}^{H}\left(\boldsymbol{\Theta}\boldsymbol{x}^{k} - \boldsymbol{y}\right) = \cdots = \sum_{j=0}^{k}\boldsymbol{\Theta}^{H}\left(\boldsymbol{y} - \boldsymbol{\Theta}\boldsymbol{x}^{j}\right) - \frac{1}{\delta}\boldsymbol{x}^{k+1} \tag{12}$$

Let $\boldsymbol{v}^{k} = \sum_{j=0}^{k-1}\boldsymbol{\Theta}^{H}\left(\boldsymbol{y} - \boldsymbol{\Theta}\boldsymbol{x}^{j}\right)$. Therefore, the following relationship can be obtained

$$\boldsymbol{v}^{k+1} = \boldsymbol{v}^{k} + \boldsymbol{\Theta}^{H}\left(\boldsymbol{y} - \boldsymbol{\Theta}\boldsymbol{x}^{k}\right) \tag{12}$$

and

$$\boldsymbol{p}^{k} = \boldsymbol{v}^{k} - \frac{1}{\delta}\boldsymbol{x}^{k} \tag{13}$$

Substituting $J = \mu \| \boldsymbol{x} \|_1$ into Eq. (9) and regardless of the constant term, then Eq. (9) can be written as

$$\boldsymbol{x}^{k+1} = \underset{\boldsymbol{x} \in C^{N}}{\operatorname{argmin}}\, \mu\|\boldsymbol{x}\|_1 + \frac{1}{2\delta}\left\|\boldsymbol{x} - \delta\left(\boldsymbol{p}^{k} - \boldsymbol{\Theta}^{H}\left(\boldsymbol{\Theta}\boldsymbol{x}^{k} - \boldsymbol{y}\right) + \frac{1}{\delta}\boldsymbol{x}^{k}\right)\right\|^2 \tag{14}$$

Substituting Eqs. (12) and (13) into Eq. (14), we have

$$\boldsymbol{x}^{k+1} = \underset{\boldsymbol{x} \in C^{N}}{\operatorname{argmin}}\, \mu\|\boldsymbol{x}\|_1 + \frac{1}{2\delta}\left\|\boldsymbol{x} - \delta\boldsymbol{v}^{k+1}\right\|^2 \tag{15}$$

Then the optimization problem (15) can be solved with the complex soft-threshold function $\boldsymbol{x}^{k+1} = \delta \operatorname{soft}_{\mu}(\boldsymbol{v}^{k+1})$. The complex soft-threshold function is defined as [9]:

$$\operatorname{soft}(u,a) = \frac{\max\left\{|u| - a, 0\right\}}{\max\left\{|u| - a, 0\right\} + a}u \tag{16}$$

According to Eq. (12), we have the complex linearized Bregman iteration(CLBI) algorithm.

$$\begin{cases} \boldsymbol{v}^{k+1} = \boldsymbol{v}^k + \boldsymbol{\Theta}^H(\boldsymbol{y} - \boldsymbol{\Theta}\boldsymbol{x}^k) \\ \boldsymbol{x}^{k+1} = \delta \mathrm{soft}_\mu(\boldsymbol{v}^{k+1}) \end{cases} \tag{17}$$

3 Simulation and analysis

In this paper, all experiments were implemented in MATLAB R2008b platform and run on a personal computer with a Inter(R) Core(TM) E7500 2.93GHz CPU and 2GB memory. We present three results of experiments to demonstrate convergence, effectiveness, and advantage of this new algorithm.

To show these properties, we consider complex signal $s = a\exp(j\omega t)$. The magnitudes a are generated from uniform distribution $\mathrm{U}(-1,1)$. There are three spikes in frequency domain. The frequencies ω are defined as: $\omega_i = k_i 2\pi/N\,(i=1,2,3)$, where k_i are chosen from $[1,2,\cdots,N]$ randomly and N=256 denotes the signal dimension. In all experiments, $\boldsymbol{\Phi}$ is the $(N/2)\times N$ Gaussian matrix and $\boldsymbol{\Psi}$ is the $N\times N$ Fourier basis $\boldsymbol{\Psi} = \exp\left(-j2\pi \boldsymbol{t}_n^T \boldsymbol{f}_d\right), \left(\boldsymbol{t}_n = (1,2,\cdots,N)\right), \boldsymbol{f}_d = \left(1/N, 2/N, \cdots, (N-1)/N, 1\right)$.

Notice that the signal is sparse in frequency domain, so we can apply our CLBI algorithm to recover the coefficient vector $\boldsymbol{x}$ of the original signal from the measured signal. Then we can obtain the original signal from $\boldsymbol{s} = \boldsymbol{\Psi}\boldsymbol{x}$. The stopping criterion is $\|\boldsymbol{\Theta}\boldsymbol{x}^k - \boldsymbol{y}\|/\|\boldsymbol{y}\| \le 10^{-5}$. The relative reconstruction error is defined as $\|\boldsymbol{\Psi}\boldsymbol{x}^k - \boldsymbol{s}\|/\|\boldsymbol{s}\|$.

Firstly, the effectiveness of the proposed algorithm is tested. Figure 1 shows the original signal in time domain. From Figure 2, we can see that the proposed algorithm reconstructs three true frequency spikes in Fourier domain. Figure 3 shows the reconstructed versus original signal in time domain, with 2.87×10^{-5} relative reconstruction error. One close-up of Figure 3 is shown in Figure 4, obviously, the proposed CLBI algorithm reconstructs the original complex signal accurately. Secondly, we demonstrate the convergence of the proposed algorithm. Figure 5 shows the logarithm of the relative reconstruction error with different iterations. From Figure 5, it can be seen that the relative reconstruction error of CLBI algorithm decreases gradually with the number of iterations. Therefore, the proposed CLBI algorithm for the complex CS reconstruction problem indeed converges. In order to verify the advantage of reconstruction speed, runtime of the real-valued LBI algorithm and the proposed CLBI algorithm is compared. Figure 6 shows runtime of the CLBI algorithm is reduced dramatically, when signal sequence is longer.

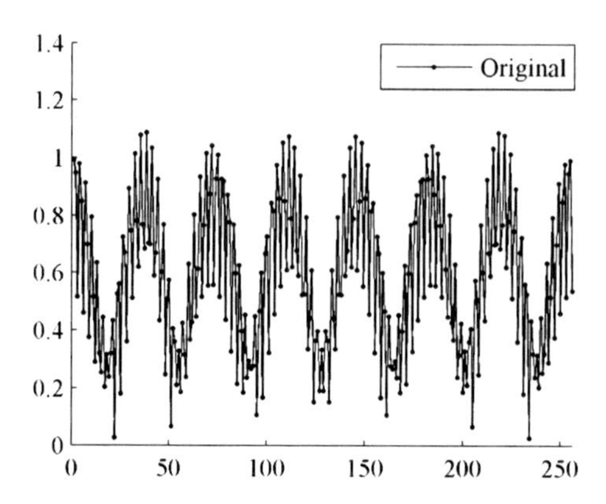

Figure 1: Original signal in time domain.

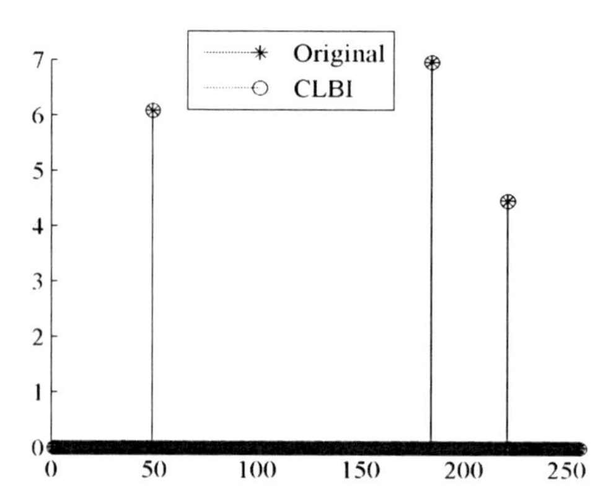

Figure 2: Reconstructed versus original signal in Fourier domain.

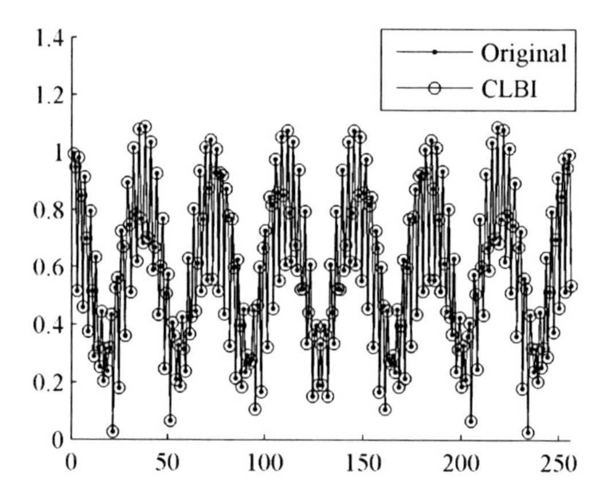

Figure 3: Reconstructed signal result in time domain.

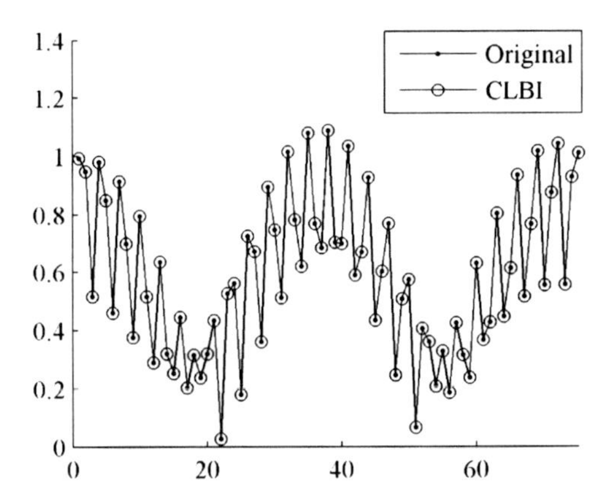

Figure 4: One close-up of Figure 3.

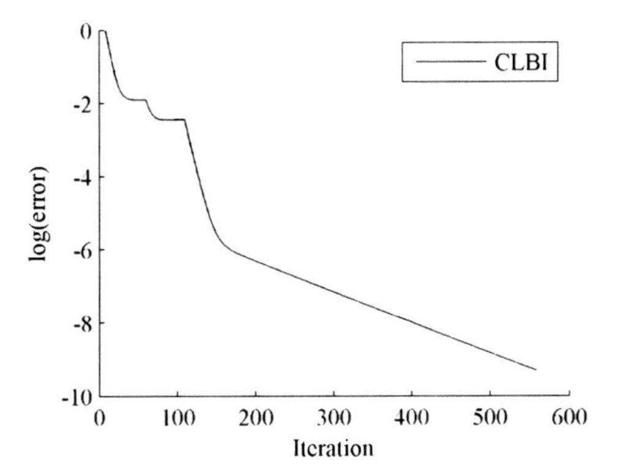

Figure 5: The logarithm of the relative reconstruction error.

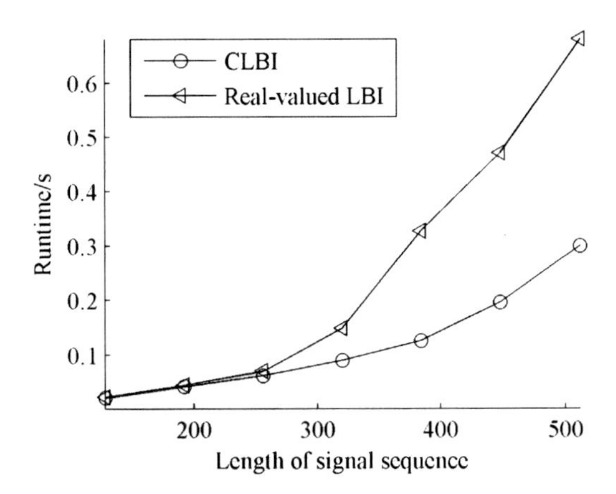

Figure 6: Runtime comparison.

4 Conclusions

In this paper, we have presented a novel algorithm for complex-valued signal recovery problem via the complex mathematical framework in CS. On the base of theoretical analysis, the convergence, effectiveness and reconstruction speed of the algorithm are simulated and analyzed. The results show that the proposed algorithm can not only be used to complex-valued signal, but also it has good features such as less computational burden, better convergent and reconstructed performance.

References

[1] Donoho, D.L., Compressed sensing. IEEE Transactions on Information Theory, 52(4), pp. 1289–1306, 2006.

[2] Candès, E.J., Romberg, J.K., Tao, T., Robust uncertainty principles: exact signal reconstruction from highly incomplete frequency information. IEEE Transactions on Information Theory, 52(2), pp. 489–509, 2006.

[3] Candès, E.J., Tao, T., Near optimal signal recovery from random projections: universal encoding strategies. IEEE Transactions on Information Theory, 52(12), pp. 5406–5425, 2006.

[4] Ender, J.H.G., On compressive sending applied to radar. Signal Processing, 90(5), pp. 1402–1414, 2010.

[5] Shechtman, Y., Beck, A., Eldar, Y.C., GESPAR: efficient phase retrieval of sparse signal. IEEE Transactions on Signal Processing, 62(4), pp. 928–938, 2014.

[6] Yin, W., Osher, S., Goldfarb, D., Darbon, J., Bregman iterative algorithms for l1-minimization with applications to compressed sensing. SIAM Journal on Imaging Sciences, 1(1), pp. 143–168, 2008.

[7] Bregman, L.M., The relaxation method of finding the common point of convex sets and its application to the solution of problems in convex programming. USSR Computational Mathematics and Mathematical Physics, 7(3), pp. 200–217, 1967.

[8] Erdogan, A.T., Kizilkale, C., Fast and low complexity blind equalization via subgradient projections. IEEE Transactions on Signal Processing, 53(7), pp. 2513–2524, 2005.

[9] Wright, S.J., Nowak, R.D., Figueiredo, M.A.T., Sparse reconstruction by separable approximation. IEEE Transactions on Signal Processing, 57(7), pp. 2479–2493, 2009.

Measurement of step voltage and touch voltage of grounding grid by adopting the method of short-range current auxiliary electrodes

Dan Cai-xian, Deng Chang-zheng, Zhao Xi-wu, Ren Yi-jing,
Zhou Yu-xin, Wu Yu-shan, Chai Lu
College of Electrical Engineering & New Energy,
China Three Gorges University, Yichang, China

Abstract

The influences of testing error which the distance of current auxiliary electrodes act on are computed and analyzed by CDEGS software for decreasing the distance of current auxiliary electrodes in the testing circuit for measurement of step voltage and touch voltage of large-sized grounding grid. According to the problem that the distance of single current auxiliary electrode is longer, the scheme including double current auxiliary electrodes is advanced, and the determining principle of the distance of current auxiliary electrode in the case of horizontal two layers soil model is summarized. Step voltage and touch voltage of EHV electric substation grounding grid are measured in the field, and testing results better verify the measuring method of short-range current auxiliary electrodes.

Keywords: large-sized grounding grid, step voltage, touch voltage, simulation computation, distance of current electrodes, field test.

1 Introduction

Due to influence of many factors, the testing of the step voltage and the contact voltage is always being as a technical problem obsessing electrical sector [1-3]. Large-sized grounding grid on this paper specifically means the substation grounding grid (the level of voltage is 500kv and above). Compared with general grounding grid, large-sized grounding grid cover larger area, and their structure is complex, also the connecting equipments are various, further more, surface potential distribution owing to fault current is uneven. Therefore it will be very

www.witpress.com, ISSN 1743-3533 (on-line)
doi:10.2495/SEEEI140521

difficult to measure step voltage and contact voltage of large-sized grounding grid point by point, so the area where the maximum appears most probably or the personal intensive place is chosen to have focus testing in general[2]. Diagonal of large-sized grounding grid usually remains a long distance (over 300m), for which the distance of current auxiliary electrodes will be more than 1500m designed by *Measure Guide for Characteristic* Parameter of Grounding Sets (DL/T475–2006) (Guide for short) [3].

On the premise of ensuring accuracy of test, properly reducing the distance of current auxiliary electrodes can significantly reduce test cost and improve testing efficiency. This paper is aimed at shortening the distance of current auxiliary electrodes, and analyzing the influence of test error to step voltage and contact voltage caused by it. Then the selecting principle of the distance of current auxiliary electrode in large-sized grounding grid test has been put forward and tested in grounding grid of extra high voltage substation.

2 The influence of test error caused by the distance of current auxiliary electrodes

The test of step voltage and contact voltage needs to build a testing circuit artificially in order to inject current to the grounding grid, and finally achieves the goal to simulating the actual situation of short-circuit current. The distance between current auxiliary electrode and edge of grounding grid is the main factor causing the error: the longer the distance is, the closer it is to the real short-circuit situation. In order to keep the test error small enough and the arranging work is not so hard at the same time, according to the characteristics of the large-sized grounding grid and different soil model [4,5], CDEGS software is used to calculate and analyze the influence of test result caused by the distance of current auxiliary electrodes. And on the basis of these calculation and analysis, engineer can choose a right the distance of current auxiliary electrodes.

2.1 Uneven soil model

The real grounding grid is set in the uneven soil. According to the method of ground model setting, the soil is usually simply solved as double grounding grid which is layering vertically. The ratio of upper and lower ground resistivity and the influence of the upper soil thickness are discussed as follow.

2.1.1 Single current electrode arrangement

In the horizontal stratification double ground model, the ratio of upper and lower ground resistivity is presented as K, the refractive index. As Figure 1 shows, the range of K is −1 to 1, K>0 means the lower resistivity is higher, K=0 means even soil, and K<0 means upper is higher.

$$K = \frac{\rho_2 - \rho_1}{\rho_1 + \rho_2} \tag{1}$$

in the type: ρ_1 – upper ground resistivity, $\Omega \cdot$ m;
ρ_2 – lower ground resistivity, $\Omega \cdot$ m.

Calculate the effect to the experiment of upper layer thickness, calculate the relationship of stride potential test error and the upper layer thickness under different resistivity. According to Figures 1 and 2, (1) $K>0$, εk increases first and then decrease alongside the increase of h, $K<0$, decreases and then increases alongside the increase of h. (2) εk is gradually close to the number in the even ground alongside the increase of H under all the condition of K.

All in all, in order to ensure the step potential measurement error and the contact potentials were within 5% considering the soil thickness influence, when $K>0.6$, S equals to 4 times of the grounding grid diagonal length: when $0.4<K\leq0.6$, S equals to 3 times of the grounding grid diagonal length[6].

2.1.2 Double current auxiliary electrodes

When $K=0.6$ or 0.8, the distance of current auxiliary electrodes S needs to come to $3D$, even $4D$. However, due to the restriction of the grounding grid surrounding, this could not be reached. Therefore, the double current auxiliary electrodes come out to decrease the single current auxiliary electrode distance. The method is shown as follows.

When K = 0.8 or 0.6, the maximum step potential of different current electrode distance, contact potential and test error are shown as follow. According to the former test error standard, K=0.6 or 0.8, the distance of current auxiliary electrodes could be twice of the grounding grid diagonal length.

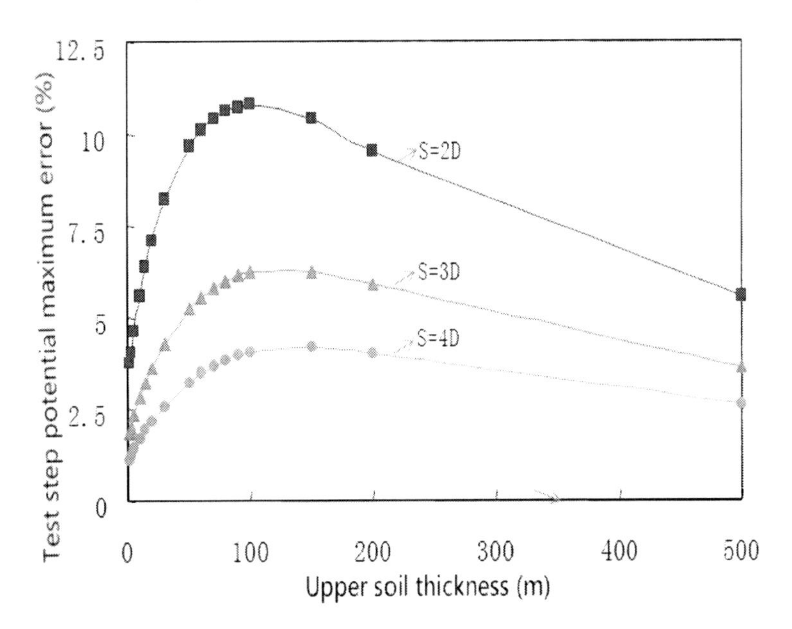

Figure 1: The relationship between the test error of step potential and the thickness of upper soil (S=2D).

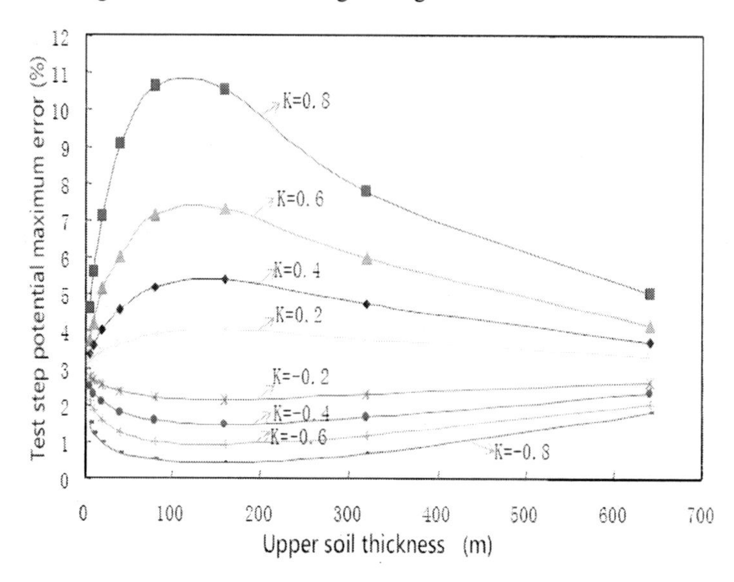

Figure 2: The relationship between the test error of step potential and the thickness of upper soil (K=0.8).

Table 1: The maximum test values and error of maximum test values and error of step potential, touch potential (K=0.8).

The distance of current auxiliary electrode	Step potential (Real value = 469.84 V/m)		Touch potential (Real value = 685.72V)	
	Test value (V/m)	Relative error ε_k (%)	Test value (V)	Relative error ε_j (%)
$S = 2D$	471.0	+0.25	684.29	−0.21
$S = 1.5D$	471.98	+0.46	753.526	+9.9
$S = 1D$	473.48	+0.77	749.53	+9.31

Under the double current auxiliary electrodes, contact potential has bigger test error compared with step potential. Using two current electrode test solutions would be able to solve some of the grounding grid at a single current electrode, and step potential of error is small, but contact potential error is a big problem at the same time.

3 The field test

The Short Current Auxiliary Electrode method was carried out in the field test of step voltage (potential) and the contact voltage (potential) on a 500kv substation grounding grid in Hubei.

3.1 Testing scheme

Test power source supply with the measurement device of YDGH-60/800 grounding resistance of grounding grid, current auxiliary electrode using 12 root 1.5m long φ50mm round steel ground pegs, the actual current amplitude remains within 45A ± 5% range in test.

The grounding grid of this transformer substation shows "L" shape, its largest diagonal according to 420m consideration. Select 5 measuring points measured soil resistivity, CDEGS software is adopted to establish two layers soil model which horizontal layered: the upper layer soil resistivity is 79.99Ω • m and thickness of 0.73m; the under soil resistivity of 112.26Ω • m, the refraction coefficient K = 0.17, select the distance of current auxiliary electrode as twice as the diagonal length of the grounding grid, namely current auxiliary electrode arrangement in 850m away from the fixed edge (southwest corner) of the grounding grid. In the case of S = 4D, current auxiliary electrode arrangement in 1760m away from the edge (southwest corner). In the case of S = 2D, the current auxiliary electrode layout in near the midpoint of the path. In the case of S = 3D case, the current auxiliary electrode layout the path of 1250m.

3.2 The test results of step voltage

Measurement of Step voltage and touch voltage adopting the potential probe layout method which from "guideline" and literature [7]. Test results are shown in Figures 3 and 4. The test results have been imputed to 60kA, so that it is easy to compare in the same short-circuit current conditions.

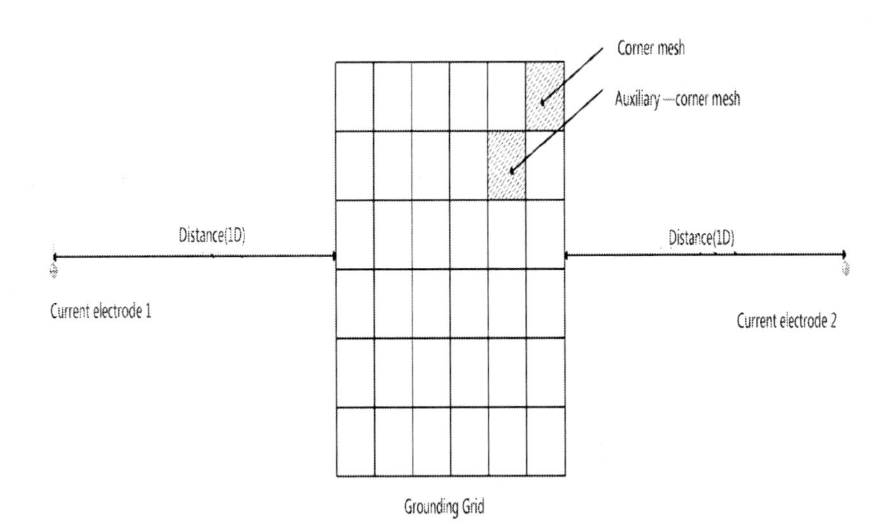

Figure 3: The arrangement graph of two current auxiliary electrodes test circuit.

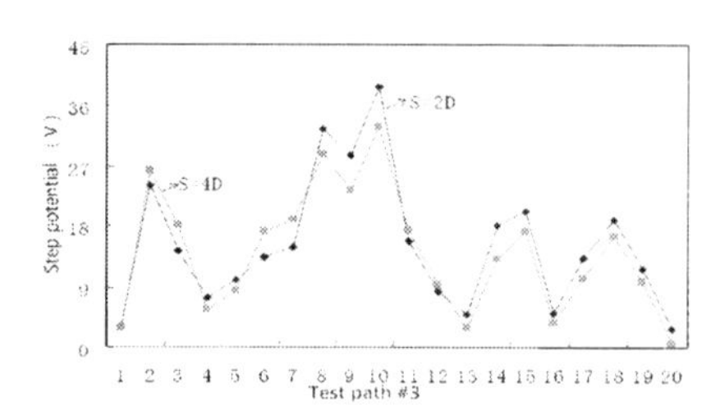

Figure 4: Test path #1.

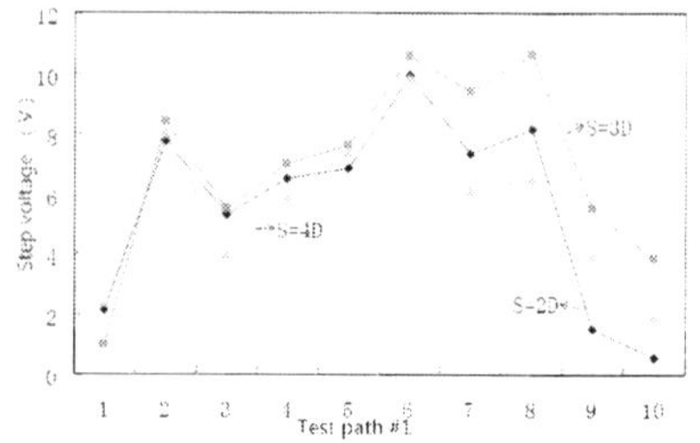

Figure 5: Test path #2.

Table 2: Test results of touch potential (*V*).

Test location	Electrode form	*S*=2D	*S*=4D
	Iron electrode	108.50	130.34
	Disk electrode	121.16	119.93
C-phase breaker	Iron electrode	118.81	158.79
	Disk electrode	128.07	102.70

The following conclusions are drawn from our analysis: the step potential (voltage) which is measured in the case of $S=2D$ is consistent with test results which is measured in the case of $S=4D$ or $S=3D$. It shows that selecting $S = 2D$ step can meet engineering testing requirements of electric potential (voltage).

3.3 The test results of touch voltage

Select #505327 A-phase, C-phase grounding switch as measuring point, because the cement pavement and grass nearby, these two cases in every group use disk electrode and iron solder electrode, using a wet rag wrapped up disk electrode and pressed on 20kg of weight, and pour some water to make good contact with the current auxiliary electrodes and cement pavement in the vicinity of the electrodes[8]. The test results are shown in Table 2.

From Table 2 shows that:

(1) The touch potential(voltage) which is measured in the case of *S*=2D is consistent with test results which is measured in the case of $S=4D$. Therefore, choose *S* = 2D can meets touch potential (voltage) of engineering testing requirements;
(2) Due to the test position close to the injecting point of measuring current, ground potential rise of grounding system are larger, the touch potential (voltage) is large.

4 Conclusion

(1) The selecting principle is verified by theoretical calculation and field measurement on the distance of current auxiliary electrodes: in horizontal

two layer soil model, when $K \leq 0.4$, we use one current auxiliary electrode arrangement, $S = 2*D$; when $0.4 \leq K \leq 0.6$ we use one current auxiliary electrode arrangement, $S = 3*D$; when $K > 0.6$, we use single current auxiliary electrode arrangement, $S = 4*D$, or two current auxiliary electrodes arrangement, $S = 2*D$.

(2) The Short Current Auxiliary Electrode method proposed in this paper can reduce the distance of current auxiliary electrodes to a certain extent, and significantly reduce the workload of arranging wire, improve test efficiency. But the selecting principle of the distance current auxiliary electrodes is still under further research and validation with the actual measured data and theoretical calculation.

References

[1] Liang Jian-wei, A test of 500-kv substation grounding device. *Electric Power Technology of Shanxi*, **14(1)**, pp. 60–62, 1994.

[2] Deng Yu-rong, He Jin-liang, Ma Yu-lin, Method of evaluation for condition of grounding system in substations, **(5)**, pp. 1–4, 2006.

[3] China Electricity Council, Measure Guide for Characteristic Parameter of Grounding Sets (DL/T475–2006), 2006.

[4] He Jin-liang, Zeng Rong, *Grounding Power System*. Science Press, Beijing, 2007.

[5] Chen Xian-lu, Liu Yu-gen, Huang Yong, *Earthing*. Chong Qing University Press, Chong Qing, 2001.

[6] Xie Guang-run, *Grounding Power System*. China Water Power Press, Beijing, 1991.

[7] IEEE-SA Standards Board, IEEE Guide for Safety in AC Substation Grounding, IEEE Std 80–2000.

[8] R. Kosztaluk, D. Mukhedkar, Y. Gervais, Field measurements of touch and step voltages. *IEEE Transactions on Power Apparatus and Systems*, **PAS-103**, p. 11, 1984.

Mechanical fault diagnosis for high voltage circuit breakers based on coil current

Xu Cheng[1], Yonggang Guan[2], Xinxia Peng[2], Kai Gao[3], Yihe Liu[4], Yu Guo[5]
[1]*State Grid Research Electric Power Research Institute, Beijing, China*
[2]*The State Key Lab of Control and Simulation of Power Systems and Generation Equipments, Department of Electrical Engineering, Tsinghua University, Beijing, China*
[3]*State Grid Shanghai Electric Power Research Institute, Shanghai, China*
[4]*State Grid Shanghai Municipal Electric Power Company, Shanghai, China*
[5]*State Grid Shanghai Pudong Electric Power Supply Company, Shanghai, China*

Abstract

A new method of mechanical fault diagnosis of high voltage circuit breaker, based on coil current, is presented. In this paper, a series tests were performed on a 12kV vacuum circuit breaker, while in different fault conditions. At the same time, coil current signals were recorded during closing and tripping operation. For diagnostic purposes, seven characteristic parameters were extracted from the coil current waveforms. The classification method, Support Vector Machine (SVM), was employed to diagnose the condition of circuit breaker, normal or faulty. The results showed that the coil current signals can be used to diagnose the mechanical condition of circuit breaker.
Keywords: circuit breaker, mechanism fault diagnosis, coil current, support vector machine.

1 Introduction

Circuit breakers are key protect and control equipment of the power system. If there are some fault in the power system, circuit breakers can quickly break the

WIT Transactions on Engineering Sciences, Vol. 107, © 2015 WIT Press
www.witpress.com, ISSN 1743-3533 (on-line)
doi:10.2495/SEEEII140531

circuit to limit the damage extent to the system. Considering the important role circuit breakers play, it is essential to assess the normal or faulty condition of circuit breakers. Some international researches on fault statistics emphasize that about 45% of major and 40% of minor failures in high voltage circuit breakers are mechanical issues [1–3].

At present, most of the researches on mechanical failures of circuit breakers are based on vibration signals [4, 5]. The vibration signals are transient ones and difficult to deal with. The vibration sensors are sensitive to its position and not convenient to install. Furthermore, the analysis methods of vibration signals are very complex. The coil current was mostly applied to analyze the electromagnet condition of the operating mechanism. In recent years, some researchers begin to pay close attention to using the coil current to evaluate the mechanical condition of circuit breakers [6, 7]. The coil current signals are easy-to-access, easy-to-measure, and non-invasive. The analyses of the current signals are very convenient.

In this paper, coil current was employed to diagnose the healthy or faulty mechanical condition of a 12kV vacuum circuit breaker with spring drive mechanism. Six different conditions were analyzed, such as normal, high operating voltage, low operating voltage, the longer travel of movable iron armature (the loose failure of electromagnet core), the stagnation of electromagnet, and the failure of buffer. The result of this study reveals a useful and practical method to diagnose the condition of circuit breakers.

2 Coil current signal

The tests were performed on a 12kV vacuum circuit breaker with spring drive mechanism under different conditions. This section introduces the characteristic parameters of coil current signal.

2.1 The operating process of circuit breakers with spring mechanism

The operation of circuit breaker begins when the coil receives the command signal. This signal creates coil current in the coil and electromagnet force upon the plunger. When the force exceeds both gravity and friction, the plunger begins to move. The motion of the plunger generates a back electromotive force, which can reduce the coil current. The plunger continues to move until it strikes the trip latch and activates the movement of spring mechanism. The stored energy in the spring will then be released to move the transmission mechanical parts and the contacts. When the plunger keeps at the maximum travel, the coil current reaches its maximum value. After a short time delay, the coil is disconnected from the supply and the coil current decays fast to zero. The iron plunger falls down to rest at its starting position.

2.2 Characteristic parameters of circuit breaker coil current

The tests were performed on a 12kV vacuum circuit breaker with spring drive mechanism under different conditions. This section introduces the characteristic parameters of coil current signal.

A typical coil current waveform of the circuit breaker recorded under normal condition is shown in Figure 1.

- T_0 is the moment when the coil is energized. Then the coil current begins to increase exponentially. The iron plunger is in stationary condition.
- T_1 is the moment that coil current reaches its first peak value.
- T_2 is the moment that electromagnet core impact trip units.
- T_3 is the auxiliary contact operation time. T_4 is the moment that coil current decreases to zero.
- I_1, I_2, I_3, which are also important characteristic values, are the magnitude of coil current at the particular time.

In this study, 30 sets of data were recorded during closing or tripping operation when the circuit breaker was in normal condition. The mean values of the seven characteristic parameters mentioned above in the closing operation are shown in Table 1. The coil current signal during the tripping operation is the same as it in the closing operation. In the following parts, we use coil current recorded in the closing operation to support our research.

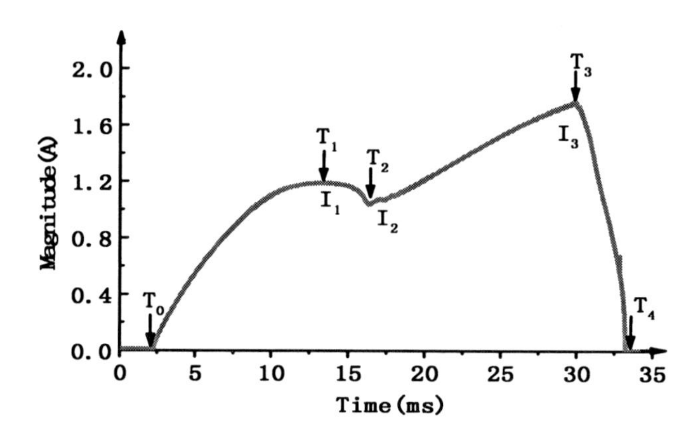

Figure 1: The typical coil current waveform.

Table 1: The mean values and variations of characteristic parameters.

Parameters	Mean Value	Maximum-Positive Deviation	Maximum-Negative Deviation
T_1(ms)	10.10	0.14	−0.17
T_2(ms)	14.48	0.07	−0.07
T_3(ms)	36.79	0.267	−0.34
T_4(ms)	41.69	0.266	−0.55
I_1(A)	1.58	0.021	−0.017
I_2(A)	1.14	0.016	−0.0169
I_3(A)	2.33	0.022	−0.037

The variations of the seven parameters are very small in the normal state. This makes it conducive to distinguish different conditions of circuit breaker.

3 Fault simulation tests

In this part, the simulations of five faults are introduced. 30 sets of data were gathered when the circuit breaker was in each abnormal condition.

3.1 The fault of coil supply voltage

The circuit breaker should be operated under the rated supply voltage, 220V DC. For some reasons, the supply voltage may fluctuate. The variations of the supply voltage can result in operation failure. For a higher voltage, the coil current may be too large to burn down the coil. For a lower voltage, the circuit breaker may not operate.

In order to investigate this kind of fault, 110% and 85% of the rated voltage were supplied to the coil in the closing operation.

3.2 The loose of electromagnet core

The operation of circuit breaker starts at the movement of the electromagnet core. If the screw or bolt is loose, the travel of electromagnet will increase and the closing or tripping operation may not be completed. To simulate this failure, we screwed off the fixing bolt, shown in Figure 2.

3.3 The stagnation of electromagnetic

Most of time through its whole life, the circuit breaker will keep in static state. So it is possible that the electromagnet components are aging and corrosion, resulting in stagnation fault. In serious case, the moving part of the electromagnet may refuse to move. Some heavy things were put to the bolt under movable iron core to simulate this fault, as shown in Figure 3.

The transmission mechanism transfer the energy stored in the spring to the moving contact. When the circuit breaker finishes the closing or opening operation, a buffer is needed to absorb surplus energy to protect the contacts and other parts.

Figure 2: The simulation of longer travel fault.

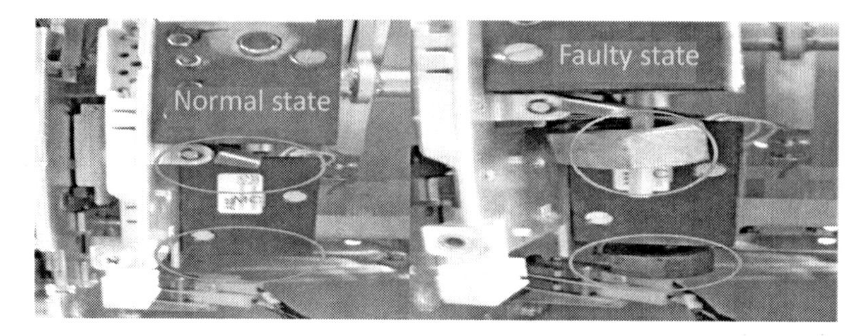

Figure 3: The simulation of the stagnation fault.

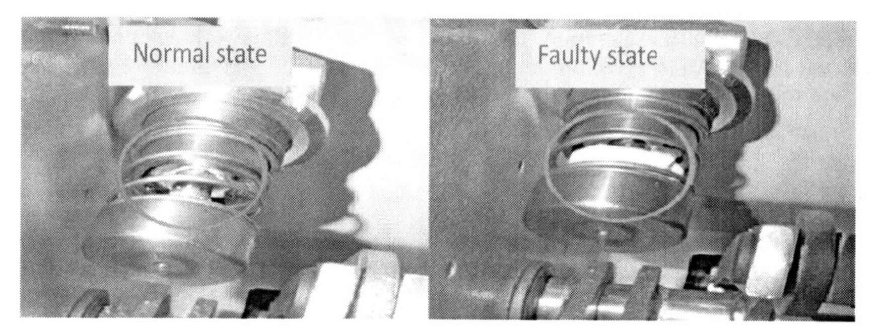

Figure 4: The simulation of the fault of buffer.

Table 2: The mean values of recoding data.

Parameters	Failures				
	110%	85%	Loose	Stagnation	Buffer
T_1(ms)	9.52	11.27	9.71	12.68	9.95
T_2(ms)	13.35	16.67	15.47	17.94	14.46
T_3(ms)	35.81	38.72	37.76	39.98	34.31
T_4(ms)	41.34	42.86	42.63	44.88	39.61
I_1(A)	1.675	1.376	1.65	1.66	1.55
I_2(A)	1.24	0.943	1.1	1.2	1.13
I_3(A)	2.55	1.911	2.3	2.27	2.28

In actual operation, the reset buffer spring may be fatigue even broken after many operations, result in stagnation fault, and reduce the buffer distance. Figure 4 shows the simulation method of this failure of buffer. A rubber was added to the buffer rod and to reduce the buffer distance.

3.4 The mean values of characteristic parameters of other conditions

Table 2 shows the mean values of seven parameters while the circuit breaker is in each faulty condition. Figure 5 shows the differences between coil current waveforms while the circuit breaker is in six different conditions.

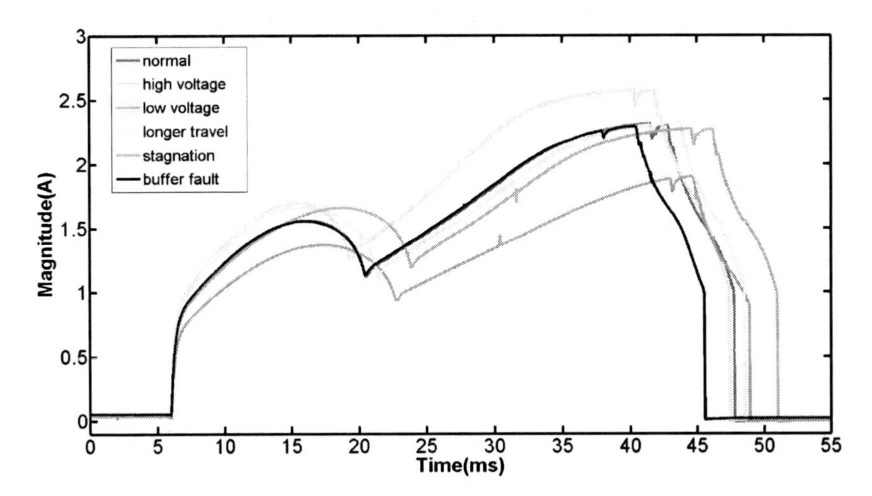

Figure 5: The comparison of coil current while in different condition.

It can be seen in Table 2 that at least two parameters in each condition differ greatly from others.

It can be seen from Figure 5 that when the mechanical condition of circuit breaker changes, the coil current waveform will also change. That means the coil current signal can be used to distinguish the normal or faulty mechanical condition of circuit breakers.

4 Condition assessment algorithm

The method, Support Vector Machine (SVM), was used to classify the different mechanical conditions of circuit breaker. An SVM classifier minimizes the generalization error by optimizing the tradeoff between the number of training error and the so-called Vapnik-Chervonenkis (VC) dimension, which is a new concept of complexity measure. SVM is very suitable for handling small sample data classification problem.

In this study, 30 tests were carried out while the circuit breaker was in every condition. Top 20 sets of data were used for training samples, while the last 10 sets of data were used to test the classification effect. The classification results are shown in Figure 6. In Figure 6, label "1" stands for "normal state," label "2" stands for "high supply voltage," label "3" stands for "low supply voltage," label "4" stands for "the loose failure of electromagnet core," label "5" stands for "the stagnation of electromagnetic," label "6" stands for "the failure of buffer".

As can be seen from Figure 6, the accuracy of classification is 100%, that is the real classification and the predict classification are all coincide. That means the differences of characteristic parameters of coil current signal are very big while the circuit breaker mechanism is in different condition, and can be used to assess the condition of the circuit breaker correctly.

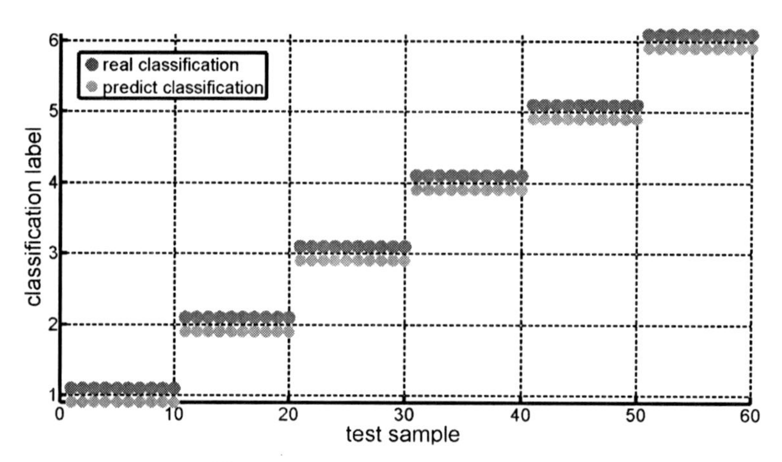

Figure 6: The classification result.

5 Conclusions

1. The coil current waveform of circuit breaker with spring operating mechanism is highly repeatable under the same condition.
2. Characteristics of the coil current waveform can be described through the peak and valley time and current values, which are the characteristic parameters, T_0, T_1, T_2, T_3, T_4, I_1, I_2, and I_3.
3. The characteristic parameters can be used to judge the condition of mechanism. The classification accuracy is 100% using the support vector machine method.
4. This method, based on coil current signal, is very simple to diagnose the mechanical condition of circuit breaker, and convenient to the field application.

Acknowledgment

This project was supported by the National High-tech R&D Program (863 Program) (2011AA05A121), State Grid Beijing Electric Power Company (520223135031), and State Grid Shanghai Municipal Electric Power Company (5209401350BR).

References

[1] CIGRE Working Group 13.06, Final report of the second international enquiry on high voltage circuit breaker failures and defects in service, Paris, August/September 1994.

[2] *IEEE Guide for Diagnostics and Failure Investigation of Power Circuit Breakers*, IEEE Std. C37.10-1995, September 1996.

[3] G.J. Paoletti, M. Baier, Failure contributors of my electrical equipment and condition assessment program development. *IEEE Transactions on Industry Applications*, **38**, pp. 1668–1676, December 2002.
[4] N. Natti, M. Kezunovic, Assessing circuit breaker performance using condition-based data and Bayesian approach. *Electric Power System Research*, **81**, pp. 1796–1804, April 2011.
[5] X. Huang, H. Xia, Design of an on-line monitoring system of mechanical characteristics of high voltage circuit breakers. In: *Proceedings of International Conference on Electronic Communication and Control*, pp. 3646–3649, 2011.
[6] J. Pan, Z. Wang, D. Lubkeman, Condition based failure rate modeling for electric network components. In: *Proceedings of IEEE/Power Engineering Society Power System Conference and Exposition*, pp. 1–6, March 2009.
[7] A.A. Razi-Kazemi, M. Vakilian, K. Niayesh, M. Lehtonen, Circuit breaker automated failure tracking based on coil current signature. *IEEE Transactions on Power Delivery*, **29**, pp. 283–290, February 2014.

Theory innovation and application for equipment fault diagnosis

Ren Xin[1,2], Xiaohu Chen[3], Yifang Yang[2], Zhang kai[2]
[1]*Graduate School 13 Team, National Defense University, Beijing, China*
[2]*Medical Protection Laboratory,*
Naval Medical Research Institute, Shanghai, China
[3]*Department of Military Logistics and Science and*
Technology Equipment, National Defense University, Beijing, China

Abstract

The fault diagnosis of equipment system becomes more and more complex, and the traditional grey correlation analysis model cannot meet the needs of equipment fault diagnosis. In order to solve the problem, we establish improved optimization model based on weighted degree of grey incidence. In this paper, firstly, traditional analysis model will be compared with improved optimization model on theory. In addition, we take a certain equipment fault diagnosis system for the case study. As a result, improved optimization model has strong anti-noise interference capability and resolution definition than traditional grey correlation analysis model. The comparison results on theory and case study indicate that the proposed optimization model can be effectively and feasibly applied on the complex information equipment system fault analysis.
Keywords: equipment, fault diagnosis, weighted degree of grey incidence.

1 Introduction

As we all know, along with the rapid social development and transformation of war form, more and more high and new technology constantly emerges, and equipment is becoming very complex. Equipment failure diagnosis becomes so much harder than ever before that traditional failure analysis method or model sometimes cannot effectively meet the needs of the reality [1] Innovation or improvement of the existing traditional model is required to make for the accurate analysis of complex information equipment failure, and it can increase the

www.witpress.com, ISSN 1743-3533 (on-line)
doi:10.2495/SEEEII140541

efficiency of equipment maintenance and affordable effectiveness, which provides guarantees to win local wars under information conditions. In this paper we compare two equipment fault analysis models and put forward the advantages of improved optimization model.

2 Traditional grey correlation analysis theory and improved weighted optimization model

2.1 Traditional grey correlation analysis theory

Grey correlation degree is defined as the main factors and secondary factors influence on the analysis results of measurement in overall and each development process of one thing [2]. This method does not need a larger sample, in the process of analysis, the deviation degree of simulation curve and the actual result reflects the correlation degree between the different factors in the system, and the primary and secondary factors affecting system process can be more accurately obtained according to the correlation degree, that is to say, the more similar the geometric shape is, the greater the degree of correlation will be.

2.2 Optimization to improve the correlation analysis model

Traditional grey correlation analysis method considers the arithmetic average of grey relational coefficient calculated simply point by point as the grey correlation degree, which cannot analyze the effect of each time correlation coefficient on the result of system analysis, and the partial correlation tendency will be missing. Grey entropy optimization model develops on the basis of grey correlation analysis theory, and it can improve the result of the correlation degree analysis and makes up for the deficiency of the traditional grey correlation analysis method. Specific analysis steps can be listed as follows:

(i) Establishment of mapping. *Definition:* $R_j \to P_j$ $P_k = \dfrac{w_k \gamma(x_0(k), x_i(k)}{\sum_{k=1}^{n} w_k \gamma(x_0(k), x_i(k)}$

this is the weighted grey correlation coefficient distribution map, P_k is the distribution density value of weighted grey correlation coefficient, and $\sum_{k=1}^{n} P_k = 1$

(ii) Correlation coefficient. Grey correlation coefficient of grey correlation theory represents the main factors influence to each sequence point, and the correlation coefficient at each point needs to remain balance. First of all, to establish the main sequence $X_0 = [x_0(1), x_0(2), ..., x_0(n)]$ and compared sequence $X_i = [x_1(1), x_1(2), ..., x_i(n)]$ $(i = 1, 2, \cdots, m)$, the correlation coefficient is defined as follow: $\varepsilon_{ik} = \dfrac{\Delta_{\min} + \rho\Delta_{\max}}{|X_0(k) + X_i(k) + \rho\Delta_{\max}|}$, and among them,

$\Delta_{\min} = \min_i \min_k |X_0(k) - X_i(k)|$, $\Delta_{\max} = \max_i \max_k |X_0(k) - X_i(k)|$, $\Delta_{\max} = \max_i \max_k |X_0(k) - X_i(k)|$, in addition, weighted grey correlation degree: $\varepsilon_{ik}(X_0, X_i) = \sum_{k=1}^{n} w_k \varepsilon_{ik}(x_0(k), x_i(k))$, w_k is the weight on the k point, and $\sum_{k=1}^{n} w_k = 1$. Definition to distinguish coefficient: $\rho \in (0,1)$, and it usually takes 0.5, in order to obtain good resolution, we need to regulate its value according to correlation degree, and it is defined as:

$$\rho = \begin{cases} \varepsilon \le \rho < 1.5\varepsilon & \Delta_{\max} > 3\Delta \\ 1.5\varepsilon \le \rho \le 2\varepsilon & \Delta_{\max} < 3\Delta \end{cases}$$

(iii) Determination of weight. The specific process that we adopt grey entropy analysis method to ascertain the weight of correlation coefficient is as follows: define functions: $H_{\otimes}(R_i) = -\sum_{k=1}^{n} p_k \ln p_k$, and this is weighted grey relation entropy of X_i, $\max H_{\otimes}(R_i) = -\sum_{k=1}^{n} p_k \ln p_k$, $s.t.$ $p_k \ge 0$, $s.t.$ $\sum_{k=1}^{n} p_k = 1$ In addition, the calculation results of weight L under maximum constraints are shown in the following step: $L = \Gamma^{-1} A$,

$$\Gamma = \begin{bmatrix} \gamma_1 & -\gamma_2 & & & \\ & \gamma_2 & -\gamma_3 & & \\ & & & & \\ & & & \gamma_{n-1} & -\gamma_n \\ 1 & 1 & 1 & 1 & 1 \end{bmatrix}, \quad L = \begin{bmatrix} w_1 \\ w_2 \\ \\ w_{n-1} \\ w_n \end{bmatrix}, \quad A = \begin{bmatrix} 0 \\ 0 \\ \\ 0 \\ 1 \end{bmatrix} \tag{1}$$

3 Construction of fault diagnosis system

Now, the complex equipment system fault analysis system is established based on the above improved grey entropy correlation analysis model, as shown in Figure 1, through the figure we can conclude that the fault diagnosis system is mainly composed of three parts [3], and they complete the data transmission with the use of serial port communication protocol: the first step is to obtain running parameters of analysis system; the second step is mainly to obtain the data for scientific processing and improve optimization weighted correlation

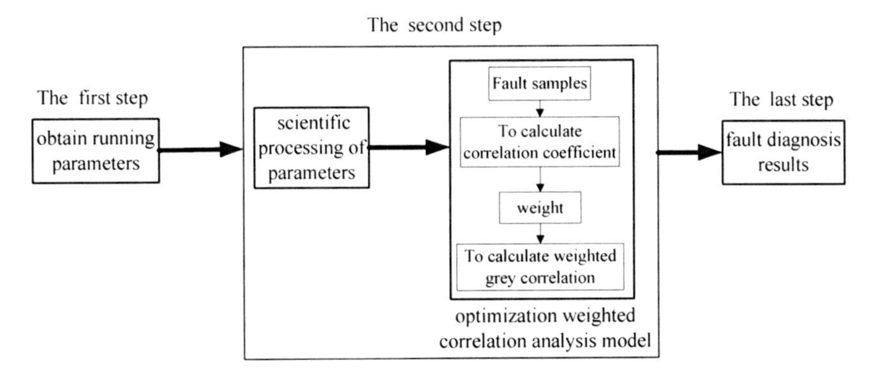

Figure 1: Structure graph of fault diagnosis system.

analysis model to calculate; and the last step is mainly to analyze the system fault diagnosis results, etc.

4 Application example

This paper takes a complex information equipment system as an example, and the implementation method is as follows: to set state standard state vector based on equipment system representation parameters, and determine each state vector under test through the correlation comparison to analyze the equipment failure mode [4]. Firstly, a certain equipment system state pattern vector is established: $G=\left[G(1) \quad G(2) \quad G(3) \quad \cdots \quad G(8)\right]$, the system has eight sets, and five kinds of failure mode are, respectively, G_1, G_2, G_3, G_4, and G_5.

$$G=\begin{bmatrix} g_1 \\ g_2 \\ g_3 \\ g_4 \\ g_5 \end{bmatrix}=\begin{bmatrix} g_1(1) & g_1(2) & \cdots & g_1(8) \\ g_2(1) & g_2(2) & \cdots & g_2(8) \\ g_3(1) & g_3(2) & \cdots & g_3(8) \\ g_4(1) & g_4(2) & \cdots & g_4(8) \\ g_5(1) & g_5(2) & \cdots & g_5(8) \end{bmatrix} \tag{2}$$

In addition, $G_0=\left[g_0(1) \quad g_0(2) \quad g_0(3) \quad \cdots \quad g_0(8)\right]$ is the failure mode vector that is to be tested. We set up the equipment system fault training sample sets, 0.5 is the normal operation value, 1.0 is the upper limit shutdown value, 0.0 is the lower limit shutdown value, 0.75 and 0.25 are, respectively, the upper and lower alarm state [5], and the values of failure mode vector are displayed in Table 1.

In the process of fault diagnosis analysis, the system state data will be collected: 0.5, 0.75, 0.25, 0.75, 0.5, 0.5, 0.5, 0.5. The analysis results of improved grey correlation are respectively as follows: $\varepsilon(x_0, x_1) = 0.5369$, $\varepsilon(x_0, x_2) = 0.5674$, $\varepsilon(x_0,$

$x_3) = 0.6612$, $\varepsilon(x_0, x_4) = 0.6107$, $\varepsilon(x_0, x_5) = 0.4969$, through the calculation results we can see that weighted correlation of x_0 and x_3 reaches maximum, and corresponding symptom set G_3 is the same as the system state data, so we judge G_3 for system failure. In addition, the correlation coefficient and weight calculation results are listed in Table 2. In order to verify the system ability to distinguish fault, we enter the state data offline: 0.50, 0.45, 0.50, 0.75, 0.35, 0.25, 0.45, 0.55, and respectively use traditional and improved optimization grey correlation model to make fault analysis of complex equipment system. After normalized processing of correlation, a correlation scatter diagram is shown in Figure 2. On the whole, two kinds of system fault correlation sorting is consistent, but the distance of two fault with improved weighted correlation model is further apart, so it is clear that improved optimization model owns higher resolution.

Table 1: Equipment system fault training sample sets.

	M_1	M_2	M_3	M_4	M_5	M_6	M_7	M_8
G_1	0.5	0.25	0. 75	0. 5	0.5	0.75	0.5	0.75
G_2	0.5	0.5	0.75	0.25	0.5	0.25	0.5	0.5
G_3	0.75	0.5	0.25	0.5	0.5	0.5	0.75	0.5
G_4	0.5	0.75	0.5	0.25	0.75	0.5	0.25	0.5
G_5	0.25	0.5	0.75	0.25	0.5	0.75	0.25	0.5

Table 2: Correlation coefficient and weight vector.

Correlation coefficient				
$r(x_0, x_1)$	$r(x_0, x_2)$	$r(x_0, x_3)$	$r(x_0, x_4)$	$r(x_0, x_5)$
1.0000	1.0000	0.5000	1.0000	0.5000
0.3333	0.5000	0.5000	1.0000	0.5000
0.3333	0.3333	1.0000	0.5000	0.3333
0.5000	0.3333	0.5000	0.3333	0.3333
1.0000	1.0000	1.0000	0.5000	1.0000
0.5000	0.5000	1.0000	1.0000	0.5000
1.0000	1.0000	0.5000	0.5000	0.5000
0.5263	0.9091	0.9091	0.9091	0.9091
Weight vector				

$w(x_0, x_1)$	$w(x_0, x_2)$	$w(x_0, x_3)$	$w(x_0, x_4)$	$w(x_0, x_5)$
0.0671	0.0709	0.1653	0.0763	0.1242
0.2013	0.1418	0.1653	0.0763	0.1242
0.2013	0.2128	0.0826	0.1527	0.1863
0.1342	0.2128	0.1653	0.2290	0.1863
0.0671	0.0709	0.0826	0.1527	0.0621
0.1342	0.1418	0.0826	0.0763	0.1242
0.0671	0.0709	0.1653	0.1527	0.1242
0.1275	0.0780	0.0909	0.0840	0.0683

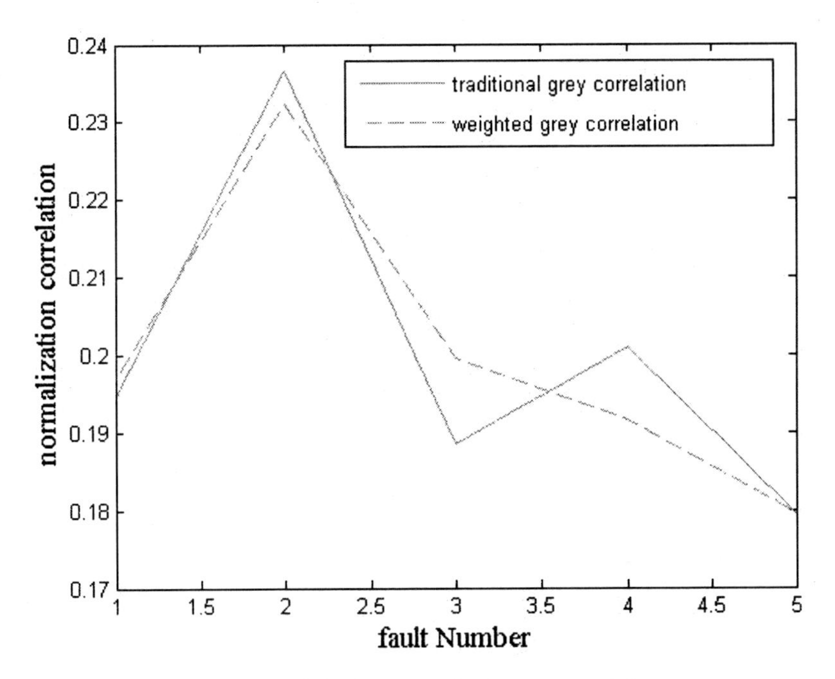

Figure 2: Linked scatter plot for two models.

5 Conclusions

Traditional grey correlation fault analysis model can overcome the problem of small fault data samples, but it exists many limitations and should be improved. Comparision analysis of theory is implemented between the traditional model and improved optimization, and then through a certain case we conclude that improved model has very high resolution and get good use effect, so the grey entropy optimization model can be effectively applied to the fault analysis of complex equipment system.

References

[1] Zhang Wei, Liang Wei, Equipment safeguard countermeasures study under complicated electromagnetic environment. *Journal of Equipment Command Institute of Technology*, **19(4)**, pp. 1–4, 2008.

[2] Marks RJ, Intelligence: computational versus artificial. *IEEE Transactions on Neural Networks*, **4(5)**, pp. 737–739, 1993.

[3] Xu Xiao-tao, Tian Cheng, Zhu Xue-wei, IETM application in information equipment support research. *Journal of Defense Technology*, **(9)**, pp. 22–29, 2008.

[4] Friend AD, Scuhgart HH, Running WS, Physiology-based gap model of forest dynamics. *Ecology*, **(3)**, pp. 792–797, 1993.

[5] Daisuke Y, Li GD, Kozo M, On the generalization of grey relation-al analysis. *Journal of Grey System*, **9(1)**, pp. 23–34, 2006.

Designing and optimizing the pulse shaping based unitary transformation for U-S OFDM

Daobin Wang, Lihua Yuan, Jingli Lei, Xiaoxiao Li, Jianming Shang
School of Science, Lanzhou University of Technology, Lanzhou, PR China

Abstract

In this paper, we proposed a pulse shaping based scheme for unitary-spreading orthogonal frequency division multiplexing system (U-S OFDM), which is an expanded version of coherent optical DFT-spreading OFDM built through replacing the spreading DFTs by a pulse shaping process. This system exploits the modified form of Nyquist pulse as each subcarrier's time-domain waveform which has a more rapid decay in time domain. Due to reducing the fields overlap between subcarriers, the proposed scheme can provide a lower PAPR and achieve good nonlinear interference mitigation. Theoretical results showed that our approach can obviously improve the signal quality for a system with more than 25600 ps/nm of chromatic dispersion generated by 1600km of standard fiber, compared with conventional DFT-spreading OFDM.
Keywords: coherent optical fiber transmission, orthogonal frequency division multiplexing, unitary spread, nonlinear inter-subcarrier interference.

1 Introduction

For future elastic optical network, orthogonal frequency division multiplexing (OFDM) is a very key enabling technology, in which the spectrum width of a sub-carrier corresponds to a frequency slot [1]. Except for the excellent robustness against chromatic dispersion (CD) and polarization mode dispersion (PMD), the compact arrangement of subcarrier's spectrums offer a high spectral efficiency in the optical OFDM system. The higher order modulation and electronic signal processing can also be implemented more flexible and easier [2].

However, one inherent drawback of an optical OFDM system is its sensitivity to fiber nonlinearity. In order to combat or alleviate this drawback, various techniques have been proposed. For example, several methods have been

www.witpress.com, ISSN 1743-3533 (on-line)
doi:10.2495/SEEEI140551

investigated to reduce the high peak to average power ratio (PAPR) of OFDM signals, such as clipping, pre-coding, partial transmission technique, selective mapping, and optical phase modulator [3]. In addition to PAPR reduction techniques, further investigations have been carried out to improve nonlinear tolerance by employing an electronic pre- and post-compensation scheme [4], RF-pilot, back-propagation or both [5, 6]. In Ref. [7], discrete Fourier transform spread (DFT-S) OFDM, which was initially proposed in 3GPP Long Term Evolution, has migrated to optical communication. DFT-precoding prior to the conventional OFDM transmitter entails a low PAPR of the OFDM signal resulting in higher nonlinear tolerance. Another interesting technique is the coherent optical single carrier frequency division multiplexing (CO-SCFDM) [8], which can be viewed as special case of the DFT-spread OFDM with only one sub-band.

More recently, the DFT-spread OFDM has been extended to a unitary-spread generalized modulation format, called as the unitary-spread OFDM [9], in which the (de)spreading (I)DFTs are replaced by more general unitary transformations. The authors of this paper have proposed one unitary transformation by having the generalized (de)spreading performed by means of N-point Hadamard matrices and they refer to the resulting system as Wavelet Spread (WAV-S) OFDM. Through numerical simulation, they found that the nonlinear tolerance of WAV-S OFDM turned out to be inferior to that of DFT-S OFDM but still exceeded that of plain OFDM. In this letter, we carefully examined the signal generating process of U-S OFDM system and proposed a modified unitary transformation for it, which is based on pulse shaping. Moreover, the simple but effective optimization schemes were also proposed in order to further reduce the PAPR. Numerical simulations showed that our proposed scheme can obviously improve the nonlinear tolerance for U-S OFDM system and outperform the conventional DFT-S OFDM.

2 System setup and principle

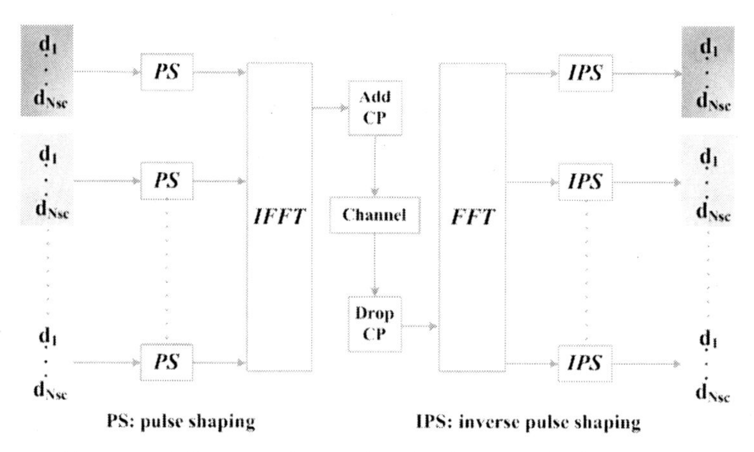

Figure 1: Conceptual diagram of the pulse shaping based unitary transformation.

The modulation and demodulation processes of the optical U-S OFDM system based on pulse shaping are schematically plotted in Figure 1. The high-speed serial data to be transmitted is divided up into M bands with N_{sc} subcarriers per band. Each group of subcarriers will first go through a unitary spreading operation and then the N point IFFT performed on the outputs of all unitary spreading will convert the frequency domain signal to time domain. Assuming $\{d_k^p; k = 0,1, \ \ ,N_{sc}-1\}$ is the *k*th data symbol to be transmitted in the *p*th sub-band $\{p = 0,1, \ \ ,M-1\}$ as shown in Figure 1. After the unitary spreading, the output of *p*th sub-band can be written as:

$$x_l^p = \sum_{k=0}^{N_{sc}-1} (\mathrm{UT})_{lk}^p \, d_k^p \tag{1}$$

where $(\mathrm{UT})_{lk}^p$ denotes the *l*,*k*-element of the *p*th unitary transformation matrix $(\mathrm{UT})^p$ and x_l^p represents the *l*th output of *p*th sub-band. The matrix $(\mathrm{UT})^p$ must satisfy $((\mathrm{UT})^p)^{-1} = ((\mathrm{UT})^p)^{\dagger}$ in order to ensure unitary operation, where † denotes the conjugate-transpose. The total number of the outputs after M blocks of unitary spreading is N, which can be equal or greater than $M \times N_{sc}$. As mentioned above, the authors of [9] took the unitary transformation as the Hadamard matrix, which contains only +1s or −1s, and have proposed the WAV-S OFDM. But, because there is no optimization, the nonlinear tolerance of WAV-S OFDM turned out to be inferior to that of DFT-S OFDM. In this work, being different with the scheme based on Hadamard matrix, we design a new unitary transformation for U-S OFDM, which replaces the spreading DFTs by a pulse shaping matrix and can be expressed as:

$$(\mathrm{UT})^p = (\mathrm{PS})^p \quad (\mathrm{PS})_{lk}^p = c_l \exp\left(-j2\pi \frac{lk}{N_{sc}}\right) \tag{2}$$

where $(\mathrm{PS})_{lk}^p$ is the *l*,*k*-element of the *p*th pulse shaping matrix $(\mathrm{PS})^p$ and c_l is the *l*th Fourier coefficient of the pulse waveform of the subcarrier. If the sin*c* function with rectangular spectral shape is chosen as the time-domain waveform of each subcarrier, the Fourier coefficient c_l will always be equal to one and our proposed scheme recovers to the conventional DFT-S OFDM [7].

Because the sin*c* function has an infinite nonzero range and is impractical for implementation, so we adopted a modified form of Nyquist pulse in this paper, which is similar to a sin*c* function but has non-rectangular spectrum. Let $\lambda = \ln 2/\alpha B$, its impulse and frequency responses can be defined as [10]:

$$r(t)=\frac{1}{T_p}\operatorname{sinc}\left(\frac{t}{T_p}\right)\frac{4\lambda\pi t\sin\left(\pi\alpha t/T_p\right)+2\lambda^2\cos\left(\pi\alpha t/T_p\right)-\lambda^2}{\left(2\pi t\right)^2+\lambda^2} \tag{3}$$

$$R(f)=\begin{cases}1 & |f|\le B(1-\alpha)\\ \exp\left\{\lambda\left[B(1-\alpha)-|f|\right]\right\} & B(1-\alpha)<|f|\le B\\ 1-\exp\left\{\lambda\left[|f|-B(1+\alpha)\right]\right\} & B<|f|<B(1+\alpha)\\ 0 & |f|\ge B(1+\alpha)\end{cases} \tag{4}$$

where T_p is the time duration of the main lobe of this pulse, B is the Nyquist frequency and α is known as a roll-off factor. Compared with the sinc pulse, the tail of this modified Nyquist pulse decays more rapidly, which allows quick truncation and minimizes the resulting inter-symbol interference (ISI) at the receiver. More importantly, the time-domain waveform with the rapid decay can decrease the power coupling between different subcarriers and obtain the reduced PAPR for U-S OFDM symbol. The cost to obtain the reduced PAPR is excess bandwidth but that may be a fair trade to obtain higher nonlinear transmission performance for coherent optical systems.

Performing N point IFFT on the outputs of pulse shaping process generates the time-domain samples of U-S OFDM system as indicated by:

$$S_m=\sum_{n=0}^{N-1}x_n\cdot\exp\left(j\frac{2\pi mn}{N}\right) \tag{5}$$

where S_m is the mth time-domain sample in one U-S OFDM symbol period and $0\le m\le N-1$. x_n is determined according to Eqs. (1) and (2).

From Eqs. (3) and (4), we can find that the pulse shaping based unitary transformation has two adjustable parameters, one is roll-off factor α and the other is the time duration T_p of modified Nyquist pulse. There is a close relationship between the PAPR and the time duration of each subcarrier's waveform. The longer time duration of each subcarrier's waveform will make the subcarrier energies superimpose and combine in time domain, which can increase the PAPR and excite FWM among the individual subcarriers due to fiber nonlinearities. So, one can optimize the investigated system's nonlinear transmission performance through reducing the time duration T_p of each subcarrier's waveform. Another optimization approach depends on the roll-off factor α, which is the measure of the excess bandwidth of the Nyquist pulse. Modifying the roll-off factor α can also reduce the fields overlap between different subcarriers and enhance the system's performance.

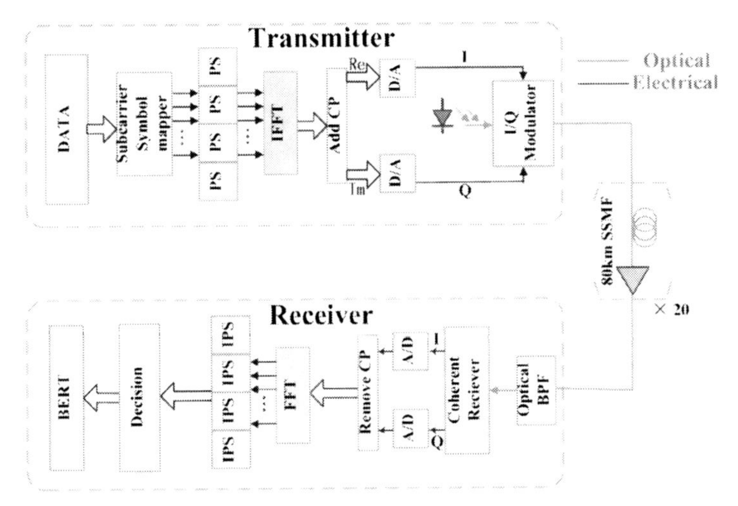

Figure 2: Coherent optical U-S OFDM system based on pulse shaping.

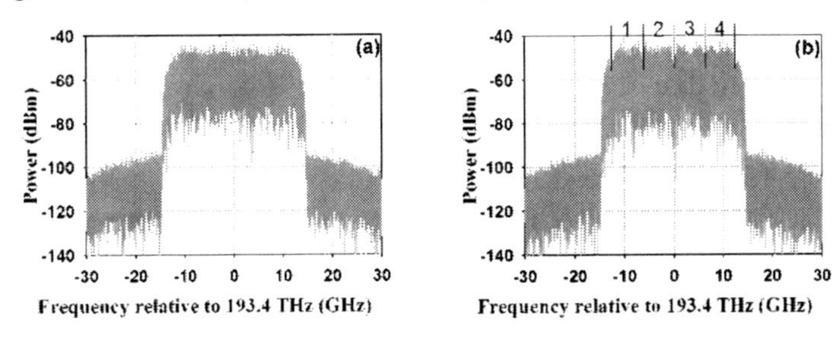

Figure 3: Spectrums for 100 Gb/s PDM DFT-S OFDM and U-S OFDM signals: (a) DFT-S OFDM; (b) pulse shaping based U-S OFDM.

After the complex U-S OFDM signals are generated, the real and imaginary components are used to drive the Mach–Zehnder modulators (MZM). One of the MZ outputs is delayed by 90 degree in order to generate the optical In-phase (I) and Quadrature (Q) components. These are then summed and injected into the fiber. In the receiver, the U-S OFDM signal is first down-converted to baseband with an optical coherent detection scheme, and then demodulated by performing FFT and baseband signal processing to recover the data, as displayed in Figure 1.

3 Simulation and numerical results

The schematic setup of the coherent optical U-S OFDM system has been shown in Figure 2. A numerical simulation based on the open-source software Optilux is con-ducted to identify the transmission performance of a PDM U-S OFDM system. On the transmitter side, a pseudorandom binary sequence of length 2×2^{17} is generated and mapped by a 4-QAM encoder. The information stream is further parsed into 1024 low-speed parallel data subcarriers. These subcarri-ers are divided into four sub-bands, and then each sub-band is processed by a matrix manipulation module described above. In each sub-band, eight pilot subcarriers are used for channel estimation. After generating the time series by an inverse fast Fourier transform processor, cyclic prefix is added to ensure a correct data recovery. The MZ modulator is used to convert electrical signals to optical signals. The laser line-width is set at zero, with adjustable launch power. The frequency of the carrier wave is set at 193.4 THz. The optical channel consists of 20 spans, each including 80 km standard single-mode fiber (SSMF) and dispersion compensating fiber (DCF). The SSMF fiber has a loss of 0.2 dB/km, a dispersion parameter of 16 ps/nm/km, and a nonlinearity coefficient of 1.31 W^{-1} km^{-1}. The optical amplifiers, with noise figure F = 6 dB, compensated for the fiber loss in each span. On the receiver side, the local oscillator laser is assumed to be perfectly aligned with power set at 10 dBm and line-width also equals to zero. The I/Q components of the U-S OFDM signal are recovered by a 2×4 90^0 optical hybrid and two pairs of photo-detectors. Each photo-detector has a responsivity of 1 A/W and is noiseless to show the noise and distortion due to the optical amplifiers and fiber nonlinearity. The converted U-S OFDM RF signal is demodulated using fast Fourier transform processor and the inverse matrix manipulation of (2). The obtained signals are fed into a 4-QAM decoder. The Q factor is extracted from the constellation using the method described in Ref. [4].

We compare nonlinearity performances between four systems: (1) the optimized U-S OFDM based on pulse shaping, where the T_p is reduced to one-half of its original length. This will enlarge the number of the outputs after unitary spreading to its double size and the IFFT/FFT size is 2048; (2) the U-S OFDM based on pulse shaping but without optimization, where the T_p remains unchanged; (3) the DFT-S OFDM; (4) the conventional OFDM. For fair comparison, the IFFT/FFT sizes of (2)–(4) are also set to 2048 points, which includes the 1024 modulated subcarriers and 1024 zero padding. Two different dispersion maps are considered in our investigation. One is 100 Gb/s transmission link without optical dispersion compensation and each span only includes 80 km standard single-mode fiber. Another is the transmission link with periodic dispersion compensation and the residual dispersion per span is 100 ps/nm. Electrical dispersion equalizer and polarization de-multiplexing at the receiver are enabled in both cases.

Figure 3(a) and (b) shows the spectrums of 100 Gb/s DFT-S OFDM and pulse shaping based U-S OFDM. As can be seen, we can clearly distinguish between all four sub-bands in Figure 3(b). For U-S OFDM system, the spectrum shape is determined by the number of sub-bands and the pulse waveform of each subcarrier, which is a modified form of Nyquist pulse in our work. According to Eq. (4), its frequency response is not a rectangular function and there is a gradual change at low frequency part. When transforming to the optical spectrum, the magnitude of low-frequency part of U-S OFDM spectrum appears some descending, compared with the high-frequency part. Therefore, the boundary between sub-bands can be easily identified in Figure 3(b).

Next, the OSNR sensitivities in back to back transmission have been measured for 100 Gb/s PDM U-S OFDM systems and the results are shown in Figure 4. The theoretical values of Q factor are calculated as following:

$$Q = 10\log 10\left\{\left(50\text{GHz}/\text{Bitrate}\right)\cdot 10^{\left(\text{OSNR}/10\right)}\right\} \tag{6}$$

We find the OSNR sensitivity of pulse shaping based U-S OFDM, if not optimized, is inferior to CO-OFDM and DFT-S OFDM system by ~0.5dB. After optimized, the difference between the OSNR sensitivities of our proposed scheme and CO-OFDM can reduce to only about 0.1dB. Finally, we investigated the nonlinear transmission performance of the long haul U-S OFDM systems. Figure 5(a) shows the electrical Q versus the launch power into each fiber span for 100 Gb/s PDM U-S OFDM and CO-OFDM systems. These results show that if the launch power into SMF is 2 dBm, our approach can improve the Q factor by 2.1 dB for a transmission link with more than 25,600 ps/nm of chromatic dispersion generated by 1600km of standard fiber, compared with the DFT-S OFDM system.

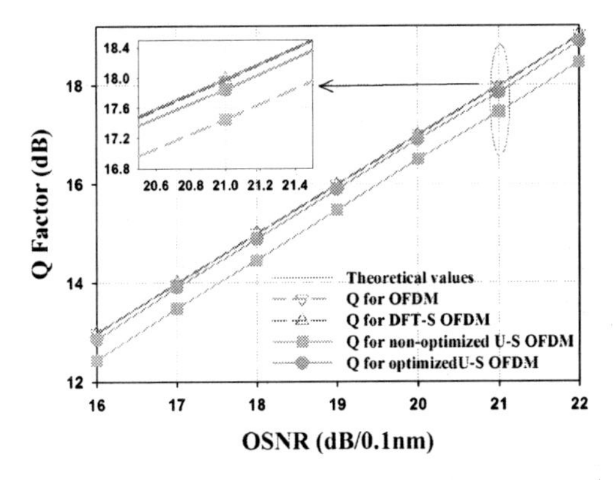

Figure 4: OSNR sensitivity in back to back transmission.

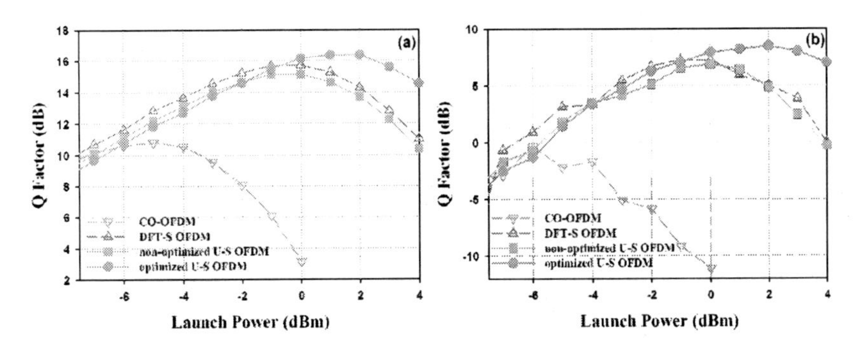

Figure 5: Signal quality versus optical launch power after 1600 km transmission. (a) the fiber link without dispersion management. (b) the fiber link with dispersion management.

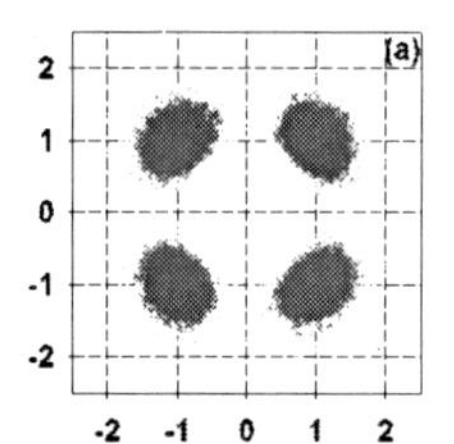

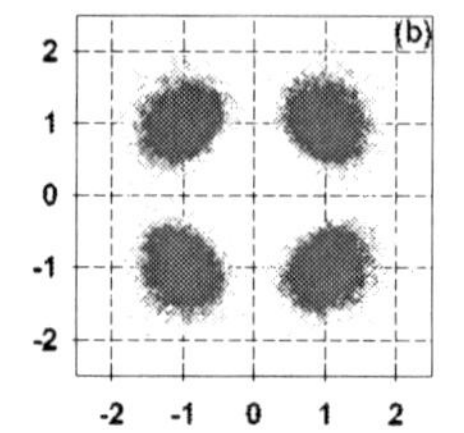

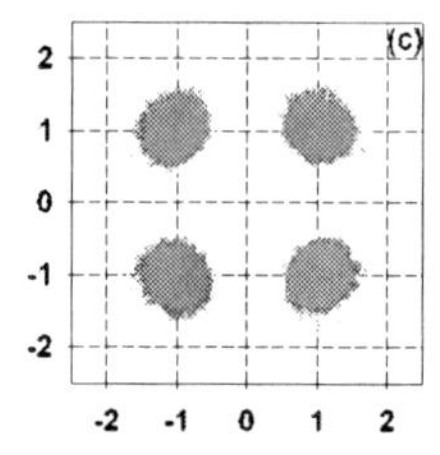

Figure 6: Constellation diagrams for U-S OFDM at 2 dBm. (a) DFT-S OFDM, (b) the non-optimized U-S OFDM based on pulse shaping, (c) the optimized U-S OFDM based on pulse shaping.

If using optical dispersion compensation, the Q factor can be increased by 3.5 dB at the optimum power for a fiber link with periodic dispersion compensation, as shown in Figure 5(b). The corresponding constellations of Figure 5(a) have been shown in Figure 6. The reduced diffusion presented in the constellation points of U-S OFDM signals demonstrated that the nonlinear inter-carrier interferences between subcarriers have been eliminated substantially.

4 Conclusions

In summary, the pulse shaping based unitary transformation is designed and optimized for U-S OFDM to combat nonlinear impairment in optical media. Through adopting the waveform with small pulse width, this system can remove the over-lapping between the adjacent subcarriers and reduce the occurrence probability of FWM effectively. Simulation results show that our proposed scheme, if optimized, can outperform the conventional DFT-S OFDM in the long-haul transmission system.

Acknowledgment

This work was supported in part by the National Natural Science Foundation of China (Nos. 61367007, 61167005), in part by the Natural Science Fund of Gansu Province of China (Nos. 1112RJZA017, 1112RJZA018) and in part by the Research Fund for the Doctoral Program of Lanzhou University of Technology.

References

[1] G. Zhang, M.D. Leenheer, A. Morea, B. Mukherjee, A survey on OFDM-based elastic core optical networking. *IEEE Communications Surveys and Tutorials*, **15(1)**, pp. 65–87, 2013.

[2] W. Shieh, C. Athaudage, Coherent optical orthogonal frequency division multiplexing. *Electronics Letters*, **42(10)**, pp. 587–589, 2006.

[3] A.J. Lowery, L.B. Du, Optical orthogonal division multiplexing for long haul optical communications: a review of the first five years. *Optical Fiber Technology*, **17(5)**, pp. 421–438, 2011.

[4] A.J. Lowery, Fiber nonlinearity pre- and post-compensation for long-haul optical links using OFDM. *Optics Express*, **15(20)**, pp. 12965–12970, 2007.
[5] A. Diaz, A. Napoli, S. Adhikari, Z. Maalej, A.P. Lobato Polo, M. Kuschnerov, J. Prat, Analysis of back propagation and RF pilot tone based nonlinearity compensation for a 9 × 224 GB/s POLMUX-16QAM system. In: *Proceedings of 37th OFC*, Los Angeles, CA, March 2012, paper OTh3C.5, 2012.
[6] L.B. Du, A.J. Lowery, Pilot-based cross-phase modulation compensation for coherent optical orthogonal frequency division multiplexing long-haul optical communications systems. *Optics Letters*, **36(9)**, pp. 1647–1649, 2011.
[7] Y. Tang, W. Shieh, B.S. Krongold DFT-spread OFDM for fiber nonlinearity mitigation. *IEEE Photonics Technology Letters*, **22(16)**, pp. 1250–1252, 2010.
[8] F. Zhang, C.C. Yang, X. Fang, T.T. Zhang, Z.Y. Chen, Nonlinear performance of multi-granularity orthogonal transmission systems with frequency division multiplexing. *Optics Express*, **21(5)**, pp. 6115–6130, 2013.
[9] G. Shulkind, M. Nazarathy, An analytical study of the improved nonlinear tolerance of DFT-spread OFDM and its unitary-spread OFDM generalization. *Optics Express*, **20(23)**, pp. 25884–25901, 2012.
[10] A. Assalini, A.M. Tonello, Improved Nyquist pulses. *IEEE Communications Letters*, **8(2)**, pp. 87–89, 2004.

MPPT control for direct driven vertical axis wind turbine generation system with permanent magnet synchronous generator

Aihua Wu[1,2], Buhui Zhao[1], Jingfeng Mao[2], Hairong Zhu[2]
[1]*School of Electrical and Information Engineering, Jiangsu University, Zhenjiang, Jiangsu, China*
[2]*School of Mechanical Engineering, Nantong University, Nantong, Jiangsu, China*

Abstract

Direct driven vertical axis wind turbine (VAWT) with permanent magnet synchronous generator (PMSG) is a new type wind power generation system, and has been focused in recent years. It has good potential for commercial application because of its excellent performances. In this paper, an H-type Darrieus VAWT generation system and its power control method are studied. Firstly, the general configuration of a VAWT generation system is described, and the mathematical model is established. Secondly, in order to improve the wind energy conversion efficiency and to realize maximum power point tracking (MPPT), an optimum tip speed ratio (TSR) control algorithm for the rated wind speed operating region is discussed. Then, the generator angular velocity closed-loop control equation and the power close-loop control system structure are worked out to implement maximum wind power absorption. Finally, the system numerical simulation model is developed using MATLAB/SIMULINK software. Simulation results demonstrate the validity of proposed power control method for the maximum power point tracking.
Keywords: vertical axis wind turbine, wind power generation systems, power control, maximum power point tracking, boost conversion.

1 Introduction

Nowadays, wind power generation has become the most competitive renewable energy according to the technical maturity and economic feasibility. It has been

www.witpress.com, ISSN 1743-3533 (on-line)
doi:10.2495/SEEEII140561

paid wide and sustained attention in the world. Compared to the commercial large-scale horizontal axis wind turbine (HAWT) generation system, the vertical axis wind turbine (VAWT) generation system has lots of particular strengths, such as simpler blade manufacturing processes, lower noise level, higher security, smaller installation size requirements, lower construction cost, no upwind device, suitable for application in low wind speed areas and densely populated area etc. [1, 2]. Hence, particular interest has been focused in recent years on VAWT generation system. And these new type VAWT generation systems are always adopted as distributed generation equipments, which supply power for urban public lighting, outdoor advertising, communication base stations, highway monitoring, rural off-grid power plants and other areas of distributed power consumer.

The VAWT generation systems are usually designed for small or medium-scale systems, using direct driven permanent magnet synchronous generator (PMSG) and fixed pitch structure so as to reduce unit costs and improve transmission efficiency. However, in order to further increase wind energy utilization efficiency, research and design advance MPPT technique to maximize the efficiency of wind energy utilization within the rated wind speed operating range is always of great importance.

In this paper, aimed to a practical application of H-type Darrieus VAWT permanent magnet direct driven off-grid wind power generation system, a MPPT control strategy for VAWT is studied. According to the VAWT natural aerodynamic mechanical properties in rated wind speed operating range, the optimum tip speed ratio (TSR) control principle and angular velocity control equation are given. Simulation experiments are carried out, and validity of the proposed control method is proven.

2 Wind power generation system configuration

Figure 1 shows a general configuration of a VAWT generation system. It consists of H-type Darrieus VAWT, PMSG, three-phase uncontrolled rectifier, Boost converter, and customer side power converter.

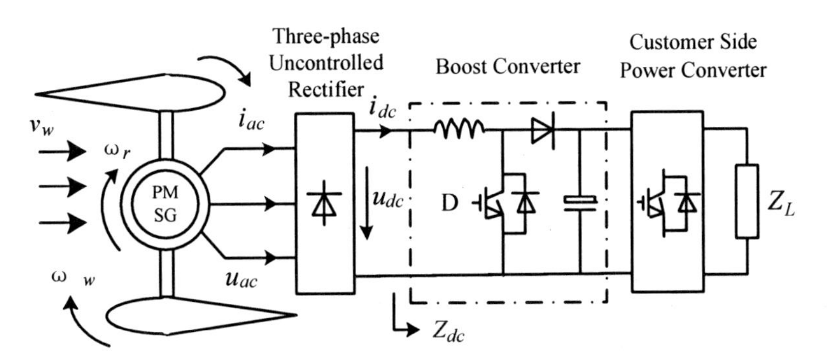

Figure 1: Configuration of a VAWT generation system.

In Figure 1, v_w is the wind speed; ω_w is the turbine angular velocity; ω_r is the generator angular velocity; u_{ac} and i_{ac} are the generator output phase voltage and current, respectively; u_{dc} and i_{dc} are the rectifier output DC voltage and current, respectively; Z_{dc} and Z_L are equivalent reactance of the Boost converter and the customer side load, respectively.

3 System mathematical model

3.1 Wind turbine model

According to aerodynamics theory, vertical axis wind turbine aerodynamic power output is [3]

$$P_a = 0.5C_p(\lambda, \beta)\rho A v_w^{\ 3} \tag{1}$$

where ρ is the atmospheric density; A is the swept area; $C_P(\lambda, \beta)$ is the aerodynamic power coefficient; β is the pitch angle; λ is the tip speed ratio (TSR), which the equation as indicated in Eq. (2).

$$\lambda = \omega_w R_H / v_w \tag{2}$$

where R_H is the radius of H-type VAWT.

Figure 2 shows the relationship curve between $C_P(\lambda,\beta)$ and λ, where β equal zero degree. It can be noticed that the optimum value of aerodynamic power coefficient (C_{p_opt}) equals 0.45 for λ equals 6.6.

From Eq. (2), it can be seen that if the turbine angular velocity is regulated in accordance with wind speed, which the ratio is always consistent with the optimal tip speed ratio λ_{opt}, then the maximum power from a wind turbine can be obtained as

$$P_{a\ \mathrm{opt}} = 0.5C_{p\max}\rho A\left(R_H\omega_w\right)^3 / \lambda_{\mathrm{opt}}^{\ 3} \tag{3}$$

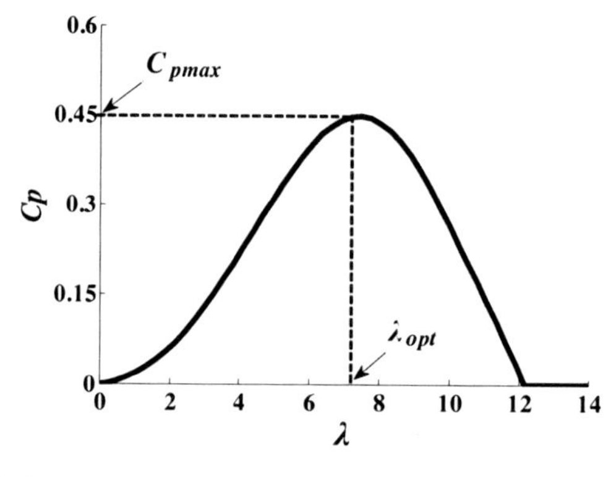

Figure 2: Relationship between power coefficient and tip speed ratio.

Therefore, the available mechanical torque can be expressed as

$$T_{a\ \mathrm{opt}} = P_{a\ \mathrm{opt}} / \omega_w \tag{4}$$

3.2 Power electronic converter model

The PMSG is modelled by the following voltage equations in the rotor *dq*-axes reference frame

$$\begin{aligned} u_d &= -R_s i_d - L_d \dot{i}_d + L_q n_p \omega_r i_q \\ u_q &= -R_s i_q - L_q \dot{i}_q - L_d n_p \omega_r i_d + n_p \omega_r \psi_f \end{aligned} \tag{5}$$

where u_d, u_q, i_d, i_q, L_d, L_q are the stator voltages, currents and inductances in the *dq* axes, respectively; R_s is the stator resistance; n_p is the PMSG pole pairs; ψ_f is the rotor permanent magnet flux linkage.

For the surface mount type PMSG, the electromagnetic torque is

$$T_e = 1.5 n_p \psi_f i_{sq} \tag{6}$$

If using three-phase power diode uncontrolled rectifier connected to the PMSG generator output terminal, the DC voltage output of uncontrolled rectifier is [4]

$$u_{\mathrm{dc}} = 3\sqrt{6} u_{\mathrm{ac}} / \pi \tag{7}$$

where

$$u_{\mathrm{ac}} = \sqrt{\left(u_{\mathrm{sd}}{}^2 + u_{\mathrm{sq}}{}^2\right)/2} \tag{8}$$

From Figure 1, DC current output of uncontrolled rectifier is

$$i_{\mathrm{dc}} = u_{\mathrm{dc}} / Z_{\mathrm{dc}} \tag{9}$$

where Z_{dc} is equivalent dynamic reactance as the condition of the Boost converter output current continuous, and its expression is

$$Z_{\mathrm{dc}} = (1 - D)^2 R_L \tag{10}$$

where D is the Boost converter PWM duty cycle.

According to power conservation theory and Eq. (7), the inductor current i_{dc} is formed as follows

$$i_{\mathrm{dc}} = \pi i_{\mathrm{ac}} / \sqrt{6} \tag{11}$$

The inductor current i_{dc} represented by Eq. (11) also can be written in terms of the dq-axes current components as follows [5]

$$i_{\rm dc} = \sqrt{3}\pi i_{\rm sq} / 6 \tag{12}$$

Incorporating Eq. (12) for Eq. (6), we have

$$T_e = k_{\rm dc} i_{\rm dc} \tag{13}$$

where $k_{\rm dc} = 3\sqrt{3} n_p \psi_f / \pi$.

3.3 Electromechanical coupling model

The quantitative relation linking the VAWT angular velocity and the PMSG angular velocity can be described as follows

$$\omega_w = G_m \omega_r \tag{14}$$

where G_m is the gear ratio of the main shaft. For direct driven system, we have $G_m = 1$.

Thus, from Figure 1 and Eq. (14), we can conclude the kinematical equation of direct driven vertical axis wind turbine with permanent magnet synchronous generator as follows

$$J_r \dot{\omega}_r = T_a - k_{\rm dc} i_{\rm ac} - B_r \omega_r \tag{15}$$

where J_r is the system inertia; B_r is the system friction coefficient.

From Eqs. (13) – (15), we can see that the effective regulation of the generator electromagnetic torque by controlling the DC side current of three-phase rectifier can influence turbine angular velocity, which leads to the passive change of tip speed ratio and aerodynamic power coefficient. By this means, we can realize the purpose of maximum wind power absorption.

4 MPPT control method

In order to capture maximum aerodynamic power, the controller must drive the turbine rotating at the optimal tip speed ratio state. So we can get the PMSG reference angular velocity from Eq. (2), this is

$$\omega_r^* (\text{II}) = \omega_{w\,\rm opt} = \lambda_{\rm opt} v_w / R_H \tag{16}$$

Accordingly, the MPPT control system structure can be designed as Figure 3.

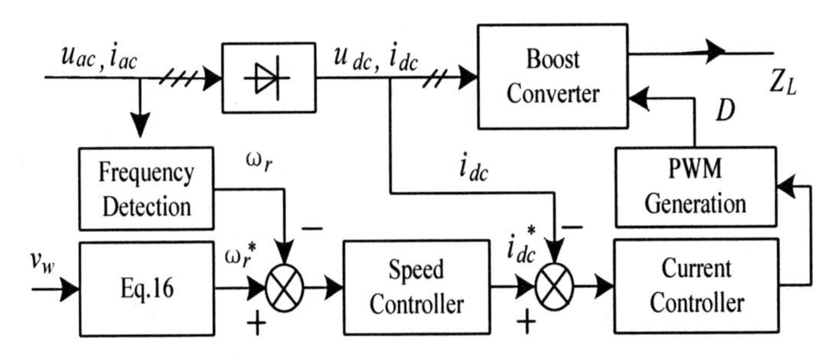

Figure 3: MPPT control system structure.

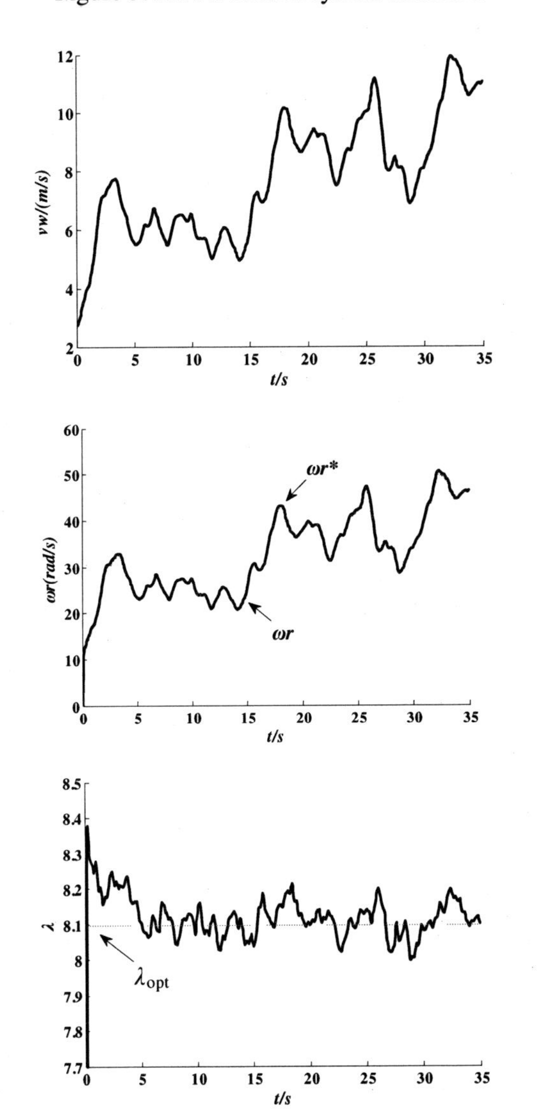

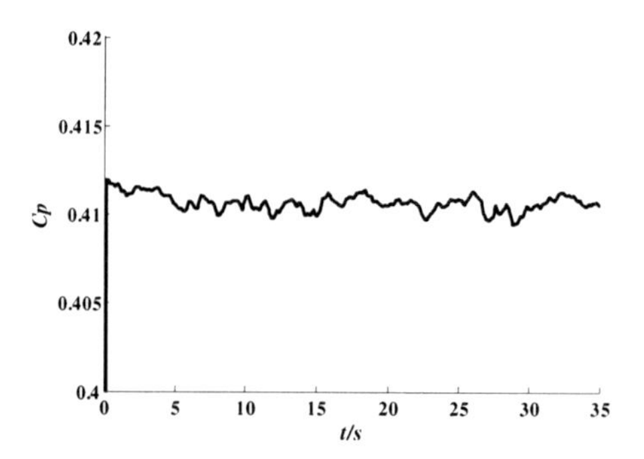

Figure 4: MPPT control simulation waveforms. (a) wind speed. (b) angular velocity. (c) tip speed ratio. (d) aerodynamic power coefficient.

5 Simulation results

The PMSG parameters are given as follows: $P_N = 5.5$KW, $n_N = 500$r/min, $n_p = 6$, $\psi_a = 0.35 \times 10^{-3}$ Wb, $U_L = 137$V, $I_N = 23.3$A. The VAWT parameters are given as follows: $R_T = 1.93$m, $H_T = 3.04$m, $A = 11.722\text{m}^2$, $v_{wN} = 12$m/s, $C_{P\max} = 0.42$, $\lambda_{\text{opt}} = 8.1$, $\rho = 1.293\text{kg/m}^3$, $J_r = 1.5\text{kg.m}^2$, $B_r = 0.02$ N·s/rad. The MPPT controller uses PID regulator, and its control parameters are designed as $K_p = 22$, $K_i = 10$, $K_d = 0.2$.

Figure 4 shows the MPPT control system simulation waveforms of VAWT angular velocity, tip speed ratio, and aerodynamic power coefficient under different wind speed changing conditions.

Figure 4 shows that the PMSG generation reference angular velocity is changed in accordance with the field wind speed, and their relationship is consistent with Eq. (16). The PMSG actual angular velocity tracks the reference angular velocity rapidly, which makes the VAWT tip speed ratio always fluctuating around the optimal value λ_{opt}. It ensures the aerodynamic power coefficient is very close to $C_{p\max}$, and means the wind power system realizes MPPT generation.

6 Conclusions

In this paper a direct driven VAWT generation system using PMSG and its MPPT control method are studied. According to the generation system configuration and MPPT working characteristics within rated wind speed region, the TSR-based closed-loop control system is designed. The numerical simulation model is built by MATLAB/SIMULINK software, and simulation results have confirmed the effectiveness of the proposed system.

Acknowledgment

This work was supported by the National Natural Science Foundation of China (61004053, 61273151, 51307089), Natural Science Foundation of Jiangsu Province

(BK20141238), Postgraduate Research Innovation Program of Jiangsu Province (CXLX13_681) and Qing Lan Project of Jiangsu Province.

References

[1] Sandra Eriksson, Hans Bernhoff, Mats Leijon, Evaluation of different turbine concepts for wind power. *Renewable and Sustainable Energy Reviews*, **12(5)**, pp. 1419–1434, 2008.

[2] Ming-Fa Tsai, Wei-Chieh Hsu, Tai-Wei Wu, Jui-Kum Wang, Design and implementation of an FPGA based digital control IC of maximum power point tracking charger for vertical-axis wind turbine generators. *International Conference on Power Electronics and Drive Systems*, November 2–5, Taipei, Taiwan, pp. 764–769, 2009.

[3] Jaohindy P., Garde F., Bastide A., Aerodynamic and mechanical system modeling of a vertical axis wind turbine. *2011 International Conference on Electrical and Control Engineering*, September 16–18, Yichang, China, pp. 5189–5192, 2011.

[4] Amei K., Takayasu Y., Ohji T., Sakui M., A maximum power control of wind generator system using a permanent magnet synchronous generator and a boost chopper circuit. In: *Proceedings of the Power Conversion Conference*, April 2–5, Osaka, Japan, pp. 1447–1452, 2002.

[5] Dalala Z.M., Zahid Z.U., Wensong Yu, Yonughoon Cho, Jih-Sheng Lai, Design and analysis of an MPPT technique for small-scale wind energy conversion systems. *IEEE Transactions on Energy Conversion*, **28(3)**, pp. 756–767, 2013.

Fault ride-through capability enhancement of grid interfacing photovoltaic system by an improved flux-coupling type SFCL

Lei Chen, Feng Zheng, Changhong Deng, Shichun Li, Miao Li
School of Electrical Engineering, Wuhan University, Wuhan, China

Abstract

At present, the application of renewable energy technologies has attracted lots of attentions around the world. Since most of the renewable energy generation units are connected to electric power systems through the inverters, it is important to ensure their operational robustness against short-circuit faults. In this paper, our research group introduces an improved flux-coupling type superconducting fault current limiter (SFCL) to enhance the fault ride-through (FRT) capability of a grid interfacing photovoltaic system. The SFCL's structural principle is presented, and its influence mechanism to the photovoltaic system's transient characteristics is discussed. Furthermore, the detailed simulation model is built in MATLAB/SIMULINK, and the system's performance behaviors under a three-phase ground fault are imitated. The demonstrated results show that, employing the SFCL can availably limit the fault current, improve the voltage sag, and guarantee the photovoltaic system's FRT capability.
Keywords: fault ride-through capability, improved flux-coupling type SFCL, photovoltaic system, transient simulation.

1 Introduction

The number of renewable energy sources connected to the public grid is increasing significantly due to the deregulation of the electric power distribution industry and to environmental problems [1]. Among these renewable sources, photovoltaic (PV) technology can be regarded as a main candidate in the future of electricity generation, and it is a promising clean energy source which does not require rotational generators [2]. Actually, one of the crucial issues regarding the development of a grid interfacing photovoltaic system is related to its fault ride-through (FRT) capability.

WIT Transactions on Engineering Sciences, Vol. 107, © 2015 WIT Press
www.witpress.com, ISSN 1743-3533 (on-line)
doi:10.2495/SEEEI140571

In general, a PV generation unit is connected to the power grid through the inverter, and this grid-interfacing PV system may stop the operation or be in unstable operation simultaneously due to inverter characteristic in PCS in case of transient disturbances or short-circuit fault [3]. However, according to the standards, the PV system (in particular for the one with large-rated capacity) should stay connected and contribute to the grid in case of severe grid voltage disturbance since the disconnection may further degrade voltage restoration during and after fault conditions [4]. In order to enhance the PV system's FRT capability, a few solutions have been suggested [5]. In view of that a fault current limiter (FCL) can directly affect the increased short-circuit current and decreased terminal voltage, its application may bring more positive effects.

In this paper, our research group introduces an improved flux-coupling type superconducting fault current limiter (SFCL) to enhance the fault ride-through (FRT) capability of a grid interfacing photovoltaic system. This paper is organized in this manner. Section 2 introduces the SFCL's structural principle, and discusses its influence mechanism to the photovoltaic system's transient characteristics. In Section 3, the detailed simulation model is built in MATLAB/SIMULINK, and the system's performance behaviors under a three-phase ground fault are imitated. In Section 4, conclusions are summarized and next steps are suggested.

2 Theoretical analysis

2.1 Structural principle of the improved flux-coupling type SFCL

The schematic configuration of the modified flux-coupling type SFCL is shown in Figure 1. This SFCL is mainly composed of a coupling transformer (CT), a controlled switch S_1 and a superconducting coil (SC). The switch S_1 and the superconducting coil are respectively connected in series with the CT's primary and secondary windings, which are wound in reverse directions. The metal oxide arrester (MOA), which can be used to suppress switching overvoltage, is connected in parallel with the CT's primary winding. L_1, L_2 are the winding self-inductances, and M is the mutual inductance. In addition, Z_s is the circuit impedance and S_{load} is the circuit load. R_{SC}/R_{moa} is recorded as the SC/MOA's normal-state resistance.

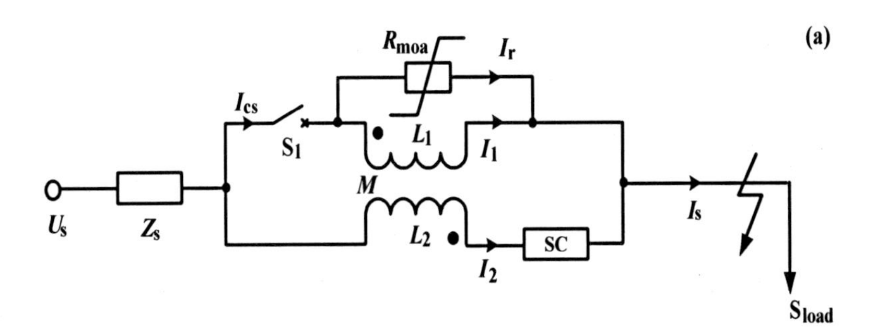

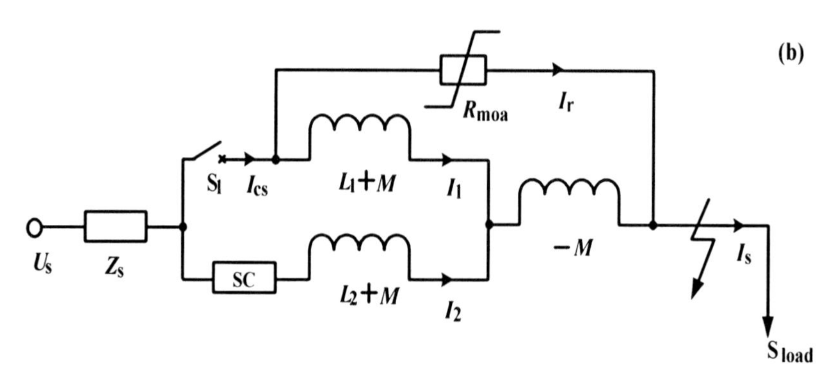

Figure 1: Structure of the improved flux-coupling type SFCL. (a) main connection, (b) equivalent circuit.

Since it is convenient to analyze the CT's characteristics by using the equivalent circuit whose parameters are expressed in terms of the mutual-inductance and self-inductances of the windings, the SFCL's impedance characteristic can be studied more clearly. According to the CT's equivalent circuit, the SFCL's electrical equivalent structure is shown in Figure 1(b), where the series and parallel connections among the coupling windings, SC, MOA and S_1 become more intuitive.

In normal (no fault) condition, the switch S_1 is closed and the SC is maintained in the zero-resistance/superconducting state. The SFCL's impedance is determined by the CT's operating impedance, which can be calculated as:

$$\begin{aligned} Z_{CT} &= j\omega[(L_1+M)//(L_2+M)-M] \\ &= j\omega(L_1L_2-M^2)/(L_1+L_2+2M) \end{aligned} \tag{1}$$

Supposing that the coupling coefficient k and transformation ratio n can be respectively expressed as $k = M/\sqrt{L_1L_2}$ and $n = \sqrt{L_1/L_2}$, the CT's operating impedance will be $Z_{CT} = j\omega L_2(1-k^2)n^2/(n^2+2kn+1)$. In the case that an iron core is used to maximize the coupling, k will be approximated to 1 and $Z_{CT} \approx 0$. The non-inductive coupling is achieved, and the MOA is "short-circuited." Consequently, the SFCL will not affect the main circuit.

After the fault happens, S_1 will be opened rapidly. Hence, a freewheeling circuit consisting of L_1 and R_{moa} will be formed, so as to restrain the switching overvoltage. Once the overvoltage is eliminated, the freewheeling circuit will be interrupted owing to the MOA's high resistance effect. Further, since the flux between the CT's two windings can no longer cancel out each other, the non-inductive coupling will be destroyed, and the superconducting coil will as well quench to its high-resistance state. Right now the SFCL will play the role, and its impedance is calculated as $Z_{SFCL} = [R_{SC} + j\omega L_2 + (kn\omega L_2)^2/(R_{moa} + n^2 j\omega L_2)]$. In view of $R_{moa} \gg n^2\omega L_2$, $Z_{SFCL} \approx R_{SC} + j\omega L_2$ can be obtained.

Compared with the original flux-coupling type SFCL [6, 7], the improved SFCL is a resistive-inductive type (hybrid type) SFCL, which absorbs the merits of the two types of SFCLs and can theoretically enhance the transient performance of a power system more efficiently [8].

2.2 Enhancement of a grid interfacing photovoltaic system by the SFCL

As shown in Figure 2, it indicates the schematic diagram of a grid interfacing photovoltaic system with the SFCL. From this figure, the PV generation unit is connected to the input side of a boost converter and its output side is connected to the DC link capacitor. Further, a three-phase voltage source inverter (VSI) is adopted to maintain constant DC voltage and supply sinusoidal current to the main grid. For the boost converter, the maximum power point tracking (MPPT) control is used to ensure the PV system's operating efficiency.

The PV system's steady state active and reactive power in dq reference frame can be written as:

$$\begin{aligned} P &= 3(E_q I_q + E_d I_d)/2 \\ Q &= 3(E_q I_d - E_d I_q)/2 \end{aligned} \tag{2}$$

where E_d and E_q are the d and q components of the system voltage, and I_d and I_q are the d and q components of the PV inverter output current. Since E_q is constant, and E_d is equal to 0 at synchronous speed, Eq. (2) can be changed to $P = 3E_qI_q/2$, $Q = 3E_qI_d/2$. When the fault happens, the PV inverter's output current I_d will depend on its rated capacity, and the output current I_q can be calculated as:

$$I_q = \sqrt{{I_{max}}^2 - I_d^2} \tag{3}$$

As the SFCL is installed before the transformer, it can help to limit the fault current contributed by the PV system, and meantime improve the voltage sag. Note that, regarding the voltage compensation caused by the SFCL, the desired voltage boosting level should be considered. If the voltage dip is too serious, the SFCL can undertake part compensation which is determined by its design parameters, and the remaining voltage compensation can be provided through the PV converter as injected reactive current to fulfill grid code requirements.

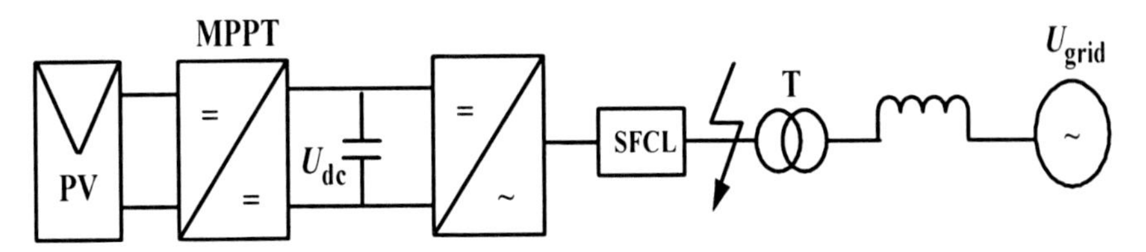

Figure 2: A grid interfacing photovoltaic system with the SFCL.

Table 1: Main parameters of the PV system integrated with the SFCL.

Improved Flux-Coupling Type SFCL	
Primary/secondary/mutual inductance	20 mH/20 mH/19.98 mH
Superconducting coil R_{sc}	2 Ω
Demonstrated PV Generation System	
PV generation's rated capacity/voltage	350kW/619V
DC voltage link	1 kV
Temperature/Radiation	25°C/1000 W/m^2
DG transformer	0.38 kV/35 kV, $Y_g - Y_g$

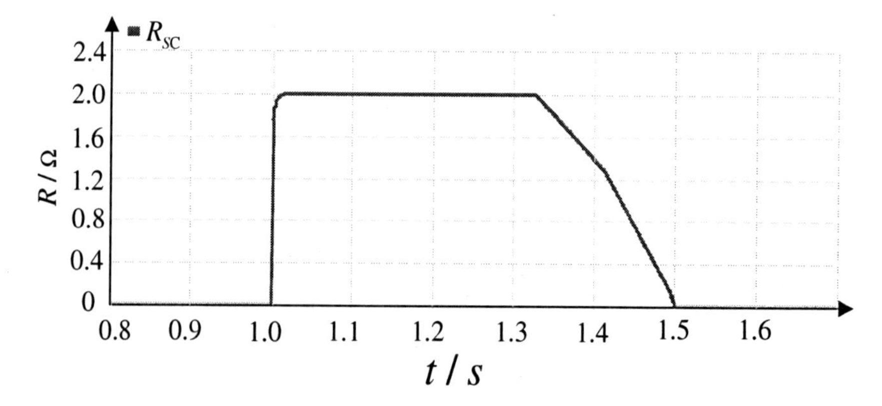

Figure 3: Quench and recovery of the imitated superconducting coil.

3 Simulation analysis

To verify the aforementioned idea's effectiveness and quantitatively access the SFCL's effects, the simulation model corresponding to Figure 2 is built in MATLAB/SIMULINK, and parts of parameters are indicated as Table 1. In regards to the modeling of the inverter control, details can be found in Ref. [9]. The fault conditions are set as that, a three-phase grounded fault happens at t =1 s, and the fault duration/resistance is 0.1 s/0.5 Ω. Figure 3 shows the quench and recovery model of the superconducting coil. The SC will quench to the normal state within 4 ms, and after the fault, it does not recover the superconducting state at once due to the thermally accumulated joule energy. The recovery time is less than 0.5 s, so as to match up the auto-reclosing's operation.

As shown in Figures 4 and 5, they indicate the SFCL's influence on the PV system's output voltage and current (near the VSI side). From the figures, employing the SFCL will limit the fault current to a lower level, and maintain the voltage as much as possible (90% of the nominal voltage). As a result, the PV generation system's FRT capability can be well enhanced.

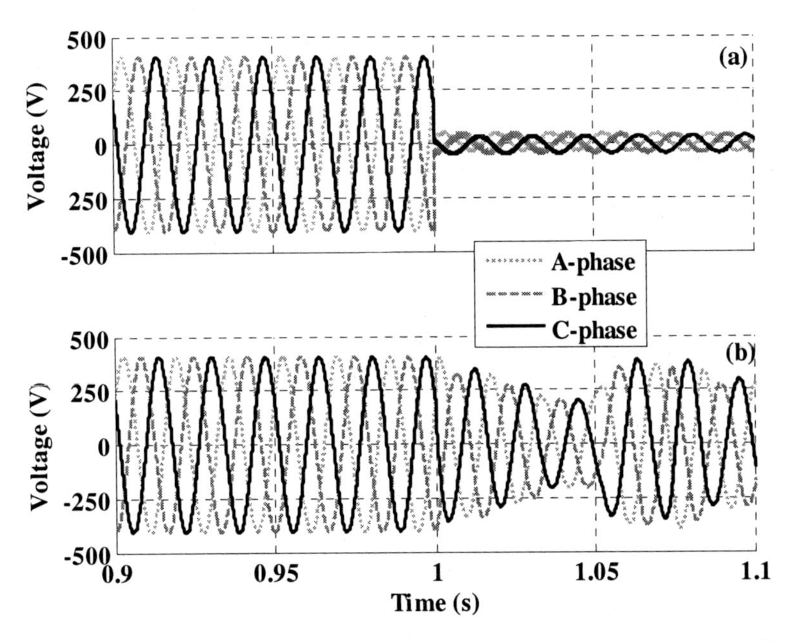

Figure 4: Voltage curves of the PV system. (a) without SFCL, (b) with the SFCL.

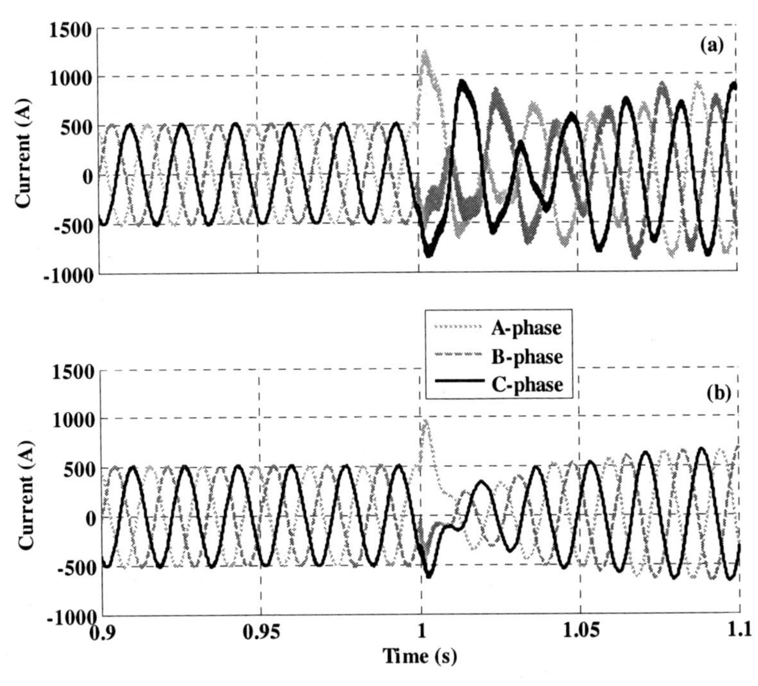

Figure 5: Current curves of the PV system. (a) without SFCL, (b) with the SFCL.

4 Conclusions

In this paper, aiming at the application of an improved flux-coupling type SFCL in the grid interfacing photovoltaic system, related theoretical analysis and simulation verification are conducted. From the results, the SFCL's positive effects on enhancing the PV system's FRT capability can be confirmed. In the near future, a real prototype of the SFCL will be made, and the experimental results will be reported in later articles.

Acknowledgment

The research work was supported in part by the Wuhan Planning Projects of Science and Technology (2013072304010827, 2013072304020824), Fundamental Research Funds for the Central Universities (2042014kf0011), and Natural Science Foundation of Hubei Province of China (2014CFB706).

References

[1] Jaume Miret, Miguel Castilla, Antonio Camacho, Lu´ıs Garc´ıa de Vicu˜na, Jos´e Matas, Control scheme for photovoltaic three-phase inverters to minimize peak currents during unbalanced grid-voltage sags. *IEEE Transactions on Power Electronics*, **27(10)**, pp. 4262–4271, 2012.
[2] C. Liu, K.T. Chau, X. Zhang, An efficient wind-photovoltaic hybrid generation system using doubly excited permanent-magnet brushless machine. *IEEE Transactions on Industrial Electronics*, **57(3)**, pp. 831–839, 2010.
[3] Hussam Alatrash, AdjeMensah, EvlynMark, Ghaith Haddad, Johan Enslin, Generator emulation controls for photovoltaic inverters. *IEEE Transactions on Smart Grid*, **3(2)**, pp. 996–1011, 2012.
[4] Kim, S.-T., Kang, B.-K., Bae, S.-H., Park, J.-W., Application of SMES and grid code compliance to wind/photovoltaic generation system. *IEEE Transactions on Applied Superconductivity*, **23(3)**, p. 5000804, 2013.
[5] Mohamed Shawky El Moursi, Weidong Xiao, Jim L. Kirtley Jr., Fault ride through capability for grid interfacing large scale PV power plants. *IET Generation, Transmission & Distribution*, **7(9)**, pp. 1027–1036, 2013.
[6] Lei Chen, Yuejin Tang, Zhi Li, Li Ren, Jing Shi, Shijie Cheng, Current limiting characteristics of a novel flux-coupling type superconducting fault current limiter. *IEEE Transactions on Applied Superconductivity*, **20(3)**, pp. 1143–1146, 2010.
[7] Li Ren, Yuejin Tang, Zhi Li, Lei Chen, Jing Shi, Fengshun Jiao, Jingdong Li, Techno-economic evaluation of a novel flux-coupling type superconducting fault current limiter. *IEEE Transactions on Applied Superconductivity*, **20(3)**, pp. 1242–1245, 2010.
[8] L. Chen, Y.J. Tang, J. Shi, N. Chen, M. Song, S.J. Cheng, Y. Hu, X.S. Chen, Influence of a voltage compensation type active superconducting

fault current limiter on the transient stability of power system. *Physica C*, **469**, pp. 1760–1764, 2009.

[9] Majid Taghizadehl, Javad Sadehl, Ebadollah Kamyab, Protection of grid connected photovoltaic system during voltage sag. In: *The International Conference on Advanced Power System Automation and Protection*, pp. 2030–2035, 2011.

Detection and parametric inversion of rough surface shallow buried target based on support vector machine

Qiyuan Zou, Qinghe Zhang, Fei Xu
College of Science, Three Gorges University, Yichang, China

Abstract

We studied the inverse scattering problem of 2-D metal target below a rough surface in this paper. Random rough surface is generated through Monte Carlo method. Relative parameters are estimated by means of a regression technique based on the support vector machine (SVM). We used the radar cross-section (RCS) as feature value, after a proper training procedure, the proposed method is able to reconstruct the geometric parameters of the buried target. Numerical results are provided for the validation of the approach.
Keywords: rough surface, buried target, support vector machine (SVM), parametric inversion.

1 Introduction

Inverse scattering is one of the most challenging problems in electromagnetic field due to its nonlinear and probabilistic characteristics. However, it is still gaining a considerable amount of attention because of its numerous applications in civil areas as well as military areas, a lot of numerical analysis methods are used to deal with inverse scattering problem.

Recent years, artificial intelligence technology like artificial neural network (ANN) is widely used in inverse scattering research [1, 2], such as half space buried target, conductive target. Since ANN is based on empirical risk minimization criterion, its generalization ability is poor for untrained samples, it has local optimum and too much learning problem as well. With the development of research, new artificial intelligence technology-support vector machines (SVM) is applied to electromagnetic inverse scattering. SVM is based on the structural risk minimization principle and statistical learning theory, the solution of which is

WIT Transactions on Engineering Sciences, Vol. 107, © 2015 WIT Press
www.witpress.com, ISSN 1743-3533 (on-line)
doi:10.2495/SEEEII140581

global optimal. With FEM, FDTD served as forward algorithm, some scholars used SVM to deal with inverse scattering problem of half space buried target and dielectric cylinder [3, 4] in frequency domain.

In this paper, we performed parametric inversion of the target below rough surface through electromagnetic simulation. Scattering field is calculated by method of moment (MoM), with RCS served as the input of SVM and geometric parameters as the output, we estimated the electromagnetic inverse scattering network to conduct parametric inversion. During numerical simulation, different observations were set to gain samples, we analyze the influence on the inversion result of different samples, achieve the estimation of buried target radius and depth on condition of more observation point detection.

2 Scattering property of rough surface buried target

The problem is shown in Figure 1. Random rough surface generated by Monte Carlo method [5] divides the region into two parts (air–soil). The air part is free space and the soil part is filled by dielectric medium, 2-D metal cylinder is buried in soil part. A tapered wave with HH polarized uniform illuminating the air–soil interface in the x–z plane. Parameters of different area are denoted with different subscript. Unit normal vector of surface and target are n_s, n_o while time-harmonic factor $e^{-j\omega t}$ is dropped. Under the illumination of the incident wave, there will be induced current J_s and magnetic current K_s on the surface and induced current J_o on the target. Field equations including integration of both induced current and magnetic current, both surface inside and outside and the metal surface should be considered to obtain the surface integral equations [6].

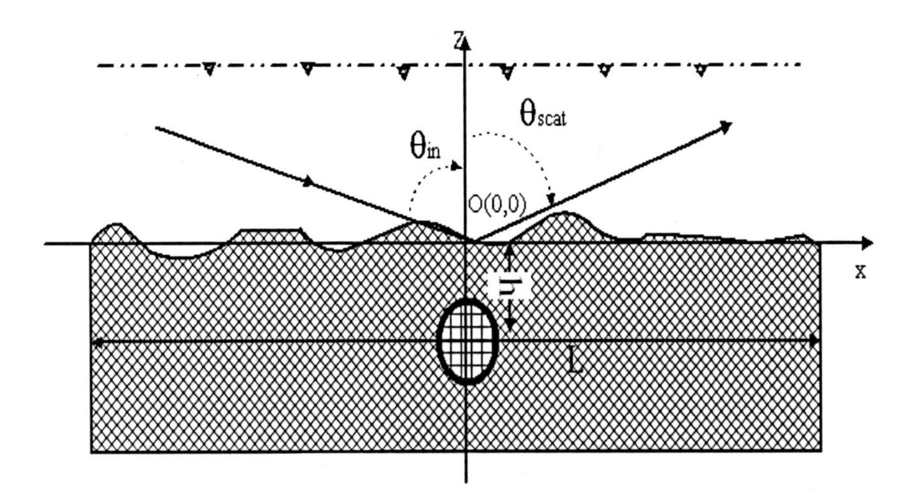

Figure 1: Random rough surface generated by Monte Carlo method.

Radiation field produced by J_s, K_s, J_o as follows:

$$\begin{aligned} E_{s1} &= ik_1\eta_1 A_1 + \frac{\eta_1}{k_1}\nabla(\nabla\cdot A_1) - \nabla\times F_1 \\ E_{s2} &= ik_2\eta_2 A_2 + \frac{\eta_2}{k_2}\nabla(\nabla\cdot A_2) - \nabla\times F_2 \\ E_{so} &= ik_2\eta_2 A_{2o} \end{aligned} \tag{1}$$

Among them, vector potential A and F represent the convolution of the source current and the Green function. Apply the equivalence principle at the surface and the target, get the equations:

$$\begin{aligned} \bar{K}_1 &= (E_{s1} + E_i)\times\hat{n}_s \\ \bar{K}_2 &= (E_{s2} + E_{so})\times(-\hat{n}_s) \\ \hat{n}_o &\times(E_{s2} + E_{so}) = 0 \end{aligned} \tag{2}$$

In combination with Eq. (1), Eq. (2) can be transformed into Eq. (3):

$$\begin{aligned} & K_t(\bar{r}) - ik_1\eta_1 A_{t1}(\bar{r}) + \left[\frac{\partial F_{x1}(\bar{r})}{\partial z} - \frac{\partial F_{z1}(\bar{r})}{\partial x}\right] = E_i(\bar{r}) \ \ (\bar{r}\in f^+) \\ & K_t(\bar{r}) + ik_2\eta_2 A_{t2}(\bar{r}) + \left[\frac{\partial F_{x2}(\bar{r})}{\partial z} - \frac{\partial F_{z2}(\bar{r})}{\partial x}\right] = 0 \qquad (\bar{r}\in f^-) \\ & ik_2\eta_2 A_{t2o}(\bar{r}) + ik_2\eta_2 A_{t2}(\bar{r}) + \left[\frac{\partial F_{x2}(\bar{r})}{\partial z} - \frac{\partial F_{z2}(\bar{r})}{\partial x}\right] = 0\ (\bar{r}\in o) \end{aligned} \tag{3}$$

the first two of Eq. (3) are integral equations about rough surface (about J_s, K_s) and last one is equation about target surface (about J_o).

Subdivide the rough surface and target surface, J_s, K_s, J_o denoted with discrete pulse basis function. We transformed Eq. (3) into matrix equation by point matching method.

$$\begin{bmatrix} \overline{\overline{Z}}_J^{ss} & \overline{\overline{Z}}_K^{ss} & 0 \\ \overline{\overline{Z}}_J^{ss} & \overline{\overline{Z}}_K^{ss} & \overline{\overline{Z}}_{J_o}^{os} \\ \overline{\overline{Z}}_J^{so} & \overline{\overline{Z}}_K^{so} & \overline{\overline{Z}}_{J_o}^{oo} \end{bmatrix} \begin{bmatrix} \bar{J}_s \\ \bar{K}_s \\ \bar{J}_o \end{bmatrix} = \begin{bmatrix} \bar{E}_i \\ 0 \\ 0 \end{bmatrix} \tag{4}$$

Superscripts of the block matrix in the equation indicate the relationship between the resource and field, while S denotes rough surface and OS denotes the target respectively. To solve Eq. (4) we can get current value, then it is possible for us to calculate the scattering field according to the current value. If incident wave is a tapered wave uniform, the RCS can be calculated as:

$$\sigma(\theta_s) = \frac{\left|\varphi_s^N(\theta_s)\right|^2}{g\sqrt{\frac{\pi}{2}}\cos\theta_i[1 - \frac{1+2\tan^2\theta_i}{2(k_1 g\cos\theta_i)^2}]} \tag{5}$$

3 SVM theory used for regression estimation

Basic principle of support vector regression [7] can be stated as follows. Given training samples $(\{X_i, y_i\}; i = 1, \cdots n; X_i \in R^n; y_i \in R)$. X_i is an n-dimension vector which is called Eigen value, used as the input of the SVM (RCS in this paper). y_i is called target value that used as output (size and depth of target in this paper). The goal is to find a function F that have a maximum fixed error ε for all the samples belonging to the training set and that is as smooth as possible, in order to reduce the effects on the estimated values due to the perturbation of the input data. In the SVM regression, with help of Lagrange multiplier method, the unknown function is constructed by linearly combining the results of a nonlinear transformation of the input samples:

$$F(X) = \sum_{i=1}^{N} (\alpha_i - \alpha_i^*) K(X_i, X) \tag{6}$$

where N is the number of support vector, α_i, α_i^* are unknown quantities called Lagrange factor and must be chosen to minimize the distance between the values predicted by the function and the known samples. Nonlinear transformation function K is a kernel function, in same literature, several kernel functions have been considered such as polynomial kernels, radial kernel function (RBF) [8].

4 Numerical results and analysis

4.1 Numerical simulation of the rough surface buried target scattering field

In order to assess the effectiveness of the SVM approach, several numerical simulations have been performed. Frequency of the incident wave f = 300MHz, incident angle $\theta_i = 30°$, length of the Gaussian rough surface $L = 10\lambda$, region below the rough surface is filled by dielectric substance of known dielectric property ($\varepsilon_2 = 2 - 0.2j$ $\mu_2 = \mu_0$), while μ_0 is the magnetic permeability of the vacuum.

Tapered plane wave [9] is used as incident wave to eliminate the edge effect caused by interception of limited size of the rough surface.

$$E_i = \exp\left[ik_0(x\sin\theta_i - z\cos\theta_i)(1+w)\right]\exp(-t) \tag{7}$$

while

$$t = \frac{(x + z\tan\theta_i)^2}{g^2},\quad w = \frac{2t-1}{(k_0 g\cos\theta_i)^2},\quad g = \frac{L}{4} \tag{8}$$

Figure 2 shows the variation of the double radio scattering cross section (RCS) when the target radius changes. As we can see from Figure 2, under the condition of other parameters remain unchanged, RCS turned up a peak value to the direction of the mirror and the peak value did not change distinctly as the target radius changed. But RCS increased distinctly as the radius increased at other directions. Reason of this phenomenon is that the interaction between the rough surface and the buried target enhances when the target radius increases.

Figure 3 shows the RCS for different target depth. We can also find that RCS turned up a peak value to the direction of the mirror and the peak value did not

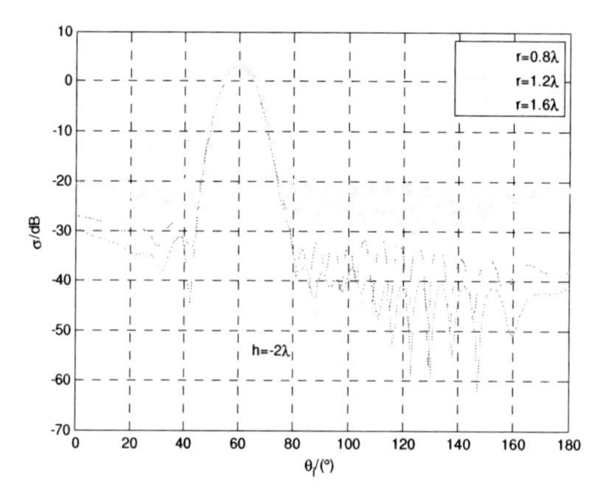

Figure 2: RCS for different target radius.

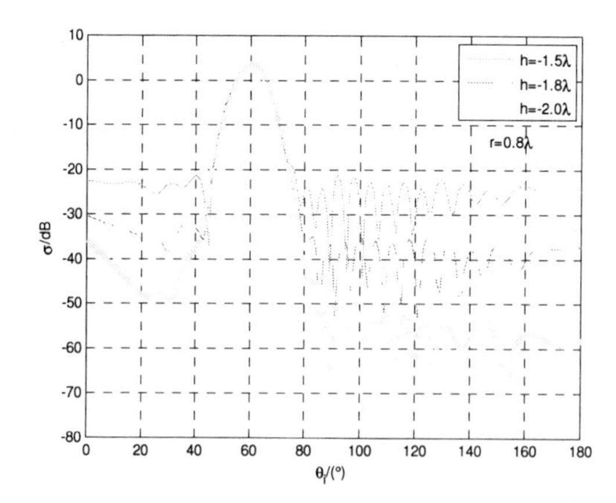

Figure 3: RCS for different target depth.

change distinctly as the target depth changed. But RCS decreased distinctly as the depth increased at other directions. This is mainly because the interaction between the rough surface and the buried target weaken when the target depth increase.

4.2 Detection and parametric inversion of target beneath rough surface

Five observation points of different direction ($\theta_s = i\pi/6$; I = 1, 2, 3, 4, 5) above rough surface were used to measure the scattering field. After gaining scattering data, we conducted parametric inversion with the help of SVM. We calculate 50 different rough surfaces to avoid the randomness of the rough surface. Training samples and testing samples are represented as (i) and (ii):

(i) $r_i = 0.8\lambda + (i-1)\times 0.02\lambda, i = 1,2\cdots 10$
$h_i = -1.2\lambda + (i-1)\times 0.06\lambda, i = 1,2\cdots 10$

(ii) $r_i = 0.802\lambda + (i-1)\times 0.02\lambda, i = 1,2\cdots 9$
$h_i = -1.206\lambda + (i-1)\times 0.06\lambda, i = 1,2\cdots 9$

As shown in Figures 4 and 5, correlation coefficient (R^2) between the retrieved and reference of the target parameters are 0.99, very close to 1. Our retrieval result is very encouraging. Compare root-mean-square error (RMSE) in two figures, Figure 4 is better than 5, this suggests that inversion of the radius is better than that of depth in selected data range. In addition, inversion result of depth from −1.7 to −1.6 in Figure 5 is scattered, this is probably because the RCS fluctuates acutely when the depth reaches -1.6λ .

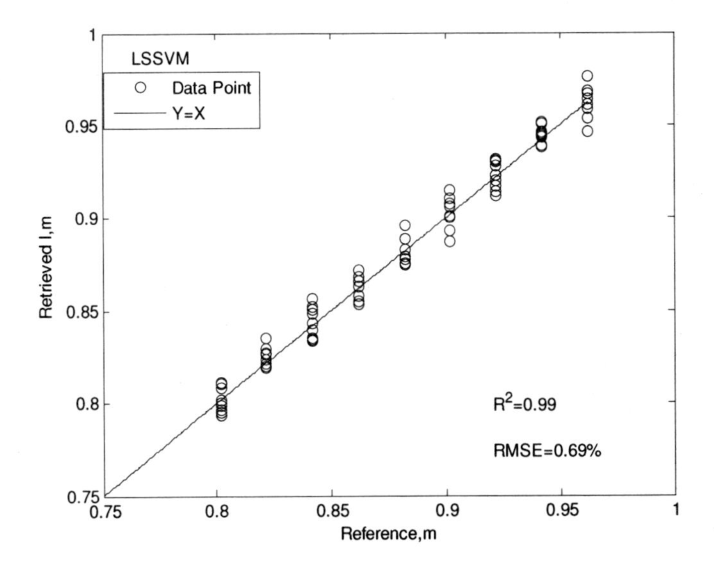

Figure 4: Inversion of the target radius.

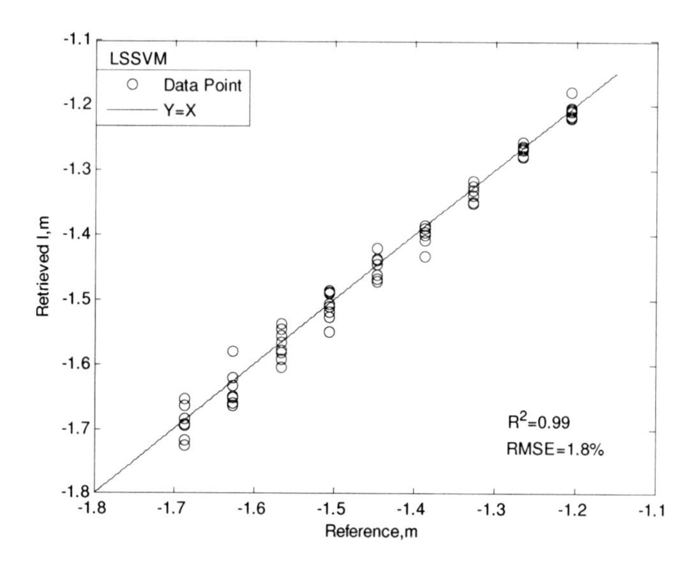

Figure 5: Inversion of the target depth.

5 Conclusions

In the proposed passage, we calculated the RCS of a 2-dimension metal target beneath a dielectric rough surface by method of moment (MoM) and discussed the effect of the target radius and depth on the scattering field. Then with help of SVM, we studied the inverse scattering problem of the buried target. RCS of the target gained at different observations were used as the input of SVM, target radius and depth as the output, we achieved the estimation of the target parameters. Accuracy of the estimation also proves the validity of the algorithm we used to calculate the scattering field.

Acknowledgment

The research work was supported by National Natural Science Foundation of China under Grant No. 61179025 and Natural Science Foundation of Hubei Provincial under Grant No. D20111201.

References

[1] Mydur R, Michalski KA, A neural-network approach to the electromagnetic imaging of elliptic conducting cylinder. *Microwave and Optical Technology Letters*, **28(5)**, pp. 303–306, 2001.
[2] Rekanos IT, On-line inverse scattering of conduction cylinders using radial basis-function neural networks. *Microwave and Optical Technology Letters*, **28(6)**, pp. 378–380, 2001.

[3] Zhang QH, Xiao BX, Zhu GQ, A new solution of real-time electromagnetic inverse scattering. *Chinese Journal of Geophysics*, **49(5)**, pp. 1546–1551, 2006 (in Chinese).
[4] Zhang QH, Inverse scattering by dielectric circular cylinder based on BP neural networks. *Chinese Journal of Radio Science*, **25**, pp. 398–402, 2010.
[5] Leung Tsang, Chi H. Chan, Kyung Pak, Haresh Sangani, Monte-Carlo simulations of large-scale problems of random rough surface scattering and applications to grazing incidence with the BMIA/canonical grid method. *IEEE Transactions on Antennas and Propagation*, **43(8)**, August 1995.
[6] Peterson AF, Ray SL, Mittra R, *Computational Methods for Electromagnetic*. IEEE Press, New York, NY, pp. 37–94, 1997.
[7] Vapnik V, Golowich S, Smola AJ, Support vector method for function approximation, regression estimation, and signal processing. *Advance in Neural Information Processing Systems*, **9**, pp. 281–287, 1996.
[8] Smola AJ, Scholkopf B, *A Tutorial on Support Vector Regression*. Royal Holloway College Press, London, 1998.
[9] Tsang L, Kong J, Ding KH, *Scattering of Electromagnetic Waves: Numerical Simulations*. Wily-Interscience, New York, NY, pp. 118–124, 2001.

Analysis for DC bias of risk and its suppress measures in Sichuan power grid

Wei Wei, Xiaobin Liang, Wei Zhen, Cangyang Chen
Sichuan Electric Power Research Institute, China

Abstract

At first the analysis of suppress measures for the main transformer including adding blocking capacitance at neutral point, current-limiting resistor, reverse flow-injection equipment are given, followed by the comparison of three method to suppress the DC bias and practical range. Based on the damage of DC bias caused by Yi bin-Jin hua UHV direct current system, this paper has the discussion on the feasibility and impact of adding blocking capacitance to suppress the DC bias taken by Fang-Mountain power plant. In this part, combined blocking capacitance configuration parameters provided by manufacturers, do analyze the reliability of inhibiting biasing current visa adding blocking capacitance at neutral point which measure taken by Fang-Mountain power plant at main transformer, as well as the effects on the surrounding station. And then in electromagnetic and electromechanical transient two simulation environment, analyze short circuit current flows through Fang-Mountain power plant's two main transformer neutral point grounding branch with 500kV/220kV side bus in single-phase fault mode, two-phase short circuit to ground fault mode, and check whether the related electrical equipment can meet the operational requirements. In the later part, the impact of Fang-Mountain power plant main transformer adding blocking capacitance on transmission line protection, protection of main transformer, generator protection are taken in analysis. Results showed that, to ground potential in biasing environment is not at the point of non-extreme points-site, adding blocking capacitance is effective for suppressing native bias current, without deteriorating risk of DC bias in around-site.

Keywords: DC bias, blocking device, short-circuit current, relay protection.

WIT Transactions on Engineering Sciences, Vol. 107, © 2015 WIT Press
www.witpress.com, ISSN 1743-3533 (on-line)
doi:10.2495/SEEEII140591

1 Introduction

This paper firstly analyzes the suppress measures for the main transformer including adding blocking capacitance at neutral point, resistor, reverse flow-injection equipment, followed by the comparison of three methods to suppress the DC bias and practical range [1-4]. Based on the damage of DC bias caused by Yi bin-Jin hua UHV direct current system, this paper has the discussion on the feasibility and impact of adding blocking capacitance to suppress the DC bias taken by Fang-Mountain power plant. In this part, combined blocking capacitance configuration parameters provided by manufacturers, do analyze the reliability of inhibiting biasing current visa adding blocking capacitance at neutral point which measure taken by Fang-Mountain power plant at main transformer, as well as the effects on the surrounding station. And then in electromagnetic and electromechanical transient two simulation environment, analyze short circuit current flows through Fang-Mountain power plant's two main transformer neutral point grounding branch with 500kV/220kV side bus in single-phase fault mode, two-phase short circuit to ground fault mode, and check whether the related electrical equipment can meet the operational requirements. In the later part, the impact of Fang-Mountain power plant main transformer adding blocking capacitance on transmission line protection, protection of main transformer, generator protection are taken in analysis. Results showed that, to ground potential in biasing environment is not at the point of non-extreme points-site, adding blocking capacitance is effective for suppressing native bias current, without deteriorating risk of DC bias in around-site. Research results will do good to mitigate risk of DC bias in Sichuan province, and also can improve the safe and stable operation of margin of power grid [5-6].

2 Comparison and analysis of measures suppressing DC bias

At present, there are three extensive DC bias suppression measures including blocking capacitance, current limiting resistor, and reverse flow-injection equipment. This section will introduce the basic principle of the three methods, followed by comparison of various methods' advantages and disadvantages, also the application scopes are given.

2.1 Blocking capacitance

Schematic diagram of the device as shown in Figure 1, the device body is formed by capacitor bank, mechanical bypass switch and rectifier bypass loop system in parallel, and it is connected between the transformer neutral point and the ground. In the case of no DC current through transformer neutral point, mechanical bypass switch K3 is in the closed position and the neutral point of transformer is in the metallic ground state. When the detected DC current through transformer neutral point exceeds the limit, the mechanical bypass switch K3 is turned to off position which makes the capacitor group get into the circuit to block DC current.

Bias current can be completely eliminated at the installation site by adding blocking capacitance. Furthermore, as the installed capacitor reactance on neutral

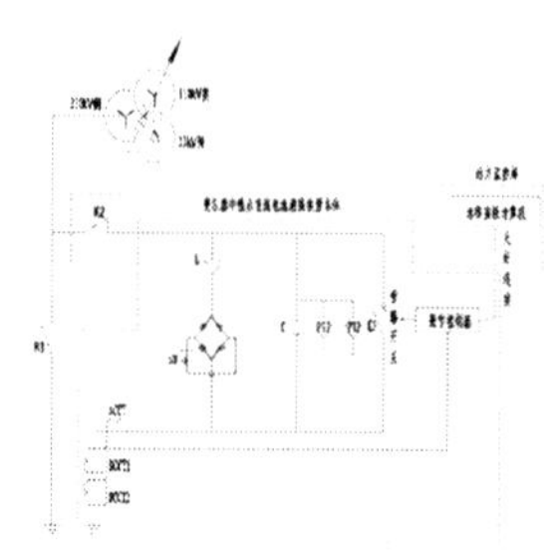

Figure 1: Schematic of DC blocking device installing.

point is often very small, and at the same time the neutral point can be grounding reliably, system grounding impedance had no effect on operation of relay protection which has been put into operating and the no-load transformer. The insulating property is not strict and then there is no need to take special measures to strengthen insulating property. For these reasons, the cost of installation and equipment t is relatively low.

2.2 Current – limiting resistor

In DC power transmission system, the earth is often used to be the return circuit. If AC system of the transformer neutral grounding point does also exist, there will be a DC current in the transformer of the transmission line flow. The value of the current is related to the potential difference between the grounding grids for transformer. And the current will cause the transformer core to be asymmetric saturation. Low frequency noise will be resulting in, and even can cause loss of transformer core, fuel tank and internal structural members increasing which will lead to overheat. Current return through the neutral will also cause electrochemical corrosion in the vicinity which will have negative effects on relay protection device.

2.3 Reverse flow injection device

Device assembly and connection as shown in Figure 2. DC-reverse injection equipment grounding terminal is the compensation grounding electrode arranged at the outside independently, and output terminal of DC generator is connected with current limiting reactor. The other end of current limiting reactor is connected with the transformer neutral point which can be used to limit the AC current into DC generator. The DC current monitor and control devices of transformer neutral point is designed to measure DC current through transformer neutral point, and then according to the DC current value and direction, DC generating device will start to generate the corresponding reverse DC current with correct direction and value automatically. By the way we can achieve the purpose of limiting transformer neutral DC current. The device is very flexible which can generate different direction and value injection current dynamically according to the DC current through neutral point. But the device is complex and

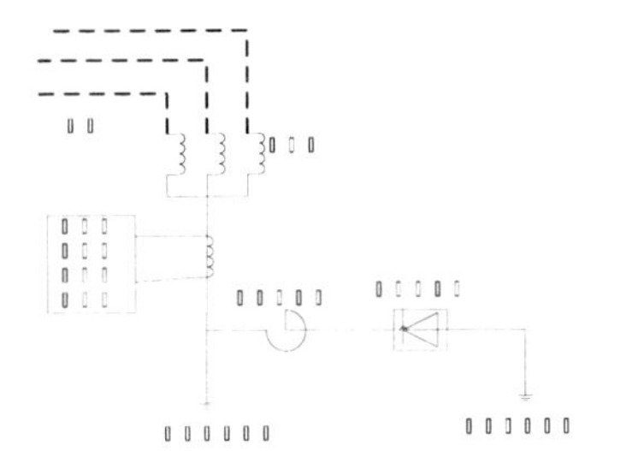

Figure 2: Method of injecting reverse current at neutral point.

expensive, also the feasibility of the system need to be verified. In addition, injecting current should consider the influence of branch current generating by grounding-grid, substation lightning line and other equipment. The reasonable control of the compensation capacity is the difficulty in this method. The method is applicable to DC smaller occasions. As its compensation adjustment control process is complex, overcompensation current should be avoided.

3 Feasibility analysis of using blocking capacitance to suppress DC bias in Fang-Mountain power plant

According to the operating arrangement, there is a transformer of neutral point grounding operation in each No. I and No. bus II of Fang-Mountain power plant. There are three transformers of 220kV in Fang-Mountain power plant. The standby transformer is in neutral point grounding operation, and one of other two main transformers is in neutral grounding operation. According to the above operation mode, one-for-two structure of DC blocking device is used at neutral point of No. 1 and No. 2 main transformers in which DC blocking device can be used for No. 1 main transformer neutral point grounding device, also can be connected to the No. 2 main transformer neutral point for grounding, but at the same time only one is allowed main to make use of the device to ground. Schematic diagram of one-for-two structure as shown in Figure 1.

3.1 Feasibility of suppression for DC bias

Network topological graph near Fang-Mountain power plant is shown in Figure 3. In normal operating mode, main transformer of Fang-Mountain power plant is in single neutral grounding operation mode. With DC blocking device, main transformer of Fang-Mountain power plant will affect Luzhou power station and the surrounding station. DC magnetic bias model of Yibin and Luzhou as the simulating object, the effect of current value and direction through station nearby caused by adding blocking device in Fang-Mountain with case which the DC grounding current is 5000A. The simulation results are shown in Table 1.

The start-up standby transformer of Fang-Mountain power plant is in neutral grounding mode. After adding blocking device, DC current through neutral point

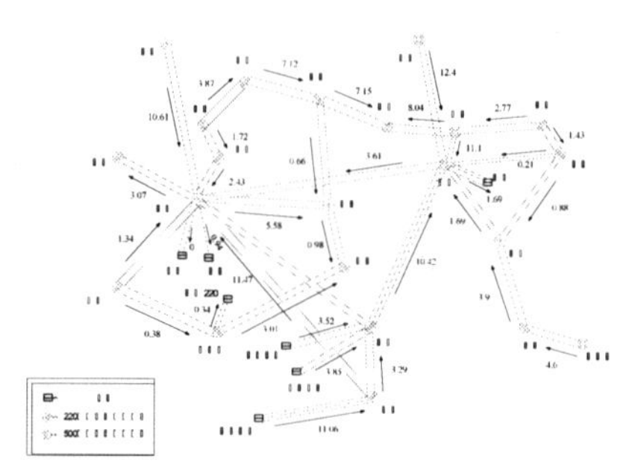

Figure 3: Power wiring diagram near Fang-Mountain power plant.

Table 1: The value of DC current in observation stations before and after installing blocking device.

Station	Before Installation	After Installation
Luzhou	−8.38	−9.98
Xufu	−4.78	−4.84
Naxi	−5.05	−6.07
Yangqiao	0.97	0.85
Yuguan	−0.95	−1.31
Linguang	4.56	6.27
Gaoshi	−2.1	−1.56
Fang-Mountain	−2.81	0.00

of single main transformer in Luzhou station increases 1.6A. Also DC current through neutral point of main transformer in Xufu, Naxi, Yuguang, and Linzhuang station increases, but the increasing amount is relatively average and there is no station's current exceeding standard. On the other hand, Yang qiao, Gaoshi station main transformer's DC current through neutral point decreases slightly.

The following neutral point of single main transformer in Fang-Mountain power plant is also in neutral grounding mode. The result of simulation analysis for DC current in observation stations before and after installing blocking device is shown in Table 2:

The main transformer of Fang-Mountain power plant is in neutral grounding mode. After adding blocking device, DC current through neutral point of main transformer in Luzhou station increases 6.28A. Also DC current through neutral point of main transformer in Xufu, Naxi, Yuguang, and Linzhuang station increases. On the other hand, Yang qiao, Gaoshi station main transformer's DC current through neutral point decreases slightly.

As can be seen from above simulation analysis, Fang-mountain power plant after adding blocking device can completely inhibit DC bias risk. Although the bias current will be transferred to other nearby stations, the amount of DC current transfer is relatively small which will not cause run the risk of other stations.

Table 2: The value of DC current in observation stations before and after installing blocking device in single main transformer.

Station	Before Installation	After Installation
Luzhou	−3.7	−9.98
Xufu	−4.17	−4.84
Naxi	−4.42	−6.07
Yangqiao	1.01	0.85
Yuguan	−1.77	−2.31
Linguang	3.22	6.27
Gaoshi	−1.75	−1.56
Fang-Mountain	−18.61	0.00

With the analysis of the reasons we can see that before adding blocking capacitor, ground potential of Fang-Mountain power plant is the midpoint for all stations. And then when Fang-Mountain power plant is in DC-blocked situation, there is no big change in the distribution of ground potential which will lead the uniform distribution. However, if ground potential of Fang-Mountain power plant is in extreme position (highest or lowest), adding blocking devices will make other stations' neutral bias current increase significantly. For this reason, there is a need to adding blocking device in large area which will cause larger investment of reliability and devices. By comparing Tables 1 and 2, as leakage reactance of start-up standby transformer is larger than main transformer, the application of start-up standby transformer can increase zero-sequence impedance and reduce the main transformer's neutral bias current. And therefore bias current of stations without blocking device can be suppressed by temporary application of start-up standby transformer.

3.2 Feasibility analysis of using resistor in Fang-Mountain power plant

Figure 4 shows current and voltage of these two main transformers' neutral point with blocking device.

It can be seen when Lu Hung 500kV line occurs two-phase ground short circuit with DC isolation device, there will be a peak voltage of 100V at No. 1 main transformer neutral point and the peak current is about 2.05kA without big changes. For No. 2 main transformer, the neutral point voltage peak maintenance at 7.4kV not exceeding safe operating range.

According to the network topology of Sichuan power grid, this section will study the feasible range of blocking capacitor, and presents the change trend of neutral point's current and voltage within the fault mode as blocking capacitor changes. By the simulation calculation, short-circuit current of 220kV bus in single-phase short circuit mode and neutral point voltage changes as shown in Figure 5.

We can see from Figure 5, as long as the guarantee of the blocking capacitor is not less than 1415μF, Fang-Mountain power plant 220kV bus single-phase short circuit will not damage No. 1 and No. 2 main transformer neutral point insulation level.

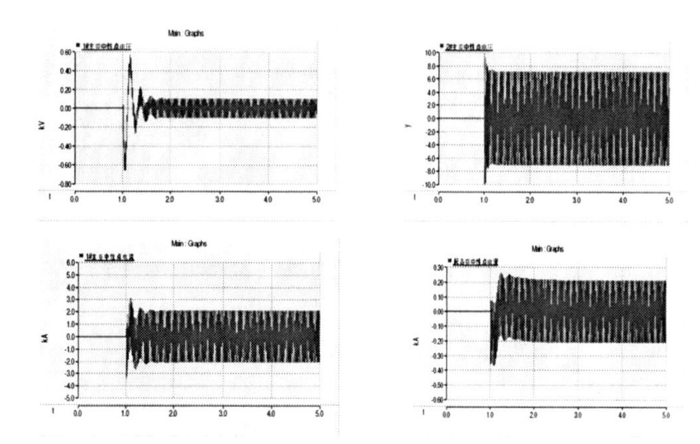

Figure 4: Simulation results with blocking device. (a) neutral point voltage of No. 1 and No. 2 main transformer. (b) neutral point current of No. 1 and kai standby transformer.

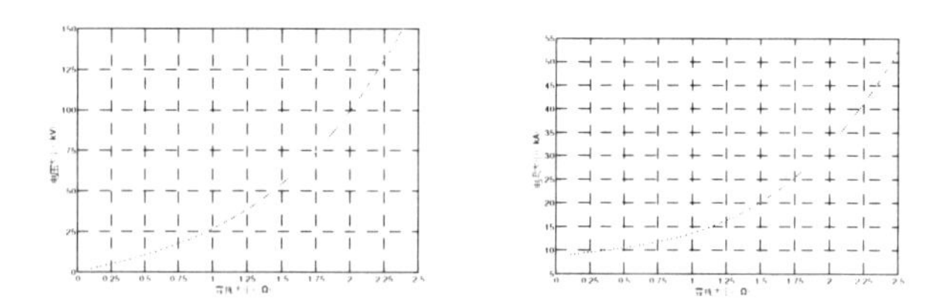

Figure 5: Short-circuit current of 220kV bus in single-phase short circuit mode and neutral point voltage changes with blocking capacitor.

3.3 Impact of Fangshan transformer adding blocking capacitance on relay protection

(1) Phase-to-phase power line distance protection

Due to that the change of main transformer neutral point grounding method does not have an impact on sequence diagram of two-phase short circuit and three-phase short circuit, so the change of main transformer grounding method does not have an impact on phase-to-phase power line distance protection.

(2) Grounding distance protection

According to the fault voltage and fault current before and after the Fang-Mountain power plant main transformer adding blocking capacitance, the change of fault current in different fault point ,when single phase short circuit to grounding fault occurring at Fang-Mountain power plant 220kV bus , fault current changes the biggest. Taking the protection method of Fang-Mountain side of Fang-Mountain-to Luzhou line for example, the incensement of the short circuit current of fault phase and zero sequence current improves the sensitivity of the protection method. As a result, the reliability of the reaction is significantly enhanced. And taking the protection method of Luzhou side of Fang-Mountain to

Luzhou line for example, before the installation of capacitor in the main transformer neutral point, the distance between Luzhou and measurement fault point of the grounded distance protection is 2.0639Ω. After the installation, the measured value is about 2.0643Ω, only changed by 0.02% compared with the original value. So the change of main transformer grounding method practically nearly has no impact on the grounding distance protection of line protection. Note: Luzhou side:grounding distance I, 1.3 Ω; grounding distance II, 6.8Ω; grounding distance III, 85 Ω.

(3) Zero sequence current protection
Because the change of Fang-Mountain power plant main transformer neutral point grounding method means the change of zero sequence equivalent network. As a result, zero sequence current and zero sequence impedance must change when the grounding fault occurred. But due to the small capacitive reactance of capacitor that installing at the main transformer, the change of zero sequence impedance and zero sequence current must be small. Considering different point and method of fault, when two phase grounding fault occurring at the point of Fang-Mountain 220kV bus, zero sequence current get the biggest. According to the calculated result, the value of zero sequence current at the short circuit point before main transformer neutral point installing capacitor is 13840 A. After installing the capacitor, the value is 13859 A and it has increased 0.14% than before. For zero sequence current protection, the incensement of zero sequence current during the fault improves the sensitivity of the protection. And as a result, the reliability of the reaction is significantly enhanced. But due to the great impact on zero sequence equivalent network of plan B, the original protective set value need to be checked out in order to make sure that the protective operation is correct.

(4) Analysis of effect on main transformer protection
Due to that the capacitor is installed in the neutral point of the Fang-Mountain power plant, the protection for the transformers on the Fang-Mountain side, for the lines between Fang Mountain and Luzhou, for the 220kV lines nearby between Luzhou and Gaoshi, and for the 500kV lines between Luzhou and Honggou all should be taken into consideration. There are three kinds of fault: single phase grounding short circuit, two phase grounding short circuit, and three-phase grounding short circuit. Only the computation results will be analyzed here while three kinds of fault, the specific computation results will not be listed, only the computation results will be analyzed here.

(5) Main transformer differential protection
For main transformer differential protection, the change of main transformer neutral point grounding method has no impact on the relationship of value between differential current and brake current in differential protection. And it has no impact on differential protection.

(6) Main transformer negative sequence current protection
As the change of main transformer neutral point grounding method has no impact on positive sequence equivalent network and negative sequence equivalent network, it will not affect negative sequence overcurrent protection.

(7) Zero sequence overcurrent protection
Due to that the zero sequence protection is based on main transformer neutral point current, the change of main transformer neutral point grounding method must have a direct impact on zero sequence overcurrent protection. According to the calculated result, take the inside fault of main transformer as an example, the value of zero sequence still gets the biggest change when two phase grounding short circuit fault occurring at the outgoing side of 220kV line Fang-Mountain power plant. The neutral point current (3I0) is 19372 A with the capacitor installed. In the case of the fault is external fault of main transformer, single phase short circuit fault occurs at Yuguan 220kV bus. The main transformer neutral point zero sequence current is 7632 A. After installing capacitor, the current is 7656 A. Because the main transformer of Fang-Mountain power plant belongs to the user, the setting value is set by user and there is no data in protect department of provincial dispatching center. According to the setting value of zero sequence current provided by Fang-Mountain power plant (zero sequence I: 4.4 kA; zero sequence II: 0.99 kA), no matter whether the capacitor is installed at the main transformer neutral point and the zero sequence is over the protective setting value.

But due to the cooperative relationship between the time setting value of main transformer protection and neighboring line protection, the change has no impact on the operation of protection.

(8) Analysis of effect on generator protection
Because the main transformer neutral grounding point is at HV side and the generator side (low voltage side) winding is delta connection, zero sequence does not circulate. As a result, there is no impact on the generator zero sequence equivalent networks. Also the main transformer neutral point grounding method has no impact on the positive sequence equivalent network and negative sequence equivalent network of generator side. So no matter whether the main transformer neutral point grounding to earth through capacitor or not will have no impact on generator protection.

4 Conclusion

At first the analysis of suppress measures for the main transformer including adding blocking capacitance at neutral point, current-limiting resistor, reverse flow-injection equipment are given, followed by the comparison of three method to suppress the DC bias and practical range. Based on the damage of DC bias caused by Yibin-Jinhua UHV direct current system, this paper have the discussion on the feasibility and impact of adding blocking capacitance to suppress the DC bias taken by Fang-Mountain power plant. In this part, combined blocking capacitance configuration parameters provided by manufacturers, do analyze the reliability of inhibiting biasing current visa adding blocking capacitance at neutral point which measure taken by Fang-Mountain power plant at main transformer, as well as the effects on the surrounding station. And then in electromagnetic and electromechanical transient two simulation environment, analyze short circuit current flows through Fang-Mountain power plant's two main transformer neutral point grounding branch with 500kV/220kV side bus in single-phase fault mode, two-phase short circuit to ground fault mode, and check whether the related electrical equipment can meet the operational requirements. In the later part, the impact of Fang-Mountain power plant main transformer adding blocking capacitance on transmission line protection, protection of main transformer, generator protection are taken in analysis. Results showed that, to ground potential in biasing environment is not at the point of non-extreme points-site, adding blocking capacitance is effective for suppressing native bias current, without deteriorating risk of DC bias in around stations.

References

[1] Xue Xiangdang, Guo Hui, Zheng Yunxiang, The study of harmful effects of geomagnetically induced current on power transformers. In: *Proceedings of the EPSA*, **11(2)**, pp. 13–19, 1999.

[2] Zhong Lianhong, Lu Peijun, Qiu Zhicheng, The influence of current of Dc earthing electrode on directly grounded transformer. *High Voltage Engineering*, **29(8)**, pp. 12–17, 2003.

[3] Zhang Yanbing, *Research on Algorithm and Simulation of Power Grid Geomagnetically Induced Currents*. North China Electric Power University, Beijing, 2005.

[4] Wang Hongliang, *Research on Ground Electrode Current Operation of HVDC Monopolar Ground Return Mode*. SouthWest JiaoTong University, Cheng Du, 2007.

[5] Li Xiaoping, Wen Xxishan, Chen Cixuan, Simulating analysis of exciting current of single phase transformer on DC bias. *High Voltage Engineering*, **11(9)**, pp. 93–99, 2005.

[6] Hao Zhiguo, Yu Yang, Zhang Baohui, Earth surface potential distribution of HVDC operation under monopole ground return mode. *Electric Power Automation Equipment*, **4(06)**, pp. 39–43, 2009.

Predicting the risk of DC-bias in Sichuan Province during ±800kV Binjin UHVDC transmission system joint debugging based on the finite element method

Wei Wei[1], Xiaobin Liang[1], Wei Zhen[1], Xiaoxu Wang[2]
[1] *Sichuan Electric Power Research Institute, Chengdu 610072, China;*
[2] *Chongqing University*

Abstract

This paper analyses the research methods of DC bias, compares the advantages and disadvantages of various domestic and international analysis method, and puts forward that the multi physics modeling method based on FEM is the most reliable and most accurate method of analyzing DC bias. This paper simulates the phenomenon of DC bias in Sichuan province during Binjin DC system joint debugging based on the finite element method. The result shows that the method's error is less than 10 percent in simulating earth potential distribution, main transformer neutral current and DC current in AC systems under the condition of ignoring the local geological structure of the local area. Through this research, we can accurately predict the risk of DC bias of multiple DC transmission lines in Sichuan province, which runs in various earth return running mode such as unipolar maintenance, unipolar latching, and bipolar asymmetry. And the study makes suppression measures as well, which plays a very important role in the safe and stable operation of power equipment.
Keywords: DC bias, FEM, multi-physics, unipolar latching.

1 Overview of FEM

Finite Element Method (FEM) originated in the mechanics calculation. Winslow, Chari, and Silvester et al. adapted it to the computation of electromagnetic field. The application has become an important turning point and has been widely used in every aspect of Electrical Engineering [1]. FEM, from the mathematical point of view, is a numerical method to approximate the solution of mathematical boundary

value problems and is the optimal approximation to the original function in solving interval under certain conditions [2]. FEM divides the closed field that belongs to the continuous function characterized by partial differential equation into a limited number of small regions, and uses the selected approximate function instead of each region to obtain the approximate numerical value of the field.

This research will use the ANSYS software developed by ANSYS Company in American to calculate and analyze the ground potential distribution and the electromagnetic characteristics of 500kV three groups of self-coupling transformer's iron core in the bias magnetic environment [3]. ANSYS software is a large-scale field analysis software based on Finite Element Theory and is widely used in engineering field. It is powerful, and can be used to calculate the electric, magnetic, thermal, power, and other fields. Besides, it has its own knack in electromagnetic field calculation. Therefore, it is reasonable to use the software to analyze the dynamic characteristics of main transformer's earth potential under DC bias conditions, the neutral point current of main transformer and dynamic characteristics under the influence of main transformer under the influence of magnetic bias.

2 The working principle and implement steps of ANSYS

The algorithm of ANSYS is based on Finite Element Method, which is used to simulate the real physical system by using mathematical approximation method [4]. By utilizing both simple and interacting elements (unit), we can use a limited number of unknown variables to approximate the real system which has unlimited number of unknown variables. When ANSYS is used in electromagnetic field analysis, the calculation parameters are usually expressed in freedom, which is known as nodal freedom. Compared with those of finite element analysis software, ANSYS has more advantages, such as a more comprehensive application, a larger computing capacity, a more convenient modelling and a more clear interface.

3 Simulate the phenomenon of DC-bias in Sichuan province during Binjin DC transmission system joint debugging based on ANSYS

3.1 Introduction of Binjin DC

Binjin DC arises from Sichuan Yibin converter station, and stops at Jinhua converter station [5]. It goes through five provinces: Sichuan, Guizhou, Hunan, Jiangxi, Zhejiang. The line length is 1653 km. The DC is an important part from west to east. The construction project for the implementation of the national strategy from West to East has great significance to meet the needs of electric power in Sichuan, and to ease the power shortage of Zhejiang. The project is designed to ±800kV bipolar, rated capacity 8000MW, planning 500kV AC outlet 8 returns. DC near zone grid is shown in Figure 1.

The statistics shows that Binjin DC has 11,500kV substations including power plants and 18,220kV substations. The straight-line distance is not more than 150km. All stations are present very close to DC bias risk.

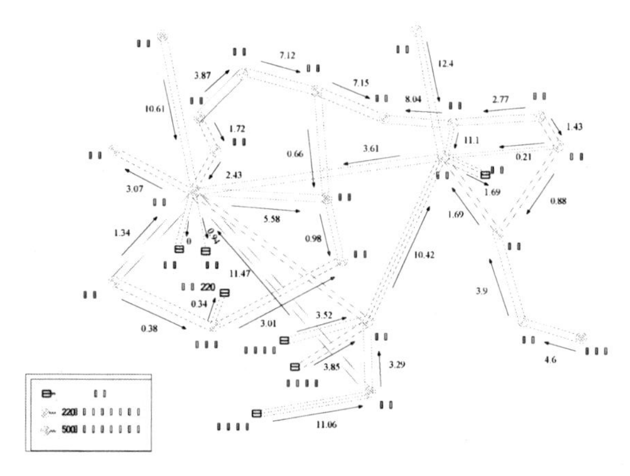

Figure 1: Grid connection of yibin converter station near zone grid.

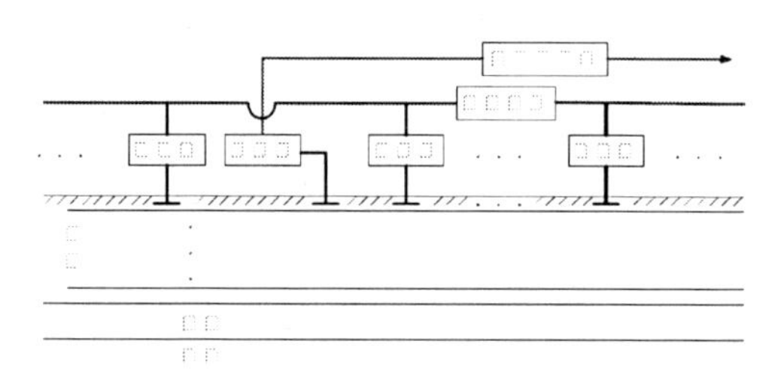

Figure 2: Diagram of unipolar earth-return operation mode.

3.2 DC bias simulation model based on ANSYS

The computational model consists of three parts [6]: the hierarchical structure of soil, Binjin DC system, Southern part of the AC system is shown in Figure 2. The model assumes Yibin to be the sending end. If it is the end of the receiving end, it would take the opposite flow calculations. The experiment is calculated in accordance with the order of the Binjin DC system during the entire experiment simulated. Calculating the distribution of different sizes into the earth entirely electric current, and the location-based GPS Yibin, Luzhou two 220kV and the substations above to determine the specific location of the station maximizes the stations simulate real topography, and calculate the various stations main transformer neutral current size. Finally, the simulation results and the actual measurement results were compared to verify the correctness and validity of the method, and lay the foundation for the development of subsequent suppression measures.

3.2.1 The hierarchical model of ground soil

In order to description of the ground soil stratification model more accurately, the simulation uses a four-layered soil model, soil resistivity of each layer are shown in Table 1.

Table 1: Soil model.

Layer	Resistivity (Ω m)	Depth (km)
Humus soil	200	0.5
Soil layer	1000	4
The original rock	6000	10
The inner earth mantle	2000	100

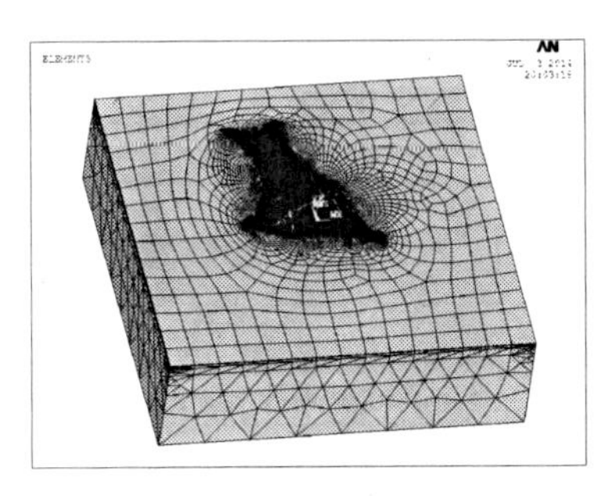

Figure 3: The discretization grid of analysis model.

The earth resistance of 500kV and 220kV substations respectively takes 0.2Ω and 0.3Ω.

3.2.2 AC systems

In order to study the extent substations affected by the DC bias, the AC systems we studied was all 220kV and 500kV substations and AC transmission lines, in Yibin, south Sichuan and Luzhou area, including the square area that the length and width are both 400km in the center of the grounding pole. Location of the observation point is determined by the coordinates: set DC grounding pole as coordinate origin, positive *X*-axis is from west to east in the horizontal direction, positive *Y*-axis is from south to north in the horizontal direction, positive *Z*-axis vertically downward from the surface. The substations' position coordinates in Luzhou and Yibin are shown in Table 2.

3.2.3 Earth electrode of DC system

Shuanglong converter station and Fulong converter station share a common earth electrode. The earth electrode is located in Da Sha Ba Village, Yibin City. The straight line distance from the earth electrode to Fulong converter station is about 72km, and about 80km to Shuanglong converter station.

3.3 DC bias inversion simulation of Binjin UHV system

3.3.1 Simulation analysis of earth potential distribution

In ANSYS analysis software, we discredited a 400*400 km^2 area earth field model and the model of power transmission line distributed in the ground. There are 334, 532 grids, as shown in Figure 3. Among them, the intensive portion of the grids is substations and transmission lines in Yibin and Luzhou.

At the earth current, respectively, −1000A, −3000A, and 5000A cases, the amplitude of surface potential is calculated as shown in Tables 2 and 3. And the computing results of surface potential are shown in Figures 4–6.

Table 2: The Surface potential of observation points, when earth current is −1000A (Unit:V).

Site Label	500kV Substation	Site Voltage
1	Luzhou	−13.03
2	Honggou	−10.09
3	Xufu	−15.14
4	Shuanglong (Yibin)	−12.36
5	Fulong	−12.88
6	Muchuan	−10.05
7	Left Xiangjiaba (power plant)	−11.34
8	Right Xiangjiaba (power plant)	−11.65
9	Left Xiluodu (power plant)	−10.14
10	Xinping (power plant)	−18.23
11	Rongzhou (power plant)	−20.70
220kV Substation		
12	Yuguan	−12.63
13	Linzhuang	−10.84
14	Yangqiao	−11.69
15	Fangshan (standby transformer)	−13.53
16	Naxi	−12.86
17	Zhendong	−11.94
18	Fengchongwan	−9.14
19	Gaoshi	−14.92
20	Baisha	−13.92
21	Ziyan	−12.37
22	Pingshan	−11.53
23	Chengnan	−14.55
24	Gongxian	−13.40
25	Fengshou	−15.71
26	Beijingba	−15.14
27	Xinping 220 (power plant)	−18.56
28	Longtou	−29.36
29	Jiangnan	15.74
30	DC Earth electrode	−377.00

Table 3: The surface potential of observation points, when earth current is 5000A (Unit: V).

Site Label	500kV Substation	Site Voltage
1	Luzhou	65.15
2	Honggou	50.43
3	Xufu	75.72
4	Shuanglong (Yibin)	61.82
5	Fulong	64.40
6	Muchuan	50.27
7	Left Xiangjiaba (power plant)	56.68
8	Right Xiangjiaba (power plant)	58.25
9	Left Xiluodu (power plant)	50.70
10	Xinping (power plant)	91.14
11	Rongzhou (power plant)	103.50
220kV Substation		
12	Yuguan	63.13
13	Linzhuang	54.22
14	Yangqiao	58.44
15	Fangshan (standby transformer)	67.63
16	Naxi	64.28
17	Zhendong	59.70
18	Fengchongwan	45.72
19	Gaoshi	74.60
20	Baisha	69.62
21	Ziyan	61.84
22	Pingshan	57.64
23	Chengnan	72.76
24	Gongxian	67.00
25	Fengshou	78.55
26	Beijingba	75.70
27	Xinping 220 (power plant)	92.78
28	Longtou	146.80
29	Jiangnan	78.70
30	DC Earth electrode	1885.00

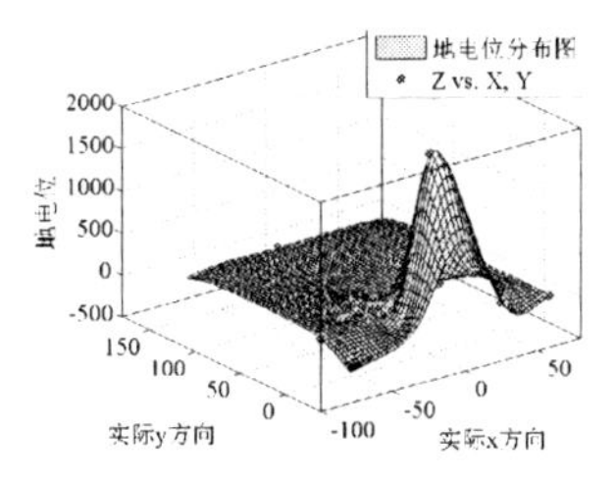

Figure 4: Earth ground potential distribution near the earth electrode, when earth current is 5000A.

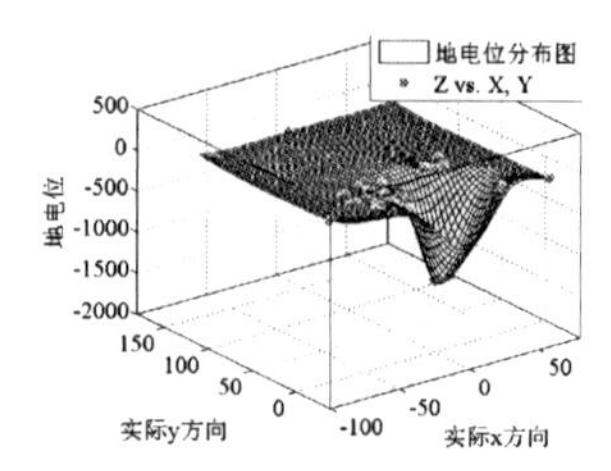

Figure 5: Earth ground potential distribution near the earth electrode, when earth current is −3000A.

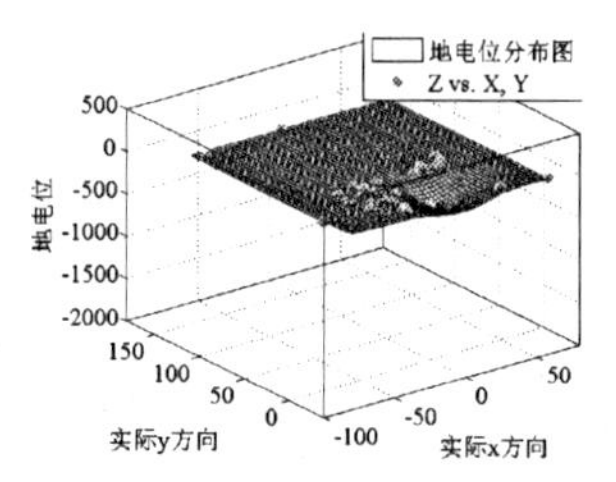

Figure 6: Earth ground potential distribution near the earth electrode, when earth current is −1000A.

3.3.2 Simulation analysis of main transformer neutral current and DC current in AC systems

The DC's computing result of each substation's main transformer neutral is shown in Table 4:

Table 4: Main transformer neutral point DC current under different bias current (Unit: A).

500kV Substation	Single Transformer Neutral Point Current		
	Earth Current: −1000A	Earth Current: −3000A	Earth Current: −5000A
Luzhou	2.06	5.03	−8.38
Honggou	−2.07	−6.2	10.33
Xufu	0.96	2.87	−4.78
Shuanglong	−0.14	−0.41	0.68
Fulong	−0.02	−0.05	0.09
Muchuan	−1.77	−5.3	8.83
Left Xiangjiaba	−1.16	−3.49	5.81
Right Xiangjiaba	−0.65	−1.94	3.23
Left Xiluodu	−0.61	−1.83	3.05
Xinping	0.31	0.94	−1.57
Rongzhou	0	0	0.00
220kV			
Yuguan	0.31	0.53	−0.95
Linzhaung	−1.31	−2.34	4.56

Yangqiao	−0.19	−0.58	0.27
Fangshan	0.56	1.69	−2.81
Naxi	1.01	3.03	−5.05
Zhendong	0.23	0.7	−1.17
Fengchongwan	−1.53	−4.6	7.67
Gaoshi	0.3	0.9	−2.1
Baisha	0.17	0.52	−0.87
Ziyan	−1.3	−3.91	6.51
Pingshan	−1.62	−4.87	8.11
Chengnan	−0.24	−0.62	0.92
Gongxian	0.6	1.8	−3
Fengshou	1.57	4.72	−7.86
Beijingba	−0.96	−2.88	4.8
Xinping 220	0.11	0.34	−0.57
Longou	1.02	2.56	−5.6
Jiangnan	−0.52	−1.35	0.19

Table 5: Comparison of measured value and simulation value, when earth current is 5000A.

Substation name	Measured Value (A)	Simulation Value (A)
Luzhou substation	−8.3	−8.38
Xufu substation	−4.1	−4.78
Naxi substation	−4.01	−5.05
Yangqiao substation	−0.24	0.27
Yuguan substation	−0.64	−0.95
Linzhuang substation	3	4.56
Gaoshi substation	−2.7	−2.1
Longtou substation	−4.7	−5.6
Jiangnan substation	−0.03	0.19
Chengnan substation	0.6	0.92
Shuanglong substation	0.23	0.68
Fangshan substation	−3.1	−2.81

In order to verify the correctness of the simulation results, we compare the 5000A earth current's simulation results with the measurement results, as shown in Table 5:

The results obtained from the above results the following analysis:

(1) The model built in this paper can accurately reflect the ground potential distribution under bias magnetic environment and the main transformer neutral current of the substation grounding near earth electrode. And the error does not exceed 10 percent.

(2) Although four-layer soil model can reduce model error caused by uneven distribution of soil, but the error of soil resistivity still exists.

(3) The actual location of earth electrode and each substation have a certain bias, which have great influence on the calculation results.

6 Conclusion

This paper simulates the phenomenon of DC bias in Sichuan province during Binjin DC system joint debugging based on the finite element method. Through comparing the simulation results with measurement results, we can conclude that the method's error is less than ten percent in simulating earth potential distribution, main transformer neutral current and DC current in AC systems under the condition of ignoring the local geological structure of the local area. Through this research, we can accurately predict the risk of DC bias of multiple DC transmission lines in Sichuan province, which run in various earth-return running mode such as unipolar maintenance, unipolar latching and bipolar asymmetry. And the study makes suppression measures as well, which plays a positive role in the safe and stable operation of power equipment.

References

[1] Xue Xiangdang, Guo Hui, Zheng Yunxiang, et al., The research of Geomagnetic induced current's damage on power transformer. *Proceedings of the Chinese Society of Universities*, **11(2)**, pp. 13–19, 1999.

[2] Zhong Lianhong, Lu Peijun, Chou Zhicheng, et al., The DC grounding current effect to the neutral grounded transformer. *High Voltage Engineering*, **29(8)**, pp. 12–17, 2003.

[3] Zhang Yanbing, *Research on Algorithm and Simulation of Power Grid Geomagnetically Induced Currents*. North China Electric Power University, Beijing, 2005.

[4] Wang Hongliang, Research on Ground Electrode Current Operation of HVDC Monopolar Ground Return Mode. Master Thesis, Southwest Jiao Tong University, 2007.

[5] Li Xiaoping, Wen Xishan, Chen Cixuan, Simulating analysis of exciting current of single phase transformer DC magnetic bias. *High Voltage Engineering*, **11(9)**, pp. 93–99, 2005.

[6] Hao Zhiguo, Yu Yang, Zhang Baohui, et al., Rules of earth surface potential distributions of HVDC monopole ground operation mode. *Electric Power Automation Equipment*, **4 (06)**, pp. 39–43, 2009.

Research on excitation characteristics change of large transformers under DC bias effects

Wei Wei[1], Xiaobin Liang[1], Hongtu Zhang[2], Wei Zhen[1]
[1]*Sichuan Electric Power Research Institute, China*
[2]*Sichuan Electric Power Dispatching Center, China*

Abstract

The paper is focused on the main transformer core saturation caused by DC bias. Based on the electromagnetic coupling principle, the differential equations of the excitation current, flux and excitation voltage are established to describe the coupling relationship associated with electric and magnetic circuits. By using Fourier series, explicit quantified solutions of the flux are obtained in terms of the excitation current containing both AC and DC components. In the end, based on the finite element theory, simulations are performed with multiple-physical fields simulation software, ANSYS, to verify the above theoretical deductions. The simulation results show that linear superposition relationship of DC and AC flux generated by the bias current are not satisfied. And the core flux generated by DC and AC bias current simultaneously is much smaller than the sum of the core flux generated by DC and AC bias current separately. These results provide an effective technical support and theoretical guidance for the large-scale primary transformer DC bias tolerance evaluation.

Keywords: DC bias, magnetization curve, saturation, Fourier series, multiple-physical fields.

1 Introduction

With the fast development of smart grid, ultra-high voltage direct current (UHVDC) technology is progressing rapidly in China. Although long-distance UHVDC transmission efficiency can be greatly improved, it may cause a series of system security and stability risks. Especially for single-stage high-power earth-loop operation and bipolar asymmetric operation, there will be DC magnetic bias near the earth electrode. The DC bias may lead to the transfer characteristics

WIT Transactions on Engineering Sciences, Vol. 107, © 2015 WIT Press
www.witpress.com, ISSN 1743-3533 (on-line)
doi:10.2495/SEEEII140611

distortion of the main transformer accompanied by a variety of adverse effects such as noise, vibration, heat, and loss, which would produce very negative impacts for the power grid systems and the transformers. Previous studies show that the transfer characteristics of main transformers have much to do with the magnetization curve of the core [1-3]. Extreme saturation characteristics of main transformers with DC bias and high remanence is the key point to improve the excitation characteristics. While at present, few research works on extreme saturation have been reported, since the magnetization curve of the core generated by DC and AC simultaneously could not be obtained easily by AC no-load test. Aiming at these problems, first differential equations of excitation current, flux and excitation voltage are established in this paper based on electromagnetic coupling. The coupling between electric and magnetic circuits is taken into consideration in these equations [4-5]. Then explicit quantified solutions of the flux are calculated in terms of the excitation current containing both AC and DC components using Fourier series. Last, based on the finite element theory, simulations are performed with multiple-physical fields simulation software ANSYS to verify the derivations above. The simulation results show that linear superposition relationship associated with DC and AC flux generated by the bias current are not satisfied. Usually the rated flux could be generated by a very small alternating excitation current (smaller than 2A). If at the same time certain DC current entry the transformer, for example 10–20 A, the generated DC flux is much smaller than the rated flux. That is, the core flux under AC and DC simultaneously is much smaller than that under AC and DC separately [6].

2 Linear model of magnetic saturation characteristics

For simplification, this paper ignores the core hysteresis loop, and replacing it with the average magnetization curve. The core nonlinear saturation characteristics is simplified and represented by a broken line as shown in Figure 1. Line OA represents the magnetization curve where the transformer is working in linear region and the slop of line OA is denoted by k1. Line AB describes the curve where the transformer is disturbed and working in saturation region, while the slop is k2. By simplifying the magnetization curve as polygonal line, the linear model of magnetic saturation characteristics is built. The excitation current waveform obtained from the linear model keeps the main features of the initial excitation current and could describe the transformer's working states well.

3 Working states of transformers with different excitation currents

The excitation current curves with AC over-excitation are shown in Figure 2. When the transformer is energized with AC over-excitation, the excitation current distortion increases with core flux density rising. Considering the nonlinear excitation characteristics, parts of the magnetic curves which lie in the linear region transferred into the saturation region. Finally the total excitation current curves become symmetrical vertex sharp waves and the excitation current contains only odd components.

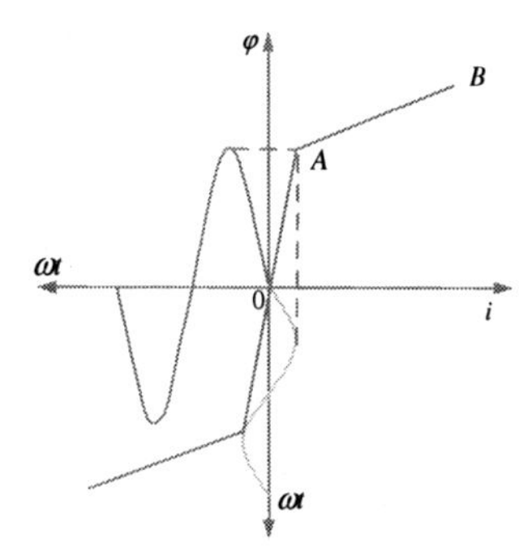

Figure 1: Rated excitation current curves of transformers.

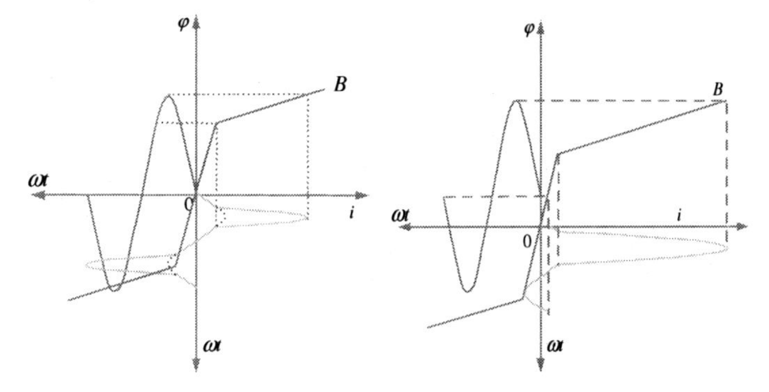

Figure 2: Excitation current curves with AC over-excitation. Figure 3: Excitation current curves with DC bias.

Figure 3 shows the excitation current curves with DC bias. The total flux density is the sum of DC and AC excitation flux. Half cycle of excitation current with the same direction of DC bias is greatly increased, while the other half is decreased relatively. So the excitation current curves become very asymmetric and contain not only odd components but also even components. The core is working in the saturation region during the half period when the excitation current is in the same direction with DC bias.

4 Excitation current calculation under DC bias effects

According to Ampere's law, the magnetomotive force balance equation could be described as follow:

$$N_1 i_1 + N_2 I_{\mathrm{dc}} = Hl \tag{1}$$

Assume that the core magnetization curve could be fitted by hyperbolic function as $H = xsh(yB)$, where x and y are parameters of the core magnetization curve

and usually are larger than 1 according to the former research works. Then Eq. (1) could be simplified as:

$$NN_1 i_1 + N_2 I_{\mathrm{dc}} = lx[sh(K\frac{y}{A}\varphi_0)ch(K\frac{y}{A}\varphi_m \sin\omega t) + ch(K\frac{y}{A}\varphi_0)sh(K\frac{y}{A}\varphi_m \sin\omega t)] \tag{2}$$

Let $m = K\frac{y}{A}\varphi_m = yB_m$, representing the utilization rate of the core and it is a design constant related to the transformer's working state. With the Fourier series expansion of $sh(K\frac{y}{h}\varphi_m \sin\omega t)$ and $ch(K\frac{y}{h}\varphi_m \sin\omega t)$, Eq. (2) could be rewritten as:

$$N_2 I_{\mathrm{dc}} = lxsh(K\frac{y}{A}\varphi_0)a_0(m) \tag{3}$$

$$N_1 i_1 = lc\left[\frac{N_2 I_{\mathrm{dc}}}{lxa_0}\left(\sum_{n=1}^{\infty} a_{2n}(m)\cos(2n\omega t)\right) + \sqrt{1+(\frac{N_2 I_{\mathrm{dc}}}{lxa_0})^2}\left(\sum_{n=1}^{\infty} a_{2n+1}(m)\sin((2n+1)\omega t)\right)\right] \tag{4}$$

$$a_{2n+1}(m) = \frac{\omega}{\pi}\int_0^T [sh(K\frac{y}{A}\varphi_m \sin\omega t)\sin((2n+1)\omega t)]\mathrm{d}t = \frac{y\varphi_m K}{2A(2n+1)}(a_{2n}(m) + a_{2(n+1)}(m)) \tag{5}$$

$$a_{2n}(m) = \frac{\omega}{\pi}\int_0^T [ch(K\frac{y}{A}\varphi_m \sin\omega t)\cos(2n\omega t)]\mathrm{d}t = -\frac{y\varphi_m K}{4An}(a_{2n-1}(m) + a_{2n+1}(m)) \tag{6}$$

$$a_0(m) = \frac{\omega}{2\pi}\int_0^T ch(K\frac{y}{A}\varphi_m \sin\omega t)\mathrm{d}t \tag{7}$$

where T is the time period of the waveforms in terms of power frequency.

When I_{dc}=0, the root mean square value of field ampere-turns is:

$$N_1 i_{ef0} = lx\sqrt{a_1^2(m) + a_3^2(m) + a_5^2(m) + \ldots\ldots} \tag{8}$$

While $I_{\mathrm{dc}} \neq 0$, the root mean square value of field ampere-turns is:

$$N_1 i_{ef0} = lx\sqrt{\begin{array}{l}(a_1^2(m) + a_3^2(m) + \ldots) + \\ (\frac{N_2 I_{\mathrm{dc}}}{lxa_0(m)})^2[a_1^2(m) + a_2^2(m) + \ldots]\end{array}} \tag{9}$$

Rewrite Eq. (3) and φ_0 is obtained:

$$\varphi_0 = \frac{A}{Ky}sh^{-1}[\frac{N_2 I_{\mathrm{dc}}}{lxa_0(m)}] \tag{10}$$

Let $H = f(B)$, which is the magnetization curve. Then the relationship between excitation current and flux could be described as $\frac{Ni}{l} = f(\frac{\varphi}{A}K)$. Rewrite this equation and φ is obtained: $\varphi = \frac{A}{K} f^{-1}(\frac{Ni}{l})$. Comparing the expression of φ with Eq. (10), it is obvious that φ_0 and I_{dc} not only follow the magnetization curve but also differ by $a_0(m)$ which could be expressed as:

$$a_0(m) = \frac{\omega}{2\pi}\int_0^T ch(K\frac{y}{A}\varphi_m \sin\omega t)dt = \frac{\omega}{2\pi}\int_0^T (1 + \frac{(K\frac{y}{A}\varphi_m)^2 \sin^2\omega t}{2!} + \ldots)dt \tag{11}$$

The flux density of core is not located in saturated region with the transformer under AC and DC bias simultaneously, that is, $a_0(m) = \frac{\omega}{2\pi}\int_0^T 1dt = 1$. This shows, given a DC flux φ_0, the current average is exactly the corresponding current i_{00} in terms of magnetization curve, as shown in Figure 4(a), and there is no distortion in excitation current curve. When the core flux density average is just located in saturated region of the magnetization curve, for example, the core flux density is 1.75T, then

$$a_0(m) = \frac{\omega}{2\pi}\int_0^T (1 + \frac{(K\frac{y}{A}\varphi_m)^2 \sin^2\omega t}{2!})dt = 1 + \frac{(K\frac{y}{A}\varphi_m)^2}{4} = 1 + 0.76y^2 \tag{12}$$

While if the core flux density is located in highly saturated region of the magnetization curve, for example, the core flux density is 2.5T, then

$$a_0(m) = \frac{\omega}{2\pi}\int_0^T (1 + \frac{(K\frac{y}{A}\varphi_m)^4 \sin^4\omega t}{4!})dt = \frac{\omega}{2\pi}(\frac{T}{2} + \frac{T}{4} + \frac{T}{8})\frac{(K\frac{y}{A}\varphi_m)^4}{4!} = 1.424y^4 \tag{13}$$

This shows that there would be half cycle of the current waveform located in saturated region, as shown in Figure 4(b). The current average i_0 (or I_{dc}) is $1+0.76y^2$ times larger than the current i_{00} corresponding to DC flux. In total, because of the nonlinear magnetic core material, the relationship between DC flux φ_0 and excitation current average i_0 is not the same as the relationship between φ and i in terms of magnetization curve.

The above analysis shows that because of the nonlinear magnetic properties of electrical steel sheet, the total flux containing DC flux would lead to the current distortion and there would be a current average while the DC component flowing through the transformer. And the DC flux corresponding to this current average is φ_0 instead of φ_{i0}. So DC flux φ_0 generated by AC and DC simultaneously is much smaller than that generated by DC bias alone. Additionally considering the saturation magnetization of the electrical steel core is usually a certain value, so the total flux is limited to a certain value. Therefore the current average would keep

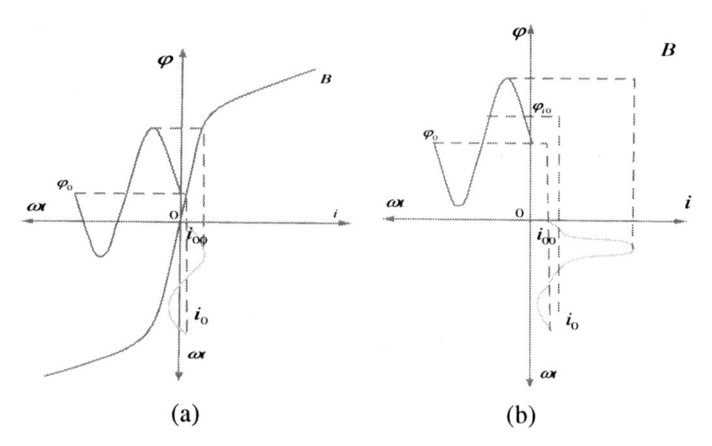

Figure 4: Magnetization curve under DC bias.

increasing because of the current distortion. But the totally flux or DC component of the total flux corresponding to this distorted current waveform needed is not too large. What's more, the transformer is usually designed to work at the knee point of the magnetization curve, that is to say the AC flux is already large enough. Thus the actual DC flux is not too large considering the fact that nowadays the property of the electrical steel sheet used in transformers is good enough. In total, although there would be DC component flowing through the transformer windings for some certain reason in the power system, the DC flux in the core that may be generated would not be too large.

Rewrite Eq. (4) and then the following formula is obtained:

$$\frac{N_2 I_{\mathrm{dc}}}{lxa_0(m)} = \frac{U_2 I_{\mathrm{dc}}}{4.44 fAB_m lxa_0(m)} = \frac{\delta S I_{\mathrm{dc}}}{V_{fe} ma_0(m) I_2} \tag{14}$$

where V_{fe} is the volume of core and $V_{\mathrm{fe}}=Al$. S is transformer capacity and $S=U_2I_2$. U_2 and I_2 denote the rated voltage and rated current of the primary side respectively. δ is a constant corresponding to core magnetization curve and $\delta = \dfrac{y}{4.44 fx}$. Eq. (14) shows that DC magnetic bias is influenced by transformer capacity, parameters of the core, property of core materials, core laminations slot and the load.

5 Analysis of excitation characteristics of transformers under DC magnetic bias with ANSYS

It could be summarized from Section 4 that with the effects of DC flux, DC component leads to the half-cycle saturation of AC excitation current and results in current average. With the AC and DC magnetic bias simultaneously, DC

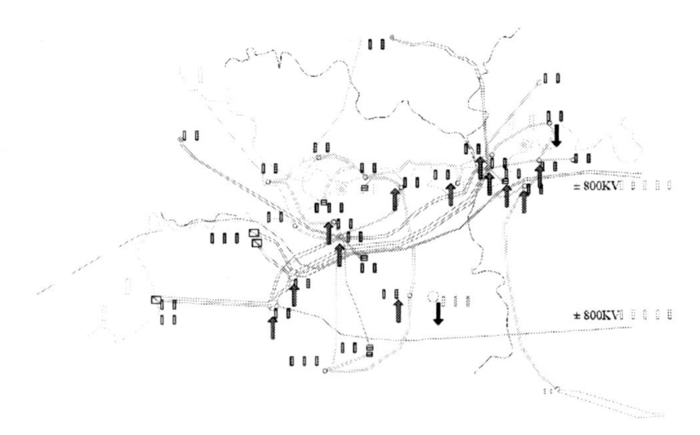

Figure 5: Current flow schematic of the 220kV and above substation in Yibin and Luzhou power system.

current value is associated with both DC flux and the work state or rated flux of the core. And the relationship between them is not directly associated with the primary magnetization curve. Therefore the DC flux generated by the some external reasons is not too large in the transformer. And it's not the same as the corresponding flux generated by DC bias alone.

This section takes the main transformer in Luzhou as an example. By using ANSYS, a detailed analysis is performed on the core flux of the 500kV main transformer in Luzhou substation under DC magnetic bias. And the excitation characteristics and the distribution of flux are obtained.

Magnetic induction and core flux of the main transformer core are calculated considering the neutral direct current of the main substation in Luzhou is 0A, −8.3A, −15A, and −25A separately. And the results are shown in Figure 6.

Figure 6 and Table 1 show that the core magnetic saturation point (or knee point) approximately is 1.4T. For the main transformer of the 500kV substation in Luzhou, the effects of the DC bias are listed as follows:

(1) Both magnetic induction and core flux of the main transformer would increase because of DC bias flowing through the main transformer's neutral line.
(2) Considering the transformer under rated condition is usually working in the region around the knee point, i.e. 1.4 Tesla, so even small DC bias may lead to the core saturation. As shown in Table 1, when the DC bias is -15A and -25A, the core flux growth has exhibited nonlinear characteristics because of the core saturation.
(3) The core flux increases from 0.515 to 0.538 with an increasing rate 4.5%, while the DC bias is changing from 0A to −25A. The simulation results are consistent with figure 5(b), i.e. DC flux under DC bias effects is much smaller than the flux generated by direct current alone. So even with a relatively large direct current, the saturation flux would not be too large. And the actual DC bias tolerance may be larger than 0.7 percent of the rated current (the standards used in the national power grid corporation).

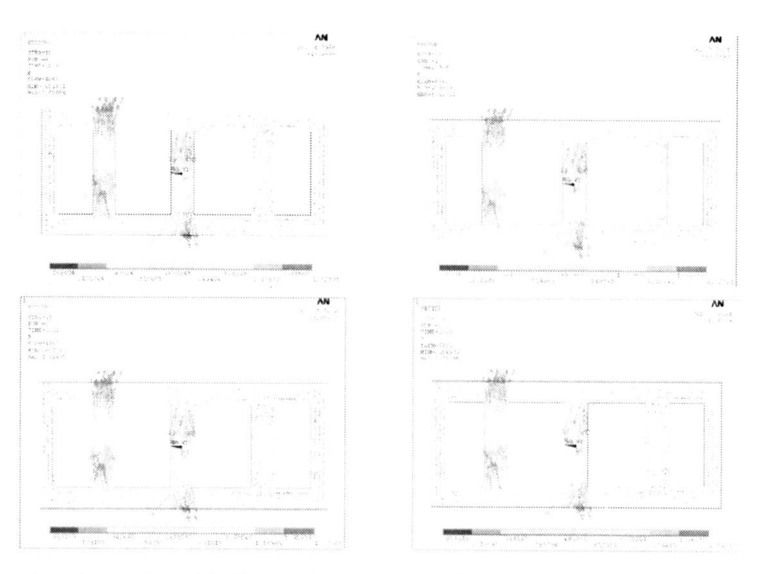

Figure 6: Magnetic field distribution of the main transformer core. (a) magnetic induction with the direct current is 0A. (b) magnetic induction with the direct current is –8.1A. (c) magnetic induction with the direct current is –15A. (d) magnetic induction with the direct current is –25A.

Table 1: Magnetic induction and flux of the main transformer under DC bias in luzhou substation.

DC Bias	Magnetic Induction of Core (T)	Flux (Wb)
0	1.36	0.515
–8.1	1.39	0.523
–15	1.41	0.534
–25	1.42	0.538

6 Conclusions

Differential equations of excitation current, flux and excitation voltage are established based on electromagnetic coupling. The coupling relationship between electric and magnetic circuits is taken into consideration in these equations. Explicit quantified solutions of the flux are calculated in terms of the excitation current containing both AC and DC components using Fourier series. According to simulation results, linear superposition relationship associated with DC and AC flux generated by the bias current are not satisfied. Excitation characteristics of large transformers under DC bias effects are analyzed. With the effects of DC flux, DC component leads to the half-cycle saturation of AC excitation current and results in current average. DC current value is associated with both DC flux and the work state or rated flux of the core. And the relationship between them is not directly associated with the primary magnetization curve. Simulations are performed with ANSYS and the derivations above are verified.

References

[1] Xue Xiangdang, Guo Hui, Zhwnf Yunxiang, et al., The study of harmful effects of geomagnetically induced current on power transformers. *Proceedings of the EPSA*, **11(2)**, pp. 13–19, 1999.
[2] Zhong Lianhong, Lu Peijun, Qiu Zhicheng, et al., The influence of current of DC earthing elcctrode on directly grounded transformer. *High Voltage Engineering*, **29(8)**, pp. 12–17, 2003.
[3] Zhang Yanbing, Studies of Arithmetic and Simulation on GIC of Power Grid. Master Degree Thesis. North China Electric Power University, Beijing, 2005.
[4] Wang Hongliang, Research on Ground Electrode Current of High Voltage Direct Current System in Ground Return. Master Degree Thesis. Southwest Jiaotong University, 2004.
[5] Li Xiaoping, Wen Xishan, Cheng Cixuan, Simulating analysis of exciting current of single phase transformer on DC bias. *High Voltage Engineering*, **31(9)**, pp. 8–10, 2005.
[6] Hao Zhiguo, Yu Yang, Zhang Baohui, et al., Earth surface potential distribution of HVDC operation under monopole ground return mode. *Electric Power Automation Equipment*, **29(06)**, pp. 10–14, 2009.

The vibration characteristics of large power transformer under DC bias

Xiaobin Liang, Wei Wei, Wei Zhen, Lijie Ding, Hua Zhang
Sichuan Electric Power Research Institute, China;

Abstract

The DC bias has significant impact on the vibration of power transformer. We describe the flux distribution of iron core when the power transformer is under DC bias condition. Then the magnetostrictive model is built, and the mode shapes and resonance frequencies of core are calculated. In order to study the impact of DC bias effect to vibration of iron core, the dynamic analysis of iron core is presented, and the vibration simulated results of iron core with different direct current are compared. It is found that, the direct current caused by DC bias will accelerate the saturation of iron core. As the direct current increase, the core vibration will increase significantly. The relation between core vibration and direct current (caused by DC bias) is nonlinear, and it is determined by the nonlinear characteristics of dynamic response equation and magnetic material.
Keywords: DC bias, magnetostrictive, core vibration, power transformer.

1 Introduction

In recent years, the ultra-high voltage direct current (UHVDC) transmission technology has developed very fast, this technology has improved the transmission efficiency significantly [1]. However, it also caused some new risks, and the DC bias is one of them [2]. When the DC transmission lines operate in single-stage high-power earth-loop operation and bipolar asymmetric operation, the DC bias is significant near the earth electrode. The direct current will cause electric field in earth, so the electric potential in the earth is different. The absolute value of electric potential near to the electrode will be higher than infinite region. If there are grounding lines in AC system, the direct current will enter the AC system by the grounding lines. For example, if the power transformer is neutral grounded, this will form direct current in the windings.

WIT Transactions on Engineering Sciences, Vol. 107, © 2015 WIT Press
www.witpress.com, ISSN 1743-3533 (on-line)
doi:10.2495/SEEEI1140621

Researchers have done many studies on the impact of DC bias on power transformer. In Ref. [3], the author studied the three-phase five limbs transformer, and gave a series of mathematic transformation and numerical nonlinear algebraic equations to solve this coupled problem. In Ref. [4], Michael and Hermann investigated the coupling vibration of large power transformer by using finite element method. However, only a few papers present the transformer vibration caused by magnetostrictive. This paper will study the vibration characteristics of power transformer under DC bias.

2 Electromagnetic behavior of transformer core under DC bias

2.1 Flux distribution and magnetic flux density

One three-phase five-limb transformer is studied in this paper. The size and all the main parameters of its iron core are given in Table 1.

The sinusoidal waveform voltages and DC bias voltage have been used to supply this power transformer. Based on this theory, voltage in each phase can be described as following:

$$\begin{bmatrix} U_A(\theta) \\ U_B(\theta) \\ U_C(\theta) \end{bmatrix} = \begin{bmatrix} U_{\max} \sin(2\pi ft) + U_{\mathrm{dc}} \\ U_{\max} \sin\left(2\pi ft + \frac{2\pi}{3}\right) + U_{\mathrm{dc}} \\ U_{\max} \sin\left(2\pi ft + \frac{2\pi}{3}\right) + U_{\mathrm{dc}} \end{bmatrix} \tag{1}$$

In which, $U_i(\theta)$ is the voltage in phase i, U_{max} is the amplitude of AC voltage, f is the electrical frequency, and U_{dc} is the DC bias voltage in each phase. We use

Table 1: Structure parameters of iron core.

Item	Value
Number of phases	3
Number of limbs	5
Core length (mm)	5500
Core thickness (mm)	1250
Core high (mm)	3820
Winding turns of primary side	556
Winding turns of secondary side	245
Rated voltage of primary side (kV)	500
Rated voltage of secondary side (kV)	220

ANSYS software to build the magnetic field model of transformer core, and solve this differential equation of transient magnetic field:

$$\sigma \frac{\partial A}{\partial t} + \nabla \times \left(\frac{\Delta \times A}{\mu_r \mu_0} \right) = \mathrm{J} \tag{2}$$

In which, μ_r is relative permeability, μ_0 is vacuum permeability, A is magnetic vector potential, J is current density, and σ is conductivity. By solving differential equations of the magnetic field, the magnetic field distribution can be known, and flux distribution of iron core is shown as follows:

The power transformer is powered by the voltage in Eq. (1), and the secondary winding is no-load. The flux distribution of iron core is presented in Figure 1. The saturation point of this transformer is about 1.4T, and we can found that the core is saturated on both sides of the middle column. In order to compare the saturation effect of iron core when there are different DC bias currents, we supply this transformer with additional direct current. So, the direct current with amplitude of 0A, −8.1A, −12A, and −25A are added in each phase, respectively. It is found that: the flux distribution in iron core don't change, which is shown in Figure 1. However, the average magnetic flux density and flux have changed, and these are presented in Table 2.

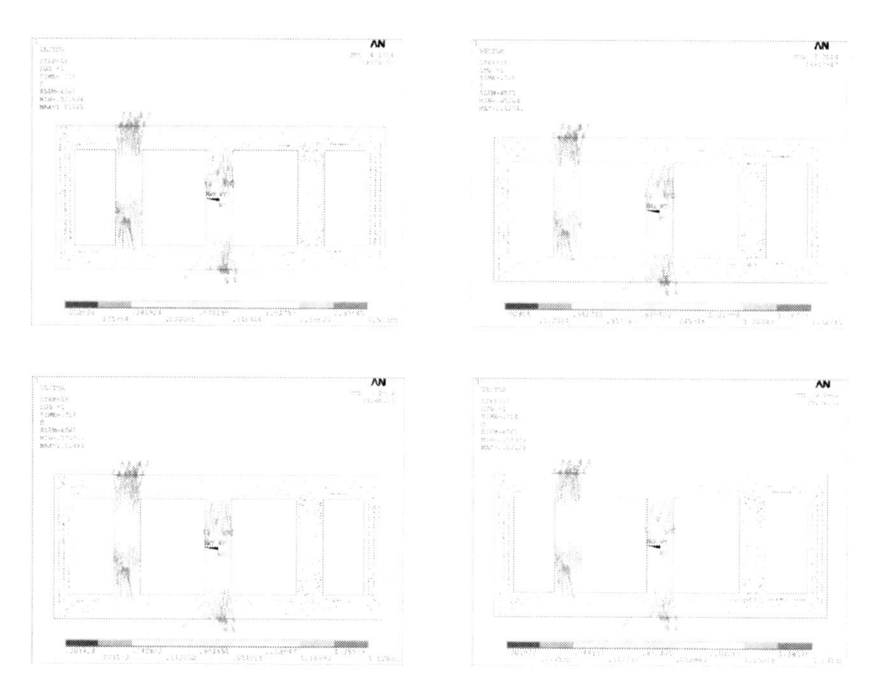

Figure 1: The Flux distribution and flux density of transformer core with different direct current. (a) I_{dc} = 0A. (b) I_{dc} = −8.1A. (c) I_{dc} = −12A. (d) I_{dc} = −25A.

Table 2: The magnetic flux density and flux of core under DC bias.

DC Bias (A)	The Magnetic Flux Density in Core (T)	The Flux in Core (Wb)
0	1.36	0.515
–8.1	1.39	0.523
–15	1.41	0.534
–25	1.42	0.538

In Table 2, it is found that: the direct current in transformer will increase the magnetic flux density in iron core, and it will increase the flux in iron core as well. When the transformer works normally, the iron core is close to saturated. So, the iron core will be saturated even the direct current is small in each phase. For example, when the direct current is –12A and –25A, the transformer has been deep saturated, and the incensement of flux in core presents nonlinear characteristic. When the direct current increases from 0A to –25A, the flux in core increases from 0.515Wb to 0.538Wb, and this pointed out that the direct current will cause the saturation of iron core, and because the nonlinear characteristic of iron core, the incensement of flux caused by direct current is quite small.

2.2 Magnetostrictive of the iron core

Magnetostrictive effect describes the change in dimensions of a material due to a change in its magnetization. This phenomenon is a manifestation of magnetoelastic coupling, which is exhibited by all magnetic materials to some extent. In resent research, the vibration caused by magnetostrictive is considered as one of the main vibration source. This paper has introduced the flux distribution of iron core under different DC bias condition, in order to study the vibration characteristic of iron core we have to build the vibration model of iron core under the DC bias condition. In the alternating magnetic field, the force density in iron core can be described by this equation:

$$\mathrm{F} = J \times B - \frac{1}{2} H^2 \nabla u + \frac{1}{2} \nabla \left(H^2 \tau \frac{\partial u}{\partial \tau} \right) \tag{3}$$

In which, F is force density, J is current density, B is magnetic flux density, H is magnetic field intensity, u is permeability of iron core, τ is density of iron core.

In Eq. (3), the first term $J \times B$ is Lorentz force, the second term $\frac{1}{2} \nabla \left(H^2 \tau \frac{\partial u}{\partial \tau} \right)$ is volume force, and the third term $\frac{1}{2} \nabla \left(H^2 \tau \frac{\partial u}{\partial \tau} \right)$ is magnetostrictive force. The magnetostrictive force is influenced by magnetic field intensity and is also influenced by the material parameters of iron core. In Ref. [5], the author has proposed a mathematical expression for magnetostrictive, and the magnetostriction component along any direction can be calculated as a nonlinear function of the magnetization using Eq. (4).

$$\lambda_i = \frac{3}{2}\lambda_s\left(\alpha_i^2 - \frac{1}{3}\right) = \frac{3}{2}\lambda_s\left[\left(\frac{M_i}{M_s}\right)^2 - \frac{1}{3}\right] \tag{4}$$

In which, λ_i is magnetostriction along the direction *i*, λ_s is magnetostriction constant, α_i is the magnetization direction cosine, M_i is magnetization along the direction *i*, and M_s is the saturation magnetization of the material.

3 Mechanical behavior of iron core

In Ref. [6], author has pointed out that the vibration of power transformer is mainly caused by magnetostrictive. So in this paper, in order to exclude the impact of other vibration sources (Lorentz force in winding for example), the transformer works under no-load condition, and the vibration of magnetostrictive in iron core is studied.

3.1 Modes and resonant frequency of iron core

In modal analysis as reported in Ref. [7], the resonant frequencies and mode shapes can be obtained by solving the following matrix equation derived from finite element method:

$$([K] - \omega_i^2[M])\{\varphi\}_i = \{0\} \tag{5}$$

The effects of windings have been taken into account in the model by using the lumped mass method. The materials used in the mechanical model are presented in Table 3.

Due to the complex structure of the motor system, more than 20 mode shapes are possible with the resonant frequencies less than 500 Hz. Table 4 gives the list of core's mode shapes and their resonant frequencies. Among these frequencies, the core deforms for mode 1, mode 5, mode 12, and mode 27 are shown in Figure 2.

Table 3: Mechanical properties of iron core.

Material	Density (kg/m^3)	Young's Modulus (GPa)	Poisson's Ratio
Si-steel	7700	200	0.3

Table 4: Resonant frequency of transformer iron core.

Mode Order	Frequency (Hz)	Mode Order	Frequency (Hz)
1	35	16	278
2	61	17	286
3	84	18	305
4	92	19	321
5	127	20	382
6	129	21	387
7	144	22	390
8	156	23	400
9	178	24	420
10	192	25	435
11	208	26	473
12	245	27	478
13	254	28	486
14	270	29	493
15	277		

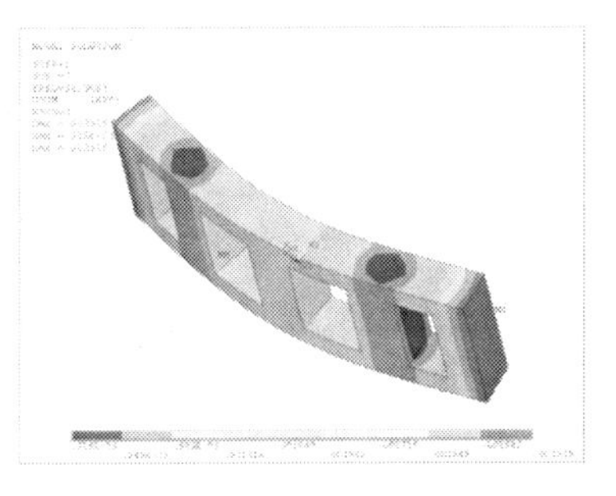

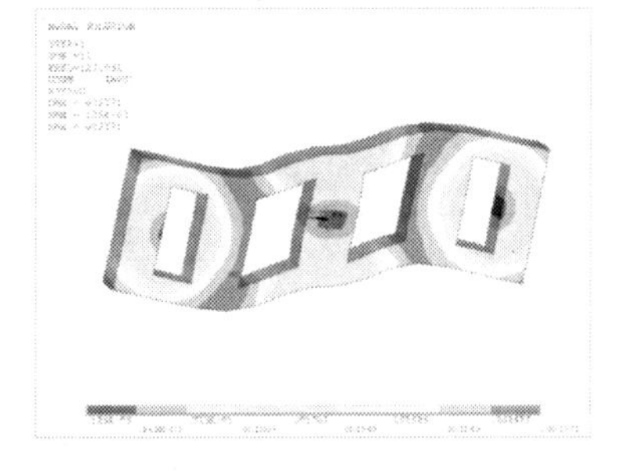

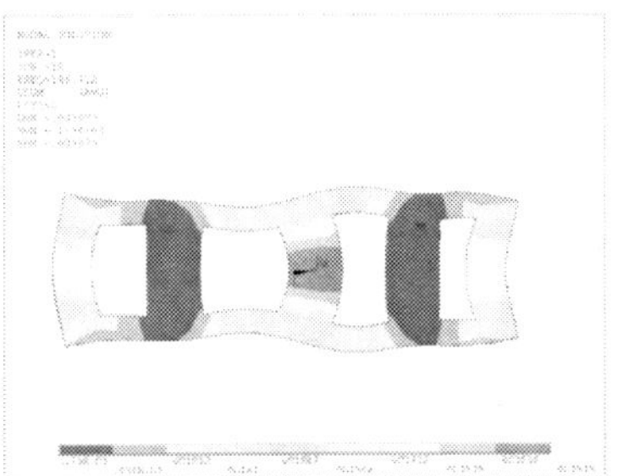

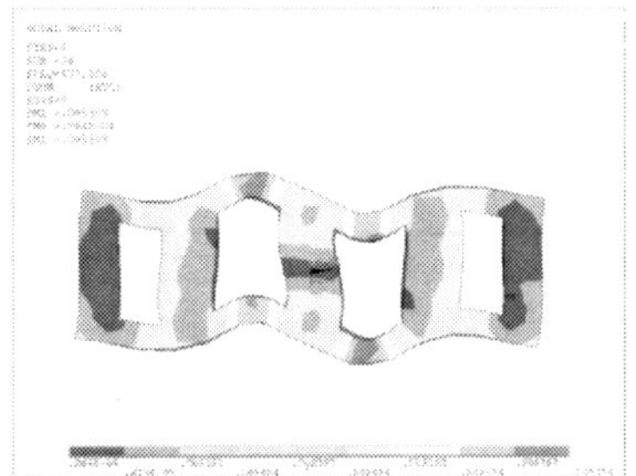

Figure 2: Iron core deformations for different mode shape. (a) mode 1, (b) mode 5, (c) mode 12, (d) mode 27.

For mode orders from 1 to 4 (low frequency), the deformation of iron core is mainly the bending of the whole core. With the incensement of the frequency, the deformation of iron core occurs in yoke and in the limbs.

When electromagnetic forces of different frequencies are applied on the stator teeth, all the mode shapes will appear at their corresponding resonant frequency, and we will study the dynamic response of iron core.

3.2 Dynamic analysis of iron core

The mechanical behavior of the iron core is simulated in no-load conditions with 3-phase supplied by sine wave voltage and different direct current in each phase. The top point on the outer surface of core is chosen to simulate its vibration behavior. The dynamic response of iron core is calculated by using mode superposition method. This method is based on the modal analysis, and sums up the factored mode shapes. The dynamic equation is expressed as follows.

$$[M]\sum_{i=1}^{n}\{\varphi_i\}\ddot{y}_i + [C]\sum_{i=1}^{n}\{\varphi_i\}\dot{y}_i + [K])\sum_{i=1}^{n}\{\varphi_i\}y_i = \{F\} \tag{6}$$

In which, [M] is mass matrix, [C] is damp matrix, [K] is elasticity matrix, Φ_i is mode shape, and y_i is deformation.

In order to compare the vibration behavior of iron core in different direct current condition, the power transformer is supplied by the same sine wave voltage, and is also supplied by different direct current in each phase. We set the direct current is 0A, –8.3A, –12A, and –25A. The vibrations of iron core are presented in Figure 3.

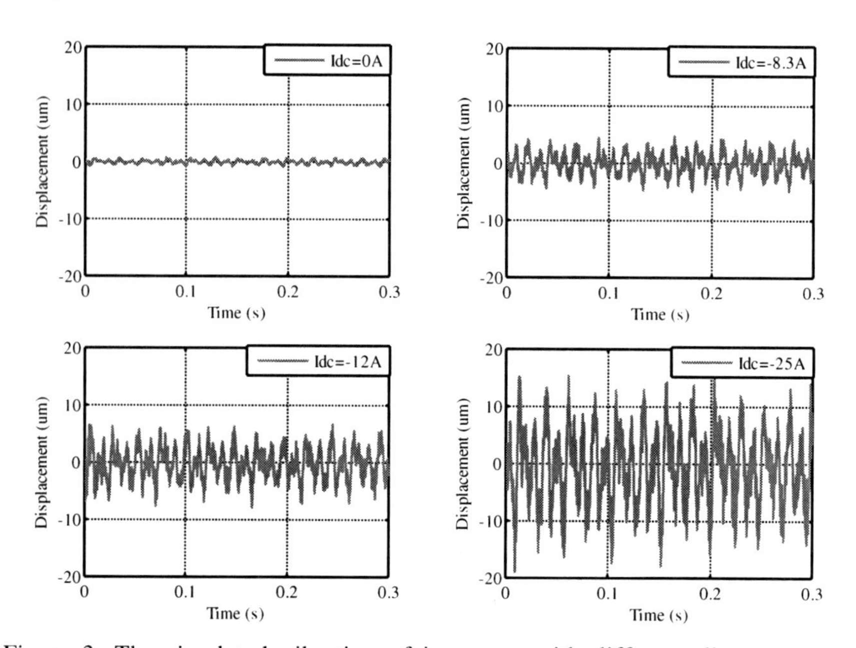

Figure 3: The simulated vibration of iron core with different direct current. (a) I_{dc} = 0A, (b) I_{dc} = –8.3, (c) I_{dc} = –12A, (d) I_{dc} = –25.

In Figure 3, we found that: if there is no direct current in winding of transformer, the vibration caused by magnetostrictive in iron core is very small, and the amplitude is less than 0.4μm. With the incensement of the direct current in each phase, the magnetostricitve vibration becomes bigger. When the direct current in each phase is −8.3A, the amplitude of vibration is about 4μm, which is 10 times bigger than result of no direct current condition. When the direct current in each phase is −12A, the amplitude of vibration is 7μm. When the direct current in each phase is −25A, the amplitude of vibration is 16μm. So, the impact of direct current caused by DC bias to the magnetostrictive vibration in iron core is very important. At the same time, we found that the impact of direct current to vibration is nonlinear, and this is determined by the nonlinear characteristics of magnetic material and dynamic response equation.

4 Conclusion

In this paper, we firstly describe the flux distribution of iron core when the power transformer is under DC bias condition. It is pointed out that the main vibration source is caused by magnetostrictive effect in no-load condition. So, the magnetostrictive model of iron core is built, the mode shapes and resonance frequencies of core are calculated. In order to study the impact of DC bias effect to vibration of iron core, the dynamic analysis of iron core is presented, and the vibration simulated results of iron core with different direct current are compared. It is found that, the direct current caused by DC bias will accelerate the saturation of iron core. As the direct current increase, the core vibration will increase significantly. The relation between core vibration and direct current (caused by DC bias) is nonlinear, and it is determined by the nonlinear characteristics of dynamic response equation and magnetic material.

References

[1] Wang Mingxin, Zhang Qiang, Analysis on influence of ground electrode current in HVDC on ac power network. *Power System Technology*, **29(3)**, pp. 9–14, 2005.

[2] Mao Xiao-ming, Wu Xiaochen, Analysis on operational problems south China ac-dc hybrid power grid. *Power System Technology*, **28(2)**, pp. 7–13, 2004.

[3] Li Xiao-ping, Wen Xi-shan, DC bias computation study on three-phase five limbs transformer. *Proceedings of the CSEE*, **30(1)**, pp. 127–131, 2010.

[4] E. Michael, L. Hermann, Investigation of load noise generation of large power transformer by means of coupled 3D FEM analysis. *Compel*, **26(3)**, pp. 788–799, 2007.

[5] S. Chikazumi, *Physics of Ferromagnetism*. Oxford University Press, New York, NY, 1997.

[6] X.M. Zhong, X.H. Yao, Q. Han, G.C. Liang, Experimental research on vibration in amorphous alloy transformer core. *Science Technology and Engineering*, **9(17)**, pp. 4934–4939, 2009.
[7] C.Y. Wu, C. Pollock, Analysis and reduction of vibration and acoustic noise in the switched reluctance drive. *IEEE Transactions on Industry Applications*, **31(1)**, pp. 91–98, 1995.

Research on nonlinear target tracking algorithm based on particle filters

Liu Kai, Liang XiaoGeng
Guidance and Control Department,
Airborne Missile Academy, Luoyang, Henan, China

Abstract

Particle filter (PF) is presented to solve the nonlinear filter and non-Gaussian problems, while the algorithms of Kalman filter and extended Kalman filter (EKF) within the Gaussian background leads to the filter precision decrease and divergence phenomenon. As a nonlinear filter algorithm based on Bayesian estimation, particle filter has original advantage at treating the parameter estimation and state filtering aspects of nonlinear non-Gaussian time-varying systems, but it takes a lot of time due to larger number of particles. Thereby extended Kalman particle filter (EKPF) is presented to solve the lower the real-time performance resulting from high computational complexity. The simulation results show that the PF approach outperforms the EKF algorithm under strong nonlinear and non-Gaussian environment, and EKPF gives better performance than EKF in solving high computation complexity.
Keywords: nonlinear target tracking, extended Kalman filter, particle filter, IEPF.

1 Introduction

Precision guide is one of the key technologies of modern weapons systems, including target detection, guidance information processing, precise positioning, and navigation, guidance algorithm design, advanced flight control system design, and so on. In the terminal guidance of air to air missiles, since the guidance information that can be measured from seeker is very limited, the lack of information in precision guide becomes more serious [1, 2]. Therefore, how to extract accurate information has become a critical core issue in precision guide.

Guidance information estimation are calculating the relative motion state based on the guidance information from seeker and the navigation information of

WIT Transactions on Engineering Sciences, Vol. 107, © 2015 WIT Press
www.witpress.com, ISSN 1743-3533 (on-line)
doi:10.2495/SEEEI140631

missile, such as relative distance and relative speed between missile and target, angle of sight, sight angular rate, and so on. The main problems that guidance information estimation encounters are: serious nonlinear circuit of the relative kinematic, more information needed for more advanced guidance law, and poor observability due to insufficient target detection information. Because of these problems, the estimation accuracy of guidance information extraction has been widely appreciated. Currently the mainly used guidance information extraction algorithm includes extended Kalman filter, tracking filter, etc. [3].

In this paper, guidance filtering based on particle filter algorithm is presented against the main problems encountered in guidance information estimation. Particle filter is a filtering method based on random sampling, mainly to solve nonlinear and non-Gaussian problems. Essence of particle filter is a recursive Bayes filter realization form getting rid of the problem that Gaussian distribution must be satisfied when solving nonlinear filtering problem. Thus particle filter is applicable to non-Gaussian nonlinear conditions.

2 Discussed problems relative motion between missile and target

Guidance information estimation are calculating the relative motion state based on the guidance information from seeker and the navigation information of missile, such as relative speed between missile and target, angle of sight, sight angular rate, and so on.

Considered the plane intercept, the missile and target relative motion is shown in Figure 1. The missile and target relative motion can be obtained as follow [4].

The discussed model is given as follows [4].

$$R\dot{q} = v_{\mathrm{m}} \sin(q - \theta_{\mathrm{m}}) - v_{\mathrm{t}} \sin(q - \theta_{\mathrm{t}})$$

$$\dot{R} = -v_{\mathrm{m}} \cos(q - \theta_{\mathrm{m}}) + v_{\mathrm{t}} \cos(q - \theta_{\mathrm{t}})$$

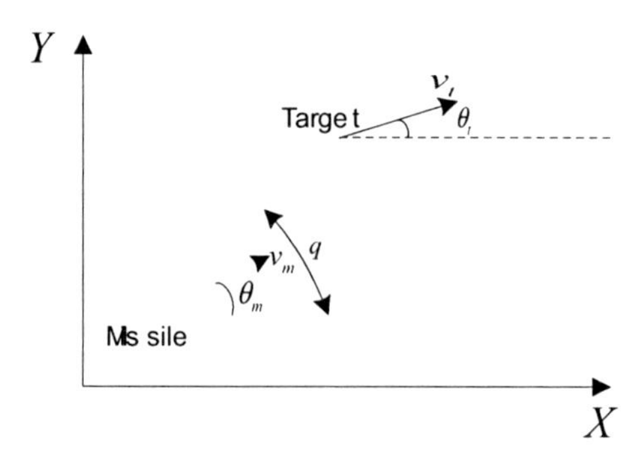

Figure 1: The plane intercept kinematic relations.

In the equations, R is the relative distance, $\dot{R}$ as the relative velocity, q as the line of sight angle, $\dot{q}$ as the angular velocity of the line of sight, v_{m} as the missile speed, θ_{m} as the trajectory angle, v_{t} as the target speed, and θ_{t} as the target speed dip.

The seeker sight angular rate output is only needed in the traditional proportional navigation. And with the improvement of the precision guide, there has been variable guidance laws, such as sliding mode variable structure guidance law, optimal guidance law, etc., which not only need the angular rate information, but also the relative position of missile projects, as well as the relative speed of the target acceleration and other information. Thus guidance filtering algorithms are needed to get the relative motion between target and missile.

3 The nonlinear filtering algorithm

3.1 Particle filter

Particle filter is a filtering method based on random sampling mainly aiming to solve nonlinear non-Gaussian problems [5]. Particle filter is a realization form of recursive Bayes filtering essence, which a series of random samples of the state space are used to approximate the system state posterior probability density function at each moment (particle) state to approximate the posterior probability density function, thereby obtaining the state estimation. The main idea is to use these sample points and weights to get the state minimum variance estimator, which is a statistical method for simulation-based filtering.

The basic steps of particle filter algorithm are as follow:

(1) Initialization: at the time $k=0$, confirm the prior probability density function $p(x_0)$ of target state representation, and extract N sample points $\left\{x_0^i, i=1,\cdots,N\right\}$, and give each particle initial weights $\left\{\lambda_0^i, i=1,\cdots,N\right\}$, $\lambda_0^i = 1/N$,

(2) Prediction that are importance sampling, and weight calculation: when $k=1$, N sample points x_k^i are extracted from transferred prior density function $p\left(x_k \mid x'_{k-1}\right)$, and the weight ($p\left(z_k \mid x_k^i\right)$ is the likelihood density function) is calculated by the formula $\lambda_k^i = \lambda_{k-1}^i p\left(z_k \mid x_k^i\right)$,

(3) Weight normalization: $\lambda_k^i = \lambda_k^i \Big/ \sum_{i=1}^{N} \lambda_k^i$,

(4) Resampling: first, defines the effective sampling scale $N_{\mathrm{eff}} = \dfrac{N}{1+\mathrm{Var}(\lambda_k^i)}$. If $N_{\mathrm{eff}} < N_{\mathrm{th}}$, the state set $\left\{x_k^i, i=1,\cdots,N\right\}$ is

resampled and reconstructed, which is similar to the distribution $p(x_k \mid z_k)$, and then the weight is reset $\lambda_k^i = 1/N$. Otherwise, no re-sampling;

(4) States update: $x_k = \sum_{i=1}^{N} x_k^i \lambda_k^i$.

Random sample particles with no clear format in the particle filter, are not subject to the constraints of model linearity and Gaussian assumption. Particle filter using these random samples can be approximated by the mathematical expectation of any arbitrary function, resulting in approximate optimal numerical solution based on physical model. Therefore, particle filter is suitable for strong non-linear and non-Gaussian situation with high accuracy, fast convergence etc.

3.2 Particle filter based on EKF

In the particle filter algorithm, EKF is used to update particles, and the approximate posterior probability density will be determined as the importance density function, namely

$$q\left(x_k^i \mid x_k^{i-1}, z_k\right) = N\left(\hat{x}_k^i, \hat{P}_k^i\right)$$

Then new particles are generated from the importance density function, and resampling is done after weight update which is the extended particle filter based on EKF [6]. The basic steps of extended particle filter based on EKF are as follows:

(1) Initialization: at the time $k = 0$, confirm the prior probability density function $p(x_0)$ of target state representation, and extract N sample points $\{x_0^i, i = 1, \cdots, N\}$, and give each particle initial weights $\{\lambda_0^i, i = 1, \cdots, N\}$, $\lambda_0^i = 1/N$,

(2) Update particles with EKF:

$$\left[\left\{x_k^i, \hat{P}_k^i\right\}_{i=1}^{N}\right] = \mathrm{EKF}\left[\left\{x_{k-1}^i, \hat{P}_k^i\right\}_{i=1}^{N}, z_k\right]$$

$$x_{k|k-1}^i = f\left(x_{k-1}^i\right)$$

$$P_{k|k-1}^i = F_k^i P_{k-1}\left(F_k^i\right)^T + F_k^i Q_k \left(F_k^i\right)^T$$

$$K_k = P^i_{k|k-1} H^T_k \left(H_k P^i_{k|k-1} H^T_k + R_k \right)^{-1}$$

$$\hat{x}^i_k = \hat{x}^i_{k-1} + K_k \left[Z_k - h\left(\hat{x}^i_{k-1} \right) \right]$$

$$\hat{P}^i_k = \left[I - K_k H_k \right] P^i_{k|k-1}$$

$$x^i_k \sim q\left(x^i_k \mid x^{i-1}_k, z_k \right) = N\left(\hat{x}^i_k, \hat{P}^i_k \right)$$

(3) Weight calculation:

$$\lambda^i_k = \lambda^i_{k-1} \frac{p\left(z_k \mid \hat{x}^i_k \right) p\left(\hat{x}^i_k \mid x^i_{k-1} \right)}{p\left(\hat{x}^i_k \mid x^i_{k-1}, z_{1:k} \right)}$$

(4) Weight normalization: $\lambda^i_k = \lambda^i_k \Big/ \sum_{i=1}^{N} \lambda^i_k$,

(5) Resampling: first, defines the effective sampling scale $N_{\text{eff}} = \dfrac{N}{1 + \text{Var}(\lambda^i_k)}$. If $N_{\text{eff}} < N_{\text{th}}$, the state set $\left\{ x^i_k, i = 1, \cdots, N \right\}$ is resampled and reconstructed, which is similar to the distribution $p\left(x_k \mid z_k \right)$, and then the weight is reset $\lambda^i_k = 1/N$. Otherwise, no re-sampling;

(6) States update $x_k = \sum_{i=1}^{N} x^i_k \lambda^i_k$.

4 Guidance filter design

4.1 Guided filter state equations

Requiring tracking large maneuvering targets, singer model of maneuvering target is chosen to estimate the relative motion state. The model is essentially a zero-mean first-order time-dependent model [5], namely:

$$\dot{a}(t) = -\alpha a(t) + \omega(t)$$

α is the frequency of the motor equal to the reciprocal of the time constant, $\omega(t)$ is white noise with zero mean and $\sigma_\omega = 2\alpha\sigma^2_a$ variance, and σ^2_a is the variance for the target acceleration. According to the target maneuvering model,

$x = [r_x, r_y, v_x, v_y, a_{tx}, a_{ty}]^T$ is chosen as the state variables. Then the filter state equations are set up in Cartesian coordinates as

$$\begin{cases} \dot{r}_x = v_x \\ \dot{r}_y = v_y \\ \dot{v}_x = a_{tx} - a_{mx} \\ \dot{v}_y = a_{ty} - a_{my} \\ \dot{a}_{tx} = -\alpha a_{tx} + w_x(t) \\ \dot{a}_{ty} = -\alpha a_{ty} + w_y(t) \end{cases}$$

In the equations, r_x, r_y are the components of the relative distance vector in the Cartesian coordinate system, v_x, v_y as the components of the relative velocity vector in the Cartesian coordinate system, a_{tx}, a_{ty} as the components of the target acceleration vector in the Cartesian coordinate system, a_{mx}, a_{my} as the components of the missile acceleration vector in the Cartesian coordinate system, w_x, w_y as the system process noise.

4.2 Target tracking measurement equations

According to Figure 1, assume that the measured information contains the target azimuth angle, the relative distance and etc. The target azimuth angle θ and the relative distance R are selected as the observation vector. The measurement equations are set up as follows, in which v_k is the system measurement noise.

$$Z_k = h(x_k) + v_k; \quad h(x_k) = \begin{bmatrix} \sqrt{x^2 + y^2} \\ \arctan\left(\frac{y}{x}\right) \end{bmatrix}$$

5 Simulation analysis

In order to verify the effectiveness of the filtering algorithms for guidance, the simulation model is established, and the simulation parameters are as follows: the target initial position is in (5000 m, 5000 m), the initial velocity is (20 m/s, 0 m/s) the sampling interval is T = 0.01 s, the simulation end time is set to 20 s, the sampling particle number of particle filter algorithm is set 500, while the number of EKPF is set 100.

Figures 2–5 show the position and speed error in X direction and Y direction, respectively, using EKF PF and EKPF. The result shows that particle filter is able to improve the target tracking accuracy effectively in nonlinear non-Gaussian condition, while the particle filter based on EKF is little worse than PF, but better than EKF.

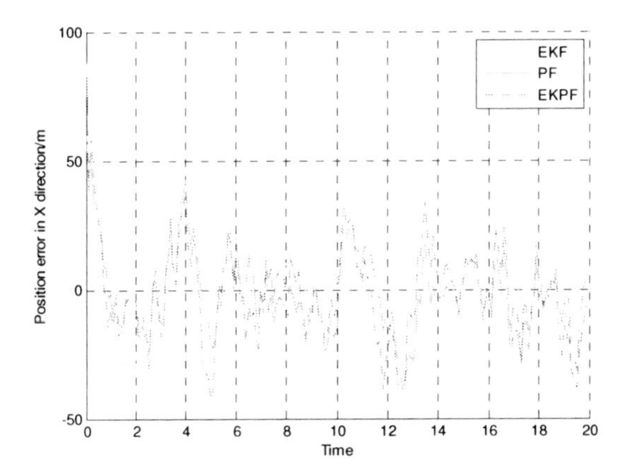

Figure 2: The position error in X direction.

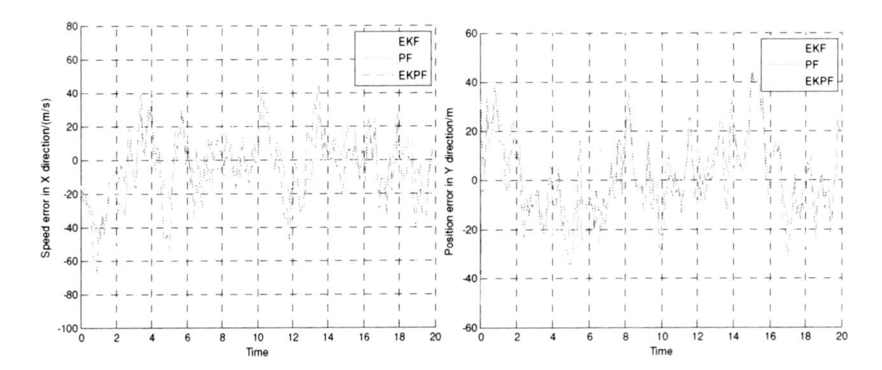

Figure 3: The speed error in X direction. Figure 4: The position error in Y direction.

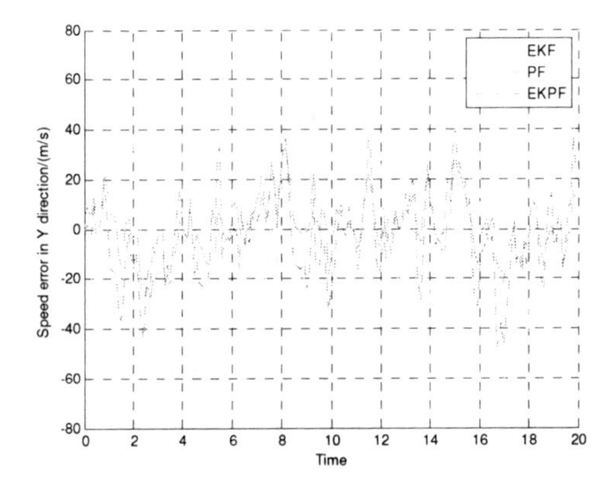

Figure 5: The speed error in Y direction.

Table 1: Filter accuracy comparison between EKF, PF, and EKPF.

	EKF	PF	EKPF
RMSEX (m)	20.86	18.99	17.86
RMSEVx (m/s)	29.45	15.72	18.65
RMSEY (m)	23.00	19.51	18.54
RMSEVy (m/s)	33.33	12.95	17.78
Time used in simulation (s)	0.38	20.38	11.18

The average root mean square error of target tracking is defined as

$$\mathrm{RMSE} = \sqrt{\frac{1}{\mathrm{Totaltime}} \sum_{j=1}^{\mathrm{Totaltime}} \left(X(i) - \hat{X}(i) \right)^2}$$

Table 1 shows the filter accuracy comparison and the time used in simulation between EKF, PF, and EKPF. The result shows that particle filter have better accuracy than EKF and particle filter based on EKF, especially in the speed error, but employ most time. And also, particle filter based on EKF is almost the same accuracy to particle filter, but employ half time of particle filter. Although particle filter based on EKF takes up a lot of calculation, it is all the same an effectual method to improve the enormous-calculation situation of particle filter.

6 Conclusions

In this paper, EKF, PF, and EKPF are presented to solve the nonlinear problem in target tracking, which are verified and compared by simulation. The simulation results show that particle filter gives better performance than EKF and particle filter based on EKF in nonlinear and non-Gaussian environment, existing problem in enormous calculation. Particle filter based on EKF is an effective way to improve the enormous-calculation situation of particle filter, meanwhile assuring the target tracking accuracy. Thus it can be concluded that the improved algorithm of particle filter is the development direction of particle filter in future.

Acknowledgment

The research work was supported by National Natural Science Foundation of China under Grant No. 61065009 and Natural Science Foundation of Qinghai Provincial under Grant No. 2011-z-756.

References

[1] Liu Xingtang, *Missile Guidance Control System Analysis Design and Simulation*. Northwestern Polytechnical University Press, Xian, 2006.

[2] C.C. Lefas, Using roll-angle measurements to track aircraft maneuvers. *IEEE Transactions on Aerospace and Electronic Systems*, **20(6)**, pp. 672–681, 1984.
[3] Venkat Durbha, S.N. Balakrishnan, Target interception with cost-based observer. *AIAA Guidance, Navigation, and Control Conference and Exhibit*, Keystone, Colorado, AIAA 2006–6218, August 21–24, 2006.
[4] Paul Zarchan, *Tactical and Strategic Missile Guidance*. American Institute of Aeronautics and Astronautics, 2012.
[5] N.J. Gordon, D.J. Salmond, A.F.M. Smith, A novel approach to nonlinear/ nongaussian Bayesian state estimation. *IEE Proceedings on Radar and Signal Processing*, **140(2)**, p. 1072113, 1993.
[6] Hu Shiqiang, Jing Zhongliang, *Principle and Application of Particle Filter*. Science Press, Beijing, 2010.

Sliding mode control for spacecraft proximity operations

Yue Chi, Wei Huo
The Seventh Research Division, Science and Technology on Aircraft Control Laboratory, Beihang University, Beijing, PR China

Abstract

Since the middle of the last century, the spacecraft, a carrier for human beings to explore and use the resources of outer space, has attracted more and more attention of the scholars. As an important part of space technology, space rendezvous and docking technology also attracted many scholars to research. In this paper, the high precision control problem of a pursuer spacecraft approaching a space target in proximity missions is investigated. The difficulties of this control problem include strong nonlinear, strong coupling, uncertainty, and external disturbance. By choosing proper sliding mode surfaces, a novel sliding mode controller is presented in presence of model uncertainties and unknown external disturbances. It is proved that the sliding mode surfaces converge to zero in finite time and the relative position and attitude errors of the closed-loop system asymptotically converge to zero. The validity of the proposed control strategy is demonstrated by numerical simulation results.
Keywords: space rendezvous and docking, sliding mode control, spacecraft control, proximity operations, modeling uncertainties, interpolation.

1 Introduction

Space rendezvous and docking has become an important research topic due to the continuous increase of orbit activity such as space station supply, repair, and automated inspection. Various studies have been carried out on the nonlinear control of spacecraft relative position and attitude, especially for spacecraft proximity operations, which is of the most importance in the process of rendezvous and docking.

Z. Zhong developed an attitude tracking sliding mode controller and a relative position tracking non-singular terminal sliding mode controller [1], and

WIT Transactions on Engineering Sciences, Vol. 107, © 2015 WIT Press
www.witpress.com, ISSN 1743-3533 (on-line)
doi:10.2495/SEEEI140641

proved that the closed-loop system is asymptotically stable in finite time. However, the upper bound of the disturbance was assumed to be known in the prior. F. Yong provided introduction of non-singular terminal sliding mode and comparison between non-singular terminal sliding mode and conventional sliding mode [2]. Y. Shuanghe developed a new form of terminal sliding mode for the trajectory tracking of robotic manipulators [3]. Y. Liang listed two fast terminal sliding modes and a non-singular fast terminal sliding mode [4]. Further, it has been verified that the fast sliding modes have better performance than the conventional one. P. Haizhou and K. Vikram used the vectorized formalism modeling position and attitude model of leader and follower spacecrafts in the research of formation flying [5]. The controller was designed to deal with unknown masses and inertial matrices of both the leader and follower spacecrafts. In view of the absence of linear velocity and angular velocity [6], a high-pass filter was introduced to the former controller to estimate the velocities using position and attitude measurements, but the uncertainties of the model were not considered as the previous research [5]. Taking perturbations due to gravitation variations, solar radiation, and atmospheric drag into account, R. Kristiansen derived a nonlinear model of relative position and attitude of leader and follower spacecrafts and presented a passivity-based PD+ controller [7], a sliding surface controller and an integrator backstopping controller based on the model [8]. But uncertainties of the model and the external disturbance were not considered.

In this paper, a sliding mode controller is developed to deal with coupled relative position and attitude dynamical equations with time-varying disturbances and parameter uncertainties, including deviation of the acting point of control force from the mass center of the pursuer, unknown mass of the pursuer, and unknown inertial matrices of both the pursuer spacecraft and the space target.

2 Modeling

2.1 Pursuer and target dynamics

The coordinate systems used in this paper are inertial frame $\mathcal{N}=\{O_{\mathcal{N}},i_{\mathcal{N}},j_{\mathcal{N}},k_{\mathcal{N}}\}$, pursuer body-fixed frame $\mathcal{P}=\{O_{\mathcal{P}},i_{\mathcal{P}}, j_{\mathcal{P}},k_{\mathcal{P}}\}$ and target body-fixed frame $\mathcal{T}=\{O_{\mathcal{T}},i_{\mathcal{T}},j_{\mathcal{T}},k_{\mathcal{T}}\}$ as shown in Figure 1. Point P is the desired pursuer proximity position along the direction of target docking port. The pursuer position and attitude models expressed in frame $\mathcal{P}$ are given as follows [9]:

$$\begin{aligned}
&\dot{\boldsymbol{r}}=\boldsymbol{v}-\boldsymbol{S}(\boldsymbol{\omega})\boldsymbol{r},\ \ m\dot{\boldsymbol{v}}+m\boldsymbol{S}(\boldsymbol{\omega})\boldsymbol{v}=\boldsymbol{f}_c+\boldsymbol{d}_f\\
&\dot{\boldsymbol{\sigma}}=[(1-\boldsymbol{\sigma}^{\mathrm{T}}\boldsymbol{\sigma})\mathrm{I}_3+2\boldsymbol{S}(\boldsymbol{\sigma})+2\boldsymbol{\sigma}\boldsymbol{\sigma}^{\mathrm{T}}]\boldsymbol{\omega}/4\\
&\boldsymbol{J}\dot{\boldsymbol{\omega}}+\boldsymbol{S}(\boldsymbol{\omega})\boldsymbol{J}\boldsymbol{\omega}=\boldsymbol{S}(\boldsymbol{\rho}_c)\boldsymbol{f}_c+\boldsymbol{\tau}_c+\boldsymbol{d}_\tau
\end{aligned}\tag{1}$$

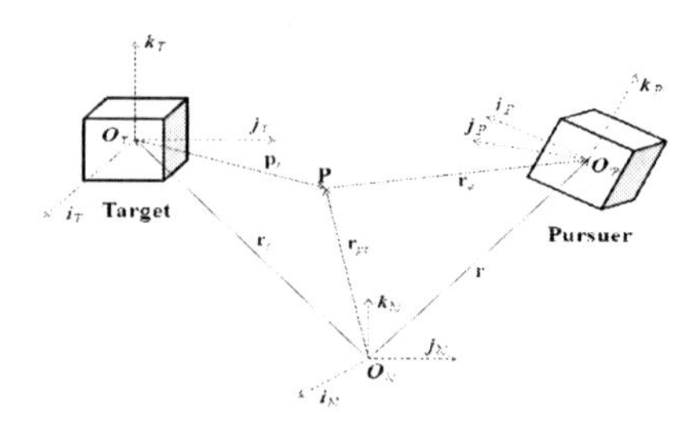

Figure 1: Definition of frames and vectors.

where $\boldsymbol{r} \in \mathbb{R}^3$ and $\boldsymbol{v} \in \mathbb{R}^3$ are position and velocity of mass center $O_{\mathcal{P}}$, $\boldsymbol{\sigma} \in \mathbb{R}^3$ and $\boldsymbol{\omega} \in \mathbb{R}^3$ are the Modified Rodrigues Parameters (MRP) describing attitude and the angular velocity of the pursuer with respect to frame $\mathcal{N}$, m, and $\boldsymbol{J} \in \mathbb{R}^{3\times3}$ are mass and inertial matrix, $\boldsymbol{f}_c$ and $\boldsymbol{\tau}_c$ are control force and control torque applied to the pursuer, $\boldsymbol{\rho}_c$ is deviation of the acting point of $\boldsymbol{f}_c$ from mass center of the pursuer, $\boldsymbol{d}_f$ and $\boldsymbol{d}_\tau$ are disturbance force and disturbance torque, $\boldsymbol{S}(\boldsymbol{a})$ is the skew symmetric matrix derived from vector $\boldsymbol{a} \in \mathbb{R}^3$, such that for any $\boldsymbol{b} \in \mathbb{R}^3$, $\boldsymbol{a} \times \boldsymbol{b} = \boldsymbol{S}(\boldsymbol{a})\boldsymbol{b}$. The target position and attitude models expressed in frame $\mathcal{T}$ are given as follows [9]:

$$\begin{aligned} &\dot{\boldsymbol{r}}_t = \boldsymbol{v}_t - \boldsymbol{S}(\boldsymbol{\omega}_t)\boldsymbol{r}_t, \quad m_t\dot{\boldsymbol{v}}_t + m_t\boldsymbol{S}(\boldsymbol{\omega}_t)\boldsymbol{v}_t = \boldsymbol{0} \\ &\dot{\boldsymbol{\sigma}}_t = [(1-\boldsymbol{\sigma}_t^{\mathrm{T}}\boldsymbol{\sigma}_t)\mathrm{I}_3 + 2S(\boldsymbol{\sigma}_t) + 2\boldsymbol{\sigma}_t\boldsymbol{\sigma}_t^{\mathrm{T}}]\boldsymbol{\omega}_t/4 \\ &\boldsymbol{J}_t\dot{\boldsymbol{\omega}}_t + \boldsymbol{S}(\boldsymbol{\omega}_t)\boldsymbol{J}_t\boldsymbol{\omega}_t = \boldsymbol{0} \end{aligned} \tag{2}$$

where $\boldsymbol{r}_t \in \mathbb{R}^3$ and $\boldsymbol{v}_t \in \mathbb{R}^3$ are position and velocity of mass center $O_{\mathcal{T}}$, $\boldsymbol{\sigma}_t \in \mathbb{R}^3$, and $\boldsymbol{\omega}_t \in \mathbb{R}^3$ are MRP and angular velocity of the target with respect to frame $\mathcal{N}$. m_t and $\boldsymbol{J}_t \in \mathbb{R}^{3\times3}$ are mass and inertial matrix, respectively.

2.2 Relative motion dynamics

From Figure 1, the position and velocity of point P in frame $\mathcal{T}$ are[10]:

$$\boldsymbol{r}_{pt} = \boldsymbol{r}_t + \boldsymbol{p}_t, \quad \boldsymbol{v}_{pt} = \boldsymbol{v}_t + \boldsymbol{S}(\boldsymbol{\omega}_t)\boldsymbol{p}_t \tag{3}$$

The pursuer position, velocity, and angular velocity relative to the target spacecraft expressed in frame $\mathcal{P}$ as:

$$\boldsymbol{r}_e = \boldsymbol{r} - \boldsymbol{R}\boldsymbol{r}_{pt}, \quad \boldsymbol{v}_e = \boldsymbol{v} - \boldsymbol{R}\boldsymbol{v}_{pt}, \quad \boldsymbol{\omega}_e = \boldsymbol{\omega} - \boldsymbol{R}\boldsymbol{\omega}_t \tag{4}$$

where $\boldsymbol{R} = -(4(1-\boldsymbol{\sigma}_e^{\mathrm{T}}\boldsymbol{\sigma}_e)S(\boldsymbol{\sigma}_e) - 8S(\boldsymbol{\sigma}_e)^{\mathrm{T}}S(\boldsymbol{\sigma}_e))/(1+\boldsymbol{\sigma}_e^{\mathrm{T}}\boldsymbol{\sigma}_e)^2 + \mathbf{I}_3$, $\mathbf{I}_3$ is the rotation matrix from $\mathcal{T}$ to $\mathcal{P}$. By substituting Eq. (4) into Eq. (1), using Eqs. (2) and (3), the pursuer position and attitude relative to the target spacecraft in frame $\mathcal{P}$ are modeled as:

$$\dot{\boldsymbol{r}}_e = \boldsymbol{v}_e - \boldsymbol{S}(\boldsymbol{\omega})\boldsymbol{r}_e,\ \ m\dot{\boldsymbol{v}}_e = -m\boldsymbol{g} - \boldsymbol{n}_1 + \boldsymbol{f}_c + \boldsymbol{d}_f \tag{5}$$

$$\begin{aligned} \dot{\boldsymbol{\sigma}}_e &= \boldsymbol{G}(\boldsymbol{\sigma}_e)\boldsymbol{\omega}_e \\ \boldsymbol{J}\dot{\boldsymbol{\omega}}_e &= -\boldsymbol{S}(\boldsymbol{\omega})\boldsymbol{J}\boldsymbol{\omega} + \boldsymbol{J}\boldsymbol{S}(\boldsymbol{\omega}_e)\boldsymbol{\omega} + \boldsymbol{n}_2 + \boldsymbol{S}(\boldsymbol{\rho}_c)\boldsymbol{f}_c + \boldsymbol{\tau}_c + \boldsymbol{d}_\tau \end{aligned} \tag{6}$$

where $\boldsymbol{r}_e$ and $\boldsymbol{v}_e$ are relative position and velocity, $\boldsymbol{\omega}_e$ is relative angular velocity, $\boldsymbol{\sigma}_e = (\boldsymbol{\sigma}_t(\boldsymbol{\sigma}^T\boldsymbol{\sigma} - 1) + \boldsymbol{\sigma}(1-\boldsymbol{\sigma}_t^T\boldsymbol{\sigma}_t) + 2$

$\boldsymbol{S}(\boldsymbol{\sigma}_t)\boldsymbol{\sigma})/(1+\boldsymbol{\sigma}^T\boldsymbol{\sigma}\boldsymbol{\sigma}_t^T\boldsymbol{\sigma}_t + 2\boldsymbol{\sigma}_t^T\boldsymbol{\sigma})$ is the MRP describing relative attitude, $\boldsymbol{g}=\boldsymbol{S}(\boldsymbol{\omega})\boldsymbol{v}_e + \boldsymbol{S}(\boldsymbol{\omega}-\boldsymbol{\omega}_e)^2\boldsymbol{R}\boldsymbol{p}_t$, $\boldsymbol{G}(\boldsymbol{\sigma}_e)=[(1-\boldsymbol{\sigma}_e^{\mathrm{T}}\boldsymbol{\sigma}_e)\mathrm{I}_3 + 2$, $\boldsymbol{S}(\boldsymbol{\sigma}_e) + 2\boldsymbol{\sigma}_e\boldsymbol{\sigma}_e^{\mathrm{T}}]/4$, $\boldsymbol{n}_1 = m\boldsymbol{R}\boldsymbol{S}(\boldsymbol{p}_t)\boldsymbol{J}_t^{-1}\boldsymbol{S}(\boldsymbol{R}^{\mathrm{T}}(\boldsymbol{\omega}-\boldsymbol{\omega}_e))\boldsymbol{J}_t\boldsymbol{R}^{\mathrm{T}}(\boldsymbol{\omega}-\boldsymbol{\omega}_e)$,

$\boldsymbol{n}_2 = \boldsymbol{J}\boldsymbol{R}\boldsymbol{J}_t^{-1}\boldsymbol{S}(\boldsymbol{R}^{\mathrm{T}}(\boldsymbol{\omega}-\boldsymbol{\omega}_e))\boldsymbol{J}_t\boldsymbol{R}^{\mathrm{T}}(\boldsymbol{\omega}-\boldsymbol{\omega}_e)$, $\boldsymbol{R}$ is the rotation matrix from $\mathcal{T}$ to $\mathcal{P}$, $\boldsymbol{p}_t$ is shown in Figure 1.

3 Sliding mode controller

3.1 Control objective

In this paper, we assume that $m, \boldsymbol{J}, \boldsymbol{J}_t, \boldsymbol{\rho}_c$ are unknown, $\hat{\boldsymbol{m}}, \hat{\boldsymbol{J}}, \hat{\boldsymbol{J}}_t, \hat{\boldsymbol{\rho}}_c$ are their estimates, respectively. $\hat{\boldsymbol{J}}_t$ is a positive diagonal matrix, $\boldsymbol{d}_f$ and $\boldsymbol{d}_t$ are unknown and meet the bounded conditions, $\|\boldsymbol{d}_f\| \le \rho_f$ and $\|\boldsymbol{d}_\tau\| \le \rho_\tau$. $\hat{\rho}_f$, $\hat{\rho}_\tau$ are estimates of ρ_f and ρ_τ , $\tilde{m} = m - \hat{m}$, $\tilde{\boldsymbol{\rho}}_c = \boldsymbol{\rho}_c - \hat{\boldsymbol{\rho}}_c$, $\tilde{\rho}_f = \rho_f - \hat{\rho}_f$, $\tilde{\rho}_\tau = \rho_\tau - \hat{\rho}_\tau$, $\tilde{\boldsymbol{J}} = \boldsymbol{J} - \hat{\boldsymbol{J}}$, $\tilde{\boldsymbol{J}}_t = \boldsymbol{J}_t - \hat{\boldsymbol{J}}_t$ and meet $|\tilde{m}| \le \bar{m}$, $|\tilde{\rho}_{ci}| \le \bar{\rho}_{ci}$, $i = 1,2,3$, $\bar{\boldsymbol{\rho}}_c = [\bar{\rho}_{c1}, \bar{\rho}_{c2}, \bar{\rho}_{c3}]^{\mathrm{T}}$, $|\tilde{\rho}_f| \le \bar{\rho}_f$, $|\tilde{\rho}_\tau| \le \bar{\rho}_\tau$, $\tilde{\boldsymbol{J}} = [\tilde{j}_{ij}] \in \mathbb{R}^{3\times3}$, $\tilde{j}_{ij} = \tilde{j}_{ji}$, $i \ne j$, $|\tilde{j}_{ij}| \le \bar{j}_{ij}$, $\bar{\boldsymbol{J}} = [\bar{j}_{ij}] \in \mathbb{R}^{3\times3}$, $\tilde{\boldsymbol{J}}_t = [\tilde{j}_{tij}] \in \mathbb{R}^{3\times3}$, $\tilde{j}_{tij} = \tilde{j}_{tji}$, $i \ne j$, $|\tilde{j}_{tij}| \le \bar{j}_{tij}$, $\bar{\boldsymbol{J}}_t = [\bar{j}_{tij}] \in \mathbb{R}^{3\times3}$. All estimates $\{\hat{m}, \hat{\boldsymbol{J}}, \hat{\boldsymbol{J}}_t, \hat{\boldsymbol{\rho}}_c, \hat{\rho}_f, \hat{\rho}_\tau\}$ and the bounds of mismatches between the estimated and actual parameters $\{\bar{m}, \bar{\boldsymbol{J}}, \bar{\boldsymbol{J}}_t, \bar{\boldsymbol{\rho}}_c, \bar{\rho}_f, \bar{\rho}_\tau\}$ are prior known. The pursuer desired proximity distance along the direction of target docking port $\boldsymbol{p}_t$ represented in frame $\mathcal{T}$ is given in advance. Our control objective is that, in presence of modeling uncertainties $\{m, \boldsymbol{J}, \boldsymbol{J}_t, \boldsymbol{\rho}_c\}$ and unknown bounded disturbances $\{\boldsymbol{d}_f, \boldsymbol{d}_\tau\}$, design sliding mode controllers $\boldsymbol{f}_c$ and $\boldsymbol{\tau}_c$ with accessible measurement information $\{\boldsymbol{r}_e, \boldsymbol{v}_e, \boldsymbol{\sigma}_e, \boldsymbol{\omega}_e, \boldsymbol{v}, \boldsymbol{\sigma}, \boldsymbol{\omega}\}$ such that $\{\mathrm{r}_e, \mathrm{v}_e, \mathrm{s}_e, \mathrm{w}_e\} \to 0$ as $t \to \infty$.

3.2 Sliding mode controller

Select sliding surface

$$s_1 = v_e + k_1 \mathrm{diag}(\mathrm{sgn}(r_e))|r_e|^{r_1}, s_2 = \omega_e + k_2 \mathrm{diag}(\mathrm{sgn}(\sigma_e))|\sigma_e|^{r_2} \quad (7)$$

where k_1, $k_2 > 0$, $1 < r_1, r_2 < 2$, $\mathrm{sgn}(a) = (\mathrm{sgn}(a_1), \mathrm{sgn}(a_2), \quad \mathrm{sgn}(a_3))^{\mathrm{T}}$, $\mathrm{diag}(a) = \mathrm{diag}(a_1, a_2, a_3)$, $|a|^r = (|a_1|^r, |a_2|^r, |a_3|^r)^{\mathrm{T}}$, $\mathrm{sgn}(a_1) = 0$, $a_1 = 0$.

Lemma 1: the derivation of Eq. (7) are

$$\dot{s}_1 = \dot{v}_e + r_1 k_1 \mathrm{diag}(|r_e|^{r_1 - 1})\dot{r}_e, \ \dot{s}_2 = \dot{\omega}_e + r_2 k_2 \mathrm{diag}(|\sigma_e|^{r_2 - 1})\dot{\sigma}_e \quad (8)$$

Proof: Eq. (7) can be expressed as

$$s_{1i} = \begin{cases} v_{ei} + k_1 r_{ei}^{r_1}, & r_{ei} > 0 \\ v_{ei}, & r_{ei} = 0 \\ v_{ei} - k_1(-r_{ei})^{r_1}, & r_{ei} < 0 \end{cases}, s_{2i} = \begin{cases} \omega_{ei} + k_2 \sigma_{ei}^{r_2}, & \sigma_{ei} > 0 \\ \omega_{ei}, & \sigma_{ei} = 0 \\ \omega_{ei} - k_2(-\sigma_{ei})^{r_2}, & \sigma_{ei} < 0 \end{cases}$$

$$i = 1, 2, 3 \quad (9)$$

it is apparent that Eq. (9) is continuous when $r_{ei} > 0$, $r_{ei} < 0$ or $\sigma_{ei} > 0$, $\sigma_{ei} < 0$, $i = 1, 2, 3$. Actually it is still continuous at $r_{ei} = 0$, $\sigma_{ei} = 0$, because we have the partial derivatives at $r_{ei} = 0$, $\sigma_{ei} = 0$. Therefore, the derivation of Eq. (7) can be expressed as Eq. (8).

We begin by defining a linear operator $L_1 : \mathbb{R}^3 \to \mathbb{R}^{3\times6}$ acting on vector $a = [a_1, a_2, a_3]^T$ as

$$L_1(a) \triangleq \begin{bmatrix} a_1 & 0 & 0 & 0 & a_3 & a_2 \\ 0 & a_2 & 0 & a_3 & 0 & a_1 \\ 0 & 0 & a_3 & a_2 & a_1 & 0 \end{bmatrix}$$

such that $Ja = L_1(a)\theta_j$, $\theta_j \triangleq [j_{11}, j_{22}, j_{33}, j_{23}, j_{13}, j_{12}]^{\mathrm{T}}$.

From Lemma 1 and Eqs. (5) and (6)

$$\begin{aligned} m\dot{s}_1 &= m\dot{v}_e + r_1 k_1 m \mathrm{diag}(|r_e|^{r_1 - 1})\dot{r}_e \\ &= f_c + d_f - n_1 + m(-g + r_1 k_1 \mathrm{diag}(|r_e|^{r_1 - 1}) \cdot \text{..........} \quad \text{.. (10)} \\ &(v_e - S(\omega)r_e)) \triangleq f_c + d_f - n_1 + mq_1 \end{aligned}$$

$$
\begin{aligned}
\boldsymbol{J}\dot{\boldsymbol{s}}_2 &= \boldsymbol{J}\dot{\boldsymbol{\omega}}_e + r_2 k_2 \boldsymbol{J}\mathrm{diag}(|\boldsymbol{\sigma}_e|^{r_2-1})\dot{\boldsymbol{\sigma}}_e \\
&= (-\boldsymbol{S}(\boldsymbol{\omega})\boldsymbol{L}_1(\boldsymbol{\omega}) + \boldsymbol{L}_1(\boldsymbol{S}(\boldsymbol{\omega}_e)\boldsymbol{\omega}) + \boldsymbol{L}_1(r_2 k_2 \mathrm{diag} \\
&\quad (|\boldsymbol{\sigma}_e|^{r_2-1})\boldsymbol{G}(\boldsymbol{\sigma}_e)\boldsymbol{\omega}_e))\boldsymbol{\theta}_j + \boldsymbol{n}_2 + \boldsymbol{S}(\boldsymbol{\rho}_c)\boldsymbol{f}_c + \boldsymbol{\tau}_c + \boldsymbol{d}_\tau \\
&\triangleq \boldsymbol{Q}_2\boldsymbol{\theta}_j + \boldsymbol{n}_2 - \boldsymbol{S}(\boldsymbol{f}_c)\boldsymbol{\rho}_c + \boldsymbol{\tau}_c + \boldsymbol{d}_\tau
\end{aligned} \tag{11}
$$

The controllers are designed as

$$
\begin{aligned}
\boldsymbol{f}_c &= -\hat{\rho}_f \,\mathrm{sgn}(\boldsymbol{s}_1) - \hat{m}\boldsymbol{q}_1 + \boldsymbol{d}_1 \\
\boldsymbol{\tau}_c &= -\hat{\rho}_\tau \,\mathrm{sgn}(\boldsymbol{s}_2) - \boldsymbol{Q}_2\hat{\boldsymbol{\theta}}_j + \boldsymbol{S}(\boldsymbol{f}_c)\hat{\boldsymbol{\rho}}_c + \boldsymbol{d}_2
\end{aligned} \tag{12}
$$

where $\boldsymbol{d}_1$ and $\boldsymbol{d}_2$ are vectors to be designed later. $\hat{\boldsymbol{\theta}}_j$ are estimates of $\boldsymbol{\theta}_j$ calculated by using $\hat{\boldsymbol{J}}$. The mismatches between the actual and estimated parameters are defined as $\tilde{m} = m - \hat{m}$, $\tilde{\rho}_f = \rho_f - \hat{\rho}_f$, $\tilde{\boldsymbol{\theta}}_j = \boldsymbol{\theta}_j - \hat{\boldsymbol{\theta}}_j$, $\tilde{\boldsymbol{\rho}}_c = \boldsymbol{\rho}_c - \hat{\boldsymbol{\rho}}_c$, and $\tilde{\rho}_\tau = \rho_\tau - \hat{\rho}_\tau$.

Lemma 3 [11]: assume that a continuous, positive-definite function $v(t)$ satisfies the following differential inequality

$$
\dot{v}(t) \le -\alpha v^{\eta}(t),\ \forall t \ge t_0,\ v(t_0) \ge 0
$$

where $\alpha > 0$, $0 < \eta < 1$ are constants. Then, for any given t_0, $v(t)$ satisfies the inequality: $v^{1-\eta}(t) \le v^{1-\eta}(t_0) - \alpha(1-\eta)(t - t_0)$, $t_0 \le t \le t_r$, and equality: $v(t) = 0$, $\forall t \ge t_r$, where $t_r = t_0 + v^{1-\eta}(t_0)/\alpha(1-\eta)$. Denote symmetric matrix $\boldsymbol{J} = [j_{11}, j_{12}, j_{13}; j_{12}, j_{22}, j_{23}; j_{13}, j_{23}, j_{33}]$ and define $\boldsymbol{x} \triangleq \boldsymbol{R}^{\mathrm{T}}(\boldsymbol{\omega} - \boldsymbol{\omega}_e)$.

$$
\begin{aligned}
\boldsymbol{n}_1 &= m\boldsymbol{R}\boldsymbol{S}(\boldsymbol{p}_t)\boldsymbol{J}_t^{-1}\boldsymbol{S}(\boldsymbol{x})\boldsymbol{J}_t\boldsymbol{x} \\
&= (\tilde{m} + \hat{m})\boldsymbol{R}\boldsymbol{S}(\boldsymbol{p}_t)(\tilde{\boldsymbol{J}}_t + \hat{\boldsymbol{J}}_t)^{-1}\boldsymbol{S}(\boldsymbol{x})(\tilde{\boldsymbol{J}}_t + \hat{\boldsymbol{J}}_t)\boldsymbol{x}
\end{aligned} \tag{13}
$$

$$
\begin{aligned}
\boldsymbol{n}_2 &= \boldsymbol{J}\boldsymbol{R}\boldsymbol{J}_t^{-1}\boldsymbol{S}(\boldsymbol{x})\boldsymbol{J}_t\boldsymbol{x} \\
&= (\tilde{\boldsymbol{J}} + \hat{\boldsymbol{J}})\boldsymbol{R}(\tilde{\boldsymbol{J}}_t + \hat{\boldsymbol{J}}_t)^{-1}\boldsymbol{S}(\boldsymbol{x})(\tilde{\boldsymbol{J}}_t + \hat{\boldsymbol{J}}_t)\boldsymbol{x}
\end{aligned} \tag{14}
$$

Besides, $(\hat{\boldsymbol{J}}_t + \tilde{\boldsymbol{J}}_t)^{-1} = \hat{\boldsymbol{J}}_t^{-1}(\mathbf{I}_3 + \tilde{\boldsymbol{J}}\hat{\boldsymbol{J}}_t^{-1})^{-1} = \hat{\boldsymbol{J}}_t^{-1}[\mathbf{I}_3 + (\mathbf{I}_3 + \tilde{\boldsymbol{J}}\hat{\boldsymbol{J}}_t^{-1})^{-1} - \mathbf{I}_3] = \hat{\boldsymbol{J}}_t^{-1}(\mathbf{I}_3 + \boldsymbol{H})$, where $\boldsymbol{H} = (\mathbf{I}_3 + \tilde{\boldsymbol{J}}\hat{\boldsymbol{J}}_t^{-1})^{-1} - \mathbf{I}_3 = [h_{ij}]$ denote $\hat{j}_m = \min\{\hat{\boldsymbol{J}}_{tii}\}(i = 1,2,3)$, $\overline{j}_m = \max(\overline{j}_{tij})(i, j = 1,2,3)$, when the upper bound of the uncertainties meets $\overline{j}_m < \hat{j}_m/12$, we can get a non-negative matrix $\overline{\boldsymbol{H}} = [\overline{h}_{ij}] \in \mathbb{R}^{3\times3}$, $\overline{h}_{ij} = 3\overline{j}_m/(\hat{j}_m - 3\overline{j}_m)$ so that $|\overline{h}_{ij}| \le h_{ij}(i, j = 1,2,3)$ [12].

Select a non-negative function

$$v_1 = \boldsymbol{s}_1^{\mathrm{T}} m \boldsymbol{s}_1 / 2 \tag{15}$$

Computing its derivation, from Eq. (10), (12) and $\|\boldsymbol{a}\| \leq \|\boldsymbol{a}\|_1$,

$$\begin{aligned}
\dot{v}_1 &= \boldsymbol{s}_1^{\mathrm{T}} m \dot{\boldsymbol{s}}_1 = \boldsymbol{s}_1^{\mathrm{T}} \boldsymbol{f}_c + \boldsymbol{s}_1^{\mathrm{T}} \boldsymbol{d}_f - \boldsymbol{s}_1^{\mathrm{T}} \boldsymbol{n}_1 + m \boldsymbol{s}_1^{\mathrm{T}} \boldsymbol{q}_1 \\
&= -\hat{\rho}_f \boldsymbol{s}_1^{\mathrm{T}} \operatorname{sgn}(\boldsymbol{s}_1) - \hat{m} \boldsymbol{s}_1^{\mathrm{T}} \boldsymbol{q}_1 + \boldsymbol{s}_1^{\mathrm{T}} (\boldsymbol{d}_1 + \boldsymbol{d}_f - \boldsymbol{n}_1) + m \boldsymbol{s}_1^{\mathrm{T}} \boldsymbol{q}_1 \\
&\leq -\hat{\rho}_f \|\boldsymbol{s}_1\|_1 + \tilde{m} \boldsymbol{s}_1^{\mathrm{T}} \boldsymbol{q}_1 - \boldsymbol{s}_1^{\mathrm{T}} \boldsymbol{n}_1 + \boldsymbol{s}_1^{\mathrm{T}} \boldsymbol{d}_1 + \rho_f \|\boldsymbol{s}_1\|_1 \\
&= \tilde{\rho}_f \|\boldsymbol{s}_1\|_1 + \tilde{m} \boldsymbol{s}_1^{\mathrm{T}} \boldsymbol{q}_1 - \boldsymbol{s}_1^{\mathrm{T}} \boldsymbol{n}_1 + \boldsymbol{s}_1^{\mathrm{T}} \boldsymbol{d}_1
\end{aligned} \tag{16}$$

For $\boldsymbol{C} = [c_{ij}]$, define $\underline{\boldsymbol{C}} = [|\, c_{ij} \,|]$, select

$$\begin{aligned}
\boldsymbol{d}_1 = &-\bar{\rho}_f \operatorname{sgn}(\boldsymbol{s}_1) - \operatorname{diag}(\operatorname{sgn}(\boldsymbol{s}_1))(\bar{m} \underline{\boldsymbol{q}}_1 + k_3 |\boldsymbol{s}_1|^{r_3}) - \operatorname{diag}(\operatorname{sgn} \\
&(\boldsymbol{s}_1))((\bar{m} + \hat{m}) \boldsymbol{R} \boldsymbol{S}(\boldsymbol{p}_t) \hat{\boldsymbol{J}}_t^{-1} (\mathbf{I}_3 + \bar{\boldsymbol{H}}) \boldsymbol{S}(\boldsymbol{x})(\bar{\boldsymbol{J}}_t + \hat{\boldsymbol{J}}_t) \boldsymbol{x})
\end{aligned} \tag{17}$$

where $k_3 > 0$, $0 < r_3 < 1$. Substituting Eq. (17) into Eq. (16) and using Lemma 2 lead to following expression:

$$\begin{aligned}
\dot{v}_1 &\leq -k_3 \boldsymbol{s}_1^{\mathrm{T}} \operatorname{diag}(\operatorname{sgn}(\boldsymbol{s}_1)) |\boldsymbol{s}_1|^{r_3} \\
&\leq -k_3 (|s_{11}|^2 + |s_{12}|^2 + |s_{13}|^2)^{(r_3+1)/2} = -k_3 (2 v_1 / m)^{(r_3+1)/2}
\end{aligned} \tag{18}$$

According to the Lemma 3, the system Eq. (5) will reach $\boldsymbol{s}_1 = 0$, within a finite time $t_{r1} = (m^{(r_3+1)/2} (2 v_1(0))^{(1-r_3)/2}) / k_3 (1 - r_3)$. When confined to $\boldsymbol{s}_1 = 0$, substituting Eq. (7) into Eq. (5), we have

$$\dot{\boldsymbol{r}}_e = -k_1 \operatorname{diag}(\operatorname{sgn}(\boldsymbol{r}_e)) |\boldsymbol{r}_e|^{r_1} - \boldsymbol{S}(\boldsymbol{\omega}) \boldsymbol{r}_e \tag{19}$$

Select a non-negative function $v_2 = \boldsymbol{r}_e^{\mathrm{T}} \boldsymbol{r}_e / 2$, computing the derivation of v_2 and using Eq. (19), we have

$$\dot{v}_2 = -k_1 \boldsymbol{r}_e^{\mathrm{T}} \operatorname{diag}(\operatorname{sgn}(\boldsymbol{r}_e)) |\boldsymbol{r}_e|^{r_1} = -k_1 \left(\|\boldsymbol{r}_e\|^{1+r_1} \right) \leq 0 \qquad (20$$

Besides,

$$\int_{t_0}^{\infty} \|\boldsymbol{r}_e\|^{1+r_1} \mathrm{d}t = -(1/k_1) \int_{t_0}^{\infty} \dot{v}_2 \mathrm{d}t \leq v_2(t_0)/k_1 < \infty, \forall t_0 \geq 0$$

which implies that $\boldsymbol{r}_e \in L_{1+r_1}$. Choose a non-negative function $v_3 = \boldsymbol{\omega}_t^{\mathrm{T}} \boldsymbol{J}_t \boldsymbol{\omega}_t / 2$, in view of Eq. (2) and taking the time derivation of v_3 lead to:

$$\dot{v}_3 = \boldsymbol{\omega}_t^{\mathrm{T}} \boldsymbol{J}_t \dot{\boldsymbol{\omega}}_t = \boldsymbol{\omega}_t^{\mathrm{T}} (-\boldsymbol{S}(\boldsymbol{\omega}_t) \boldsymbol{J}_t \boldsymbol{\omega}_t) = \mathbf{0} \tag{21}$$

it follows that v_3 is a constant, which implies that $\boldsymbol{\omega}_t \in L_\infty$. From Eq. (7) and $0 \le \boldsymbol{\sigma}_e^{\mathrm{T}} \boldsymbol{\sigma}_e \le 1$, we know that $\boldsymbol{\sigma}_e, \boldsymbol{\omega}_e \in L_\infty$. From Eqs (4) and (19), we know that $\boldsymbol{\omega} \in L_\infty$, $\dot{\boldsymbol{r}}_e \in L_\infty$. From Babalat Lemma [13], we conclude that $\boldsymbol{r}_e \to 0$ as $t \to \infty$. From Eq. (7), we know that $\boldsymbol{v}_e \to 0$ as $t \to \infty$. Select a non-negative function $v_4 = \boldsymbol{s}_2^{\mathrm{T}} \boldsymbol{J} \boldsymbol{s}_2 / 2$, computing its derivation, from Eqs. (11), (12) and $\|\boldsymbol{a}\| \le \|\boldsymbol{a}\|_1$,

$$\begin{aligned} \dot{v}_4 &= \boldsymbol{s}_2^{\mathrm{T}} \boldsymbol{J} \dot{\boldsymbol{s}}_2 = \boldsymbol{s}_2^{\mathrm{T}} \boldsymbol{Q}_2 \boldsymbol{\theta}_j + \boldsymbol{s}_2^{\mathrm{T}} \boldsymbol{n}_2 - \boldsymbol{s}_2^{\mathrm{T}} \boldsymbol{S}(\boldsymbol{f}_c) \boldsymbol{\rho}_c + \boldsymbol{s}_2^{\mathrm{T}} \boldsymbol{\tau}_c + \boldsymbol{s}_2^{\mathrm{T}} \boldsymbol{d}_\tau \\ &\le \boldsymbol{s}_2^{\mathrm{T}} \boldsymbol{Q}_2 \tilde{\boldsymbol{\theta}}_j + \boldsymbol{s}_2^{\mathrm{T}} \boldsymbol{n}_2 - \boldsymbol{s}_2^{\mathrm{T}} \boldsymbol{S}(\boldsymbol{f}_c) \tilde{\boldsymbol{\rho}}_c + \tilde{\rho}_\tau \|\boldsymbol{s}_2\|_1 + \boldsymbol{s}_2^{\mathrm{T}} \boldsymbol{d}_2 \end{aligned} \tag{22}$$

Select

$$\begin{aligned} \boldsymbol{d}_2 = &-\bar{\rho}_\tau \mathrm{sgn}(\boldsymbol{s}_2) - \mathrm{diag}(\mathrm{sgn}(\boldsymbol{s}_2))(\underline{\boldsymbol{S}}(\boldsymbol{f}_c)\bar{\boldsymbol{\rho}}_c + \underline{\boldsymbol{Q}}_2 \bar{\boldsymbol{\theta}}_j \\ &+ k_4 |\boldsymbol{s}_2|^{r_4} + (\bar{\boldsymbol{J}} + \hat{\boldsymbol{J}}) \boldsymbol{R} \hat{\boldsymbol{J}}_t^{-1} (\mathbf{I}_3 + \bar{\boldsymbol{H}}) \boldsymbol{S}(\boldsymbol{x})(\bar{\boldsymbol{J}}_t + \hat{\boldsymbol{J}}_t)\boldsymbol{x}) \end{aligned} \tag{23}$$

where $k_4 > 0$, $0 < r_4 < 1$, $\bar{\boldsymbol{\theta}}_j$ is the bound of $\tilde{\boldsymbol{\theta}}_j$, which is calculated by using $\bar{\boldsymbol{J}}$. Substituting Eq. (23) into Eq. (22) and using Lemma 2 lead to following expression:

$$\begin{aligned} \dot{v}_4 &\le -k_4 \boldsymbol{s}_2^{\mathrm{T}} \mathrm{diag}(\mathrm{sgn}(\boldsymbol{s}_2)) |\boldsymbol{s}_2|^{r_4} \\ &\le -k_4 (|s_{21}|^2 + |s_{22}|^2 + |s_{23}|^2)^{(r_4+1)/2} \\ &\le -k_4 (2 v_4 / \lambda_{\max}(\boldsymbol{J}))^{(r_4+1)/2} \end{aligned} \tag{24}$$

When confined to $s_2 = 0$, substituting Eq. (7) into Eq. (6), we have

$$\dot{\boldsymbol{\sigma}}_e = -k_2 \boldsymbol{G}(\boldsymbol{\sigma}_e) \mathrm{diag}(\mathrm{sgn}(\boldsymbol{\sigma}_e)) |\boldsymbol{\sigma}_e|^{r_2} \tag{25}$$

Select a non-negative function $v_5 = \boldsymbol{\sigma}_e^{\mathrm{T}} \boldsymbol{\sigma}_e / 2$, computing the derivation of v_5, from Eq. (25) and $\boldsymbol{\sigma}_e^{\mathrm{T}} \boldsymbol{G}(\boldsymbol{\sigma}_e) = (1 + \boldsymbol{\sigma}_e^{\mathrm{T}} \boldsymbol{\sigma}_e) \boldsymbol{\sigma}_e^{\mathrm{T}} / 4$,

$$\begin{aligned} \dot{v}_5 &= -k_2 \boldsymbol{\sigma}_e^{\mathrm{T}} \boldsymbol{G}(\boldsymbol{\sigma}_e) \mathrm{diag}(\mathrm{sgn}(\boldsymbol{\sigma}_e)) |\boldsymbol{\sigma}_e|^{r_2} \\ &\le -k_2 \boldsymbol{\sigma}_e^{\mathrm{T}} \mathrm{diag}(|\boldsymbol{\sigma}_e|^{r_2}) \mathrm{sgn}(\boldsymbol{\sigma}_e) / 4 \\ &= -k_2 \left(|\boldsymbol{\sigma}_{e1}|^{1+r_2} + |\boldsymbol{\sigma}_{e2}|^{1+r_2} + |\boldsymbol{\sigma}_{e3}|^{1+r_2} \right) / 4 \end{aligned} \tag{26}$$

since $\dot{v}_5$ is negative definite, from Lyapunov's theorem [13], we conclude that $\boldsymbol{\sigma}_e \to 0$ as $t \to \infty$. From Eq. (7), we know that $\boldsymbol{\omega}_e \to 0$ as $t \to \infty$.

4 Numerical simulation

In this section, a simulation scenario is considered to show the effect of the proposed controllers. The target position in frame $\mathcal{T}$ is given by $\boldsymbol{p}_t = [0,5,0]^{\mathrm{T}}(\mathrm{m})$. Choose controller parameters $k_1 = 0.05$, $k_2 = 0.2$, $k_3 = 20$, $k_4 = 100$, $r_1 = 1.6$, $r_2 = 1.05$, $r_3 = 0.6$, and $r_4 = 0.95$. Estimates are chosen as $\hat{\rho}_f = \hat{\rho}_\tau = \sqrt{27} \times 10^{-2}$, $\hat{m} = 52.38(\mathrm{kg})$, $\hat{\boldsymbol{\rho}}_c = [0.027, 0.018, 0.0225]^{\mathrm{T}}(\mathrm{m})$,

$$\hat{\boldsymbol{J}}_t = \mathrm{diag}(3300, 3100, 2400)(\mathrm{kg\,m^2})$$

$$\hat{\boldsymbol{J}} = \begin{bmatrix} 538.47 & -20.25 & -46.35 \\ -20.25 & 381.96 & -24.3 \\ -46.35 & -24.3 & 237.24 \end{bmatrix} (\mathrm{kg\,m^2})$$

Bounds of mismatches between the estimated and actual parameters are chosen as $\bar{m} = 11.64(\mathrm{kg})$, $\bar{\rho}_f = \bar{\rho}_\tau = 0.2 \times \sqrt{27} \times 10^{-2}$, $\bar{\boldsymbol{\rho}}_c = [0.006, 0.004, 0.005]^{\mathrm{T}}(\mathrm{m})$,

$$\bar{\boldsymbol{J}} = \begin{bmatrix} 119.66 & 4.5 & 10.3 \\ 4.5 & 84.88 & 5.4 \\ 10.3 & 4.5 & 52.72 \end{bmatrix}, \quad \bar{\boldsymbol{J}}_t = \begin{bmatrix} 100 & 160 & 160 \\ 160 & 100 & 160 \\ 160 & 160 & 100 \end{bmatrix} (\mathrm{kg\,m^2})$$

In the simulation, parameters of the pursuer and the target are as follows: $m = 58.2(\mathrm{kg})$, $\boldsymbol{\rho}_c = [0.03, 0.02, 0.025]^{\mathrm{T}}(\mathrm{m})$,

$$\boldsymbol{J} = \begin{bmatrix} 598.3 & -22.5 & -51.5 \\ -22.5 & 424.4 & -27 \\ -51.5 & -27 & 263.6 \end{bmatrix} (\mathrm{kg\,m^2}) \; \boldsymbol{J}_t = \begin{bmatrix} 3336.3 & -135.4 & -154.2 \\ -135.4 & 3184.5 & -148.5 \\ -154.2 & -148.5 & 2423.7 \end{bmatrix} (\mathrm{kg\,m^2}),$$

$$\boldsymbol{d}_f = \begin{bmatrix} 1 + \sin(\pi t/125) + \sin(\pi t/200) \\ 1 + \sin(\pi t/125) + \sin(\pi t/250) \\ 1 + \cos(\pi t/125) + \cos(\pi t/250) \end{bmatrix} \cdot 10^{-2}(\mathrm{N}),$$

$$\boldsymbol{d}_\tau = \begin{bmatrix} 1 + \sin(\pi t/125) + \sin(\pi t/200) \\ 1 + \sin(\pi t/125) + \sin(\pi t/250) \\ 1 + \cos(\pi t/125) + \cos(\pi t/250) \end{bmatrix} \cdot 10^{-2}(\mathrm{Nm})$$

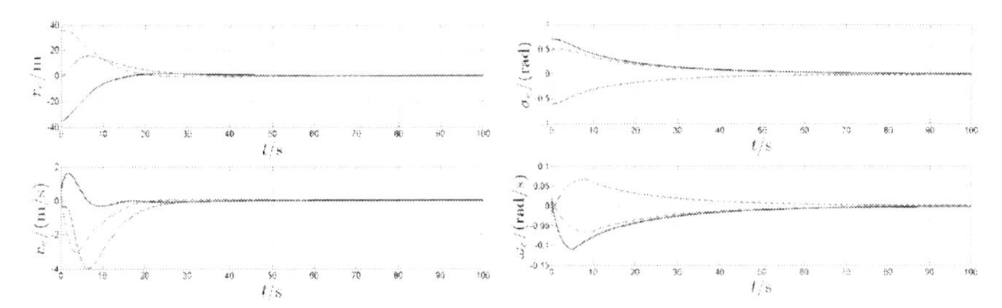

Figure 2: Relative position and velocity.

Figure 3: Relative attitude and angular velocity.

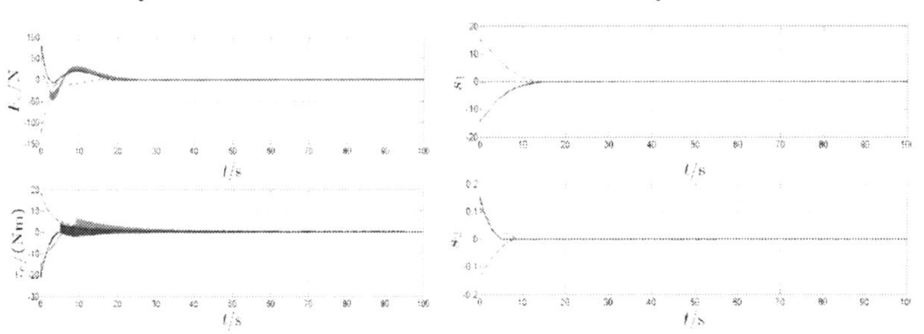

Figure 4: Control force and control torque.

Figure 5: Time history of sliding mode surface.

The initial state of pursuer is set to $\boldsymbol{r}=[1,1,1]^{\mathrm{T}}\times 7.078\times 10^{6}$ (m), $\boldsymbol{v}=[0,0,0]^{\mathrm{T}}$ (m/s), $\boldsymbol{\sigma}=[0,0,0]^{\mathrm{T}}$ (rad), $\boldsymbol{\omega}=[0,0,0]^{\mathrm{T}}$ (rad/s). The initial relative position, velocity, attitude, and angular velocity are set to $\boldsymbol{r}_e=[50/\sqrt{2},0,-50/\sqrt{2}]^{\mathrm{T}}$ (m), $\boldsymbol{v}_e=[0.5,-0.5,\ 0.5]^{\mathrm{T}}$ (m/s), $\boldsymbol{\sigma}_e=[0.5,-0.6,0.7]^{\mathrm{T}}$ (rad), and $\boldsymbol{\omega}_e=[0.02,-0.02,0.02]^{\mathrm{T}}$ (rad/s). Results of the numerical simulation are provided in Figures 2–5. Figures 2 and 3 depict the control effects under the proposed control force $\boldsymbol{f}_c$ and control torque $\boldsymbol{\tau}_c$. Figure 3 shows the time history of $\boldsymbol{f}_c$ and $\boldsymbol{\tau}_c$. Figure 4 shows the time history of sliding mode surface. From the results we conclude that the proposed sliding mode controllers effectively control the pursuer to track the target motion and attitude in spite of external disturbances and model uncertainties.

5 Conclusions

The control problem of space autonomous rendezvous and docking operation missions is studied in this paper. Asymptotical convergence of relative position and attitude errors are proved via Lyapunov analysis for proposed sliding mode controllers. The control law shows acceptable accuracy and robustness in the presence of unmodeled disturbances, uncertain inertia parameters and mass. We provide numerical simulation studies to demonstrate the performance of the developed control law.

Acknowledgments

This work was supported by the National Key Program of Natural Science Foundation of China under grant 61134005 and the National Basic Research Program of China (973 Program) under grant 2012CB821204.

References

[1] Z. Zhong, Z. Huibo, C. Baowen, et al., Sliding mode control for satellite proximity operations with coupled attitude and orbit dynamics. *2nd International Conference on Intelligent Control and Information Processing (ICICIP)*, July 25–28, 2011.
[2] F. Yong, Y. Xinghuo, M. Zhihong, Non-singular terminal sliding mode control of rigid manipulators. *Automatica*, **38(12)**, pp. 2159–2167, 2002.
[3] Y. Shuanghe, Y. Xinghuo, S. Bijan, et al., Continuous finite-time control for robotic manipulators with terminal sliding mode. *Automatica*, **41(11)**, pp. 1957–1964, 2005.
[4] Y. Liang, Y. Jianying, Nonsingular fast terminal sliding-mode control for nonlinear dynamical systems. *International Journal of Robust and Nonlinear Control*, **21(16)**, pp. 1865–1879, 2011.
[5] P. Haizhou, K. Vikram, Adaptive nonlinear control for spacecraft formation flying with coupled translational and attitude dynamics. *Proceedings of the 40th IEEE Conference on Decision and Control*, 2001.
[6] W. Hong, P. Haizhou, K. Vikram, Output feedback control for spacecraft formation flying with coupled translation and attitude dynamics. *American Control Conference*, June 8–10, 2005.
[7] R. Kristiansen, E.I. Grotli, P.J. Nicklasson, et al., A model of relative translation and rotation in leader-follower spacecraft formations. *Modeling, Identification and Control*, **28(1), pp.** 3–13, 2007.
[8] R. Kristiansen, P.J. Nicklasson, J.T. Gravdahl, Spacecraft coordination control in 6DOF: Integrator backstepping vs passivity-based control. *Automatica*, **44(11)**, pp. 2896–2901, 2008.
[9] H. Peter, *Spacecraft Attitude Dynamics*. Courier Dover Publications, 2012.
[10] Y. Liang, Y. Jianying, Robust finite-time convergence of chaotic systems via adaptive terminal sliding mode scheme. *Communications in Nonlinear Science and Numerical Simulation*, **16(6)**, pp. 2405–2413, 2011.
[11] W. Yuye, F. Yong, Y. Xinghuo, et al., Terminal sliding mode control of MIMO linear systems with unmatched uncertainties. *The 29th Annual Conference of the IEEE of Industrial Electronics Society*, 2003.
[12] S. Guang, H. Wei, Direct-adaptive fuzzy predictive control of satellite attitude. *Acta Automatica Sinica*, 1151–1159p, 2010.
[13] H.K. Khalil, J. Grizzle, *Nonlinear Systems*. Prentice Hall, Upper Saddle River, 2002.

The stock decision of repairable aviation spares

Yuan Liu[1], Yun-xiang Chen[2], Bing-xiang Wang[1],
Dian-cheng Zhang[1]
[1]*The Unit of PLA, Xi'an, China*
[2]*Institute of Equipment Management and Safety Engineering, Air Force Engineering University, Xi'an, China*

Abstract

In response to the material requirements and reduce the cost, in view of the stock optimization problem of repairable aviation spares, the calculation method of aviation spares turnaround time is given, and on this basis, based on availability and inventory cost of the inventory optimization model are established, and using the marginal analysis method to solve the model. Finally through the concrete example analysis, proved the validity of the model.
Keywords: repairable, aviation spares, turnaround time, availability, cost.

1 Introduction

Because the repairable parts can repair many times, the price is high, the reserves of repairable parts optimization problem has attracted much attention. Although the repairable parts have small proportion in the total items of the supply chain, but the cost of a security in the supply chain security occupies a large proportion of the total cost. Once if the inventory quantity is insufficient, it will make the aircraft cannot timely repair or maintenance work is not completed as scheduled. On the contrary, if too many, will cause inventory costs increase, capital backlog, and other issues.

Difficulty in funds tension and material source, especially some imported spare parts is conditioned by the foreign party, it is particularly important to study repairable parts inventory optimization.

At present, there are a lot of research findings on repairable parts storage [1–3], but the turnaround time for repairable spares were not given specific in-depth calculation method. Therefore, to solve this problem, set up a turnaround time of air material spare parts optimization model, it is very important and practical for equipment to realize precise guarantees.

WIT Transactions on Engineering Sciences, Vol. 107, © 2015 WIT Press
www.witpress.com, ISSN 1743-3533 (on-line)
doi:10.2495/SEEEI140651

2 Repairable turnaround time

Repairable spares should be reused by repairing. Therefore, according to the repairability, scrap rate, repairable spares within the procurement cycle operating modes, as shown in Figure 1.

Turnaround time is the average time that the repairable unit is removed from the equipment to been repaired by departments. Turnaround time of repairable components, as shown in Figure 2.

Assume that failure part by immediate repair strategy, and disassembly time and transit time should be neglected. So repairable spare parts turnaround time is equal to the repair time.

Because of limited maintenance resources, faults need to be queued for repair. So repair time is that failure enter maintenance department awaiting repair until the repair is completed. Assume that the interval time of spare parts arriving and service times submit to exponentially distributed, the turnaround time of repairable spares can be solved by queuing theory. And repairable spares quantity is limited. Thus, the system can be viewed as M|M|U|K [4] service systems, namely, there are u-parallel maintenance group, in the system, and the number of needing to be repaired are not better than k repairable spares. Figure 3 is a repairable spares state transition diagram.

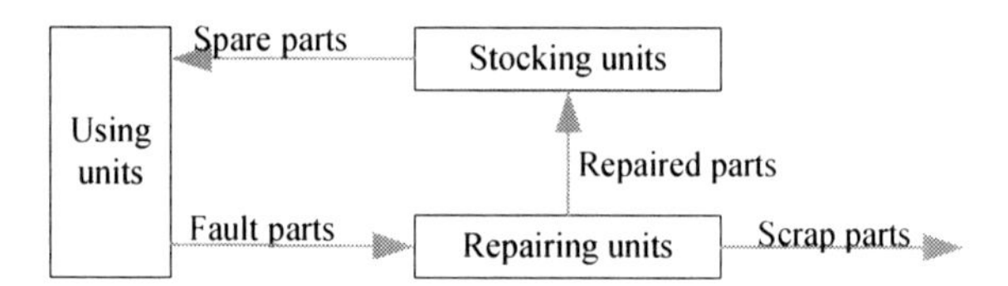

Figure 1: Operation mode of repairable spares.

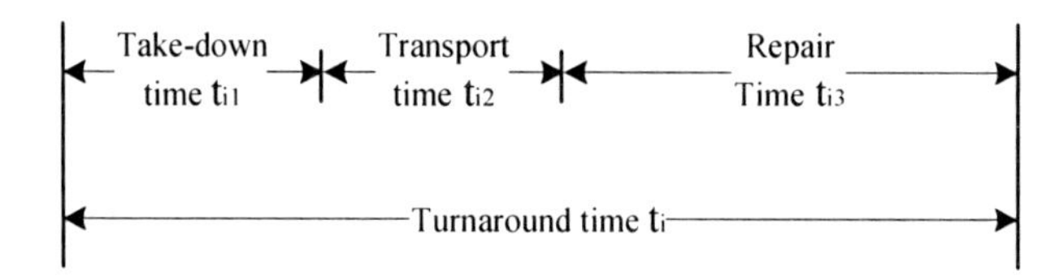

Figure 2: Turnaround time of repairable components.

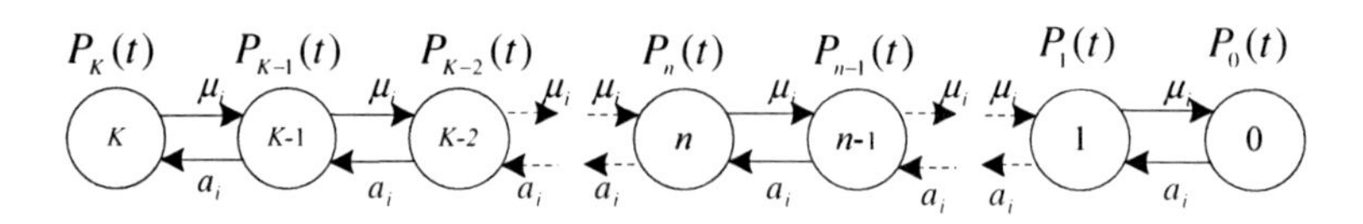

Figure 3: Repairable spares state transition diagram.

In the figure, circles with characters represent the number of needing repair, $P_n(t)$ above circle represent the probability of n failures.

Its conversion process can be described as: at any given time t, n repairable spares with class i need to be repaired in the system, The next moment, one possible situation, after a failure was repaired, it would come into reserving or working. System state changes into n-1 failure. The change strength is fix rate μ_i, indicated by a line with an arrow marked with ui in the figure. The other situation, working part fails, the failure being sent into the service. System state changes into n + 1 failure. change strength is demand rate a_i, indicated by a line with an arrow marked with a_i in the figure.

Based on state transition diagrams and Kolmogorov's equation, State limit probability of n failures in the management system is given, the probability formula can be expressed as [4–6]:

Let $\rho = \frac{a}{\mu}$, $\rho_C = \frac{a}{U\mu}$

Then

$$P_n = \begin{cases} \frac{1}{n!}\rho^n P_0 & 1 \le n \le U \\ \frac{1}{U!U^{n-M}}\rho^n P_0 & U \le n \le K \end{cases} \tag{1}$$

$$P_0 = \begin{cases} \left[\sum_{n=0}^{U-1} \frac{1}{n!}\rho^n + \frac{\rho^U}{U!} + \frac{1-\rho_U^{K-U+1}}{1-\rho_U}\right] & \rho_U \ne 1 \\ \left[\sum_{n=0}^{U-1} \frac{1}{n!}U^U + \frac{U^U}{U!}(K-M+1)\right]^{-1} & \rho_U = 1 \end{cases} \tag{2}$$

Suppose the number of the failures is N_q, which is need to wait for repairing in the system, the expected value of N_q can be expressed as:

$$E[N_q] = \sum_{n=U}^{K}(n-U)P_n = \sum_{n=U}^{K}\frac{n-U}{U!U^{n-2}}\rho^n P_0 \tag{3}$$

Suppose the number of the failures is N_s, which is repairing and need to wait for repairing in the system, the expected value of N_s can be expressed

$$E[N_s] = E[N_q] + M + P_0\sum_{n=0}^{U-1}\frac{(n-U)\rho^n}{n!} \tag{4}$$

Suppose the time from failures arriving at repair sectors to it leaving is T_s, the expected value of T_s equals parts turnaround time t_i, can be expressed as:

$$t_i = E\,[T_s] = \frac{E\,[N_s]}{a} \tag{5}$$

When $t_i \leqq 1/a$, namely, the turnaround time is less than or equal to the time between failures (MTBF), then you need reserve just a spare parts.

When $t_i > 1/a$, namely, the turnaround time is greater than the time between failures (MTBF), in order to meet spare parts requirements during the turn-around period, have to stock a certain number of spares.

Turnaround time of spare parts more than $1/a$ probability is:

$$P\left\{T_s > \frac{1}{a}\right\} = \begin{cases} U_1 e^{-\mu \cdot \frac{1}{a}} + U_2 e^{-\frac{U}{a}(1-\rho U)} & \rho \neq U-1 \\ [1 + P_d \cdot \frac{\mu}{a}] e^{-\mu \cdot \frac{1}{a}} & \rho = U-1 \end{cases} \tag{6}$$

In the formula, P_d is the probability of faults needing to wait for maintenance. It can be showed as:

$$P_d = \sum_{n=U}^{K} P_n \tag{7}$$

3 The inventory optimization model of repairable aviation spares

Repairable spares should be reused by repairing within the procurement cycle $[0, T]$. Therefore, in the $[0, T]$ the parts reserves should be the demand for parts of meeting the Support requirements in turnaround time.

There are two main type about Spare part optimization goals: first, support capability index; the second is inventory cost index. Support capability index is measured by aircraft availability mainly.

3.1 Aircraft availability model

Aircraft availability is the probability which the plane was in good condition at any time. When one or more of the key equipment failure, the parts for changing failures need be required by repair demand from material unit, the demand could be met immediately, may also arise out of stock. If a shortage occurs, maintenance department must wait for material unit to restock, during this period, the aircraft is in fault state.

Therefore, the aircraft availability could be seen as the percentage of expectation of the planes, the planes is not grounded because of the shortage of spare parts in a random point in time [7–9]. Assuming any spare parts shortages at a time will cause the grounded aircraft and spare parts requirements subject to value a of Poisson process, then:

$$A = \prod_{i=1}^{I} \left\{1 - EBO_i / (NZ_i)\right\}^{Z_i} \tag{8}$$

In the formula: Z_i is the number of fixing parts i in the one plane, and n is the number of aircraft. EBO_i is the number of the expected shortages of the parts i.

Assuming that the spare parts requirement obeys the Poisson process of Failure rate for λ.

The equation for the number of expected stock:

$$\text{EBO}(S_i) = \begin{cases} \sum_{x=S+1}^{\infty} (x - S_i) P(x \mid aT), & t_i \geq T \\ \sum_{x=S+1}^{\infty} (x - S_i) P(x \mid at_i), & t_i < T \end{cases} \tag{9}$$

Among them, the X_i is the demand for equipment i; S_i is the reserves of equipment i; $P(X_i = x)$ is the probability of additional supply of spare parts.

$$P(x) = (\lambda T)^x \, \mathrm{e}^{-\lambda T} / x! \quad x = 0, 1, 2, \cdots \tag{10}$$

3.2 The establishment of cost model

Spare parts cost includes: spare parts purchase cost, ordering cost, storage cost and shortage cost [2]. The spare parts cost is expensive, so the purchase cost is the largest proportion of total cost, about 60–70% of total cost. Therefore, here we mainly discuss how the purchase cost of spare parts influence spare parts reserve quantity decision. Reserve quantity of spare parts S is denoted as $S = (s_1, s_2, \ldots, s_i)$. Spare parts unit price C is denoted as $C = (c_1, c_2, \ldots, c_i)$. The purchase cost of spare parts can be expressed as:

$$S \times C^T = E_m \tag{11}$$

3.3 Optimization model of spare parts reserve quantity

In wartime, the main purpose is to win the battle, everything should be subordinated to the military benefit led principle.

Usually, it is mainly to complete flight training and on duty. The successful completion of the work need a certain aircraft using availability as security, and security costs will become the key consideration. According to the different goal needs, we can establish two optimization model of spare parts reserve quantity.

Model 1: Based on aircraft availability as objective function, security cost as constraint conditions.

$$\max A \quad \text{s.t.} \quad E \leq E_m \tag{12}$$

Model 2: Based on security cost as objective function, aircraft availability as constraint conditions.

$$\min E \quad \text{s.t.} \quad A \geq A_0 \tag{13}$$

In the formula: A is aircraft availability, A_0 is the lowest aircraft availability; E is the spare parts security cost, E_m is spare parts security total cost quota.

4 Algorithm analysis

The model belongs to nonlinear programming problem. Because of more types of aviation spare parts, planning on using a general solution method is more complex process, so this paper uses the method of marginal analysis [10] to optimized or solved. The marginal effect analysis is an evolutionary optimization technique, for the allocation of scarce resources to obtain maximum benefit. It can be considered by weighing analysis of the marginal unit of cost and benefit, in order to achieve the rational use of resources. With the availability model as an example, spare parts allocation optimization flow-process diagram based on marginal analysis, as shown in Figure 4.

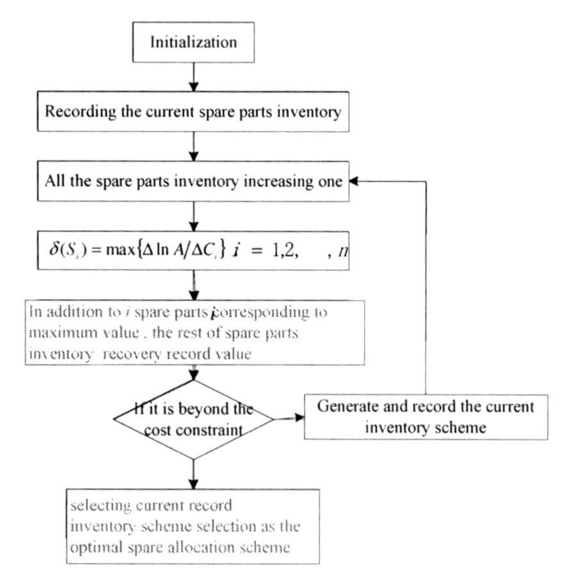

Figure 4: Critical chain.

5 The example analysis

Some flight group have 10 aircraft, to simplify the problem, assuming that the equipment is made up of two main parts, equipment stand-alone installation number is 1, and obedience of the Poisson distribution. According to the consumption of the mission 3 years data of spare parts, calculating the spare parts plan that spare parts cost is no more than 250 thousands yuan and aircraft availability is not less than 0.95. The relevant parameters of these two parts as shown in Table 1.

According to the data analysis in Table 1, using marginal analysis we can obtain several combinations of spare parts 1 and 2 reserves, when the total security costs is not more than 250 thousands yuan, and availability is not less than 0.95, as shown in Table 2.

Table 1: The calculated values by the model of spares storage level for an example.

Spare name	1		2	
Average annual demand	10		50	
Mean time to repair	0.1		0.08	
Spare unit price/10 Kyuan	5		1	
s	EBO (s)	v	EBO (s)	v
0	1.000		4.000	
1	0.368	0.126	3.018	0.982
2	0.023	0.053	2.110	0.908
3	0.004	0.016	1.348	0.762
4	0.001	0.004	0.782	0.567
5	0.000		0.410	0.371
6	0.000		0.195	0.215
7	0.000		0.085	0.111
8	0.000		0.034	0.051
9	0.000		0.012	0.021
10	0.000		0.004	0.008

Table 2: The projects of spares storage in limiting bound.

s_1	s_2	A/%	E/10 Kyuan
1	7	95.5	12
2	7	98.9	17
2	8	99.4	18
2	9	99.7	19
3	9	99.8	24
3	10	99.9	25

Table 3: Compare projects at the same cost.

s_1	s_2	A (%)	E/10 Kyuan
5	0	60	25
4	5	95.8	25
3	10	99.9	25

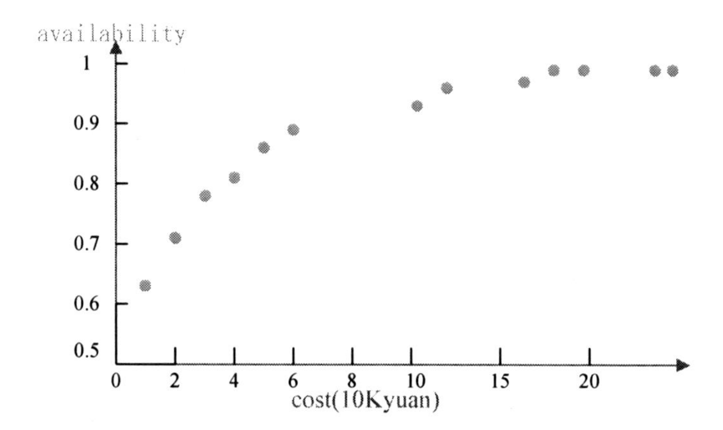

Figure 5: The relation between the cost and the availability.

In order to illustrate the effectiveness of the method, there will be 250 thousands yuan in total cost, we compares availability and total cost of possible combinations. As shown in Table 3.

Through the contrast analysis, we can see clearly that third reserve scheme is the optimal solution, and the marginal analysis method calculates the consistent results. Thus, the reserve of spare parts combination Table 2 is the optimal reserve plan. The optimal relation curve between the availability and the cost as shown in Figure 5.

The discrete points on the curve in the figure, is the highest availability in the specified cost conditions, with the same availability under the lowest cost significance.

6 Conclusion

In order to realize precise equipment support, this paper researched the reserve problem about repairable aircraft spare parts which possess a large proportion of aircraft material support total fee. It firstly introduced the operation mode about repairable aircraft spare parts in the supply security system. According to the operation characteristics, it analyzed repairable aircraft spare parts turnover time in the procurement cycle. Secondly, it established the spare parts reserve optimization model which requirement is followed Poisson distribution under immediately repair strategies, proposed the model solving method based on marginal analysis. Finally, through the analysis of specific examples, the model is validated and further analysis.

References

[1] Li Shu-guang, Zhao Yan-Jun, Xu Cheng, The study of requirement forecasting and decision-marking marking methodology of equipment's maintenance spares-parts. Acta Armamentarii, (32)7, pp. 901–904, 2011.
[2] Han Xin-cai, Air Materiel Management Engineering. Blue Sky Publishers, Beijing, 2, pp. 136–211, 2003.
[3] Liu Xiao-chun, Huang Ai-jun, Ma Fang, et al., Requirement forecast of equipment maintenance parts based on exponential smoothing method. Equipment Environmental Engineering, (9)6, pp. 109–110, 2012.
[4] Hao Jie-zhong, Yang Jian-jun, Yang Ruo-peng, The Strategy Analysis of Equipment Technology Support. National Defence Industry Publishers, Beijing, pp. 34–40, 2006.
[5] Zhu Yi-fei, Huang Guo-c, Markovian decision programming model study on spare parts inventory. Journal of Air Force Engineering University, 2(2), pp. 91–94, 2001.
[6] Zhou Jiang-hua, Xiao Gang, Miao Yu-hong, Analysis of optimal spares supply programming. Journal of Mechanical Strength, 26(3), pp. 270–273, 2004.
[7] Craig C. Sherbrooke, Optimal Inventory Modeling of Systems Multi-Echelon Techniques (2nd ed.). Kluwer Academic Publishers, Norwell, MA, 2004.
[8] Du Jun-gang, Du Xin, He Ya-qun, Premium allocation of spare inventory in two-echelon supply system based on aircraft operation availability. Ordnance Industry Automation, 28(1), pp. 39–44, 2009.
[9] Yang Ping-lv, Bao Lei, Optimization model of inventory spares based on METRIC. Ship Electronic Engineering, (30)12, pp. 161–163, 2010.
[10] Nie Tao, Sheng Wen, Research on two-echelon supply support optimizing for repairable spare parts of K: N system. System Engineering and Electronics, (32)7, pp. 1452–1454, 2010.

A C-band broadband and miniaturized substrate integrated waveguide circulator

Shuai Zhu, Wei Tian, Liang Chen,
Xiaoguang Wang, Longjiang Deng
School of Microelectronics and Solid-State Electronics, University of Electronic Science and Technology of China, Chengdu, China

Abstract

The substrate integrated waveguide (SIW) has been studied and applied in the microwave and millimeter wave field. In order to obtain broadband and miniaturized SIW circulator, a C-band T-junction SIW circulator is designed by High Frequency Structure Simulator (HFSS) and processed on the printed circuit broad (PCB) with the size of 18 mm*18 mm in this paper. Further, the circulator is measured by vector network analyzer with a 35.7% bandwidth at the −20dB return loss points, insertion loss better than 1.15dB. The broad bandwidth and small size indicating such device is an excellent candidate for C-band circulator applications. Dual circulation property of a ferrite junction is obtained and a tentative analysis is given.

Keywords: circulator, dual circulation property, ferrite, C-band, substrate integrated waveguide (SIW).

1 Introduction

A new technology, called substrate integrated waveguide (SIW) [1] has attracted much attention of researchers, which shows the advantages of both traditional waveguides and microstrips, such as low cost, high Q factor, high power capacity, low weight, and compact size. SIW technology also enables a possibility that all the passive and active components can be integrated completely on the same substrate, components such as filter, power divider, antenna, amplifier, and circulator have been developed using SIW transmission lines [2, 3].

Among these devices, the SIW ferrite junction circulator has potential applications in integrated communication and radar system. This kind of Y-junction

WIT Transactions on Engineering Sciences, Vol. 107, © 2015 WIT Press
www.witpress.com, ISSN 1743-3533 (on-line)
doi:10.2495/SEEEI1140661

circulator has been proposed in Refs. [4–6]. SIW usually needs to be transited to the microstrip lines in order to integrate with external components. For the transition of microstrip-to-SIW, some expressions for the variation of impedances are proposed in Ref. [7], such as exponential, triangular, and Klopfenstein, and they all have small reflection.

In this paper, a C-band SIW T-junction circulator is presented and demonstrated. A linear tapered microstrip is adopted to realize the impedance matching between SIW and 50Ω microstrip. Compared with previous work in Ref. [6], the circulator presented in this paper owns the broader bandwidth, less insertion loss and the shorter microstrip length as well as smaller size, the agreement of the simulation and experimental results is also better. Moreover, the T-junction SIW circulator is first obtained by rotating microstrip line based on Y-junction in this paper.

2 Design of circulator

2.1 Design rules and modes of SIW

The configuration of SIW is illustrated in Figure 1, which can be considered as a special rectangular waveguide with periodical via-hole forming the narrow walls. The energy leaking between via-holes can be ignored if the distance between the via-holes and their diameter satisfy the following relationships:

$$s / \lambda_c < 0.25 \tag{1}$$

$$s / d \leq 2 \tag{2}$$

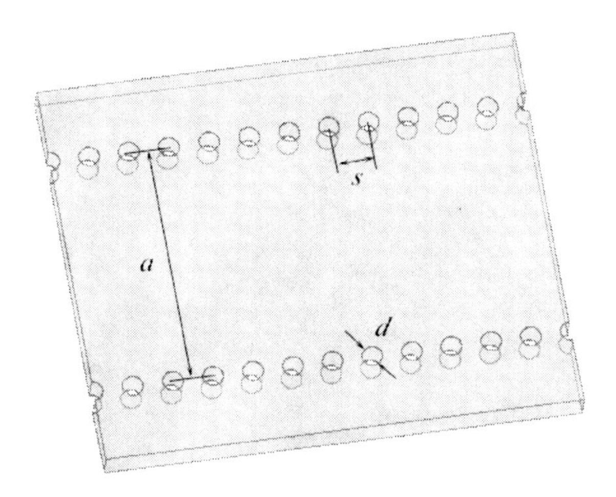

Figure 1: Configuration of SIW.

where s and d are, respectively, the period length and the diameter of the via-hole, and λ_c is the cutoff wavelength [8]. TE_{10} mode is the fundamental mode of SIW and only TE modes can be excited in this periodical structure, the detailed discussion has been given in Ref. [9].

2.2 Materials selection and circulator design

For the design of circulator, ferrite and substrate materials are selected strictly. Proper saturation magnetization and low dielectric loss materials should be considered firstly. In this paper, the nickel–zinc ferrite is chosen with saturation magnetization $4\pi Ms$ = 1700 Gauss and the relative dielectric constant ε_f is 14.5, the normalized magnetization p is calculated by

$$p = \gamma 4\pi Ms / f_0 \tag{3}$$

where γ is gyromagnetic coefficient and f_0 is center frequency. Usually, circulator can acquire broad bandwidth when the value of p is between 0.4 and 0.8. The ferrite cylinder is designed to be at the same height with the substrate. The radius of the ferrite cylinder R_f is calculated according to Eq. (4)

$$R_f = \frac{1.84\lambda}{2\pi\left(\varepsilon_f \mu_e\right)^{\frac{1}{2}}} \tag{4}$$

where λ is the operating wavelength in the free space, μ_e is the effective magnetic permeability[10].

In order to diminish the size, the substrate material with high dielectric constant is prime choice. In the paper, the printed circuit broad (PCB) Taconic CER-10 (tm) is chosen as substrate material, whose height is 0.762 mm, relative dielectric constant ε_r is 10. The size of SIW is designed as s = 0.7 mm, d = 0.5 mm and the width a is calculated by

$$a_{\text{eff}} = a - 1.08\frac{d^2}{s} + 0.1\frac{d^2}{a} \tag{5}$$

where a_{eff} is the equivalent width of the rectangular waveguide [9]. A tapered microstrip line with the length of l_t is used for impedance match, and $w_{\_50\Omega}$ and $w_{\_SIW}$ are, respectively, its widths of two ends.

3 Results and discussions

In our study, the commercial software High Frequency Structure Simulator (HFSS) is chosen to design the circulator model. When $w_{\_50}$ = 0.7 mm, $w_{\_SIW}$ = 3.3 mm, l_t = 4.8 mm, R_f = 3.81 mm, s = 0.7 mm, a = 10.6 mm, d = 0.5 mm, and h = 0.762 mm,

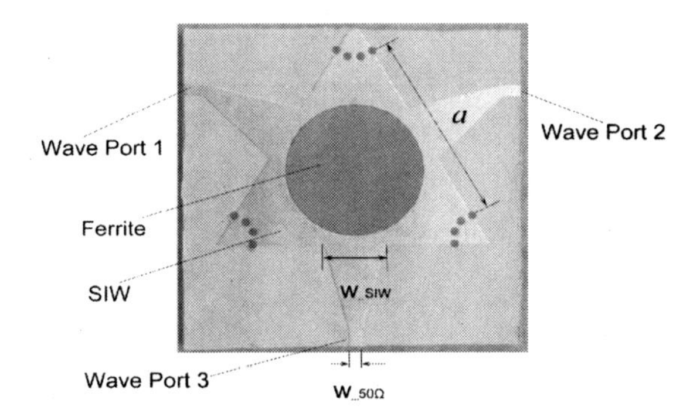

Figure 2: T-junction SIW circulator.

a good working performance is acquired. Then the circulator is manufactured and measured as shown in Figure 2.

The circulator was measured by vector network analyzer at room temperature. In our experiment, the cylindrical permanent magnet was used to provide biasing magnetic field, and whose location above the ferrite was fixed by manual tuning. The comparison of S-parameters between simulation results and measured results are illustrated in Figures 3 and 4. A 37.8% bandwidth of return loss S_{11} at −20dB with the insertion loss better than 0.4dB is obtained by HFSS simulation. Similarity, a 35.7% bandwidth of return loss S_{11} with the insertion loss better than 1.15dB is measured. The isolation S_{31} of simulation and measured results has the same trend of frequency response, while the measured results are worse than simulation.

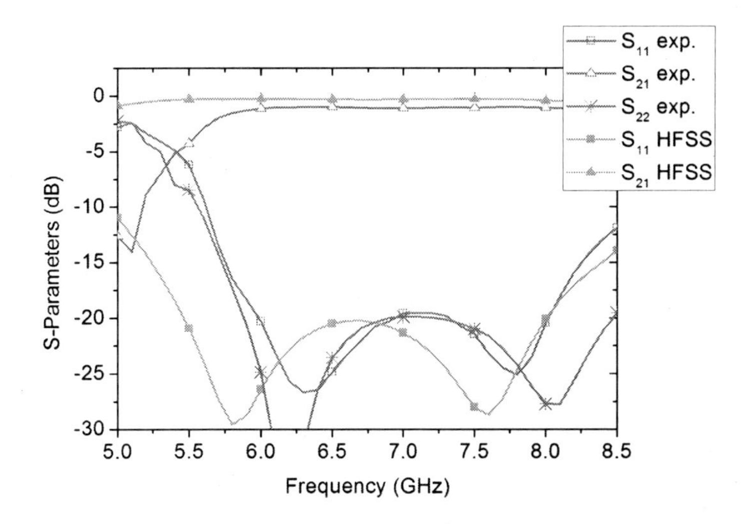

Figure 3: Comparison of S-parameters between simulation results and experimental results.

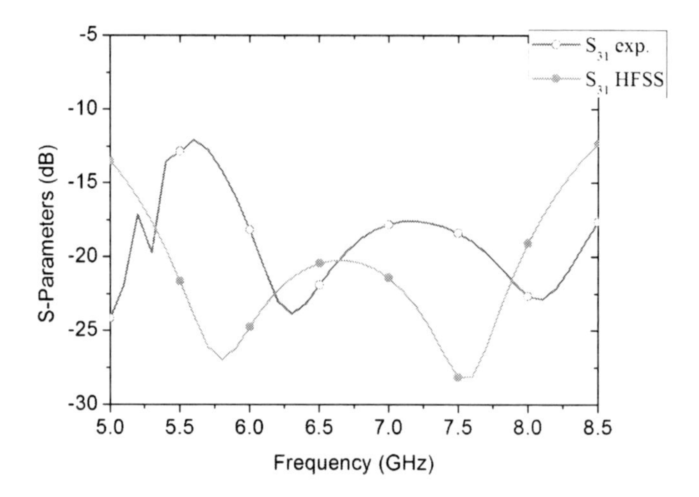

Figure 4: Comparison of S_{31} between simulation results and experimental results.

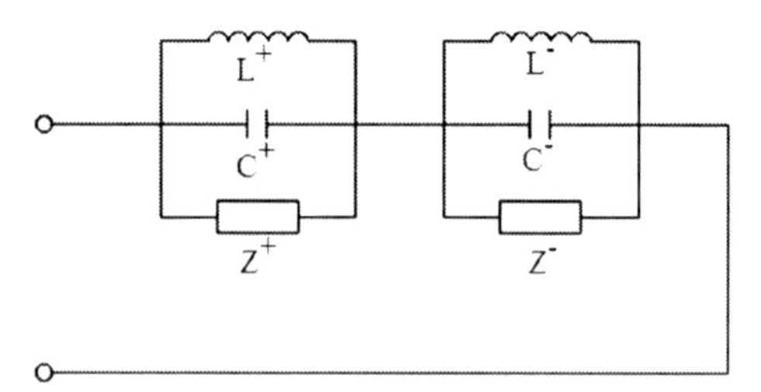

Figure 5: The equivalent circuit of ferrite junction.

The operating bandwidth is broad due to the dual circulation properties of ferrite. The ferrite cylinder can be equivalent a series circuit which consists of two parallel resonance circuits as shown in Figure 5. Two resonant frequencies are degenerate and only when ferrite is magnetized by biasing magnetic field, two resonance frequencies separate. The farther two resonance frequencies separate, the broader bandwidth will be. In order to obtain favorable dual circulation properties, the impedance match of circulator circuit is needed. In this paper, a tapered microstrip transformer at the terminal of SIW is employed and optimized, when its length approach a quarter-wavelength, the impedance match is satisfied.

Some reasons can be explained why the measured insertion loss is worse than simulation. First, the resonance line width ΔH of ferrite was neglected during the simulation which contributes much of the insertion loss of the circulator, the ΔH is bigger, the more loss will be. In addition, the ferrite at low saturation leaded to the loss increasing significantly during the tests. Last, the poor contact and fabrication inaccuracy also made insertion loss worse.

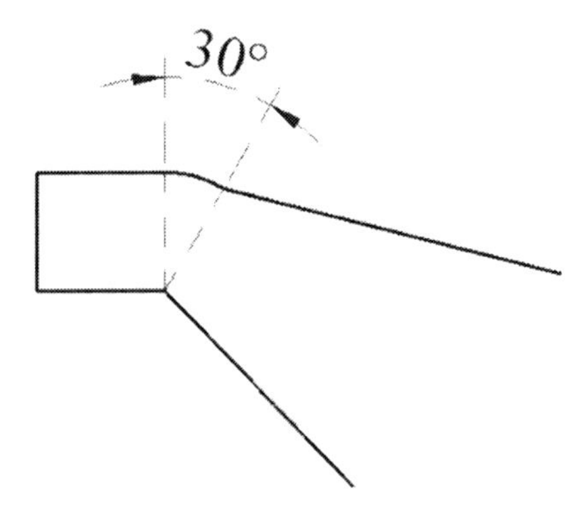

Figure 6: Microstrip line in wave port 1 with 30 degree rotation.

The experiment results verify simulation design well, while there is also a deficiency that measured results skew to high-frequency compared to simulation results. The main reason is the biasing magnetic field is not strong enough to make ferrite saturated which causes the frequency deviation. Moreover, the actual processing size a is 10.1 mm, not the designed value 10.6 mm. With the reduction of a, the propagation constant skews to high-frequency.

Coincidence degree of S_{11} and S_{22} shows the good consistency of port 1 and port 2. T-junction is better to meet the layout requirement in some circuit design, but it is also more difficult to design due to 3 ports are not exactly same. However, in this paper, this problem is solved well. A change of Y-junction to T-junction is realized through a 30 degree rotation of fan shaped microstrip in port 1 and port 2 as shown in Figure 6. By adjusting the width and length of microstrip lines, we get an outstanding consistency of 3 ports.

4 Conclusions

This paper presents a C-band broadband and miniaturized T-junction SIW circulator. The device is manufactured and measured, showing good agreement between simulation and experiment results. In order to obtain broad operating bandwidth, dual circulation property of ferrite is utilized. Bandwidths between 35–40% of return loss at −20dB and the insertion loss better than 1.15dB are achieved. Further, the dual circulation property is discussed briefly. Compared with [6] that the return loss S_{11} exceeds −20dB in all the given frequency, a 26.1% bandwidth at −20dB isolation S_{31} and the insertion loss about 2dB, the T-junction SIW circulator presented in this paper has a pronounced improvement both in performance and size.

References

[1] L. Yan, W. Hong, K. Wu, T.J. Cui, Investigations on the propagation characteristics of the substrate integrated waveguide based on the method

of lines. *IEE Proceedings on Microwaves, Antennas and Propagation*, 152(19), pp. 35–42, 2005.
[2] Lukasz Szydlowski, Adam Lamecki, Michal Mrozowski, Design of microwave lossy filter based on substrate. *IEEE Microwave and Wireless Components Letters*, 21(5), pp. 249–251, May 2011.
[3] Zhebin Wang, Sulav Adhikari, David Dousset, Chan-Wang Park, Ke Wu, Substrate integrated waveguide (SIW) power amplifier using CBCPW-to-SIW transition for matching network. *2012 IEEE MTT-S International Microwave Symposium Digest*, pp. 1–3, June 2012.
[4] Bouchra Rahali, Mohammed Feham, Design of K-band substrate integrated waveguidecoupler, circulator and power divider. *International Journal of Information and Electronics Engineering*, 4(1), January 2014.
[5] K. Wu, W. D´Orazio, J. Helszain, A substrate integrated wave-guide degree-2 circulator. *IEEE Microwave and Wireless Components Letters*, 14(5), pp. 207–209, May 2004.
[6] Wenquan Che, Xiao Jing Ji, Edward K.N. Yung, Miniaturized planar ferrite junction circulator in the form of substrate-integrated waveguide. *International Journal of RF and Microwave Computer-Aided Engineering*, 18, pp. 8–13, January 2008.
[7] D.M. Pozar, Microwave Engineering. University of Massachussetts at Amherst, Proyecto Final de Carerra, pp. 256–267, February 1998.
[8] D. Deslandes, K. Wu, Accurate modeling, wave mechanisms, and design considerations of a substrate integrated waveguide. *IEEE Transactions on Microwave Theory and Techniques*, 54, pp. 2516–2526, June 2006.
[9] F. Xu, K. Wu, Guided-wave and leakage characteristics of substrate integrated waveguide. *IEEE Transactions on Microwave Theory and Techniques*, 53, pp. 66–73, January 2005.
[10] H. Bosma, On Stripline Y-circulation at UHF*. *IEEE Transactions on Microwave Theory and Techniques*, pp. 61–73, 1964.

The combination method for evidences and its application on patient diagnosis

Wang Ping[1–3], Zhu Xuemei[1,3]
[1]*Key Laboratory of Intelligent Computing and Signal Processing, Ministry of Education, Anhui University, China*
[2]*Science Computing and Intelligent Information Processing of Guangxi Higher Education Key Laboratory, Guangxi Teachers Education University, Nanning, China*
[3]*Xihua University, Chengdu, China*

Abstract

This paper proposes a new method for the fusion of non-independent evidence. The method eliminates the influence of relevance between evidences with decreasing the basic probability assignment of certainty and increasing the uncertainty of non-independent evidences according to correlation coefficient $|r_{ij}|$, and then the problem which relevant evidences support hypothesis repeatedly can be coped well. There is an example to illustrate how the method deal with the fusion of relevant evidences, the result shows that the method is better to combine the relevant evidences, and then it can be used for data fusion of relevant information.

Keywords: evidence theory, combination rule, correlation coefficient.

1 Introduction

Dempster–Shafer (D–S) evidence theory is an efficient algorithm to deal with uncertain, incomplete and vague information in data fusion. It was put forward by Dempster [1], and then developed by Shafer [2], so it was called D–S evidence theory. The theory can be used to reason from uncertain information. Because D–S evidence theory can provide a combined framework [3], and it can combine cumulative evidence for changing prior opinions if only there is new evidence [4], it can be suitable for taking into account the difference between knowledge types and get better conclusion [5].

WIT Transactions on Engineering Sciences, Vol. 107, © 2015 WIT Press
www.witpress.com, ISSN 1743-3533 (on-line)
doi:10.2495/SEEEI140671

For classical D–S evidence theory, evidence independence is required; it made it difficult to apply the evidence theory. Many researchers try to resolve these problems. Wang et al. proposed the combined fuzzy logic/D–S evidence theory method [6]. Parikh et al. proposed a new method based on the fact that the use of predictive accuracy for basic probability assignment can improve performance over that provided by traditional basic probability assignment methods [4]. Jones et al. noted that it is necessary to ensure the validity of basic probability assignment [7]; but they did not give a method to do. In this paper, we put forward an improvement method of D–S theory based on the correlation coefficient between two evidences, and analyzed the actual influence of the method to patient diagnosis.

2 The classic D–S evidence theory and combination rule

Ω is the space of hypothesis called a frame of discernment. If the number of elements in Ω is N, then the elements of power set of Ω (2^{Ω}) are 2^N. Each element of power set is corresponding to a hypothesis (subset) of value of Ω_N. A subset that contains at least two elements of Ω is called a compound hypothesis.

D–S evidence theory defined the mass function M on 2^{Ω} as:

M: $2^{\Omega} \rightarrow [0, 1]$

For each subset A (hypothesis) in Ω, let it corresponding to a decimal fraction $M \in [0, 1]$, with the following property:

(1) $M(\emptyset) = 0$, $\emptyset$ is called empty set or impossible case

(2) $\sum_{A \subseteq \Omega} M(A) = 1$

So we call function M basic probability assignment function on 2^{Ω}, $M(A)$ is the basic probability assignment (BPA) of A.

The D–S evidence theory defined two functions: belief (BEL) and plausibility (PL), they are both derived from the mass function M. The belief function of hypothesis is defined as:

BEL: $2^{\Omega} \rightarrow [0, 1]$

$$BEL(A) = \sum_{B \subseteq A} M(B), \quad \forall A \subseteq \Omega \quad A \neq \varphi$$

BEL(A) represents the total probability of all subset of A; showing the whole belief degree or reliability of A.

Then: BEL($\emptyset$) = 0 and BEL(Ω) = 1.

Define the Plausibility Function of hypothesis: $2^{\Omega} \rightarrow [0,1]$

$$PL(A) = 1 - BEL(\bar{A}) = \sum_{B \cap A \neq \varphi} M(B), \quad \forall A \subseteq \Omega$$

where $\bar{A} = \Omega - A$

We can use BEL(A) and PL(A) to measure the uncertainty of hypothesis A. Because BEL(A) and PL(A) are the lower limit and upper limit of P(A) separately, that is:

$$BEL(A) \leq P(A) \leq PL(A)$$

The combination rule of D–S evidence theory provides a method to combine two evidences [4]. Suppose M_1 and M_2 are two independent basic probability assignment on 2^{Ω}, BEL_1 and BEL_2 are two belief functions on 2^{Ω}, M_1 and M_2 are corresponding basic probability assignment separately, focal elements are $A_1, \ldots, A_k$ and $B_1, \ldots, B_r$, further more:

$$K_1 = \sum_{\substack{i,j \\ A_i \cap B_j = \varphi}} M_1(A_i) M_2(B_j) < 1$$

The combination rule of D–S theory is defined as:

$$M(C) = \begin{cases} \dfrac{\sum_{\substack{i,j \\ A_i \cap B_j = C}} M_1(A_i) M_2(B_j)}{1 - K_1} & \forall C \subset U \quad C \neq \varphi \\ 0 & C = \varphi \end{cases}$$

Provided that $A, B \subseteq U$, the belief interval of A and B are separately:

$$\mathrm{EI}_1(A) = [\mathrm{BEL}_1(A), \mathrm{PL}_1(A)]$$

And:

$$\mathrm{EI}_2(B) = [\mathrm{BEL}_2(B), \mathrm{PL}_2(B)]$$

So the belief interval after combination is:

$$\begin{aligned} &\mathrm{EI}_1(A) \oplus \mathrm{EI}_2(B) \\ &= [1 - K_2(1 - \mathrm{BEL}_1(A))(1 - \mathrm{BEL}_2(B)),\ K_2 \mathrm{PL}_1(A) \mathrm{PL}_2(B)] \end{aligned}$$

where

$$K_2 = \left\{1 - \left[\mathrm{BEL}_1(A)\mathrm{BEL}_2(\overline{B})\mathrm{BEL}_1(\overline{A})\mathrm{BEL}_2(B)\right]\right\}^{-1}$$

D–S combination evidence theory makes decision based on two evidences. For the combination of multi-evidence, the conclusion can be inferred from the combination of every two evidences.

3 The combination problem of non-independent evidence

When the evidence are non-independent each other, D–S combination rule may lead to false results. Let's imagine a scene which a patient sees doctor, the patient says that he feels his chest uncomfortable, and the doctor auscultates his chest with Stethoscope first, and he thinks the patient may be arrhythmia. The doctor direct an intern, due to the authority of doctor, the intern checks the patient and has the same diagnosis as him. The doctor suggests the patient takes an examination with Electrocardiogram (ECG), and ECG report shows that the patient may be coronary thrombosis, and then there are three evidences: Intern, Doctor, Electrocardiogram. Suppose that the frame of discernment Ω is {A: coronary thrombosis, B: arrhythmia}, their diagnosis results can be expressed as follow:

Doctor: $M_D(B) = 0.8$, $M_D(\Omega) = 0.2$
Intern: $M_I(B) = 0.8$, $M_I(\Omega) = 0.2$
ECG: $M_E(A) = 0.9$, $M_E(\Omega) = 0.1$

Then we will fuse these evidences with the classic D–S combination rule. First we combine the diagnoses of Doctor and Intern, the combination m can be processed as Table 1.

The result after fusion is:

$$M_1(B) = M_D(B) \oplus M_I(B) = 0.96$$

$$M_1(\Omega) = M_E(\Omega) \oplus M_W(\Omega) = 0.04$$

Then combine above result with the ECG report. It is shown as Table 2.

And then the final fusion result is:

$$M(A) = 0.036;\ M(B) = 0.096;\ M(\Omega) = 0.004;\ M(\emptyset) = 0.$$

The disagreement coefficient of evidence is: $K = 0.864$

Standardized the focal element (divided by $(1-K)$), the fusion result is:

$$M(A) = 0.036/0.136 = 0.27;\ M(B) = 0.096/0.136 = 0.70;\ M(\Omega) = 0.004/0.136 = 0.03.$$

So the belief interval of coronary thrombosis is [0.27, 0.30], and the belief interval of arrhythmia is [0.70, 0.73]. The conclusion shows that the patient is arrhythmia, the ECG report is voted down by the diagnosis of two doctor, obviously it is unreasonable. With the DS combination theory, the result of fusion is always decided by majority supporters.

Table 1: Fusion of doctor and intern.

	$M_I(B).=.0.8$	$M_I(\Omega).=.0.2$
$M_D(B).= 0.8$	$(B)0.64$	$(B)0.16$
$M_D(\Omega) = 0.2$	$(B)0.16$	$(\Omega)0.04$

Table 2: Fusion of result of first step and ECG.

	$M_1(B) = 0.96$	$M_1(\Omega) = 0.04$
$M_E(A) = 0.9$	$(\emptyset)0.864$	$(A)0.036$
$M_E(\Omega) = 0.1$	$(B)0.096$	$(\Omega)0.004$

4 The combination rule for relevant

We propose a new combination method that based on D–S theory, which eliminates the relevance between evidences according to their correlation coefficient. The new method may draw a more reliable conclusion than previous.

The method relies on hypothesis as follows:

1) There are many evidences for focal element A, they are 1,2…N, the basic probability assignment of each evidence is: $M_i(A)$, $i \in N$.
2) With the experiment result or prediction, the correlation coefficient of each two of evidences r_{ij} can be decided.
3) Eliminate the relevance according to the correlation coefficient of evidence for one of the two evidences. The basic probability assignment after eliminated relevance is $(1-|r_{ij}|)\,M_i(A)$
4) Because the probability assignment of certainty is decreased, the probability assignment of uncertainty is increased, the incremental part is $|r_{ij}|M_i(A)$.

The correlation coefficient r_{ij} can be calculated with the historic data of two evidence source. Suppose there are two evidences X and Y, their basic probability evaluation to same focal element in N times are (x_1, y_1), (x_2, y_2), …, (x_n, y_n), separately, then:

$$r_{ij} = \frac{n\sum_{i=1}^{n} x_i y_i - \sum_{i=1}^{n} x_i \sum_{i=1}^{n} y_i}{\sqrt{[n\sum_{i=1}^{n} x_i^2 - (\sum_{i=1}^{n} x_i)^2][n\sum_{i=1}^{n} y_i^2 - (\sum_{i=1}^{n} y_i)^2]}}$$

where $\sum_{i=1}^{n} x_i, \quad \bar{y} = \frac{1}{n}\sum_{i=1}^{n} y_i$

If the belief function and plausibility function are separately:

$\mathrm{BEL}_i(A)$ and $\mathrm{PL}_i(A)$, $i \in N$

Then the new combination rule is:

$$M(C) = \begin{cases} \dfrac{\sum_{\substack{i,j \\ A_i \cap B_j = C}} (1-|r_{ij}|)M_1(A_i)M_2(B_j)}{1-K_1} & \forall C \subset U \quad C \neq \varphi \\ 0 & C = \varphi \end{cases}$$

where $K_1 = \sum_{\substack{i,j \\ A_i \cap B_j = \varphi}} (1-|r_{ij}|)M_1(A_i)M_2(B_j) < 1$

The confidence interval after combination is:

$$\begin{aligned} &\mathrm{EI}_1(A) \oplus \mathrm{EI}_2(B) \\ &= [1-K_2(1-\mathrm{BEL}_1(A))(1-\mathrm{BEL}_2(B)), K_2\mathrm{PL}_1(A)\mathrm{PL}_2(B)] \end{aligned}$$

where $K_2 = \left\{1-\left[\mathrm{BEL}_1(A)\mathrm{BEL}_2(\bar{B})\mathrm{BEL}_1(\bar{A})\mathrm{BEL}_2(B)\right]\right\}^{-1}$

5 The application of new method for patient diagnosis

With the above algorithm, we need to decide the correlation coefficients between each of evidence. Because the intern is directed by the doctor, he will believe the diagnosis of the doctor in most of cases. According to their experimental results of diagnosis, we infer correlation coefficients between the intern and the doctor is $r_{DI} = 0.9$. The ECG report is from the instrument, it can be thought as an independent evidence with the doctor or intern, then the correlation coefficient between ECG report and doctor or intern is 0, that is, $r_{IE} = r_{DE} = 0$.

The first step, the diagnosis of the doctor and the intern can be fused. With the new combination rule, we eliminate the relevance according to the correlation coefficient of evidence for one of the two combination evidences, then:

The intern: $M_I(B) = (1-|r_{DI}|)\times 0.8 = 0.08$

$$M_I(\Omega) = 0.2.+ .|r_{DI}|\times 0.8 = 0.92$$

The fusion result is shown in Table 3.

The result after fusion is:

$$M_1(B) = M_D(B) \oplus M_I(B) = 0.816$$

$$M_1(\Omega) = M_E(\Omega) \oplus M_W(\Omega) = 0.184$$

Then combine above result with the ECG report. Because $r_{IE} = r_{DE} = 0$, the two evidence can be combined with original data. It is shown in Table 4.

And then the final fusion result is:

$$M(A) = 0.1656;\ M(B) = 0.0816;\ M(\Omega) = 0.184;\ M(\emptyset) = 0.$$

The disagreement coefficient of evidence is: $K = 0.7344$.

Standardized the focal element (divided by coefficient (1-K)), the fusion result is: $M(A) = 0.1656/0.2656 = 0.624$; $M(B) = 0.307$; $M(\Omega) = 0.069$.

Table 3: Fusion of doctor and intern.

	$M_I(B) = 0.08$	$M_I(\Omega) = 0.92$
$M_D(B) = 0.8$	$(B)0.064$	$(B)0.736$
$M_D(\Omega) = 0.2$	$(B)0.016$	$(\Omega)0.184$

Table 4: Fusion of result of first step and ECG.

	$M_1(B) = 0.816$	$M_1(\Omega) = 0.184$
$M_E(A) = 0.9$	$(\emptyset)0.7344$	$(A)0.1656$
$M_E(\Omega) = 0.1$	$(B)0.0816$	$(\Omega)0.0184$

The final conclusion shows that the belief interval of coronary thrombosis is [0.624, 0.693], and the belief interval of arrhythmia is [0.307, 0.376]. The fusion conclusion shows that the patient is more probably coronary thrombosis.

6 Conclusion

The paper propose the new combination rule to eliminate the relevance between evidences with correlation coefficient, and the influence of similar evidence can be decreased. As the example shown, the fusion conclusion is more credible.

Acknowledgment

The project was supported by Science Computing and Intelligent Information Processing of Guangxi higher education key laboratory (GXSCIIP201202), and Chunhui Program of Ministry of Education (Z2012029), and partially Supported by Scientific Research Fund of Sichuan Provincial Education Department of China (09ZZ029), and partially supported by research fund of key laboratory of signal and information processing of Xihua University (szjj2012–015).

References

[1] Dempster, A.P., Upper and lower probabilities induced by a multi-valued mapping. *The Annals of Statistics*, **28**, pp. 325–339, 1967.

[2] Shafer, G., *A Mathematical Theory of Evidence*. Princeton University Press, Princeton, NJ, 1976.

[3] Fabre, S., Appriou, A., Briottet, X., Presentation and description of two classification methods using data fusion based on sensor management. *Information Fusion*, **2(1)**, pp. 49–71, 2001.

[4] Parikh, C.R., Pont, M.J., et al., Application of Dempster–Shafer theory in condition monitoring applications: a case study. *Pattern Recognition Letters*, **22(6–7)**, pp. 777–785, 2001.

[5] Valérie Kaftandjian, Olivier Dupuis, Daniel Babot, Yue Min Zhu, Uncertainty modelling using Dempster–Shafer theory for improving detection of weld defects. *Pattern Recognition Letters*, **24(1–3)**, pp. 547–564, 2003.

[6] Wang, P., Propes, N., Khiripet, N., Li, Y., Vachtsevanos, G., An integrated approach to machine fault diagnosis. *IEEE Annual Textile Fiber and Film Industry Technical Conference*, Atlanta, May 4–6, pp. 59–65, 1999.

[7] Jones, R.W., Lowe, A., Harrison, M.J.A., Framework for intelligent medical diagnosis using the theory of evidence. *Knowledge-Based System*, **15**, pp. 77–84, 2002.

Research on electromagnetic shielding problems of equipment support under the condition of information war

Ren Xin[1,2], Jiwen Cui[3], Yifang Yang[2], Zhang Kai[2]
[1]*Graduate School, National Defense University, Beijing, China*
[2]*Medical Protection Laboratory,*
Naval Medical Research Institute, Shanghai, China
[3]*Department of Military Logistics and Science and*
Technology Equipment, National Defense University, Beijing, China

Abstract

Electromagnetic environment is more and more complex in information war-field, especially the electromagnetic pulse weapons or high power microwave weapons can form strong electromagnetic pulse environment, which would pose a serious threat to military equipment. In order to win the information-based war, people should pay great attention to the research on the electromagnetic shielding problem in equipment support. In this paper, we reveal the influence of electromagnetic environment on each element of equipment support by analyzing the physical essence of electromagnetic environment and mechanism. Finally, we put forward some measures and suggestions how to strengthen electromagnetic shielding of equipment support under the condition of information war, and these electromagnetic shielding measures would be effectively used in wartime and daily equipment support training.

Keywords: information warfare, electromagnetic environment, equipment support, electromagnetic shielding.

1 Introduction

With the extensive application of information technology in the field of military, local information warfare under high-tech conditions has played on the stage as a new form of combat [1]. In order to meet the needs of information war develop-

WIT Transactions on Engineering Sciences, Vol. 107, © 2015 WIT Press
www.witpress.com, ISSN 1743-3533 (on-line)
doi:10.2495/SEEEI140681

ment, high and new technology taking the information technology as the core has been applied in weapon system, which increasingly promotes the development of weapon equipment information. Actually, the information superiority is essentially electromagnetic dominance. The complex electromagnetic environment on the information battlefield brings great effects on weapon equipment performance, so equipment support training should be vigorously carried out under complicated electromagnetic environment to improve equipment support ability of information war. To study the electromagnetic environment effect of weapons equipment and protective countermeasures will be an extremely urgent and realistic task in the current weapon equipment development [2].

2 Electromagnetic environment effect and mechanism on the information battlefield

Various electromagnetic activities exist in the whole operational process on the information battlefield, and the complex battlefield electromagnetic environment mainly contains the natural environment and artificial environment factors. The effect of electromagnetic hazard source on weapon equipment is called "Electromagnetic environment effect" [3]. The mechanism of electromagnetic environment effect is given a brief introduction as follows:

1. Electromagnetic frequency interference. Interference signals are generated by instantaneous change of electromagnetic pulse, and they cause electromagnetic frequency interference to make information equipment malfunction or function failure by entering the amplification circuit.
2. Thermal effect. Heating effect of electrostatic discharge or high power electromagnetic pulse will make electronic equipment components overheating, and it can lead to the deterioration or failure of electronic equipment performance.
3. Surge effect. Pulse current is induced on the shielding shell, just like surge flowing on the shell, and the cumulative effect can reduce the equipment reliability. Especially when gaps and holes appear, surge will enter inner-system and cause the damage of sensitive device.
4. Strong electric field effect. Strong electric field formed by the electromagnetic hazard sources can breakdown the gate oxide of components and medium between the metallization line, which cause circuit failure and make components unable to work normally.
5. Magnetic effect. Strong current caused by electromagnetic pulse can produce strong magnetic field, and the electromagnetic energy is directly coupled to the inner-system resulting in the failure of electronic devices [4].

3 The effect of electromagnetic environment on equipment support

Equipment support is collectively called support measures and organization command activities ensuring weapon equipment to successfully complete the mission, and it is an important part of equipment work. In the process of equipment support, the elements of equipment support may get electromagnetic signals they do not want to receive and the high power microwave and electromagnetic pulse will bring serious electromagnetic interference to equipment support activities. The formation of the electromagnetic interference must have three basic elements as follows: the electromagnetic interference sources, coupling channels and interference objects. The relationship of the three elements will be shown in Figure 1.

The main way of electromagnetic interference effect on Equipment support elements is embodied in the following respects.

1. The effect of complex electromagnetic environment on the equipment support command. In order to make scientific and accurate equipment support command decisions, radio communication technology is usually used to collect a lot of relevant information from various information sources on the future information battlefield. But radio communication will be seriously affected by the complex electromagnetic environment. The phenomenon about data transmission of continuity and mistake usually takes place, and the rear echelon cannot achieve normal and effective contact with forward support forces by radio equipment leading to limited information channel or data suspension of equipment support, which will reduce the stability of equipment support information system. A decrease in the ability of equipment support command can easily lead that the decision-maker makes wrong decisions due to battlefield situation misjudge, and it will cause great influence on the process of war.

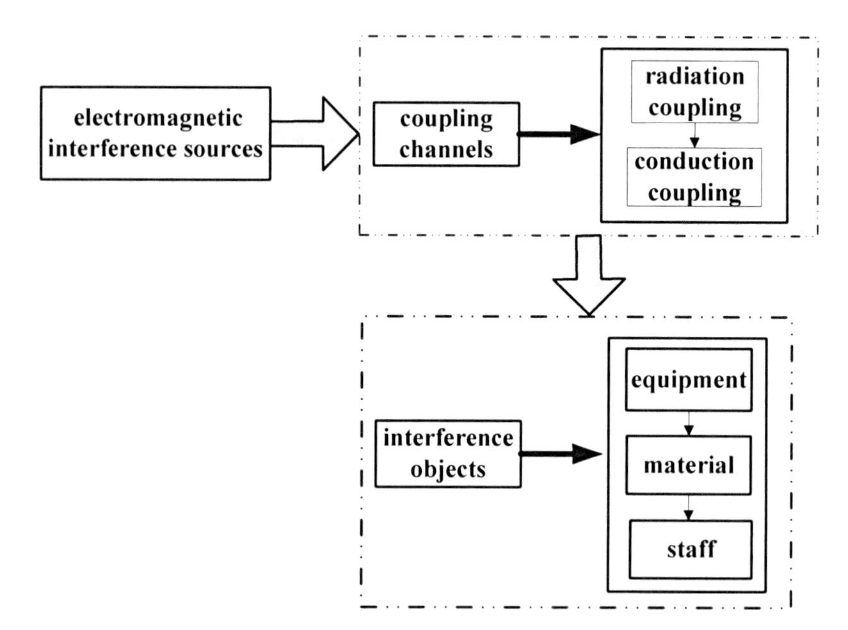

Figure 1: Relationship of the three formation elements.

2. The effect of complex electromagnetic environment on equipment material supply and management. Firstly, because communication command system mainly relies on electromagnetic activities in war, the equip commanders cannot get the condition of equipment consumption and troops demand resulting in equipment support "fog." In addition, the high technical scout system leads to the increase of equipment support supply force transparency [5]. The protection ability of equipment support supply force is weak and it is easily attacked by the enemy during the war, which cannot realize the continuous and efficient supply of equipment support. Again, because a large number of high power microwave and electromagnetic pulse weapons are widely used in information war, equipment is badly damaged. The threat of complex electromagnetic environment to storage and maintenance of weapons and ammunition is widening constantly on the battlefield, and non-combat consumption of equipment material is significantly increasing. These factors bring greater difficulty to equipment material management for equipment command authority.
3. The effect of complex electromagnetic environment on equipment maintenance support. Electromagnetic interference is imposed on maintenance support system by the microelectronics technology, and operation and storage information in the system database is destroyed, which leads that equipment maintenance personnel can only rely on experience and intuition to rescue the damaged equipment on the battlefield using manual work with heavy workload and low efficiency. Aiming at the electromagnetic signal source of equipment maintenance support forces, many means such as electromagnetic suppression, interference and destruction are used. The equipment maintenance power support and its self-survival will be very difficult, which causes to disconnect between equipment maintenance support and operations. Finally, the equipment maintenance support cannot be continuously implemented on the battlefield.

4 How to carry out electromagnetic shielding of equipment support

4.1 Strengthen the electromagnetic shielding theory study of equipment

In order to understand the connotation and composition of the complex electromagnetic environment, we must firstly study the basic theory such as the electromagnetic wave, electromagnetic radiation, and electromagnetic spectrum, and learn how to make electronic war. In addition, we should study the performance characteristics of information weapon equipment and failure mechanism of weapon equipment to grasp equipment electromagnetic characteristics and variation rule of equipment support under the complex electromagnetic environment. Only In this way can we effectively improve the training pertinence of equipment support. Finally, the operational characteristics and rules should be researched in complex electromagnetic environment, and the knowledge of understanding or using frequency, frequency protection and anti-interference also should be learned. We must strengthen the security of equipment and information system and effectively master the equipment support demand of different object in different operational stage and direction.

4.2 Master the electromagnetic shielding technology of equipment

The electromagnetic shielding technology can play an important role on equipment support under information battlefield environment conditions, and the electromagnetic shielding technology of equipment usually can be mainly listed as follows:

1. Ground handling. Taking the earth as a potential reference point, the electronic equipment is connected with the earth in appropriate ways. The influence of external electromagnetic field is effectively suppressed and a large amount of charge accumulated in the surface of the machine casing by electrostatic induction will be released, which can improve the stability of the electronic equipment circuit system to ensure the normal operation of equipment.
2. Technology of electromagnetic shielding. Its main principle is that electromagnetic radiation environment and electronic devices sensitive to electromagnetic pulse is isolated on the space, and the effect of electromagnetic pulse field on the equipment and system will be reduced. By controlling the leakage of internal radiation electromagnetic energy and preventing external radiation from entering certain area, the purpose of equipment electromagnetic shielding of is ultimately realized.
3. Other emergency shielding measures and technology. Filtering technology is an example, it prevents electromagnetic oscillation from entering the equipment along any external connecting line, and the filtering effect is achieved by absorbing the undesirable frequency components.

4.3 Electromagnetic shielding training of equipment support

People should fully understand the effect of complex electromagnetic environment on equipment support and actively explore new training content and method of equipment support, because it is very important to improve the battle effectiveness.

1. Shielding training of equipment support command under complicated electromagnetic environment. The scientific decision is provided for equipment support command relying on the modern electromagnetic information technology. Equipment commanders carry out the deduction and validation of training programs with the use of simulation system in order to predict the possible results of different program, and they choose the best solution to achieve qualitative leap of equipment support command decision. In addition, equipment commanders must strictly differentiate the non-scientific electromagnetic information and false

electromagnetic phenomenon and capture the essence of things through the phenomenon. Doing so can improve the ability of observation, identification, analysis and comparison and to obtain useful electromagnetic information. Equipment command and coordination is the key to satisfy the electromagnetic information requirement of equipment command decision under the complex electromagnetic environment in time of war.

2. Equipment training of material supply and management under complicated electromagnetic environment. In order to improve the ability of equipment material supply and management on the battlefield, a variety of high and new techniques such as multi-band camouflage net, compound camouflage coatings and corner reflector are adapted [6]. Firstly, because the shielding ability of equipment support force is limited, equipment shielding system should constantly improve camouflage protection ability in the war and avoid the soft kill and hard damage of electromagnetic wave to ensure that the equipment material supply is safe and reliable. In addition, the effective electromagnetic shielding measures should be taken to improve the electromagnetic shielding ability of equipment material management in the main place for the safety of the equipment material management power, and the safety protection system of modern equipment material will be established.
3. Equipment maintenance support training under complicated electromagnetic environment. At the first, the complex electromagnetic environment of equipment maintenance support training is to be constructed in order to explore the regulation of equipment maintenance support on the basis of integration of existing equipment under complicated electromagnetic environment. The integrated complex electromagnetic environment will be created by taking advantage of real weapon equipment, all kinds of signal simulator, computer simulation technology and distributed interactive simulation technology to flexibly simulate electromagnetic situation

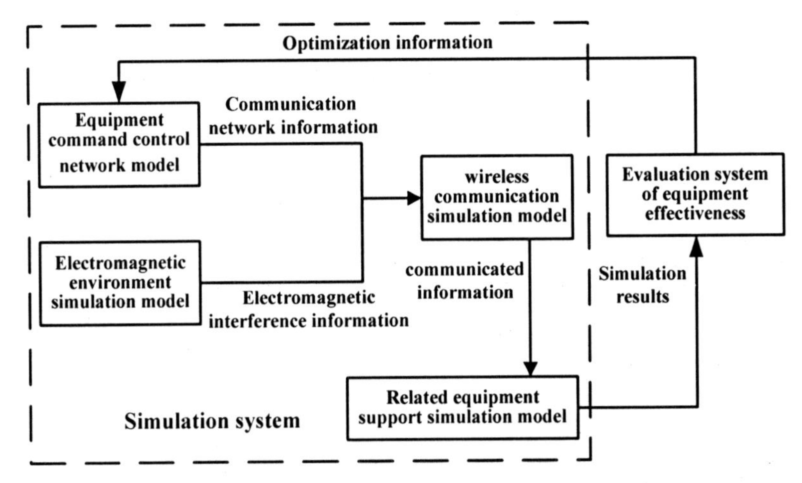

Figure 2: The operating diagram of simulation model for equipment support.

on the implementation of equipment maintenance support. The repair place is built to effectively prevent electromagnetic interference, and the targeted measures are taken to improve their protective ability by strengthening defense ability training of equipment survival, and in order to ensure the accuracy and safety of equipment maintenance through innovation of equipment maintenance support method. Besides, three-dimensional simulation of virtual battlefield can be established in computer software environment and the virtual equipment shall be placed in the electromagnetic environment in order to carry out collaborative training online in accordance with the operational mode. The operating diagram of simulation model for equipment support is shown in Figure 2.

5 Conclusions

The complicated electromagnetic environment has inevitably existed on the battlefield, and weapon equipment has become increasingly information-based and electromagnetic sensitization. Because the effect of complex electromagnetic environment on weapon effectiveness and equipment support is gradually increasing, the electromagnetic compatibility of the equipment should be improved to enhance the battlefield adaptability, and study the electromagnetic environment effect of weapon equipment and countermeasures to and has become a research topic is urgently. This paper discusses mechanism and influence on equipment support of complex electromagnetic environment, and puts forward some targeted countermeasures and suggestions, which will have some reference significance for improving the ability of equipment support under complex electromagnetic environment in the future.

References

[1] Ding Hong-Bao, Lin Xiang-Guo, Wang Wen-Ming, The difficulties of medical service support and countermeasures in local war under complicated electromagnetic environment. *Defense Health Forum*, **16(4)**, p. 201, 2007.
[2] Huang Tao, Liu Chang-Yong, Zhang Guo-Yun, A brief analysis of problems and countermeasures about equipment support training under complex electromagnetic environment. *Research on Equipment of Academic*, **(1)**, pp. 12–13, 2008.
[3] Robert D. Goldblum, Electromagnetic environmental effects (E3) within the military. ITEM 2000.
[4] Manuel W. Wik, et al., The threat of intentional electromagnetic interference. In: *Proc. Asia Pacific CEEM*, Shanghai, China, 2000.
[5] He Hong, et al., *Electromagnetic Compatibility and Electromagnetic Interference*. National Defense Industry Press, Beijing, 2007.
[6] Cai Ji-Wei, et al., Study on command and control modeling of equipment support under the complex electromagnetic environment. *Journal of Hebei University of Science and Technology*, **32(12)**, pp. 92–95, 2011.

Electrical impedance tomography system used in pulmonary function based on FPGA

Hou Hailing[1,2], Wang Huaxiang[2], Chen Xiaoyan[1]
[1]*College of Electronic Information and Automation, Tianjin University of Science and Technology, Tianjin, China*
[2]*School of Electrical Engineering and Automation, Tianjin University, Tianjin, China*

Abstract

In electrical impedance tomography system, DC polarization voltage has effect on measurement signals which exists at the interface between the metal electrodes and the skin or the electrolyte solutions. To solve the problem, a band-pass filter is designed and implemented. The measured waveforms show that the DC polarization voltage is filtered out effectively and signal-to-noise ratio of the system is improved. Multiperiod undersampling technique is applied to data acquisition of multifrequency EIT system whose frequency ranges from 10KHz to 10MHz. Numerical experiments show that the combination of the undersampling technique and digital phase-sensitive detector can maximize signal-to-noise ratio in the presence of Gaussian wideband noise by Matlab simulation. Finally, conjugate gradient algorithm is adopted to image media distribution in the statistic experiments and get better imaging results.

Keywords: electrical impedance tomography, polarization voltage, band-pass filter, undersampling, digital phase-sensitive.

1 Introduction

Medical research has shown that the tissues of human body have different impedance characteristics [1], and some pathological phenomena and biological activities may cause changes in tissue impedance. Therefore biological tissue impedance carries rich pathological and physiological information. In 1985, Brown and others first discussed the potential applications of EIT (Electrical

www.witpress.com, ISSN 1743-3533 (on-line)
doi:10.2495/SEEEI140691

Impedance Tomography) in medical field, and proposed that it can be applied to lung ventilation image monitoring [2].

In EIT, alternating current or voltage is applied to an object by means of excitation electrodes placed on body surface. Meanwhile, by measuring the voltage (or current) of electrodes, the corresponding electrical impedance and its changes can be calculated. Then according to different application purposes, some electrical properties of the tissue or organs, such as impedance, admittance, permittivity, can be acquired which are associated with the state of human physiology, pathology [3]. This information not only can reflect the anatomical structure, but more importantly, functional image be expected to obtain.

EIT technique has an attractive application prospect for its advantages such as non-invasion, non-radiate, quick-response and low-cost. While, in EIT system, the current field which is built in human tissue is a 'soft field' and the EIT inverse problem is poorly conditioned and ill-posed. For example, a slight variation in measured voltage has a large effect on imaging results. So improving the signal-to-noise ratio of the hardware circuit is the premise of stable imaging of EIT system. In addition, medical studies have learned that the real part and imaginary part of electrical impedance of human tissue all contain abundant physiological and pathological information. However, the imaginary part is not easy to extract, which magnitude increases with the rising of excitation frequency. Clinical EIT imaging systems have limited to collect data at frequencies of less than 100kHz [4, 5], with the exception of Halter's design which had a bandwidth of 10MHz for breast imaging. With the increase of frequency, the result of the measurement will be greatly influenced due to the existence of the parasitic impedance that corrupts the signals being measured at high frequencies. Besides that, the EIT inverse problem is poorly conditioned and ill-posed, thus requiring that the measured data have a high degree of precision. It is difficult to choose the analog-digital converter with high resolution, high conversion rate and low cost.

Experiments found that DC polarization voltage has great impact on useful signals which exists at the interface between the metal electrodes and the skin or the electrolyte solutions. In this design, an active band-pass filter can effectually attenuate the DC signal, which make the system SNR be further improved.

In this hardware circuit, multi-period undersampling technique is adopted to gather high frequency signals. Numerical experiments proved that multi-period undersampling technique can be well applied to EIT system, and DPSD technique has good inhibition effect on Gauss white noise.

2 Overall structure of the system

This EIT system hardware is mainly made up of three parts: sensing electrode array, data collection system and computer image reconstruction unit. System block diagram is shown in Figure 1.

Data collection system chooses Xilinx's FPGA – Spartan-3 as the control core, and PicoBlaze, the embedded microprocessor soft core, is responsible for data collection and process control. Spartan-3's DDS (Direct Digital Synthesizer) IP core generates sine wave digital signal, which is then converted to analog sine

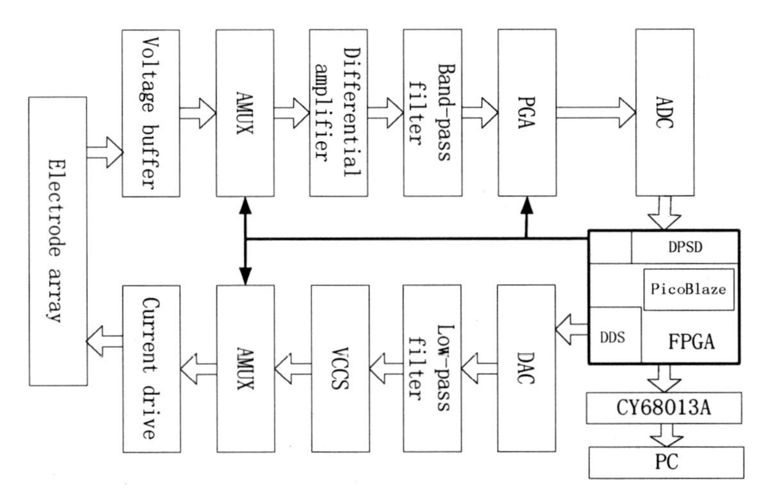

Figure 1: System block diagram.

wave signal by a 14-bit digital to analog converter – DAC904. High-order harmonics contained in the analog sine wave are attenuated by a second order low-pass filter. Voltage controlled current source circuit converts voltage signal to current signal which is built with current feedback amplifier – AD844. Current stimulation signal is loaded into the test tissue after flowing through current drive circuit.

FPGA controls analog multiplexers, scans electrode array and collects test signals. The weak measurement signal is amplified by high CMRR, differential amplifier – AD8129 which can suppress common-mode signals in a great extent.

Through the observation to the oscillator, measurement signal contains both high frequency noise and DC polarization voltage signal. An active band-pass filter is designed to suppress DC offset signal which may corrupt useful signal. EIT system is a kind of system with large dynamic range. So, if the gain of amplifier circuits is fixed, the small signal is easily polluted by the noise. In order to make full use of ADC's input range, programmable gain amplifier – THS7001 is adopted to be post-amplifier circuit. Analog measurement signals are converted to digital signals by ADC, and then the real part and imaginary part of the electrical bioimpedance are calculated through digital phase sensitive demodulation unit implemented by FPGA. Then, the results are sent to a computer through USB2.0 chip – CY68013A. Finally, the conductivity and permittivity distribution inside the body are imaged in 2-D by the reconstruction algorithm using the knowledge of the applied excitation, the measured data and the electrode geometry.

3 Elimination of polarization voltage

In actual measurement, electric charges will be generated at the interface between the metal electrodes and the skin or the electrolyte solutions, and then produce a certain potential difference which is known as polarization voltage. The polarization voltage is from several millivolt to dozens of millivolt which exhibits direct form. The magnitude of polarization voltage is related to the current flowing through electrodes, contact impedance between electrodes and skin [5]. It is inevitable to corrupt the useful signals because the measurement signals which contain polarization voltage will be beyond the scope of ADC measurements for a 2.5V DC voltage bias, as shown in Figure 2.

In experiments which excitation frequency is 100KHz, measure waveform observed is shown in Figure 2 after signals are amplified by a differential amplifier. In Figure 2, electrode 1 and 2 are exciting electrodes, and measure the voltages of electrodes 3–8 in order. It can be seen clearly that polarization voltage corrupts the useful signals. Measurement waveforms of all electrode couples are not in the same axis and the polarization voltages are diverse from each other, as shown in Figure 2(a). And in Figure 2(b), the top or the bottom of the measurement waveforms are eliminated which cannot be used for calculating the results.

To eliminate the impact of DC polarization voltage on measurement, a band-pass filter is designed which is made up by a second-order high-pass filter and a second-order low-pass filter. The filter can greatly suppress DC voltage and high frequency noise contained in measurement signals verified through simulation and observed through an oscilloscope.

3.1 Design of the high-pass filter

By the above analysis, a high-pass filter is needed to remove the DC polarization voltage in order to avoid signal distortion. A second order voltage-controlled high-pass filter is designed as shown in Figure 3(a).

The transfer function of the second order voltage-controlled high-pass filter is

$$A_{\mathrm{u}}(s)=\frac{(sCR)^2 A_{\mathrm{up}}(s)}{1+(3-A_{\mathrm{up}})sCR+(sCR)^2} \tag{1}$$

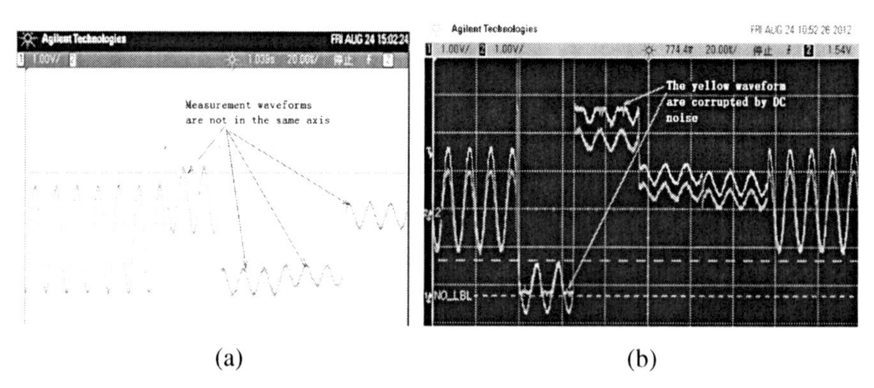

(a) (b)

Figure 2: Measurement signals affected by DC polarization voltage.

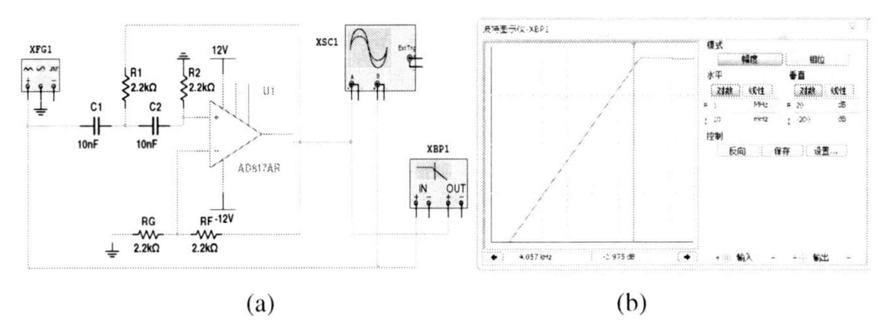

(a) (b)

Figure 3: (a) Second order voltage controlled high-pass filter. (b) Amplitude-frequency characteristic.

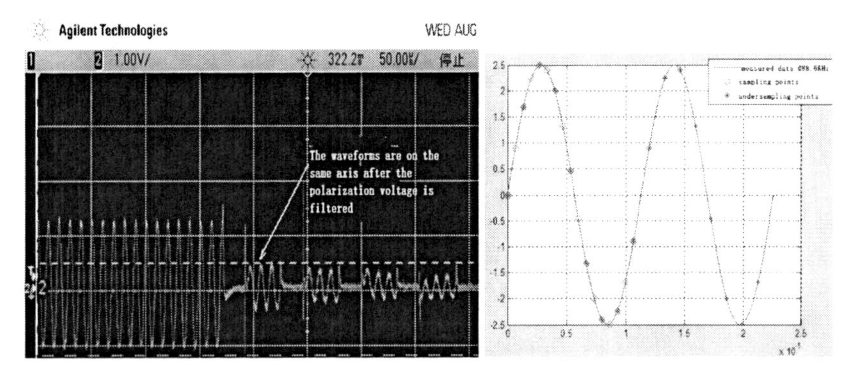

Figure 4: Waveform without DC offset voltage. Figure 5: Undersampling scheme.

where A_{up} is the passband gain, RF = R1, $A_{up} = 1 + R_F/R_1 = 2$; Q-factor $Q = 1/(3-A_{up}) = 1$; $R = 2.2\text{k}\Omega$, $C = 0.01\mu\text{F}$, characteristic frequency $f_0 = 1/(2\pi RC) = 7284\text{Hz}$, time constant $\tau = RC = 22\mu\text{s}$.

Frequency response formula is

$$\dot{A}_u = \frac{A_{up}}{1-(\frac{f_0}{f})^2 + j\frac{1}{Q}(\frac{f_0}{f})} \tag{2}$$

Amplitude-frequency curve is shown in Figure 3(b) simulated by MultiSim. When $f << f_0$, the slope of the curve is + 40dB/dec. After the voltage signals are processed through the filter, the actual curve is observed, as shown in Figure 4. As can be seen from this figure, DC polarization voltage contained in measurement signals is removed and all signals are on the same axis. The measured waveform is observed when exciting with intermittent sine waves which adopt direct digital frequency synthesizer (DDFS) technique [6]. This intermittent sine waves keep zero for a certain period of time, which can extend the time that signals pass through 0 volt. This keeping zero switch strategy combining with digital phase sensitive demodulation technique can improve the SNR of the system.

4 The combination of multi-period undersampling technique and digital phase sensitive demodulation

4.1 The multi-period undersampling technique

As described above, the real part and imaginary part of tissue impedance contain abundant physiological and pathological information. Jossinet [7] suggested that the measurement signals could improve diagnostic value when the excitation frequency exceeds 1MHz. This is because the imaginary part information can be highlighted when the excitation frequency reaches a few MHz. Meanwhile, according to Nyquist sampling theorem, in order to recover the original signal, sampling frequency F_s is at least twice the highest frequency fm of the measured signals, in the engineering practice, F_s often is 5–10 times of fm. For example, if excitation frequency is 10MHz, F_s should be 50–100MHz. It is difficult to choose analog to digital converter with more than 16-bit resolution, high conversion rate, and low cost. To solve the bandwidth issue, multi-period undersampling technique is adopted. The undersampling method is similar to the sampling oscilloscope capturing periodic waveforms of frequencies much higher than the sampling rate of the scope. The samples are obtained over multiple cycles of the sinusoid. The number of sampling points per cycle is not an integer, so the sampling positions inside each cycle are different. With the appropriate choice of sampling frequency, the samples obtained are equivalent to those obtained by sampling at a higher rate over one cycle of the sinusoid [8]. This property is demonstrated in Figure 5, where the actual sampling points spread over two cycles.

4.2 Digital phase sensitive demodulation (DPSD)

Phase-sensitive detection is a very powerful technique for measuring the amplitude and phase of a signal which is contaminated with, or even buried in Gaussian noise or other non-coherent interference. The digital phase sensitive demodulation unit implemented with FPGA is based on match filter theory. A_s is well known, for a particular signal contaminated with Gaussian noise, the filter which gives the best SNR [9] improvement is the signal's matched filter.

Define in-phase reference signal V_{ci}, quadrature reference signal V_{si} and measurement signal V_i.

$$V_{ci} = \cos(2\pi i F_{sig} / F_s) \tag{3}$$

$$V_{si} = \sin(2\pi i F_{sig} / F_s) \tag{4}$$

$$V_i = A\cos(2\pi i F_{sig} / F_s + \varphi) \tag{5}$$

where F_s is sampling frequency, F_{sig} is signal frequency, A is amplitude, φ is the initial phase. After multiply accumulation, the results are

$$R=\sum_{i=0}^{N-1}V_{ci}V_i=\sum_{i=0}^{N-1}\cos(2\pi iF_{sig}/F_s)A\cos(2\pi iF_{sig}/F_s+\varphi)=\frac{1}{2}NA\cos\varphi \quad (6)$$

$$I=\sum_{i=0}^{N-1}V_{si}V_i=\sum_{i=0}^{N-1}\sin(2\pi iF_{sig}/F_s)A\cos(2\pi iF_{sig}/F_s+\varphi)=\frac{1}{2}NA\sin\varphi \quad (7)$$

In Formula (6) and (7), $N = F_s/F_{sig}$, the amplitude and phase can be easily calculated from formulas (6) and (7).

5 Static experiment results

In the experiments, an perspex salt bath with 300mm diameter is used, a concentration of 0.7% saline simulates human tissue, and 16 electrodes are stimulated and measured with 100KHz frequency signal, as shown in Figure 6. Conjugate gradient algorithm [10–12] is adopted to image media distribution in the experiments, and the results of imaging are shown in Figure 7(a).

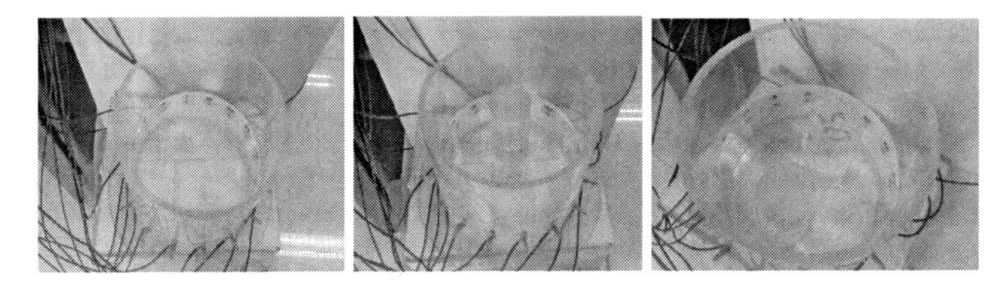

Figure 6: Experimental salt bath. (a) 1 empty, (b) one bar, (c) two bars.

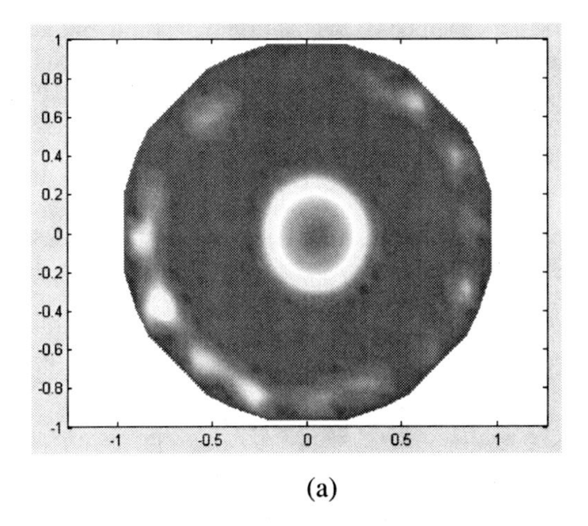

(a)

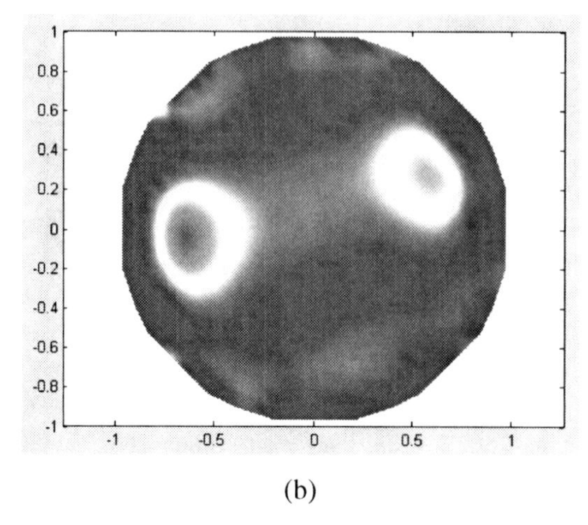

(b)

Figure 7: Reconstructed images. (a) one bar, (b) two bars.

6 Conclusions

In this paper, a band-pass filter is designed and effectively suppresses DC polarization voltage. Thereby the SNR is improved further. The system adopts undersampling technique combining DPSD which can give the best SNR improvement for Gaussian noise, and can acquire signal with higher frequency by using ADC with lower conversion rate.

Acknowledgment

The research work was supported by National Natural Science Foundation of China under Grant No. 61301246 and Natural Science Foundation of Tianjin under Grant No. 12JCYBJC19300.

References

[1] A.M. Dijkstra, B.H. Brown, A.D. Leathard, et al., Review clinical application of electrical impedance tomography. *Journal of Medical Engineering and Technology*, **(17)3**, pp. 89–98.

[2] B.H. Brown, et al., Applied potential tomography: possible clinical applications. *Clinical Physics and Physiological Measurement*, **6(2)**, pp. 109–121, 1985.

[3] He Yongbo, Wang Huaxiang, Ma Min, A high precision electrical impedance tomography system for human body. *Journal of Electronic Measurement and Instrument*, **20(2)**, pp. 48–51, 2006.

[4] R. Halter, A. Hartov, K.D. Paulsen, Design and implementation of a high frequency electrical tomography system. *Physiological Measurement*, **25**, pp. 379–390, 2004.

[5] Wang Sanqiang, He Wei, Shi Jian, New design of pre-amplification circuit for EEG signals. *Journal of Chongqing University*, **29(6)**, pp. 51–53, 2006.
[6] Cui Ziqiang, Wang Huaxiang, Improvements on real-time performance of electrical capacitance tomography. *Chinese Journal of Scientific Instrument*, **31(9)**, pp. 1939–1945, 2010.
[7] J. Jossinet, The impedivity of freshly excised human breast tissue. *Physiological Measurement*, **19**, pp. 61–75, 1998.
[8] Ning Liu, G.J. Saulnier, J.C. Newell, A multichannel synthesizer and voltmeter for electrical impedance tomography. In: *Proceedings of the 25' Annual International Conference of the IEEE EMBS Cancun*, Mexico, September 17–21, 2003.
[9] R. Smith, I. Freeston, B. Brown, et al., Design of a phase-sensitive detector to maximize signal-to-noise ratio in the presence of Gaussian wideband noise. *Measurement Science and Technology*, **3**, pp. 1054–1062, 1992.
[10] M. Hanke, Conjugate gradient type methods for ill-posed problems. Scientific and Technical, Longman, Harlow, 1995.
[11] Rajen Manicon Murugan, An Improved Electrical Impedance Tomography (EIT) Algorithm for the Detection and Diagnosis of Early Stages of Breast Cancer, PhD thesis, University of Manitoba, 1999.
[12] Xie Kexin, Han Jian, Lin Youlian, *Optimization Methods*. Tianjin University Press, Tianjin, 2004.

Research of the voltage-source three-phase three-level PWM rectifier

Wang Shuo, Huang Mei, Niu Liyong
School of Electrical Engineering,
Beijing Jiaotong University, Beijing, China

Abstract

The neutral-point clamped three-phase three-level PWM rectifier, possessed of the advantages as small volume, simple control method and the two-way flow of energy, is the earliest, most used multilevel topology structure. This paper gives its mathematical model, a double closed loop control strategy based on SVPWM modulation. Based on the inherent neutral-point unbalance of the three-level PWM rectifier, a precise control method based on the variable ratio factor is selected. Finally based on Matlab/Simulink, this paper constructs simulation model of three-level PWM rectifier and the simulation and experimental results verify the correctness and feasibility of the proposed control scheme.
Keywords: three-phase voltage-source rectifier, balance of neutral point potential, TMS320F28335.

1 Introduction

Nowadays, Neural-Point-Clamped three-level PWM rectifier which is more and more researched and applied could be adapted to the situation like high ac voltage and large capacity [1]. Comparing with two-level rectifier, the three-level PWM rectifier has the advantages as follow [2–4]:

The peak voltage that each main power switch tube takes is only half of the value of two-level PWM rectifier. It is able to generate input current which has better sine wave even when the switching frequency is not too high. Under the same switching frequency and controlling condition, the harmonic waves of the three-level PWM rectifier input current is much less than the two-level rectifier.

The neutral-point voltage balancing problem of three-level NPC rectifier has been widely recognized in literature. Various strategies have been presented, and

WIT Transactions on Engineering Sciences, Vol. 107, © 2015 WIT Press
www.witpress.com, ISSN 1743-3533 (on-line)
doi:10.2495/SEEEI140701

successful operation has been demonstrated with a dc-link voltage balance maintained. This paper proposed a precise control method based on the variable ratio factor and the simulative results prove the validity and effectiveness of the proposed control method.

2 Voltage control method and mathematical model of three-level PWM

The main circuit of three-level neutral point clamped (NPC) PWM rectifier is shown in Figure 1, which consists of 12 switches, 6 clamping diodes, two output capacitor (C1, C2). States of Switch $S1$ and $S3$ are complementary, so are the states of switches $S2$ and $S4$. Switches $S1$ and $S4$ cannot conduct at the same time, and then three-level modulation of the AC input voltage is obtained. As to the different combination of four switch states per phase, there are three different levels:

$$V_{aN}\begin{cases} U_{dc1} & S1,S2: \text{on s3,s4: off} \\ 0 & S2,S3: \text{on s1,s4: off} \\ U_{dc2} & S3,S4: \text{on s1,s2: off} \end{cases} \tag{1}$$

In the three-phase static coordinates (a, b, c), according to the Kirchhoff's law, the state equation is:

$$\begin{cases} u_{sa} = L_s \dfrac{d_{ia}}{dt} + i_a R_s + \left(S_{1a} - \dfrac{S_{1a}+S_{1b}+S_{1c}}{3}\right)U_{dc1} - \left(S_{2a} - \dfrac{S_{2a}+S_{2b}+S_{2c}}{3}\right)U_{dc2} \\ u_{sb} = L_s \dfrac{d_{ib}}{dt} + i_b R_s + \left(S_{1b} - \dfrac{S_{1a}+S_{1b}+S_{1c}}{3}\right)U_{dc1} - \left(S_{2b} - \dfrac{S_{2a}+S_{2b}+S_{2c}}{3}\right)U_{dc2} \\ u_{sc} = L_s \dfrac{d_{ic}}{dt} + i_c R_s + \left(S_{1c} - \dfrac{S_{1a}+S_{1b}+S_{1c}}{3}\right)U_{dc1} - \left(S_{2c} - \dfrac{S_{2a}+S_{2b}+S_{2c}}{3}\right)U_{dc2} \end{cases} \tag{2}$$

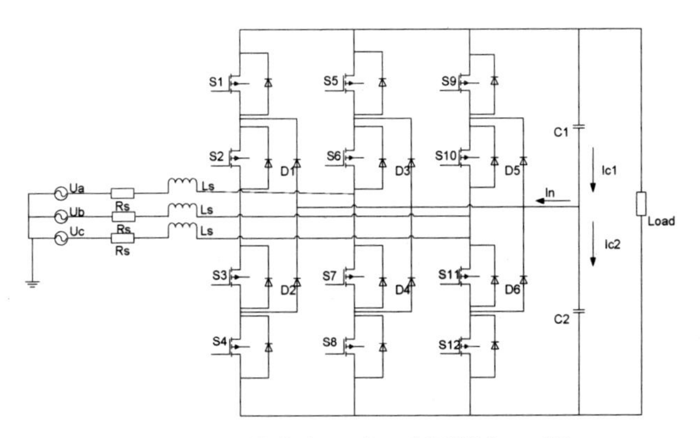

Figure 1: NPC three-level PWM rectifier.

$$\begin{cases} C_d \dfrac{dU_{dc1}}{dt} = S_{1a}i_a + S_{1b}i_b + S_{1c}i_c - i_L \\ C_d \dfrac{dU_{dc2}}{dt} = S_{2a}i_a + S_{2b}i_b + S_{2c}i_c - i_L \end{cases} \tag{3}$$

Through Clark transformation and park transformation, the mathematical model of three-level PWM rectifier on two-phase synchronous rotating (d, q) coordinate is as follows:

$$\begin{cases} L_s \dfrac{d_{id}}{dt} - \omega L_s i_q + R_s i_d = u_{sd} - S_{d1}U_{dc1} + S_{d2}U_{dc2} \\ L_s \dfrac{d_{id}}{dt} - \omega L_s i_d + R_s i_q = u_{sq} - S_{q1}U_{dc1} + S_{q2}U_{dc2} \end{cases} \tag{4}$$

$$\begin{cases} C_d \dfrac{dU_{dc1}}{dt} = \dfrac{3}{2}\left(i_{d1}S_{d1} + i_{q1}S_{q1}\right) - i_L \\ C_d \dfrac{dU_{dc2}}{dt} = \dfrac{3}{2}\left(i_{d2}S_{d2} + i_{q2}S_{q2}\right) - i_L \end{cases} \tag{5}$$

Figure 2 shows the model of three-level PWM rectifier on (d, q) coordinate.

This paper puts forward a double closed-loop control method which consists of outer voltage loop and inner current loop based on three-phase voltage source PWM rectifier. The outer voltage loop controls DC side voltage of PWM rectifier and inner current loop control AC current to realize operate in the unity power factor.

As formula (4), assuming that u_d, u_q as rectifier front voltage of AC side, then

$$\begin{cases} u_d = u_{sd} + \omega L_s i_q - L_s \dfrac{di_d}{dt} - R_s i_d \\ uq = u_{sq} + \omega L_s i_d - L_s \dfrac{di_q}{dt} - R_s i_q \end{cases} \tag{6}$$

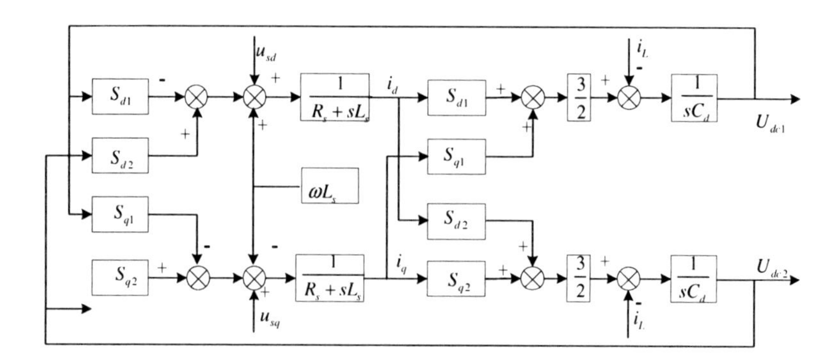

Figure 2: Model of three-level PWM rectifier on (d, q) coordinate.

As Eq. (6) shows, i_d and i_q, the grid current in (d, q) coordinates, are intercoupling which brings inconvenience to the current controller design. For this, using feedforward decoupling control strategy to realize the adjustment without error of i_d and i_q with two PI regulators. Therefore, the control equation of space command vector voltage V^* is as follows:

$$\begin{cases} v_d^* = u_{sd} + \omega L_s i_q - \left(K_p + \dfrac{K_1}{s}\right)\left(i_d^* - i_d\right) \\ v_q^* = u_{sq} + \omega L_s i_d - \left(K_p + \dfrac{K_1}{s}\right)\left(i_q^* - i_q\right) \end{cases} \tag{7}$$

where v_d and v_q are components of space command vector voltage V^* in (d, q) coordinates, i_d and i_q are components of current vector of AC side (d, q) coordinates, i_d^* and i_q^* are gird current instructions, K_p is proportional gain and K_i is integral gain.

Formula (7) introduces $\omega L_s i_q$, $\omega L_s i_d$ and u_{sd}, u_{sq} as feedforward compensation, which is called current feedforward decoupling. The current state equation after decoupling is as follows:

$$\begin{cases} \left(K_p + \dfrac{K_1}{s}\right)\left(i_d^* - i_d\right) = L_s \dfrac{di_d}{dt} + R_s i_d \\ \left(K_p + \dfrac{K_1}{s}\right)\left(i_q^* - i_q\right) = L_s \dfrac{di_d}{dt} + R_s i_q \end{cases} \tag{8}$$

Obviously, this control algorithm of current feedforward decoupling succeeds to achieve the i_d, i_q decoupling control, and i_d, i_q equation turns into a simple first-order system which simplifies the design of the control system. When the rectifier is running in the unit power factor, the referenced reactive current i_q^* should be 0 and the referenced Active current i_d^* can be calculated from the voltage loop:

$$i_d^* = \left(K_{ip} + \frac{K_{il}}{T}\right)\left(U_{dc}^* - U_{dc}\right) \tag{9}$$

This design of the outer voltage loop can make the DC side voltage track the command voltage and then get the referenced active current i_d^*. Figure 3 shows the principle of double closed-loop control system.

3 The method of neutral point potential balance

Based on the assumptions of $C_1 = C_2 = C_s$, $\Delta U = U_{dc1} = U_{dc2}$, according to the Kirchhoff's current law, the neutral current equation is:

$$i_n = i_{c1} - i_{c2} = 2C_s \frac{d\left(U_{dc1} - U_{dc2}\right)}{dt} = -2C_s \frac{d(\Delta U_{dc})}{dt} \tag{10}$$

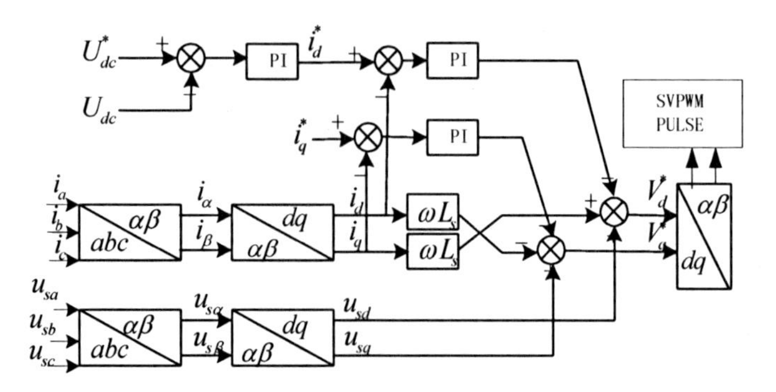

Figure 3: Block diagram of double closed-loop control system.

According to Eq. (9), it can be seen that the unbalance of neutral point potential is due to that the current flowing in and out of neutral point exists, which makes one capacitor charge and the other discharge, when the neutral point current flow through the dc-link capacitors, the electric charge stored in the two capacitors will definitely vary so as to cause the change of the neutral point potential. As this paper is based on the SVPWM modulation [5, 6], it can be found that the zero vector and large vectors have no effect on neutral point potential. Furthermore, the neutral point current flow through small vectors and medium vectors which definitely can cause the unbalance of neutral point potential. However, there is only one switch state of medium vectors and each small vector has two switching states, so controlling the time of each small vector can make the neutral point potential balance. This paper introduces the concept of ratio factor ρ:

$$\rho = \frac{\left(U_{dc1} - U_{dc2}\right)}{v} = \frac{\Delta U_{dc}}{v} \tag{11}$$

Without changing total time of small vector, the ratio factor can change the control time of each small vector which can make the neutral point potential balance.

4 Simulation and experimental results

In order to verify the operation of the proposed rectifier and control method, the system of Figure 4 has been simulated in MATLAB/SIMULINK environment and the experimental verifications are carried on a 20kVA three-level PWM rectifier prototype based on MOSFET as well. A fully digital control system with digital signal processor (DSP) TMS320F28335 and complex programmable logic device (CPLD) XC95144 is designed to fulfill the control algorithms. The main parameters used for simulation are defined as follows: power source voltage, U_s = 220V, supply fundamental frequency, f= 50Hz, Switching frequency, f_s = 20kHZ, filter line inductance, L =1mH, Resistance of reactors, R_s =0.35Ω, DC link capacitor, C = 4700 μF, Load resistance, R=24Ω, DC link voltage, U^*_{DC} =660V.

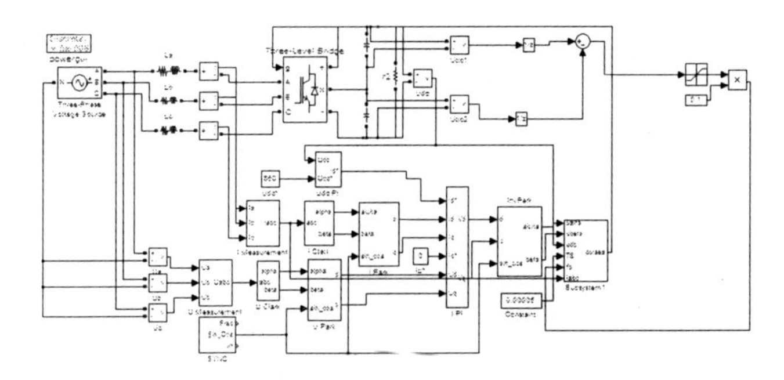

Figure 4: Simulation structure of three-level rectifier.

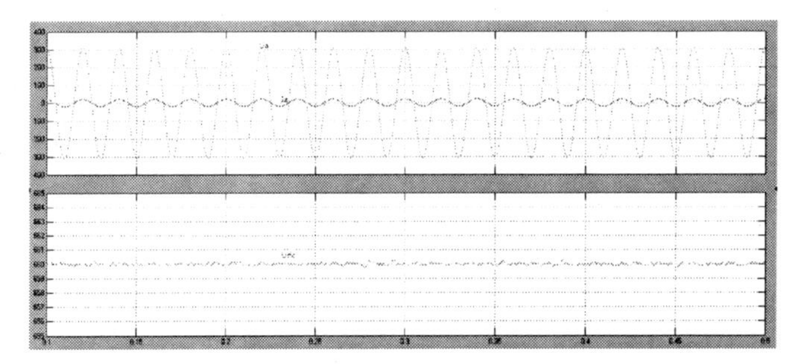

Figure 5: Voltage and current waveform of grid side and voltage waveform of DC side.

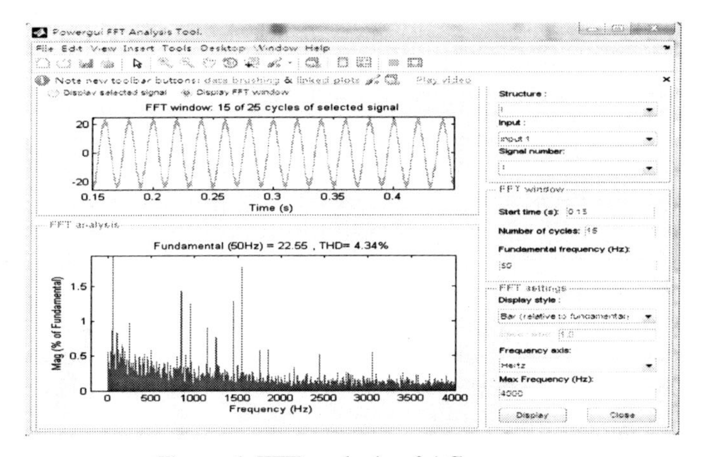

Figure 6: FFT analysis of AC current.

Figures 5–7 are the simulation results. Figure 5 is voltage and current waveform of grid side and voltage waveform of DC side. It can be seen that the rectifier's voltage and current waveforms keep the same phase angle to realize unit power factor working, and the DC side voltage U_{DC} remained in the 660V. Figure 6 gives the harmonic spectrum of the input current. The spectrum

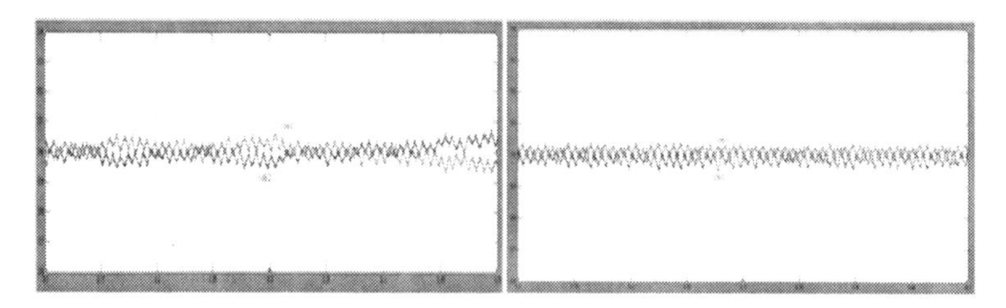

Figure 7: Comparison of DC voltage waveform.

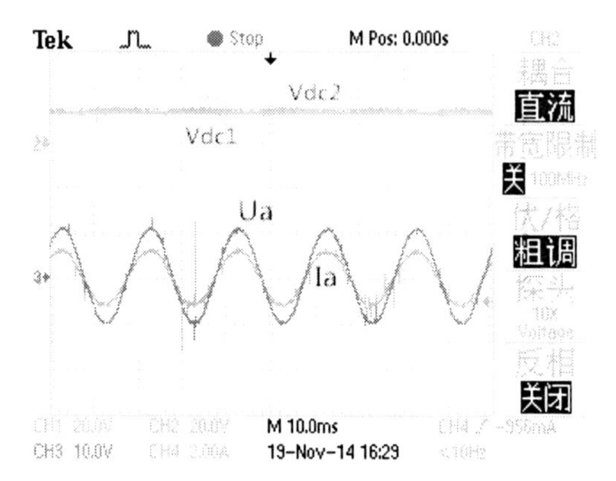

Figure 8: Current and voltage of the grid side and DC voltage waveforms under low input voltage condition.

illustrates that the THD is equal to 4.34%. Therefore, the three-level PWM rectifier can work in unit power factor with this strategy. Figure 7 is the two capacitors' voltage, before and after neutral point potential balance control. It can be concluded that the neutral-point potential balance is obtained well. Figure 8 is the experimental result of steady state which represents the current and voltage of the grid side and DC voltage waveforms under low input voltage condition. It can be found that, the input current and the output voltage of the rectifier are stable, and the neutral-point voltage balance is obtained well. The simulation results and the experimental results show well steady and dynamic performance of the rectifier using the proposed control scheme base on the proportional plus resonant controller.

5 Conclusion

This paper proposed a new PWM control algorithm based on the SVPWM for a neutral-point-clamped three-level PWM rectifier. Pure sinusoidal waveforms can be achieved in grid currents with a unity power factor by controlling the input current. Finally, the control strategy is simulated in MATLAB / SIMULINK environment and is implemented in the there-level PWM rectifier experimental system based on DSP. The simulation and experimental results are consistent with the theoretical analysis and verify the excellent performances of the proposed PWM control method for three-level rectifiers. The PWM control strategy presented in this paper is worth to be applied especially in high-power medium-voltage circumstance.

References

[1] A. Nabae, I. Takahashi, H. Akagi, A new neutral-point clamped PWM inverter. *IEEE Transactions on Industry Applications*, **IA-17(5)**, pp. 518–523, 1981.

[2] Y.C. Zhang, Z.M. Zhao, M. Eltawil, L.Q. Yuan, Performance evaluation of three control strategies for three-level neutral point clamped PWM rectifier. In: *Proceedings of IEEE APEC'08*, pp. 259–264, 2008.

[3] Jin Hongyuan, Zou Yunpin, Lin Lei, Research on the technology of the neutral-point voltage balance and dual-loop control scheme for three-level PWM rectifier. In: *Proceedings of the CSEE*, **26**, pp. 64–68, 2006.

[4] A.R. Beig, V.T. Ranganathan, Space vector based bus clamped PWM algorithms for three level inverters: implementation, performance analysis and application considerations. *IEEE Proc. of APEC03*, New York, NY, IEEE, pp. 569–575, 2003.

[5] Thomas Nathenas, Georgios Adamidis, A new approach for SVPWM of a three-level inverter-induction motor fed-neutral point balancing algorithm. *Simulation Modeling Practice and Theory*, **(29)**, pp. 1–17, 2012.

[6] Han Yanqing, Wang Yong, Zhao Bo, SVPWM current control based on indeterminate frequency hysteresis. *Power Electronics*, **47(2)**, pp. 79–81, 2013.

Design and implementation of 3-axis servo platform driving system based on magnetic encoder

Yangzhi Guo
School of Electrical and Electronic Engineering,
Hubei University of Technology, Wuhan, China

Abstract

Nowadays, most 3-axis servo platform drivers are developed in open-loop control, obviously, it's quite difficult to guarantee the workload position and orientation accuracy. Compared with the platform drivers using optical encoder feedback, this paper introduced a driving method which is based on non-contact magnetic encoder feedback. Brushless dc motors were used as the executors, and the average motor currents were detected in real time, while, at the same time, deviation angles of motor shafts in the platform were detected by the magnetic encoders. With the development of an actual 3-axis servo platform prototype, recommended control system schema has had proved quite a good dynamic performance and stability in platform driving.
Keywords: platform driver, brushless dc motor, magnetic encoder.

1 Introduction

With the development of technology, the UAV aerial photography [1] and vehicle stability servo platform video processing applications are widely used. High precision servo control system is the basis to realize the platform active vibration cancellation capability, it's also the key factor to reduce the complexity of platform structure and improve the precision of platform action response. The high-frequency vibration of the platform is mainly from a variety of vibration through the base conduction, which is characterized by a high frequency, low amplitude. While, the low-frequency vibration is mainly from the attitude jitters of the platform generated during exercise in accordance with instructions.

WIT Transactions on Engineering Sciences, Vol. 107, © 2015 WIT Press
www.witpress.com, ISSN 1743-3533 (on-line)
doi:10.2495/SEEEI140711

Most traditional 3-axis servo platform driving control system is an open-loop structure, posture vibration compensation of the platform can only be achieved to a certain extent, and its ability to eliminate vibration is extremely limited. In literature [1, 2], vane-servomotor was employed as the execution unit, but such motor has some disadvantages such as low response, small output torque, and limited movement stroke. While, stepper motor is used as execution unit in literature [3–5]. It is relatively in large size and weight, and easy to lose step and overshoot, these will also reduce the accuracy of the control system. With respect to vane-servomotor and stepper motor, brushless dc motor has a faster response, larger pull-in torque [6], higher power density, which offers control system much better sensitivity and bigger response bandwidth that represents stronger active vibration absorption potentials.

Currently, the majority closed-loop control system of the servo platform driver [7] uses an optical encoder motor feedback, which complicates the installation structure, intolerable vibration and shock, the shortcomings of the relatively low output accuracy, that increases the difficulty of platform driver and control system design. Non-contact magnetic encoder has a simpler structure, impact resistance, insensitive to vibration, higher precision output data, and is much more suitable for the servo platform driving system development.

2 The principle of servo platform balance control

Servo platform contains three brushless dc motors, each level of the load are mounted directly on a certain motor output shaft. Three orthogonal axes of the output shaft synthesize the compensation movement of the three axes (X, Y, and Z)

Figure 1: 3-Axis servo platform.

in the reference coordinate system, shown as Figure 1. In order to maintain a stable posture of the load, the force of the platform must be in balance. The algebraic sum of the motor drive torque and the gravity torque, vibration, and other disturbing torques should be zero.

For any motor shaft, ignore the quality of the connecting rod and the existence of external disturbance torque, following formula could be derived according to torque balance equation:

$$\begin{cases} T_x + mgL\sin\theta_x + ma_x\cos\theta_x = 0 \\ T_y + mgL\sin\theta_y + ma_y\cos\theta_y = 0 \\ T_z + ma_z\cos\theta_z = 0 \end{cases} \tag{1}$$

wherein θ as a center of the load relative to the motor shaft angle, m for the quality of motor drive shaft load, L for the connecting rod length, T as the motor output shaft torque, a is the acceleration of the motor axis.

Due to the existence of external vibration interference, reciprocating load jitters occur, which need to be compensated for by the angle of the motor output. The stability of the workload orientation is achieved by the adjustment of θ which compensates the tiny vibration displacements, and leads to an almost-constant-rated position and orientation.

The differential of formula (1) is:

It's straightforward that the orientation stability of the 3-axis servo Platform is not only subjected to the motor shaft angle θ, but also to the small deviation of the motor shaft angle $\Delta\theta$, just as Figure.2.

$$\begin{cases} \Delta T_x = -mgL\cos\theta_x\cdot\Delta\theta_x + ma_x\sin\theta_x\cdot\Delta\theta_x \\ \Delta T_y = -mgL\cos\theta_y\cdot\Delta\theta_y + ma_y\sin\theta_y\cdot\Delta\theta_y \\ \Delta T_z = ma_z\sin\theta_z\cdot\Delta\theta_z \end{cases} \tag{2}$$

In order to maintain a non-shaking stable platform orientation, it is necessary to adjust the motor torque output by a small torque ΔT in real time regarding both the motor shaft angle θ and its small deviation $\Delta\theta$. $\Delta\theta$ are measured by a

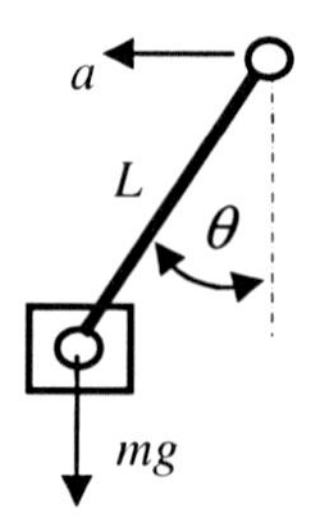

Figure 2: Moment equilibrium diagram.

set of high precision non-contact magnetic encoders, a and θ are derived from the real-time feedback and the quaternions of gyroscope posture and settlement, ΔT is represented by brushless dc motor drive current adjustment.

By the differential equation of quaternions, the relationship between the angular velocity gyroscope output value and the quaternions is as follows [8, 9]:

$$\begin{pmatrix} q_0 \\ q_1 \\ q_2 \\ q_3 \end{pmatrix}_{t+\Delta t} = \begin{pmatrix} q_0 \\ q_1 \\ q_2 \\ q_3 \end{pmatrix}_t + \frac{\Delta t}{2} \begin{pmatrix} -a_x \cdot q_1 - a_y \cdot q_2 - a_z \cdot q_3 \\ +a_x \cdot q_0 - a_y \cdot q_3 + a_z \cdot q_2 \\ +a_x \cdot q_3 + a_y \cdot q_0 - a_z \cdot q_1 \\ -a_x \cdot q_2 + a_y \cdot q_1 + a_z \cdot q_0 \end{pmatrix} \tag{3}$$

Considering the transformation relationship between quaternions and Euler angle, we have:

$$\beta = \mathrm{ArcTan}\left[\frac{2(q_2 \cdot q_3 + q_0 \cdot q_1)}{q_0^2 - q_1^2 - q_2^2 + q_3^2}\right] \tag{4}$$

$$\varphi = -\mathrm{ArcSin}\left[q_0^2 - q_1^2 - q_2^2 + q_3^2\right] \tag{5}$$

$$\phi = \mathrm{ArcTan}\left[\frac{2(q_1 \cdot q_2 + q_0 \cdot q_3)}{q_0^2 + q_1^2 - q_2^2 - q_3^2}\right] \tag{6}$$

Wherein, q_0, q_1, q_2, and q_3 are four parameters of quaternions. a_x, a_y, and a_z, are the three gyroscope output values respectively in the direction of *X*, *Y*, and *Z*. β, φ, and ϕ are three Euler angles in direction of the pitch, roll, and heading respectively.

By the speed-torque properties of brushless dc motor, the relationship between electromagnetic torque and the driving current should be subjected to [10]:

$$T_e = \frac{e_A \cdot i_A + e_B \cdot i_B + e_c \cdot i_c}{\Omega} \tag{7}$$

where T_e represents the electromagnetic torque, e_A, e_B, and e_C are the three reverse electromotive force of phase A, B, and C, respectively. Ω is mechanical angular velocity for the motor.

3 Hardware design of the system

To test and verify the feasibility of the above theoretical analysis, a servo platform prototype has had been developed. We choose magnetic encoder AS5145B as the motor position feedback and its schema is shown in Figure 3.

Figure 3: AS5145B and magnet structure.

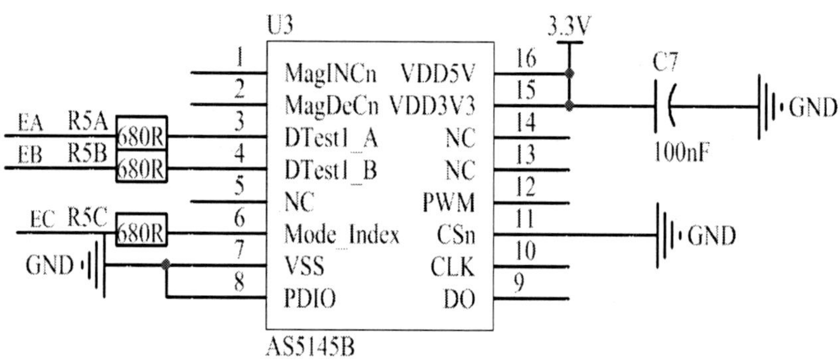

Figure 4: AS5145B circuit diagram.

Figure 4 is the magnetic encoder circuit. AS5145B can sense the induction magnetic field changes on the surface of the chip [11–13], then it converts the changes into 12 bits binary code. The output can be in the form of incremental PWM. Through the use of the fourfold frequency technology, AS5145B can measure the precision of the deviation angle as follows:

$$\theta = 360° / 4096 = 0.087° \quad (8)$$

At the same time, to simplify the circuit design, we chose the BLDC drive chip L6234D to drive the motor. The L6234D is a triple half bridge to drive a brushless dc motor. It utilizes the SENSE terminals to connect a resistor to the ground for the motor average current sampling. Its circuit diagram is shown in Figure 5.

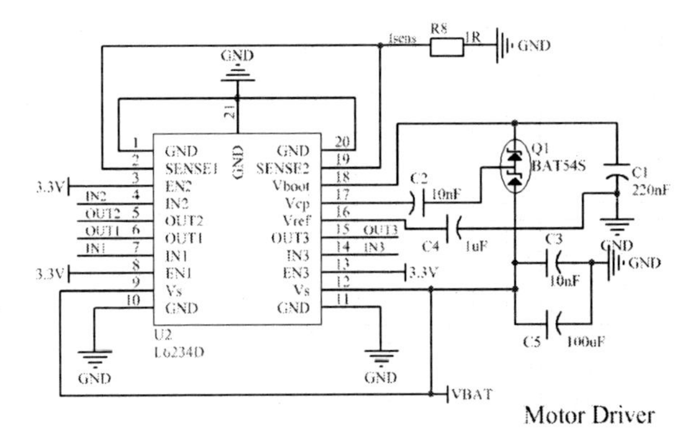

Figure 5: BLDC drive circuit diagram.

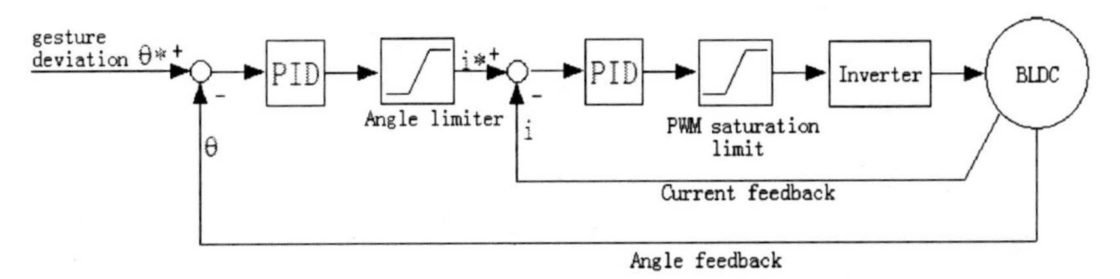

Figure 6: The controller structure.

4 The design of the controller

Compared with traditional open-loop control system, the improved closed-loop drive control system block diagram is shown in Figure 6.
The PID controller with incremental PID formula as follows [7]:

$$\Delta u(t) = K_P\left(e(t) - e(t-1)\right) + K_I e(t) + K_D\left(e(t) - 2e(t-1) + e(t-2)\right) \quad (9)$$

In order to improve the system stability and reduce the system maintenance cost, multiple MCU control strategy of each motor is controlled by independent MCU.

5 Experimental result

Figure 1 is the recommended prototype 3-axis servo platform. Two photos were taken 50cm away in front of a cup of flower. Figure 7(a) is the photo taken by a camera in the hand. Because of the inevitability shaking of the hand, the picture is obviously fuzzy and vague. Figure 7(b) was taken with the same camera that installed in recommended 3-axis servo platform that was hold in the same hand. Deliberately shaking was applied upon the 3-axis servo platform, however, it's straightforward that the photo is surely to be much more clear and the picture is much better.

Figure 7: The comparison of a photo before and after using the platform. (a) photo by a camera before using the platform. (b) photo by a camera after using the platform.

6 Conclusions

A driving system of a 3-axis servo platform has been developed with the help of new magnetic encoder and brushless dc motor. Experiments has proved that recommended control schema and driving hardware has the ability to eliminated platform vibration effectively, meanwhile, miniature, possibly low cost and high precision could be achieved at the same time by the schema put forward in this paper.

References

[1] Yu Bao-yi, Pei Hai-long, Design and implementation of pan-tilt control system in UAV. *Automation and Instrumentation*, **27(9)**, pp. 48–51, 2012.

[2] Li Xiang-qing, Sun Xiu-xia, Peng Jian-liang, Motion compensation based gimbal controller design for small UAV. *Systems Engineering and Electronics*, **33(2)**, pp. 376–379, 2011.

[3] Liu Jun-cheng, Design of the vision table based on DSP and tyro. *Control and Automation*, **24(5)**, pp. 143–144, 2008.

[4] Wu Yi-fei, Li Sheng, Cai Hua, Design and implementation of pan-tilt control system based on MSP430 MCU. *Control and Automation*, **22(20)**, pp. 90–93, 2006.

[5] Shi Lei, Han Bao-ling, Luo Qing-sheng, Guo Zhen, Design and realization of the rotational station control system for intelligent ball-robot. *Machinery Design and Manufacture*, **3**, pp. 148–150, 2011.

[6] Wang Fang, Li Hong, Development of control technology of brushless direct current motor. *Journal of Beijing Institute of Machinery*, **22(1)**, pp. 38–42, 2007.

[7] Cao Ying, Cheng Lei, Fang Kang-ling, Design and implementation of robot's console with servo and drive integration. *Computing Technology and Automation*, **27(4)**, pp. 31–34, 2008.

[8] Wang Ya-feng, Liu Hua-ping, Sun Fu-chun, Zhang You-an, All-attitude navigation and control in SINS based on error quaternions. *Journal of Chinese Inertial Technology*, **15(4)**, pp. 390–393, 2007.
[9] Shi Wen-ming, Xu Bin, Chen Li-min, Realization of the quaternion-based Kalman Filter for strapdown AHRS. *Techniques of Automation and Applications*, **24(11)**, pp. 6–8, 2005.
[10] Xia Chang-liang, Brushless dc motor control system. *Science Press*, 2009.
[11] Zhang Li-hua, Wu Hong-xing, Zheng Ji-gui, Xu Ren-heng, Technology of position servo magnetic encoder based on industrial robot. *Micromotors*, **46(10)**, pp. 56–60, 2013.
[12] Yu Fei, Zhao Ji-min, Luo Xiang, Analysis and research of the algorithm of magnetic rotary encoder. *Small and Special Electrical Machines*, **39(10)**, pp. 20–22, 2011.
[13] Wu Zhong, Lv Xu-ming, Design of angular velocity and position observer for servo motors with magnetic encoders. *Proceedings of the Chinese society for Electrical Engineering*, **31(9)**, pp. 82–87, 2011.

The double closed loop control simulation of cascade STATCOM for harmonic suppression

Ting Zhang, Jun Liu
School of Electrical and Electronic Engineering,
Hubei University of Technology, Wuhan, China

Abstract

In view of the harmonic problems in STATCOM, this paper discusses the control strategy of voltage and current harmonic suppression of double closed loop. The outer voltage loop consists of: the second notch filter and PI control algorithm, in order to curb the second ripple of voltage. The inner current loop is composed of embedding repetitive control algorithm in PI controller with feed forward compensation, in order to improve the ability of anti-interference of harmonic and accelerate the ability of the response of the current loop. In this paper, it is contrasted that double closed loop used PI controller and current loop embedded in repetitive controller, voltage loop embedded in the second order notch filter for harmonic suppression ability by setting up the STATCOM device's simulation model. Finally, it is to prove the correctness and effectiveness with the algorithm of the control strategy.
Keywords: STATCOM, double closed loop control, second order notch filter, repetitive controller.

1 Introduction

It is essential to improve the voltage utilization efficiency, to obtain pollution-free electric energy and to stabilize the voltage. However the voltage stability is affected by many factors, such as power generation and transmission and distribution sector, the nonlinear element in power network system. They will cause the fluctuation of grid voltage and power loss outage. Voltage fluctuation will generate harmonics which will increase consumption of reactive power in the transmission and distribution. Therefore, it is very significance that researched on

www.witpress.com, ISSN 1743-3533 (on-line)
doi:10.2495/SEEEI140721

reactive compensation can curb harmonic suppression for improving dynamic stability of power system in grid [1].

With the development of modern power electronic technology, voltage stability one way to stabilize the voltage is dynamic reactive power compensation in power grid. Static synchronous compensator (STATCOM) is an advanced method [2] in the reactive power compensation device with the desired reactor, small capacitor and low cost. If they cannot be suppressed in the voltage loop, and the output of voltage loop current equals loop followed input. The periodic two secondary voltage ripple and current harmonics will be generated in STATCOM.

In order to try to suppress periodic two voltage harmonics in the voltage loop, the two second notch filter is applied to suppress the two harmonic voltage. For the current loop, when the input is periodic sine wave, PI control cannot realize tracking without error. Repetitive control [3] can be introduced to further reduce the error. Repetitive control based on the internal model principle of waveform use the repetitive disturbance to correct the cycle of the output waveform. It can achieve relative ideal steady index without sampling multiple variables, controlling the speed and complicating the algorithm. In this paper, it can suppress the harmonics by embedding repetitive control in current loop and the second order notch filter in voltage loop.

It is particularly discussed the design of PI's double closed loop controller [4] and the method of the second order notch filter [5] and repetitive controller in this paper. The model is used to stimulate the analysis of the ability of to suppress the harmonics with the embedded second order notch filter and repetitive.

2 Design of double closed loop control system in the STATCOM

Single-phase single link STATCOM device model is shown in Figure 1. It assumes that converter is noless and storage none worse for the wear in the ideal case. Then according to the instantaneous power DC energy conservation is $u_{ab}i_s = u_{dc}i_{dc}$ [6, 7].

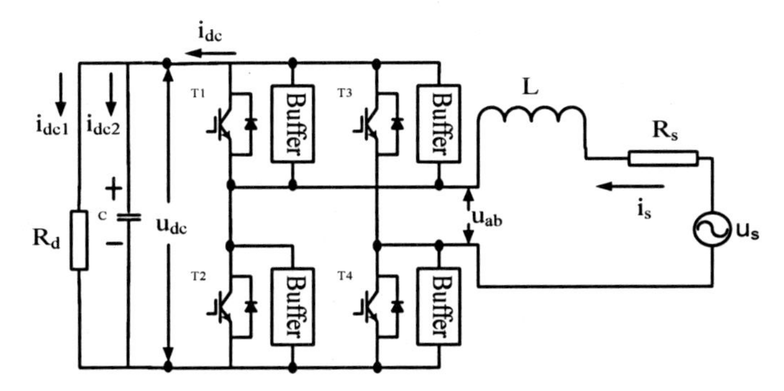

Figure 1: Single-phase single link STATCOM device model.

Assuming that i_s and u_s of network side in the converter is that pure sine wave. Direct voltage is constant u_{dc}, ignoring high-frequency component in u_{ab}. R_s is considered as the equivalent resistance of filter inductance L. C represents DC capacitor, R_d as load resistor. w as the network side frequency. And instantaneous available converter DC side current value is

$$i_{dc} = \frac{U_{ab} i_s}{U_{dc}} = \frac{U_{ab} \sin\left(wt - \delta_{ab}\right) I_s \sin wt}{U_{dc}} = I_{dc} \cos \delta_{ab} - I_{dc} \cos\left(2wt - \delta_{ab}\right) \quad (1)$$

It is shown that the converter DC side current contains not only the DC component, but also two harmonic component as given in Eq. (1). The two harmonic component, multiplied by two times resonance branch, link bus capacitor and DC side current load producing a pulsating voltage by the main component of the two harmonic in DC side. Double closed loop control strategy is adopted in this paper and the control system block diagram is shown in Figure 2:

It consists of a voltage regulator $G_v(s)$, the current loop regulator $G_i(s)$, the instantaneous reactive power component values sin(wt), the PWM link model $G_{PWM}(s)$, object to the current $G_{ii}(s)$ and voltage controlled loop object $G_{vv}(s)$. Current loop controls AC current to the given value constantly, and the voltage loop controls DC capacitor voltage constantly [6]. If the inductor current is regarded as a controlled current loop object, the object controlled transfer function $G_{ii}(s)$ can be written as $G_{ii}\left(s\right) = \dfrac{U_S\left(s\right) - U_{ab}\left(s\right)}{I_S\left(s\right)} = \dfrac{1}{Ls + R}$. If the DC capacitor C can be regarded as the controlled object of voltage loop, I_{dc1} represented the load current disturbance, the DC output arm side current is $I_{dc}(s)$, and the transfer function of the object $G_{vv}(s)$ is $G_{vv}\left(s\right) = \dfrac{U_{dc}\left(s\right)}{I_{dc}\left(s\right) - I_{dc1}\left(s\right)} = \dfrac{1}{sC}$. It is assumed that the voltage and current PI regulator is represented, respectively, in Figure 2.

$$\begin{cases} G_v\left(s\right) = K_{vp} + \dfrac{1}{T_v s} \\ G_i\left(s\right) = K_{ip} + \dfrac{1}{T_i s} \end{cases} \quad (2)$$

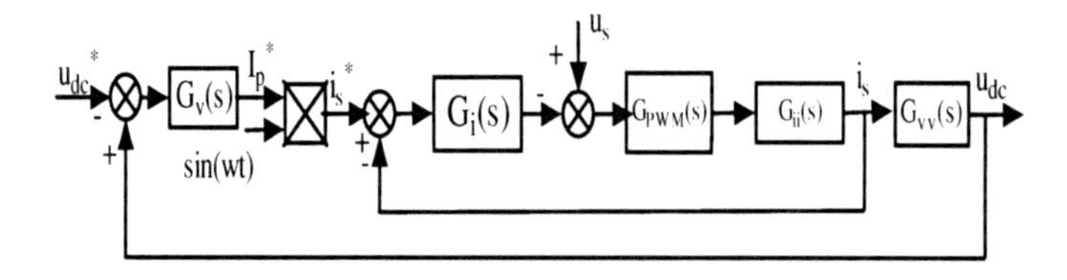

Figure 2: The double closed loop control system block diagram.

Proportional integral coefficient and time constant of the voltage and current loop PI regulator, respectively, are K_{vp}, K_{ip}, T_v, T_i. PWM link model can be expressed as $G_{pwm}(s) = \frac{K_{pwm}}{sT_s/2+1}$ where T_s represents the switch cycle and K_{PWM} is voltage gain of the converter. The current closed loop transfer function is given by:

$$G_{ic}(s) = \frac{G_i(s)G_o(s)}{1+G_i(s)G_o(s)} = \frac{K_{ip}K_{pwm}/R}{\frac{S^2 T_i T_s}{2} + ST_s + K_{ip}K_{pwm}/R} \tag{3}$$

The voltage closed loop transfer function is described by:

$$G_{vc}(s) = \frac{G_v G_{ic} G_{vv}}{1+G_v G_{ic} G_{vv}} \tag{4}$$

3 The design of the second order notch filter

There are a lot of methods to eliminate the voltage second times ripple. For one thing, the common method is the addition of a low-pass filter in the voltage loop, but the requirement of low pass filter [8] with a cut-off frequency is very low, resulting from phase lag, so the stability of the control system is reduced. For another, the appropriate digital filters the two times harmonic. Sufficient attenuation can be provided by the second order notch filter [8, 9] in the 2 harmonic, meanwhile the low frequency phase lag is small, suitable for using on in this occasion. Control block diagram of the two order digital filter added, is shown in Figure 3.

The simulation of the second order notch filter transfer function $H(s)$ is mentioned below:

$$H(s) = \frac{s^2 + \Omega_o^2}{s^2 + Bs + \Omega_o^2} \tag{5}$$

$H(s)$ can be transformed into second order digital notch filter [8, 9] by using Tustin transform method. When the analog notch consists of frequency 100H,

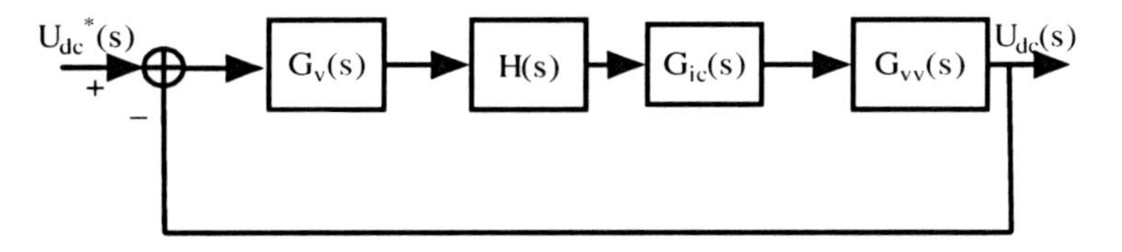

Figure 3: Control block diagram of two order the filter.

3dB bandwidth 5Hz and the sampling frequency 3kHz, the transfer function of the second order digital notch filter is as follows:

$$H(z)=\frac{0.9974z^2-1.984z+0.9974}{z^2-1.984z+0.9948} \tag{6}$$

4 Design of current loop embedded in repetitive controller

Repetitive control [10–16] is a waveform control technique based on the principle of endometrial. The current loop is embedded in the block diagram design of the repetitive controller which is shown in Figure 4.

Figure 5 is shown as specific structure diagram of repetitive controller $G_{re}(z)$, in addition the transfer function expression of $G_{re}(z)$ as shown:

$$G_{re}(z)=\frac{K_r Q(z) z^{-N} F(z)}{1-Q(z)z^{-N}} \tag{7}$$

$N = f_s/f$, switching frequency of PWM inverter is f_s in formula, and f is the fundamental frequency of power grid. k_r is regarded as the controller gain. Compensation filter is represented by $F(z)$. $Q(z)$ is low pass filter. Generally, to improve system stability $Q(z) = (z + 2 + z^{-1})/4$. The design of the repetitive controller reference literature [11–15].

Choosing control parameters $k_r = 1$, $F(z) = 1.25z^2-1.25z + 3.75$ and using a zero order holder to transform, closed current loop transfer function can be gained:

$$G_{ic1}(s)=\frac{(1+G_{re}(s)G_{ic}(s))}{1+(1+G_{re}(s))G_{ic}(s)} \tag{8}$$

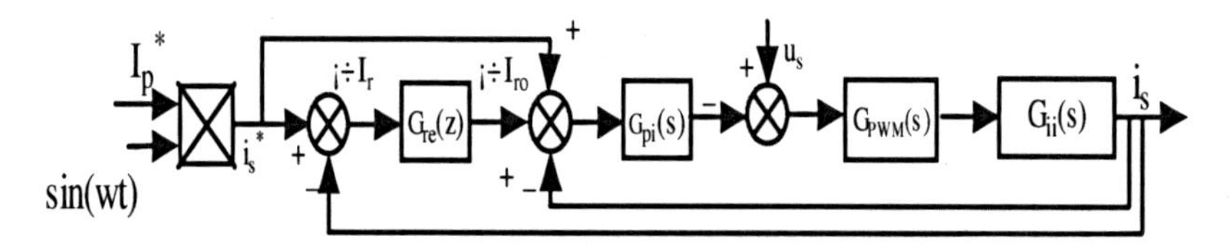

Figure 4: Current loop embedded in block diagram of the repetitive controller.

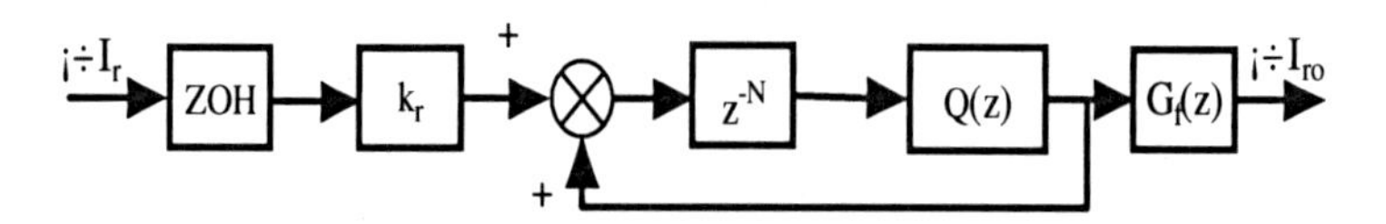

Figure 5: Control diagram of repetitive controller.

5 Analysis of STATCOM simulation

In order to verify the correctness of the above theory analysis, the simulation model [13–16] is built. The main parameters are as follow:

When the current loop adopts feed forward compensation PI control voltage loop used the PI controller, the high harmonic component and serious waveform distortion can be shown in Figure 6. The total harmonic current THD is 5.91%.

After introduced into the second order digital notch filter, the waveform of STATCOM output voltage and the analyses about the spectrum of the network side current is as shown in Figure 7. It is indicated that the grid side current is

Parameters	Numerical	Parameters	Numerical
Grid voltage (v)	100	The DC capacitor (μf)	850
DC voltage reference (v)	400	The load resistance (Ω)	20
AC inductance (mH)	4.1	The switching frequency (kHz)	3

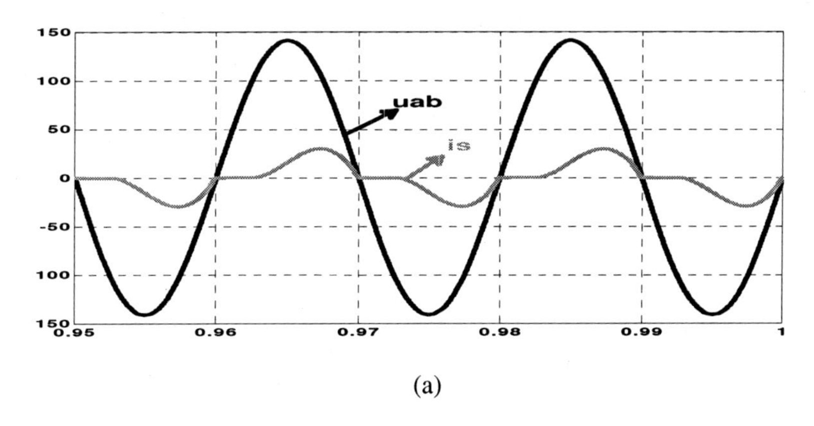

(a)

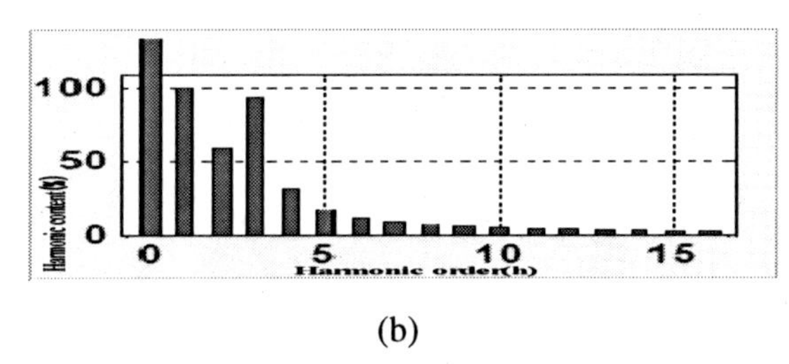

(b)

Figure 6: The waveform of STATCOM output voltage and grid side current and spectral analysis of the network side current. (a) the waveform of STATCOM output voltage and grid side current, (b) spectral analysis of the network side current.

diminutively affected by the harmonic voltage, the waveforms changing smoothly, two order ripple component reducing by 80%.

The total harmonic distortion content of the current is 4.1%. It is 1.81% lower than that of PI control. But the distortion still exists.

After introduced into repetitive control, the waveform of STATCOM output voltage and the analyses about the spectrum of the network side current is shown in Figure 8. It is indicated that harmonic suppression ability is greatly improved. Total harmonic distortion (THD) is reduced to 1.8%. The quality of the network side current waveform is significantly improved.

When the second order digital notch filter associates with repetitive control, the network side current waveform followed is as shown Figure 9. It can be seen that the tracking control error obtained favorable and the error is based on achieving zero.

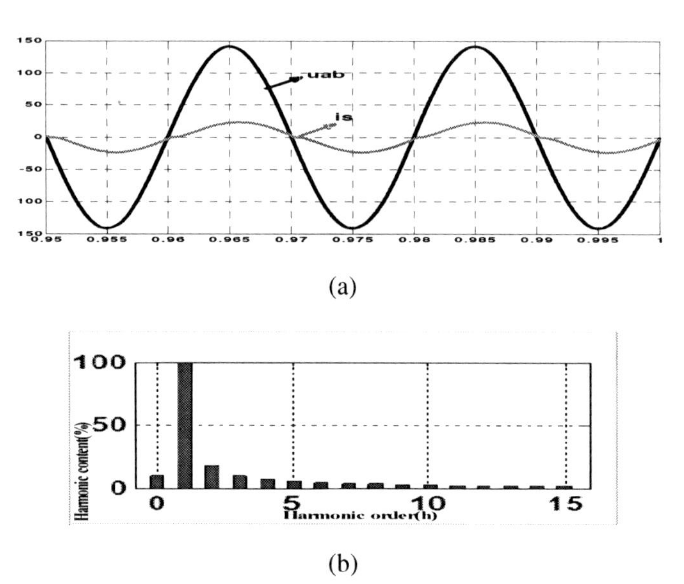

Figure 7: The waveform of STATCOM output voltage and grid side current and spectral analysis of the network side current with the two order digital analysis. (a) The waveform of STATCOM output voltage and grid side current, (b) spectral analysis of the network side current.

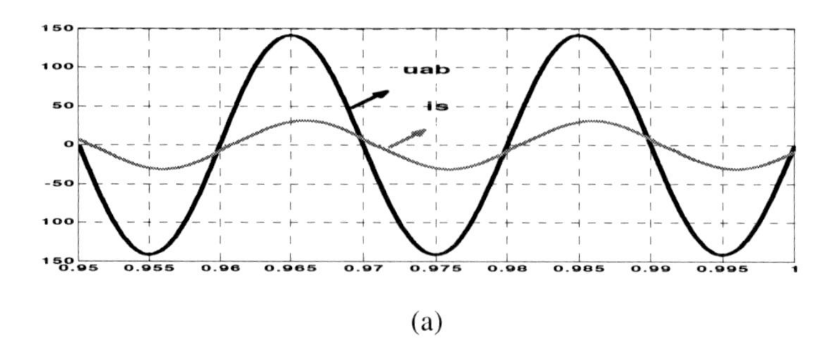

(a)

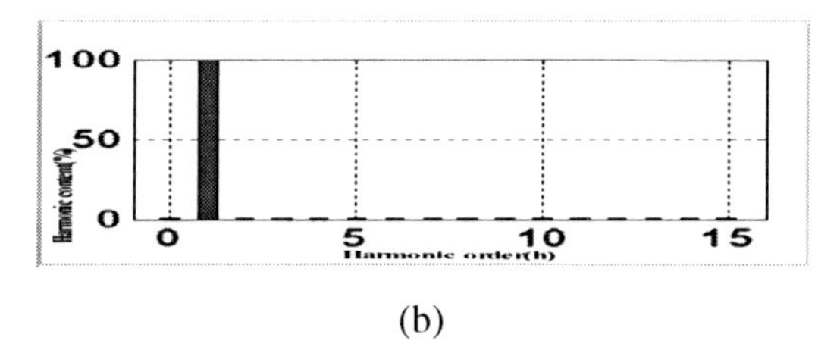

(b)

Figure 8: The waveform of STATCOM output voltage and grid side current and spectral analysis of the network side current with repetitive control. (a) The waveform of STATCOM output voltage and grid side current, (b) spectral analysis of the network side current.

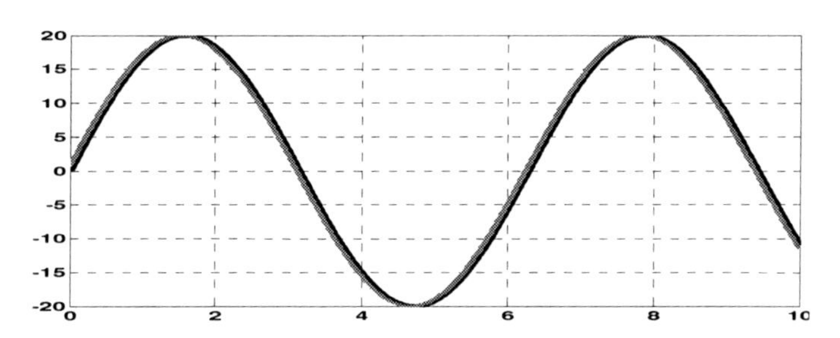

Figure 9: The current tracking waveform of STATCOM network side.

6 Conclusions

This paper has discussed the design of PI double closed loop controller, and has discussed dentally the design method of two order notch filter and repetitive controller. It is contrasted that double closed loop using PI controller and current loop embedded in repetitive controller, voltage loop embedded in the second order notch filter for harmonic suppression ability through building the STATCOM device simulation model.

Compared with the general feed forward compensation of PI control algorithm, the simulation results show that the method that the second order notch filters associates with repetitive controller is greatly reducing the harmonic pollution of power network, and to improving the rapidity and stability of the system.

References

[1] Xiao Xiangning, *Analysis and Control of Power Quality*. China Power Press, Beijing, 2004.

[2] Liu Wen-hua, Song Qiang, ±50 Mvar STATCOM based on chain circuit converter employing IGCT's. *Proceedings of the CSEE*, **28(15)**, pp. 55–60, 2008.

[3] Liu Zhao, *Research on Key Techniques in Cascade Multilevel STATCOM for Wind Power Generation*. Huazhong University of Science and Technology, Wuhan, 2012 (in Chinese).

[4] Hebo, *Control Theory and Control Engineering*. Huazhong University of Science and Technology, Wuhan, 2012 (in Chinese).
[5] Guo Weinong, Chen Jian, Study on digital dual-loop control for inverters based on state-observer. *Proceedings of the CSEE*, **22(9)**, pp. 64–68, 2002 (in Chinese).
[6] ChengJian, *Power Electronics and Power Electronic Converter and Control Technology*. Higher Education Press, Beijing, 2002.
[7] Peng Li, *Research on Control Technique for PWM Inverters Based on State-Space Theory*. Huazhong University of Science and Technology, Wuhan, 2004 (in Chinese).
[8] Wei Wenhui, Song Qiang, Teng Letian, et al., Research on anti-fault dynamic control of static synchronous compensator using cascade multilevel inverters. *Proceedings of the CSEE*, **25(4)**, pp. 19–24, 2005 (in Chinese).
[9] Wu XH, Panda SK, Xu JX, DC link voltage and supply-side current harmonics minimization of three phase PWM boost rectifiers using frequency domain based repetitive current controllers. *IEEE Transactions on Power Electronics*, **23(4)**, pp. 1987–1997, 2008.
[10] Gao Ji-lei, Research on harmonic current elimination method of single-phase PWM rectifiers. *Proceedings of the CSEE*, **21(30)**, 32–39, 2010 (in Chinese).
[11] HuangHui, The *Research of Single-Phase PWM Rectifier and Control Strategy*. Huazhong University of Science and Technology, Wuhan, 2012 (in Chinese).
[12] Zhang B, Wang DW, Zhou KL, et al., Linear phase lead compensation repetitive control of a CVCF PWM inverter. *IEEE Transactions on Industrial Electronics*, **55(4)**, pp. 1595–1602, 2008.
[13] Gong Li, Kang Yong, Research on key techniques of cascaded H-bridge multilevel STATCOM. *Journal of Electrician Technique*, **26(10)**, pp. 217–223, 2011.
[14] Zhang Kai, Peng Li, Xiong Jian, et al., State-feedback-with-integral control plus repetitive control for PWM inverters. *Proceedings of the CSEE*, **26(10)**, pp. 56–62, 2006 (in Chinese).
[15] Zhang B, Wang DW, Zhou KL, et al., Linear phase lead compensation repetitive control of a CVCF PWM inverter. *IEEE Transactions on Industrial Electronics*, **55(4)**, pp. 1595–1602, 2008.
[16] Longman RW, Iterative learning control and repetitive control for engineering practice. *International Journal of Control*, **73(10)**, pp. 930–954, 2000.

Based on the PID and repetitive control of four quadrant research of PWM rectifier

Jian Pan, Yuyang Li, Xiaolei Zhang
School of Electrical and Electronic Engineering,
Hubei University of Technology, Wuhan, China

Abstract

We propose and demonstrate a novel control strategy based on the Proportion Integration Differentiation (PID) controller and the repetitive controller, due to it can eliminate the harmonic in the input current on high-power condition which exists in tradition PID controllers. The feature of the proposed control strategy can not only inherits the good characteristic of dynamic response of PID controller, but also can suppresses the unwanted harmonic wave. The improved digital phase-stop-loop (DPLL) can decrease the delayed error caused by the formation of orthogonal virtually coordinate system using two integral units in a traditional method. The function of DPLL can realize the unit power factor and improve the current waveform on the Alternating Current (AC) side. By using the digital band-stop-filter, we can eliminate the second harmonic on the Direct Current (DC) bus. We simulate the proposed method with MATLAB, and it shows a good feature of feasible and effective.

Keywords: four-quadrant rectifier, repetitive controller, digital phase-lock-loop, band-stop-filter.

1 Introduction

The topology structure of Pulsed Wave Modulation (PWM) four quadrant rectifier has been widely used in Uninterruptible Power System (UPS), traction locomotive and other occasions, due to its various advantages such as the bidirectional power flow, network side input sinusoidal current waveform, fast dynamic response performance, and high power factor. However, there are two key problems of four quadrant rectifier to achieve high power factor and energy feedback, which also reflect in the comprehensive control of the grid side current and stabilize the bus voltage. One problem is how to obtain good sinusoidal

WIT Transactions on Engineering Sciences, Vol. 107, © 2015 WIT Press
www.witpress.com, ISSN 1743-3533 (on-line)
doi:10.2495/SEEEI1140731

current and high power factor and the other is how to realize more stable DC voltage in the bus at the network side. The traditional PID control cannot eliminate the grid side current harmonic when PWM four quadrant rectifier harmonic pollution occurs, particularly in the existence of a certain degree of distortion of the grid side current. Therefore, a novel control strategy of grid side current harmonics suppression has attracted considerable interest. In order to improve the control effect, various schemes have been proposed to improve the control algorithm of PWM four quadrant rectifier. In Ref. [1], a control method based on current prediction has been proposed to obtain high quality input currents in the lower switching losses. And in Ref. [2] a genetic algorithm is introduced to improve the control precision. Using variable step size adaptive control algorithm proposed in Ref. [3], it significantly reduces the delay system of digital control, and improves the system stability margin. Ref. [4] demonstrates a repetitive control method, which can eliminate the error tracking network side current and improve the system's stability. In our work, a new novel control strategy based on the PID controller and the repetitive controller has been proposed, improving the dynamic response and harmonic suppression. In addition, in order to obtain a stable voltage outer loop voltage reference value, we also design a notch filter which can effectively improve the DC side voltage quality. Finally, in order to make the four quadrant to achieve high power factor, we use an improved DPLL which made of an improved integral order instead of general construction of orthogonal signal with an improved phase link. The method can effectively reduce the control delay and improve the performance of the system.

2 Modeling method of four quadrant rectifier

Topological structure of PWM four quadrant rectifier is shown in Figure 1.

$$\text{Define: } S_A = \begin{cases} 1 & S_1\text{open}, S_2\text{close} \\ 0 & S_1\text{close}, S_2\text{open} \end{cases}, \qquad S_B = \begin{cases} 1 & S_3\text{open}, S_4\text{close} \\ 0 & S_3\text{close}, S_4\text{open} \end{cases} \tag{1}$$

Then, we can deduce the state equation of four quadrant rectifier from Figure 1.

The relationship and the change rule between the grid side current i_S and bus voltage u_{dc} can be clearly illustrated from Formula (2). By choosing the appropriate control strategy for the grid side current i_s and bus voltage u_{dc} control, in the desired results can be achieved.

$$\begin{cases} L\dfrac{di_s}{dt} = u_s - i_s R - (S_A - S_B)u_{dc} \\ C\dfrac{du_{dc}}{dt} = (S_A - S_B)i_s - \dfrac{u_{dc}}{R_L} \end{cases} \tag{2}$$

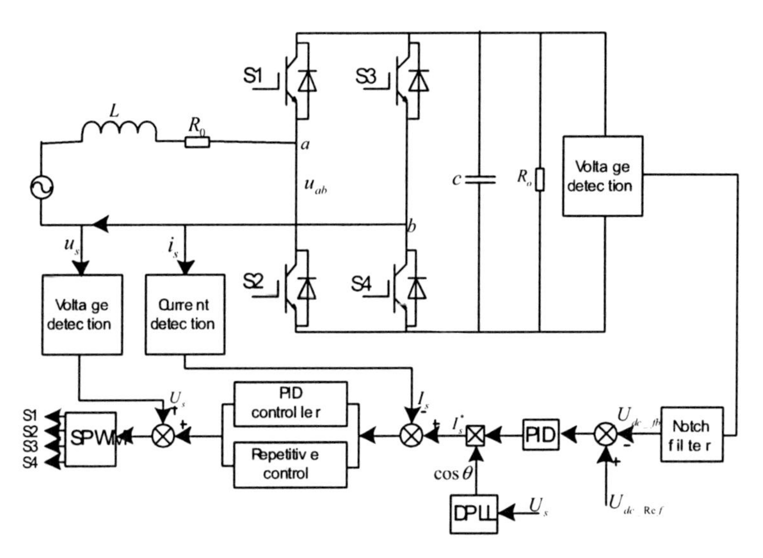

Figure 1: The four quadrant rectifier topology and control strategy frame.

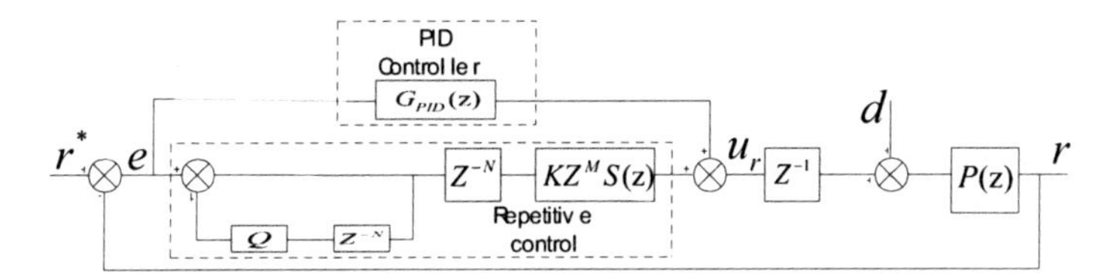

Figure 2: The four quadrant rectifier current inner loop control block diagram.

3 Compound control based on PID and repetitive control

PID control is an empirical control algorithm; it has been widely used in the simulation and control of sine wave rectifier system. The proposed algorithm has a fast dynamic response and robustness [5, 6]. However, when it applied to the sine wave inverter digital, the output characteristics is bad. In recent years, repetitive control is one of the most developed a new control algorithm; this algorithm has excellent steady-state output characteristics and robustness. Due to its own defects, there is always a period of output delay control, so the dynamic response is not ideal. Furthermore, we propose to put the two control algorithm combined each other, form a new kind of control scheme. A concrete control method is shown in Figure 2:

As shown in Figure 2, the dashed box represent PID control and repetitive control, respectively. Repetitive control by cycle delay link and compensator $KZ^M S(z)$, $P(z)$ as the control object transfer function, d as the disturbance signal. In this paper, using the sampling chip AD7606, and a high-speed asynchronous sampling mode, the switching frequency 10K sampling frequency is much greater than 2K, this is to eliminate the switching frequency interference to the control system, improve the division phase compensation and the regulating ability. Therefore, the Z^{-N} by Z^{-40} to Z^{-200}. In addition, the need for a positive feedback cycle delay values Q in the design value is usually a slightly less than 1 constants, or in order to facilitate the phase compensation, also can put the design into a low pass filter, low pass filter using $Q(z^2, z, z^{-2})$, such as calculation formula (3). Among them $a+b=2$. In the previous scheme, $a=b=1$, Q namely a zero phase low-pass filter.

$$Q(z^2, z, z^{-2}) = \frac{az^2 + bz^{-2} + 2}{4} \tag{3}$$

Compensator $KZ^M S(z)$ gain usually take a constant less than 1, this paper is 0.75, in order to control the compensation intensity, thereby affecting the harmonic suppression ability. $S(z)$ designed as a two order low pass filter, to ensure that the system in low frequency to obtain the unit gain, and realize the effective attenuation in high frequency. Therefore, this paper designed compensator (4) is shown as a formula.

$$KZ^M S(z) = \frac{0.0983295z^{22} + 0.196659z^{21} + 0.0983295z^{20}}{z^2 - 0.7478z + 0.272215} \tag{4}$$

4 The design of single-phase digital phase-locked loop

The single-phase PLL is mainly by the phase loop (PD), a loop filter (LF) and a voltage controlled oscillator (VCO) is composed of three parts, the majority of the phase-locked loop by the discriminator design difference [7]. The main function of the single-phase PLL phase link is the realization of the virtual coordinate system; produce a delayed input signal of 90 degrees orthogonal signal [8]. There are many ways to generate orthogonal signals, including the use of more is a delay processing of the input signal frequency. Methods currently used two first-order generalized integrator to delay is relatively common, as shown in Figure 3, but this method also has an obvious shortcomings, that is, in the digital realization,

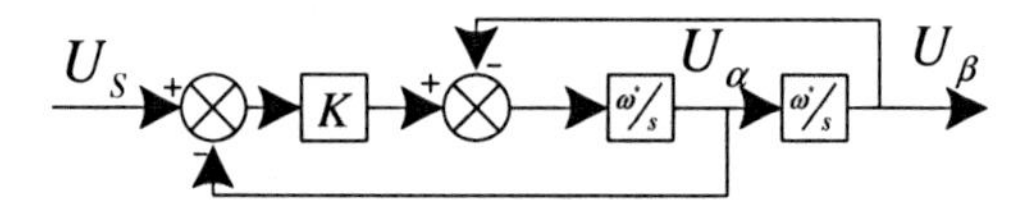

Figure 3: The traditional two order general integral phase detector based on phase link.

through two integral, will cause a delay of sample time, the strict control on the delay should not be ignored, and will consume additional hardware resources. Therefore, this paper proposes a new phase segment, with the input signal through a general integral get another orthogonal signal; effectively improve the problem, structure diagram as shown in Figure 4.

Figure 4 shows the phase transfer function for link: $G(S) = \frac{\omega * - s}{\omega * + s}$, its corresponding Bode is shown in Figure 5. From the Bode diagram can be seen 90 degrees lag in 50Hz. Figure 6 is a phase-locked control diagram based on Figure 4 phase link.

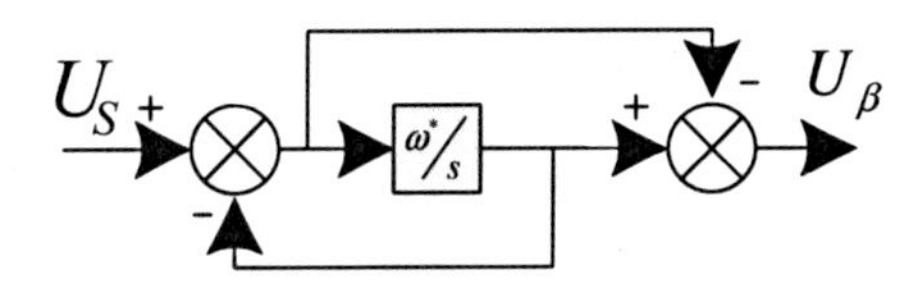

Figure 4: The improved phase link.

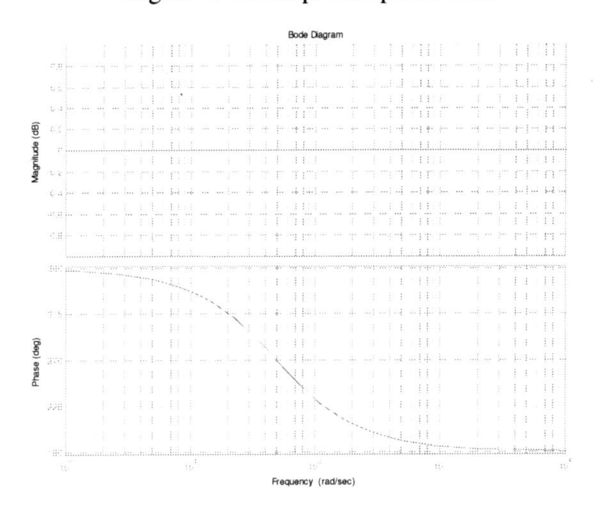

Figure 5: The improved phase link Bode diagram.

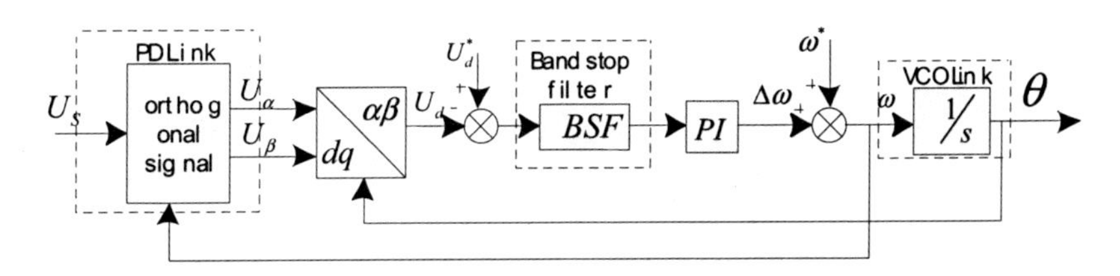

Figure 6: The improved phase link PLL control block diagram.

5 Notch filter design

By literature [3] is available, the output DC bus voltage containing the AC component. The AC component will bring third harmonic current loop for composite control. This third harmonic influence PWM four quadrant rectifier current control effect [8]. Therefore, in this paper, in order to eliminate the two octave fundamental frequency, the design of a notch filter $\omega_0 = 100\text{Hz}$, the bandwidth of 3dB (B=10Hz), the sampling frequency is 2KHz notch filter. Discrete digital notch filter transfer function as shown in formula (5):

$$G(s) = \frac{s^2 + \omega_0^2}{s^2 + Bs + \omega_0^2} \xrightarrow{s=\frac{1}{T_s}\frac{1-z^{-1}}{1+z^{-1}}} G(z) = \frac{0.995 - 1.98z^{-1} + 0.9975z^{-2}}{1 - 1.99z^{-1} + 0.99z^{-2}} \quad (5)$$

6 The simulation results and analysis

In the MATLAB simulation model of PWM rectifier. Make L=1.5mH, a capacitance C=37600uF and resistance R=0.04, the input voltage is 200V, the intermediate bus voltage 540V. Voltage outer-loop transfer function of PID is $G(S) = 1.1565 + \frac{35.5}{S}$. The inner-loop transfer function is $G(S) = 0.4 + \frac{300}{S}$. The simulation results are shown in Figures 7 and 8.

According to the simulation, when the grid side voltage is the fundamental, 3 and 5 harmonics, only THD control when the input current THD is 10.54%, PID+ THD3.73% for the repetitive control. According to the two groups before and after the experiment, found that the current PID control embedded repetitive control can significantly inhibit the harmonic.

Figure 9 is a fundamental input signal only containing 50Hz, Figure 10 the input signal into 3, 5 harmonic in including fundamental wave, square wave PLL output signal, sine wave input signal is phase locked. The simulation waveform

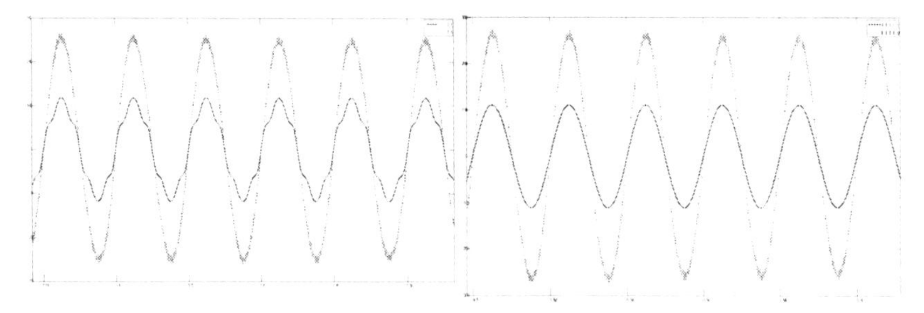

Figure 7: The grid side with harmonic adopts PID control.

Figure 8: The grid side with harmonic adopts PID and repetitive control.

Figure 9: The grid side voltage only contains fundamental.

Figure 10: The grid side voltage contains 3,5 harmonic.

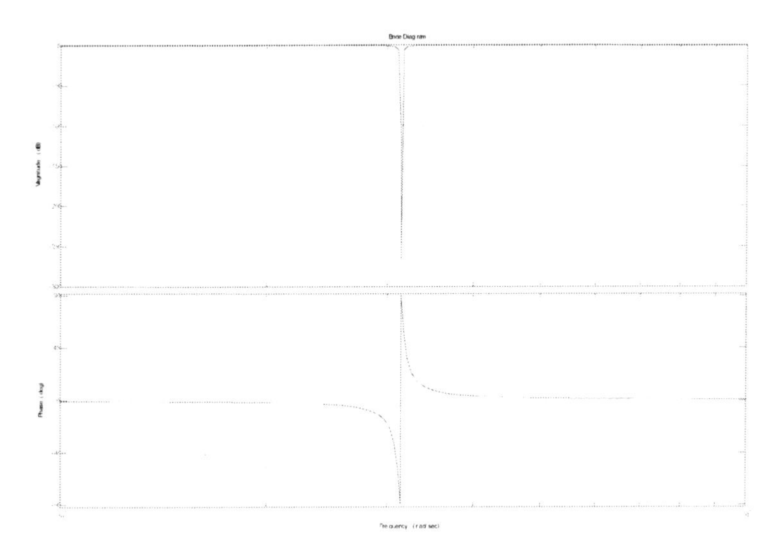

Figure 11: The input signal contains only fundamental.

can be seen synchronous phase output signal at the 2 cycle can quickly track the input signal, phase lock speed is fast, and the input signal and the phase output angle error for the 0 basic, high precision of PLL. When the input signal is subjected to harmonic pollution, the presence of multiple zero crossing phase output signal, can quickly track the input signal. Having strong anti-interference ability[9-12].

Figure 11 presents type (10) corresponding to the Bode diagram, we can see that, the harmonic filter on 100Hz has obvious inhibition effect. Figure 12 is the MATLAB simulation output DC bus voltage fluctuations in depression before and after the wave filter, wave front ripple reaches 1.34%, filtering after only 0.149%, to achieve the desired effect of filter.

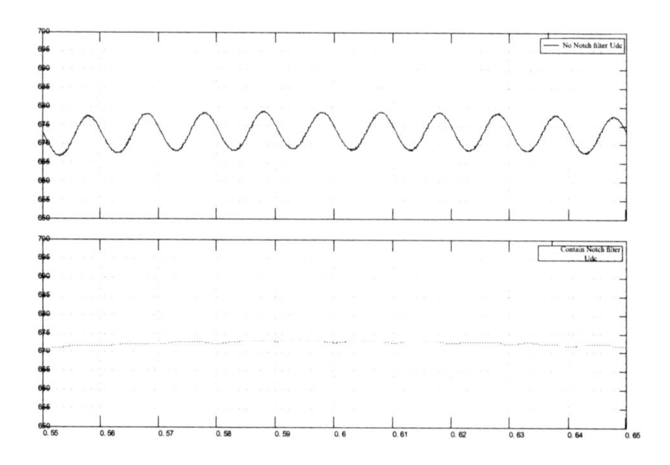

Figure 12: The input signal contains only 3,5 harmonic.

7 Conclusions

A novel control strategy based on compound control has been proposed and demonstrated. By using the algorithm of the inner current loop control and PID embedded repetitive control, instead of traditional controller, we can obtain that the algorithm can quickly and accurately track the given value to control effect, realizing no static error and effectively suppressing harmonics. Compared with the traditional phase link which consists of two first-order general integral phase link, the improved phase link we proposed can effectively reduce the control time delay, increase the phase speed, enhance the ability of anti-interference, and decrease the consumption of hardware resource. At the same time, the proposed notch filter can effectively eliminate the bus voltage two harmonic component, suppress the three harmonic input current, and improved input current waveform. The system shows a good dynamic performance through the simulation of the improved algorithm. And it can make a significant role in the engineering applications.

References

[1] Pericle Zanchetta, David B. Gerry, Vito Giuseppe Monopoli, et al., Predictive current control for multilevel active rectifiers with reduced switching frequency. *IEEE Transactions on Industrial Electronics*, 2008.

[2] F. Jafari, A. Dastfan, Optimization of single-phase PWM rectifier performance by using the genetic algorithm. *Proceedings of ICREPQ*, Granda, Spain, 2010.

[3] Wu Zhen-Xing, Zou Yun-Ping, Zhang Yun, Adaptive predictive controller of supply current applied in single-phase PWM rectifier. *Journal of Electrician Technique*, 2010.

[4] Gao Ji-Lei, Zhang Ya-Jing, Lin Fei, Research on harmonic current elimination method of single-phase PWM rectifiers. *Chinese CSEE*, 2010.

[5] Ho MT, Chung H, An integrated inverter with maximum power tracking for grid-connected PV systems. *IEEE Transactions on Power Electronics*, 2005.
[6] Sivakumar S, Parsons T, Modeling, analysis and control of bidirectional power flow in grid connected inverter systems. *Power Conversion Conference*, Osaka, Japan, 2002.
[7] Yang Liu, Liu Hui-Jin, Chen Yun-Ping, A new algorithm for random harmonic current detection in three-phase four-wine system. *Proceedings of the CSEE*, 2005.
[8] Zhang Zhong, Zhang Jin-Ping, Li Guo-Feng, AC drive traction system of high power HXD2 freight electric locomotive. *Electric Drive for Locomotives*, 2008.
[9] Zhao Qing-Lin, Guo Xiao-Qiang, Wu Wei-Yang, Research on control strategy for single-phase grid-connected inverter. *Proceedings of the CSEE*, 2007.
[10] Deng H, Oruganti R, Srinivasan D, Adaptive digital control for UPS inverter applications with compensation of time delay. *The Nineteenth Annual IEEE Applied Power Electronics Conference*, Piscataway, USA, 2004.
[11] Griñó R, Cardoner R, Costa-Castelló R, Digital repetitive control of a three-phase four-wire shunt active filter. *IEEE Transactions on Industrial Electronics*, 2007.
[12] García-Cerrada A, Pinzón-Ardila O, Feliu-Batlle V, Application of a repetitive controller for a three-phase active power filter. *IEEE Transactions on Power Electronics*, 2007.

Non-rigid structure from motion in trajectory space based on SDP

Yaming Wang, Lingling Tong, Zhang Zhang
School of Information Science and Technology, Zhejiang Sci-Tech University, Hangzhou, China

Abstract

This paper addresses the problem of non-rigid structure from motion (NRSFM) in trajectory space. We use a common trajectory basis, Discrete Cosine Transform (DCT) basis to increase the number of constraints, making the estimation of the unknowns more stable. Parameters should be optimized after the factorization of the measurement matrix. The selection of optimization method affects the accuracy of the structure reconstruction. This paper proposes a trace-minimization approach based on SDP (semi-definite programming). Combining the orthonormality constraint and trace minimization constraint, the Levenberg–Marquard (LM) algorithm is utilized for further optimization. The reconstruction results on the real image sequences of Yoga show that our approach runs reliably, and improves the accuracy of NRSFM effectively.
Keywords: NRSFM, trajectory basis, SDP.

1 Introduction

NRSFM is a process of recovering the time varying 3D coordinates of points on a non-rigid object from their 2D places in an image sequence. At present, NRSFM has two major approaches: trajectory basis method and shape basis method. Tomasi [1] first proposed factorization approaches for recovering rigid structure, and the methods were extended to manage non-rigidity in the seminal paper by Bregler et al. [2]. The core is the observed shapes can be represented as a linear combination of a compact set of basis shapes. But the shape basis of a mouth smiling movement, for example, cannot be recycled to compactly represent a person walking. So, instead of representation by a shape space, Akhter et al. [3] proposed to represent the time-varying structure of a non-rigid object as a linear

WIT Transactions on Engineering Sciences, Vol. 107, © 2015 WIT Press
www.witpress.com, ISSN 1743-3533 (on-line)
doi:10.2495/SEEEI140741

combination of a set of basis trajectories. The primary advantage of trajectory space is that DCT basis can be predefined to close to many real trajectories, which reducing the number of unknowns and improving the stability in estimation.

In order to further improve the accuracy of NRSFM, we try to use SDP method and LM algorithm to further optimize. This paper bases on the predefined DCT basis to recovering the motion and structure of the non-rigid object. For the 2D signal of $N \times N$ sample points, the DCT formula can be depicted as follows.

$$F(u,v) = c(u)c(v)\sum_{i=0}^{N-1}\sum_{j=0}^{N-1} f(i,j)\cos[\frac{(i+0.5)u\pi}{N}]\cos[\frac{(j+0.5)v\pi}{N}] \tag{1}$$

where $c(u)$ and $c(v)$ are the coefficients,

$$c(u) = \begin{cases} \sqrt{\frac{1}{N}}, u = 0 \\ \sqrt{\frac{2}{N}}, u \neq 0 \end{cases}, \quad c(v) = \begin{cases} \sqrt{\frac{1}{N}}, v = 0 \\ \sqrt{\frac{2}{N}}, v \neq 0 \end{cases} \tag{2}$$

2 Discussed problems

In fact, 3D reconstruction for non-rigid motion is equivalent to the decomposition of measure matrix W. Namely, decompose the matrix W into the rotation matrix R of camera and the structure matrix S of non-rigid object. Then estimate the corresponding unknown parameters by a series of constraints. Last use optimization method to recovering S of non-rigid object.

The measured 2D trajectories are included in measurement matrix W, containing the location of N image points across M frames.

$$W = \begin{bmatrix} X \\ Y \end{bmatrix} = \begin{bmatrix} x_{11} & . & . & . & x_{1N} \\ y_{11} & . & . & . & y_{1N} \\ . & . & . & . & . \\ . & . & . & . & . \\ . & . & . & . & . \\ x_{M1} & . & . & . & x_{MN} \\ y_{M1} & . & . & . & y_{MN} \end{bmatrix}_{2M \times N} \tag{3}$$

The measurement matrix W can be decomposed as $W = RS$, where $R = blkdiag(R_1, R_2, \cdots, R_M) \in R^{2M \times 3M}$, R_i $(i = 1, 2, \cdots, M)$ is an orthogonal projection matrix.

The structure matrix S can be decomposed into the trajectory basis matrix Θ and the coefficient matrix A [3], $S_{3M\times N}=\Theta_{3M\times 3K}\mathrm{A}_{3K\times N}$, and define $\Lambda = R\Theta$, we get:

$$W = RS = R\Theta\mathrm{A} = \Lambda\mathrm{A} \tag{4}$$

Factorize W with SVD method, we can get:

$$W = \hat{\Lambda}\hat{\mathrm{A}} \tag{5}$$

Generally, however, the matrix $\hat{\Lambda}$ and $\hat{\mathrm{A}}$ will not be equal to Λ and A, respectively, because SVD is not unique. Any non-singular orthogonal matrix [4] $Q \in R^{3K\times 3K}$ can be inserted between $\hat{\Lambda}$ and $\hat{\mathrm{A}}$, and get a new valid decomposition $W = \hat{\Lambda}\hat{\mathrm{A}} = \hat{\Lambda}QQ^{-1}\hat{\mathrm{A}} = \Lambda\mathrm{A}$. This matrix Q is called correction matrix.

According to Akhter et al. [3], we define the first, $K+1$ st and $2K+1$ st columns of the correction matrix Q as Q_k, then we can get the matrix R:

$$\hat{\Lambda}Q_k = \begin{bmatrix} \theta_{11}R_1 \\ . \\ . \\ . \\ \theta_{M1}R_M \end{bmatrix} \tag{6}$$

$$\hat{\Lambda}_{2i-1:2i}\,Q_k(\hat{\Lambda}_{2i-1:2i}\,Q_k)^T = \hat{\Lambda}_{2i-1:2i}\,Q_kQ_k^T\,\hat{\Lambda}^T{}_{2i-1:2i} = \theta_{i,1}^2 I_{2\times 2}, i=1,2,...,M \tag{7}$$

where $\hat{\Lambda}_{2i-1:2i} \in R^{2\times 3K}$ denotes the two rows of matrix $\hat{\Lambda}$ at positions and $2i-1$ and $2i$. Once Q_k is obtained, we can get the matrix R and S.

3 The solution based on SDP

3.1 A linear equation system with G_k

Set a gram matrix $G_k = Q_kQ_k^T \in R^{3K\times 3K}$, according to formula (7), we can get linear equations of G_k as follows:

$$\hat{\Lambda}_{2i-1}\,G_k\,\hat{\Lambda}_{2i-1}^T = \theta_{i,1}^2$$

$$\hat{\Lambda}_{2i} G_k \hat{\Lambda}_{2i}^T = \theta_{i,1}^2$$

$$\hat{\Lambda}_{2i-1} G_k \hat{\Lambda}_{2i}^T = 0$$

$$\hat{\Lambda}_{2i} G_k \hat{\Lambda}_{2i-1}^T = 0, i = 1, 2, \cdots M \tag{8}$$

Xiao et al. [5] found that due to the inherent ambiguity of orthogonal constraint, we can't obtain the only solution of the matrix Q. But Akhter et al. [6] showed that the inherent ambiguity will not necessarily lead to a fuzzy shape. Later, they proved that only using the constraint can recover the only structure. But they did not put forward any method to solve Q. According to Yuchao Dai [7], G_k is the solution space of matrix C, and C can be got directly from the input image data. Denote $vec(*)$ as the operator of vectorization, and $g_k = vec(G_k)$. Use $vec(EXF^T) = (F \otimes E)vec(X)$, we update the formula (8) as follows:

$$\begin{bmatrix} \hat{\Lambda}_{2i-1} \otimes \hat{\Lambda}_{2i-1} \\ \hat{\Lambda}_{2i} \otimes \hat{\Lambda}_{2i} \\ \hat{\Lambda}_{2i-1} \otimes \hat{\Lambda}_{2i} \\ \hat{\Lambda}_{2i} \otimes \hat{\Lambda}_{2i-1} \end{bmatrix} g_K = C_i g_K = \begin{bmatrix} \theta_{i,1}^2 \\ \theta_{i,1}^2 \\ 0 \\ 0 \end{bmatrix} = D_i \tag{9}$$

Stacking Eq. (9) from $i = 1$ to $i = M$, we get:

$$Cvec(G_k) = Cg_k = D \tag{10}$$

where matrix $C = [C_1^T, C_2^T, ..., C_M^T]^T$ and $D = [D_1^T, D_2^T, ..., D_M^T]^T$ are known. This is a linear system of equations about vector g_k, which has $9k^2$ unknowns. G_k is a symmetric matrix, G_k has $(3k)(3k+1)/2$ independent variables. In order to find out g_k, the required minimum frames of video sequence must be satisfied $F \geq (3k)(3k+1)/8$.

According to Dai Yu-chao [7], under non-degenerate and noise-free conditions, any correct solution of G_k must lie in the intersection of the solution space of matrix C and a rank-3 positive semi-definite matrix cone, G_k belongs to:

$$\{\mathrm{rank}(G_k) = 3\} \cap \{G_k \succeq 0\} \cap \{Cvec(G_k) = D\} \tag{11}$$

3.2 Solve matrix G_k with SDP

According to Dai Yu-chao [7], since rank-function itself is not very numerically stably and rank-minimization is an NP-hard problem in general. So relax the rank-function to a nuclear-norm minimization form, that is, $\min \| G_k \|_*$. Moreover, because $G_k = Q_k Q_k^T$ is a symmetric positive definite matrix, the nuclear norm is simply its trace. Thus we have $\| G_k \|_* = trace(G_k)$. Then we arrive at the following trace-minimization to solve G_k.

$$\begin{aligned} \min \quad & \text{trace}(G_k), \text{s.t.} \\ & C\text{vec}(G_k) = D \\ & G_k \succeq 0 \end{aligned} \tag{12}$$

The trace-minimization problem of G_k is a standard of SDP. Once G_k is calculated, then Q_k is extracted by using Cholesky Factorization. To further improve the accuracy, we use LM algorithm to optimize the initial value obtained by Cholesky Factorization and satisfy the following minimum:

$$\min_{Q_k} \sum_{i=1}^{F} [\left(2\frac{\hat{\Lambda}_{2i-1} Q_k Q_k^T \hat{\Lambda}_{2i}}{\hat{\Lambda}_{2i} Q_k Q_k^T \hat{\Lambda}_{2i}} \right)^2 + \left(1 - \frac{\hat{\Lambda}_{2i} Q_k Q_k^T \hat{\Lambda}_{2i}}{\hat{\Lambda}_{2i-1} Q_k Q_k^T \hat{\Lambda}_{2i-1}} \right)^2 + trace(Q_k Q_k^T)] \tag{13}$$

3.3 Compute the structure matrix S

According to formula (6), once Q_k is known, rotation matrix R can be estimated. And use the pseudo inverse method to calculate S. Since $\left(\Lambda^T \Lambda\right)^{-1} \Lambda^T \Lambda = E$, $W = \Lambda \mathrm{A}$, we get coefficient matrix A as follows:

$$\mathrm{A} = (\Lambda^T \Lambda)^{-1} \Lambda^T W \tag{14}$$

At last, we get the structure matrix $S = \Theta \mathrm{A} = \Theta(\Lambda^T \Lambda)^{-1} \Lambda^T W$.

4 Experimental verification

The experiments chose the real image sequence of Yoga, which come from http://cvlab.lums.edu.pk/nrsfm. This paper gives result comparison between Akhter's, which are shown in Figures 1 and 2.

Compare Figure 1 and Figure 2, when $K = 4$, 7, and 8, the proposed method is lightly better than Akhter's, but when $K = 12$, the reconstruction precision of our proposed method is increased significantly.

As we can summarize from Figure 2 and Table 1, the proposed method improves the accuracy of reconstruction. The method is proved to be effective.

Figure 1: Reconstruction using Akhter proposed method with different values of K. (a) $K = 4$, (b) $K = 7$, (c) $K = 8$, (d) $K = 12$.

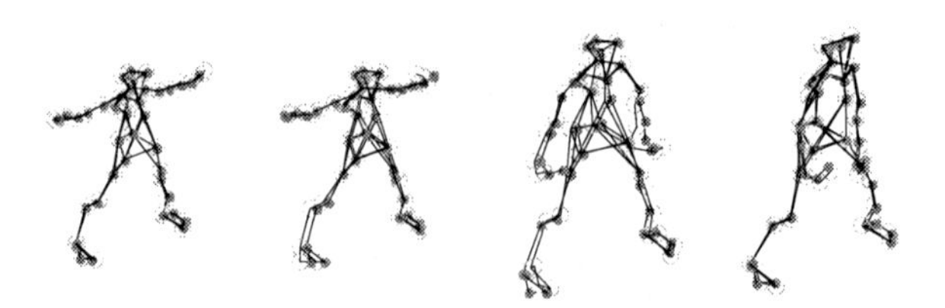

Figure 2: Reconstruction using our proposed method with different values of K. (a) $K=4$, (b) $K=7$, (c) $K=8$, (d) $K=12$.

Table 1: The comparison of the reconstruction error using two approaches.

Experiment Approaches	Reconstruction Error of Structure
Akhter's LM algorithm	0.1622
The proposed method	0.1352

5 Conclusions

This paper uses the SDP to solve the trace-minimization problem of the matrix, and combines the LM algorithm to further improve. The reconstruction of the Yoga sequence shows that the proposed method improves the precision. And the recovered structure is very close to the original structure. But for dramatic movement video sequence, our proposed approach can't improve the accuracy of the reconstruction. So, we need to further improve this method.

Acknowledgment

The work is supported by National Natural Science Foundation of China under Grant No. 61272311.

References

[1] C. Tomasi, T. Kanade, Shape and motion from image streams under orthography: a factorization method. *International Journal of Computer Vision*, **9(2)**, pp. 137–154, 1992.

[2] C. Bregler, A. Hertzmann, H. Biermann, Recovering non-rigid 3D shape from image streams. *IEEE Conference on Computer Vision and Pattern Recognition*, **2**, pp. 690–696, 2000.

[3] I. Akhter, Y. Sheikh, S. Khan, T. Kanade, Nonrigid structure from motion in trajectory space. *Advances in Neural Information Processing System*, pp. 41–48, 2008.

[4] L. Torresani, A. Hertzmann, C. Bregler, Nonrigid structure from-motion: estimating shape and motion with hierarchical priors. *IEEE Transactions on Pattern Analysis and Machine Intelligence*, **30(5)**, pp. 878–892, 2008.

[5] J. Xiao, J.X. Chai, T. Kanade, A closed-form solution to non-rigid shape and motion recovery. *International Journal of Computer Vision*, **67(2)**, pp. 573–587, 2004.
[6] I. Akhter, Y. Sheikh, S. Khan, In defense of orthonormality constraints for nonrigid structure from motion. *IEEE Conference on Computer Vision and Pattern Recognition*, pp. 1534–1541, 2009.
[7] Dai Yu-chao, Li Hong-dong, He Ming-yi, A simple prior-free method for non-rigid structure-from-motion factorization. *IEEE Conference on Computer Vision and Pattern Recognition*, pp. 2018–2025, 2012.

Determination method of optimal confidence of wind power for economic dispatch

Haixiang Zong, Ying Wang, Guoqiang Yang, Kaifeng Zhang
Key Laboratory of Measurement and Control of CSE,
School of Automation, Southeast University, Nanjing, China

Abstract

In the economic dispatch problem considering wind power uncertainty, different confidences of wind power are corresponding to different allocations of spinning reserve, which further influences the overall economically operation of the power system. This paper proposes a determination method of optimal confidence of wind power. Within the confidence intervals, the spinning reserve is used to satisfy the vibration of wind power as the economic cost. Outside the confidence intervals, the wind curtailment and load shedding are considered as the risk cost. An economic dispatch model is established, and the objective of the optimization is the overall cost in order to figure out the optimal confidence. An improved particle swarm algorithm is presented to solve the optimization problem. Simulation results indicate that the overall cost is the best under the optimal confidence.
Keywords: optimal confidence, wind power, risk cost, particle swarm algorithm.

1 Introduction

Recently, large-scale wind power integration has brought a great challenge to power dispatching. It's risky to make dispatch schedule according to the forecasting curve of wind power as the wind power is unpredictable precisely. Therefore, economic dispatch considering the uncertainty of wind power has led a significant research direction in power dispatching.

The confidence of wind power is used to describe the uncertainty of wind power, it is the reflection of the probability of wind power in some certain range, which is called as confidence interval. When the output distribution of forecasting

WIT Transactions on Engineering Sciences, Vol. 107, © 2015 WIT Press
www.witpress.com, ISSN 1743-3533 (on-line)
doi:10.2495/SEEEI140751

wind power is fixed, the higher the confidence is, the larger the confidence interval is. On the one hand, there must be larger amounts of spinning reserve to satisfy the vibration of wind power, which means the economic cost is higher. On the other hand, the risk cost considering wind curtailment and load shedding is lower outside the confidence interval. On the contrary, when the confidence is lower, the economic cost is lower with a higher risk cost. Therefore, the overall cost is related to the confidence of wind power, and the minimum cost must exist under some certain confidence. That's the reason why we have to find an optimal confidence.

Nowadays, some studies have been made about power dispatching considering the confidence of wind power. In Refs. [1, 2] different forecasting functions of wind power have been used to describe the prediction error of wind power, Chance Constrained Programming model has been established to solve the problem, the constraints are made based on the confidence. It can't ensure the reliable operation of the system under extreme circumstances. In Refs. [3, 4] the uncertain problem has been converted to a certain problem considering the confidence of wind power, however, the confidence is chosen according to the experience, which is no thesis basis. At the same time, many literatures have researched the relationship between economy and risk. In Refs. [5, 6] the objective functions of traditional economic dispatch only consider operating costs of thermal power, no attention has been paid to the risk cost for uncertainty of wind power. In Ref. [7] optimal confidence of wind power has been mentioned to balance the economic cost and risk cost. In Refs. [8, 9] the risk has been quantified to punishment cost in the objective function, which can achieve the balance of economy and risk.

The paper is organized as follows. In the next section, we propose the problem formulation. In Section 3, an economic dispatch problem model is established and the constraints are given. Section 4 is the case study of the optimization of the problem. Finally, we conclude our paper in Section 5.

2 Problem formulation

Once the confidence of wind power is determined, the outputs of wind farm can be divided to two parts considering that the forecasting error of wind power meet the normal distribution, just as Figure 1 shows.

In Figure 1, $p_{w\max}$ is the capacity of wind power, $p_{wt\max}$ and $p_{wt\min}$ are the upper and lower limits of the confidence interval of wind power respectively.

One is that within the confidence interval, according to the operation conditions of the units in the grid, there must be an optimal allocation of spinning reserve which refers to the optimal generation cost of the thermal plants to satisfy the vibration of wind power. The part can be defined as economic cost.

The other is the shaded area in the figure which refers to wind curtailment and load shedding, respectively. The difference between the upper limit of confidence interval and the capacity of wind power is defined as risk of wind curtailment. Meanwhile, the difference between zero and the lower limit of confidence interval is defined as risk of load shedding. Both of the risk make up the risk cost.

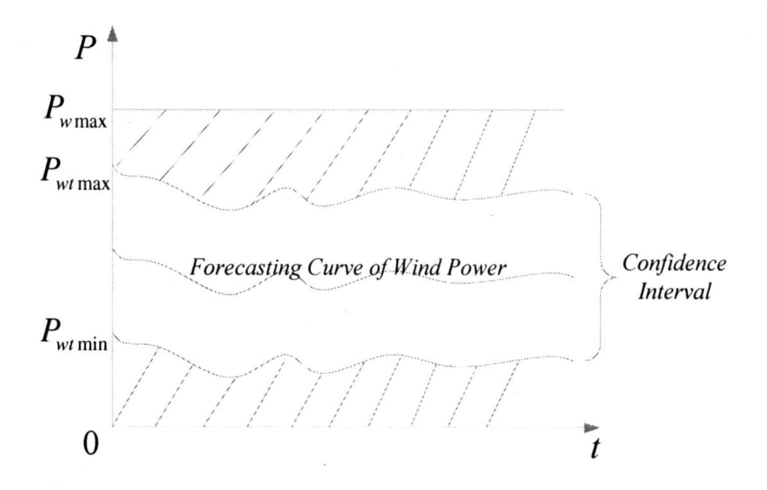

Figure 1: Overall cost considering the confidence of wind power.

Therefore, the problem can be converted to calculate the optimal overall cost at some certain confidence. It can be formulated as:

$$\begin{cases} F = \min(F_{\text{eco}} + F_{\text{risk}}) \\ \text{s.t.} \quad \mu \in [0,100] \end{cases} \tag{1}$$

where F_{eco} and F_{risk} mean the economic cost and the risk cost, respectively, μ represents the confidence of wind power.

3 Optimization model

The objective function can be defined as follows:

$$F = F_{\text{eco}} + F_{\text{risk}} = \min\left(\sum_{t=1}^{T}\sum_{i=1}^{I} a_i p_{i,t}^2 + b_i p_{i,t} + c_i\right) + \sum_{t=1}^{T} \beta \cdot p_{wlt} + \sum_{t=1}^{T} \gamma \cdot p_{lct} \tag{2}$$

where a_i, b_i, c_i are fuel coefficients of unit, T and I indicate time and amount of fossil power plant respectively. $p_{i,t}$ represents the generation schedule of i unit in t period. β is the loss coefficient of wind curtailment whose unit is $\$/\text{MW}\cdot\text{h}$ and p_{wlt} is amount of wind curtailment; γ is the loss coefficient of load shedding whose unit is $\$/\text{MW}\cdot\text{h}$ and p_{lct} is amount of load shedding.

The constraints are shown as follows:

1. Power balance constraints:

$$\sum_{i=1}^{I} p_{i,t} + p_{wt} = p_{lt} \quad t = 1, \quad T \tag{3}$$

where $p_{i,t}$ is the planned output of the unit i at time t. p_{wt} is the forecasting output of wind power at time t, p_{lt} is the forecasting load at time t.

2. Generation limits of thermal plants:

$$p_{i\min} \leq p_{i,t} \leq p_{i\max} \tag{4}$$

where $p_{i\min}$ and $p_{i\max}$ are the lower and upper limit of real power generation of unit i.

3. Unit ramping up constraints:

$$-\Delta_{i,d} T_{60} \leq p_{i,t} - p_{i,t-1} \leq \Delta_{i,u} T_{60} \tag{5}$$

where $\Delta_{i,d}$ and $\Delta_{i,u}$ are the ramp-down and ramp-up rate limits of unit i. The time of unit ramping up is one hour.

4. Wind curtailment and load shedding constraints:

$$p_{wlt} = p_{w\max} - p_{wt\max} \tag{6}$$

$$p_{lct} = p_{wt\min} \tag{7}$$

where $p_{w\max}$ is the capacity of wind power, $p_{wt\max}$ and $p_{wt\min}$ are the upper and lower limit of the confidence interval of wind power.

5. Spinning reserve constraints:

$$\sum_{i=1}^{n} \min\left(p_{i\max} - p_{i,t}, \Delta_{i,u} T_{15}\right) \geq p_{wt} - p_{wt\min} \tag{8}$$

$$\sum_{i=1}^{n} \min\left(p_{i,t} - p_{i\min}, \Delta_{i,\mathrm{d}} T_{15}\right) \geq p_{wt\max} - p_{wt} \tag{9}$$

4 Case study

In this paper, we assumed a system consisting of six thermal power units and only one wind farm. The data of thermal power units is in Table 1. The wind farm installed capacity was 200MW. The results of forecasting wind power and load are shown in Tables 2 and 3, respectively. An improved particle swarm algorithm was presented to solve the optimization problem [11]. The population size in PSO algorithm was 40, acceleration factors were $c1$=2, $c2$=2. The inertial factor was $w = w_{\max} - \left(w_{\max} - w_{\min}\right) k / \mathrm{K}_{\max}$, $w_{\max} = 0.9$, $w_{\min} = 0.4$, k represented the current iteration number and $K_{\max}$ represented the max iteration number 100. The loss coefficient of wind abandoning was $\beta = 80$ $/MW·h and the loss coefficient of load shedding was $\gamma = 120\ \$/\mathrm{MW \cdot h}$. We performed calculations using the algorithm presented above. Tables 1–3 are listed in the appendix. The simulation results are shown in Figures 2 and 3.

As shown in Figure 2, the economic cost decreases with lower confidence, since the spinning reserve required is much less when confidence goes down. The line becomes smooth at the point of 70%, then 70% is the optimal confidence. The rise is not obvious in the chart, we make a comparison with the condition that the loss coefficient of wind curtailment was $\beta = 120\ \$/\mathrm{MW \cdot h}$ and the loss coefficient of load shedding was $\gamma = 140\ \$/\mathrm{MW \cdot h}$. The simulation results are shown in Figure 4.

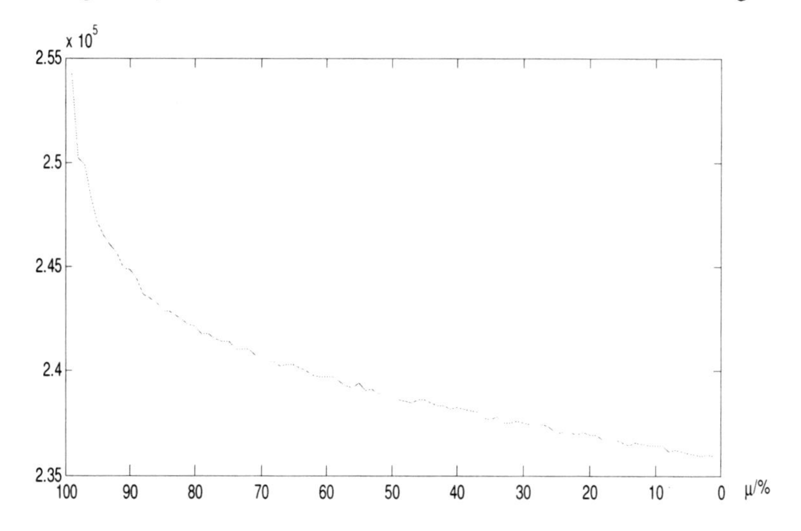

Figure 2: The economic cost under the confidence ranging from 100% to 0%. μ represents confidence.

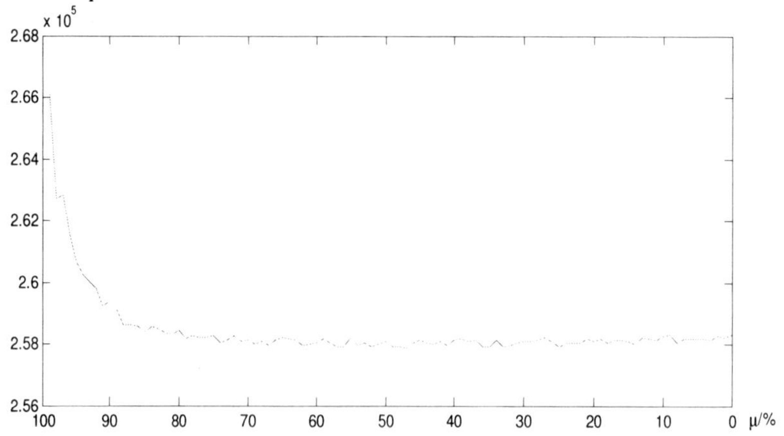

Figure 3: The overall cost under the confidence ranging from 100% to 0%. μ represents confidence.

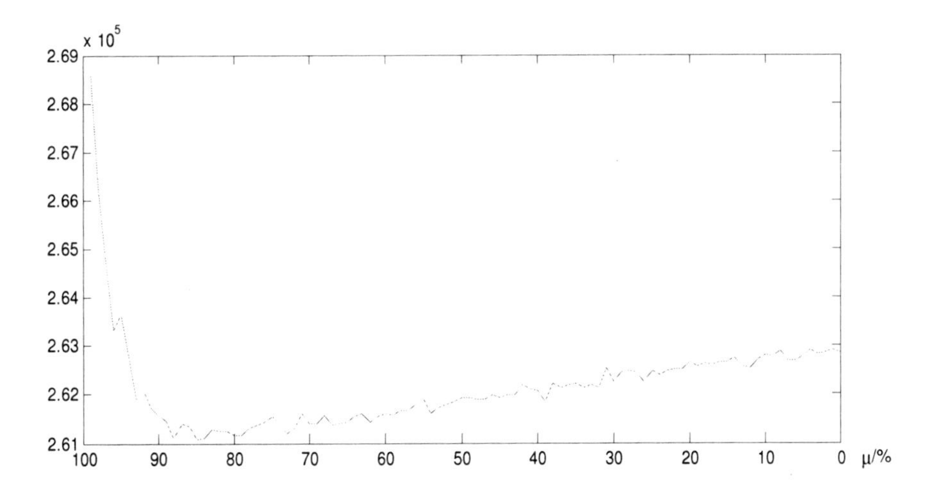

Figure 4: The overall cost under the confidence ranging from 100% to 0%.

The rise is clearer in this chart, and the optimal confidence becomes 85%, that's because the optimal confidence is related to the loss coefficient of wind curtailment and the loss coefficient of load shedding, both of which are determined by the dispatch department.

As shown in the figures, the overall costs under optimal confidence are better than the others, which indicates better allocation of the spinning reserve and appropriate risk costs. Therefore, the optimal confidence can make suggestions for the optimal output interval of wind power.

5 Conclusions

Based on the concept of confidence, this paper modifies the objective function of traditional economic dispatch model, in order to determine the system's optimal confidence interval. Results show that the proposed method is more economic and theoretical. The optimal confidence in this paper confirms the optimal confidence interval of wind power, which provides valuable information for setting the upper and lower limits for wind farms.

Acknowledgments

This work is supported by National Natural Science Foundation of China (Nos. 51177019 and 51477157) and State Grid Corporation of China (Research of the Key Technique for Large-scale New Energy Integration Dispatching Plan and Security Correction based on Probability Analysis).

References

[1] Shen Zhou, Yang Wei, Zhong Hai-bo, Teng Bai-an, Power system containing wind farm optimization scheduling with multi-objective based on chance-constrained and random simulation. *Journal of Electric Power*, **28(1)**, pp. 44–49, 2013.

[2] Sun Yuan-zhang, Wu Jun, Li Guo-jie, He Jian, Dynamic economic dispatch considering wind power penetration based on wind speed forecasting and stochastic programming. *Proceedings of the CSEE*, **29(4)**, pp. 41–47, 2009.

[3] Lin Haiming, Liu Tianqi, Li Xingyuan, Optimal active power flow considering uncertainties of wind power output and load. *Power System Technology*, **37(6)**, 1584–1589, 2013.

[4] Zhang Haifeng, Gao Feng, Wu Jiang, Liu Kun, A dynamic economic dispatching model for power grid containing wind power generation system. *Power System Technology*, **37(5)**, pp. 1298–1303, 2013.

[5] Liu Dewei, Guo Jianbo, Huang Yuehui, Wang Weisheng, Dynamic economic dispatch of wind integrated power system based on wind power probabilistic forecasting and operation risk constraints. *Proceedings of the CSEE*, **33(16)**, pp. 9–15, 2013.

[6] Zhou Wei, Sun Hui, Gu Hong, Ma Qian, Chen Xiaodong, Dynamic economic dispatch of wind integrated power systems based on risk reserve constraints. *Proceedings of the CSEE*, **32(1)**, pp. 47–55, 2012.

[7] Dong Xiaotian, Yan Zheng, Feng Donghan, et al., Power system economic dispatch considering penalty cost of wind farm output. *Power System Technology*, **36(8)**, pp. 76–80, 2012.

[8] Wang Zhaoxu, Zhou Ming, Wu Di, Xu Shenzhuan, Research on dynamic economic dispatch for grid-connected wind power system. *Modern Electric Power*, **28(2)**, pp. 37–42, 2011.

[9] Xia Shu, Zhou Ming, Li Geng-yin, Dynamic economic dispatch of power system containing large-scale wind farm. *Power System Protection and Control*, **39(13)**, pp. 71–77, 2011.

[10] F. Bouffard, F.D. Galiana, Stochastic security for operations planning with significant wind power generation. *IEEE Transactions on Power Systems*, **23(2)**, pp. 306–316, 2008.

[11] Liu Yong, Hou Zhi-jian, Jiang Chuan-wen, Economic dispatch by a modified particle swarm optimization algorithm. *Relay*, **34(20)**, pp. 24–27, 2006.

Appendix

Table 1: The Data of Thermal Power Units.

Units	a ($/MW^2)	b ($/MW)	c ($)	p_{max} (MW)	p_{min} (MW)	Δ_u (MW/h)	Δ_d (MW/h)
1	0.0070	7.0	240	120	50	50	90
2	0.0095	10.0	200	200	50	50	90
3	0.0090	8.5	220	300	80	65	100
4	0.0090	11.0	200	150	50	50	90
5	0.0080	10.5	220	200	50	50	90
6	0.0075	12.0	190	120	50	50	90

Table 2: The Results of Forecasting Wind Power.

Period	Wind (MW)	Period	Wind (MW)	Period	Wind (MW)
1	120.0	9	94.5	17	133.4
2	150.0	10	112.4	18	108.5
3	100.0	11	126.7	19	100.0
4	80.0	12	130.5	20	120.0
5	60.0	13	133.2	21	140.0
6	50.0	14	129.5	22	150.0
7	70.6	15	147.2	23	150.0
8	80.5	16	140.7	24	160.0

Table 3: The Results of Forecasting Load.

Period	Load (MW)	Period	Load (MW)	Period	Load (MW)
1	500.0	9	1094.6	17	842.0
2	550.0	10	1178.8	18	926.2
3	580.0	11	1220.9	19	1010.4
4	650.0	12	1263.0	20	950.0
5	730.0	13	1178.8	21	850.0
6	800.0	14	1094.6	22	730.0
7	968.3	15	1010.4	23	600.0
8	1010.4	16	884.1	24	550.0

Study on the relationship between pressure and displacement at the top part of men's socks

He Nan Dong, Rui Dan
College of Quartermaster Technology,
Jilin University, Changchun, China

Abstract

The study focused on the top part of men's socks, probing into the relationship between pressure on the lower leg exerted by the top part of socks and the displacement on the leg under stress. The displacement on the leg under stress was simulated using finite element method and the functional relationship between pressure and displacement at any cross-sectional point of the lower leg where the top part of socks located was obtained through the Lagrange quadratic interpolation, analyzing the multiple relation between pressure and displacement in detail. A novel conceive on appraisal of pressure comfort through the relationship between pressure and human body surface displacement was proposed by this paper, with which method subjectivity and uncertainty by subjective assessment on pressure comfort can be avoided, providing a theoretical reference for optimal design of the top part of man's socks.

Keywords: top part of socks, pressure, displacement, finite element, comfort.

1 Introduction

The research scope of clothing comfort mainly includes heat and moisture factor, touch factor and pressure factor [1]. Among them, pressure comfort is important index to determine whether a garment is comfortable for the human body or not. Traditional research method to evaluate clothing pressure is by means of man's subjective feelings after the participants wearing clothing [2–5]. Due to the differences between individuals, serious shortcomings in subjective evaluation makes subjective evaluation unable to estimate pressure comfort objectively [6]. A certain amount of inward displacement perpendicular to the surface of human body will emerge because of the clothing pressure on human body. In 1975,

www.witpress.com, ISSN 1743-3533 (on-line)
doi:10.2495/SEEEII140761

Verillo pointed out that displacement less than 0.001 mm could effectively cause generation of pressure under ideal conditions [7]. In this study, the tiny displacement of human body under clothing pressure was obtained through the ANSYS finite element software to discuss the functional relation between pressure and displacement.

Sock is one of the important accessories in people's daily life. Early studies on socks mainly concentrated on thermal-wet comfort [8] and the health care efficacy [9, 10], while there were few achievements of researches on pressure of socks [11, 12]. This study focused on the top part of men's socks, with cross-sectional structure of the leg where the top part of socks located, the participants and experimental socks used in this paper consistent with that mentioned in literature [13]. The displacement of human body surface was calculated when the top part of socks was exerting pressure to the lower legs through finite element software ANSYS10.0 and the relationship between pressure and displacement was investigated deeply.

2 Research on distribution trend of stress and displacement at the top part of socks

In a literature [13], 3-D body scanning was performed on 50 males meeting the requirements of the test with non-contact 3-D body scanner, together with cross section of the region 6 cm above the level of medial ankle bulge of the participants' right foot intercepted, then the standard curve model of the cross section of human lower leg where the top of man's socks located was obtained with Matlab. According to division of angle, this cross-section curve was divided into 72 points [13] as shown in Figure 1 (Units: mm).

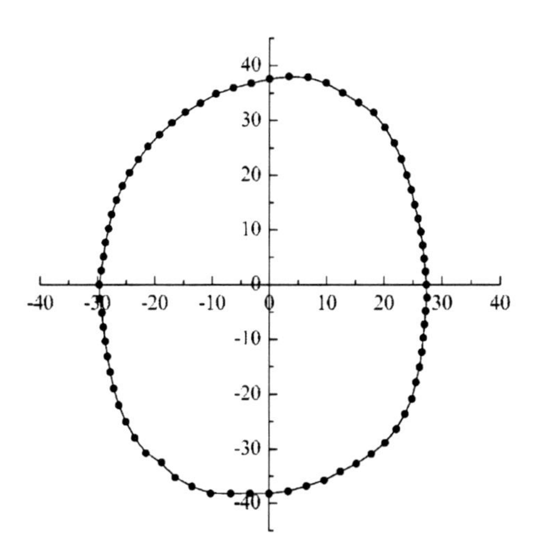

Figure 1: Grown form of the cross section of human lower leg.

Pressure test was performed on 72 cross-sectional points of lower leg when the participants were wearing six types of experimental socks with AMI3037S-5 clothing contact pressure tester. Standard figures of pressure and displacement of 72 cross-sectional points of lower leg were achieved with simulation of finite element software ANSYS as shown in Figures 2 and 3.

As Figure 2 shows, the maximal pressure values of the six types of socks appear near 60° and 265°. According to the standard curve of leg cross-section at the top part of socks, the radius of curvature near the two points of human leg are relatively small. According to the law of Laplace $P = T/R$, under the premise of uniform distribution of socks' welts elasticity modulus, pressure is inversely proportional to radius of curvature, that is to say, pressure is bigger where the radius of curvature is smaller [14]. The smallest pressure value appears near 0°, because radius of curvature of this point is biggest. When sock' welt is exerting pressure to the leg, the general situation of pressure distribution is as follows: [0°, 60°], [150°, 180°], and [210°, 270°] are the pressure-increasing regions, [60°, 150°], [180°, 210°], and [270°, 360°] are the pressure-decreasing regions.

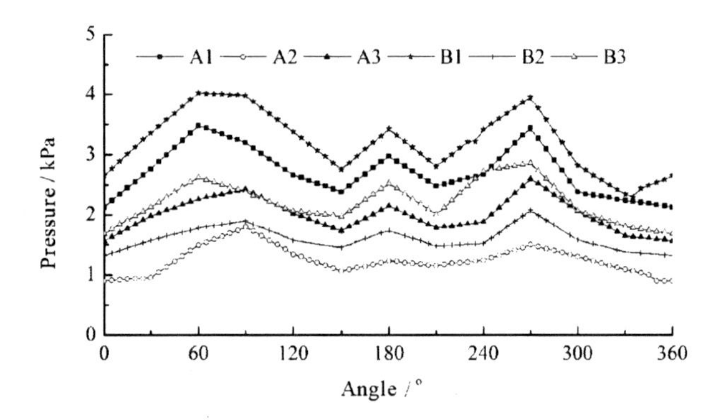

Figure 2: Pressure value of 72 points.

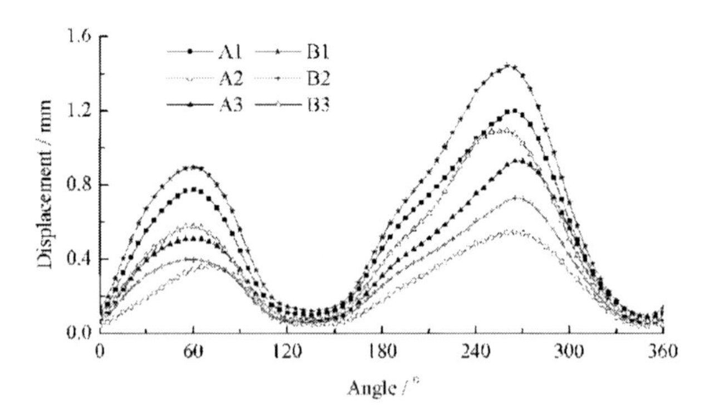

Figure 3: Displacement value of 72 points.

As Figure 2 shows, the pressure values of 60° and 265° are almost the same. However, Figure 3 shows that of 265° is obviously bigger than that of 60°, which is because under the circumstance that curvature radius of the two points are approximate, point 265° is further away from tibia and fibula. The general situation of displacement distribution is as follows: [0°, 60°] and [135°, 265°] are the displacement-increasing regions, [60°, 135°] and [265°, 360°] are the displacement-decreasing regions.

3 Calculation of functional relations between pressure and displacement

3.1 Calculation of functional relations between pressure and displacement of the known points

In this paper, fitted curves on pressure and displacement of 72 known points were made for six types of socks, respectively, using SPSS 13.0. Only seven turning points of pressure or displacement of 0°, 60°, 90°, 150°, 180°, 210°, and 270° are selected to analyze and illustrate the functional relation between pressure and displacement (Figure 4).

Figure 4 shows the fitted curve of pressure and displacement on 0° when six types of socks exert pressure to the lower leg. Degree of fitting of the curve is 0.999, indicating good fitting effect. The quadratic curve equation between pressure and displacement is:

$$Y = -0.004X^2 + 0.066X - 0.012 \quad (1)$$

In the same manner, the quadratic curve equations and fitting curves of the rest six points could also be made, whose degree of fitting of the curves are all more

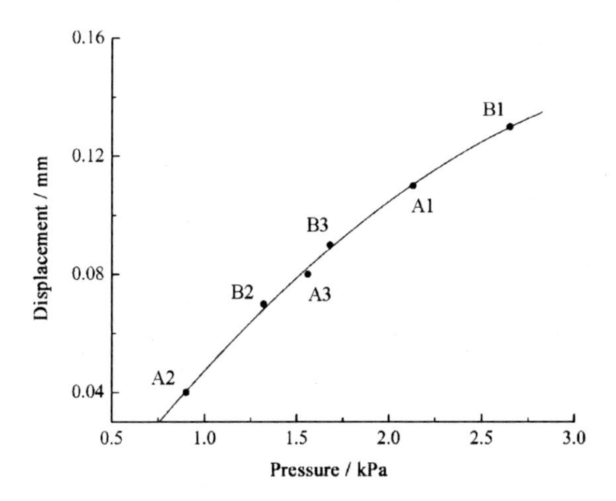

Figure 4: Fitted curve of pressure and displacement on 0°.

than 0.9, indicating good fitting effects. The quadratic curve equations are as follows:

$$60°,\ Y = 0.006X^2 + 0.185X + 0.053 \tag{2}$$

$$90°,\ Y = 0.011X^2 + 0.07X + 0.103 \tag{3}$$

$$150°,\ Y = 0.002X^2 + 0.046X + 0.005 \tag{4}$$

$$180°,\ Y = 0.004X^2 + 0.139X + 0.001 \tag{5}$$

$$210°,\ Y = -0.006X^2 + 0.348X - 0.066 \tag{6}$$

$$270°,\ Y = -0.003X^2 + 0.36X + 0.003 \tag{7}$$

3.2 Functional relations between pressure and displacement of random points by interpolation

The functional relations between pressure and displacement of the 72 known points of the cross-section of the lower leg could be obtained using SPSS. However, only knowing the functional relations between pressure and displacement of the tested points is far from being desired in the practical application. Here, the functional relations between pressure and displacement of random points of the cross-section of the lower leg can be predicted by Lagrange quadratic interpolation.

Lagrange quadratic polynomial is:

$$L_2(x) = \frac{(x-x_1)(x-x_2)}{(x_0-x_1)(x_0-x_2)}y_0 + \frac{(x-x_0)(x-x_2)}{(x_1-x_0)(x_1-x_2)}y_1 + \frac{(x-x_0)(x-x_1)}{(x_2-x_0)(x_2-x_1)}y_2 \tag{8}$$

Taking sock A1 as an example, the functional relation between pressure and displacement of 1° is calculated by Lagrange linear polynomial and Lagrange quadratic polynomial respectively, the detailed procedure is as follows:

To decrease truncation error in the process of interpolation calculation, the nodes selected should be as close to interpolation point x as possible. In formulas (8), $L(x)$ is interpolating function, x is interpolation point, x_0, x_1, x_2 are interpolation nodes nearest interpolation point x. For sock A1, the figures of the pressure and displacement of some points nearest point 1° are shown in Table 1.

Table 1: Part of pressure and displacement datum for sock A1.

x Angle (°)	0	5	10	15	20
Y_p Pressure (kPa)	2.13	2.24	2.35	2.45	2.55
Y_d Displacement (mm)	0.109	0.158	0.237	0.325	0.409

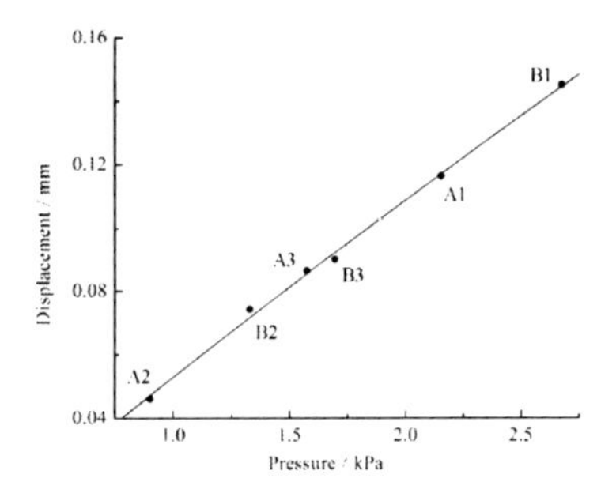

Figure 5: Fitted curve of pressure and displacement on 1°.

The three interpolation nodes nearest 1°selected are $x_0 = 0°$, $x_1 = 5°$, $x_2 = 10°$, respectively. The figures of the pressure and displacement in Table 1 are substituted into Lagrange interpolation formulas (8). The figures of the pressure and displacement of point 1° are obtained.

Lagrange quadratic polynomial: $Y_{P_2} = 2.152 \quad Y_{d_2} = 0.1164$.

To test the accuracy of the results, stress testing is performed at point 1° of sock A1 using pressure tester AMI3037S-5 mentioned above, the pressure got is 2.15kPa, which is then input into finite element method of ANSYS to simulate analysis, the figure of displacement got is 0.1094mm. Comparing the predicted figure with the tested figure, we find the accuracy of quadratic interpolation is almost the same as predicted figure by interpolating function.

In the same manner, the figures of pressure and displacement of point 1° of the rest five types of socks can be obtained using Lagrange quadratic polynomial.

A2: $Y_P = 0.902\text{kpa} \quad Y_d = 0.0461\text{mm}$ A3: $Y_P = 1.574\text{kpa} \quad Y_d = 0.0864\text{mm}$

B1: $Y_P = 2.674\text{kpa} \quad Y_d = 0.145\text{mm}$ B2: $Y_P = 1.328\text{kpa} \quad Y_d = 0.0743\text{mm}$

B3: $Y_P = 1.696\text{kpa} \quad Y_d = 0.0938\text{mm}$

The fitted curve of pressure and displacement on 1° is made using SPSS just as Figure 5 shows.

The quadratic curve equation is:

$$Y = -0.004X^2 + 0.068X - 0.012 \tag{9}$$

Degree of fitting of the curve is 0.999, indicating good fitting effect. With the same method, the quadratic curve equations between pressure and displacement of random cross sectional points of lower leg where the top part of socks is located can be obtained.

4 Multiple relationship of pressure and displacement

The experiments above illustrate that the pressure exerted and displacement produced by sock B1 are both the biggest among the six tested socks. Therefore, sock B1 is chosen as the basis to probe into the multiple relationship of pressure at the same cross-sectional point and displacement between the rest five socks and sock B1. In this research, 12 angles of 0°, 60°, 120°, 180°, 240°, 300° is selected to make analysis on the multiple relationship of pressure and displacement (Table 2).

As seen from Table 2, the multiple relationships of pressure and displacement match basically at the same cross-sectional point of the leg for different kinds of socks. Taking angle 0° as an example, the pressure exerted to the leg by sock B1 is 1.244 times of that of A1 while the displacement produced by sock B1 is 1.24 times of that of A1, which indicated the multiple relationship of pressure and displacement is almost the same. Table 2 showed that the rest angles represented the similar tendency. Because of the difficulties when measuring the displacement in the practical procedure, the changing tendency of displacement could be approximately inferred according to the multiple relationship of pressure and displacement, which may provide theoretical principles of judging pressure comfort to some extent.

Table 2: Multiple relation of pressure and displacement.

Angle		B1/A1	B1/A2	B1/A3	B1/B2	B1/B3
0°	Pressure (kPa)	1.244	2.944	1.699	2.008	1.577
	Displacement (mm)	1.24	3.109	1.67	1.94	1.536
60°	Pressure (kPa)	1.155	2.698	1.779	2.258	1.534
	Displacement (mm)	1.16	2.607	1.762	2.255	1.556
120°	Pressure (kPa)	1.271	2.522	1.673	2.153	1.641
	Displacement (mm)	1.275	2.465	1.668	2.147	1.654
180°	Pressure (kPa)	1.155	2.789	1.595	1.971	1.361
	Displacement (mm)	1.138	2.909	1.685	2.013	1.374
240°	Pressure (kPa)	1.281	2.758	1.819	2.25	1.257
	Displacement (mm)	1.248	2.706	1.738	2.17	1.266
300°	Pressure (kPa)	1.185	2.169	1.369	1.785	1.362
	Displacement (mm)	1.161	2.075	1.246	1.674	1.382

5 Conclusion

The research probed into the pressure exerted on the lower leg by the top part of men's socks, simulating the body surface displacement of the leg under pressure when dressed by using finite element method of ANSYS and obtaining the functional relationship between pressure and displacement at any point of cross section of the lower leg where top part of socks was located through interpolation, and finally the multiple relation between pressure and displacement was analyzed.

The research proposed a new concept of estimating relationship between pressure and human body surface displacement under pressure to judge pressure comfort. Subjectivity and uncertainty created once by subjective feelings about pressure comfort after dressing is avoided with this method and how to analysis pressure comfort qualitatively using human body surface displacement under pressure remains a future direction.

References

[1] Li, Y., Clothing comfortable and its application. *Textile Asia*, **29(7)**, pp. 29–33, 1998.

[2] Maccracken, R., Joy, K.I., Free-form deformation with lattice of arbitrary topology. *ACM Computer Graphics*, **30(3)**, pp. 181–188, 1996.

[3] Nakahashi, M., Morooka, H., Morooka H., et al., Sen'i Seihin Shohi Kagaku, Effect of clothing pressure on front and back of lower leg on compressive feeling. *Journal of the Japan Research Association for Textile End-Uses*, **40(10)**, pp. 661–668, 1999.

[4] Kawabata, H., Tanaka, Y., Sakai, T., et al., Sen'i Gakkaishi, Measurement of garment pressure: pressure estimation from local strain of fabric. *Journal of the Society of Fiber Science and Technology*, Japan, **44(3)**, pp. 142–148, 1988.

[5] Makabe, H., Momota, H., Mitsuno, T., Ueta, K., Effect of covered area at the waist on clothing pressure. *Seni-Gakkaishi*, **49(10)**, pp. 513–521, 1993.

[6] Tae Jin Kang, Chung Hee Park, Youngmin Jun, et al., Development of a tool to evaluate the comfort of a baseball cap from objective pressure measurement: (1) Holding power and pressure distribution. *Textile Research Journal*, **77(9)**, pp. 653–660, 2007.

[7] Verillo R.T., Cutaneous sensations. Experimental sensory psychology. Scott Foresman, Glenview, pp. 1–10, 1975.

[8] Uchiyama, S., Tsuchida, K., Harada, T., The transfer properties of moisture and heat through clothing materials, Part 2: Wear sensation of socks and analysis of the microclimate within clothing on the simulator. *Journal of the Textile Machinery Society of Japan*, **35(5)**, pp. 210–218, 1982.

[9] Sano, J., Hygienic studies on summer socks. *Journal of Home Economics*, **10**, pp. 240–245, 1959.

[10] Mizunoue, Y., Hygienic of socks. *I-seikatsu Kenkyu*, **4(3)**, pp. 46–51, 1977.

[11] Emmanuel, A., A device for objectively measuring the force to stretch socks. *Journal of the Textile Institute*, **71(1)**, p. 175, 1980.

[12] Merrit, E., Measuring fabric tension in socks while testing for size. *Knitting Technique*, **11(4)**, pp. 284–286, 1989.
[13] Dan Rui, Fan Xuerong, Chen Dong-sheng, Wang Qiang, Study on the pressure at top part of man's socks using finite element method. *Journal of Textile Research*, **32(1)**, pp. 105–110, 2011.
[14] Denton, M.J., Fit, stretch and comfort. *Textiles*, **3**, pp. 12–17, 1972.

Research on dynamic management and model of collaborative sharing of enterprise multi-source information resources based on cloud computing

Yicheng Yu, Wei Chen, Fu Fang
College of Economy and Management,
Shanghai Institute of Technology, Shanghai, China

Abstract

The paper aims to promote the enterprise cloud computing application of multi-source information resources sharing, collaborative problem solving multi-source information resources. Take the enterprise as the discussion object, through the analysis of cloud computing company multi-source information resources and its service model, model to explore its coordinating the resources sharing and dynamic management method. Then, based on theoretical analysis, graphical hierarchical topology, argument deduction form conclusion. The paper first introduced the cloud computing and cloud classification and service mode, multi-source information resources and multi-source information resources as a service, and then analyzes the sharing mode of multi-source information resource sharing model, relates to the outline, service platform construction, content and service settings, finally described the multi-source cloud sharing information resources dynamic management, etc.

Keywords: cloud computing, service mode, multi-source information resource, resource sharing model, dynamic management.

1 Introduction

The rapid development of information technology and its application called continuous innovation network computing mode.

A new model of computing have emerged one after another, following the distributed computing, parallel computing, grid computing, utility computing

WIT Transactions on Engineering Sciences, Vol. 107, © 2015 WIT Press
www.witpress.com, ISSN 1743-3533 (on-line)
doi:10.2495/SEEEII140771

after the IT field raises the network rookie: cloud computing. Cloud computing is the development of Parallel Computing, Distributed Computing and Grid Computing, is an Internet based super computing model and information resource integration using the commercial realization. The calculation model is a through Internet in the way of providing service dynamic scalable virtual, allowing enterprises access to needed service related [1, 2] in poorly understood the situation through Internet. Recently, the sharing of information resources in cloud computing model and its application has been for people to look on.

2 Classification and service model of cloud computing

2.1 Cloud characteristics and classification

The cloud extended over the scale of calculation, and reliability, versatility, economy, on-demand service and virtual features already for the world attention and become a hot research at home and abroad in the field of current network, some international mainstream IT companies: Google, Microsoft, Amazon, IBM, Oracle [3] according to this characteristics have launched a cloud computing platform, cloud computing related information application and service, etc. According to the differences of cloud computing application opening can be divided into: Public Cloud or External Cloud and Private Cloud or enterprise Cloud, Mixtued Cloud and extension of community type [4].

2.2 Cloud service model

The main service of cloud computing include: Infrastructure as a Service:IaaS, Platform as a Service:PaaS, Data as a Service:DaaS, Software as a Service:SaaS. Through the organic combination of them, many cloud services can be based on the application into them, then can obtain the network cloud services convenient. Recently, people from the resource management perspective leads: Supervision and Management as a Service [5]:MaaS, Human as a Service: HaaS [6], etc.

3 Multi-source information resources and services

3.1 Multi-source information resources

Cloud computing environment with a variety of distribution is often mixed in the information resources in different geographical location, different spatial scale is huge, involving data resources, physical resources, model resources, knowledge resources and computing resources[7], their combination and together, they formed the multi-source information resources in Ref. [8]. Figure 1 shows the information resources under the cloud computing environment. Source information resource service to service providers, two basic elements of service enterprise users, the former including infrastructure service providers, platform providers, application software provider and multi-source information resource service provider.

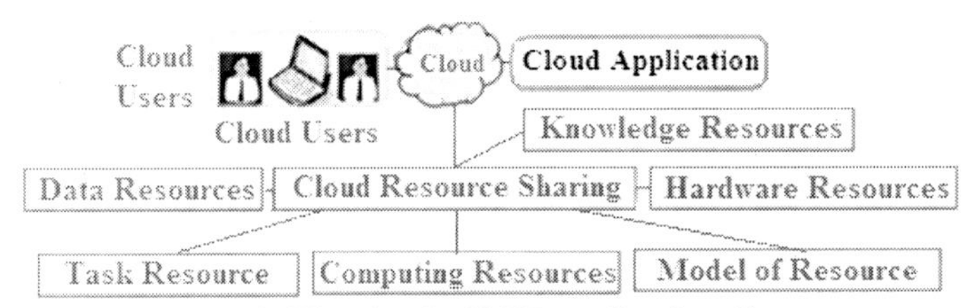

Figure 1: The cloud of MIR application diagram.

3.2 Multi-source information resources as a service

Cloud computing provides feasible methods and ways for the implementation of information resources of cloud services, calls the people of multi-source information resources as a service inquiry. Ref. [9] leads to the information resources of cloud knowledge as a service:KaaS and the cloud of multi-source information resources as a service:MIRaaS, then the actual existence of multi-source information resources (MIR) for the analysis, leads to a new model of cloud service: cloud multi-source information resources as a service [9, 10] (Multi-Source Iinformation Resource as a Service, MIRaaS).

MIRaaS is the cloud service providers use the powerful cloud computing technology will be distributed in different regions of the information resource integration and re deployment, multi-source information resources construction of cloud. MIRaaS involves multi-source information resources of cloud service providers, cloud infrastructure service provider, platform cloud service providers, application software cloud service providers, human and supervision service provider, information resources of cloud service provider and information resources users of cloud services. The MIRaaS architecture consists of five layers, the structure is shown in Figure 2.

Change MIRaaS mode of management and service of the traditional information resources in a certain extent. Compared with the traditional information service mode of MIRaaS has: virtualization, diversity, service, intelligent, automation etc. Enterprise users according to the actual needs of service providers by Xiang Yun ordering required cloud of multi-source information resource service. When the user presses the ordering requirements Xiang Yun service providers pay the fees payable, to obtain the required multi information resources of cloud services. Can also be initialized processing service corresponding to the existing cloud service type, the multi-source resources organization, classification, reconstruction and optimization to form a set of user benefits for the wizard to enterprise resource management method, so that the existing multiclass cloud service system cooperative association, intelligent information service efficiently.

4 Sharing model of multi-source information resource

In order to make full use of multi-source information resources sharing and collaborative, should calculate the multi-source information processing characteristics, dynamic and multiple forms of cloud services based on the cloud, explore

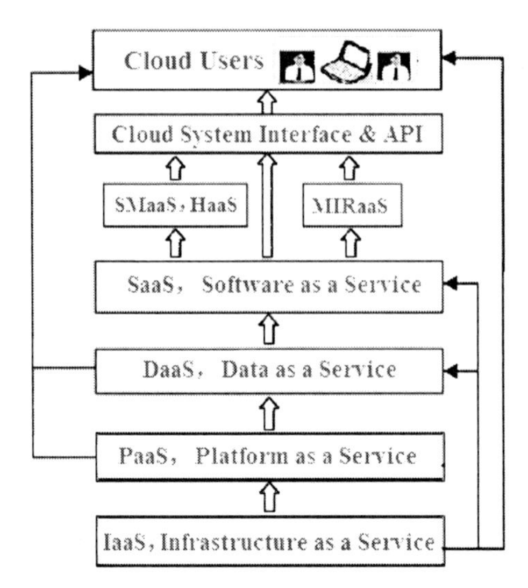

Figure 2: MIRaaS system structure diagram.

multi-source information resource sharing [11] and management model of. Cloud of multi-source information resources collaborative sharing mainly depend on the basic of cloud services (IaaS, PaaS, DaaS, MIRaaS, SaaS, etc.), construction, platform in the cloud content construction, cloud services, construction and dynamic management, integration of centralization and decentralization of the hybrid collaborative sharing mode.

4.1 Sharing pattern outline

Multi information resources sharing cloud can be composed of distribution companies in the private cloud, each private cloud into several sub company, information service institutions and other private cloud, private cloud interconnected self-managed by various institutions and enterprise cloud, can the overall management set function management mechanism. Multi-source information resource co sharing model architecture level bottom-up divided into: resource layer, middleware layer and service layer and other components, as shown in Figure 3.

1. The underlying resource layer consists of hardware resources, data resources, computing resources, knowledge resources, the resource, task resource model is composed by the hardware resources, resource management server, storage device, a cloud equipment composition. The rest of the multi-resource is the integration of body standard each information resource isomorphism or nearly isomorphic, so as to form the basic operation of collaborative sharing of multi-source information resources.
2. The middle management in resource layer and service layer, multi-source information resources is the key to the realization of collaborative sharing. This layer is mainly responsible for the enterprise cloud resources security under the foundation of management, involving numerous application task scheduling,

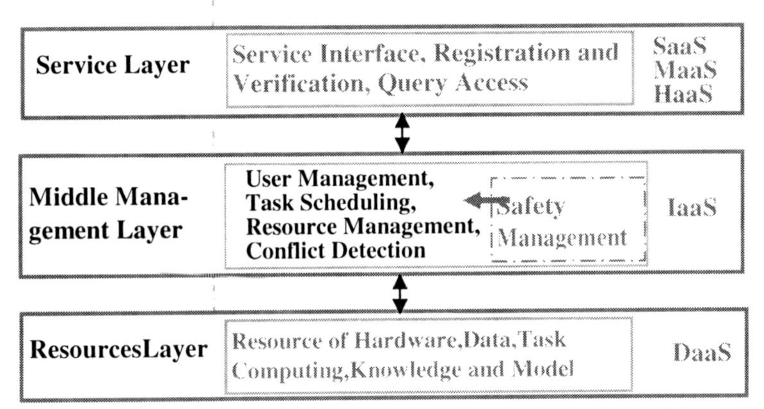

Figure 3: Collaborative-sharing of information resources model on the cloud.

efficient management, resource conflict detection processing, etc.

3. The top layer is the realization of collaborative service layer of the sharing of information resources, service interface through the verification, the platform layer effectively and reliably cloud users access resources retrieval. Enterprise cloud can establish a cloud resource index database; users can quickly search and intelligent access according to the needs of applications in the enterprise cloud resource index database, high efficiency so as to enhance the user's query accuracy and resource sharing.

4.2 The construction of service platform

Set up the cloud service platform should be consistent with the collaborative sharing of information resource requirements to the world by the application and easy access, can complete the security access, Sasuke user registration, query and a series of operations, effectively identify the user's identity and prevent the illegal operation and invasion. In order to make full use of multi-source information resources sharing and collaborative, based on sharing model under the outline of the cloud service platform should be constructed with the infrastructure for the premise, to Resource Co sharing oriented, strengthening the multi-source information resources collaborative sharing operation situation, and strive to build a modern service platform with low cost, high standard, intelligent, scalable that involves basic platform of collaborative sharing (Iaas, PaaS, Daas). Involves basic platform synergy sharing (Iaas, PaaS, Daas), public service platform: public cloud, local service platform: a private cloud and the cluster service platform: mixed cloud. Among them:

1. The public cloud is mainly supplied by third party service providers, is mainly to provide environment for all types of software facilities and network infrastructure with unified authentication, resource integration, data services, regional resource scheduling and other functions.

2. The private cloud is jointly composed of system user application system, user application platform and service providers, to provide SaaS, MIRaaS services, and implementation of the collaborative interaction and distributed resource sharing of public service platform, etc.
3. Mixed cloud is mainly for the different operating systems or provide free environment sharing platform's resources, the formation of a high concentration of information and service cluster, strengthen the high integration of resources and collaborative sharing.

4.3 Content and service settings

1. Content and service settings platform content settings. This is critical for enterprise resource sharing, should be based on cloud type categories set.
   (a) Setting of the public cloud content, the content of this cloud mainly for the enterprise public service platform, usually composed of many software, as the cluster cloud to provide various application service local service platform.
   (b) Set the private cloud content, the content of main body set up by different information associated with the public cloud content localization application platform, involving SaaS, MIRaaS and other information services, and provide the data backup routine and construction of local application of database, the effective sharing by application of public cloud computing platform integration of information resources.
   (c) Set the mixed cloud content. This is mainly reflected in the content of the cluster service platform based on the integration of information resources based construction, including the lateral resource cooperative service platform based industries and sharing of the enterprise level service platform based on the longitudinal resources.
2. The construction of the platform of services. Shared service model of cloud computing based on enterprise resources, should be based on the infrastructure and the basic platform construction as the foundation, middle management oriented, service layer as the goal, to maximize the specification system construction of sharing of information resources, the system hardware configuration for the Hadoop service system of Microsoft Windows Azure cloud computing services platform. Through the enterprise and industry cloud information service platform to carry out multi-level information sharing services, provide: IaaS, PaaS, DaaS, SaaS, and MIRaaS, and the use of interactive platform, collaborative sharing of information resources of enterprises.

5 The dynamic management of information resources sharing of multi-source cloud

Dynamic management [12] of enterprise of multi-source information resources under the cloud computing is a relatively complex problem, mode, how to effective management is concerned by the sharing of multi-source information resources industry. Multi source cloud the sharing of information resources dynamic management of information resources in the implementation of is a system of dynamic optimization and allocation, discovery and matching, expression and organization, analysis and monitoring of the management process. The cloud of multi-source information resources dynamic management system architecture is shown in Figure 4. This framework includes the dynamic expression and organization, multi-source information resource discovery and matching, optimization and distribution and results of real-time analysis and monitoring of four parts, presents each part contains the main function and relevance.

Allocation and scheduling, service mechanism and mode optimization of dynamic optimization and allocation can support multi-source information resources; also can perform real-time analysis and monitoring work on the web resource

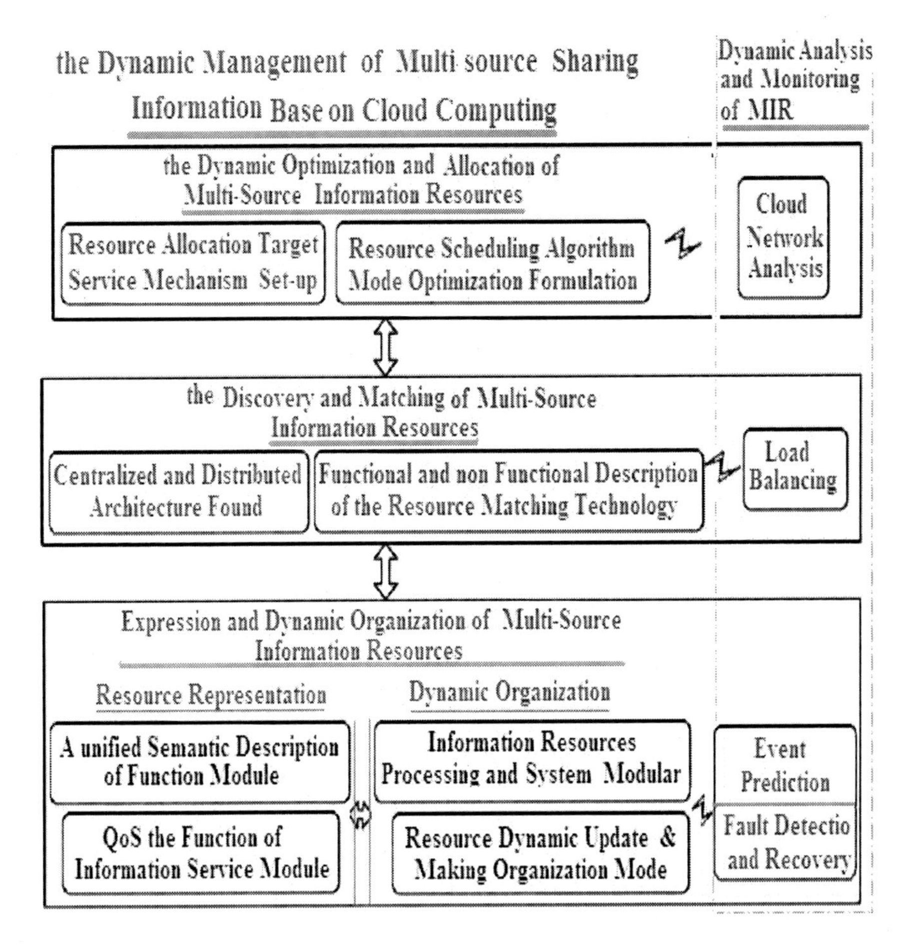

Figure 4: The cloud of MIR dynamic management system framework.

state. Resource monitoring and resource scheduling is one of the key. Resource monitoring Kona AppScale resource monitoring mechanism, using Ruby on Rails Web architecture and has a process level resources Web monitoring analysis. Resource scheduling by improved Hadoop cloud storage resource scheduling and scheduling management of cloud computation and storage resources dynamic scheduling mechanism of CSDS [1] implementation of cloud resources, to achieve the low cost and high profit rate, the maximum to meet the request of users and resources utilization. Cloud computing to extension in basic service provision, expand and enrich the application of cloud computing resource sharing the broadness and service mode, so as to realize the collaborative sharing of information resources, improve the utilization of manufacturing resources, to achieve resource efficiency.

Resource discovery, broad architecture description and resource module matching support multi-source information resource discovery and matching processing. In a cloud environment, the types of information resource are rich and great number and dynamic nature of this requires that the system has some degree of integration cooperation. To maintain local sites in autonomous systems, is also required to provide the corresponding QoS support to the complexity and heterogeneity of the underlying resources shield, put up information resource management in cloud computing various decentralized, improve the efficient management of the utilization of resources and realize the resources. Through the organization of module making mechanism, dynamic update technology and information resource transfer copy, load balancing monitoring improve multi-source information resources in multiple levels of expression and organization.

Information resources on heterogeneous platforms resources into different organizations have, each other with different control strategies, different operation rules, pricing model, load capacity and balance effect. The modules of the system in view of this situation, presents a real-time, expansibility, adaptability of utility, make the cloud information resources can be dynamic management, to provide quality information resource service for user of cloud.

6 Conclusions

Cloud computing is a product of the development of network technology, the application of multi-source information resources leading enterprises and walking synergy sharing, is the key to the future development of the information industry innovation lies. In this paper, the multi-source information resource and service sharing pattern, outline, service platform construction, content and service settings and multiple cloud sharing information resources dynamic management are also explored. However, the branch technology research is still in the initial stage, the content needs to be further perfected. The author only aiming to look forward to the application of cloud computing technology perfected.

Acknowledgment

This project was supported by: Ministry of Education (11YJA630185); the National Natural Science Foundation (61272435/F020701); Shanghai Institute of

Technology Foundation Program (SJ201101); SIT graduate education reform foundation item: 4521ZK130059086, SIT key reform fund project (1010T130029).

References

[1] IBM, IBM Introduces Ready-to-Use Cloud Computing [EB/OL]. www-03.ibm.com/press/us/en/pressrelease/22613.wss,2011-11-15

[2] Elmore AJ, Das S, Agrawal D, et al., Zephyr: Live migration in shared nothing databases for elastic cloud platforms. SIGMOD Record, 2011.

[3] Amazon web services (TM) Amazon Elastic Compute Cloud (Amazon EC2) [EB/OL]. http://aws. Amazon.com/ec2. 2011-10-24

[4] Chen Gui Fen, Wang Xu, Chen Hang, et al., Research on digital agricultural information resources sharing plan based on cloud computing. *Computer and Computing Technologies in Agriculture*. Springer, Berlin, pp. 346–354, 2011.

[5] Ai Yong, Zhang Xin, Ke Jie, Jin Guo, Ma Ji, Application of information resource view on cloud computing service. *Journal of Computational Information Systems*, **9(2)**, pp. 593–601, 2013.

[6] Xiao Jing, Wang Zhiyuan, A priority based scheduling strategy for virtual machine allocation in cloud computing environment. In: *Proceedings of the International Conference on Cloud Computing and Service Computing*, CSC.2012, Shanghai, China, pp. 50–55, 2012.

[7] Li, Keqin, Optimal load distribution for multiple heterogeneous blade servers in a cloud computing environment. *Journal of Grid Computing*, 2013.

[8] Alabbadi MM., Cloud computing for education and learning: education and learning as a service (ELaaS). In: *Proceedings of the 14th International Conference on Interactive Collaborative Learning*, pp. 589–594, 2011.

[9] Deng Zhong-Hua, Hu Wei, Zhao You-lin, Analysis on the development of knowledge base for the information resources cloud. In: *Proceedings of International Conference on Energy and Environment*, pp. 199–202, 2011.

[10] Wang Xiaoyu, The research of multi-source information resource in the cloud system and cloud service model of cloud computing. *Application Research of Computers*, **3**, pp. 784–788, 2014.

[11] Xu Dayu, Yang Shanlin, A method of multi-source information resources management under the cloud computing. *Computer Integrated Manufacturing System*, **9**, pp. 202–203, 2012.

[12] Xue Yu, Research on the model of resource scheduling optimization under the cloud computing environment. *Computer Simulation*, **5**, pp. 361–365, 2013.

Exploring and constructing a model on dynamic structure of mental accounting

Ye Zhongkai[1], Shan Xiaohong[2], Wang Ning[1]
[1]The College of Economics and Management,
Beijing University of Posts and Telecommunications, Beijing, China
[2]The College of Economics and Management,
Beijing University of Technology, Beijing, China

Abstract

With the deductive theory method this study figures out the formation process of the dynamic structure of the mental accounting system and the mechanism of its impact on consumer decisions. By analyzing the existing researches on the mental accounting structure, this study puts forward the general pattern of the static structure, named "Three Main Accounts – Multiple Sub-Accounts," with reference to the property management. This study finds that the static structure is the basis of the dynamic structure, and introduces the "Degree of Involvement" to judge the possibility of the dynamic structure. Based on "Target Representation Model," this study analyzes four stages that form the dynamic structure and try to explain how it affects consumers' decision-making process. Finally a dynamic interaction model of mental accounting system is built, which is an important supplement for mental accounting theory.
Keywords: mental accounting, property management, degree of involvement, interaction model.

1 Introduction

The notion of mental accounting, introduced by Thaler (1980 and 1985), is the set of cognitive operations used by individuals and households to organize, evaluate, and keep track of financial activities [1].

After nearly 30 years' development, this theory has been improved gradually and also promotes the theoretical breakthroughs in interdisciplinary areas, such as consumer decisions areas, financial investment areas, and so on (Zhou, 2004;

www.witpress.com, ISSN 1743-3533 (on-line)
doi:10.2495/SEEEI140781

Hyeong, 2010; He et al., 2011; Claudia, 2013). Among the researches of mental accounting on economy, most of them have focused on the theoretical application while studies related to the theoretical basis and the impact mechanism were few. This hinders the further development of mental accounting and its application. Therefore, considering individuals' mental accounting structure as an essentially basic part in the study of mental accounting structure, it will be fruitful to research into its structural elements, formation process related to the individuals' behaviors.

The previous studies about mental accounting structure have achieved certain results. Some scholars (Kahneman, Tversky, 1984; Brend, Higgins, 1998) have proposed structure formation theories, such as discrete type and continuous type theories. Some scholars (Henderson, 1992; Thaler, 1999) determined the dimensionality of structure classification, based on the source of wealth, consumption patterns and so on, which has been proved in subsequent empirical research [2]. There were also some studies focusing on the structure of specific groups of people (L.D.M., 2009; John, 2011). However, there are some problems in the previous studies.

Individual's mental accounting is unstable [3] with the simultaneous coexistence of static structure and dynamic structure. Indeed, the dimensionality of structure classification proposed by the previous studies can reflect the static structure (also called implicit structure [4]) of individual's mental accounting.

The research data were mostly obtained through questionnaires and interviews, which allowed respondents to recall their memories about their previous consumption habits and multiple consumption behaviors. However, the data may be strongly affected by researchers' experiences in designing the research and the limitations posed by these self-conducted research methods. The current dimensionality of structure classification, therefore, has no uniform standard.

Likewise, the previous studies have verified the structure formation theories [5], such as "local account" and "Target representative model," in scenario-simulated experiments. This method can prove the presence of dynamic changes of the mental account in the specific decision-making context (Kahneman, Tversky, 1984). However, there is no detailed study about the basis for the dynamic formation process and its impact on decision-making. Also, there is a lack of studies on the systematic mode of mental accounting and the relationship between static and dynamic structure.

Based on the facts above, this study will focus on the general pattern of the static structure and try to explain the formation of dynamic structure to try explaining the mechanism about how the dynamic structure influences the consumer decisions.

2 Static basis of dynamic structure

Dynamic structure cannot come out of nowhere since each consumption behavior is affected by daily consumption habit [6]. So the general model of static structure should be discussed first as it is the most direct presentation of consumption habits. Thaler defined mental accounting as individuals' psychological process involved in the coding, organization, and evaluation of financial activities. On the one hand, "coding" and "organization" reflect the psychological cognitive process and its results when individuals are managing their properties [7]. On the other hand, daily property management is often perceived when recalls or summarizes its property usage. Therefore, this study will combs static structure in the view of property management.

Property management includes income, expense, and savings. Individual determines the income sub-accounts basing on the different sources of wealth, and according to the different demands met by different expense to confirm the expense sub-accounts, and depending on the benefits and risks to ensure the savings sub-accounts. This pattern, called "Three Main Accounts - Multiple Sub-Accounts" (as shown in Figure 1), will be affected by individual's economic conditions and consumption habits and it tends to be stable when individuals' property management habits form.

As for concerned on the relationship between the various accounts, there is a relatively stable and tendentious correspondence between the income sub-accounts and the expense sub-accounts. Just like research proved that sources of income had effect on use. Furthermore, based on the consumption function ($C = f(I, A, F)$) in behavior lifecycle hypothesis, Thaler pointed out that when individual has to spend the money in savings accounts, the risk savings account precedes

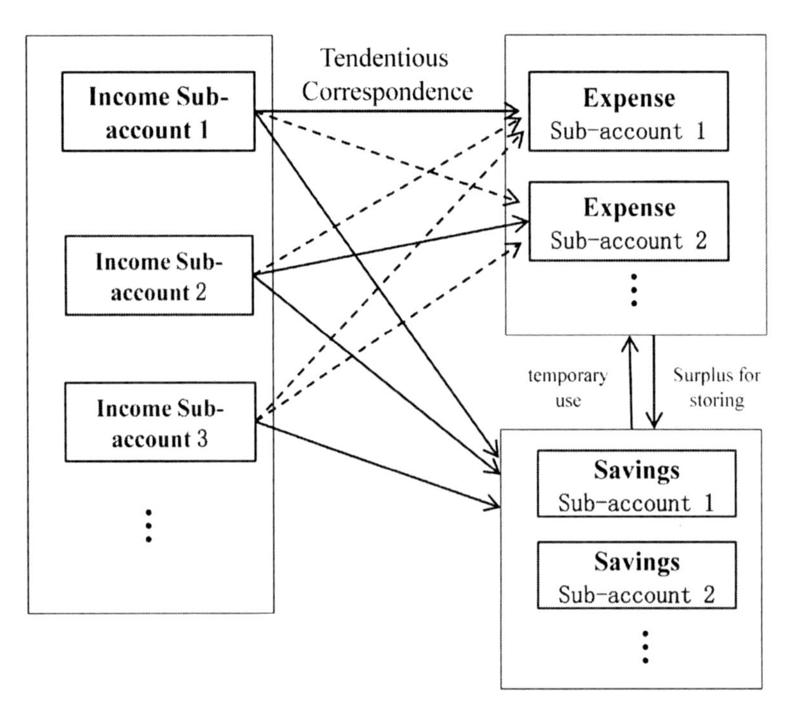

Figure 1: The general pattern of the static structure.

the security savings accounts. Correspondingly, extra funds will flow into the savings accounts when all expenses have been met.

3 Elements and formation of dynamic structure

The formation of dynamic structure is the "Valuation" process by which people try to maximize the effectiveness of mental accounts [8]. Prospect theory proposed by Kahneman has suggested that the "valuation" is based on the selected reference point to determine the gain and loss. Therefore, dynamic account is a collection of all accounts coded as "Gain." It will be elaborated from two aspects as follows.

3.1 Formation conditions of dynamic accounts

Because the dynamic changes of mental accounting structure occurs under the objective condition, the conditions affecting the dynamic structure should include "Internal Environment (IE)" and "External Environment (EE)." IE refers to the virtual environment formed by information related to the consumption purpose after the consumption demand has generated. EE refers to the real environment formed by entity factors, such as store layout, when the consumption behavior happens. Considering mental accounting as individual psychological cognitive process, when measuring the degree of influence on the structure, the IE should be the decisive factor while the EE is induction factor.

In behavioral research, the researchers used the "Degree of Involvement" to describe the degree of subjective involvement about goods [9]. So this study introduces "Degree of Involvement" as an indicator to measure the IE, which can be integrated measured not only by the degree of effort and time invested in searching goods information and brand choice, but also by the potential value and the attribute value of the product.

The level of "Degree of involvement" has impact on generation of dynamic account. The low level of "Degree of involvement" will result in direct consumption while the high level of "Degree of involvement" will trigger the dynamic change of the structure

3.2 Formation of the dynamic account and the impact on consumption decisions

The dynamic changes triggered by the high level of "Degree of involvement" consist of four stages.

Stage One: The generation of temporary sub-accounts. The consumption purposes inspired by consumer environment are the individual's needs. As individuals' needs are registered as a corresponding expense sub-account in the static structure of the mental accounts, the static structure should serve as a basis of the generation of dynamic account. Specifically, individuals with high level of "Degree of involvement" will primarily make judgments about funds from certain or several static expense sub-accounts based on their relationships with the consumption purposes [10], then confirm the corresponding temporary sub-accounts. And the higher the level of “Degree of involvement” is, the clearer of the temporary sub-accounts structure will be. Additionally, the fund size of the temporary sub-accounts could have an expected threshold under the influence of “self-control” [11].

Stage Two: The formation of the dynamic account. Based on the “Target Representation Model” [8], all the weights of the temporary sub-accounts will be determined in the dynamic account according to their own contributions to meet individual’s needs. After that, the individual’s maximum revenue dynamic account will be built with distinct composition where temporary sub-account should play a leading role and the others should play complementary roles.

Stage Three: The determination of consumption decisions. If individual consumption behaviors occurs in the objective circumstances, Soman et al. (2008) found that the inducing factors in the EE, such as discounts, rebates and so on, have a significant impact on consumer decisions. So this study speculates that the existing dynamic account will not be changed if the inducing factors are weak or inexistent. However when the inducing factors are large enough, they will drive individuals to generate new demand, or to make a wrong judgment on existing demand [12]. In this way, they could guide individuals to change consumption tendency. They could even reconstruct the existing dynamic account and lead to irrational consumption.

Stage Four: Completion of consumption and dynamic account disappears.

4 Discussion

This study finds the presence of an interactive structure consisting of both static structure and dynamic structure in the mental accounting system, and finally

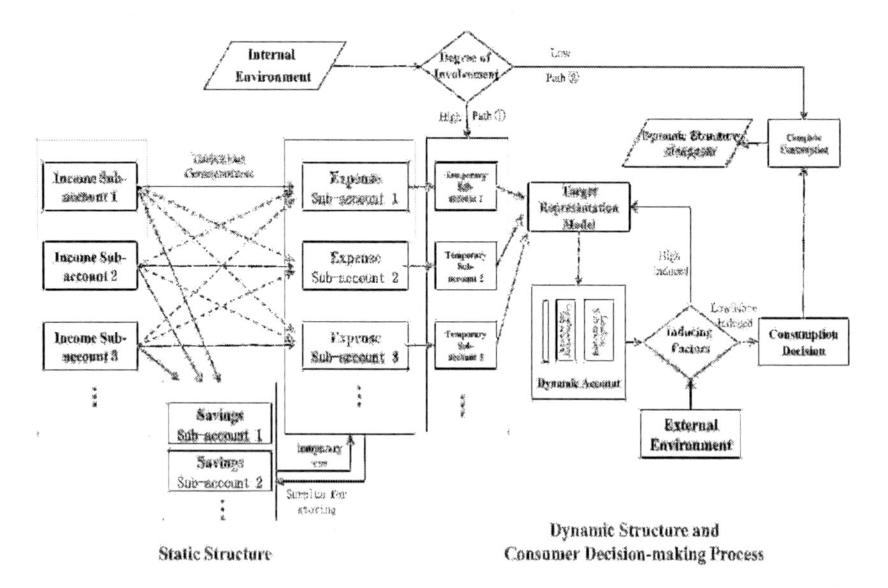

Figure 2: Dynamic interaction model of mental accounting system.

builds a Dynamic Interaction Model of Mental Accounting System (DIMMA) (as shown in Figure 2). The left side of DIMMA represents the static structure of mental accounts, and the other side is dynamic structure and consumer decision-making process. When the degree of involvement is low, individuals will consumes directly via path (2) without triggering dynamic structure. Conversely, Individuals with high degree of involvement will undergo path (1), which contains the process of producing temporary sub-accounts, determining the dynamic account and consumption decisions, and ultimately ignoring the dynamic structure after having already completed consumption. Otherwise, Individual may be affected by inducing factors and reconstruct dynamic account at the decision-making stage. Analysis shows that the DIMMA has three major characteristics. Firstly, the DIMMA is a type of kind of structure capable of self-improvement. From the view of developing the habit of property management, the frequent and similar dynamic structures will become a new static structure after Individual's multiple recall-processes. And the theory of short-term memory can also prove this viewpoint.

Secondly, the DIMMA is a systematic structure with one static structure but multiple dynamic structures at the same time. In the short term the static structure is stable and unique since it is a psychological mapping of the property management habits. However, the dynamic structure is based on the demands, which means the more demands the more dynamic structures even in the same period.

Thirdly, the DIMMA is static-vague but dynamic-distinct. It is often difficult for individuals to distinguish their psychological division of revenues and expense all the time. Even individuals may classify the same consumption behavior into different static sub-accounts at different stages. Therefore, the static structure is vague in their daily lives. However, based on the Target Representation Model, individuals will take serious consideration of the belonging, rights and other aspects of temporary sub-accounts. And the results in the dynamic structures are distinct in the short term.

5 Conclusions

In summary, this study elaborates on the general pattern of the static structure which has demonstrated the universality of the mental accounting's category dimension in the previous studies, and explains the formation process of dynamic structure of mental accounting system and the mechanism of its impact on consumer decisions, and finally builds a dynamic interaction model of mental accounting system. This study is an important supplement for mental accounting theory. Enterprises can also take advantages of the characteristic s of consumer mental accounting structure to improve their services and marketing strategies.

Acknowledgment

This work was financially supported by the Humanities and Social Sciences General Program Projects of Beijing Education Commission (No. SM20131000 5002).

References

[1] Thaler R., Mental accounting and consumer choice. *Marketing Science*, **4(3)**, pp. 199–214, 1985.

[2] Ramphal S., Mental accounting: the psychology of South African consumer behaviour, 2010.

[3] Nunes J.C., Mental accounting: flexible accounts, order effects and incommensurable entries. *Advances in Consumer Research*, **28**, p. 70, 2001.

[4] Li A., Ling W., Fang L., et al., The implicit structure of mental accounting among Chinese people. *Acta Psychologica Sinica*, **39(4)**, pp. 706–714, 2007.

[5] Li A., Ling W., The nonfungibility and mental arithmetic of mental accounting. *Psychological Science*, **(4)**, 2004.

[6] Gitman L., Joehnk M., Billingsley R., Personal financial planning. *Cengage Learning*, 2013.

[7] Thaler R.H., Mental accounting matters. *Journal of Behavioral Decision Making*, **12(3)**, pp. 183–206, 1999.

[8] Kahneman D., Tversky A., Choices, values, and frames. *American Psychologist*, **39(4)**, p. 341, 1984.

[9] Hollebeek L.D., Jaeger S.R., Brodie R.J., et al., The influence of involvement on purchase intention for new world wine. *Food Quality and Preference*, **18(8)**, pp. 1033–1049, 2007.

[10] Lu Q., Liu R., Li W., et al., Price-taker bidding strategy based on mental accounting. *IEEE International Conference on Industrial Technology*, ICIT 2008. IEEE, pp. 1–6, 2008.

[11] Jia G., Liu L., The theoretical and empirical analysis of the irrational consumption behavior. *Consumer Economics*, **22(1)**, pp. 85–88, 2006.

[12] Godek J., Murray K.B., Effects of spikes in the price of gasoline on behavioral intentions: a mental accounting explanation. *Journal of Behavioral Decision Making*, **25(3)**, pp. 295–302, 2012.

Comparison and simulation of sorting strategy in the distribution center

Chen Nan, Yin Jing
School of Mechanical-Electronic and Automobile Engineering,
Beijing University of Civil Engineering and Architecture, Beijing, China

Abstract

According to the key influence factor of the distribution center efficiency-order sorting as the research object. In this paper, the logistics distribution center of a certain electric commercial enterprise as the case background, focusing on the research of the order batching strategy, the two different order batching strategy are compared with the traditional strategy of first in first sorting. Combined with the dynamic simulation on the platform of flexi, the sorting operation time and transport moving distance from this two aspects data of different sorting strategy are provided after the model run, thus the efficiency of different sorting strategy is compared and analyzed. Then the dominant strategy for this case is determined at last. This is an effective way to reduce the sorting cost in a distribution center by optimizing the order batching strategy, it has important real value to make the distribution center improve the level of service and operation.
Keywords: distribution center, Flexsim, order sorting, order batching, simulation.

1 Introduction

Under the environment of e-commerce, variety and frequency of the delivery goods in the distribution center have extremely increased. As the core task of distribution center, sorting operation is the main source of manpower and time. Sorting operation efficiency directly affects the distribution center operating efficiency and the speed of response to customer orders, which is an important factor in evaluation of service levels. Therefore, distribution center should choose appropriate sorting equipment, in addition to more efficient sorting strategy. The key factors affect sorting efficiency including the order sorting strategy, the classification and centralized approach after sorting the goods and the goods

WIT Transactions on Engineering Sciences, Vol. 107, © 2015 WIT Press
www.witpress.com, ISSN 1743-3533 (on-line)
doi:10.2495/SEEEI140791

transportation route. As a result of the sorting operation power comes from the orders, so order sorting strategy is particularly important. In addition, in the study of the past have more research focus on the design and optimization of the transportation and walking paths. So in this paper, combined with the practical case, the order sorting strategy is emphatically considered, it makes sorting operation efficiency increasing significantly through applying the sorting strategy.

In this paper, a certain logistics distribution center of sorting operation, packaging process and related data as the case background, combined with the dynamic simulation on the platform of Flexsim, modeling the different batch sorting strategy, and through running data of picking time and moving distance are analyzed. Thus, high efficiency of sorting strategy is determined and customer orders could be quickly responded.

2 Comparative analysis of sorting strategy

Sorting operation is the process of the goods is concentrated, processed and placed from the shelves of distribution center, based on the different type and quantity customer order. The purpose of sorting is concentrating the goods in customer order correctly and rapidly. In the actual distribution center, it should take the appropriate sorting strategy to choose goods based on the actual order characteristics and requirements, so as to achieve the aim of improving the sorting efficiency.

The key factors that affect the sorting efficiency mainly including division sorting, dividing order form, order batching and order categorizing, where zone sorting and order batching are two important directions of the sorting system development. In this case, an operator is responsible for one area, therefore only division sorting strategy is considered. Aiming at the order batching, the following study carries on, the Standard of order batching general according to goods quantity, order quantity, time limit and convergence route. The above characteristic and application condition comparison of two order sorting strategies and four order batching rules as shown in Table 1. In addition, it assumptions

Table 1: Comparison of order sorting strategy and order batching rule.

Order Sorting Strategy	Order Batching Rules	Requirement Frequency	Applicable Condition
Division sorting		Periodicity or abruptness	Big orders, high timeliness
Order batching	Goods quantity	Periodicity	Big orders and stable, less types of goods but bigger in quantity
	Order quantity	Periodicity or Aperiodicity	Big orders and stable, less types of goods
	Time limit	Abruptness or Periodicity	Big orders and stable, less types of goods and smaller in quantity
	Convergence route	Periodicity	More types of goods and bigger in quantity

distribution center is sufficient supplies in this case, so the replenishment system is simplified.

It is worth that the goods after sorting also needs classified and concentrated according to the customer orders difference when using the order batching strategy, each batch of goods is classified and concentrated in the shipping area after classifying goods and packaging operation.

In conclusion, according to the actual case condition, three sorting schemes are compared in this paper: single order sorting, batch sorting according to order quantity and batch sorting according to goods quantity.

3 The simulation for different sorting strategy

3.1 Background

A certain electricity enterprise is a new generation of B2C online shopping platform under the flag of a native well-known electronics mall. It has covered the traditional home appliance, food and daily production. This enterprise has more than 10 warehouses in the country. In addition to providing local service, it could provide delivery service for other provinces and cities. The different schemes of order sorting strategy is compared and analyzed for a certain distribution center in this enterprise, and packing goods after sorting operation is completed, finally the goods is transported to the designated area to wait for shipping. This procedure achieved each function of the distribution center completely.

3.2 Simulation process analysis

1. Workflow analysis

With the customer orders provided by the electric commercial enterprise, goods from warehouse to distribution center, then packing the goods of each order, until finish goods waiting for delivery in the shipping area through transport line. The concrete operation is shown in Figure 1.

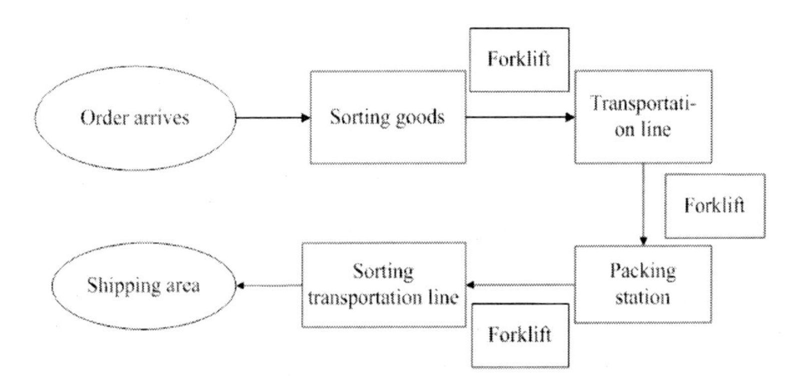

Figure 1: The operation workflow.

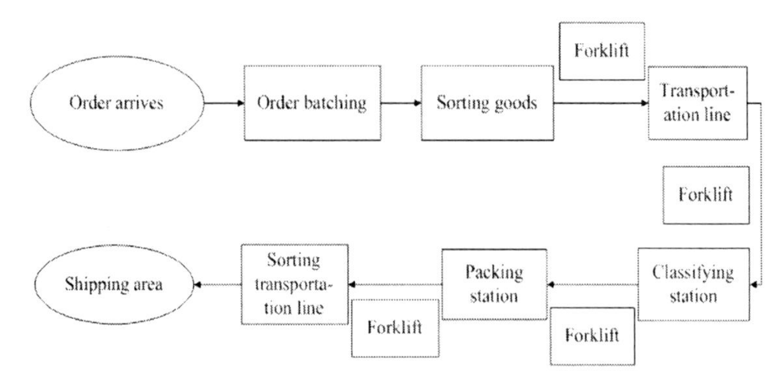

Figure 2: The operation workflow of two batch sorting strategy.

This paper discusses the other two kinds of order sorting strategy are batch sorting according to order quantity and batch sorting according to goods quantity, then sorting large orders after the order batching. The first thing is classified orders before batch sorting according to goods quantity, the next is combined each orders into the batch orders to sorting, through classifying and packing each orders, until finish goods waiting for delivery in the shipping area. The concrete operation of the above-mentioned two order sorting strategy is shown in Figure 2.

2. Assumptions

This system is aimed at one distribution center, that is, a delivery location and multiple customer orders.

The distribution center stored goods are divided into five categories, including class A consumer products, class B foods, class C books, class D dress and shoes, class E other items.

An order contains only one type of goods, it is possible one customer exists in multiple orders.

3. Parameters

The main transportation equipment is forklift, its maximum speed is 2m/s, acceleration is $1 m/s^2$, and capacity is 30 pieces. After analysis of the original order arrived data, the 500 orders in the peak-time are selected as the sample data of simulation. Obtaining the simulation required information table after adjustment of the original orders, including the order arrival time, the name of goods, the classification of goods and the quantity of goods, part of information as shown in Table 2.

4. Order sorting strategy

There are two kinds of order sorting methods, one strategy is batch sorting according to order quantity, that is, meeting 10 orders are combined into a new order to process. The other is batch sorting according to goods quantity, the quantity of single-variety goods meets a maximum of 30 pieces are combined into a new order after classified orders according to different kinds of goods.

4 Results and discussion

4.1 Dynamic simulation

By using the Flexsim7.3.0 version simulation software, the distribution center layout designed above could be displayed through 3D dynamic perspective drawing, as shown in Figure 3.

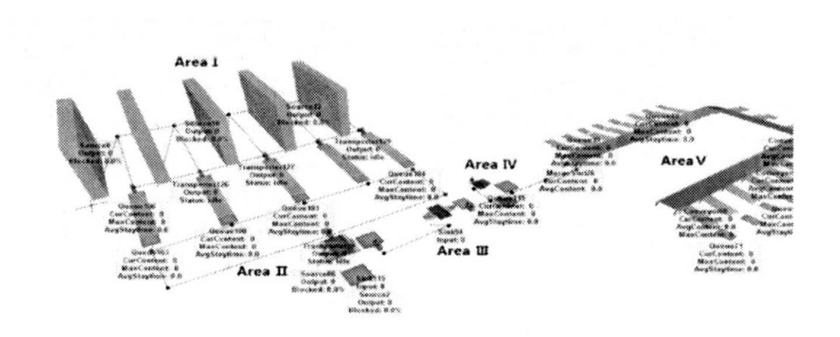

Figure 3: The distribution center layout.

Table 2: Part of information.

Arrived in List	Arrival Time (s)	Products Name	Quantity	Classification
Arrival1	34294	Barilla Spaghetti Pasta, 32 Oz	1	B
Arrival2	34315	Coffeemate Original Canister, 35.3-Ounce	1	B
Arrival3	34331	Oreo Double Stuff Sand-wich Cookie, 15.35 Oz	1	B
Arrival4	34468	Nutella Hazelnut Spread, 13 Oz	1	B
Arrival5	34530	Pringles Cheddar Cheese, 5.96 Oz	1	B
Arrival6	34572	Enfamil Newborn Baby Formula - Powder– 22.2 oz	2	B
Arrival7	34576	Earth's Best Sensitivity Baby Formula - Powder – 23.2 oz	3	B
Arrival8	34580	Doritos Tortilla Chips, Cool Ranch, 11 Oz	1	B
Arrival9	34638	Lay's Potato Chips, Party Size Classic, 13.75 Oz	1	B
Arrival10	34651	Lay's Potato Chips, Party Size Classic, 13.75 Oz	1	B

Including area I is warehouse; area II is the order arrival area; area III is classification station, it does not exist in the single order sorting strategy; area IV is packing station; area V is shipping area.

4.2 Statistical data analysis

For each sorting strategy, the speed of order sorting reflects the operation efficiency directly. Comparing and analysis the sorting speed of 500 orders in the peak-time through recording arrival time of the first order into the sorting center and completion time of order sorting. The quantity of completing order sorting changing with time as shown in Figure 4 (single order sorting, batch sorting according to order quantity and batch sorting according to goods quantity in that order), comparison and analysis as shown in Table 3.

Based on above simulation statistic table, it concludes that the distribution center can complete sorting task in the same day (before 86400s) in the context of order batching strategy, and the efficiency of batch sorting strategy according to goods quantity is better than batch sorting strategy according to order quantity, the former strategy can response to customer requirements more rapidly. In addition, taking order batching strategy can not only shorten the moving distance during sorting, but also reduce the repetition time in the process of looking for goods, as a result, the sorting efficiency can be improved. Therefore, statistical analysis of transport distance is necessary for the order batching strategy. The comparison of transport distance is shown in Table 4.

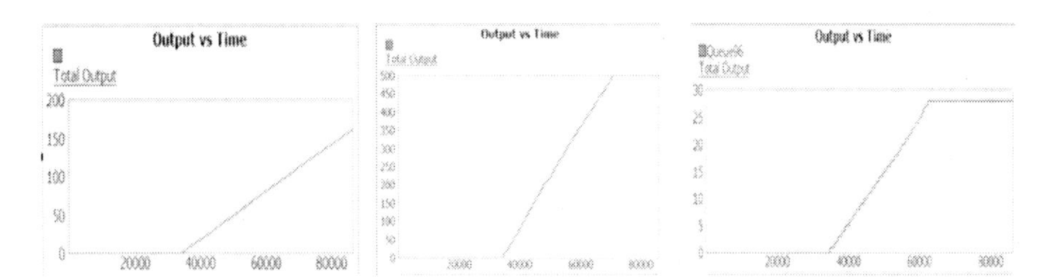

Figure 4: Order sorting quantity changing with time.

Table 3: The comparison and analysis of sorting efficiency.

Order Sorting Strategy	Completed Order Quantity	Arrival Time of the First Order (s)	Completion Time of Order Sorting (s)
Single order sorting	162	34294	86400
Batch sorting according to order quantity	500	34294	71129
Batch sorting according to goods quantity	500	34294	62305

Table 4: The comparison of transport distance.

Order Sorting Strategy	Forklift 1 Transport Distance (m)	Forklift 2 Transport Distance (m)	Forklift 3 Transport Distance (m)	Forklift 4 Transport Distance (m)	Forklift 5 Transport Distance (m)	Average Transport Distance (m)
Batch sorting according to order quantity	22504.47	16231.42	18330.09	16197.74	11659.21	16984.59
Batch sorting according to goods quantity	9598.53	5364.47	5673.15	5223.34	3037.39	5779.38

Table 4 shows that premise conditions of the same order quantity, the transport distance of the second scheme is less than the first one. Therefore, batch sorting strategy according to goods quantity has more advantages.

In conclusion, the batch sorting strategy according to goods quantity either from the perspective of the sorting speed and the efficiency of transport tools has more obvious advantage. Premise conditions of other requirements is same, the batch sorting efficiency is higher. In batch sorting scheme, batch sorting strategy according to goods quantity can make the best of transport tools, reducing the repetition distance and time in the process of looking for goods, so the efficiency of transport tools can be improved.

In addition, in the comparative study on the three sorting strategy, a truth that the utilization of transport tools is not high. But the utilization of transport tools depends on many factors, such as division sorting strategy and the order arrival time. In this paper didn't do related statement anymore.

5 Conclusions

The simulation analysis research of sorting strategy in the distribution center, it has important real value to make the distribution center improve the level of management and operation. Especially for electricity enterprise, improving the efficiency of sorting operation will make a major contribution to the whole distribution center benefit. In this paper, it is based on the specific distribution center sorting system of a certain electricity enterprise as the research object, and putting forward three different sorting strategy mainly aimed at two important factors that they include the batch sorting and the concrete batch sorting strategy. At last, combined with the dynamic simulation on the platform of Flexsim, the comparative analysis on three kinds of sorting scheme is carried on, from the statistics data of order sorting efficiency and transport tools efficiency. In conclusion, the batch sorting strategy according to goods quantity is the most excellent for this logistic system.

In fact, the sorting efficiency of distribution center influenced by multiple factors, and the sorting strategy is much more than three kinds of schemes puts forward in this paper. The suitable sorting scheme will change for the different actual system. Therefore, it needs to the simulation analysis in time based on the real situation of distribution center constantly, until the most appropriate sorting strategy is selected which it makes the efficiency optimization.

Acknowledgment

We are grateful to Beijing University of Civil Engineering and Architecture and Beijing Engineering Research Center of Monitoring for Construction Safety for providing research facilities.

References

In the key factors that influence efficiency of sorting, the batch sorting is an important direction of the distribution center system development. According to the article [1] and article [2], they are all according to the different sorting strategy, and comparing four sorting strategy after modeling analysis; article [3] and article [4] using shorten sorting operation time as the goal, they are all propose a effectively batch sorting decision method that can improve the efficiency of order sorting. This paper is aimed at different batch sorting strategy simulation modeling, then the scientific analysis and preferred has carried on. article [5] and article [6] aiming at the shortest moving distance, the order batching strategy model is established and solved; article [7] summarizes the existing single order sorting and order batching strategy, thus putting forward the optimization sorting method; article [8] compared the order batching strategy and the common method through the computer simulation, and validated the order batching strategy is more efficient; article [9] and article [10] using Flexsim software make simulation model for the yard and certain warehouse, the simulation provides an effective method of evaluation and analysis for their own system optimization scheme.

[1] Ming Gao, SanYuan Zhou, The optimization research of sorting strategy based on FLEXSIM. *Logistics Technology*, **28(9)**, pp. 90–92, 2009.
[2] XiangYing Meng, Jing Yin, The comparison and simulation distribution center sorting strategy based on Flexsim. *Beijing University of Civil Engineering and Architecture*, **29(2)**, pp. 49–52, 2013.
[3] Jia Feng, GuoQing Guo, A optimization algorithm of the order sorting method for distribution center. *Industrial Engineering*, **(5)**, pp. 123–127, 2008.
[4] Hwang H., Kim D.G., Order-batching heuristics based on cluster analysis in a low-level.
[5] Hsu C.M., Chen K.Y., Chen M.C., Batching orders in warehouses by minimizing travel distance with genetic algorithms. *Computers in Industry*, **56(2)**, pp. 169–178, 2007.
[6] ZhanLei Wang, The optimization research of order batching and sorting route for distribution center. Ji Lin University, 2013.

[7] Jia Feng, A Optimization Method for Distribution Center Order Sorting. Ji Nan University, 2008.
[8] Hong Wang, The Optimization Method for Double Area Warehouse Order Picking. Zhong Nan University, 2007.
[9] Xin Zheng, The Optimization and Simulation Research for Container Yard Based on Flexsim. Beijing Jiaotong University, 2008.
[10] ShouWen Ji, XinLei Cao, XingHua Wu, The optimization simulation research for warehouse system based on Flexsim. *Logistics Technology*, **25(10)**, pp. 72–74, 2009.

A diagnosis method for motor bearing fault based on nonlinear output frequency response functions

Changqing Xu, Chidong Qiu, Guozhu Cheng, Zhengyu Xue
Department of Marine Engineering,
Dalian Maritime University, Dalian, China

Abstract

A diagnosis method for motor bearing fault based on the nonlinear output frequency response functions (NOFRFs) is introduced in this paper. It based on that stator current and external radial flux density were acted as the input and output respectively, then the Volterra series time-domain model for motor was identified, and the NOFRFs was estimated. By analysing the eigenvalue change of each order NOFRFs, the severity of bearing fault could be inferred. The experimental motor was installed a healthy bearing or bearing whose outer raceway has single point damage respectively. Experimental result shows that the method based on NOFRFs is suitable for the identification of the bearing fault severity.

Keywords: motor, bearing, fault, Volterra series, NOFRFs.

1 Introduction

Induction motor is an electromagnetic-mechanical energy conversion device. Motor with bearing fault will cause load torque ripple [1], thereby, affect the rotor magneto-motive force (MMF) as well as the stray flux generated by the stator coil-ends [2].

The frequency-domain kernels of Volterra series, namely generalized frequency response functions (GFRFs) can reflect the frequency response characteristics of the system, including the harmonic characteristic, modulation characteristics, etc. But the GFRFs are multi-dimensional functions, and very difficult to identify. In order to simplify the analysis of nonlinear system in frequency domain, Lang and Billings introduced the concept of NOFRFs which are one-dimensional functions for frequency variable [3].

WIT Transactions on Engineering Sciences, Vol. 107, © 2015 WIT Press
www.witpress.com, ISSN 1743-3533 (on-line)
doi:10.2495/SEEEI140801

In this paper, the stator current and external magnetic flux density of squirrel-cage induction motor are analysed, and the nonlinear relationship between them is explained. Then, the basic theory of Volterra series and NOFRFs is illustrated, and the method based on them is introduced. The changing of NOFRFs can reflect the nonlinear characteristics of the system, and this can be applied to the bearing fault diagnosis. By estimating the eigenvalues of NOFRFs can effectively identify the severity of bearing fault.

2 Theoretical analysis

2.1 Stator MMF and rotor MMF

The stator MMF of the induction motor which is fed by three-phase balanced power supply can be represented as:

$$F_v^s(\theta,t) = F_s \sum_{v=1}^{\infty} \frac{1}{v} k_{dpv} \cos(\omega_s t - vp\theta) \tag{1}$$

where $v = 6n + 1$, $n = 0, \pm1, \pm2, \ldots$, F_s is the amplitude of MMF, ω_s is the angular frequency of power supply, p is the number of pole pairs, k_{dpv} is the winding coefficient of vth harmonic MMF, θ is the stator space mechanical angle.

The synthetic MMF of squirrel-cage rotor in the stator reference frame is [4]:

$$F_r^s(\theta,t) = \sum_{k=1}^{\infty} RF_u \cos((\omega_s t \mp \lambda R\omega_r t \pm \varphi)) \tag{2}$$

where F_u is amplitude of single rotor circuit MMF, R is the number of rotor bars, $\lambda = 0,1,2,\ldots$, $\omega_r = (1-s)\omega_s/p$, s is the slip and φ is the phase angle. When the bearing of motor has single point damage, the torque ripple leads to a phase modulation of the rotor MMF,

$$F_r^s(\theta,t) = \sum_{k=1}^{\infty} RF_u \cos(\omega_s t \mp \lambda R\omega_r t + \beta\cos(\omega_c t) \pm \varphi) \tag{3}$$

where $\beta = p\Gamma_c/J\omega_c^2$, it is the modulation index. $\omega_c = 2\pi f_c$, f_c is the vibration characteristic frequency of bearing fault, and Γ_c is the additional torque produced by fault which is proportional to the fault severity.

2.2 Stator current and external magnetic flux density

Assumes that the air gap of motor is uniform, the area is S, and the air gap permanent is constant, expressed by Λ_0. Then, the expression of stator voltage equation is:

$$V_m(t) = R_s I_m + N_s \frac{\mathrm{d}\Phi_m(t)}{\mathrm{d}t} = R_s I_m + N_s \frac{\mathrm{d}(F_m \Lambda_0 S)}{\mathrm{d}t} \tag{4}$$

where I_m is stator phase current, R_s is stator resistance, N_s is the effective number of winding turns. F_m is air gap MMF which is the sum of stator MMF and rotor MMF, then the stator current can be calculated.

$$F_m(\theta,t) = F_v^s \cos(\omega_s t - p\theta) + F_r^s \cos(\omega_s t \mp \lambda R\omega_r t + n\omega_c t \pm \varphi) \tag{5}$$

$$I_m = \frac{V_m}{R_s} - N_s \Lambda_0 S(\frac{dF_v^s}{dt} + \frac{dF_r^s}{dt}) \tag{6}$$

The stray flux is the result of the stator and rotor currents, and emitted from the stator coil-ends and rotor end-rings [2]. Therefore, it is associated with the stator MMF and rotor MMF. Stator MMF can generate magnetic flux density B_e^r at the end-rings of rotor bars. So the magnetic flux density of the kth end-ring of rotor bars is:

$$B_{ek}^r(\theta,t) = F_v^s(\theta,t)\Lambda(\theta,t) = F_v^s \Lambda_0 \cos\left[\omega_s t - v(1-s)\omega_s t + \frac{2\pi(k-1)pv}{R}\right] \tag{7}$$

In the same way, the magnetic flux density of the mth stator coil-end is:

$$B_{cm}^s(\theta,t) = F_r^s(\theta,t)\Lambda(\theta,t) = F_r^s \Lambda_0 \cos(\omega_s t \mp \mu R\omega_r t + n\omega_c t \pm \varphi) \tag{8}$$

The external magnetic flux density at a point P of the motor is the sum of all the magnetic flux density generated in the coil-ends of stator B_{cm}^s and rotor end-rings B_{ek}^r [5].

$$B_P = \sum B_{cm}^s + \sum B_{ek}^r \tag{9}$$

Basing on above analysis, the motor external magnetic flux density and stator current expressions have the same part F_r^s, but they have different part F_v^s. When a function which input is stator current, and output is external magnetic flux density, it is a nonlinear function.

$$B_p = a_1 I + a_2 I^2 + \cdots + a_n I^n \tag{10}$$

Since the external magnetic flux density and stator current are associated with the rotor MMF, they all vary with the load and severity of fault. Therefore, the linear coefficient a_1 only related to the position of magnetoresistive sensor. The nonlinear part is associated with the stator MMF. Due to the stator MMF is proportional to the stator current, and the amplitude of the stator current is related to the load and severity of the bearing fault. When the load is the same, the more serious the bearing fault, the greater the amplitude of stator MMF, therefore the ratio of nonlinear part will correspondingly become large.

3 Volterra series and NOFRFs

3.1 Basic concept

The discrete form of Volterra series [6] is:

$$y(n)=\sum_{k=1}^{N}\sum_{i_1=0}^{L_1-1}\cdots\sum_{i_k=0}^{L_k-1}[h_k(i_1,i_2,\cdots,i_k)]\prod_{m=1}^{k}x(n-i_m)+e(n) \tag{11}$$

where N is the highest order of the system, L_k is the memory length of the kth order of Volterra kernels, $e(n)$ is called truncation error.

The output frequency response of the nonlinear system can be expressed as:

$$Y(j\omega)=\sum_{n=1}^{N}G_n(j\omega)U_n(j\omega) \tag{12}$$

$$G_n(j\omega)=\frac{\int_{\omega=\omega_1+\cdots+\omega_n} H_n(j\omega_1,\ldots,j\omega_n)\prod_{i=1}^{n}U(j\omega_i)\mathrm{d}\sigma_{\omega n}}{\int_{\omega=\omega_1+\cdots+\omega_n}\prod_{i=1}^{n}U(j\omega_i)\mathrm{d}\sigma_{\omega n}} \tag{13}$$

where $G_n(j\omega)$ are the nonlinear output frequency response functions (NOFRFs). $U_n(j\omega)$ are the Fourier transform of $u^n(t)$, $u^n(t)$ represents the nth power of $u(t)$. The eigenvalue $Fe(n)$ of NOFRFs can reflect the nonlinear degree of system, which is defined as:

$$Fe(n)=\frac{\int_{-\infty}^{\infty}\left|G_n(j\omega)\right|\mathrm{d}\omega}{\sum_{n=1}^{N}\int_{-\infty}^{\infty}\left|G_n(j\omega)\right|\mathrm{d}\omega}(1\le n\le N) \tag{14}$$

If a system is linear in normal state, $Fe(1)$ is close to one. Assume the system has occurred nonlinear distortion, the $Fe(1)$ decreases, whereas $Fe(n > 1)$ become large.

3.2 Bearing fault diagnosis based on Volterra series and NOFRFs

Bearing fault diagnosis based on Volterra series and NOFRFs is shown as Figure 1 and summarized as follows:

1. The synchronous sampling of stator current and external radial magnetic flux density is achieved, Volterra series model for motor is then constructed, and the time-domain kernels of Volterra series are identified.

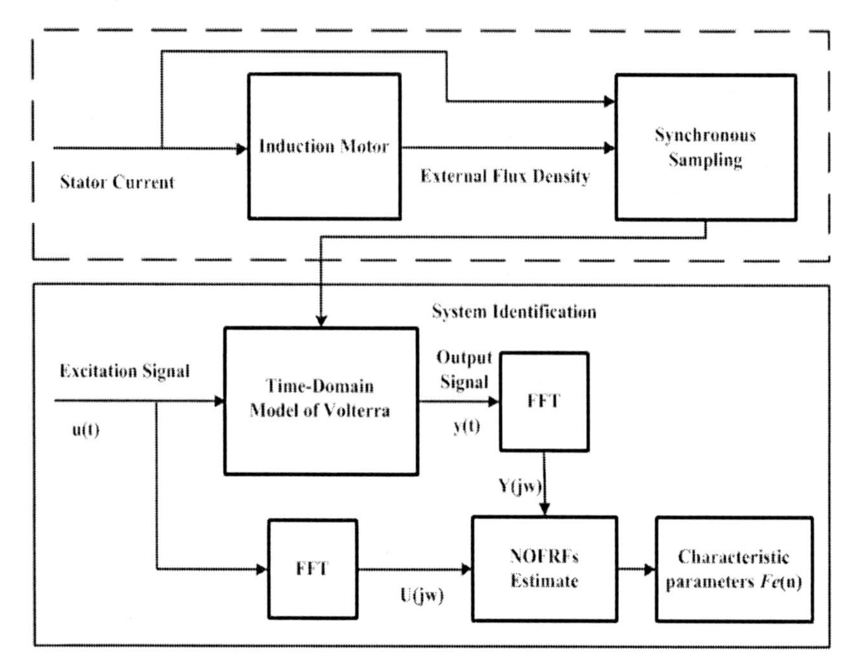

Figure 1: Bearing fault diagnosis based on Volterra series and NOFRFs.

2. Through the identified time-domain kernels, Volterra series model can be used to describe the motor. Then the excitation signals are added to the model, and the input and output of the model are implemented with FFT.
3. Estimate the NOFRFs of system and calculate the eigenvalues $Fe(n)$.

4 Experiment analysis

In order to illustrate the effectiveness of the diagnosis method, and describes the nonlinear characteristics of bearing fault. The experimental system is built and shown in Figure 2. The signal acquisition of stator current and external magnetic flux density is achieved by the Hall current sensor, magnetoresistive sensor and data acquisition card. The bearing outer raceway are artificially set a small hole and large hole respectively, as shown in Figure 3, that represents bearings with slight fault and serious fault. Since load changes affect the nonlinearity of system, the motor will connect a fixed load and use the bearings with no fault, slight fault and serious fault in outer raceway respectively. The position of the magnetic resistance sensor is fixed too. The signals of stator current and external magnetic flux density are collected. Recursive least squares algorithm is used to identify the time-domain kernels of Volterra series, and the model for induction motor is constructed. The sinusoidal signals with frequency of 50Hz and different amplitudes are used as the excitations of model, and then the NOFRFs were estimated. The NOFRFs are shown in Table 1.

As shown in Table 1, along with the increasing severity of bearing fault, the first order components of eigenvalues *Fe*(*n*) become smaller, and second order components *Fe*(2) become significantly larger, that indicates the increase in nonlinearity of motor and the severity of fault.

Table 1: Comparison of *Fe*(*n*)s under different condition.

	Fe(1)	*Fe*(2)	*Fe*(3)
No Fault	0.5060	0.0412	0.4527
Slight Fault	0.4230	0.4706	0.1063
Serious Fault	0.2429	0.6524	0.1047

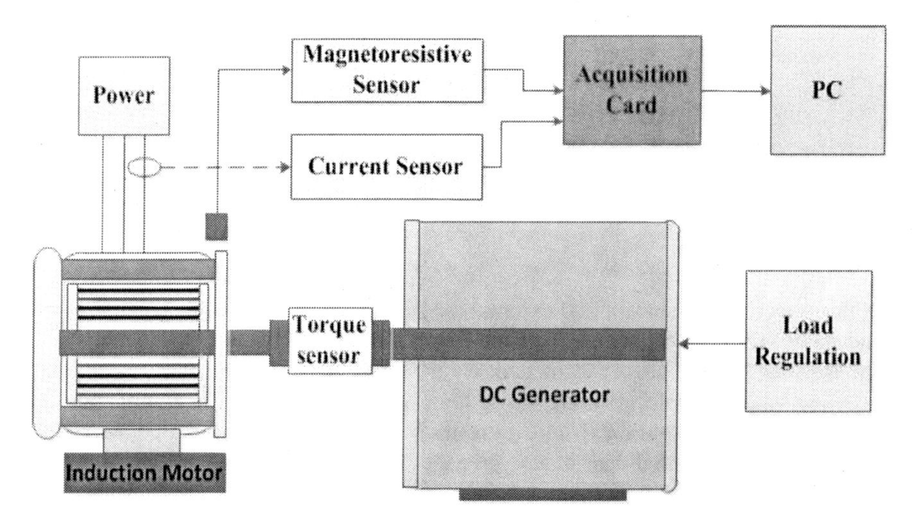

Figure 2: Experimental system for bearing fault diagnosis.

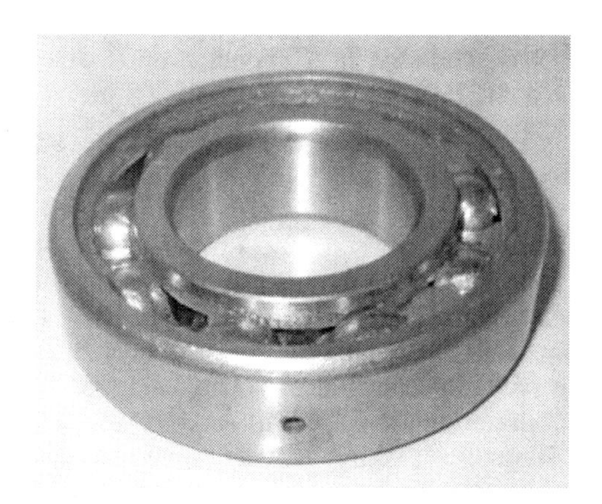

Small hole

Large hole

Figure 3: Bearing with slight and serious outer raceway fault.

5 Conclusions

In this paper, stator current and external magnetic flux density were acted as input and output of Volterra series, the time-domain model for motor was established, and then the NOFRFs were estimated. By means of the experiments on motor that installed bearings with no fault, slight fault and serious fault, the eigenvalues of NOFRFs could be obtained and compared. The results showed that NOFRFs could describe the nonlinear characteristics of motor and the severity of bearing fault.

Acknowledgment

The research work was supported by National Natural Science Foundation of China under Grant No. 51279020.

References

[1] Blödt, M., Granjon, P., Raison, B., Rostaing, G., Models for bearing damage detection in induction motors using stator current monitoring. *IEEE Transactions on Industrial Electronics*, **55(4)**, pp. 1813–1822, 2008.
[2] Henao, H., Demian, C., Capolino, G.A., A frequency-domain detection of stator winding faults in induction machines using an external flux sensor. *IEEE Transactions on Industry Application*, **39(5)**, pp. 1272–1279, 2002.
[3] Lang, Z.Q., Billings, S.A., Energy transfer properties of nonlinear systems in the frequency domain. *International Journal of Control*, **78(5)**, pp. 354–362, 2005.

[4] Joksimovic, G., Djurovic, M., Penman, J., Cage rotor MMF: winding function approach. *IEEE Power Engineering Review*, **21(4)**, pp. 64–66, 2001.
[5] Henao, H., Capolino, G.A., Martis, C., On the stray flux analysis for the detection of the three-phase induction machine faults. *Industry Application Conference, 38th IAS Annual Meeting*, pp. 1368–1373, 2003.
[6] Boyd, S., Chua, L.O., Fading memory and the problem of approximating nonlinear operators with volterra series. *IEEE Transactions on Circuits and Systems*, **32(11)**, pp. 1150–1161, 1985.

The stator current eigen frequencies induced by rotor slot harmonic and bearing fault

Guozhu Cheng, Chidong Qiu, Changqing Xu, Zhengyu Xue
Marine Engineering College, Dalian Maritime University, Dalian, China

Abstract

The spectrum analysis method of stator current for bearing fault diagnosis is mostly based on ideal motor model, and there is merely consideration the effect of internal motor harmonic. Considering the effect of rotor slot harmonics, the stator current frequency expression was newly deduced, some new current eigenfrequencies for bearing fault was discovered. The finite element simulation and experiments were, respectively, implemented for squirrel cage motor with 28 and 26 rotor bars, result shows that those proposed eigenfrequencies are subsistent and suitable for the identification of bearing fault.
Keywords: motor, bearing, fault, rotor slot harmonics.

1 Introduction

The method based on stator current spectrum analysis has the feature of low cost, non-invasion, and easy signal acquisition [1], which is one of the main methods for condition monitoring and fault diagnosis for induction motor [2]. Previous researches on motor bearing fault detection were mainly based on the ideal motor model, without considering the effect of the internal harmonics in an actual motor. It is well known that slot harmonics, a main harmonic inside the motor, widely exist in most motors. Joksimović et al.'s [3] research shows that the rotor slot harmonics of squirrel cage motor has significant effect on the rotor MMF spectrum. The emphasis of this paper is to consider the effect of rotor slot harmonics on motor. Section 2 deduces the stator current frequency expression under the normal state. Section 3 studies the frequency characteristics of bearing fault state to get a novel bearing fault eigenfrequency expression. In Section 4, experimental results based on hardware and finite element simulation verify the theoretical derivation.

WIT Transactions on Engineering Sciences, Vol. 107, © 2015 WIT Press
www.witpress.com, ISSN 1743-3533 (on-line)
doi:10.2495/SEEEI140811

2 Stator current feature induced by rotor slot harmonics

A squirrel cage rotor model is shown in Figure 1. A loop of squirrel cage rotor can be regarded as the coil pitch $\alpha = 2\pi/m$ and only one turn, where m is the number of rotor bars. One loop of a1, b1, b2, and a2 is defined as loop1, and the other loop of b1, c1, c2, and b2 is loop2. The angle difference between loop2 and loop1 is $2\pi/m$ in space phase. The current flows in these two loops with same magnitude and frequency but shifted in phase by $vp\cdot 2\pi/m$.

Rotor current is influenced by the stator flux density v^{th} space harmonic. The MMF expression produced by loop1 and loop2 are shown in work [4].

$$F_{L1}(t,\theta_r) = \sum_{k=1}^{\infty} \frac{2}{k\pi} \sin\left(\frac{k\pi}{m}\right) \cos(k\theta_r) I_{rv^{th}} \cos(s_v \omega_s t) \tag{1}$$

$$F_{L2}(t,\theta_r) = \sum_{k=1}^{\infty} \frac{2}{k\pi} \sin\left(\frac{k\pi}{m}\right) \cos k\left(\theta_r - \frac{2\pi}{m}\right) I_{rv^{th}} \cos\left(s_v \omega_s t - vp\frac{2\pi}{m}\right) \tag{2}$$

where s is the motor slip, $s_v=1 - v\,(1 - s)$, $v = 6h + 1$, $h = 0, \pm 1, \pm 2, \ldots$

Eq. (2) is simplified by trigonometric function relationship. And then, the other loops' MMF expression is obtained according to the form of loop2. The MMF of cage rotor is obtained by summing up all rotors loops' MMF.

$$F_r(t,\theta_r) = \sum_{n=0}^{m-1} \sum_{k=1}^{\infty} F_{vk} \left[\cos\left(s_v \omega_s t + k\theta_r - n(k+vp)\frac{2\pi}{m}\right) + \cos\left(s_v \omega_s t - k\theta_r + n(k-vp)\frac{2\pi}{m}\right) \right] \tag{3}$$

It is known from Eq. (3) that the corresponding rotor MMF equals the sum of m equations when k works as an arbitrary positive integer. The value of the summation would not be zero only when $k = |\beta m \pm vp|$, and can be simplified as:

$$F_{rv}^{R}(t,\theta_r) = F_{vk} \cos(s_v \omega_s t \pm (\beta m \pm vp)\theta_r) \tag{4}$$

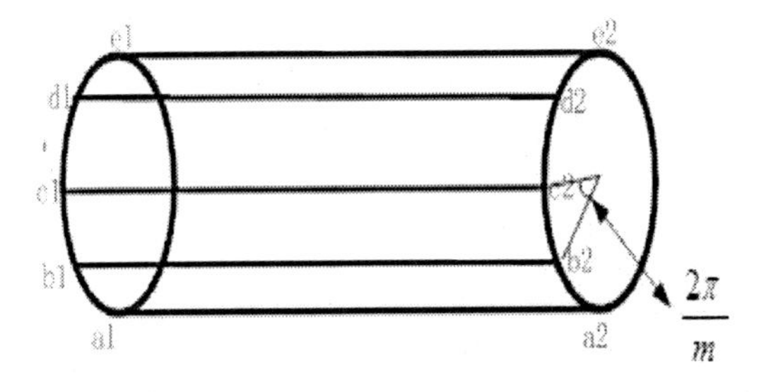

Figure 1: Squirrel cage rotor model.

where, $\beta = 0, 1, 2, \ldots$ Eq. (4) is changed into the stator reference frame by $\theta_s = \theta_r + \omega_s t\,((1 - s)/p)$.

$$F_{rv}^{S}(t,\theta_r) = F_{vk}\cos\left(\left(1 \pm \frac{\beta m}{p}(1-s)\right)\omega_s t \mp (\beta m - vp)\theta_s\right) \tag{5}$$

The air gap of normal motor can be regarded as uniform, ignoring the motor magnetic saturation, and the air gap permeance Λ can be assumed to be constant. The air gap flux density $B\,(\theta, t)$ is the product of total MMF and air gap permeance Λ.

$$B(\theta,t) = \left[F_s(\theta,t) + F_{rv}^{S}(\theta,t)\right]\Lambda \tag{6}$$

The expression of arbitrary phase flux $\Phi(t)$ is obtained by integration expression (Eq. (6)) with its structure.

$$\Phi(\theta,t) = \Phi_s\cos(\omega_s t + \phi_s) + \Phi_r\cos(\omega_s t \pm \beta m \cdot \omega_r t + \phi_r) \tag{7}$$

The relation between the flux and stator current can be given via stator voltage equation.

$$V(t) = R_s I(t) + d\phi(t)/dt = R_s I(t) + N_k d\Phi(t)/dt \tag{8}$$

where $\varphi(t)$ is flux linkage, N_k is effective winding turns, $V(\mathrm{t})$ is the motor power supply voltage. It can be found out from Eq. (8) that there is a linear relation between stator current and the differential of flux. Then, the stator current expression is

$$I(t) = I_0\sin(\omega_s t + \phi_s) + I_1\sin(\omega_s t \pm \beta m \cdot \omega_r t + \phi_r) \tag{9}$$

The instantaneous frequency expression can be obtained by taking the derivative of phase in Eq. (9). Considering the harmonic components in three-phase balanced power, stator current frequency expression is

$$f_i = | \pm n f_s \pm \beta m \cdot f_r | \tag{10}$$

where, $n = 1, 3, 5, \ldots$, $\beta = 0, 1, 2, \ldots$, m is the number of rotor bars.

Many previous studies show that the motor dynamic eccentric can cause rotational frequency component contained in stator current. According to Eq. (10), for the normal states, stator current will also contain item associated with rotational frequency, which caused by internal rotor slot harmonics in motor. The presence of rotational frequency in stator current spectrum cannot be regarded as a basis to determine whether there is a dynamic eccentricity in a motor.

3 Stator current feature induced by bearing fault

According to the bearing mechanical characteristics, if there is a single point fault in bearing, a small vibration torque of T_c whose frequency is f_c will be added when shaft pass through the fault. According to Ref. [5], the expression of rotating machinery angle with bearing fault is

$$\theta^*(t) = \frac{T_b}{J\omega_c^2}\cos(\omega_c t) + \omega_{r0}t = A\cos(\omega_c t) + \omega_r t \tag{11}$$

where J is the total inertia of motor, $A = T_b/J\omega_c^2$.

The expression of rotor MMF with fault feature can be changed into the stator reference frame by $\theta_s = \theta_r + \theta^*$, and it is shown as follows:

$$F_{rv}^S(t,\theta_r) = F_{vk}\cos(\omega_s t \pm \beta m\cdot\omega_r t \mp (\beta m \pm vp)(\theta_s - A\cos(\omega_c t))) \tag{12}$$

Air-gap eccentricity caused by bearing failure is very small, and can be neglected. Ignoring the motor magnetic saturation, the magnetic flux density is as same as the MMF waveform. An arbitrary phase flux $\Phi_b(t)$ is equal to the integral of magnetic flux density with its structure. To simplify the analysis, flux expression is given, which only contains fault feature component.

$$\Phi_b(\theta,t) = \Phi_b\cos(\omega_s t \pm \beta m\cdot\omega_r t \pm (\beta m \pm vp)\cdot A\cos(\omega_c t))) \tag{13}$$

It can be found out from stator voltage equation that there is a linear relation between stator current and the differential of flux, and the stator current expression with fault feature is as follow:

$$\begin{aligned} I_b(t) = {} & I_{m0}\sin(\omega_s t \pm \beta m\cdot\omega_r t \pm (\beta m \pm vp)\cdot A\cos(\omega_c t)) \\ & + I_{m1}\cos(\omega_s t \pm \beta m\cdot\omega_r t \pm (\beta m \pm vp)\cdot A\cos(\omega_c t) \pm \omega_c t) \end{aligned} \tag{14}$$

The instantaneous frequency expression can be obtained by taking the derivative of phase in Eq. (14). The value of A is small enough to be ignored. Considering the harmonic components in three-phase balanced power, the frequency expression of stator current with bearing fault is

$$f_{ib} = | \pm nf_s \pm \beta m\cdot f_r \pm kf_c | \tag{15}$$

It is known from Eq. (15) that the spectrum of stator current with bearing fault will appear the novel eigenfrequencies under the influence of rotor slot harmonic. Considering the number of rotor bars and the rotor rotating frequency in actual, there are some new eigenfrequencies in the medium frequency band, and it is richer than in the low frequency band.

4 Simulation and experiment

In order to simulate the torque vibration caused by bearing failure, an experimental platform is built up as shown in Figure 2. It consists of a cage induction motor driving a DC generator. The standard parameters of squirrel cage motor are 2 pole pairs, 50 Hz, 380V, 2.2 kW, 28 rotor bars, and its bearing type is 6206. The load of DC generator is two series variable resistor. One remains unchanged as the basic load. The other is controlled by the PWM switch which can control the connection or disconnection of the load circuit. The signal is generated by the control circuit. DC generator load fluctuation will cause vibration of the shaft, further cause the induction motor torque vibration. The torque vibration generated by bearing fault can be simulated like that.

The control circuit generates the PWM switch signal with 50% duty cycle and 14 Hz. The motor rotating speed is 1460 r/min in the experiment, so the rotation frequency is 24.3 Hz. The stator current data is acquired, then Matlab is used to implement spectrum analysis. The result is shown in Figure 3b.

In addition to these eigenfrequencies of $f_s \pm f_c$ and $f_s \pm 3f_c$, the $f_s \pm f_r$ also appear in Figure 3a, where f_c is the simulative failure frequency, f_s is the power frequency, and f_r is the rotation frequency. The reason for this result is that the dynamic eccentric is hardly avoided owing to motor manufacture and installation technology. Dynamic eccentricity will produce the harmonic current $f_d = f_s \pm f_r$. The experimental results prove it well. From Figure 3c, multiple frequency components from experiment results accord with the proposed expressions. Meanwhile, such as bearing failure, dynamic eccentricity will induce some new eigenfrequencies under the influence of rotor slot harmonics. The results of this experiment are consistent with the conclusion given by Mohamed [6] who studies the imbalance load of motor with the influence of rotor slot harmonics.

Table 1 is obtained from theoretical derivation and experiment results.

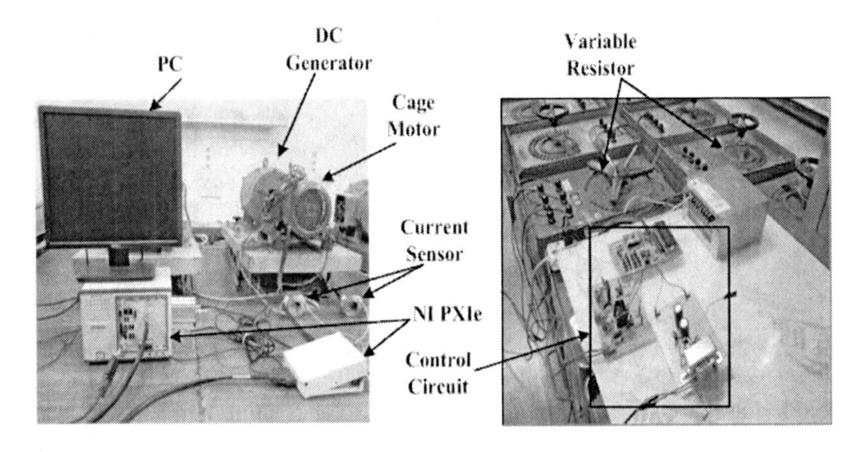

Figure 2: The experimental platform for motor bearing fault.

In order to remove the interference of the dynamic eccentricity in the actual machine and validate the analytical results better, a cage motor model is built via the finite element simulation. Its standard parameter is 2.2 kW, 380 V, 50 Hz, 26 rotor bars. The outer rail fault of bearing is simulated via injecting an oscillation torque with 87 Hz into the load torque. The motor speed is 1449 r/min in this experiment, and the rotating frequency f_r is 24.15 Hz. Power spectrum analysis is implemented by Matlab, and the result is shown in Figure 4b.

where, $n = 1, 3, 5, \ldots, \beta = 0, 1, 2, \ldots, m$ is the number of rotor bars.

It is clear in Figure 4c those eigenfrequencies, induced by rotor slot harmonic and bearing fault, exist in the stator current spectrum. Those are accord with the fault eigenfrequency expression that is derived in this paper. By comparing

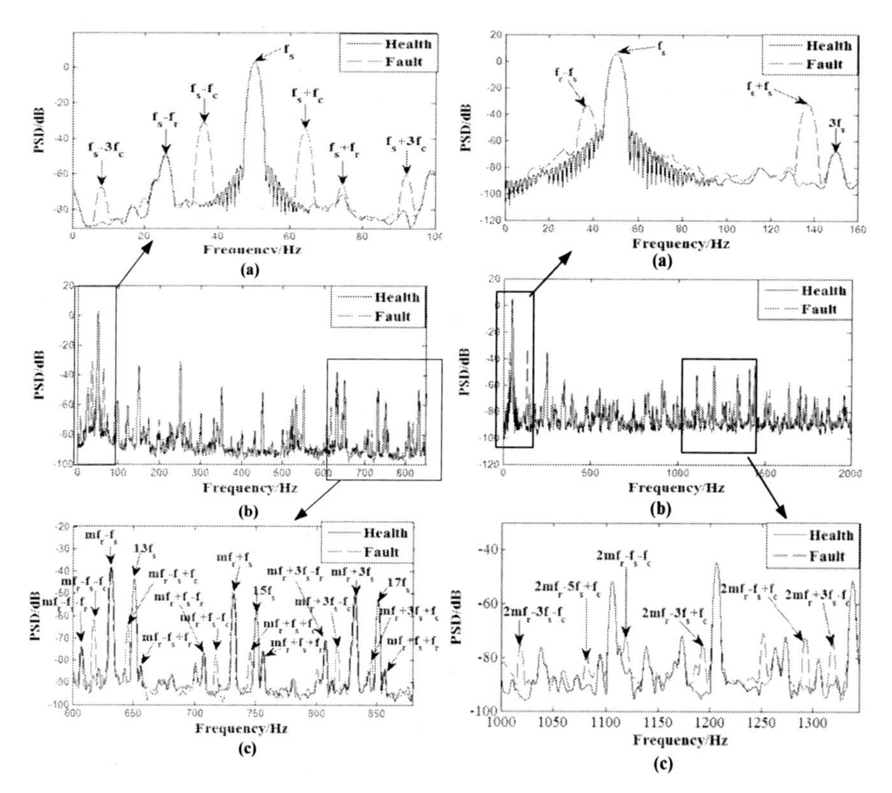

Figure 3: The stator current power spectrum for hardware experiment.

Figure 4: The stator current power spectrum for finite element simulation.

Table 1: Expression of the current frequency under different states.

The different states of motor	Normal	Bearing fault	Dynamic eccentricity
Eigenfrequencies	$nf_s \pm \beta mf_r$	$nf_s \pm \beta mf_r \pm kf_c$	$nf_s \pm \beta mf_r \pm f_r$

www.witpress.com, ISSN 1743-3533 (on-line)

Figure 4c with Figure 4a, it can be clearly found that the eigenfrequency in medium frequency band is richer than that in low frequency band. Therefore, when bearing fault detection is implemented for actual motor, the reliability of fault diagnosis can be improved well via extracting those eigenfrequencies from medium frequency band.

5 Conclusions

This paper considered the effect of rotor slot harmonic and deduced the stator current frequency expression under normal states and bearing fault states, respectively. The stator current eigenfrequency will contain item associated with the rotational frequency and the number of rotor bars. Therefore, the presence of rotational frequency in stator current spectrum cannot be regarded as a basis to determine whether there is a dynamic eccentricity in a motor. Comparing to the previous researches, the spectrum of stator current with bearing failure will appear new eigenfrequencies within medium frequency band. The experimental results confirm that the eigenfrequency components in medium frequency band are not only existent, but also richer than in low frequency band. Therefore, the reliability of bearing fault diagnosis can be improved via an effective method that extracts the eigenfrequency from medium frequency band.

Acknowledgment

The research work was supported by National Natural Science Foundation of China under Grant No. 51279020. The support is greatly appreciated.

References

[1] W. Zhou, T.G. Habetler, R.G. Harley, Incipient bearing fault detection via motor stator current noise cancellation using Wiener filter. *IEEE Transactions on Industry Applications*, **45(4)**, pp. 1309–1317, 2009.

[2] M.E.H. Benhouzid, G.B. Kliman, What stator current processing-based technique to use for induction motor rotor faults diagnosis. *IEEE Transactions on Energy Conversion*, **18(2)**, pp. 238–244, 2003.

[3] G. Joksimovic, M. Durovic, J. Penman, Cage rotor MMF-winding function approach. *IEEE Power Engineering Review Letters*, **21(4)**, pp. 64–66, 2001.

[4] G. Joksimovic, M. Durovic, A. Obradovic, Skew and linear rise of MMF across slot modeling winding function approach. *IEEE Transactions on Energy Conversion*, **14**, pp. 315–320, 1999.

[5] M. Blodt, P. Granjon, B. Raison, et al., Models for bearing damage detection in induction motors using stator current monitoring. *IEEE Transactions on Industrial Electronics*, **55(4)**, pp. 1813–1822, 2008.

[6] S. Mohamed, B. Khmais, C. Abdelkader, Stator current analysis of a squirrel cage motor running under mechanical unbalance condition. *International Multi-Conference on Systems, Signals & Devices (SSD)*, Hammamet, Tunisia, 2013.

Author Index

WIT Transactions on Engineering Sciences, Vol. 107, © 2015 WIT Press
www.witpress.com, ISSN 1743-3533 (on-line)
doi:10.2495/SEEEII14index1

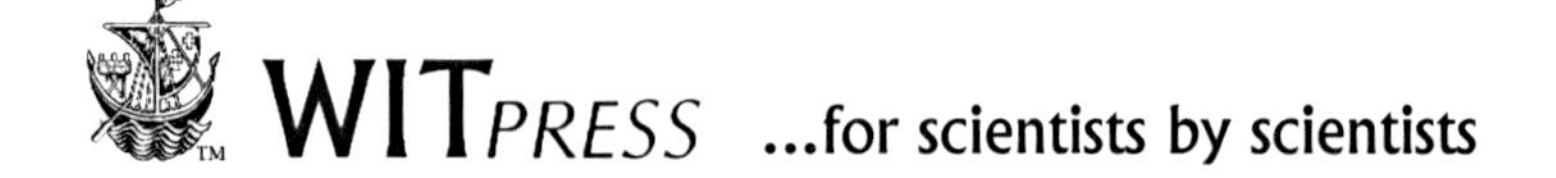
WIT PRESS ...for scientists by scientists

CPSIA information can be obtained at www.ICGtesting.com
Printed in the USA
BVOW04*0342210715

408355BV00001BB/1/P

9 781784 661175